SPINGLER

MARINE BIOLOGY

MARINE BIOLOGY

Second Edition

Herbert H. Webber
Western Washington University

Harold V. Thurman
Mt. San Antonio College

HarperCollins*Publishers*

Sponsoring Editor: Glyn Davies
Project Coordination, Text Design: Proof Positive/Farrowlyne Associates, Inc.
Cover Design: MNO Production Services, Inc.
Cover Photo: Alex Kerstitch/Sea of Cortez Enterprises
Photo Research: Karen Koblik
Production: Michael Weinstein
Compositor: The Clarinda Company
Printer and Binder: R.R. Donnelley & Sons Company
Cover Printer: New England Book Components

Acknowledgments for the copyrighted materials not credited on the page where they appear are listed in the Acknowledgments section beginning on page 413. This section is considered an extension of the copyright page.

Marine Biology, Second Edition

Library of Congress Cataloging-in-Publication Data

Webber, Herbert H.
Marine biology / Herbert H. Webber, Harold V. Thurman.—2nd ed.
p. cm.
Thurman's name appears first on the earlier edition.
Includes bibliographical references and index.
ISBN 0-673-39913-3
1. Marine biology. I. Thurman, Harold V. II. Title.
QH91.W43 1991
574.92—dc20 90-47470
CIP

92 93 9 8 7 6 5 4 3

CONTENTS IN BRIEF

CONTENTS

PREFACE

Marine Biology, Second Edition is an introductory text, written for students in the first two years of college. While a college-level general biology course is good preparation for taking a marine biology course, no such coursework is assumed of students using this book.

The units of this textbook are divided into four main themes, or areas of exploration. First, the student is familiarized with the field of marine biology through chapters on the history and development of the field, plus an introduction to the physical components of the marine environment. Second, an in-depth examination of the diversity of marine life is presented. The emphasis in this area is on the large animals—the fish, birds, and mammals. Third, an ecological approach to marine biology is used to examine marine habitats. Fourth, human interaction with and in the marine environment is discussed, particularly in terms of the use of marine life as food, and the threat of human pollution in the oceans.

These four units are divided into 15 chapters, which allows instructors to identify the areas of interest that best fit their courses. Interspersed throughout the chapters are special features that present topics of current interest, or focus on the work of particular marine scientists. In an introductory text, these features allow students to understand the nature of detailed study, as well as familiarize them with the personalities of the scientists in the field.

Part One, Introduction and Overview, covers the definition and brief history of the evolution of marine biology as a field of study (Ch. 1). It surveys the basic nature of the marine environment (Ch. 2), including the life-styles of marine organisms and the physical characteristics of the bottom of the ocean.

Special features in Part One include a color section that highlights the nature of color in the marine environment and in marine organisms. Specifically, variation in the color of ocean water, the range of color of marine life, the absorption of color at depth by seawater, and the use of color in marine research are all documented. In the special feature in Chapter 2, the oceanographic conditions that cause El Niño are explored.

Part Two, The Biology of Marine Organisms, discusses the diversity of marine life from a systematic and evolutionary perspective. Chapter 3 examines the importance of the marine environment to the evolution of marine life, starting with the ideas that scientists have about the origin of life and the pathway to the evolution of the Monerans (i.e., the bacteria and related forms). Chapter 3 continues with a discussion of the evolutionary relationships and distribution in the marine environment of the other kingdoms of life: the Protista (protozoa, algae, and most phytoplankton), the Metazoa (the animals), the Plantae (the true plants), and the Fungi (molds and mushrooms). Chapter 4, Monera, Protista, Fungi, and Plantae, considers the bacteria, the photosynthetic marine organisms, and the microscopic consumers of the marine environment. Chapter 5, Kingdom Metazoa: The Animals, discusses the diversity of marine animals without backbones. Chapters 6 through 8 consider the marine vertebrate animals: Chapter 6—the fish; Chapter 7—the reptiles and birds; Chapter 8—the marine mammals.

Special features in Part Two include a presentation of the idea that the origin of life was on the ocean floor (Ch. 3); intraspecific aggression in sea anemones (Ch. 5); the right- and left-eyed nature of flatfish (Ch. 6); the extraordinary feeding and migration patterns of the albatross (Ch. 7); the importance of antarctic krill to whales, and the disturbance of the ocean bottom that gray whales and walruses cause from their method of feeding (Ch. 8).

Part Three, Marine Habitats, presents an ecological approach to studying marine biology. Chapter 9 introduces the ecological concepts the student needs for understanding the ecological approach. Next, the marine ecological systems are divided into four major habitats: the photic zone (Ch. 10); the deep ocean (Ch. 11); the intertidal zone (Ch. 12); and the coastal ocean (Ch. 13). In each chapter, the physical factors that affect the marine habitat under discussion are explicated. For example, waves and tides are discussed in Chapter 12, The Intertidal Zone.

Special features in Part Three include a color photo survey of the major habitats and intertidal life, marine mammals, and the hydrothermal vent community. Other special features explore the interplay between coral reefs and nutrients, and the relationship between micro and macro food chains (Ch. 9); the relationship between gulf stream rings and marine plankton (Ch. 10); deep-sea research and deep-ocean benthos (Ch. 11); the timing of grunion reproduction and the tides (Ch. 12); and the relationship between coral reefs and associated sponge populations (Ch. 13).

Part Four, The Ocean Under Stress, offers an overview of current concerns about overexploitation of marine resources and ocean pollution. Chapter 14, Food from the Sea, traces the path of overexploitation of fishery resources by humans, and introduces the concept of total allowable catch, a method of determining the yield of a fishery based on ecological data and the ability of the marine ecosystem to support the fishery. Chapter 15, Marine Pollution, examines the threat of accidental and planned discharge of the wastes of human society into the oceans. The chapter also looks at oceanic capacity to assimilate these wastes. Pollution sources including petroleum, sewage, radioactive wastes, and mercury are considered. A critique of the performance of the United States Food and Drug Administration in regard to its handling of mercury contamination of fish, and a summary of international efforts to combat marine pollution, complete the chapter.

Special features in Part Four include an examination of the loss of estuaries in the United States, due to dredging and filling, and proposals to use the deep sea for the disposal of nuclear wastes (Ch. 15).

Because the marine environment typically is so foreign to human experience, the introductory student needs a basic understanding of the important physical factors of the marine environment. *Marine Biology* discusses in detail the myriad physical factors that affect marine life. Basic material such as the nature of seawater, the effect of salt water on marine life, the distribution of light in the ocean, the effects of viscosity on floating and swimming forms, and the distribution of oxygen and nutrients are discussed in an introductory chapter (Ch. 2). Physical factors that are of particular importance to specific habitats are discussed in the appropriate chapter on ecological habitats: Tides and waves are discussed in Chapter 12, The Intertidal Zone. Oceanic circulation (vertical and horizontal) is discussed in Chapter 10, The Photic Zone.

Coastal currents, salinity, and the nature of estuaries are discussed in Chapter 13, The Coastal Zone.

Instructors of marine biology courses may wish to emphasize either a systematic approach to the diversity of marine life, or an ecological approach to marine habitats. *Marine Biology* offers a detailed treatment of both approaches. Students must have a basic conceptual appreciation of the diversity of marine life to understand the ecological approach. Likewise, students who are taught with an emphasis on the study of diversity of marine life must have some ecological appreciation if they are to understand the patterns of diversity. Whatever the emphasis of the instructor, *Marine Biology* provides adequate material for an introductory course.

We wish to thank the following reviewers who offered critical comments on the manuscript at various stages: David Campbell, Rider College; P. Kelly Williams, University of Dayton; A.L. Bratt, Calvin College; Donald Dorfman, Monmouth College; Thomas Niesen, San Francisco State University; Joseph Simon, University of South Florida; Wendell Patton, Ohio Wesleyan University; R.E. Terwilliger, Oregon Institute of Marine Biology; Patrick Williams, University of Dayton; Cynthia Groat, Bowling Green State University; Neil Crenshaw, Indian River Community College; Richard Brusca, University of Southern California; Randall Stovall, Valencia Community College; Susan Cormier, University of Louisville; William Gorham, California Lutheran University; Peter Pedersen, Cuesta College; and Harry D. Corwin, University of Pittsburgh.

We also wish to thank everyone at HarperCollins and Proof Positive/Farrowlyne Associates, Inc., for their help in bringing the manuscript to book form. We especially appreciate the efforts of the project editor, Marie Donovan, and the designer, Ann Skuran, of Proof Positive/Farrowlyne Associates.

A Special Note to the Student: Key terms to study are indicated in boldface type in each chapter, and are further defined in the Glossary at the end of the book. The end-of-chapter summaries and questions and exercises are designed to aid you in your firm understanding of the marine biology principles introduced in this text. The suggested readings will provide you with in-depth treatments and extensions of the topics discussed in the chapter, and should prove helpful in preparing term papers and reports. While metric conversions are included where necessary in each chapter, Appendix A reviews the more widely used metric conversion factors. Appendix B compiles the various scientific prefixes and suffixes you should become familiar with for further study. Appendix C outlines the major taxonomic classifications of the organisms you will encounter in this, your introduction to the field of marine biology.

H.H. Webber

H.V. Thurman

MARINE BIOLOGY

PART ONE

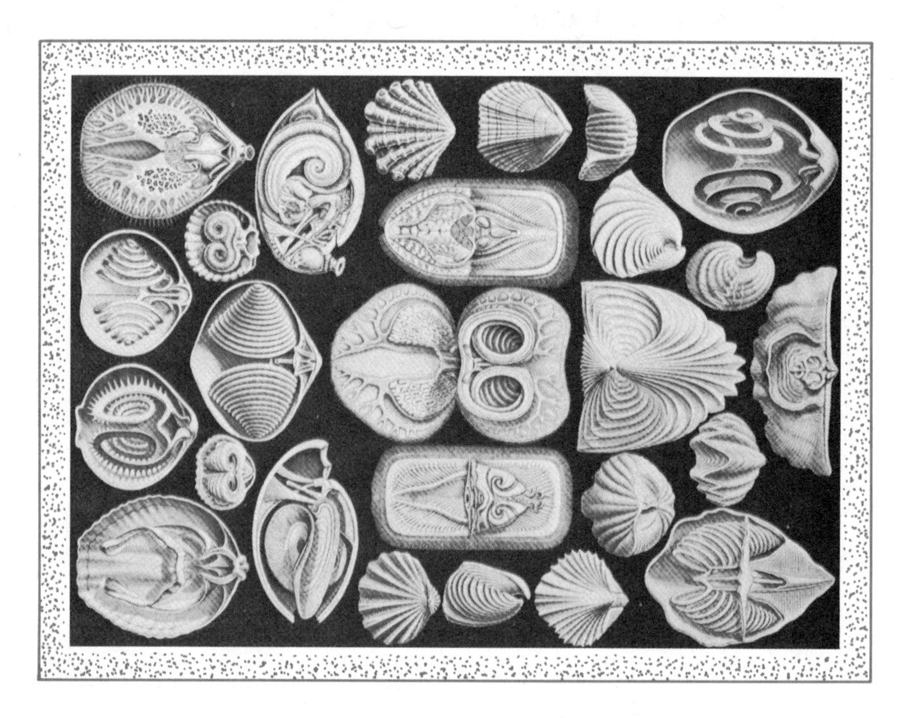

INTRODUCTION AND OVERVIEW

We start the study of marine biology in Chapter 1 by defining it, and tracing the history of the development of the field. Although human fascination with life in the oceans is ancient, it is only fairly recently that its study has become a unified science.

Why study marine biology at the introductory level? Almost 50 percent of the population of the United States lives on the fringe of land bordering the oceans. Too, most people have a basic curiosity for marine organisms and are naturally drawn to learn more about them. Besides their intrinsic interest, research studies on marine life have contributed many basic and important discoveries to our understanding of the broader field of biology. Finally, the increasing importance of marine organisms in providing food, medical, and other resources for humans adds an additional dimension to the importance of studying marine biology.

Life forms in the marine environment are faced with very different environmental conditions from those on land. Only the surface of the ocean is lighted. Photosynthesis is mostly restricted to the top 100 m (300 ft). Below 1,000 m (3,000 ft) the ocean is in total darkness. Those organisms that live on the shorelines must deal with the movement of water caused by tides and waves. Those marine organisms that live in the water column have to continually fight the effects of gravity, and so develop adaptations to either float or swim. In Chapter 2 we discuss the basic physical factors that influence the nature of marine life.

1

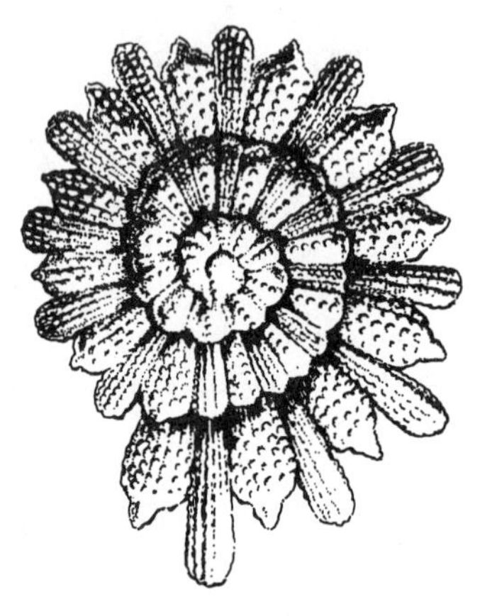

THE FIELD OF MARINE BIOLOGY

Marine biology is the study of organisms that live in the oceans. In this book, we focus on the marine organisms and their communities. Nevertheless, the subject areas of interest to marine biologists span the full dimensions of the discipline of biology. Some marine biologists specialize in biochemistry; others, in studies of the cell. Yet others focus on organisms, populations, or ecosystems. At each of these system levels, studies of life in the ocean make important contributions to our overall understanding of biology.

Biochemical studies of marine organisms have led to the discovery of many drugs not found in life on land. Cell biologists have made important discoveries about what nerves do by using jellyfish (Cnidarians), which have nerve cells but no central nervous system (i.e., brain). Even more important are studies of the giant nerve cells of squids, which have helped us understand how nerve cells work. Studies on the diversity of marine life are central to our understanding of the evolution of life on the planet. The oceans contain a much greater variety of life forms than those found either on land or in fresh water. Studies of the anatomy and development of these diverse forms help us understand the early evolution of life. Studies at the population and community level have contributed to our general understanding of ecology. The phenomenal reproductive potential of marine organisms (e.g., an oyster releases over 2 million eggs at each spawning), and the complex life history of many bottom-living species that have larval stages living free in the water column, help us understand the ecological relationships of populations.

Studies of marine ecosystems have yielded major theoretical and practical results. The recent discovery of a community of animals associated with undersea geological rift zones has given us the first example of a complex ecosystem that is not either directly or indirectly dependent on the sun as a source of energy. On the practical side, the increasing importance of world fisheries has stimulated studies on population dynamics and the capacity of the ocean's ecosystems to support species important for fisheries harvest.

How, then, should the student approach the study of marine biology? As a specific course, marine biology is relatively new to the curriculum of the undergraduate student. Only in the last 10–15 years have such courses been generally available. Before this, students wishing to study marine biology had to take separate courses in zoology, phycology, ecology, fisheries biology, etc. For a study of the physical environment of the oceans, students took a course in oceanography, generally at the graduate level.

Why, then, is a lower-division course in marine biology offered? After all, there are virtually no equivalent courses in 'land biology' or 'air biology'. The answer to this question has three parts. The first reason concerns the recent attempt to unify the studies of marine biologists into a cohesive body of knowledge. For over a hundred years marine scientists have worked more or less independently, with little communication. Those who studied the biology of oceans (biological oceanographers) typically had little communication with those who studied the sea shore and shallow waters (marine biologists). The steps taken to join research endeavors have made this undergraduate course possible.

The second reason this separate, specialized biology course is offered is because the organisms that dominate the marine environment are quite different from those that are common on land. Humans have at least a superficial understanding of the nature of life on land, since it is the environment in which we live. In many ways, however, life in the ocean is alien, unfamiliar, and inaccessible. This course will make you more familiar with the vast array of life forms in the sea.

The third reason for studying marine biology is that the biological resources of the oceans are becoming increasingly important to humans. Expansion of the cultivation and harvesting of marine species used for food requires a better understanding of the biology of marine life. Too, knowledge of the ocean's ability to dilute and decompose waste products is crucial to controlling pollution of the marine system.

A comment should be made about the scope of subjects introduced in this book. It is not possible to survey the entire field of marine biology in an introductory text. The focus of this book, therefore, is twofold: the study of organism diversity in the marine environment, and the study of the ecological communities in which different organisms are found. These aspects of marine biology are the most familiar to the beginning student. In our survey of the ecological communities of the oceans we include an introduction to the physical environment. The physical factors that shape the distribution of life in the oceans are not the same as those on land. Not only is the ocean immense (covering over 70% of the planet's surface, to an average depth of almost 4,000 m, or 12,000 ft); physical factors such as tides, waves, currents, hydrostatic pressure, salinity, and depth of light penetration are all important in creating the environment that sustains life in the oceans.

In the remainder of this chapter we will discuss in more detail the history of marine biology, and the importance of the biological resources found in the oceans.

THE HISTORY OF MARINE BIOLOGY

Early civilizations such as the Phoenicians, Greeks, and Vikings had particular interest in the marine environment. In fact, it was the Greeks who used the word *oceanus* to refer to what they thought was a wide, everflowing river that circled around a world that was flat. Early coastal civilizations used the ocean as a source of food collected from the shoreline proper, and from small fishing vessels that worked the nearshore areas. At this time, longer-distance water travel was important for trade, and travel between countries was common.

During the sixteenth century European countries started exploring the extent and location of the great oceans of the world, including the nature of the oceans' biology. During this time the science of marine biology became firmly established.

The earliest marine biologists were generalists who studied life in the oceans as part of the larger discipline of biology. Many were naturalists aboard ships that were sent on expeditions to survey the oceans. Charles Darwin sailed on the HMS *Beagle* in 1831 (Fig. 1.1). In 1839 Sir James Ross collected the first information on the natural history of Antarctica (Fig. 1.2). Thomas Henry Huxley was the naturalist on the HMS *Rattlesnake* in 1846, and G.C. Wallich sailed on the HMS *Bulldog* in 1860. These early naturalists documented life in the marine environment wherever they found it—from dredges and nets pulled behind the ship, to the life of the shoreline observed when the naturalists could leave their ships.

Not all the early marine biologists went to sea. Ernst Haeckel, one of Europe's most well-known biologists of the nineteenth century, studied marine life wherever the opportunity arose. Some of his artistic illustrations of marine life are included in this book (see Ch. 5).

In the later 1800s, researchers in marine biology polarized into specialists who studied life in the open ocean (biological oceanographers) and those who studied life on the seashore (marine biologists). It is only in the twentieth century that these two directions merged, and a synthesis of the fields developed.

The interest in studying biological oceanography developed in the 1800s because there was doubt that life existed on the deep-ocean floor. Some scientists insisted that life could not exist in this dark, high-pressure, and—at least, in their belief—oxygenless environment. To help settle this dispute, as well as to learn about the nature of life in the open ocean, the Royal Society of London recommended that funds be provided by the British government for the first major expedition specifically for the study of oceanography. This expedition was on the ship HMS *Challenger*. The *Challenger* expedition was directed to investigate the following:

1. Physical conditions of the deep sea in the great ocean basins;
2. Chemical composition of seawater at all depths in the ocean;
3. Physical and chemical characteristics of the de-

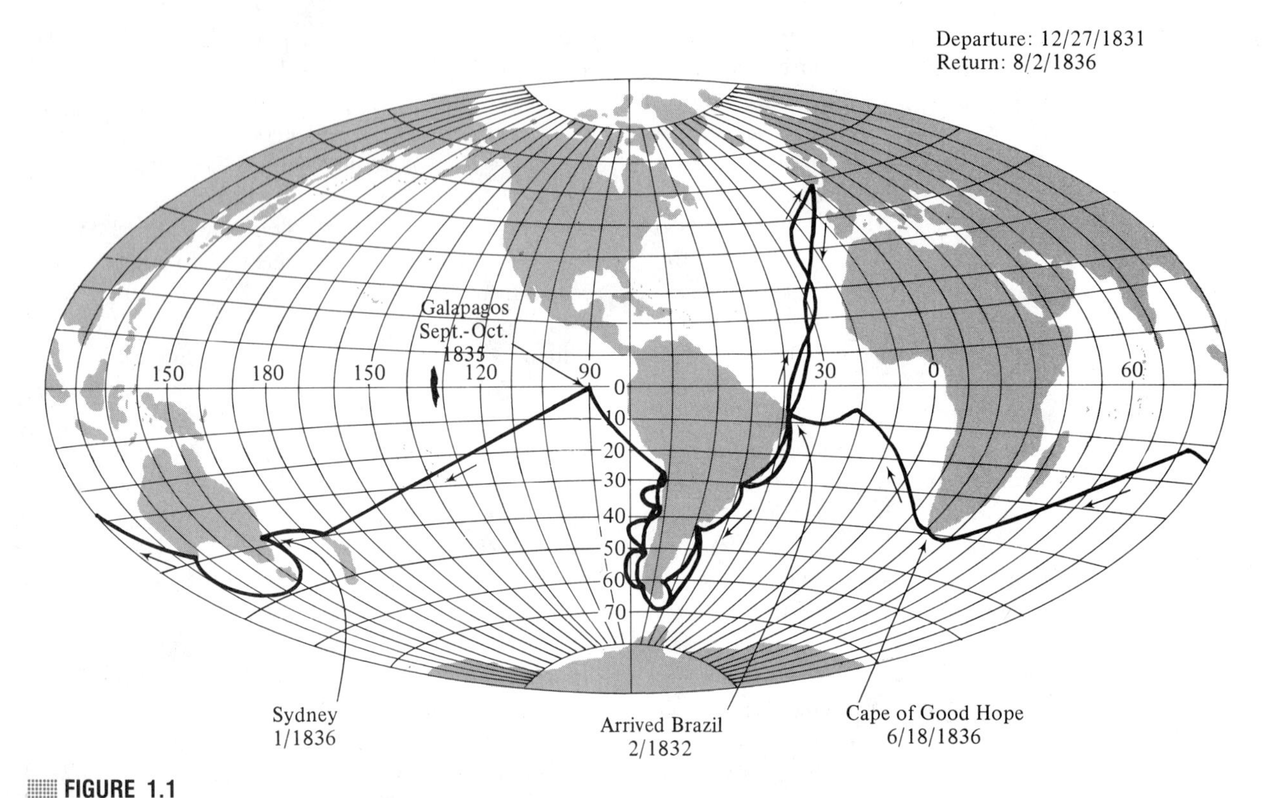

FIGURE 1.1
Voyage of the *Beagle*

Sailing as naturalist aboard the HMS *Beagle,* Charles Darwin gathered the evidence that allowed him to develop his theory of biological evolution through natural selection. (Base map courtesy of National Ocean Survey.)

posits of the sea floor and the nature of their origin; and

4. Distribution of organic life at all depths in the sea, as well as on the sea floor.

A staff of six scientists under the direction of C. Wyville Thompson left England in December 1872 aboard the HMS *Challenger* (Fig. 1.3). A 2,306-ton ship, refitted to conduct scientific investigations, the *Challenger* conducted 151 open-water trawls and 360 soundings and dredgings at deep-water locations. From its sampling of ocean life at depths down to 9 km (27,000 ft), some 4,717 new species of marine life were recovered and identified. The results of this voyage—conducted exclusively for the purpose of investigating the nature of life in the oceans—left no doubt that the oceans were inhabited throughout their depth and breadth. The famous *Challenger* reports are considered to be the basis on which the science of oceanography has been built. Today, biologists who specialize in studying the organisms of the distant and deep ocean are known as biological oceanographers.

At the same time that expeditions were being sent to study life in distant oceans, research facilities dedicated to study marine life were being established on the shore. The first was the Statione Zoologica, established at Naples, Italy in 1872. The Plymouth (England) Laboratory of the Marine Biological Association of the United Kingdom followed in 1888. In the United States, the Woods Hole (Massachusetts) Oceanographic Institute was established in 1888. These and other shore-based research facilities all have circulating sea water systems that permit the maintenance and study of marine life under laboratory conditions. Their work has given the field its major impetus to study in depth the biology of the sea shore and shallow water. Currently, there are hundreds of marine laboratory facilities around the world; virtually all maritime nations have them. In the United States, every coastal state has at least one marine laboratory. People who use the facilities of marine laboratories are generally called marine biologists.

Recently, there has been a convergence of the specializations of the traditional biological oceanographers and marine biologists. Many coastal marine laboratories support oceanography ships. Technically, 'biological oceanography' is synonymous with 'marine biology'. Clearly there is little basis to separate the two, and their continued integration is to be applauded. Introductory textbooks in marine biology such as this are possible because of integration of the work of those marine sci-

FIGURE 1.2
Ross Expedition

Members of the Sir James Clark Ross expedition to Antarctica after landing on Possession Island. The penguins seem to be at least as curious about the visitors as the members of the expedition are about them. (Reprinted by permission of the Bettman Archive.)

entists who study life on the sea shore and in shallow water, and those who study life in the open ocean and deep water.

THE OCEANS AS A RESOURCE

We have learned a great deal about the distribution of life in the ocean. Yet we still know relatively little about the sizes of various populations, and even less about the natural history of those populations that have received the most study. Of course, for most marine scientists, the primary reason for investigating the nature of life in the ocean is to better understand its ecology, and thereby be better able to appreciate and preserve it. Since oceans represent an important food resource for the world's people, ocean research is often elaborate and expensive. Thus, much research is undertaken with the aim of increasing the size of the world fishery. Special attention has been given to increasing our knowledge of populations that serve as natural open-ocean fisheries and may be adaptable to **mariculture,** the cultivation of edible or otherwise useful ocean species through natural and artificial means.

As marine biologists consider how to achieve the optimum development and use of the ocean's resources, they recognize the need to minimize the polluting effects that accompany increasing exploitation. Although more than 30 percent of the world's oil production now comes from offshore fields, such use actually is in its infancy. Careless exploitation of the physical resources of the ocean, however, can reduce their yield through pollution of the coastal ocean. Since over 90 percent of the biological resources of the ocean are thought to reside in coastal waters no more than 200 km from the shore, it is critical that pollution from shore-based and nearshore sources be prevented.

To obtain a realistic view of the ecological challenges that lie ahead, one need only consider the fact that the human population is increasing at a rate of about 10,000 individuals per hour. At this rate, the world's population will more than double in 30 years. It seems challenging enough to double the present availability of food and water. When we consider, however, that future populations will be expecting a higher standard of living—more comfort, better transportation and communication—it becomes apparent that a great deal of planning and cooperation will be required if such needs are to be met. An increasing share of the resources required to meet these needs will, of necessity, come from the oceans. Care must be taken to preserve the marine environment during their harvest.

Medical Products

Medical science also has exploited the ocean's biological resources. Products range from the vitamins A and D provided by cod-liver oil to the poison extracted from the stonefish that, in appropriate doses, is used to lower blood pressure. Antibiotics derived from marine organisms include those extracted from dinoflagellate toxins, bacteria and virus inhibitors extracted from a variety of molluscs (snails, clams, and squids), and those derived from green algae and sponges that kill penicillin-resistant staphylococcus. Biochemicals from hagfish and weeverfish venom are effective in regulating the human heartbeat. The electric eel provides a chemical (paralidoxime chloride) that is an antidote to insect poisons. Cancer therapy has been aided by holothurin extracted from sea cucumbers, which has anticarcinogenic properties. The sea worm contains bonellinin, which inhibits cancer growth. The puffer fish is the source of tetradotoxin, which can be used to relieve pain in terminal cancer cases. Snails produce chemicals that can be used to relax and contract muscles. Continued investigation into the medical uses of biochemicals derived from marine organisms can be expected to lead to the discovery of many more useful substances.

Food from the Ocean

In 1985, total world fisheries catch was 72.1 million metric tonnes, or 86.5 tons (Fig. 1.4). Of this, 40 million tonnes was from pelagic (i.e., upper-oceanic) fish stocks, and 28 million tonnes from bottom-living fish. Of this 72.2 million tonnes, not all of it was eaten directly by humans. A total of 32 percent of the catch (21 tonnes) was converted into fish meal and oil (Fig. 1.4).

FIGURE 1.3

(A) The HMS *Challenger* made the first major oceanographic voyage (Dec. 1872–May 1876). *(From the* Challenger *Report, Great Britain, 1895.)* **(B)** The route of the *Challenger.* (Base map courtesy of National Ocean Survey.)

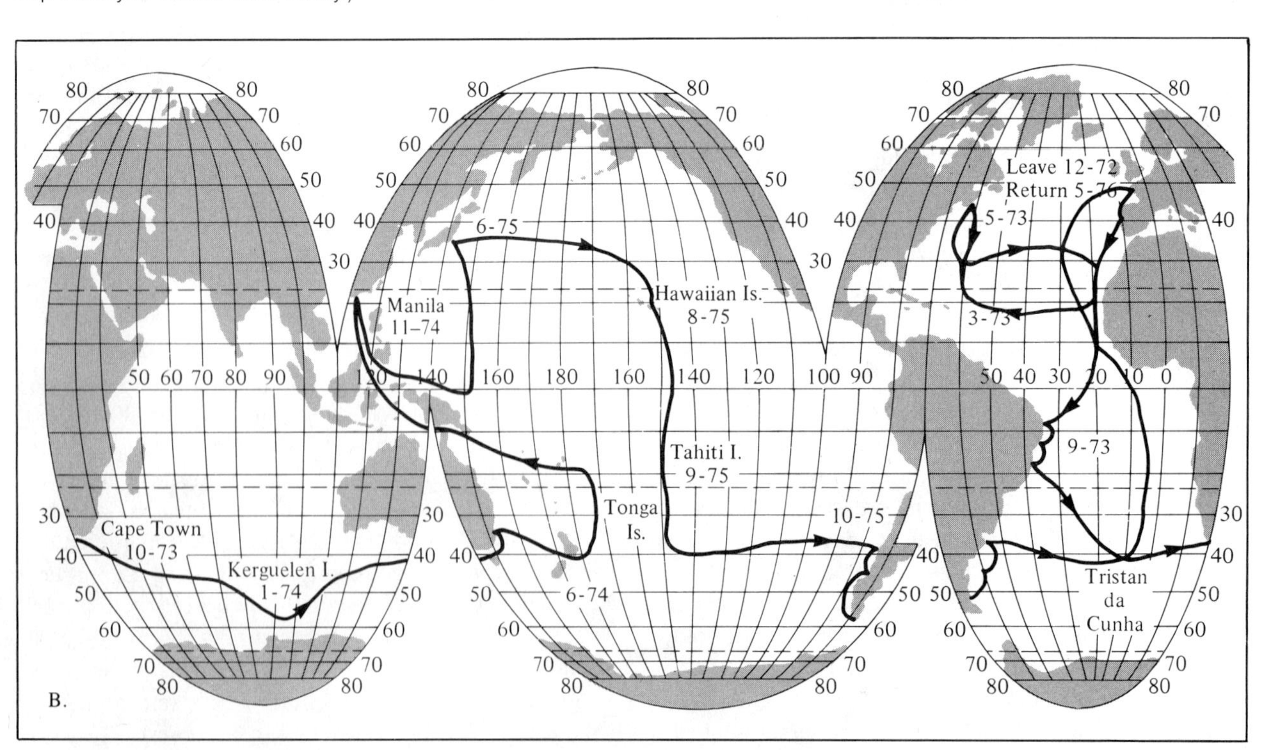

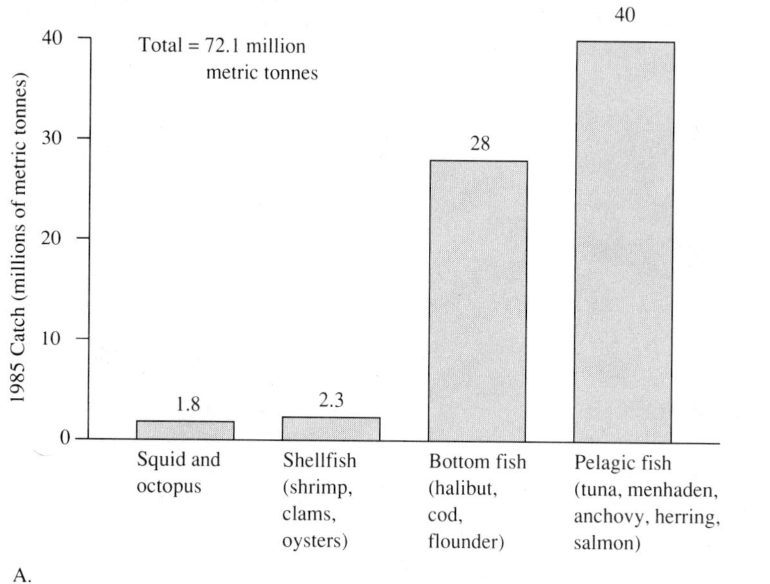

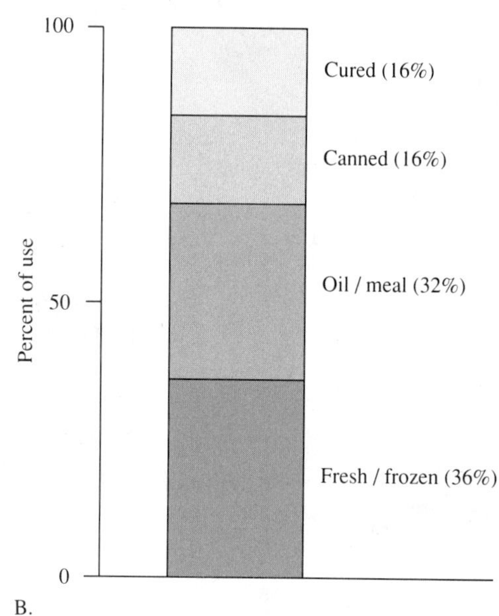

FIGURE 1.4
World Fishery Catch Statistics, 1985

(A) Catch by type and habitat. The pelagic zone is most important in total amount caught. **(B)** Utilization of the catch.

Most of this converted fish was used as food for other animals (e.g., chickens and pigs).

How important is food from the sea for human nutrition? Currently, the ocean is the source of only about 10 percent of the world's protein supply. If the fish used for feeding livestock were converted to fish protein concentrate (FPC) and eaten directly by humans, the protein made available through the marine fisheries would greatly increase. This argument sounds very convincing in light of the fact that diseases of protein deficiency account for most infant deaths and a significant proportion of adult debilitations and deaths in the developing nations of the world.

Can additional food be obtained from the oceans? During the 1966 Oceanographic Congress held in Moscow, some fishery biologists insisted that as much as 200 million metric tonnes (240 million tons) of food could be taken from the oceans each year. This prediction was overly optimistic. The annual take was then about 60 million tonnes and is presently just over 70 million tonnes (Fig. 1.4). Up until 1970 the rate of increase in the size of the world fishery averaged about 1.5 percent per year. Although there is evidence that the take may be peaking for the Atlantic and Pacific oceans, the Indian Ocean has the prospect of future increase. Present estimates of the limits of the world's conventional fishery range from 90 to 100 million tons. This estimate could possibly be increased significantly by directing our attention toward unconventional fisheries such as **krill** (small shrimplike crustaceans that are abundant in Antarctic waters) or pelagic squid.

The prospect of a large increase in the size of the world fishery appears unlikely. Some might argue that the best thing fishery biologists could do to meet the future needs of the world's population is to abandon their study of the oceans and turn their attention instead to the problem of how to slow the rate of human population growth.

If the natural open-ocean fisheries of the future are unable to keep pace with the growth in human population, mariculture may help close the gap. Mariculture is successfully practiced in the Far East, and has added to the world food supply (Fig. 1.5). The least costly projects involve algae, bivalve molluscs (clams and oysters), and some finfish. Many of these projects use food naturally available in the coastal ocean. Recent developments have added shrimp to the list of animals that can be raised artificially with commercial success. However, an operation that is a commercial success and an operation that increases the available world food supply are not necessarily the same thing. While the algae, bivalves, and finfish grow to maturity in and receive their nutrients from the natural environment, shrimps and lobsters (other animals with commercial potential) must be raised in artificial environments and artificially fed. They will be available only to those

FIGURE 1.5
Milkfish Ponds

Milkfish are harvested using a gill-net. Brackish-water fish culture techniques applied to ponds such as these in the Philippines are a successful example of marine mariculture. (FAO photo by P. Boonserm.)

who can pay very high prices, and their production will not contribute to the poorer people of the world.

Mariculture can surely help increase the world's food supply, but it is possible that growth in the mariculture industry will be matched by a decrease in natural ocean fisheries. Thus, there are realistic limits to the ability of the ocean to meet our future needs. To reduce the degree of uncertainty, we must greatly increase our knowledge of the ability of the oceans to provide a continued yield of fisheries harvest sustained over time.

SUMMARY

The history of marine biology is rooted in two areas. The great oceanographic expeditions established the basis of our knowledge of life in the open and deep oceans, while marine laboratories have promoted studies of life on the sea shore and the shallow oceans. Recent efforts to unify the results of these two initiatives have set the stage for a synthesis of the field of marine biology.

As a field of study, marine biology makes important contributions to the discipline of biology. There are a large number of unique life forms found only in the oceans, in part due to the profound difference between living in a three-dimensional water space and living on land. Marine biology also contributes to our basic understanding of biology, from refining evolutionary theory to expanding our knowledge of basic physiology and biochemistry.

The physical factors that shape living organisms in the marine environment differ from those of the land. The uniqueness of life in the oceans, coupled with the increasing importance of marine organisms as food and medical products, makes an introduction to marine biology course an important one in the undergraduate biology curriculum.

Increasingly, the medical field is looking toward marine life as a source of drugs for various diseases such as cancer, and antibiotics for infections. Even more important is the use of marine life as a source of food. An increase in the annual world harvest of just over 70 million metric tonnes is crucial to help feed the world's ever-increasing population. Most marine biologists, however, predict that annual harvests of around 100 million metric tonnes will be the most we can expect. Humans are just starting to develop the science of mariculture. Although its costs are high, we can expect significant yield increases in fishery products through these enterprises.

QUESTIONS AND EXERCISES

1. Name and describe the major accomplishments of the first major voyage undertaken exclusively for the purpose of investigating the nature of life in the oceans.
2. Describe some of the major medical benefits that have been derived from marine organisms.
3. Discuss the degree to which the protein requirements of humans are being met by marine fisheries.
4. Discuss the advantages and disadvantages of increasing the food taken from the ocean by (a) increased natural fisheries and (b) increased mariculture.
5. What is the difference between biological oceanography and marine biology?
6. Trace the history of the development of marine biology as a field of study.

REFERENCES

Bailey, H.S., Jr. 1953. The voyage of the *Challenger. Sci. Amer.* 188(5):88–94.

Banister, K. and Campbell, A., eds. 1985. *Encyclopedia of aquatic life.* New York: Facts on File.

Buzzati-Traverso, A.A. 1958. *Perspectives in marine biology.* Berkeley: University of California Press.

Cushing, D.H. 1988. *The provident sea.* New Rochelle, NY: Cambridge University Press.

Cromie, W.J. 1962. *Exploring the secrets of the sea.* Englewood Cliffs, NJ: Prentice-Hall.

Hedgpeth, J.W., ed. 1957. Treatise on marine ecology and paleoecology. Vol. I, ecology. *Memoir,* 67(1).

Mills, E.L. 1989. *Biological oceanography: An early history.* Ithaca, NY: Cornell University Press.

SUGGESTED READING

OCEANS

Lowenstein, J.M. 1989. Medicines from the sea. 22(1):72

OCEANUS

Harvesting the sea. 1979. 22(1):14–23.

Marine biomedicine. 1976. 19(2).

SEA FRONTIERS

McClintock, J. 1987. Remote sensing: Adding to our knowledge of oceans. 33(2):105–113.

Nicols, S. 1987. Krill: Food of the future. 33:12–17.

Root, M. 1988. Underwater pharmacy. 34(1):42–47.

2

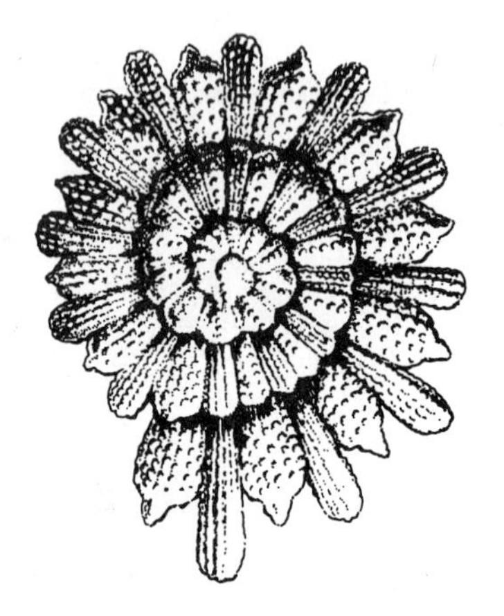

THE MARINE ENVIRONMENT

The forms of life inhabiting today's oceans are the products of millions of years of evolution. From the sunlit surface waters to the dark abyss and throughout the breadth of the oceans, various organisms have adapted to and prosper in widely different physical environments. Organisms living in the ocean are strongly influenced by salinity of ocean water, distribution of light and density stratification caused by surface heating of the ocean, and water motions. An understanding of these and other physical factors is essential for our consideration of marine ecology.

LANDFORMS OF THE OCEAN FLOOR

As investigations into the depth of the ocean have proceeded, it has become apparent that a broad shelflike feature that steepens and slopes off into deep basins has generally developed around the continents. Quite commonly, linear mountain ranges run through the deep basins. In all oceans, but especially around the margin of the Pacific Ocean, deep linear trenches up to 11 km (6.8 mi) in depth may separate the slopes from the deep-ocean basins.

Continental Margin

Extending beyond the shore around the continents is a thick wedge of sediment that in places is more than 15 km (9.3 mi) thick. The surface of this deposit can be divided into three provinces: the continental shelf, continental slope, and continental rise (Fig. 2.1).

The **continental shelf** is a gently sloping surface that extends from the shoreline to the **shelf break,** a point at which a significant increase in the rate of slope occurs. The shelf ranges in width from less than 1 km (0.62 mi) to more than 1,300 km (807 mi) off the north shore of Alaska and Siberia. The shelf break occurs at an average depth of 135 m (443 ft), but is as deep as 350 m (1,148 ft) off Antarctica, and is geologically contiguous with the continent. One can get an impression of the topography of the continental shelf by looking at the topographic features of the coastal region above sea level because virtually all of the continental shelf was exposed to subaerial erosion 18,000 years ago, when the last advance of the Pleistocene ice sheet was at its maximum. Because the continental mass is the basic source of nutrients for ocean plants, the most biologically productive waters in the ocean overlie the continental shelf.

Beyond the continental break lies the **continental slope,** where the average angle of inclination is about 4°. Around most of the Pacific Ocean, the slope extends down into deep ocean trenches to a depth as great as 11,022 m (6,840 ft) in the Marianas Trench. Where no trenches have developed, the continental slope is bounded on its seaward edge by the continental rise. One of the most impressive features of the continental slope is the **submarine canyon** (Fig. 2.2). Although some of these canyons align with the mouths of present rivers draining the continents, they are thought to be primarily the product of **turbidity currents**. It is believed these currents are initiated by disturbances such as earthquakes that cause unstable masses of sediment

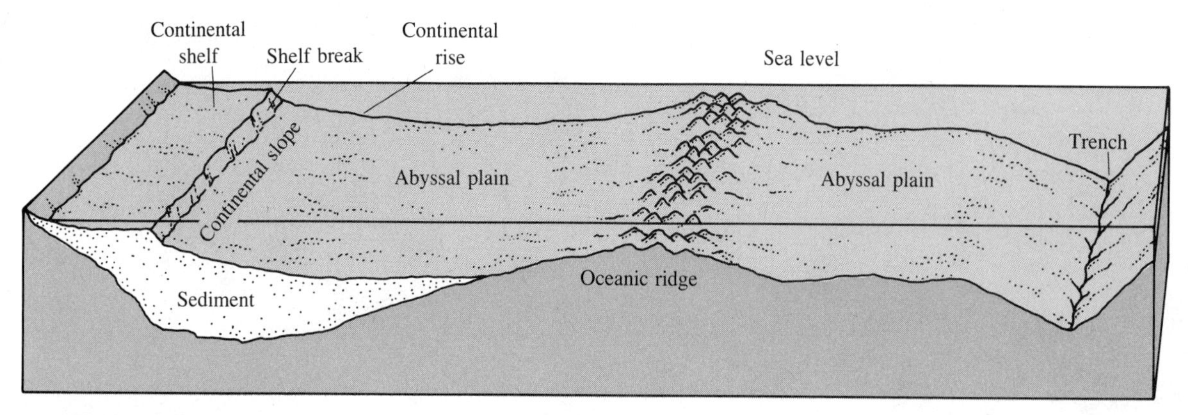

FIGURE 2.1
Physiographic Provinces of the Ocean Floor

A longitudinal profile showing submarine physiographic provinces encountered from the shoreline to the deep-ocean basin.

to break off the edge of the continental shelf, mix with ocean water, and flow as density currents down the continental slope, eroding it into canyons. This theory is supported by the fact that the remains of organisms known to live on the continental shelf are found in the deposits at the seaward end of the canyons.

The **continental rise** represents the seaward edge of the wedge of sediment deposited at the margin of the continents and slopes gently into the deep-ocean basins. Deep boundary currents that flow along the base of the continental slope distribute the sediment provided by the turbidity currents to form the rise.

Deep-Ocean Basin

Beyond the continental margin sediment wedge lies the deep-ocean basin, the main features of which are the abyssal plains and mid-ocean ridges (Fig. 2.3).

Abyssal plains are flat depositional features covering the deepest parts of the deep-ocean basins. The sediment for these deposits includes **lithogenous** particles (of rock origin) carried from the continents by ocean currents and winds; **biogenous** particles (of biological origin), primarily represented by the hard parts of microscopic plants and animals; **hydrogenous** particles (of water origin), which form deposits like manganese nodules that precipitate out of ocean water; and **cosmogenous** particles (of outer-space origin), such as tektites, which are fragments of meteorites. Extending above the abyssal plains may be volcanic peaks called seamounts if they extend more than 1 km above the ocean floor, or abyssal hills if smaller. Volcanic peaks that have been flattened by wave erosion before subsiding beneath the ocean surface are called tablemounts or guyots.

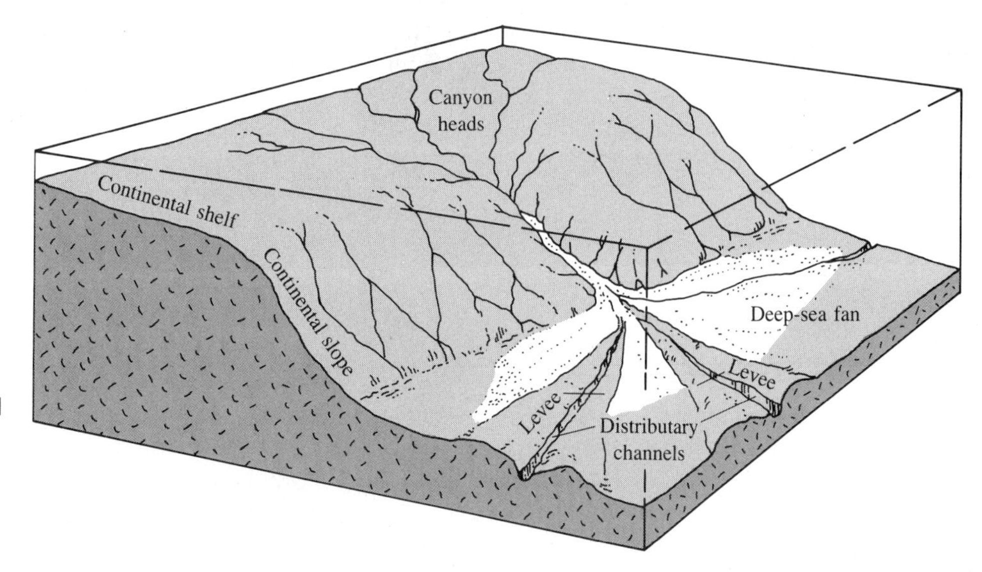

FIGURE 2.2
Submarine Canyons and Turbidity Currents

Erosional and depositional features associated with the formation of submarine canyons.

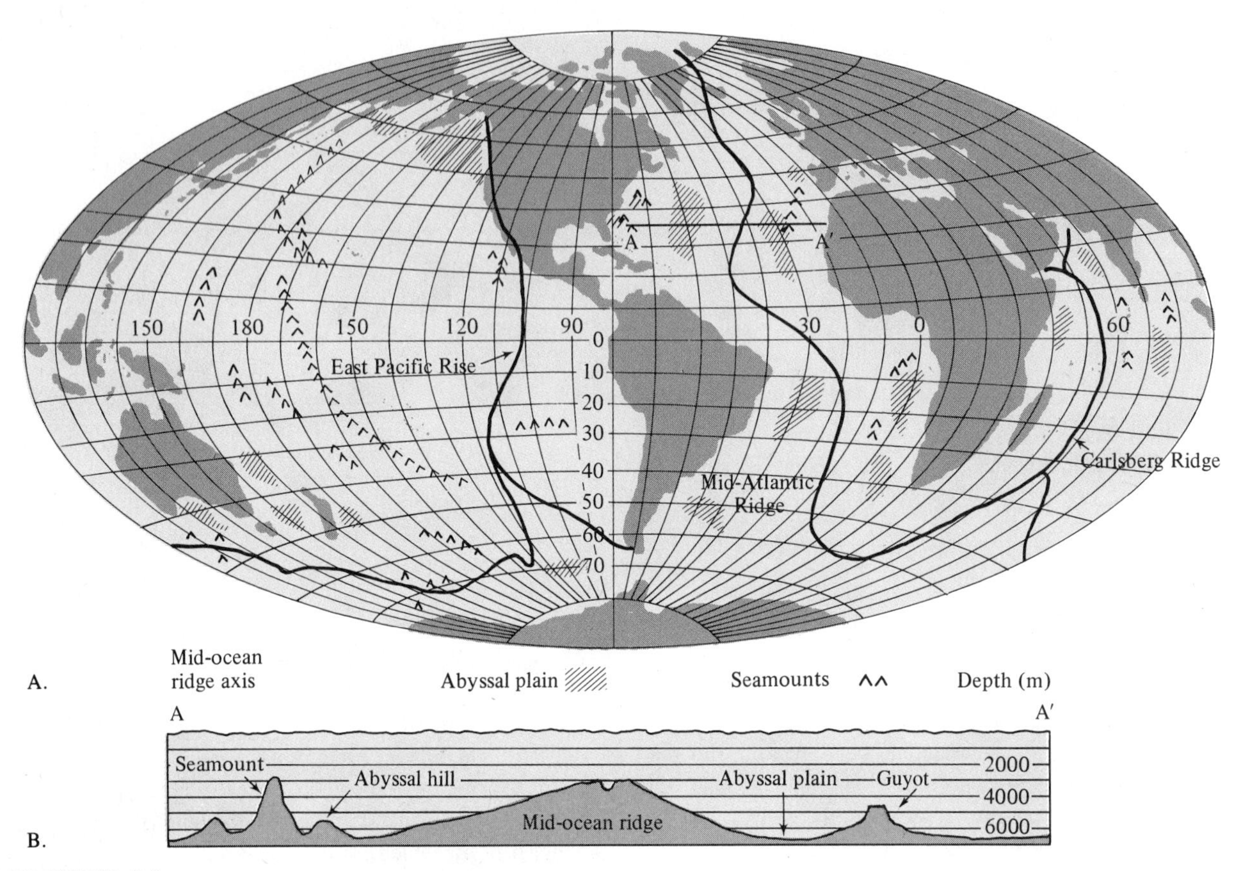

FIGURE 2.3
Mid-Ocean Ridges, Abyssal Plains, and Volcanic Peaks

(A) Some 65,000 km (40,400 mi) of oceanic mountain ranges extend across the floors of the world ocean. Flat-floored basins with thin sediment cover produce abyssal plains at the base of many mid-ocean mountain ranges. **(B)** Features of the ocean floor include volcanic peaks called *abyssal hills* (less than 1 km, or 0.62 mi, above the floor) and *seamounts* (more than 1 km). (Base map courtesy of National Ocean Survey.)

Extending across some 65,000 km (40,365 mi) of the deep-ocean basin are **mid-ocean ridges,** volcanic mountain ranges with rift valleys running along their axes. The mountains rise an average of 2.5 km (1.6 mi) above the floor of the deep-ocean basin, breaking the ocean surface occasionally to produce volcanic islands such as Iceland.

PROPERTIES OF OCEAN WATER

The Nature of Water

As the only substance that exists naturally in all three states—solid, liquid, and gas—on the earth's surface, water has some other remarkable properties that we must understand to fully appreciate the ocean environment. Water has the greatest heat capacity of any naturally occurring liquid on the earth; abnormally high melting and boiling points for a substance of its molecular mass; a high surface tension; and a great ability to dissolve other substances. Perhaps its most peculiar property is that of being less dense in its solid than in its liquid state.

All the properties listed above result from the unique shape of the water molecule. It is composed of one atom of oxygen (O) and two atoms of hydrogen (H). The hydrogen atoms are separated by an angle of 105°, as shown in Figure 2.4, resulting in the asymmetrical distribution of electrically charged particles within the molecule. The oxygen atom has a nucleus composed of eight positively charged (+) particles called protons, with an equal number of negatively charged (−) particles called electrons in a cloud surrounding the nucleus. Each hydrogen atom has one proton in its nucleus and one electron orbiting the nucleus. When the hydrogen atoms bond with the oxygen atom to produce

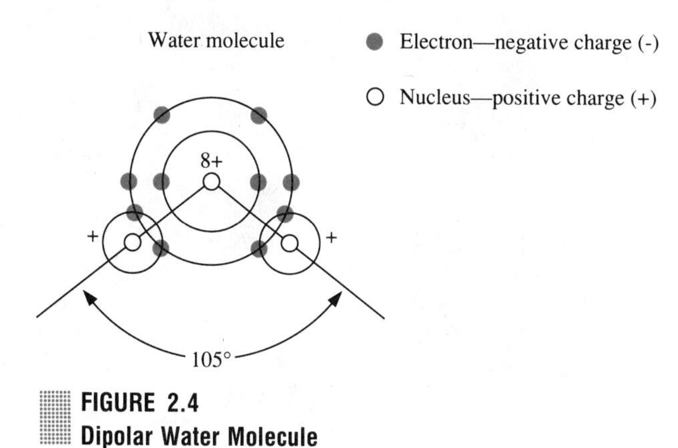

FIGURE 2.4
Dipolar Water Molecule

When two hydrogen atoms combine with an oxygen atom, they produce a water molecule, in which there is a 105° angle between the hydrogen atoms. The electrons continue to orbit their respective nuclei. Yet, since the oxygen nucleus has a larger positive charge than the hydrogen nuclei, the hydrogen electrons spend more time on the oxygen side, which causes (a) the oxygen end of the molecule to have a slight excess of negative charge, and (b) each of the hydrogen nucleii to have a slight excess of positive charge.

a water molecule, the electrons tend to stay close to the oxygen atom with its large positive charge. This makes the oxygen end of the molecule assume a slightly negative character, leaving the positively charged hydrogen nuclei somewhat exposed to give that region of the molecule a slightly positive character. The molecule is dipolar, with a slight positive charge at one end and a slight negative charge at the other.

It is this dipolar nature of the water molecule that gives water its ability to dissolve compounds composed of ions (electrically charged atoms or molecules) and dipolar molecules. It also produces the strong intermolecular bonds between water molecules called **hydrogen bonds.** It is the extra energy required to overcome the strength of the hydrogen bonds that accounts for the unusual thermal properties of water. These unusual properties are described in Table 2.1, along with their contribution to important biological and physical phenomena of the oceans.

Salinity

The properties of ocean water are determined to a great extent by the properties of pure water. It is the great capacity of water to act as a solvent that gives the ocean its 'saltiness'. Ocean water is a complex solution of salts in water. The solute, or dissolved substance, is composed primarily of ions of elements that have been dissolved from the rocks that make up the earth's crust or released into the oceans from the earth's mantle through volcanic activity.

Salinity, for the purpose of our discussion, is simply the total amount of dissolved solids in ocean water. By a more rigorous definition, however, it is the total weight of solid material dissolved in a kilogram of seawater when all the carbonate has been converted to oxide, all bromine and iodine are replaced by chlorine, and all organic matter is completely oxidized. Given this more precise definition, salinity may appear to be a very complex property, one very difficult to measure. The ocean, however, is well mixed and the relative abundance of major constituents is essentially constant. This is particularly true of the six major components that make up over 99 percent of the salinity of the oceans (Tab. 2.2).

Chemically, it is only necessary to measure the concentration of one of the major constitutents in order to determine the salinity of a given sample. The constituent that occurs in the greatest abundance and is the easiest to measure accurately is the chloride ion, Cl^-. The portion of the weight of a given sample of water that is the direct result of the presence of this ion is called **chlorinity** and is usually expressed in grams per kilogram of ocean water (g/kg) or parts per thousand (‰) (Fig. 2.5). The chloride ion accounts for 55.04 percent of the dissolved solids in any sample of ocean water. Therefore, by measuring its concentration, we can determine the total salinity in parts per thousand by the following formula:

$$\text{Salinity (‰)} = 1.80655 \times \text{Chlorinity (‰)}.$$

For example, given the average chlorinity of the ocean, 19.2 ‰, the ocean's average salinity can be calculated as

$$1.80655 \times 19.2\ \text{‰, or } 34.7\ \text{‰}.$$

The chemical determination of the chloride content in seawater can be made with a high degree of accuracy, but generally has required a great deal of time and care. Advancements in instrumentation in the field of oceanography have greatly simplified this task. Since the electrical conductivity of ocean water is known to increase with increased temperature and salinity, salinity is now most commonly determined by measuring the electrical conductivity of ocean water with a modern conductance-measuring instrument, the salinometer. Salinity can be determined to a precision of 0.003 ‰ by this method.

Because the components of salinity have a density greater than the density of water, water density increases with increased salinity. Marine scientists are greatly concerned with the relationships between water

TABLE 2.1
Properties of Water

Property	Physical and Biological Significance
Physical States. Water is the only substance that occurs as a gas, liquid, and solid within the temperature range found at the earth's surface.	As a gaseous (vapor) component of the atmosphere, water helps transfer heat from warm low latitudes to cold high latitudes. Liquid water also contributes to this process through ocean currents. Additionally, liquid water runs across the continents, dissolving minerals from the rocks and carrying them to the oceans. It is this form of water that accounts for over 85% of the mass of most marine organisms and serves as the medium in which the chemical reactions that support life occur. The freezing of water into a solid (ice) during the winter season at the higher latitudes increases surface salinity and makes possible the sinking of dense surface water. This water is the only source of oxygen for the deep ocean.
Solvent Property. Water can dissolve more substances than any other common liquid.	Ocean water carries dissolved within it the nutrients required by marine plants and the oxygen needed by animals. It is this property that has produced the 'saltiness' of the oceans.
Heat Capacity. The quantity of heat required to change the temperature of 1 g of a substance 1°C. Water has the highest heat capacity of all common liquids. The heat capacity of water is used as the unit of heat quantity, *calorie.*	Heat capacity is a major factor in making water the most important moderator of climate. It accounts for the narrow range of temperature change found at any location in the ocean. Water can gain or lose a relatively great amount of heat while undergoing a much smaller temperature change than would occur in other substances.
Latent Heat of Melting. The quantity of heat gained or lost per gram by a substance changing from a solid to a liquid or from a liquid to a solid without a change in temperature. For water, it is 80 cal—the highest of all common substances.	When ice forms, most of the heat energy lost is released to the heat-deficient atmosphere. When ice melts, the energy gained by the water is manifested as molecular energy of the liquid water. This prevents the high-latitude ocean from becoming much warmer or colder than the −1.8°C (28.7°F) freezing temperature of ocean water.
Latent Heat of Vaporization. The quantity of heat gained or lost per gram by a substance changing from a liquid to a gas or from a gas to a liquid without a change in temperature. It is higher for water, 540 cal at 100°C (212°F), than for any common substance. It is 585 cal at 20°C (68°F), at which much of the evaporation from the ocean surface occurs.	A great amount of excess heat energy is removed from the low-latitude ocean by evaporation and released through precipitation into the atmosphere at heat-deficient higher latitudes. It is this property that contributes greatly to the fact that the polar regions do not get increasingly colder and the equatorial region does not get increasingly hotter.
Surface Tension. Highest of all common liquids. Cohesive attraction of hydrogen bonds causes a molecule thick 'skin' to form on a water surface and makes capillarity possible.	Some organisms such as Halobates use this 'skin' as a walking surface. Others such as Glaucus hang from its undersurface.
Density. Mass per Unit Volume (g/cm^3). The density of ocean water is increased by decreasing the temperature and increasing salinity and pressure. For pure water, the temperature of maximum density is 4°C (39.2°F), but ocean water with a salinity greater than 24.7 ‰ will get denser as the temperature is lowered to the freezing temperature.	Plankton that stay near the surface through buoyancy and frictional resistance to sinking are greatly influenced by the effect of temperature on density. In low-density warm water, plankton must be smaller or more ornate to obtain the increased ratio of surface area to body mass necessary to remain afloat.

density, salinity, and temperature since density distribution is an important determinant of ocean circulation.

The average salinity of the world ocean is 34.7 ‰. Salinity values in the deep ocean remain very near this figure, while surface salinities range from less than 15‰ in the Baltic Sea to 40 ‰ in the Red Sea. In the major ocean basins, surface salinity is generally lowest in the high latitudes reaching values as low as 32 ‰ in the North Pacific. The high rate of evaporation in the subtropics raises surface salinities to 37.2 ‰ in the Atlantic Ocean, while increased precipitation along the equator lowers salinities to less than 35 ‰. In polar regions, freezing and thawing of sea ice, respectively, increase and decrease surface salinity on a seasonal basis.

Solar Radiation

Distribution of Light. The radiant energy from the sun that reaches the earth's surface is called **insolation.** Radiation emitted by the sun includes many types, as indicated in Figure 2.6, which shows the **electromag-**

TABLE 2.2
Ocean Salinity

Major Constituents (over 100 parts per million)		Minor Constituents (1–100 parts per million)		Trace Elements (less than 1 part per million)	
Ion	*Percentage*	*Element*	*P.P.M.*		
Chloride (Cl^-)	55.04	Bromine	65.0	Nitrogen	Iodine
Sodium (Na^+)	30.61	Carbon	28.0	Lithium	Iron
Sulfate (SO_4^{-2})	7.68	Strontium	8.0	Rubidium	Zinc
Magnesium (Mg^{+2})	3.69	Boron	4.6	Phosphorus	Molybdenum
Calcium (Ca^{+2})	1.16	Silicon	3.0		
Potassium (K^+)	1.10	Fluorine	1.0		
	99.28				

FIGURE 2.5
Composition of One Kilogram of Ocean Water

In a 1-kg (1,000-g) sample of ocean water, a salinity of 34.7 ‰ (parts per thousand) is accounted for by pure water and components of salinity.

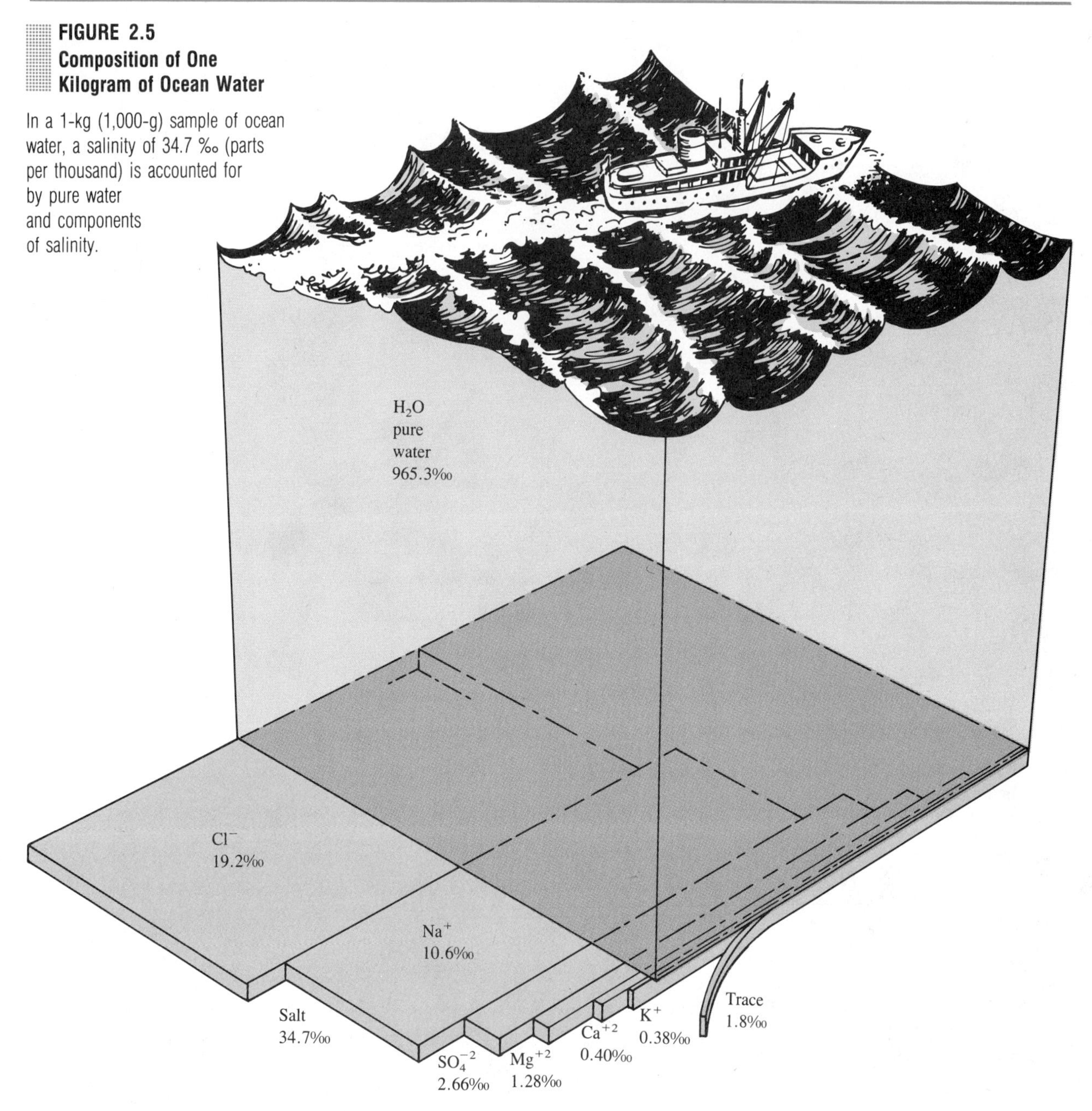

netic spectrum. Marine biologists are concerned only with the **visible spectrum** and the infrared and ultraviolet regions on either side of it. We experience the visible spectrum as light, the longer-wavelength **infrared radiation** as heat, and the shorter-wavelength **ultraviolet radiation** as the penetrating rays that cause sunburn.

Only infrared and visible light penetrate beneath the ocean surface. The infrared is absorbed by a depth of 1 m, and blue light may penetrate to 1,000 m (3,280 ft), although less than 0.1 percent of the incoming visible radiation remains below a depth of 100 m (328 ft). Throughout most of the ocean this is the maximum depth at which photosynthesis can occur.

In the clearest open tropical oceans where there is little organic matter in the water, the ocean takes on a deep blue color. The reason for this is that the small, molecular-size particles characteristic of these waters scatter the blue wavelengths of light more than any other wavelength. In nutrient-rich coastal waters, sediment from continental runoff and high concentrations of microscopic marine organisms increase the average particle size so that the scattering of green wavelengths of light becomes greater and the ocean appears green in color. The green color of chlorophyll-containing microscopic plants also contributes to this coloration. (See the special feature, "Color in the Marine Environment", in the insert at the beginning of this text.) In some coastal waters, dense populations of microscopic plants cause the water to become brownish or red in color, producing what is called a **red tide.**

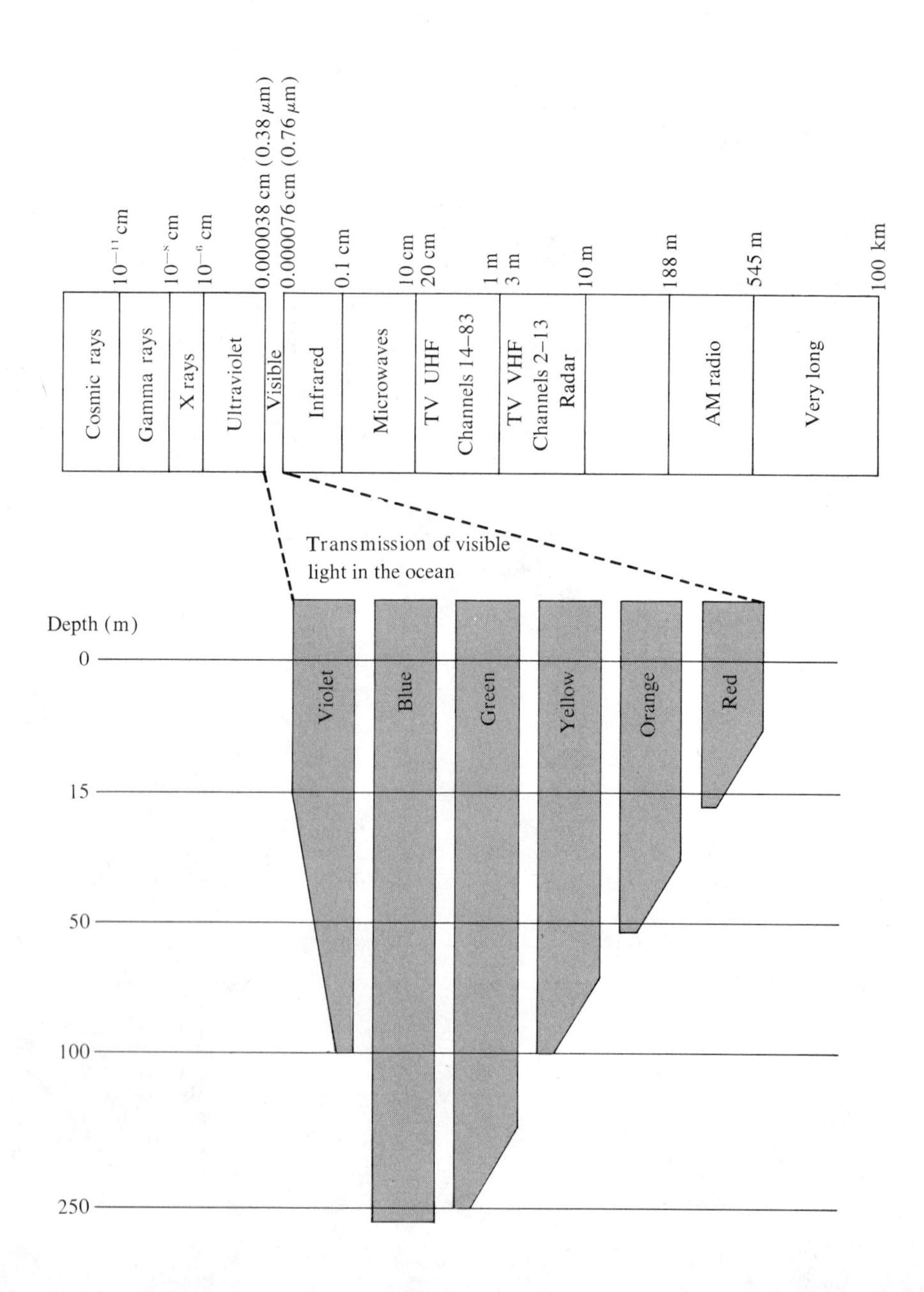

FIGURE 2.6
The Electromagnetic Spectrum and Transmission of Visible Light in Water

The wavelength of spectrum components increases from left to right. The shorter wavelengths of the spectrum's visible portion—especially blue—penetrate deeper into the ocean than do the longer wavelengths. (See further discussion of this phenomenon in the color insert at the beginning of this book.)

Density Stratification. Only about 0.1 percent of the insolation is used in plant photosynthesis. About 47 percent is absorbed by the continents and oceans after passing through the atmosphere, primarily as visible light. Only 19 percent is absorbed directly by the atmosphere. The energy absorbed by the continents and oceans is transferred as heat to the atmosphere by evaporation, conduction, and radiation to generate the earth's wind systems. Some of the energy represented by the winds is then returned through frictional stress to the ocean as the **kinetic energy** of the currents and waves. These phenomena are summarized in Figure 2.7. Currents and waves are discussed in Chapters 10 and 12, respectively.

The surface heating of the ocean affects the distribution of temperature, salinity, and density within the ocean. In the deep-ocean water below about 1,000 m, conditions are relatively uniform. Salinity varies little from 34.7 ‰, temperature is about 3°C (37.4°F), and density, which depends to a great extent on temperature and salinity distribution, is uniform. At the surface, temperatures range from about −1.8°C (−28.8°F) in the high latitudes to values above 25°C (77°F) at the equator. Surface salinity may be as low as 32.5 ‰ in the polar oceans, rising to over 37 ‰ in the subtropics, and falling to about 34.5 ‰ in equatorial regions (Fig. 2.8).

These surface values may be found to a depth of as much as 200 m (656 ft) in the **mixed layer** of the ocean, where winds mix the surface water through wave action. This mixed layer possesses a lower density than the ocean **deep water** in the low latitudes where surface temperatures are highest. Beginning at a depth of about 200 m, the high surface salinity of the subtropics and the high temperatures of the low latitudes decrease rapidly with increasing depth, while density undergoes a rapid increase. At a depth of about 1,000 m (3,280 ft), salinity, temperature, and density values have usually assumed values characteristic of the deep water. This layer of water in which rapid changes in salinity, temperature, and density occur is, gravitationally, the most stable layer in the oceans. It is impossible for the high-density deep water to mix across it with the low-density surface water. The names applied to these gradient zones where changes in salinity, temperature, and density occur rapidly are (respectively) the **halocline** (*halo*-salinity, *cline*-slope), **thermocline** (*thermo*-heat), and **pycnocline** (*pycno*-density) (Fig. 2.9). Because the salinities and temperatures found in polar surface waters are more nearly those of the deep water, no well-developed halocline, thermocline, or pycnocline is found at these latitudes, and deep water can readily mix with surface water. Surface water may sink to the bottom or deep water may rise to the surface under the natural conditions that prevail. The degree to which the pycnocline develops has a great effect on the distribution of biological productivity within the oceans, because a well-developed pycnocline prevents nutrient-rich deep water from carrying these nutrients to the surface where plants are depleting the available supply. The pycnocline is permanent in the low latitudes, while at mid-latitudes it is well developed in summer and poorly developed in winter.

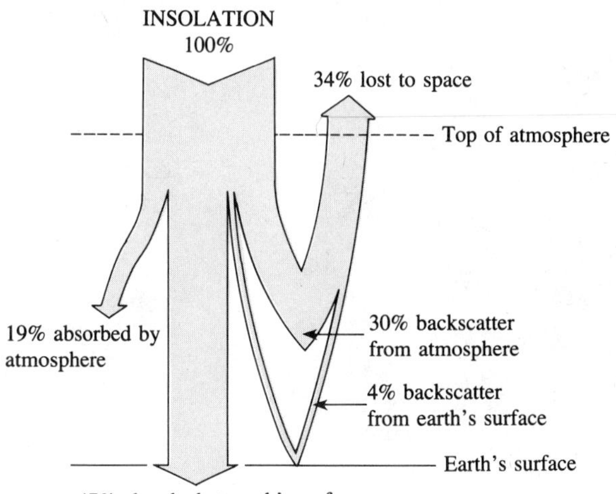

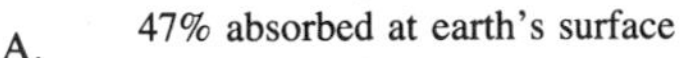

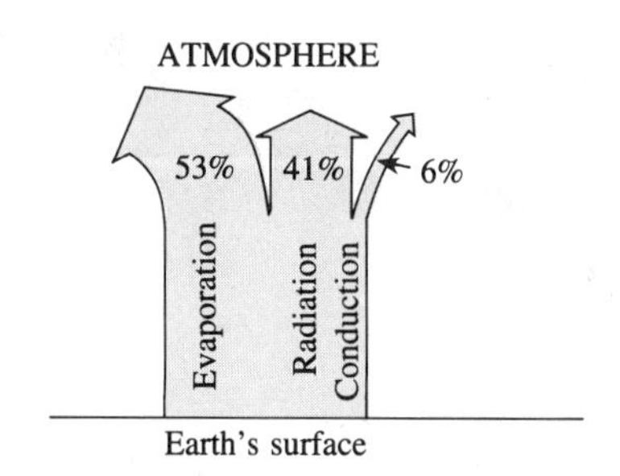

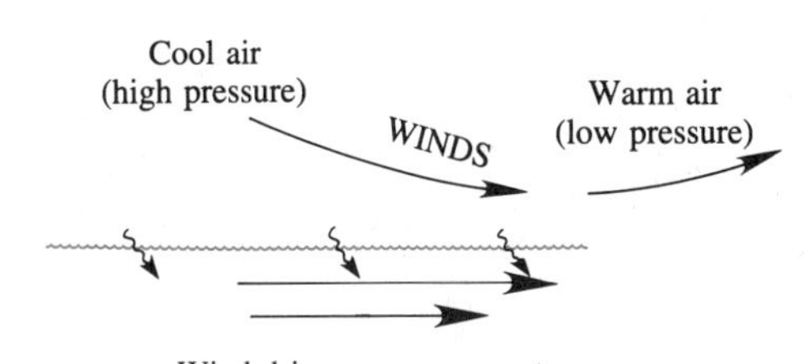

FIGURE 2.7
Effects of Surface Heating of the Ocean

(A) Absorption and backscatter of radiant energy reaching the earth's surface. **(B)** Heat absorbed by oceans and continents is returned to the atmosphere through evaporation, radiation, and conduction. **(C)** Conversion of heat energy to kinetic energy in the form of ocean waves and currents.

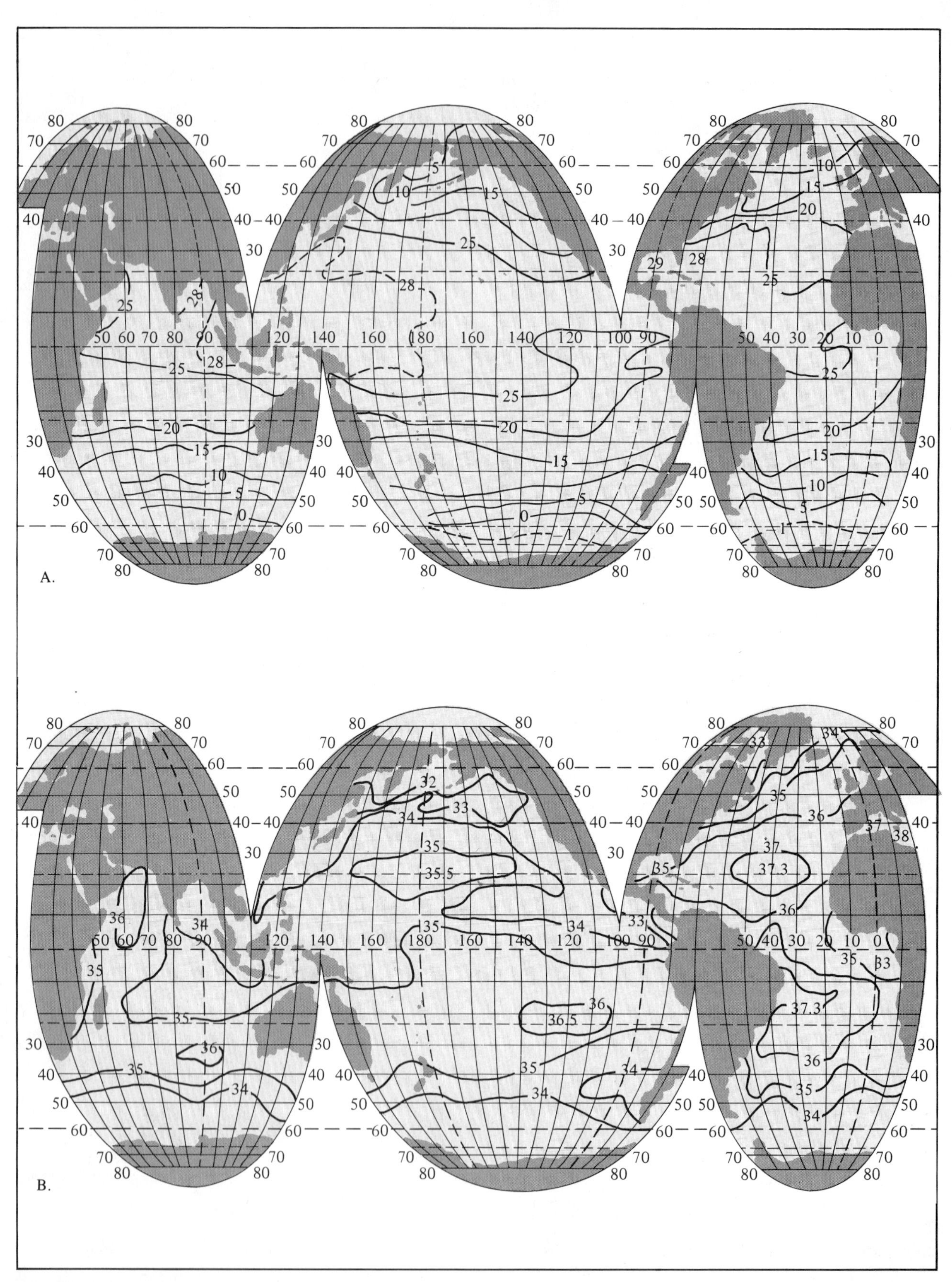

FIGURE 2.8
Surface Temperature (A), Salinity (B) of the Oceans

Temperatures are given in °C and salinity in ‰ for the month of August. (Base maps courtesy of National Ocean Survey.)

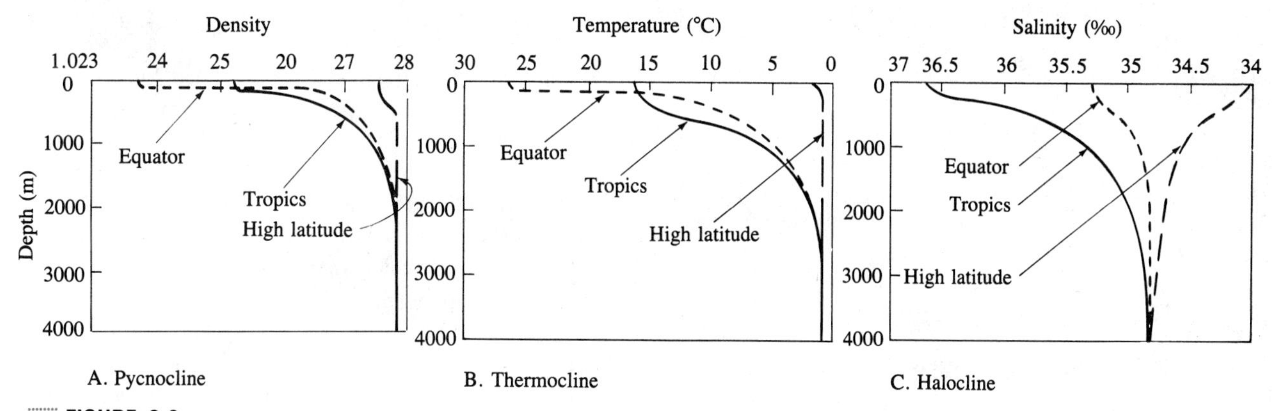

FIGURE 2.9
Vertical Profiles for Density, Temperature, and Salinity

(A) In low-latitude equatorial and tropical regions, a thin layer of low-density surface water is separated from high-density deep water by a zone of rapid density change, the pycnocline. The pycnocline is poorly developed or missing in high-latitude waters. **(B)** The gravitationally stable pycnocline is primarily the result of and coincides with a zone of rapid vertical decrease in temperature, the thermocline. The thermocline is also more highly developed in low latitudes. **(C)** A zone of rapid vertical change in salinity, the halocline, may also affect density. The halocline usually represents a decrease in salinity with increasing depth in low latitudes but the opposite condition may prevail in high-latitude areas. (After G.L. Pickard, 1963, *Descriptive physical oceanography*. Pergamon Press Ltd., Oxford, England. © 1963. Reprinted by permission of Pergamon Press Ltd.)

Sound Transmission

The thermocline has a significant effect on the pattern of sound transmission in the ocean. At a temperature of 20°C, sound travels at an average velocity of 1,450 m/s (3,250 mi/h) in ocean water, compared to 334 m/s (750 mi/h) in the atmosphere. The velocity of sound transmission in the oceans increases with increased salinity, temperature, and pressure. Salinity changes are less important than changes in temperature and pressure in changing the velocity of sound transmission, so only the effects of these latter factors are represented in Figure 2.10. Pressure increases steadily from the surface to the ocean bottom and by itself should cause the velocity of sound transmission to increase with increased depth. Throughout much of the ocean, however, there is a well-developed thermocline where temperature decreases rapidly with increasing depth. The rate of temperature decrease in the thermocline more than offsets

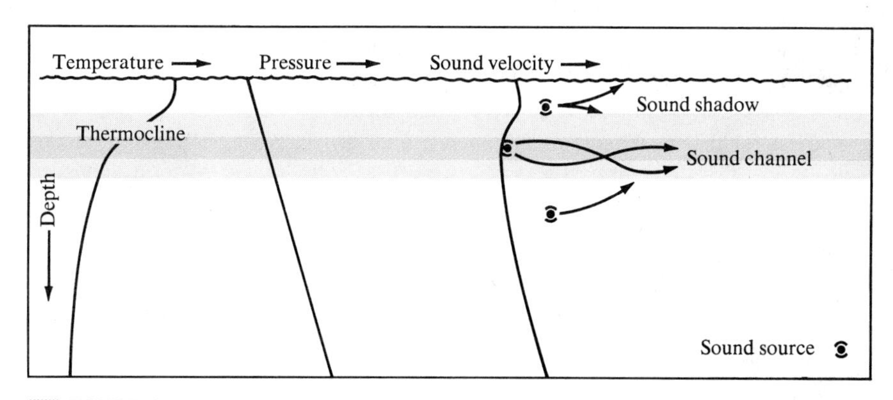

FIGURE 2.10
Transmission of Sound in the Ocean

Velocity of sound in the ocean increases with increases in the temperature and pressure (and to some extent salinity). Although pressure increases steadily with depth, which increases the velocity of sound, the rapid decrease in temperature represented by the thermocline offsets the pressure increase and produces a low-velocity sound channel. Sound is trapped here by the refraction, or bending, of sound waves.

the pressure increase from the top to the bottom of the thermocline, and a zone of low-velocity sound transmission exists at the base of the thermocline. Due to the refraction (bending) of sound waves, sound is trapped in this low-velocity zone, called a **SOFAR** (sound fixing and ranging) **channel,** and can be transmitted great distances within it. Marine animals may use this phenomenon to a much greater extent than we presently realize as a means of long-distance communication.

GENERAL CONDITIONS

All of the major divisions of the animal kingdom have members that live in the marine environment. In fact, all are thought to have originated there, and five phyla are exclusively marine. Throughout most of the open ocean, only microscopic, floating, photosynthetic cells (phytoplankton) are found, while the more complex and highly developed plants are restricted to its margins and the continents.

The ocean environment is far more stable than the terrestrial environment, and the organisms that live in the oceans generally have not developed highly specialized regulatory systems to cope with sudden changes that might occur within their environment. They are, therefore, sensitive, in different degrees, to small changes in temperature, turbidity, salinity, and so on.

Water constitutes 80 percent of the mass of **protoplasm,** the substance of life. Over 65 percent of the mass of a human being and 95 percent of that of the jellyfish is accounted for by water (Tab. 2.3). Water carries dissolved within it the gases and minerals needed by organisms and is itself one of the raw materials required for the synthesis of organic molecules. Land plants and animals have developed very complex systems and devices to distribute water throughout their bodies and prevent it from being lost. The threat of desiccation (drying out) through atmospheric exposure does not exist for the inhabitants of the open ocean, as they are abundantly supplied with water.

TABLE 2.3
Water Content of Organisms (Percent of Mass)

Organism	Water Content (%)
Human	65
Herring	67
Lobster	79
Jellyfish	95

Effects of Salinity

Marine animals are significantly affected by relatively small changes in their environmental conditions. One that is highly important is change of salinity. Some oysters that live in coastal environment are capable of withstanding a considerable range of salinity. During floods, the salinity is extremely low. On a daily basis, as the tides force ocean water into the river mouths and draw it out again, the salinity changes dramatically. The oysters, and to some extent other organisms that inhabit the coastal regions, have a tolerance for a wide range of salinity conditions; they are **euryhaline.** Other marine organisms, particularly those that inhabit the open ocean, can withstand only very small changes in salinity; these organisms are **stenohaline.**

The relative proportions of the major dissolved constituents of ocean water are very nearly the same as the relative proportions of these salts in the extracellular body fluids of all animals, particularly marine invertebrates. Even animals that have adapted themselves to life on the continents, both vertebrates and invertebrates, have these similar ratios. Although the relative amounts of dissolved materials in the body fluids of animals and in ocean water are impressively similar, the salinities of the body fluid of an animal and the water in which it lives may not be the same. Both marine and fresh-water animals have special mechanisms to insure that the concentration of salts in their body fluids are maintained at the proper level. To understand these mechanisms, we need to look at the processes of diffusion and osmosis, which allow dissolved substances and water to be transported into and out of the animal's cells.

Cell membranes are **semipermeable;** that is, they allow the passage of molecules of some substances while screening out others. Cells may take in the nutrient substances they need from the surrounding fluid medium, where they exist in high concentration by diffusion. **Diffusion** is the movement of molecules from areas of high concentration to areas of low concentration (Fig. 2.11). The nutrient compounds, to which the cell membrane is permeable, pass from the surrounding medium where they are more concentrated into the cell where they are less concentrated. The waste materials that a cell must dispose of after it has released the energy held in organic molecules through respiration are passed out of the cell by the same method. As the concentration of waste materials becomes greater inside the cell than in the external medium, they pass out of the cell into the surrounding fluid. The waste products then will be carried away by the circulating fluid that services the cells in higher animals, or by the surrounding water medium that bathes the simple one-celled organisms of the oceans. In many cases, the process of transport of nutrients and waste materials through the cell membrane by diffusion is accompanied or replaced by more complex systems of active transport of molecules through the cell membrane.

Just as molecules of solute are traveling into and out of the cell, molecules of the solvent (water) are doing likewise through a diffusionlike process called **osmosis.** When aqueous solutions of unequal salinity are separated by a water-permeable membrane that is impermeable to solute, water molecules can move through the membrane into the more concentrated solution (Fig. 2.12). Put another way, there is a net transfer of water molecules from the side of greater concentration of water molecules to the side of lesser concentration of water molecules. The two solutions can be compared in terms of their osmotic pressures. The **osmotic pressure** of a solution is the force that must be applied to keep pure water that is separated from it by a water-permeable membrane from passing through the membrane and diluting the solution. The higher the salinity of the solution, the higher its osmotic pressure will be. We thus can compare the osmotic pressure of solutions by comparing their concentrations of dissolved solids, or salinity. This can be done readily by determining their freezing point, since the addition of dissolved solids to an aqueous solution lowers its freezing point (Fig. 2.13). This effect is due to the interference of the dissolved particles with the formation of ice crystals. The lower the freezing point, the higher the osmotic pressure of a solution.

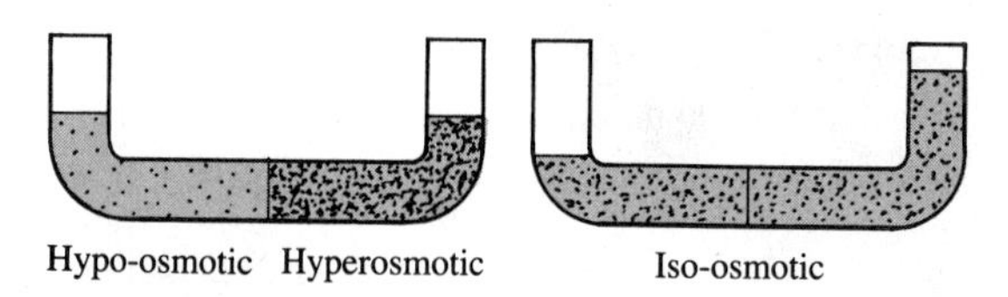

FIGURE 2.12
Osmosis

Separating two water solutions of different salinities by a membrane that allows the passage of water molecules but not molecules of dissolved substances sets the stage for osmosis to occur. Water molecules achieve a net motion by diffusion through the membrane into the more concentrated (hyperosmotic) solution from the less-concentrated (hypo-osmotic) solution. Osmosis ends when the salinity of the two solutions is the same. They are then iso-osmotic.

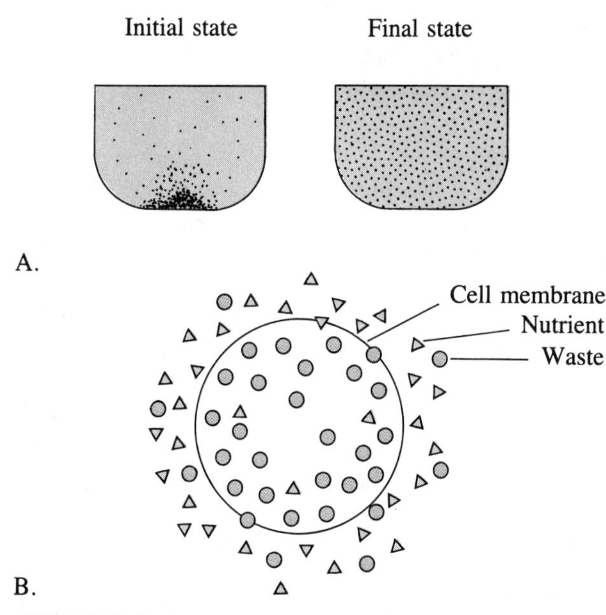

FIGURE 2.11
Diffusion

(A) If a substance soluble in water is placed in a pile on the bottom of a beaker of water, it eventually will become evenly distributed through random molecular motion. Diffusion occurs when the molecules of a substance move from areas of high concentration in the substance to areas of low concentration. **(B)** Nutrients high in concentration outside the cell, diffuse into the cell through the cell membrane. Waste is in higher concentration inside the cell and diffuses out of the cell through the cell membrane.

If the salinity of the body fluid and the external medium are equal, they are said to be **iso-osmotic** and have equal osmotic pressure. No net transfer of water will occur through the membrane. If the external fluid medium has a lower osmotic pressure than the body fluid within the cells of an organism, water will pass through the cell walls into the cells. Such an organism is **hyperosmotic** relative to the external medium. If the osmotic pressure within the cells of an organism are less than that of the external medium, water from the cells will pass through the cell membranes into the external medium. This organism is described as **hypo-osmotic** relative to the external medium.

The movement of water molecules through the semipermeable membrane in either direction (osmosis), the movement of nutrient molecules from the external medium into the cell, and the movement of molecules of waste material from the cell—all these processes are going on at the same time. During these processes, molecules of all the substances in the system are passing through the membrane in both directions, but there is a *net* transport of molecules of a given substance from the side on which they are more highly concentrated to the side where they are less concentrated.

In the case of many marine invertebrates, the body fluids and the external medium in which they are im-

FIGURE 2.13
Effect of Salinity on Osmotic Pressure and Freezing

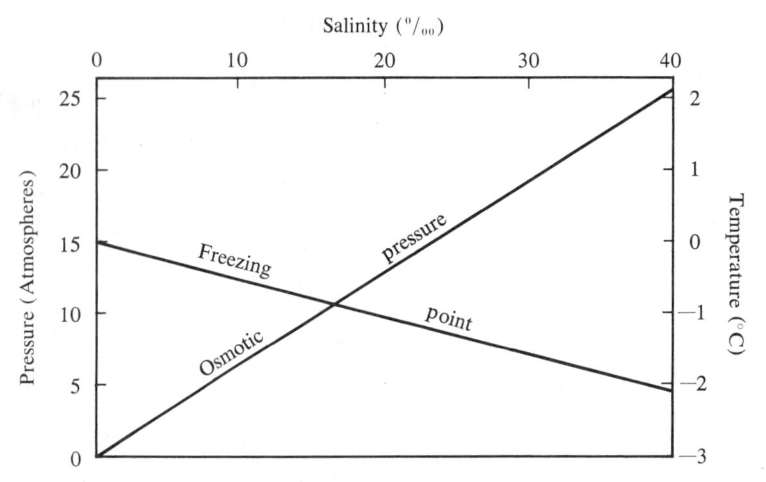

mersed are in a nearly iso-osmotic state. They do not have to develop special mechanisms to maintain their body fluids at a proper concentration. Thus, they have an advantage over their fresh-water relatives who live in a condition where their body fluids are hyperosmotic in relation to the low-salinity medium in which they live (Tab. 2.4).

Many marine fish have body fluids slightly more than one-third as saline as ocean water. This may indicate that they evolved in low-salinity coastal waters. They are, therefore, hypo-osmotic in relation to the surrounding medium. Without developing some means of regulation, the saltwater fish would be continually losing water from its body fluids to the surrounding ocean environment. This outward flow is counteracted, in part, by the ability of marine fish to drink ocean water and secrete the salts through 'chloride cells' located in the gills. Another way in which these fish retain water within their bodies, and thereby maintain a low osmotic pressure, is by discharging a very small amount of highly concentrated urine (Fig. 2.14).

Fresh-water fish are hyperosmotic in relation to the very dilute external medium in which they live. The osmotic pressure of their body fluids may be 20–30 times greater than that of the fresh water that surrounds them. Their problem is to avoid taking large quantities of water into their body cells that would eventually rupture the cell walls. In order to solve this problem, the fresh-water fish does not drink water. Also, its cells have the capacity to absorb salt and the water that is gained because its body fluids are greatly hyperosmotic is disposed of by well-developed kidneys and tubules that allow the excretion of a large volume of very dilute urine (Fig. 2.14).

The mechanisms we have been discussing explain why humans cannot drink salt water when stranded at sea. The osmotic pressure of our body fluids is about one-fourth that of ocean water. Drinking ocean water would cause a drastic loss of water from the cells of the body through the process of osmosis, causing eventual dehydration.

Availability of Nutrients

The distribution of life throughout the ocean depends primarily on the availability of nutrients required for photosynthesis. In regions where there is a large supply of nutrients, marine populations reach their greatest density. Where are these areas in which we find the greatest density of biomass? To answer this question, we must consider the sources of nutrients.

Running water that erodes the continents transports the eroded particles and dissolved inorganic compounds to the ocean basins. The dissolved nitrogen and phosphorus compounds serve as the basic nutrient supply for marine photosynthesis. Because the continents are the major sources of these inorganic nutrients, the greatest concentrations of marine life are found at their margins. The oceans contain progressively lower concentrations of marine life toward the open sea.

To make use of the available nutrients, marine **phototrophs** (organisms that carry on photosynthesis) must have sunlight as a source of energy. The ocean water represents a significant barrier to the penetration of solar radiation. In very clear water, solar energy may be detected to depth of 1 km (0.62 mi), but sufficient light intensity to support photosynthetic organisms is restricted to depths of less than 100 m (328 ft) throughout most of the world ocean. Near the coast, photosynthesis is restricted to a thinner zone because of higher concentrations of suspended sediment from river runoff and microscopic plankton due to higher rates of biological

TABLE 2.4
Freezing Points of Body Fluids in Marine and Fresh-Water Organisms (Percent Water in Parentheses)

A comparison of the freezing points of the body fluids of some marine and fresh-water organisms and the waters in which they live. The lower the freezing point of the fluid, the higher the salinity and osmotic pressure and salinity of the fluid. Marine invertebrates and sharks are essentially iso-osmotic, while most marine vertebrates are hypo-osmotic. Essentially all fresh-water forms are hyperosmotic.

Marine Animals		
	Freezing Points (°C)	
Invertebrates	Body Fluid (Percent Water)	Water (Percent Water)
Annelid worm	−1.70 (96.6)	−1.70 (96.6)
Mussel	−2.75 (94.5)	−2.12 (95.7)
Octopus	−2.15 (95.7)	−2.13 (95.7)
Lobster	−1.80 (96.4)	−1.80 (96.4)
Crab	−1.87 (96.3)	−1.87 (96.3)
Vertebrates		
Cod	−0.74 (98.5)	−1.90 (96.2)
Shark	−1.89 (97.2)	−1.90 (96.2)
Turtle	−0.50 (99.0)	−1.90 (96.2)
Seal	−0.50 (99.0)	−1.90 (96.2)
Fresh-Water Animals		
Invertebrates		
Mussel	−0.15 (99.7)	−0.02 (99.9+)
Water flea	−0.20 (99.6)	−0.02 (99.9+)
Crayfish	−0.80 (98.4)	−0.02 (99.9+)
Vertebrates		
Eel	−0.62 (98.8)	−0.02 (99.9+)
Carp	−0.50 (99.0)	−0.02 (99.9+)
Water snake	−0.50 (99.0)	−0.02 (99.9+)
Manatee	−0.50 (99.0)	−0.02 (99.9+)

productivity. The deepest confirmed instance of phototrophs sustaining themselves in the ocean is that of an alga observed at a depth of 268 m (878 ft) on San Salvador Seamount in the Bahamas.

Over the eras of geologic time, new life forms have evolved to fit into all biological niches. Many of these forms have adapted to seemingly deleterious conditions. Still, as long as the basic requirements for the production of the food supply are met, life can exist under a great range of physical conditions.

Along the continental margins, life is more abundant in some areas than others. The areas containing the greatest concentrations of biomass are characterized by low water temperatures. At low temperatures, the water holds in solution greater amounts of the important gases—oxygen and carbon dioxide—than would be found in warmer waters. A process that contributes to higher levels of production along some coasts is **coastal upwelling,** which provides a flow of deeper, cold and nutrient-rich waters to the surface. Upwelling is discussed further in Chapter 10.

Areas of high and low organic production in the ocean can be determined visually by observing the color of ocean water. Coastal waters are usually greenish in color because they contain more large particulate matter that disperses solar radiation in such a way that the wavelengths most scattered are those of green and yellow. This color is also partly due to the presence of yellow-green pigments of the abundant **phytoplankton,** phototrophs that live suspended in ocean water.

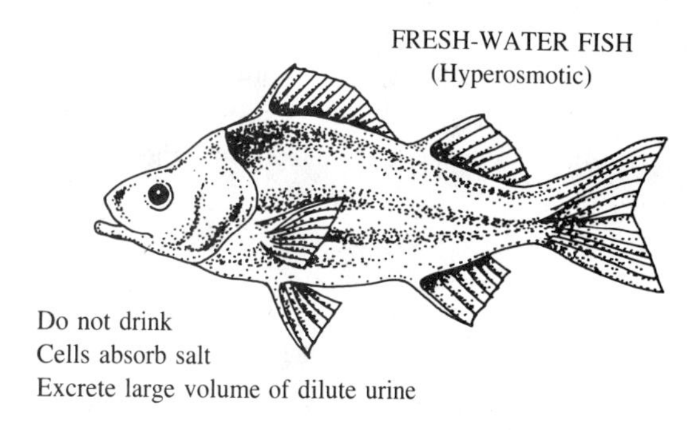

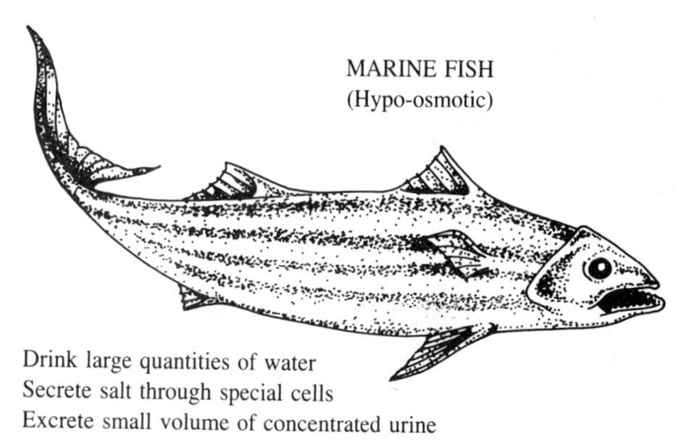

FIGURE 2.14
Fresh-Water and Marine Fish

Size and Support

Assuming the availability of nutrients, the major requirements of phytoplankton are (a) that they maintain themselves within the upper layer of ocean water, where solar radiation is available for photosynthesis, and (b) that they take in nutrients from the surrounding waters and expel waste materials. Their ability to meet these requirements is influenced by their size and shape. The ease with which phytoplankton maintain their position in the upper layers of the ocean is explained by the relationship between surface area and body mass. The greater the surface area per unit of body mass, the easier it is for an organism to maintain its position because there is more surface per unit of body mass in frictional contact with the surrounding water. Greater surface area per unit of body mass means greater frictional resistance to sinking.

Figure 2.15 shows how surface area per unit of mass increases with decreasing size. Cube A has ten times as much surface area per unit of volume as Cube B. If both cubes are composed of the same substance, their masses are proportional to their volumes, since

$$\text{Mass} = \text{Volume} \times \text{Density}.$$

If Cubes A and B were plankton, Cube A would have ten times as much frictional resistance to sinking per unit of mass as Cube B. It could stay afloat by exerting far less energy than Cube B. The relationship of size to sinking rate is expressed in **Stokes Law,** which states that the larger the particle (up to 60 μm), the faster it will sink.

For phytoplankton that depend significantly on diffusion for uptake of nutrients and disposal of wastes through the cell membrane, the amount of surface area per unit of mass is also important. The greater the surface area per unit of mass, the more efficiently these materials can be diffused through the membrane. Thus we find that individual cells in all organisms are microscopic, regardless of the size of the organism. Small size enhances the efficiency of diffusion and other active transport mechanisms that move materials across the cell membrane.

In addition to their small size, many microscopic marine organisms have other features that help them stay in the upper waters of the ocean. Diatoms, for example, commonly have long needlelike extensions from their siliceous tests (protective coverings). These hollow extensions serve to further increase the ratio of surface area to the mass of the organism, thereby creating more frictional resistance to sinking. Another way in which small organisms increase their ability to stay in the upper layers of the ocean is by producing a drop of oil stored in the organism. This decreases their overall density and increases their buoyancy.

Viscosity

We are familiar with the fact that liquids flow with varying degrees of ease, and we know that syrup flows less readily than water. This condition is indicative of the higher viscosity of syrup as compared to water. **Viscosity,** defined as internal resistance to flow, is a characteristic of all fluids.

FIGURE 2.15 Relationship of Surface Area to Body Mass

If both cubes are composed of the same substance, their masses are proportional to their volumes, since mass = volume × density. Cube *A* has ten times as much surface area per unit of volume as does Cube *B*. If Cubes *A* and *B* were plankton, Cube *A* would have ten times as much frictional resistance to sinking per unit of mass as Cube *B*. It could stay afloat by exerting far less energy than Cube *B*. If Cube *A* were a planktonic alga, it could also take in nutrients and dispose of waste through the cell wall ten times as efficiently as an alga with the dimensions of Cube *B*.

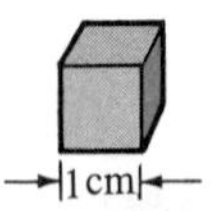

The viscosity of ocean water is affected by two variables—temperature and salinity. Viscosity of ocean water increases with an increase in salinity but is affected even more by temperature, a lowering of which increases viscosity. As a result of this temperature relationship, we find that floating plants and animals in colder waters have less need for extensions to aid them in floating than do those that occupy warmer waters. It has been observed that members of the same species of floating crustaceans will be very ornate, with featherlike appendages, where they occupy warmer waters; these appendages are missing in the colder, more viscous environments. It appears then that high viscosity would benefit the floating members of the marine biocommunity because it would make it easier for them to maintain their positions in the surface layers.

With increasing size of organisms and a change in the mode of life, viscosity ceases to enhance the ability of an organism to pursue life and becomes an obstacle instead. This is particularly true of the large organisms that swim freely in the open ocean. They must pursue prey and displace water to move ahead. The more rapidly it swims, the greater the stress that is created on the organism. Not only must water be displaced ahead of the animal, but water must move in behind it to occupy the space that it has vacated. The latter is one of the more important considerations in streamlining.

The familiar fusiform shape of the free-swimming fish and the marine mammals manifests the type of adaptations that must be achieved in streamlining an organism to move with a minimum of effort through the water. We see that a common shape achieved to meet this need is a laterally flattened body presenting a small cross section at the anterior end with a gradually tapering posterior. This form, which is characteristic of the bony fishes, reduces stress that results from movement through the water and makes it possible for this movement to be achieved with a minimum energy expenditure (Fig. 2.16).

Temperature

Of all the conditions we can measure in the ocean, there are probably no physical characteristics we have discussed that are more important to life in the oceans than temperature. It has been stated that the marine environment is much more stable than that of the land. A comparison of temperature ranges that exist in the sea to those that are found on land will serve to exemplify this condition.

The temperature range that exists in the sea is much narrower than that found on land. The minimum temperature observed in the sea is never much less than −2°C (28.4°F), and the maximum seldom exceeds 28°C (82.4°F). This compares to the continental temperature range from a low of −88°C (−127°F) to a high of about 58°C (136.4°F). The temperature found on the continents varies considerably on a daily as well as seasonal basis. Daily variations at the ocean surface rarely exceed .2° or .3°C (.36°F, .54°F), although they may be as great as 2° or 3°C (3.6°F, 5.4°F) in shallower coastal waters. Seasonal temperature variations are also small. They range from 2°C at the equator to 8°C at 35° to 45° latitude and decrease again in the higher latitudes. Seasonal variations of temperature in shallower coastal areas may be as high as 15°C.

The temperature variations characteristic of surface waters are reduced in magnitude at greater depths. In the deep oceans daily or seasonal variations in temperature are of little or no importance. Throughout the deeper parts of the ocean the temperature remains uniformly low. At the bottom of the ocean basin, at depths in excess of 1.5 km, temperatures hover around the 2° to 3°C mark regardless of latitude.

A decrease in temperature increases density, viscosity, and the capacity of water to hold gases in solution. All these changes have significant effects upon the organisms inhabiting the ocean. Where the capacity of water to contain dissolved gases is great and where so-

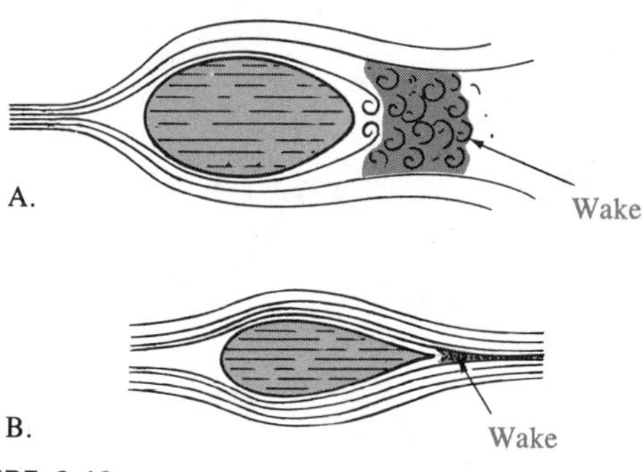

FIGURE 2.16
Streamlining

Due to the viscosity of water, any body moving rapidly through it must be designed so that it produces as little stress as possible as it displaces the water through which it moves. After the water has moved past the body, it must fill in behind the body with as little eddy action as possible. The reduction of wake is important because the wake represents a vacuum into which everything around it tends to be pulled. This pull is also exerted on the animal that created the wake, and increases the energy expenditure required to maintain a given rate of forward motion. The broad structure of body shape *A* creates a large resistance area in front, and a large wake region behind. It is poorly streamlined. The teardrop shape of body *B* is well rounded in front, and tapers gradually to the rear, producing only a small wake region. It is well streamlined.

lar energy is available, vast plant communities can develop, as they do in the high latitudes during the summer seasons. In these areas, life is abundant because of the great quantities of carbon dioxide needed by plants for photosynthesis and oxygen needed by the animals that feed upon the plants.

Floating organisms in general are larger individually in colder waters than in warmer waters, although the tropical populations appear to be characterized by a larger number of species. However, the total biomass of floating organisms in the planktonic environments of colder high latitudes greatly exceeds that of the warmer tropics. The fact that organisms in the tropics are smaller than those observed in the higher latitudes could quite possibly be related to the lower viscosity of lower-latitude waters. Because they are smaller and because they have ornate plumage, these tropical species can expose more surface area per unit of body mass (Fig. 2.17). Such plumose adaptations are strikingly absent in the larger cold-water species.

An increase in temperature increases the rate of biological activities which, according to Van Hoff's Law, doubles with each increase in temperature of 10°C (18°F). Tropical organisms apparently grow faster, have a shorter life expectancy, and reproduce earlier and more frequently than their counterparts in colder waters.

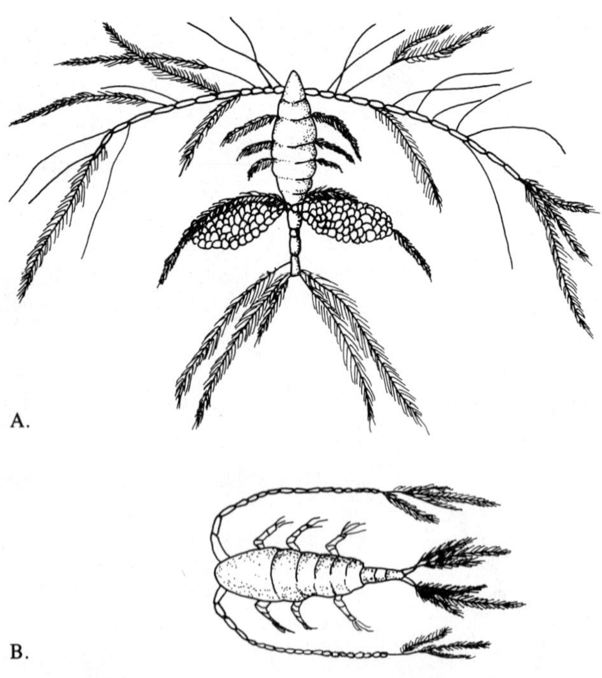

FIGURE 2.17
Plumage in Warm- and Cold-Water Species

(A) The copepod *Oithona* displays the ornate plumage characteristic of warm-water varieties. **(B)** The copepod *Calanus* displays the less ornate appendages found on temperate- and cold-water forms.

Some species of animals can live successfully only in cooler waters, and others can do so only in warmer waters. Within both groups, many species can withstand only a very small change in temperature and are called **stenothermal.** Other varieties are apparently little affected by temperature change and can withstand changes over a large range; they are classified as **eurythermal.** Stenothermal organisms are found predominantly in the open ocean and at the greater depths where dramatic changes in temperature are extremely unlikely. The reef-building corals are an exception, however. They prefer a temperature within the range of 20°–21°C (68°–70°F), but are found only in shallow coastal waters. Eurythermal organisms are more common in the shallow coastal waters in temperate latitudes, where the largest ranges of temperature occur, and in the surface waters of the open ocean.

DIVISIONS OF THE MARINE ENVIRONMENT

The marine environment can be divided into two basic parts: the ocean water itself, or the **pelagic environment;** and the ocean bottom, or the **benthic environment** (Fig. 2.18). The pelagic environment is further divided into two provinces: The **neritic province,** which extends from the shore seaward, includes all water overlying an ocean bottom less than 200 m (656 ft) in depth. Beyond this 200 m depth lies the **oceanic province.**

Pelagic Environment

Because the oceanic province includes waters ranging greatly in depth, from the ocean surface to the bottom of the deepest ocean trenches, it is further subdivided into four biozones on the basis of light penetration and water depth: The **epipelagic zone** extends from the surface to a depth of 200 m; the **mesopelagic zone,** from 200–1,000 m (656–3,280 ft); and the **bathypelagic zone,** from 1,000–4,000 m (3,280–13,120 ft). The deepest parts of the ocean below 4,000 m make up the **abyssopelagic zone.**

An important factor in determining the distribution of life in the oceanic province and boundaries of biozones within it is the availability of light. The **euphotic zone** extends from the surface to a depth where there is enough light to support photosynthesis, rarely more than 100 m (328 ft). Below this zone there are small but measurable quantities of light within the **dysphotic zone** to a depth of about 1,000 m. Below this depth, in the **aphotic zone,** there is no light at all.

Color in the Marine Environment

As we pursue our study of the marine environment, the role of light in determining the nature and distribution of life becomes clear. While other physical factors that influence marine life exist—water motions, salinity levels, and geology, to name a few—lightwaves and their scattering cause a phenomenon widely explored and appreciated by scientists and nonscientists alike: the intriguing range of colors found in the marine environment. By studying the nature of marine color, we will understand better the interaction of the physical and biological factors in the oceans.

To most of us, the dominant color we associate with the ocean is the blue of its surface waters. The majority of this surface color is due to the reflection of the blue

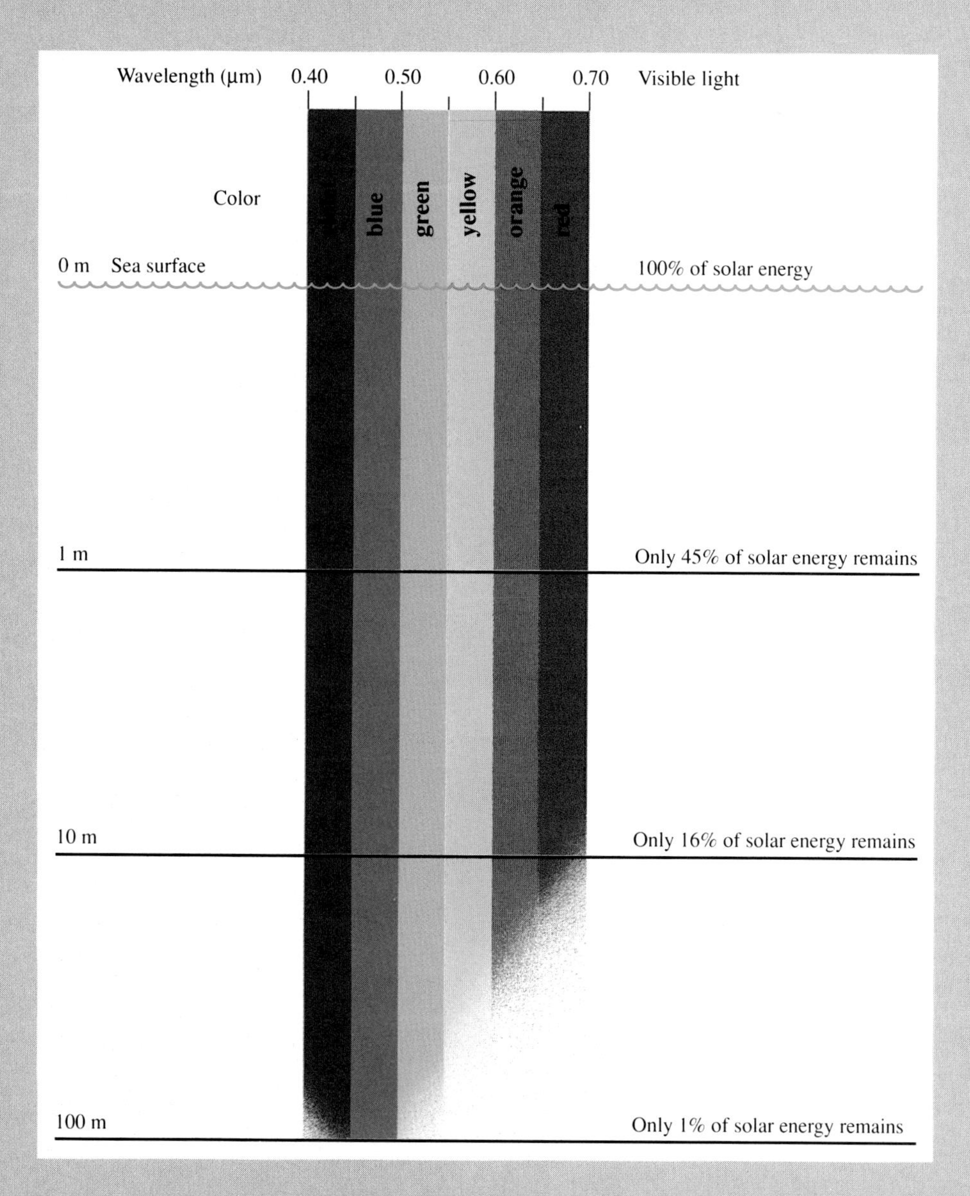

Transmission of visible light through seawater. Note that the various wavelengths of light are not absorbed at a uniform rate—reds are absorbed most rapidly with depth. The 100 m depth is used as an example here. In turbid coastal waters, light may be extinguished by 30 m; in clear oceanic areas, by as much as 250 m.

The open ocean is mostly deep blue. The intensity of blue in this Caribbean coral reef area is dependent on depth. The shallower waters over the reef are a lighter blue, whereas the deeper water in the center is a more intense blue.

Coastal water, if free of sediments and low in biological productivity, can be as blue as the open ocean. This photograph was taken in the Pacific Ocean off Baja Mexico.

color of the sky, as is evident when the blue color is obscured by a passing cloud. Other colors are readily seen, too—the vibrant reds of corals, the brilliant oranges of tropical fish. Yet, as you descend the depths, the predominant color of the water and the objects in it is blue. Colors such as red and orange disappear quickly—typically by 15 m (approximately 50 ft). This is because the various wavelengths of natural light are absorbed at different rates underwater, as the illustration shows. The red wavelengths are absorbed first, whereas the blue wavelengths penetrate deeper.

The surface color of the water is affected by suspended particles as well. In coastal areas where sediments from land sources accumulate, water color tends toward gray and gray-blue. Too, where waters are biologically productive, the color of the water is affected by the living organisms. For example, high levels of

Inland marine waters generally have high biological productivity. This location, in Washington state, has high concentrations of diatoms that give a greenish/yellowish tinge to the water.

In Alaska, some glaciers terminate directly in the ocean. The soil material carried by the ice turns the surface waters gray.

In estuaries, where river water mixes with the ocean, the water color is influenced by suspended particles. In this estuary, the river water has a high silt and clay concentration, so the resulting estuarine water is gray in color.

phytoplankton give the water a greenish or yellowish color. In areas of intense phytoplankton blooms, the color of the organism will dominate the water. Red tides—caused by blooms of dinoflagellates—are a frequent occurrence in coastal waters of the Atlantic and Pacific Oceans.

The color of marine organisms is varied. Those fortunate enough to visit the seashore at low tide will likely see the browns, greens, and reds of the prominent macroalgae. These colors are related to photosynthesis: The color of green algae is dominanted by chlorophyll; the brown and red algae also have chlorophyll, but their colors are dominated by red and brown pigments that are involved with photosynthesis in an accessory way.

Animal life in the surface waters is characterized by a lack of color. Although larger animals have pigments, the more-numerous plankton are mostly transparent. An exception is those plankton that live right at the air-water interface. These organisms are characterized by an intense blue color that results from a pigment that functions to absorb harmful ultraviolet radiation.

The red color seen here is caused by large concentrations of a reddish dinoflagellate. Some red tide blooms are due to a dinoflagellate that has a toxin that causes paralytic shellfish poisoning. Not all red tides, however, have the toxin.

Organisms that live in the deep sea—far below any light from the surface—are also brightly colored. Deep-sea fish typically evidence intense shades of red, yellow, or black.

Some animals are capable of changing their color. This is a protective measure that allows fish and invertebrates to mimic the color of their surroundings. By changing the color of special cells in their skin, these organisms can change both their color and pattern.

Marine biologists use color as a tool to communicate their findings. They use satellites in space to provide information on the physical and biological features in the ocean. Currents are mapped, temperatures taken, and the extent of biological productivity measured. The results of these satellite studies are summarized in maps, using different color shades and values to express the particular data collected.

The red soft coral on this coral reef is at a depth of 10 m (33 ft). At this depth there are still blue wavelengths but no red ones. The photograph on the left was taken with natural light, which is blue and so shows the soft coral as gray. When artificial light is used—the image shown below—the brilliant red of the soft coral appears.

This algae, found on a West Pacific coral reef, is brilliantly green because of chlorophyll. Generally, algae on coral reefs are inconspicuous because they are eaten by herbivores.

This is one of the few species of true plants that can tolerate total submergence in seawater. The eelgrass **Zostera marina** *has circumpolar distribution and is common in the shallow protected waters of the Northern Hemisphere.*

The giant kelp forests of California are formed mainly by the brown algae Macrocystis pyrifera. *Although the giant kelp contains chlorophyll, the green color is masked by a brown accessory photosynthetic pigment.*

In some brown algae the accessory pigment gives a black color. This algae is the sea palm Pterogophora *from the west coast of Washington state.*

Some accessory pigments for photosynthesis give a red color to the algae. Red algae do best in shaded conditions, either under a canopy of brown algae or at deeper depths than brown algae.

Zooplankton that live near the surface generally are lacking color. The most common animal in the sea is a copepod of the genus Calanus. *These copepods are about 1 mm in length.*

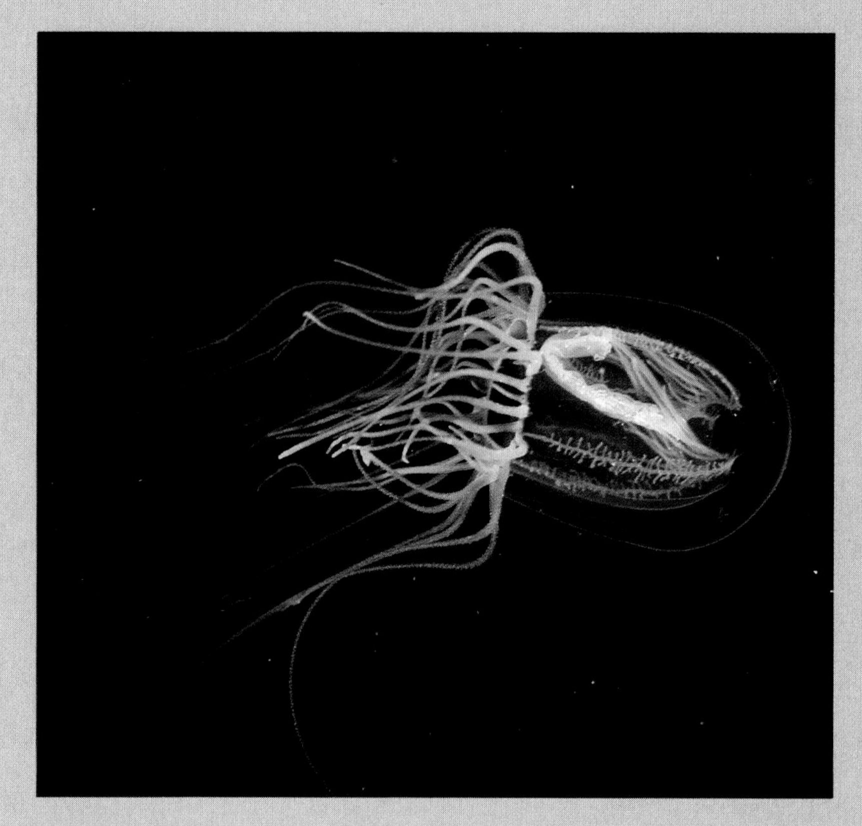

Without the artificial light of a flash, this medusae (Polyorchis) *from the Northeastern Pacific would look transparent.*

The leptocephalus larvae of the European eel live in the plankton for over a year. They are transparent and are about 10 cm (4 in) long.

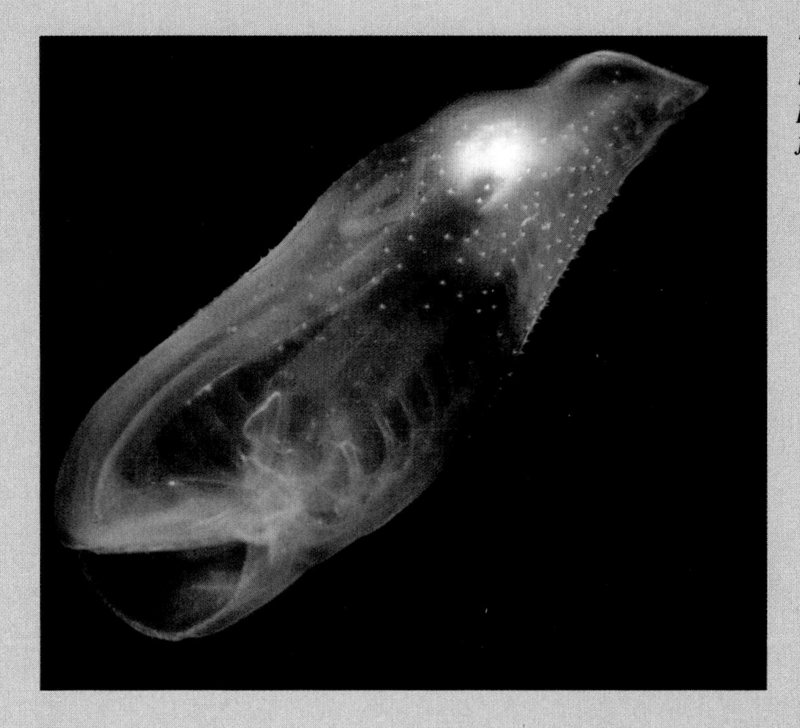

This pelagic tunicate (a salp) is transparent except for the phytoplankton in its gut that it filtered out of the sea water.

Plankton that are deeper than surface waters are still mostly devoid of color. This glass medusae (Aglantha) *is highlighted with red.*

Tunicate salps may grow in colonies. This one is long enough to encircle the diver.

This medusae is larger than the diver. Despite its size, there still is virtually no color. Its only pigment is from the batteries of nematocysts on the tentacles.

Not all plankton that live near the surface are transparent. Those that live at the interface between air and water are characterized by a deep blue or purple color. This johnny-by-the-wind (Velella) *has a band of purple that protects it from ultraviolet radiation.*

Animals that live in the deep sea below the depth of any natural light still have bright colors. This anglerfish (Melanocetus) *is bright red in artificial light.*

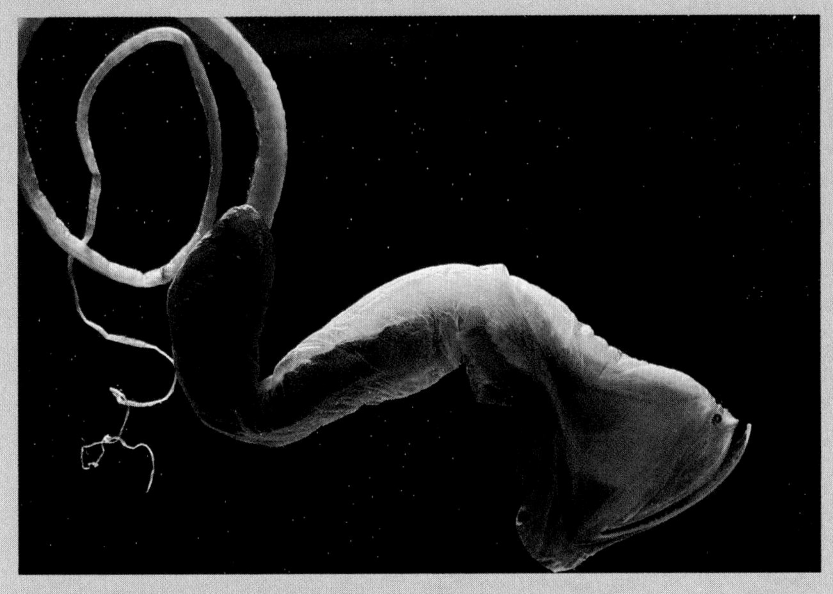

The deep-sea swallower (Saccopharynx) *is not heavily colored. The dark area is from a fish in its stomach. The swallower gets its name for being able to swallow prey almost its own size.*

The coelocanth is a rare fish found in the deep sea off the west coast of Africa. It is immensely important to biologists because it has fleshy fin lobes rather than the regular fin rays of most fish. These fleshy lobes are thought to indicate a close evolutionary affinity with land animals. Its basic color is black. The white spots are from scales lost during preservation.

Some fish can change colors to match their backgrounds. The flounder in the top photograph has a color pattern that matches the shell debris background. The flounder below has matched the sandy background.

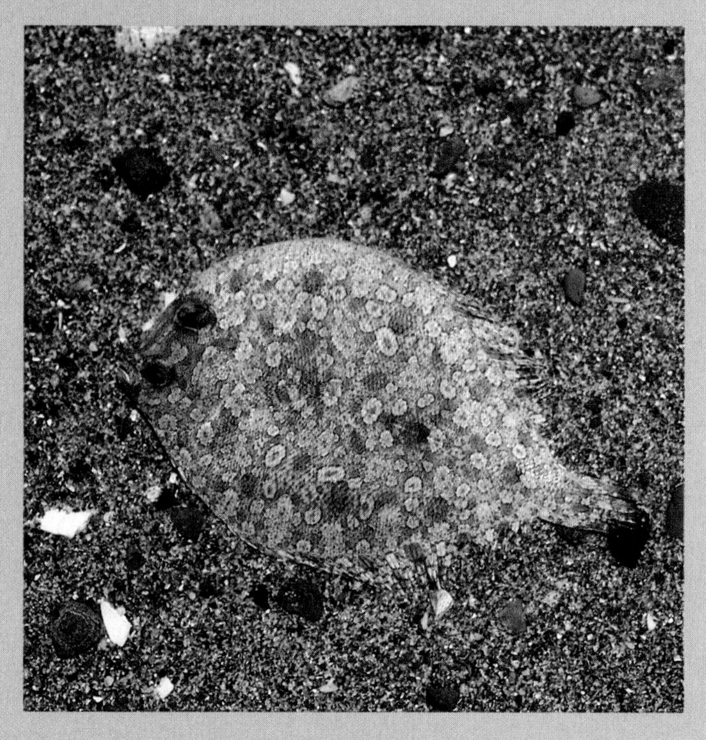

The octopus also can change colors. Probably as a fright response, the octopus blanches and takes on the light color of the background.

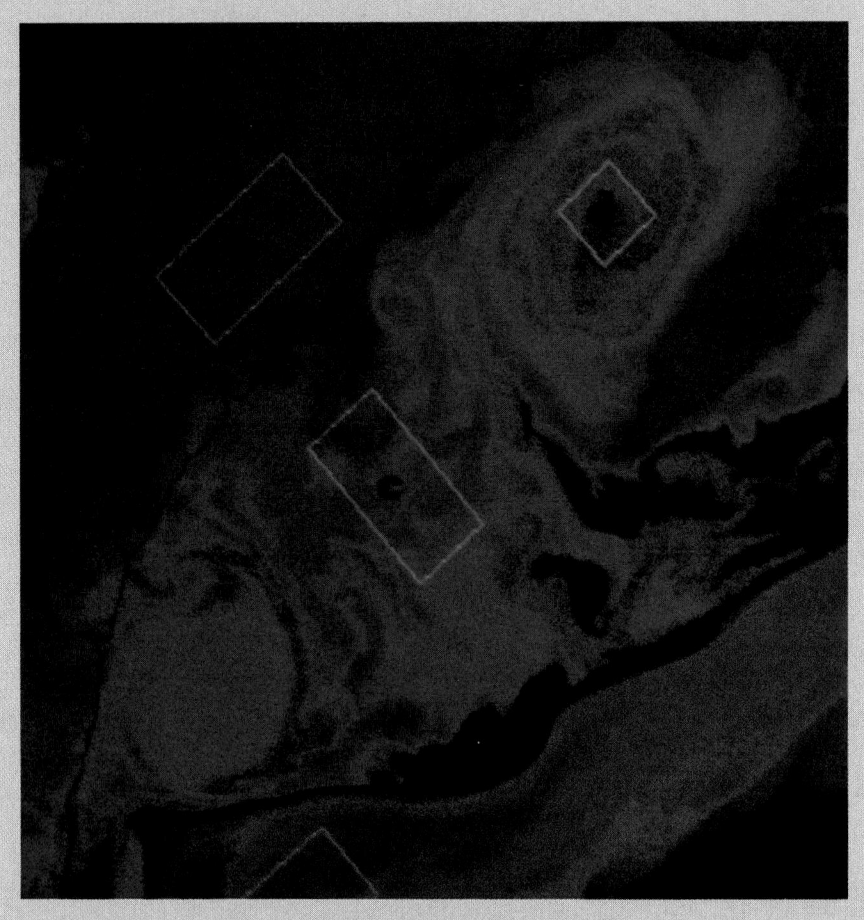

The NOAA 7 *satellite measures surface water temperature by means of a radiometer. The colors represent differences in temperature. The orange-yellow band is the warm Gulf Stream. Green is the somewhat cooler Sargasso Sea water. Progressively darker shades of blue indicate the decreasing temperatures of the water to the north. Square d is a warm-core ring. Warm-core rings are a bit of the Gulf Stream that was pinched off and persisted in the surrounding cooler water.*

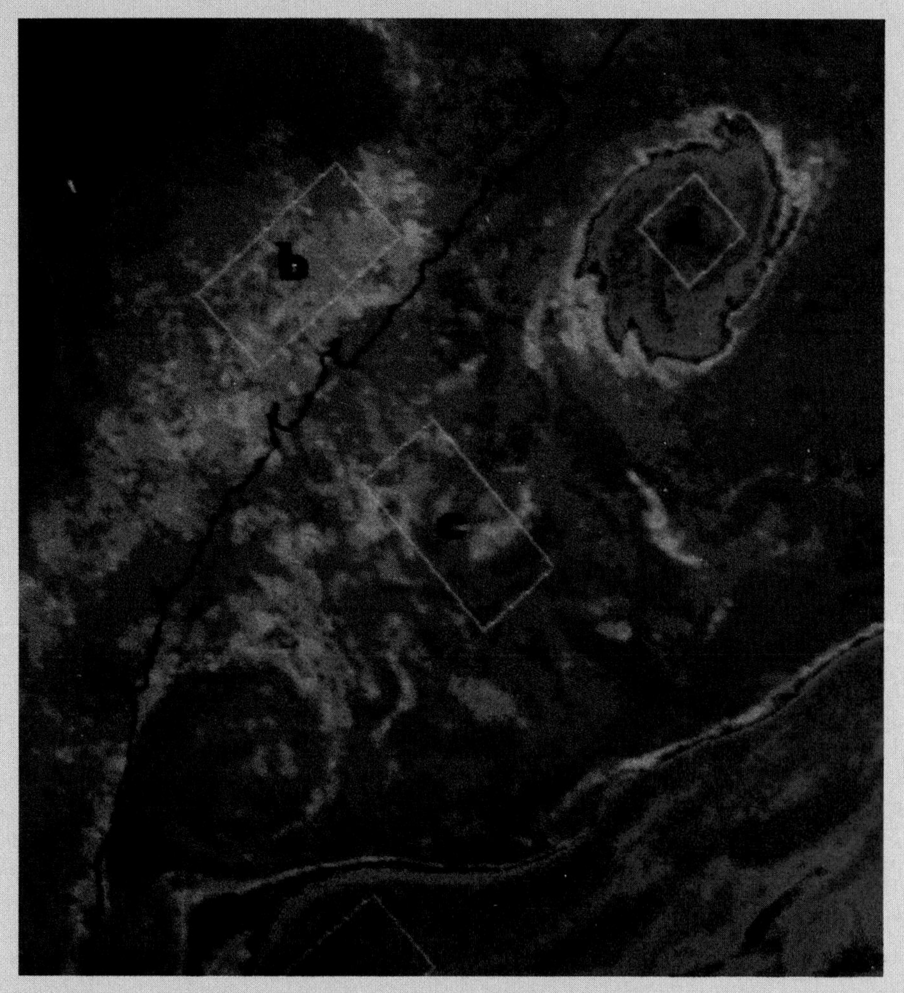

Another satellite, the Nimbus 7, *has a color scanner that can measure the concentration of chlorophyll in the water. Chlorophyll is a measure of the phytoplankton activity in the surface waters. The orange indicates high levels of chlorophyll pigment. Progressively darker shades of blue indicate decreasing levels of chlorophyll. There is a strong inverse correlation between temperature and chlorophyll concentration: Colder waters have higher phytoplankton activity. Square d, the warm-core ring, has much lower phytoplankton activity than the surrounding colder ring. The reason for the lower phytoplankton activity in colder water is not due directly to temperature. Instead, the warmer water has lower concentrations of the nutrient nitrate. The low concentration of nitrate limits the phytoplankton activity.*

The color scanner of the Nimbus *satellite has been used to map the phytoplankton activity of the world ocean from 1978 through 1981. The level of phytoplankton chlorophyll varies from less than 0.1 mg/m^3 (purple) to above 10 mg/m^3 (red). (Black indicates areas with insufficient data.) The lowest values are found in the vast reaches of open oceans. Because strong stratification of surface waters prevents nutrients in deeper water from reaching the surface, biologically these areas are deserts. Near the equator, equatorial upwelling—caused by trade winds—brings nutrient-rich water to the surface, so productivity increases. At latitudes above 40°, strong westerly winds bring surface waters with deeper nutrient-rich waters, so productivity is high. The highest levels of phytoplankton activity are near the coastlines where estuaries, coastal currents, and coastal upwelling bring high levels of nutrients to the surface.*

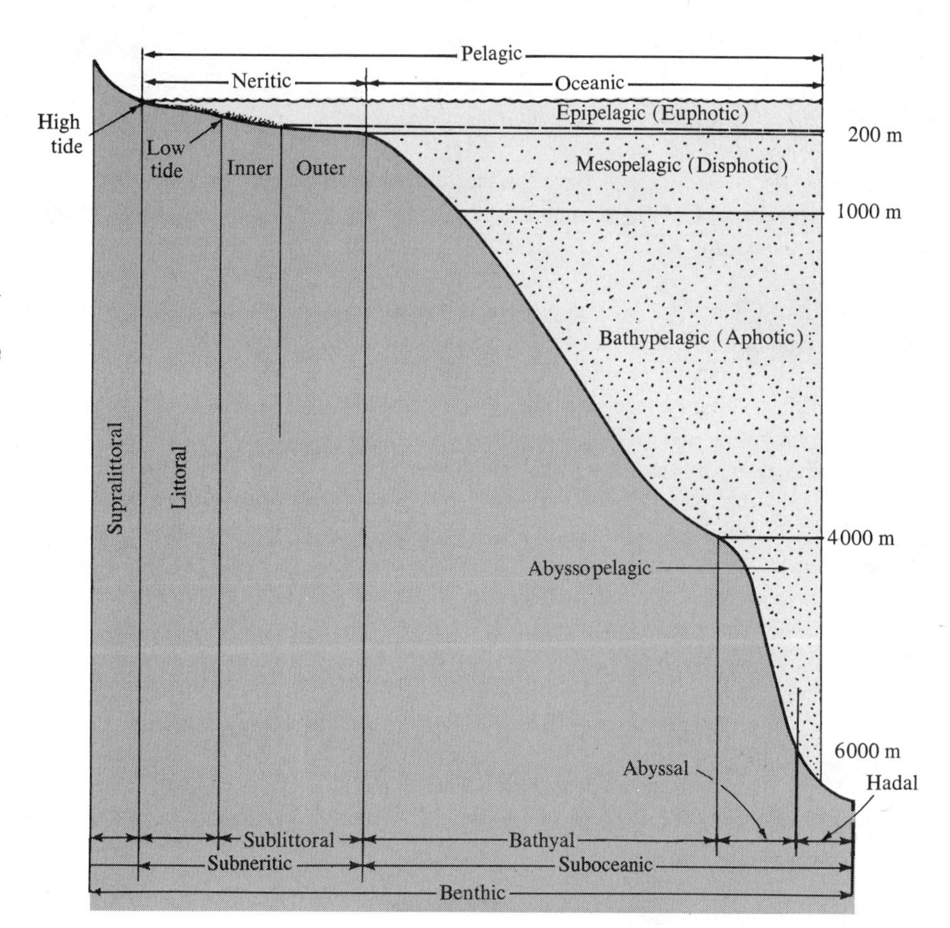

FIGURE 2.18
Biozones

The pelagic (water) environment is divided into the neritic and oceanic provinces. The oceanic province is further divided into zones based on the changing physical conditions in the ocean water with increasing depth. The benthic (bottom) environment is divided into the subneritic and suboceanic provinces. These are further subdivided into zones based on the changing physical conditions on the ocean floor with increasing depth.

The epipelagic zone is the only oceanic biozone in which there is sufficient light to support photosynthesis. The boundary between it and the mesopelagic zone (200 m) is also the approximate depth at which dissolved oxygen begins to decrease significantly. This decrease in oxygen occurs because no plants are found beneath about 150 m and because the dead organic tissue (detritus) descending from the biologically productive upper waters is undergoing decomposition by bacterial oxidation (Fig. 2.19). This decomposition below 200 m also causes an abrupt increase in the nutrient content of the water (Fig. 2.19). This depth is also the approximate bottom of the mixed layer, seasonal thermocline, and surface-water mass.

Within the mesopelagic zone, a **dissolved oxygen minimum** occurs between 700 and 1,000 m. The dissolved oxygen content increases both above and below the dissolved oxygen minimum. The intermediate water masses that move horizontally in this depth range often possess the highest content of plant nutrients in the ocean. The bases of the permanent thermocline and the top of the aphotic zone characteristically occur at 1,000 m, at the boundary between the mesopelagic and abyssopelagic zones.

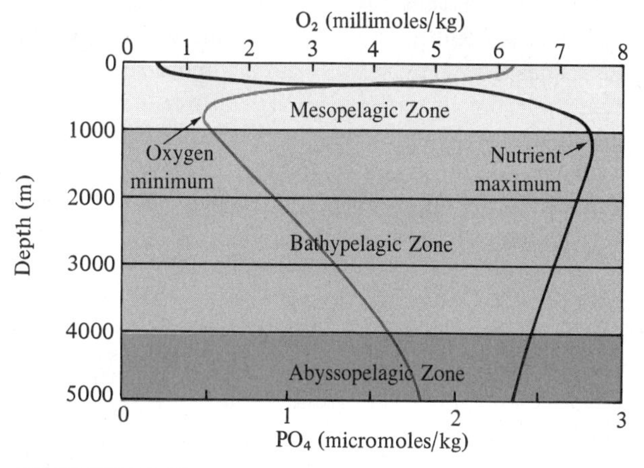

FIGURE 2.19
Distribution of Oxygen (O_2) and Plant Nutrient Phosphate (PO_4) in Water Column (mid- to low latitudes)

Oxygen is abundant in surface water due to mixing with the atmosphere and photosynthesis. Nutrient content is low in surface water because of usage by plants. Oxygen content decreases and nutrient content increases abruptly below the photosynthetic zone as a result of bacterial remineralization. An oxygen minimum and nutrient maximum are recorded at or near the base of the mesopelagic zone. Nutrient levels remain high to the ocean bottom, while oxygen content increases with depth as water masses carry oxygen into the deep ocean.

Although the mesopelagic zone contains little light, some animals there seem to be able to sense the presence of light. This discovery was made during the testing of sonar equipment by the U.S. Navy early in World War II. A sound-reflecting surface, the **deep scattering layer (DSL),** that changed depth on a daily basis was detected; the surface was located at a depth of 100–200 m during the night and sank to depths as great as 900 m during the day. After considerable investigation it was determined that this echo-creating surface was produced by masses of migrating marine life that moved closer to the surface at night and then to a greater depth during the day. It appeared to be a response to the changing intensity of solar radiation (Fig. 2.20).

The use of plankton nets and submersibles has indicated that the deep scattering layer contains layers of siphonophores, copepods, euphausids, cephalopods, and small fish. Tests using fish with air bladders suggest that they may be the primary cause of the sound reflection since a very small concentration of these fish is sufficient to produce such an effect. Fish, however, are predators and undoubtedly, if their presence is responsible for the echoes, the organisms on which they prey are responsible for the cyclic movement of the fish. The vertical movement of the smaller organisms, which respond to the changing intensity of light, draws the larger fish that prey upon them.

Other organisms have adapted to the disphotic mesopelagic zone in different ways. Some species of fish have unusually large eyes 100 times more sensitive to light than the human eye. Other fish and especially shrimp and squid are bioluminescent. Approximately 80 percent of the inhabitants in the mesopelagic zone carry light-producing **photophores,** glandular cells containing luminous bacteria surrounded by dark pigments. Some of the cells contain lenses to amplify the radiation. This cold light is produced by a chemical process involving the compound luciferin. The molecules of luciferin are excited and emit photons of light in the presence of the enzyme luciferase and oxygen. Only a 1-percent loss of energy is required to produce the illumination. The process is similar to that found in the firefly.

The aphotic bathypelagic zone and abyssopelagic zone, below the mesopelagic zone, represent more than 90 percent of the living space in the oceanic province. In this region of total darkness many totally blind fish exist. Very bizarre and small predatory species make up the fish population, and many species of shrimp that normally feed on detritus become predators at these depths, where food is scarce. The animals that live in the aphotic zones feed mostly upon each other and have developed impressive warning devices and unusual apparatus to make them more efficient predators. They have small expandable bodies, extremely large mouths relative to body size, and very efficient sets of teeth (Fig. 2.21).

The physical boundaries of the bathypelagic zone are generally set by the top and bottom of the deep-water masses. Oxygen content increases with depth in this zone, as the water masses carry oxygen from the cold surface waters where they formed to the deep ocean. The abyssopelagic zone is the realm of the bottom-water masses which commonly move in the opposite direction of the overlying deep-water masses.

Benthic Environment

The benthic, or bottom, environment can be subdivided into two large regions: the **subneritic province,** which extends from the spring high-tide shoreline to a depth of

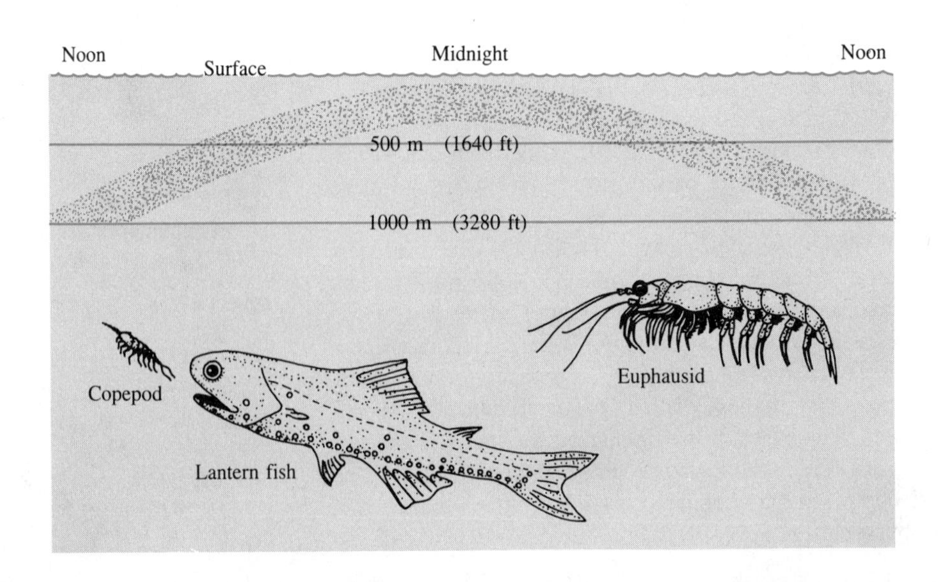

FIGURE 2.20 Deep Scattering Layer

The deep scattering layer, which scatters and reflects sonar signals within the mesopelagic zone, is caused by a migrating mass of marine organisms, among them euphausids and myctophids (lantern fish). These fish feed on smaller planktonic organisms that migrate vertically in the water column on a daily cycle.

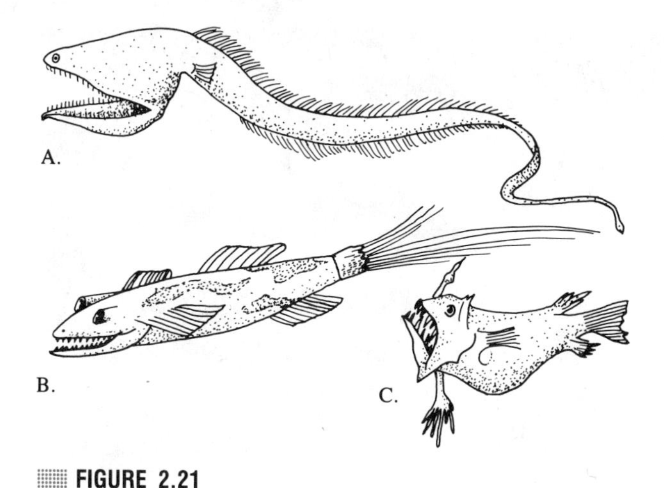

FIGURE 2.21
Deep-Sea Fish

(A) *Eurypharynx pelecanoides* (50–60 cm). **(B)** *Gigantura chuni* (12 cm). **(C)** *Linophryne bicornis* (5–8 cm).

200 m, and the **suboceanic province,** which includes all the benthic environment below 200 m (Fig. 2.18).

Above the normal high-tide line is a transitional region called the **supralittoral fringe.** This zone, commonly known as the spray zone, is covered with water only during periods of extremely high tides and when tsunami or large storm waves break on the shore. The intertidal zone is classified as the **eulittoral zone,** and the zone from normal low tide out to the 200 m depth is called the **sublittoral** (shallow subtidal) **fringe.** The inner sublittoral extends to a depth of approximately 50 m, but this seaward limit varies considerably because it is set at the depth at which no plants are found growing attached to the ocean bottom. This depth is determined primarily by the amount of solar radiation that penetrates the surface water. In areas of high turbidity the seaward limit of the inner sublittoral occurs at shallower depths due to the decreased penetration of solar radiation, while in areas of unusually clear water it may occur well below the 50 m mark. The outer sublittoral stretches from the seaward limit of the inner sublittoral to a depth of 200 m or the continental break, the seaward edge of the continental shelf.

The base, or seaward limit, of the inner sublittoral generally coincides with the base of the euphotic zone, the water zone near the ocean surface where enough light penetrates to support photosynthesis. Currents in general are stronger across the outer sublittoral bottom than along the upper continental slope of the bathyal zone. Also, sediments are coarser on the continenatal shelf than on the continental slope.

Beyond the continental break in the suboceanic system, the **bathyal zone** extends from a depth of 200 to 4,000 m and corresponds generally to the geomorphic province found beyond the continental shelf, the continental slope. From 4,000 to 6,000 m stretches the **abyssal zone,** which represents more than 80 percent of the surface area of the benethic environment and almost 57 percent of the earth's surface. Last, the **hadal zone** includes all depths below 6,000 m (19,680 ft). This zone is found only in the trenches along the margins of the continents.

Changes in the sediment type of the continental slope to a depth of about 4,000 m are subtle. Throughout much of the ocean, the calcium carbonate compensation depth occurs at about this depth, so sediments in the bathyal zone may contain considerable calcium carbonate, while those below contain little or none. In general, the deposits of neritic sediment found around the continents begin to be replaced by oceanic sediment below 4,000 m.

The ocean floor of the abyssal zone is covered by soft oceanic sediment primarily of the abyssal clay variety. The tracks and burrows of animals that live in this sediment are frequently recorded in bottom photographs (Fig. 2.22). In the hadal zone of the ocean trenches—deep, linear depressions in the deep-ocean bottom—unique faunal assemblages have developed.

MODES OF EXISTENCE IN THE OCEAN

The ways organisms live in the ocean have been divided into three fundamental life styles: planktonic, nektonic, and benthic. A brief description of each life style and a discussion of some of the organisms characteristic of each follows.

Plankton

The **plankton** include all organisms that drift with the ocean currents. Many plankters do have the ability to move but can propel themselves only weakly or only in a vertical direction, so their position in the ocean is determined primarily by ocean currents. Phototrophs that follow this drifting life style in the upper layers of the ocean are called phytoplankton. Planktonic **heterotrophs,** organisms that depend on an external food supply, are called **zooplankton** (Fig. 2.23). Most of the biomass of the earth is found adrift in the oceans as plankton.

The plankton include larger animals and plants such as jellyfish, salps, and Sargassum, which are classified as **macroplankton** and **megaplankton.** Smaller **microplankton** can be captured in fine-meshed silk plankton nets. **Nanoplankton** include the smallest plants that are too small to be filtered from the water by a silk net and must be removed by other types of micro-

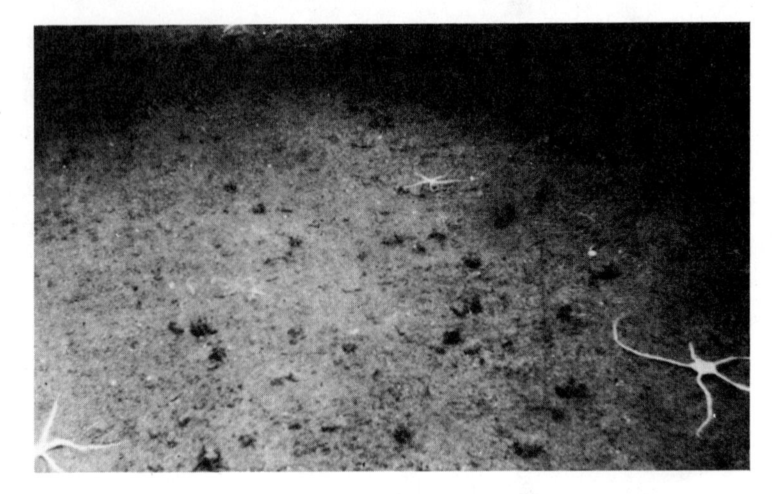

FIGURE 2.22
Benthic Organisms

Serpent stars are abundant in the deep sea, but their variety is much reduced from that in shallower waters.

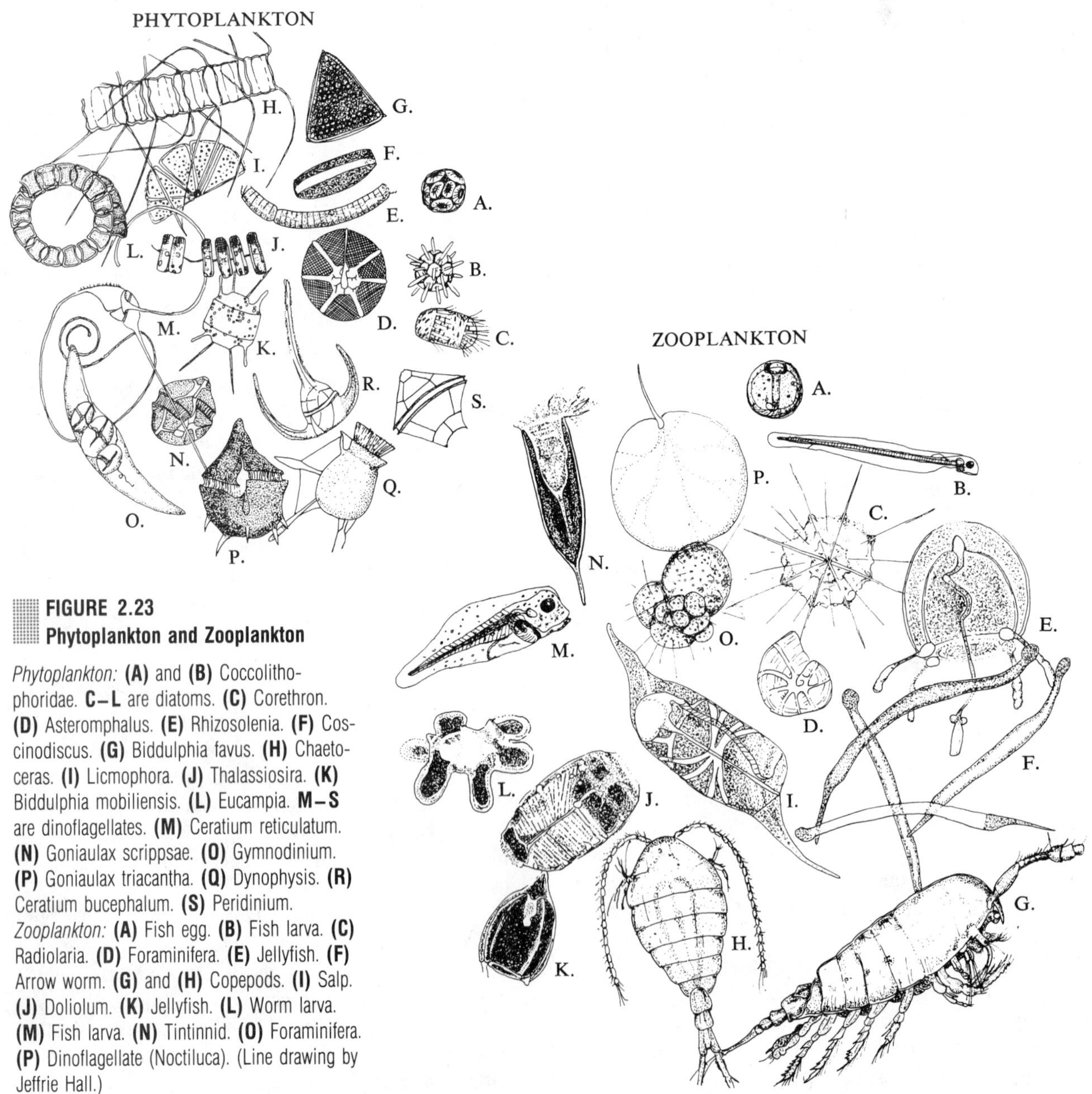

FIGURE 2.23
Phytoplankton and Zooplankton

Phytoplankton: **(A)** and **(B)** Coccolithophoridae. **C–L** are diatoms. **(C)** Corethron. **(D)** Asteromphalus. **(E)** Rhizosolenia. **(F)** Coscinodiscus. **(G)** Biddulphia favus. **(H)** Chaetoceras. **(I)** Licmophora. **(J)** Thalassiosira. **(K)** Biddulphia mobiliensis. **(L)** Eucampia. **M–S** are dinoflagellates. **(M)** Ceratium reticulatum. **(N)** Goniaulax scrippsae. **(O)** Gymnodinium. **(P)** Goniaulax triacantha. **(Q)** Dynophysis. **(R)** Ceratium bucephalum. **(S)** Peridinium. *Zooplankton:* **(A)** Fish egg. **(B)** Fish larva. **(C)** Radiolaria. **(D)** Foraminifera. **(E)** Jellyfish. **(F)** Arrow worm. **(G)** and **(H)** Copepods. **(I)** Salp. **(J)** Doliolum. **(K)** Jellyfish. **(L)** Worm larva. **(M)** Fish larva. **(N)** Tintinnid. **(O)** Foraminifera. **(P)** Dinoflagellate (Noctiluca). (Line drawing by Jeffrie Hall.)

filters. Table 2.5 shows the various size ranges of plankton and names given to each.

Plankton are also classified according to what portion of their life cycle they spend within the plankton community. Organisms such as planktonic diatoms and copepods that spend their entire life in the plankton are **holoplankton.** Other organisms, the **meroplankton,** spend a portion of their life cycle as plankton. Many of the nekton and very nearly all of the benthos make their home in the plankton community during their larval stages. As adults, benthos sink to the bottom and nekton begin to swim freely.

Nekton

The **nekton** include all animals that can move independently of the ocean currents. Nekton can determine their positions within relatively small areas of the ocean and in many cases are capable of long migrations. Included in the nekton are most adult fish and squid, marine mammals, and marine reptiles (Fig. 2.24).

The freely moving nekton, with some exceptions, are not able to move at will throughout the breadth of the ocean. They are effectively limited in their lateral range by invisible but impenetrable barriers created by gradual changes in temperature, salinity, viscosity, and availability of nutrients. Large numbers of fish frequently die because of temporary lateral shifts of masses of ocean water. Vertical range may be determined by pressure for all fish that contain swim bladders and mammals that belong to the nekton.

Fish appear to be everywhere but are generally considered to be more abundant near the continents and islands and in the colder waters. Some ocean fish, such as the salmon, ascend rivers to spawn. Many eels do just the reverse, growing to maturity in fresh water, then descending the streams to breed in the great depths of the ocean.

Benthos

The **benthos** live on or in the ocean bottom. Among the benthos, **infauna** live buried in sand, shells, or mud, and **epifauna** live attached to rocks or move over the surface of the ocean bottom. Benthos that live on the bottom but move with relative ease through the water above the ocean floor, such as certain shrimp and demersal flounder, are called **nektobenthos** (Fig. 2.25).

The littoral and shallow sublittoral are the only zones in which we find the macroscopic algae, including the greens, the browns, and the reds, attached to the bottom. This is because they are the only benthic zones

TABLE 2-5
Classification of Plankton by Size

Not all plankton are tiny. Plankton are defined by the way in which they move (drifting), rather than by size. (After Sieburth et al., 1978.)

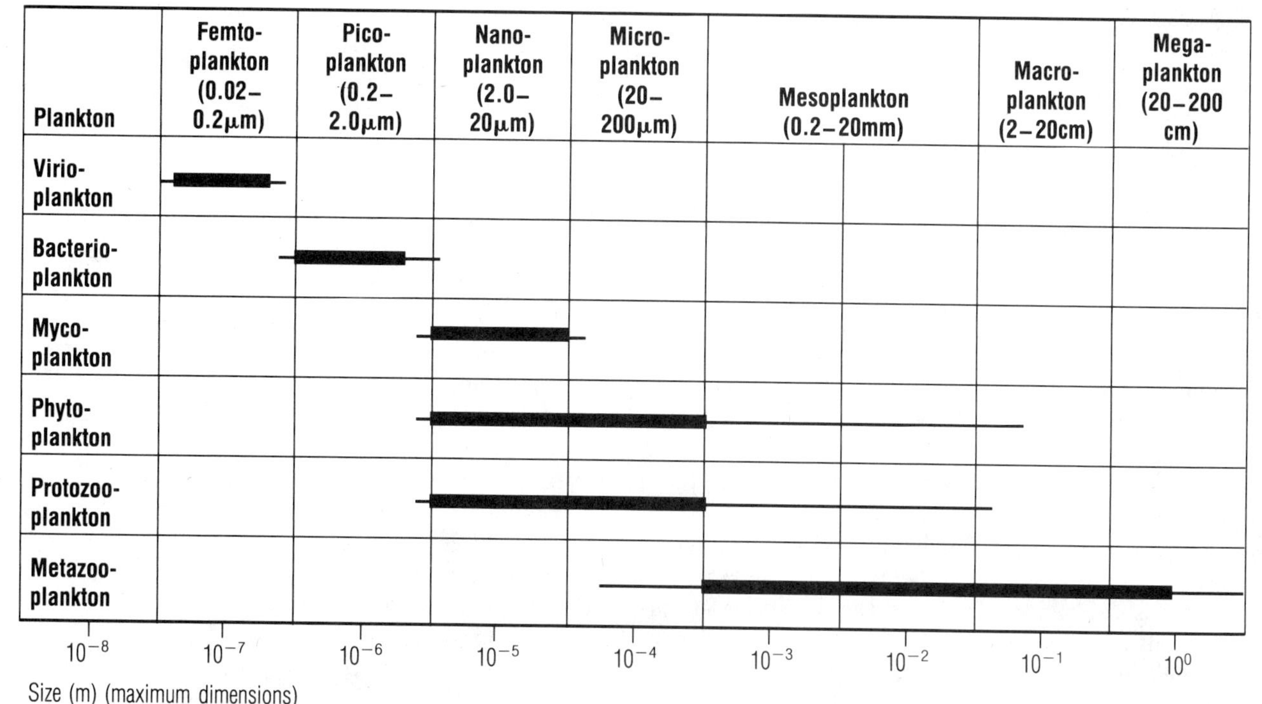

FIGURE 2.24
Nekton

(A) Bluefin tuna. **(B)** Bottlenosed dolphin. **(C)** Nurse shark. **(D)** Barracuda. **(E)** Striped bass. **(F)** Sardine. **(G)** Deep-ocean fish. **(H)** Squid. **(I)** Angler fish. **(J)** Lantern Fish. **(K)** Gulper. (Line drawing by Jeffrie Hall.)

to which light can penetrate. These nearshore zones offer a great diversity of physical and nutritive conditions. Animal species have developed in great numbers there because of the variations within the habit. Moving across the bottom from the littoral into the deeper benthic environments, it is commonly observed that an inverse relationship exists between the distance from shore and the number of benthic species that can be found. Some studies, however, have indicated an increase in species diversity towards the shelf break.

Throughout most of the benthic environment, animals live in a region of perpetual darkness where photosynthesis cannot occur. They must feed on each other or on whatever outside nutrients fall from the productive zone near the surface. The environment of the deep-sea bottom is characterized by coldness, stillness, and darkness. Under these conditions, life would be expected to move at a relatively slow pace. For those animals that move around on the bottom, streamlining is of little importance.

The organisms that live in the suboceanic benthic realm normally have quite a wide range of distribution because physical conditions for life do not vary greatly on the deep-ocean floor, even over great distances. A few species found in the littoral province and at depths of several kilometers appear to be extremely tolerant of pressure changes.

We have confined our discussion in this chapter to general conditions for life in the marine environment and the general adaptations of many populations of marine organisms to the various subenvironments. In the next few chapters, we will look at the taxonomic groupings of organisms within the ecological framework we have established here in order to understand the evolutionary as well as the ecological relationships among marine organisms.

SUMMARY

To fully appreciate the biology of the oceans, one must understand the fundamental nature of the physical marine environment. Marine geology concerns itself with a major aspect of this environment, the geology of the ocean floor. The ocean floor surrounding the continents is called the continental margin. It includes a gently sloping continental shelf that extends from the shore to an average depth of 135 m (443 ft). At its seaward edge, the shelf break, the ocean floor steepens to become the continental slope which descends into the deep ocean at an average slope of 4°. The continental rise is another gently sloping surface that leads from the base of the continental slope to the abyssal plains. The continental slope and shelf are cut by submarine canyons thought to have been cut by turbidity currents.

Mountain ranges called mid-ocean ridges and rises extend through all ocean basins and are bounded by abyssal plains, flat surfaces covered by deep-ocean sediment.

To appreciate the unique character of ocean water it is necessary to understand the unusual properties of pure water. Water has an unusually high freezing point and boiling point, latent heats of melting and vaporization, and heat capacity. It also has high surface tension and dissolves a large range of substances. All these properties result from the dipolar nature of the water molecule, which produces the strong intermolecular bond called the hydrogen bond.

The amount of dissolved solids in water is referred to as salinity and is expressed in grams per kilogram (g/kg) or parts per thousand (‰) by weight. The average salinity of ocean water is 34.7 ‰, and 99.28 percent of the dissolved solids are accounted for by six ions: chloride, sodium, sulfate, magnesium, calcium, and potassium. Salinity can be measured in many ways, but it is most often determined by making the chemical determination of chlorinity or by using a salinometer to

FIGURE 2.25
Benthos of the Intertidal and Subtidal Zones

(A) Sand dollar. **(B)** Clam. **(C)** Crab. **(D)** Abalone. **(E)** Sea urchins. **(F)** Sea anemones. **(G)** Brittle star. **(H)** Sponge. **(I)** Acorn barnacles. **(J)** Snail. **(K)** Mussels. **(L)** Gooseneck barnacles. **(M)** Sea star. **(N)** Brain coral. **(O)** Sea cucumber. **(P)** Lamp shell. **(Q)** Sea Lily. **(R)** Sea squirt. (Line drawing by Jeffrie Hall.)

measure electrical conductance. Increased salinity increases the density of ocean water.

Of the solar radiation striking the ocean's surface (insolation), only the infrared and visible light penetrate significantly beneath the ocean surface. Infrared is absorbed in the top 10 cm (4 in) and visible light is 99.9 percent absorbed by a depth of 100 m. The clear, open ocean that is biologically nonproductive is deep blue because of the scattering of blue wavelengths of light by molecular-sized particles. In biologically productive coastal waters where microscopic organisms are concentrated, the larger average particle size causes a scattering of longer, green wavelengths of light. This—along with the green pigment, chlorophyll, present in microscopic plants—produces a green color.

Because the ocean is heated from the surface by solar radiation, surface waters in the low- and mid-latitudes may be significantly warmer and more saline than the deep waters below a depth of 1,000 m (3,280 ft). In these regions, the mixed layer, which extends to a depth of about 200 m below the surface, is separated from the deep water below 1,000 m by a zone of rapidly changing temperature, the thermocline, and a zone of rapidly changing salinity, the halocline. Coinciding closely with the thermocline is a zone of rapidly changing density, the pycnocline, that separates the low-density surface water from the denser deep water. Well-developed pycnoclines can significantly inhibit biological productivity, since they prevent nutrient-rich deep water from mixing with surface water where plants use up the available nutrient supply.

The speed of sound in the ocean increases with increased temperature and pressure. Thus the decrease in temperature associated with the thermocline produces a zone of minimum sound velocity near the base of the thermocline called the SOFAR channel. Sound is trapped in this channel and marine animals may use it to communicate over long distances.

Although the relative amounts of the constituents of salinity in the ocean and in the body fluids of the organisms is often very nearly the same, the salinity of one may differ greatly from the other. If the body fluids of an organism and ocean water are separated by a semipermeable membrane that allows the passage of water molecules, problems of osmoregulation may develop. Marine invertebrates and sharks are essentially iso-osmotic, having body fluids with a salinity similar to that of ocean water. They rarely face osmoregulatory problems. Most marine vertebrates are hypo-osmotic, having body fluids with a salinity lower than that of ocean water, and tend to lose water through osmosis. Fresh-water organisms are essentially all hyperosmotic, having body fluids much higher in salinity than the water in which they live, so they must compensate with special body mechanisms for a tendency to take water into their cells through osmosis.

For life to flourish in any environment, there must be a sufficient food supply. The basic producers of food are phytoplankton, so the requirements of phytoplankton must be met if food is to be plentiful. The availability of nutrients and solar radiation make phytoplankton life possible. Since solar radiation is available only in surface water of the ocean, phytoplankton are restricted to a thin layer of surface water usually no more than 100 m deep. Nutrients derived ultimately from the continents are much more abundant near continents. Although much is yet to be learned about the distribution of life in the oceans, it appears that the biomass concentration of the oceans decreases away from the continents and with increased depth. The color of the oceans ranges from green in highly productive regions to blue in areas of low productivity.

The phytoplankton that must stay in surface water to receive sunlight and the small animals that feed on them do not have effective means of locomotion. They depend, therefore, on their small size and other adaptations to give them a high ratio of surface area per unit of body mass, and thus greater frictional resistance to sinking. Large animals that swim freely face altogether different problems and generally have streamlined bodies to reduce frictional resistance to motion.

Compared to life in colder regions, organisms living in warm water tend to be smaller, consist of a greater number of species, and constitute a much smaller total biomass. Warm-water organisms also tend to live shorter lives and reproduce earlier and more frequently than their cold-water counterparts.

The marine environment is divided into two basic parts: the pelagic (water) environment and the benthic (bottom) environment. These regions, which are further divided primarily on the basis of depth, are inhabited by organisms we can classify into three categories on the basis of life style: (1) the plankton, free-floating forms with little power of locomotion; (2) the nekton, or free swimmers; and (3) the benthos, or bottom-dwellers.

QUESTIONS AND EXERCISES

1. Draw a profile (similar to Fig. 2.1) that shows the continental shelf, continental slope, continental rise (ocean trench may replace rise in some areas—show this option), abyssal plains, mid-ocean ridge, abyssal hills, and seamounts.
2. Select three properties of water from Table 2.1 and discuss their physical or biological significance to the marine environment.
3. How is it possible to determine the total salinity of ocean water by measuring the concentration of only one component of salinity?
4. Discuss the reasons for the different colors of open ocean water far from shore and coastal water.
5. In what way does the development of a pycnocline affect the biological productivity of ocean water?
6. Describe the relationships between the speed of sound in water, temperature, and pressure, and explain how their interaction produces the SOFAR channel.
7. Define the terms *euryhaline, stenohaline, eurythermal,* and *stenothermal.* Where in the marine environment will organisms displaying a well-developed degree of each characteristic be found?
8. How are osmotic pressure, salinity, and freezing point of a solution related?
9. What problem of osmotic regulation is faced by hypotonic fish in the ocean? How have these animals adapted to solve this problem?
10. What role do the continents play in furnishing nutrients and how does the availability of nutrients affect the concentration of life in the oceans?
11. Why is biological productivity relatively low in the tropical open ocean where the penetration of sunlight is greatest?
12. Discuss the characteristics of the coastal ocean where unusually high concentrations of marine life are found.
13. What factors create the color difference between coastal waters and the less productive open-ocean water?
14. Compare the ability to resist sinking of an organism with an average linear dimension of 1 cm to that of an organism with an average linear dimension of 5 cm. Discuss some adaptations other than size used by organisms to increase their resistance to sinking.
15. Changes in water temperature significantly affect the density, viscosity of water, and ability of water to hold gases in solution. Discuss how decreased water temperature changes these variables and may affect marine life.
16. Describe how higher water temperatures in the tropics may account for the greater number of species in these regions compared to low-temperature, high-latitude areas.
17. Construct a table listing the subdivisions of the benthic and pelagic environments, and the physical factors used in defining their boundaries.
18. Describe the vertical distribution of oxygen and nutrients in the oceanic province and discuss the factors responsible for this distribution.
19. Describe the life styles of plankton, nekton, and benthos. Why is it likely that plankton account for a relatively larger percentage of the biomass of the oceans than the benthos and nekton?
20. List the subdivisions of plankton and benthos and the criteria used for assigning individual species to each.

REFERENCES

Borgese, E.M., and **Ginsburg, N.** 1988. *Ocean yearbook 7.* Chicago: The University of Chicago Press.

Coker, R.E. 1962. *This great and wide sea: An introduction to oceanography and marine biology.* New York: Harper and Row.

Hedpeth, J., and **Hinton, S.** 1961. *Common seashore life of southern California.* Healdsburg, CA: Naturegraph.

Isaacs, J.D. 1969. The nature of oceanic life. *Scientific American* 221:65–79.

May, R.M. 1988. How many species are there on earth? *Science* 241:4872, 1441–48.

Roughgarden, J., Gaines, S., and **Passingham, H.** 1988. Recruitment dynamics in complex life cycles. *Science* 241:4872, 1460–66.

Sieburth, J.M.N. 1979. *Sea microbes.* New York: Oxford University Press.

Sverdrup, H., Johnson, M., and **Fleming, R.** 1942. *The oceans.* Englewood Cliffs, NJ: Prentice-Hall.

Thorson, G. 1971. *Life in the sea.* New York: McGraw-Hill.

SUGGESTED READING

OCEANUS

Changing Climate and the Oceans. 1987. 29(4):1–100.

El-Sayed, S.Z. 1975. Biology of the southern ocean. 18(4).

The Oceans and Global Warming. 1989. 32(2):1–75.

SCIENTIFIC AMERICAN

Denton, E. 1960. The buoyancy of marine animals. 203(1):118–129.

Eastman, J.T., and **DeVries, A.L.** 1986. Antarctic fishes. 255(5):106–114.

Horn, M.H., and **Gibson, R.N.** 1988. Intertidal fishes. 258(1):64–71.

Isaacs, J.D. 1969. The nature of oceanic life. 221(3):146–165.

Stewart, R.W. 1969. The atmosphere and the ocean. 221(3):76–105.

SEA FRONTIERS

Bachand, R.G. 1985. Vision in marine animals. 31(2):68–74.

Burton R. 1977. Antarctica: Rich around the edges. 23(5):287–295.

DeVorsey, L., Jr. 1982. Where does the Gulf of Mexico end and the Atlantic Ocean begin? 28(4):231–239.

Gruber, M. 1970. Patterns of marine life. 16(4):194–205.

Hammer, R.M. 1974. Pelagic adaptations. 16(1):2–12.

Patterson, S. 1975. To be seen or not to be seen. 21(1):14–20.

Perrine, D. 1987. The strange case of the freshwater marine fishes. 33(2):114–119.

Schellenger, K. 1974. Marine life of the Galapagos. 20(6):322–332.

Thresher, R. 1975. A place to live. 21(5):258–267.

Williams, L.B., and **Williams, E.H., Jr.** 1988. Coral reef bleaching: Current crisis, future warning. 34(2):80–87.

EL NIÑO–SOUTHERN OSCILLATION EVENTS

During the 1920s, G.T. Walker identified a condition he called the Southern Oscillation (SO). It involves a periodic decrease in the pressure difference between an atmospheric high-pressure cell in the southeast Pacific and an Indo-Australian low-pressure cell (Fig. 1). This weakening of the pressure difference occurs every three–five years and is best developed in the month of January during the Southern Hemisphere summer season. It is usually accompanied by decreased strength of the trade winds and increased flow of the equatorial countercurrent that warms the surface waters of the equatorial eastern Pacific Ocean. This periodic warming of the coastal waters of Ecuador and Peru had previously been noted because it occurred along with catastrophic rains in the normally arid coastal regions of these countries. This warming of coastal waters was named El Niño (EN) (The Child) because it usually occurred around Christmas. Subsequent realization that the two events were related phenomena resulted in development of the term El Niño–Southern Oscillation (ENSO). There have been eight ENSO events since 1950.

ENSO events have been traced to the temporary decline of anchoveta and coastal bird populations in the Peruvian coastal waters during and following each event. Although other factors were involved, the anchoveta fishery collapsed during a well-developed ENSO event of 1972, and the coastal bird population that maintains the guano deposits on arid offshore islands was decimated by the decline in its food supply. Both populations are yet to recover.

Normally, the Walker Circulation (see Fig. 1) propels the southeast trade winds toward the Indo-Australian low-pressure cell, where they rise and produce high levels of precipitation. After the rising air is depleted of its moisture, it descends on the eastern Pacific high-pressure cell off the coast of South America to produce the normally arid conditions found there.

One precursor of an ENSO event is the movement of the Indo-Australian low-pressure cell to the east, beginning in October or November. In extreme cases—such as the 1982–1983 event—severe droughts can occur in the Indo-Australian region because the low-pressure cell moves so far east.

Concurrent with the eastward shift of the Indo-Australian low-pressure cell, the meteorological equator, or Intertropical Convergence Zone (ITCZ), moves south. The ITCZ is where the northeast trade winds and southeast trade winds meet and rise to produce precipitation. Its normal seasonal migration is from 10° N latitude in August to 3° N in February, but during ENSO events it may move south of the equator in the eastern Pacific. Associated with this southern shift are weak trade winds, a decrease in coastal upwelling off the Peruvian coast, and an unusually thick column of abnormally warm surface water in the eastern Pacific. These initial events are amplifications of normal seasonal fluctuations.

Figures 2A–D illustrate averaged temperature anomalies for ENSO events occurring between 1950 and 1973. Figure 2A shows the fall season (March–May) warming of the eastern Pacific that marks the start of an event. As the ENSO develops, the weakened trade winds and anomalous warmth of surface waters observed in the eastern Pacific spread toward the west (Fig. 2B). The coming of unusually warm surface waters to Kiritimati (Christmas Island; 2° N, 157° W) can be predicted by earlier observation of increased surface temperatures off the coast of Peru. The ENSO event is fully developed by January (Fig. 2C). Heavy rainfall from the southward shift of the ITCZ strikes the coast of Ecuador and Peru—which is usually dry—and spreads westward across the Pacific. The increased intensity of the

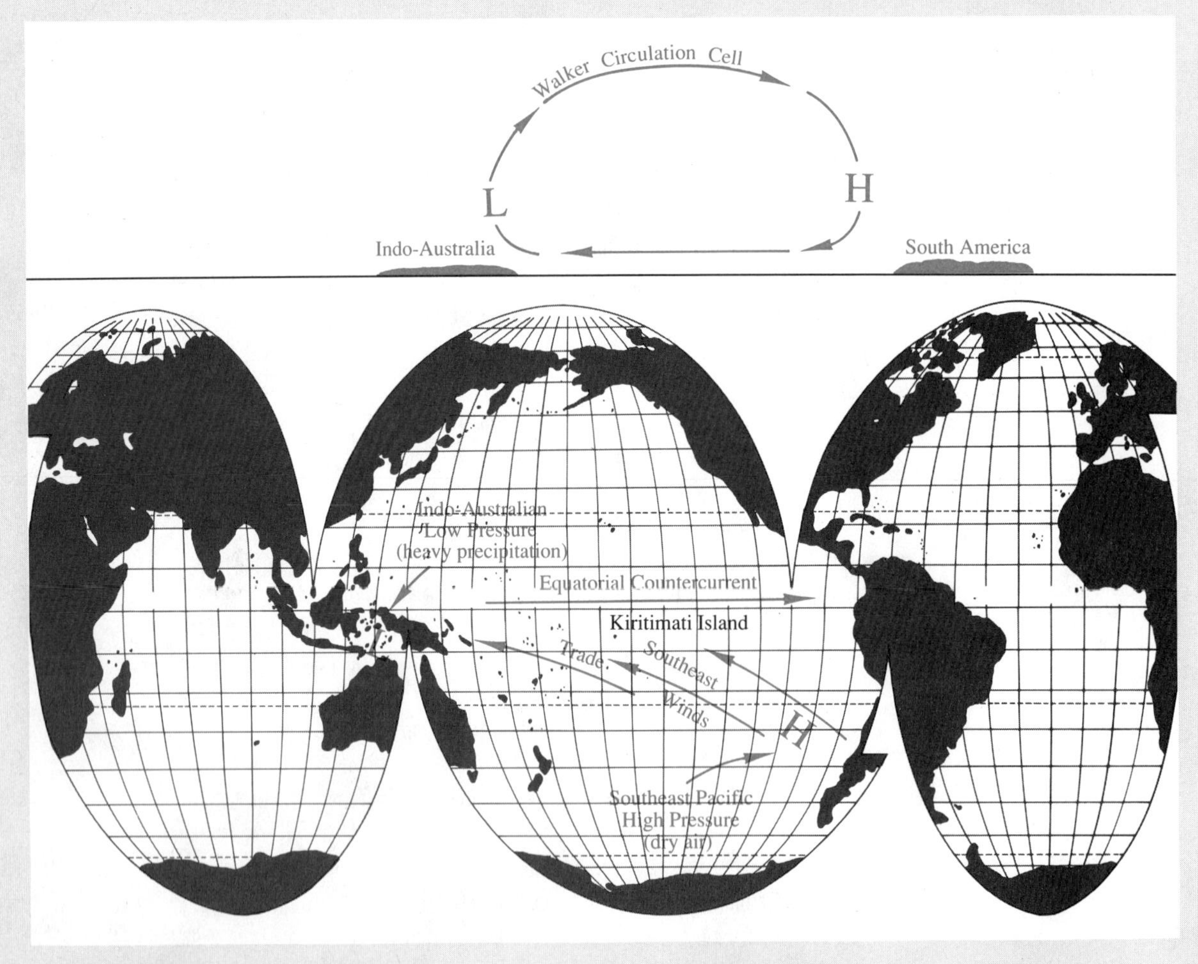

FIGURE 1
Walker Circulation

The high- and low-pressure cells producing the Walker Circulation shown are typical of the Southern Hemisphere winters. In summer, they weaken and the Southeast Trade Winds slow. This allows a larger eastward flow of warm water in the Equatorial Countercurrent. During El Niño years, winds that blow out of the west may even replace the Southeast Trade Winds.

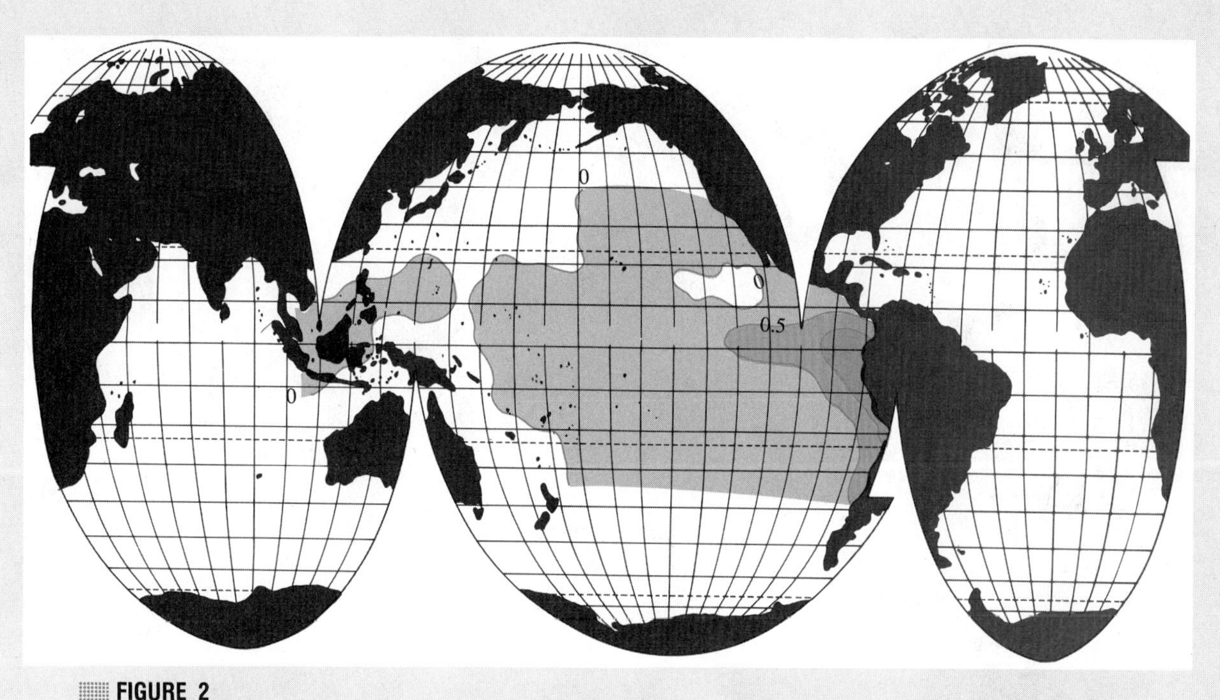

FIGURE 2
Sea Temperature Anomaly (°C) Resulting from the Averaging of ENSO Event Temperature Anomalies from 1950–1973

(A) After onset: average of March, April, and May temperature anomalies. **(B)** During development: average of August, September, and October anomalies. **(C)** Maximum development: average of December, January, and February anomalies. **(D)** Ending of event: average of May, June, and July anomalies. The typical progress of an ENSO event shows that the abnormally high surface-water temperatures first appear in the eastern Pacific off the coast of Ecuador and Peru **(A).** This condition begins to be observable by December or January along the coast of South America and is well developed during the March–May period shown. In **B** and **C** it can be seen to develop in a westerly direction along the equator, and reaches maximum development in the central Pacific by the following February. By July **(D),** conditions return to near normal, with the exception of a significant negative temperature anomaly in the eastern Pacific, called La Niña. (Courtesy of NOAA.)

- –0.5 to –1.0°C
- 0 to –0.5°C
- 0 to 0.5°C above the average surface water temperature for non-ENSO years
- 0.5 to 1.0°C above average
- 1.0 to 1.5°C above average
- More than 1.5°C above average

eastward flowing Equatorial Countercurrent causes a rise in sea level along the western coast of the Americas that progresses poleward in both hemispheres. The ENSO event ends 12–18 months after it starts, with a gradual return to normal conditions that begins in the southeastern tropical Pacific and spreads to the west (Fig. 2D). Actually, as Figure 2D shows, there is a negative temperature anomaly that exceeds 0.5°C (the temperature drops to 0.5°C below average). This period of anomalous cooling has been referred to as La Niña.

A very intense 1982–1983 ENSO event that caused a severe drought in the Indo-Australian region was unlike most events because initially it was confined to the

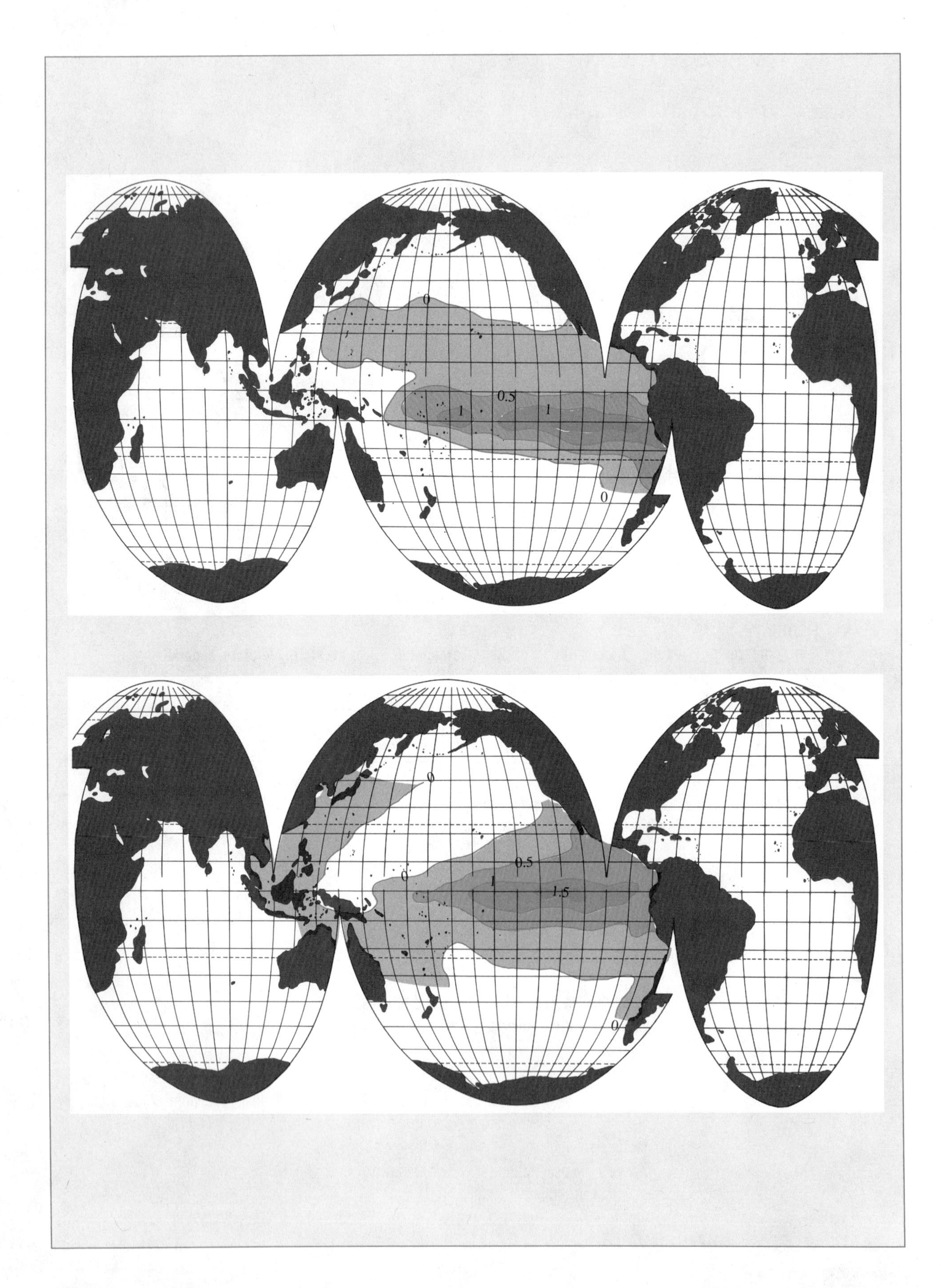
0
0.5
1
1
0
0
0.5
0
1
1.5
0

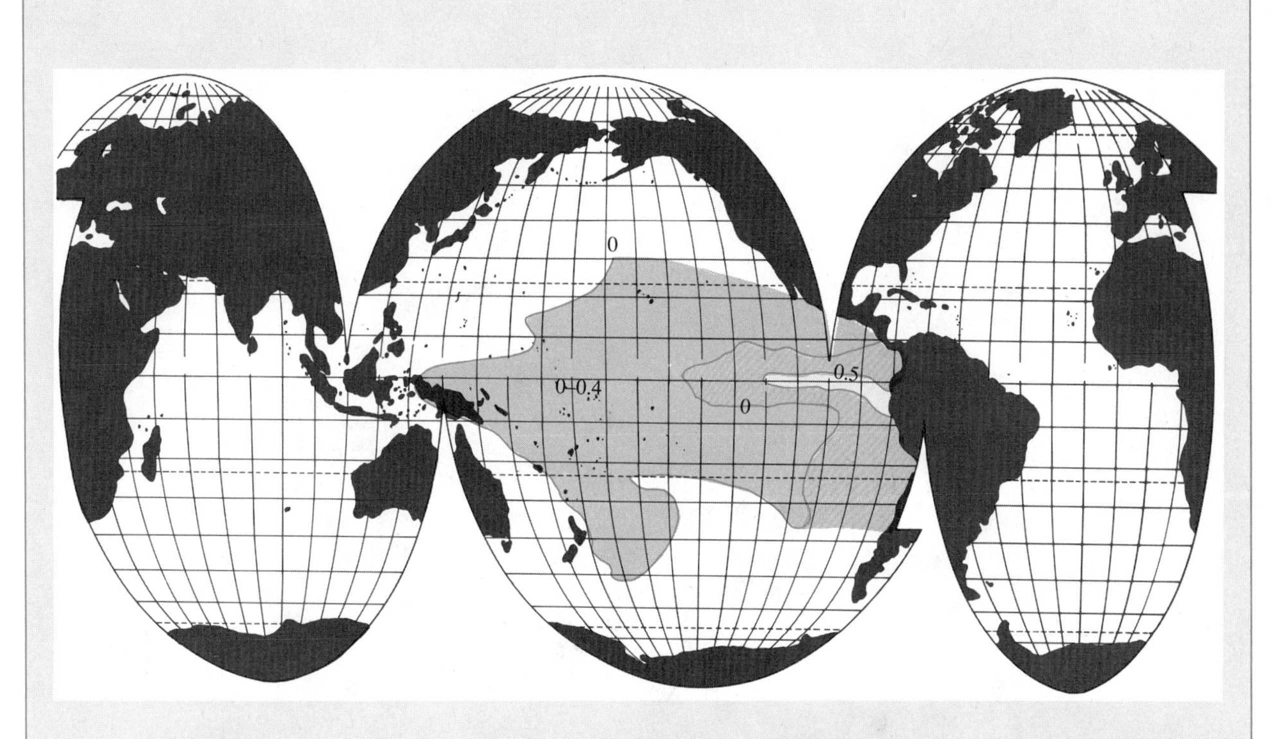

central and western tropical Pacific, then spread to the east late in its development. In November 1982 it was observed that virtually all of the 17 million adult birds that inhabit Kiritimati had abandoned their nestlings. This event indicated the severity of this ENSO in the central Pacific, as such an abandonment is not known to have occurred before. It is assumed that the spread of a thick layer of warm water over the surrounding ocean prevented the rise of nutrients into the surface waters, causing the fish to leave in search of better feeding grounds. The birds, in turn, were forced to leave in search of the fish. If the birds did not find their necessary supply of fish, a large percentage of the adults may have died in addition to the nestlings.

In addition to the climatic effects directly related to the changes in the strength of the Walker Circulation Cell, the El Niño phase is known to correspond with droughts in southeastern Africa, India, and northern South America, and to unusually heavy rainfall in the southeastern United States. The La Niña condition appears to be associated with rainfall effects that are the opposite of those caused by El Niño.

Although, historically, Las Niñas occur about every four years, none had developed during ENSO events between 1975 and 1987. However, a 1987–1988 ENSO closed out with a whopper. By June 1988, the temperatures of central tropical Pacific waters were 2°C (3.6°F) below average values for that time of year. Were the above-average temperatures of the 1980s recorded because there were no Las Niñas? When you read this, you will know if 1989 was a cool year, as might be predicted to follow the 1988 La Niña. As for the researchers who are trying to understand it all, they are still working with bits and pieces. But the pieces are getting larger.

PART TWO

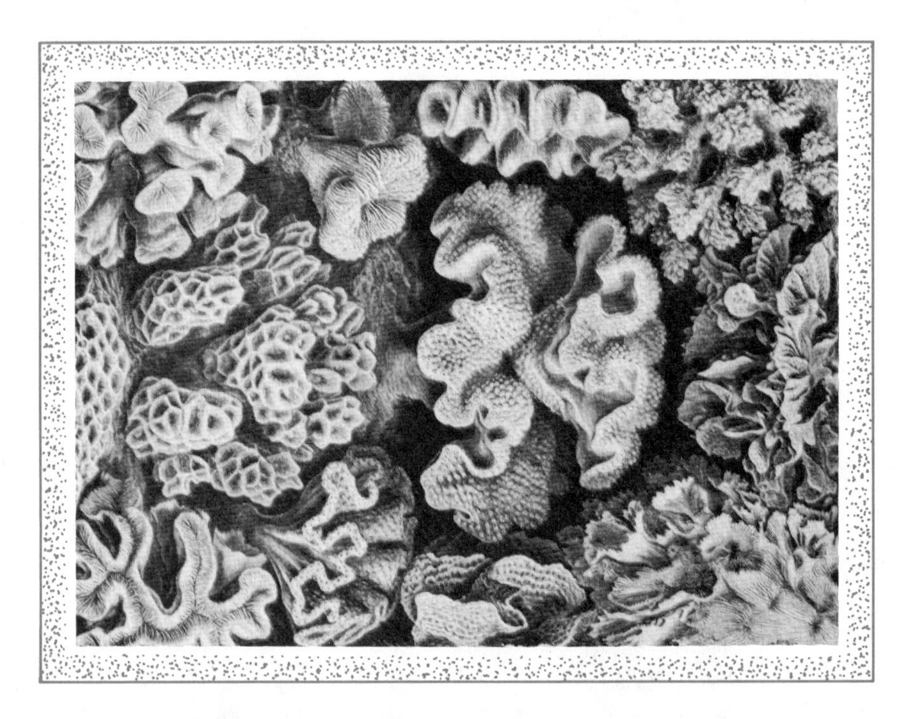

THE BIOLOGY OF MARINE ORGANISMS

Over one million species of life are distributed through the various habitats of the marine environment. How do we study this incredible diversity? First, we examine the major patterns that show how life in the sea is related. The community of organisms living in the ocean represent over 2 billion years of evolutionary history. Chapter 3 describes the evolutionary paths that life in the sea has followed.

We use the systematic approach to classify the categories of life. Five kingdoms are used to classify marine life. Our emphasis is on the animal kingdom. Four chapters are devoted to marine animals. Chapter 5 covers the invertebrate animals. This is the largest group of animal species and includes representatives from all the animal phyla. The remaining animals are those that belong to a single class of one of the animal phyla (Chordata). This class (the Vertebrata) includes the fish (Chapter 6), the reptiles and birds (Chapter 7), and the marine mammals (Chapter 8).

Although we emphasize the animals we must also consider the other types of life in the ocean. There are four other kingdoms of life, all discussed in Chapter 4. The kingdoms Plantae and Fungi have only a few species in the marine environment and their absence would not change things very much. There are two remaining kingdoms that make important contributions to the diversity of marine life. One, the Monera, includes the bacteria (both heterotrophic and photosynthetic species). Bacterial photosynthesis, however, is a small amount of the oceans' total, and the bulk of the species that are autotrophs, that is, use sunlight to produce energy-rich organic molecules, belong to the kingdom Protista. The diatoms of the water column and the algae of the coastal fringe are protists. Not all protists are autotrophs, though. The ocean's single-cell animal-like species, the Protozoa, are also protists.

Study of the organisms that live in the marine environment requires that we examine all the major groupings of life on earth. Evolutionary biologists are confident that most of the early evolution of life on the planet occurred in the sea and that life only secondarily colonized land. This is shown by the 14 out of 30 animal phyla that are totally restricted to the oceans.

Some animals have made the reverse trip. Marine mammals are derived from terrestrial forms; whales returned to the ocean only 40 million years ago. Birds and reptiles also are derived from land stocks.

3

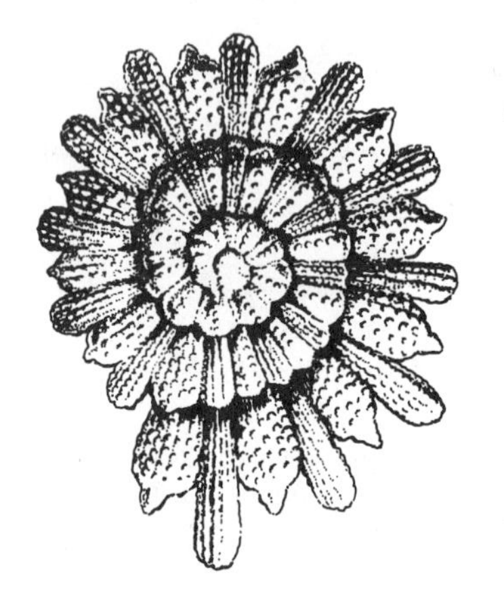

SYSTEMATICS AND EVOLUTION OF MARINE LIFE

Biologists have named at least 3 million different existing species of living organisms on earth, and estimate that there are more than 10 million additional species that remain undescribed. Added to existing species are some 75 million extinct species that have evolved and disappeared in the almost 4 billion years that life has flourished on the planet. It is through the science of sytematics and evolution that biologists find order in the diversity of life. There are two major objectives of the science of systematics and evolution. The first is to group organisms for the purpose of classification and further study. The second is to describe the evolutionary relationships between species and groups of species.

As humans, our affinities are with life on land, and there is no question that the plants and animals of the land are much more diverse than those of the ocean. For example, there are over 1 million species of insects alone. However, if you trace the evolutionary history of terrestrial species far enough, you will discover that they have their origins in the ocean. Research on the systematics and evolution of life in the oceans supports the classification system that divides living organisms into five kingdoms (Fig. 3.1). These kingdoms are the Monera, the Protista, the Plantae, the Metazoa, and the Fungi.

The story of how life started on the planet and the evolutionary pathways to existing species can now be told with reasonable certainty. It is safe to assume that the earliest steps in the evolution of life occurred in the oceans. In this chapter we will examine the systematics of living organisms and sketch how the oceans acted as a cradle for the origin and evolution of life.

The science of systematics was firmly established by Carolus Linnaeus in 1735. Linnaeus divided all living things into two major groups, or kingdoms—the plants and the animals. When the emphasis in biology was the study of the organisms of the terrestrial environment, the division of living things into plant and animals worked well. After all, the obvious organisms of that environment are the plants (Plantae) and the animals (Metazoa). The two-kingdom scheme, however, proved inadequate for classifying the organisms of the marine environment. Plants, which are the dominant photosynthetic organisms on land, contribute only a small fraction of the photosynthesis in the ocean. The dominant photosynthetic organisms of the sea are phytoplankton, or single-celled organisms. Along the shoreline, macroscopic algae are common, but these photosynthetic organisms lack the structural complexity (roots, stems, leaves) that is characteristic of plants.

There are many alternative schemes to the Linnaean two-kingdoms scheme. Some biologists argue for three, four, five, and even (up to) eight kingdoms. We use the generally accepted five-kingdom scheme in this text: Monera (bacteria), Plantae (true plants), Metazoa (multicellular animals), Protista (single-celled organisms such as protozoa and diatoms) and Fungi (molds and mushrooms). There are some 50 phyla in the five major kingdoms. The student of marine biology should be familiar with all the phyla because at least some species in each phylum are found only in the marine environment. In fact, some phyla are almost exclusively marine. Of the five kingdoms of life, only the monerans, protists, and animals are richly represented in the marine environment.

SYSTEMATICS

Classifying and naming species is the work of biologists called **systematists.** To do this, they group organisms into natural units that reflect evolutionary relationships. The data used in making these determinations are the organism's external morphology and internal anatomy, as well as a variety of biochemical and immunological data. A great amount of information also can be obtained by studying the embryonic development of the organisms.

Biological systematists work with a proscribed ordering, or **taxonomy,** in their process of identifying, naming, and classifying organisms. The hierarchy of taxonomic classifications starts with the broad category of Kingdom, and descends with ever-increasing specificity as outlined in the following:

Kingdom
Phylum
Class
Order
Family
Genus
Species

A higher-order category will include one or more of its succeeding categories in its classification. For example, the category Genus contains one (or more) species; a Family classification will include various genera (plural of genus) of different species; etc.

The smallest unit of systematic classification is the **species.** Although difficult to define precisely, a species is generally considered to be an interbreeding group of organisms (i.e., a population) that produce fertile young

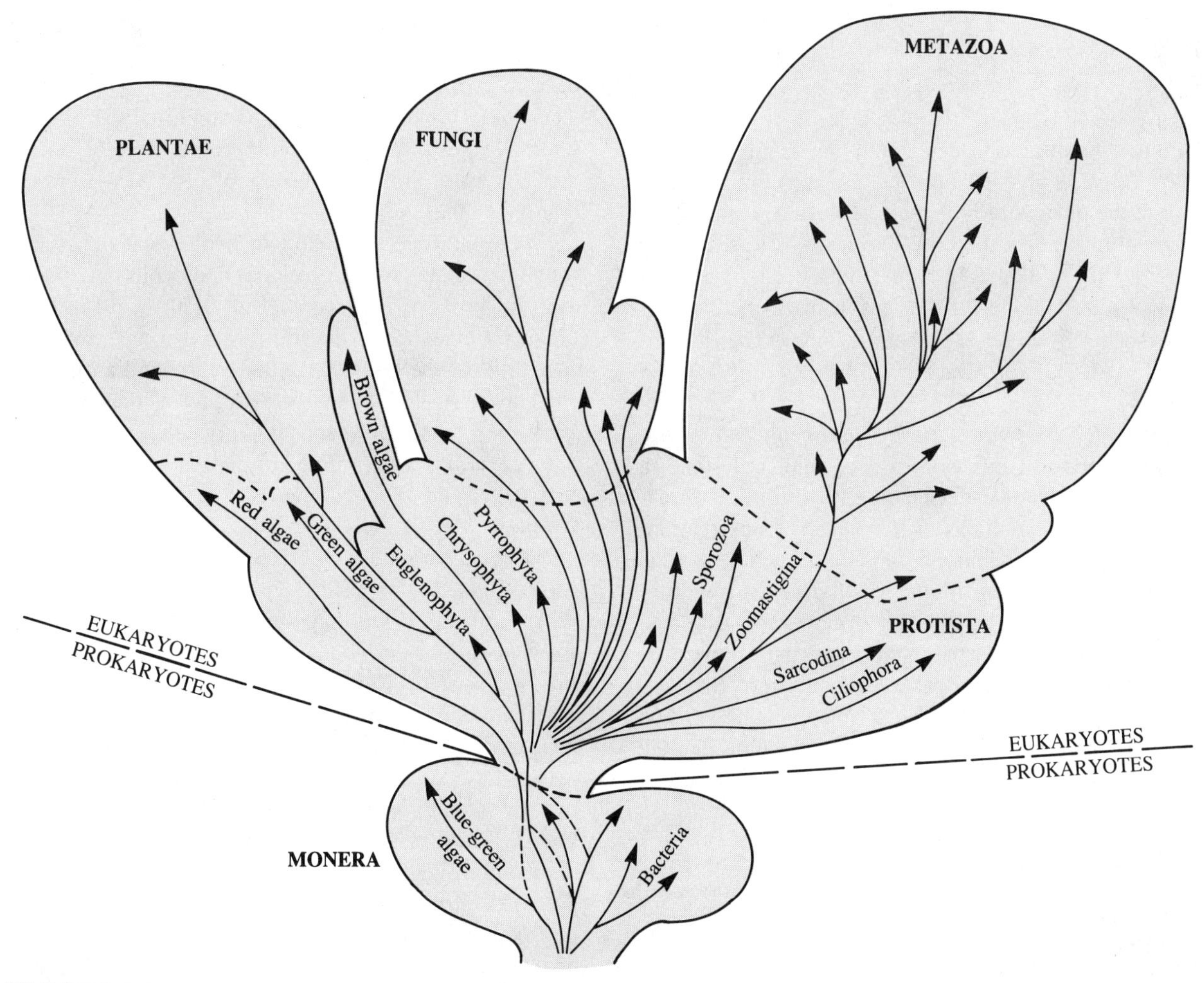

FIGURE 3.1
Probable Evolutionary Affinities of the Five Kingdoms of Life

and that cannot reproduce with any other organisms. Linnaeus developed the convention that each species is identified by a scientific name that has two parts—the genus and the species. (This double-name system is called **binomial nomenclature.**) The genus name is capitalized and both, because species names are latinized, are italicized. For example, the systematic (species) name of the common dolphin is *Delphinus delphis*.

Examples of the taxonomic classification of a common marine animal, the common dolphin, and a common marine plant, giant kelp, are given below:

COMMON DOLPHIN		GIANT KELP
Metazoa	**Kingdom**	Protista
Chordata	**Phylum**	Phaeophyta
Mammalia	**Class**	Phaeophycae
Odontocete	**Order**	Laminariales
Delphinidae	**Family**	Lessoniaceae
Delphinus	**Genus**	*Macrocystis*
delphis	**Species**	*pyrifera*

EVOLUTION

The great number of species of living things on earth today have evolved from the first primitive cells through the process of biological **evolution.** Until 1859, the widely held view in the Western world was that humans, and very likely all animals and plants, were the result of a special creator that put them on the earth in the exact form we observe them in today. On November 24, 1859, Charles Darwin published *On the Origin of Species by Means of Natural Selection, or the Preservation of Favoured Races in the Struggle for Life.* The publication of this work was a major milestone in the human quest for understanding the origins and relationships of living things. After years of observation, which included a four-year-and-eight-month voyage around the world aboard the HMS *Beagle* (see Fig. 1.1), Darwin presented a theory of how organisms had undergone change over the immense period of time represented by the geologic past. Although not yet developed as a science in Darwin's time, genetics would ultimately describe the mechanism of evolution implied by the natural selection process that Darwin envisioned. Darwin wrote,

> Well may we affirm that every part of the world is habitable. Whether lakes of brine, or those subterranean ones hidden beneath volcanic mountains—warm mineral springs—the upper reaches of the atmosphere, and even the surface of perpetual snow—all support organic beings.

The **natural selection** process described by Darwin is achieved through reproduction. The reproductive capacity of all species exceeds that necessary to maintain the population, and at some point the needs of an expanding population exceed the natural resources available to it. At this point, the population becomes stressed, and competition within the population for the available resources increases. Although the progeny of reproduction closely resemble their parents, each organism is unique. Each offspring has a complement of genes different from any organism that has come before or that will follow. According to the laws of natural selection, any individual difference, even a subtle one, that enhances the ability of the organism to survive and reproduce in the environment to which it is subjected may show up again in the genetic mix of some of its offspring. If it is truly beneficial, those offspring that possess the trait will survive and reproduce better than those that don't. The new characteristic is thereafter considered to be an **adaptation** and becomes a common characteristic of a new population. When population differences become pronounced and populations become reproductively separate, a new species is developed. Note that the environment does *not* cause the adaptation; it only selects it to be amplified through differential rates of reproduction, because those organisms that possess it are more successful in that environment than those that do not.

The evidence in support of evolutionary change is found in the studies of molecular and evolutionary biology and in the fossil record. The fossil record includes abundant evidence of the life forms that have inhabited the earth for the last 600 million years, and a diminishing chronicle of algal and bacterial forms that continues back to 3.8 billion years ago (Fig. 3.2). This record shows that life was confined to the sea until 430 million years ago, and included members of all the present-day phyla.

The history of vertebrates is well recorded. The original vertebrates were fish; their fossils are found in rocks some 500 million years old. The amphibians evolved from them and moved onto the land 400 million years ago. More complex vertebrates (reptiles, birds and mammals), well adapted to living on the land, appear in the fossil record by about 200 million years ago (Tab. 3.1). The fossil record tells us that the success of virtually every species was fleeting, and after a time period measured in millions of years species go to extinction. Since the requirements for survival in nature are constantly changing, it is unlikely that all members of a population are perfectly suited to their environment or capable of coping successfully with all the possible changes that may occur. Thus, it appears that each suc-

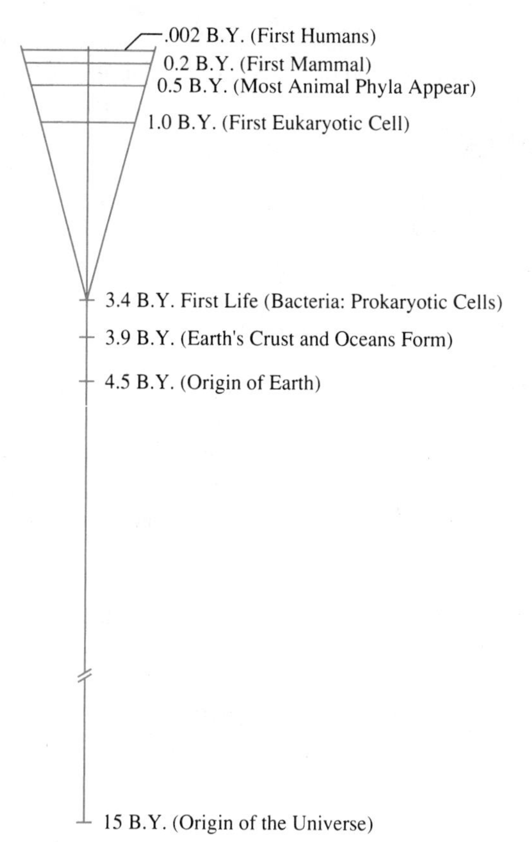

FIGURE 3.2
Major Time Events in the Evolution of Life on Earth

Note the relatively short time between the origin of the earth and the origin of life. The V-shape indicates an ever-increasing biological diversity (number of species) with time. Also note the relatively long time period between the evolution of prokaryotic cells and eukaryotic cells. (B.Y. = billion years)

cess is balanced by a subsequent failure. There is no permanent success.

The amazing aspect of Darwin's theory of evolution is that the discoveries of biological science made during the 125 years since its publication have not required major restructuring of the theory. A tenet of Darwin's theory that has been strongly supported by recent work of molecular biologists is that present-day life forms are descendants of one or a few early forms. Support for the idea of common descent has come from the discovery that all species use a similar genetic code.

One early criticism of Darwin's idea of gradual modification through descent was that changes substantial enough to produce a new species would have to occur rapidly under very stressful environmental pressure. Followers of this **saltationist** view believed new species 'leaped' into existence. The saltationist theory is based based on the 'sudden' appearance in the fossil record of markedly different species of plants and animals, and was suggested by Gregor Mendel's early genetic research with plants. While Darwin believed that traits were blended through reproduction, Mendel's work with peas indicated that crossing a smooth-surfaced pea with a rough-surfaced pea produced a rough-surfaced pea, not a blend of these characteristics.

Although biological evolution is widely accepted as a process, much remains to be learned about the mechanisms and details of the evolutionary past. The ability to observe gene structure in detail undoubtedly will lead to major advances at the molecular level. Whether this new knowledge will lead also to an increased understanding of the process of speciation and rates of evolution remains to be seen.

TABLE 3.1
Evolution of the Vertebrates As Shown by the Fossil Record (B.P. = before present)

Era	Period	Duration (million years)	Start of Period (million years B.P.)	Event
Cenozoic	Tertiary	65	2.5	Major mammal radiation
Mesozoic	Cretaceous	76	67.5	Bird radiation
	Jurassic	54	141	Dinosaurs abundant
	Triassic	30	195	First mammals
Paleozoic	Permian	55	225	Reptiles dominate
	Carboniferous	65	280	Amphibians radiate
	Devonian	50	345	Fish radiate
	Silurian	40	395	
	Ordovician	65	435	First vertebrates
	Cambrian	70	500	

EVOLUTION OF THE FIRST CELL: THE ORIGIN OF LIFE

The time scale of the evolution of life on earth is immense. We are confident that the age of the earth is 4.5 billion years. This is the age of rocks brought back from the moon, which was formed at the same time as the earth. It is hard to comprehend the time scale of a billion years. With the earth's human population at 5 billion and the annual economies of many countries exceeding a billion dollars, we have become familiar with billion as a measure of size. To put in it perspective, however, if you were to count from zero to one billion, using one second to speak each number, and counting 12 hours a day, it would take you 64 years to reach the billion mark.

The first fossil remains of living organisms are single-celled bacteria dating back 3.4 billion years (Fig. 3.2). Since the earth's crust formed and water condensed to form the oceans about 3.9 billion years ago, this means that life evolved in the 500-million-year period between 3.4 billion years ago and 3.9 billion years ago. What can scientists tell us about this process? First, based on studies of other planets in the solar system, we are sure that the early atmosphere was very different than the present one. Our atmosphere is dominated by nitrogen (N_2) and oxygen (O_2). The ancient atmosphere was dominated by ammonia (NH_3) and methane (CH_4), and also had water vapor (H_2O) and carbon dioxide (CO_2). If you pass an electrical discharge (i.e., lightning) through a mixture of these gases, a wide range of organic compounds is produced (Fig. 3.3). These include all the amino acids, sugars, fatty acids, and the molecules that make up the nucleic acids DNA and RNA. In fact, all the basic types of organic molecules found in living organisms can be produced. These kinds of experiments demonstrate that the complex organic molecules that are the building blocks of living systems could have developed in the chemical conditions of the early ocean. Under certain conditions, even molecules that form simple membranes have been produced.

The next step in the evolution of life, that of the combination of these organic molecules into a cellular form capable of reproducing itself and using the existing organic molecules of the environment as a source of energy, is total conjecture at this point. Until laboratory experimentation can produce a metabolically self-regulating cell that uses DNA as genetic information and can reproduce, we will not be able to say anything conclusive about the origin of the first cell. We can, however, confidently say some things about the conditions under which this evolution occurred. Scientists estimate that energy exchanges in the atmosphere produced a concentration of dissolved organic molecules in the ocean of around 1 percent. It is likely that the evolution of the cell occurred where concentrations were greater. Scientist Jack Corliss argues that life could have evolved at oceanic plate boundaries, where hydrogen-rich gases from the earth's interior mixed with carbon-rich gases from the atmosphere (see the special feature at the end of this chapter).

Another scientist, J. Ycas, suggests an alternative for the first cell. He proposes that the first cell might have evolved in estuaries and used the energy from the alternation between high and low salt (sodium) concentration found in estuary conditions to assist metabolism. This energy supplement from the estuary would have allowed the first cell to utilize the relatively low concentration of organic material dissolved in the water. Again, these ideas are just conjecture, and much more scientific work remains to be done in understanding the evolution of the early cell. We are sure, however, that the first cells were heterotrophs that ingested organic molecules dissolved in the ocean, and that they had the structural characteristics of the present-day Monerans (bacteria).

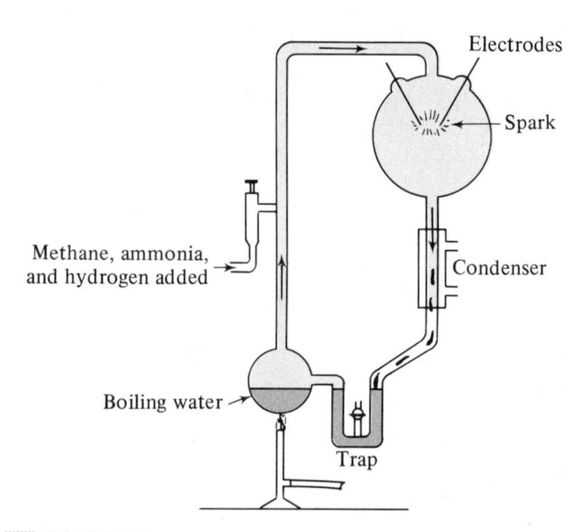

FIGURE 3.3

The apparatus used by Stanley L. Miller in the 1953 experiment that resulted in the synthesis of the basic components of life, amino acids. A mixture of water vapor, methane, ammonia, and hydrogen were subjected to an electrical spark that provided the energy for synthesis. This mixture is thought to resemble the composition of the atmosphere and oceans that existed when 'organic' molecules first formed on earth.

KINGDOM MONERA

Monerans are the organisms we commonly call bacteria. They are defined on the basis of the morphology of the cell (Fig. 3.4). The most important feature is that there is no nucleus. Instead, the genetic information (DNA) is distributed throughout the cell in one long chromosome. Other characteristics of moneran cells include the absence of cell structures such as mitochondria and chloroplasts. Moneran cells are also small (1–10 μm), whereas nonmoneran cells are generally an order of magnitude larger (10–100 μm). These differences set monerans apart from all other living organisms.

The 5,000 or so species of bacteria are classified many ways. The basic problem of bacteria taxonomy is that there is not a great deal of morphological difference between cells that are ecologically or physiologically very different. Older classifications tended to divide the monera into two major groups, or phyla—the blue-green algae and the bacteria. This reflects a time when blue-green algae were thought to be related to other algae. However, the prokaryotic cells of the so-called blue-green algae clearly indicate they are bacteria, and so modern classification calls blue-green algae cyanobacteria ('blue-green bacteria'), and considers them one of the phyla of bacteria. Lynn Margulis and Karen Schwartz in their book *Five Kingdoms* classify bacteria into 16 phyla on the basis of their function. Table 3.2 shows a simplified taxonomy of the bacteria that divides them into (a) anaerobic heterotrophic groups (the fermenting bacteria); (b) the photosynthetic bacteria (both anaerobic and aerobic); and (c) the aerobic heterotrophic bacteria.

Evolution of Monerans

As mentioned in the section on the evolution of life, the first cells are thought to have been monerans. The story of the evolution of the bacteria is a story of the response of the metabolism of the bacterial cell to the problem of obtaining an energy source. The first energy source was likely the pool of dissolved organic molecules that were formed by inorganic processes during the time period of the origin of life (see Fig. 3.2). These first bacteria were of the fermenting type. That is, they were **anaerobic** (did not use oxygen in their metabolism), and obtained energy by chemically converting the organic molecules into simpler inorganic molecules. The rate of utilization of these dissolved organic molecules probably far exceeded their rate of production and soon life on earth was faced with the first energy crisis. The first major evolution of a metabolic system in response to this energy crisis was probably the utilization of light as

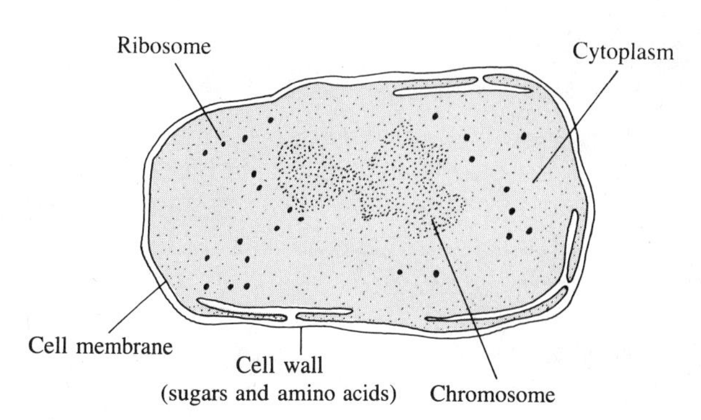

FIGURE 3.4
Prokaryotic Cell

an energy source. This first photosynthesis, however, was likely not the same as is found in present-day plants and blue-green bacteria (Tab. 3.3). The essence of a photosynthetic reaction is to find a source of hydrogen to link to an inorganic molecule to form an energy-rich organic molecule. The common photosynthesis of plants and blue-green bacteria uses water as the source of hydrogen (Tab. 3.3A). The first photosynthesis, however, likely used other molecules as a source of hydrogen. One common molecule in early oceans was hydrogen sulphide gas (H_2S). The first photosynthetic bacteria used light energy to break apart the H_2S molecule to yield hydrogen (Tab. 3.3B). The end product of this kind of photosynthesis is elemental sulphur. The modern-day green and purple sulfur bacteria still have this kind of photosynthesis.

Relying on hydrogen sulphide gas as a source of hydrogen at some point limited the expansion of bacteria on earth. A more plentiful supply of hydrogen was needed to permit further expansion of life. A virtually inexhaustible source of hydrogen is the water molecule (H_2O). The problem with using water as a source of hydrogen is that the chemical bond between the oxygen

TABLE 3.2
Taxonomy of the Bacteria

Group I. Aerobic Heterotrophic Bacteria
Pseudomonads
Nitrogen-fixing
Group II. Anaerobic Fermenting Heterotrophic Bacteria
Methane producers
Lactic-acid bacteria
Closterida type
Spirochaete type
Group III. Autotrophic Bacteria
Cyanobacteria (photosynthetic)
Green and purple sulfur (photosynthetic)
Chemoautotrophic

TABLE 3.3
Basic Metabolic Reactions Found in Monerans

(A) Photosynthesis using water as a hydrogen source. **(B)** Photosynthesis using hydrogen sulphide gas as a hydrogen source. **(C)** Anaerobic respiration. **(D)** Aerobic respiration.

(A) $$6CO_2 + 12H_2O \xrightarrow{\text{Sunlight and chlorophyll}} C_6H_{12}O_6 + 6O_2 + 6H_2O$$
Carbon dioxide + Water → Glucose + Oxygen + Water

(B) $$6CO_2 + 24H_2S + 6O_2 \longrightarrow C_6H_{12}O_6 + 24S + 18H_2O$$
Carbon dioxide + Hydrogen sulfide + Oxygen → Carbohydrate (glucose) + Sulfur + Water

(C) $$C_6H_{12}O_6 \xrightarrow{\text{Anaerobic respiration}} 2C_2H_5OH + 2CO_2 + 2ATP$$
Glucose → Ethyl alcohol + Carbon dioxide + Energy

(D) $$C_6H_{12}O_6 + 6O_2 \longrightarrow 6CO_2 + 6H_2O + 36\ ATP$$
Glucose + Oxygen → Carbon dioxide + Water + Energy

and hydrogen atoms is very strong. It took the evolution of an additional metabolic step in the already existing photosynthetic bacteria before light energy could be used to fracture the water molecule. The bacteria that developed this kind of photosynthesis are related to the blue-green bacteria of today.

Once water was available as a source of hydrogen, the energy available for the evolution and development of bacteria was virtually inexhaustible. There was, however, a devastating effect on the earth's anaerobic bacteria from this new form of photosynthesis. Since the source of the hydrogen was water (H_2O), the remaining oxygen (O_2) was released as a waste product. Oxygen started to accumulate in the atmosphere. This oxygen was toxic to the metabolism of existing life on the planet. The oxygenation of the atmosphere has been described as the most widespread and overwhelming pollution event known in the history of life on earth. However, life adapted. An additional evolutionary development in bacteria that was crucial to this adaptation was the appearance of sex. In bacteria, sex is the recombination of genetic material and is not necesarily related to reproduction. Bacteria may exchange genes at almost any time. They may incorporate genetic material from viruses, from other bacteria, or even from the DNA of a dead bacterium. This genetic plasticity means that bacteria can quickly respond to environmental change. The process of Darwinian natural selection permitted the earth's bacteria to respond to the oxygen pollution by evolving new forms that were not only tolerant of oxygen, but actually used it as part of the metabolism.

Perhaps the most important evolutionary response to the presence of oxygen in the atmosphere (and therefore dissolved in at least surface waters) was the evolution of the cytochrome-c metabolism system. When an organism breaks down an organic substrate as a source of energy, the energy is stored in the molecule adenosine triphosphate (Fig. 3.5). The number of ATP molecules produced for each molecule of organic material is a measure of the energy available. If we examine the energy yield of a molecule of the sugar glucose, we find that during anaerobic metabolism generally two molecules of ATP are produced for each molecule of sugar (Tab. 3.3C). In **aerobic** metabolism (one that uses oxygen and cytochrome-c metabolism), however, there are generally 36 molecules of ATP produced. The difference is due to the presence of the cytochrome-c system. We all know that oxygen promotes combustion and the release of energy. Essentially, the cytochrome-c, or oxidative metabolism, system channels the oxidation power of oxygen to extract the maximum energy possible from the organic molecule. The end result of the oxydative metabolism is that our glucose molecule is converted to carbon dioxide and water (Tab. 3.3D).

This is a superficial sketch of the kinds of metabolic evolution that took place as bacteria colonized the earth in the more than 2-billion-year time period from 3.4 billion years ago until at least 1 billion years ago. The important point to note is that virtually all the metabolic machinery found in organisms of the other four kingdoms evolved first in the bacteria. Simply put, the rest of evolution has been a process of packaging the metabolism evolved in the bacteria. And how much of

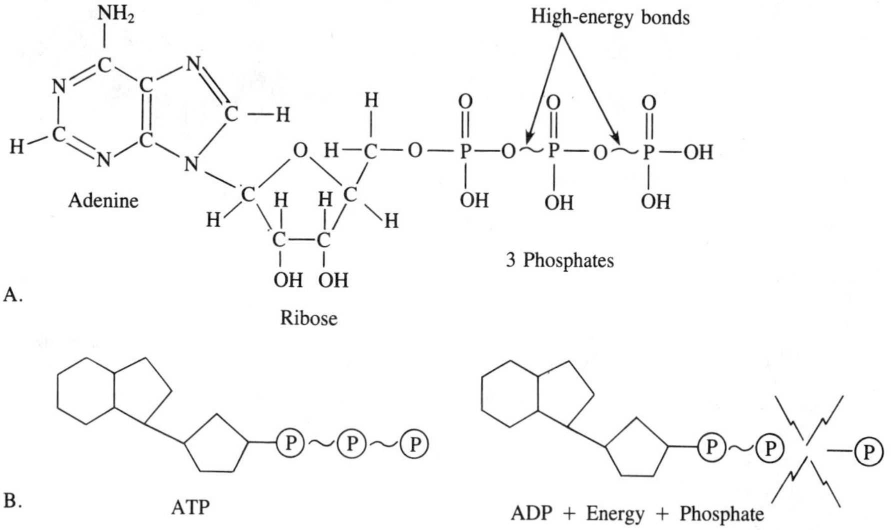

FIGURE 3.5
(A) A model of the compound adenosine triphosphate (ATP). The letters identify positions of atoms of the elements carbon (C), hydrogen (H), nitrogen (N), oxygen (O), and phosphorus (P). **(B)** Energy is released when the high-energy bond connecting the second and third phosphate groups is broken.

this evolution occurred in the ancient oceans? Probably most of it. We think that because the oceans were the most extensive bodies of water on the planet. However, bacteria are ubiquitous in all aquatic environments today, and there is no direct evidence to prove that the evolutionary steps all took place in the marine environment.

THE EUKARYOTIC CELL

If all monerans are prokaryotic cells, what, then, of the rest of life? All species belonging to the four other kingdoms (Protista, Metazoa, Plantae, and Fungi) have similar cells called eukaryotic cells. The eukaryotic cell (Fig. 3.6) is larger and structurally more complex than the prokaryotic cell. The DNA is in the form of multiple chromosomes and is enclosed in a nuclear membrane. Other differences (Tab. 3.4) include the presence of complex cell organelles—the mitochondria—for aerobic metabolism, and the chloroplasts for photosynthesis. In eukaryotic cells, cell division is a complex affair called **mitosis,** where pairs of chromosomes are separated and guided into daughter cells. Sex in eukaryotic cells involves the passage of one chromosome from each pair to each gamete (sperm or egg).

What do we know of the evolution of the eukaryotic cell? There is no direct evidence to address this question. We do know that eukaryotic cells are ancient. The first fossils of eukaryotic cells date from at least 1 billion years ago and perhaps as early as 1.6 billion

TABLE 3.4
Comparison of the Major Differences of Prokaryotic and Eukaryotic Cells

Prokaryotes	Eukaryotes
Mostly small cells (1-10 μm); All are microscopic single cells.	Mostly large cells (10-100 μm). Some are microbes; most are large, multicellular organisms.
DNA in nucleoid, not membrane-bounded. No chromosomes.	Membrane-bounded nucleus containing chromosomes made of DNA, RNA, and proteins.
Cell division direct, mostly by binary fission. Sexual systems rare; when sex does take place, genetic material is transferred from donor to recipient.	Cell division by various forms of mitosis. Sexual systems common; equal participation of both partners (male and female) in fertilization. Alternation of diploid and haploid forms by meiosis and fertilization.
Multicellular forms rare. No tissue development.	Multicellular organisms show extensive development of tissues.
Mitochondria absent; enzymes for oxidation of organic molecules bound to cell membranes (not packaged separately).	Enzymes for oxidation of 3-carbon organic acids are packaged within mitochondria.

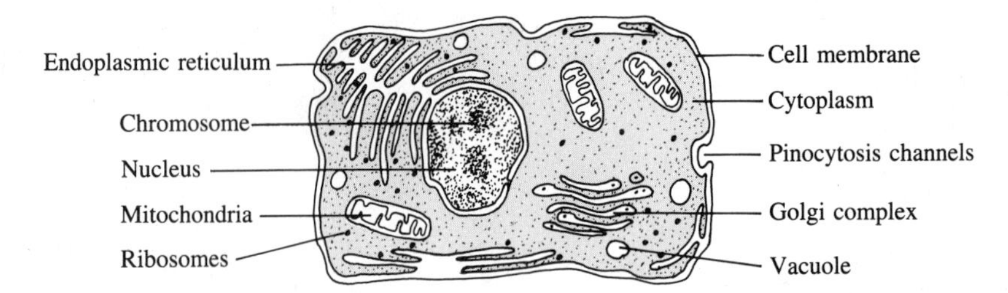

FIGURE 3.6
Eukaryotic Cell

In addition to the cell wall, cell membrane, chromosomes, ribosomes, and cytoplasm of the prokaryotic cell, the eukaryotic cell has a *nucleus* which contains the chromosomes. Food particles are ingested through *pinocytosis channels* and stored in *vacuoles.* Enzymes that release energy through respiration are concentrated in the *mitochondria,* while others are associated with the *endoplasmic reticulum.* Photosynthetic eukaryotic cells also contain *chloroplasts* within which photosynthesis occurs.

years. These first eukaryotic cells had the typical nucleus, mitochondria, and chloroplasts (Fig. 3.6). Although there is no direct evidence of the origin of eukaryotic cells, there is fascinating speculation. Some biologists argue that the eukaryotic cell evolved from a symbiosis of a number of bacterial cells living within a single membrane. This hypothesis argues that mitochondria and chloroplasts were once free-living bacteria that have become symbiotic, and that the system of microtubules found in the eukaryotic cell is also derived from a bacterial symbiont. There is evidence for this point of view. Both mitochondria and chloroplasts have their own DNA that is not part of the genetic information of the nucleus, and it is used for producing some of the protein needed when chloroplasts or mitochondria divide.

Whatever the process of the evolution of the eukaryotic cell, we know it served as the building block for evolution of the other four kingdoms of organisms.

KINGDOM PROTISTA

Protists are unicellular or colonial organisms with eukaryotic cells. Some are photosynthetic, while others are heterotrophic. When protists occur as colonial forms (i.e., the algae), there is little difference in the morphology of each cell. That is, the protists do not form different tissues. This feature distinguishes colonial protists from plants or animals.

There are some 50,000 species of protists divided into at least eight phyla (Tab. 3.5). Six of these are important in the marine environment—the Protozoa, the Chrysophyta (diatoms and coccolithophores), the Pyrrophyta (dinoflagellates), the Chlorophyta (green algae), the Phaeophyta (brown algae), and the Rhodophyta (red algae).

The evolution of the protists is not well known. By definition, once a eukaryotic cell had evolved from the monerans, the protists existed. It is likely that protista groups developed from a number of different bacteria types (Fig. 3.7). Although we have evidence of protists early in the fossil record, it was not until recently that they appeared in large numbers. Each of the diatoms, dinoflagellates, protozoa, and algae has a fossil record.

The Diatoms

Diatoms evolved only relatively recently. The first clear fossil record is from the Cretaceous period, some 140 million years ago. This makes the diatoms one of the youngest groups not only of the protists but of other kingdoms as well. Most animal groups are much older.

TABLE 3.5
Phyla of the Kingdom Protista

Phylum Protozoa	Unicellular heterotrophs including radiolarians, foraminiferans, ciliates, amoebas, and tintinnids
Phylum Euglenophyta	Mostly fresh-water, green, unicellular
Phylum Chrysophyta	Diatoms, coccolithophores, and flagellates
Phylum Pyrrophyta	Dinoflagellates
Phylum Chlorophyta	Green algae
Phylum Rhodophyta	Red algae
Phylum Phaeophyta	Brown algae
Phylum Gymnomycota	Slime molds, mostly fresh-water

FIGURE 3.7
Probable Evolutionary Relationships Between the Monera and Protista

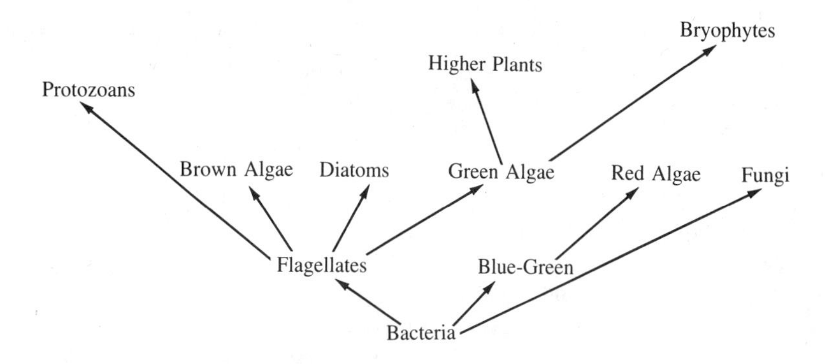

The Dinoflagellates

The origin of the dinoflagellates in the marine environment dates back to about 435 million years ago, although there are possible fossil records back to 900 million years ago. Dinoflagellates did not become common until about 200 million years ago.

The Algae

Calcareous green algae of the type important to the formation of coral reefs have a fossil history going back over 500 million years ago. Filamentous green algae are known from 400 million years ago. Red algae fossils date back to between 300 and 400 million years ago. There are no clear fossil remains of the brown algae and the evolutionary history of the group is unknown.

The Protozoa

Tintinnid protozoa are not common in the fossil record until some 140 million years ago which means they are a relatively recent group in the marine environment. Radiolarians, on the other hand, evolved earlier, and fossils are common from 400 million years ago. Foraminifera, which have a good fossil record, became common for the first time some 400 million years ago. The soft-bodied amoeba and ciliates have no clear fossil record.

The protists are the ancestors of the remaining three kingdoms. Probably the plants evolved from green algae, the animals from protozoa, and the fungi from some other protist.

KINGDOM METAZOA

The animal kingdom has over 1 million described species, divided into over 30 different phyla (Tab. 3.6). In the study of zoology, animals are traditionally divided into the vertebrates, which have a backbone, and the invertebrates. Often the invertebrates are called the 'lower animals' and the vertebrates the 'higher animals', giving the erroneous impression that animals with backbones are an evolutionary development from the invertebrates. Invertebrate phyla have a long fossil record of their own that goes back at least 500 million years.

TABLE 3.6
Phyla in the Kingdom Metazoa

Phylum	Number of Species
Porifera	10,000
Cnidaria (Coelenterata)	9,000
Ctenophora	90
Platyhelminthes	12,700
Nemertea	650
Mesozoa	50
Nematoda	10,000
Gastrotricha	400
Kinorhyncha	100
Gnathostomulida	80
Rotifera	1,800
Nematomorpha	230
Acanthocephala	500
Brachiopoda	280
Bryozoa	4,000
Entoprocta	60
Phoronida	10
Sipuncula	300
Echiura	150
Pogonophora	80
Tardigrada	400
Pentastomida	90
Mollusca	100,000
Annelida	8,700
Arthropoda	923,000
Chaetognatha	50
Hemichordata	80
Echinodermata	6,000
Chordata	39,000

Both vertebrates and invertebrates have been evolving for a long time period, and all animal groups have a rich evolutionary history filled with many examples of modern evolutionary specialization. Many species of invertebrates are just as 'high' on the evolutionary scale as the vertebrates.

Invertebrates are generally smaller than vertebrates, and loosely the two groups are referred to as 'large animals' (vertebrates) and 'small animals' (invertebrates). In the marine environment the mammals, fish, and birds make up most of the large animals (the squid and octopus are notable exceptions), but the small animals dominate in number. For example, if you explored a biologically rich rocky shore in temperate latitudes, you would find only a few invertebrate animals that exceeded a size of 10 cm, or 4 in (e.g., sea stars, sea cucumbers, and perhaps a crab or anemone). There would be many more animals in the range of 1–10 cm (e.g., barnacles, chitons, snails, crabs, anemones, flatworms, polychaete worms, brittle stars). We could make a list of perhaps 50 or so species that would fall into this size range. If we examined the animals in the size range of 1 mm–1 cm, the richness would be much greater. We would find literally hundreds of species of a great variety of phyla. As we discuss representatives of various phyla, keep this fact in mind. Because the larger species of a group are easier to collect, examine, and experiment with, they often are the ones that receive the most attention.

Let us examine the formal definition of an animal. According to Barnes (1980), "Metazoans are multicellular, motile, heterotrophic organisms that develop from embryos. The gametes never form within unicellular structures but rather are produced within multicellular sex organs or at least within surrounding somatic cells." Notice that this definition emphasizes reproduction and structure, and does not rely on our commonsensical image of an animal. When we think of animals, some features that come quickly to mind are the ability to move, a definite body shape, a predictable size, a sensory system for monitoring the environment, and a centralized nervous system (brain) to evaluate the environment and make appropriate responses. In the marine environment, however, many animal species are **sessile,** that is, they are permanently attached to or embedded in some sort of substrate. Sessile species, therefore, are capable of only very limited movement. Some animals, such as sponges, sea anemones, and jelly fish, have no central nervous system, and sponges, in fact, do not even have nerves or a sensory system. Thus our basic definition of what an animal is emphasizes two fundamental characteristics common to all animals: First, animals are multicellular, and different cell types have different structural and functional properties. Second, animals reproduce by the production of a single cell (through the fusion of an egg and sperm) that is capable of developing into all the various cell types of the adult.

Now our formal definition makes more sense. An essential part of being an animal is development through an embryo. Another important characteristic is that animals are heterotrophic; that is, they rely on other organisms as a source of food. Animals have to eat to obtain these energy-rich molecules produced by other organisms. Even though the characteristic of being a heterotroph holds for all animals, it is not diagnostic. Other organisms including bacteria, protozoa, and some dinoflagellates are also heterotrophs.

Metazoan Evolution

Evolutionary relationships of the metazoan groups have been derived from work with marine invertebrates. Most of the evolution of the animal groups occurred in the sea, and the **phylogeny** (evolutionary history) of the invertebrates is an important topic in marine biology.

Biologists think that the earliest multicellular animals evolved from one or, at the most a few, protozoan types (probably flagellated protozoa), and that the remainder of the animal phyla evolved from those early forms (Fig. 3.8). The fossil record of early animals is not good. Most phyla were well established over 500 million years ago, and the first animals probably had few hard parts suitable for fossilization. The evidence for phylogeny is derived primarily from the characteristics of early embryonic development within each group.

The phyla that became established early in the evolutionary history are referred to as the **lower phyla** and the later ones as the **higher phyla.** The terms *lower* and *higher* are generally applied to larger taxonomic groups such as phyla, classes, and orders. The phylogeny of animals also refers to groups as being either primitive or specialized. Generally, these terms describe the evolution of species within a phylum. **Primitive species** are those that show many of the characteristics of the phylum as it was in the early stages of its evolution. **Specialized species** are those that have departed to a greater or lesser degree from the general phylum characteristics. Primitive species are not always simpler than specialized species. For example, there is a type of barnacle that is parasitic on crabs that is little more than a bag of cells living off the tissues of the crab. Even though this parasite has a structural simplicity, it is a specialized species in its evolutionary history.

The lowest phylum is generally considered to be the sponges (Porifera). Sponges have little evolutionary affinity with other invertebrate phyla and are considered to be an early branch of animals that has not contributed to the evolution of other phyla. Next on the evolution-

FIGURE 3.8
Metazoan Phylogeny

This phylogenetic tree maps the probable evolutionary relationships of the Metazoa.

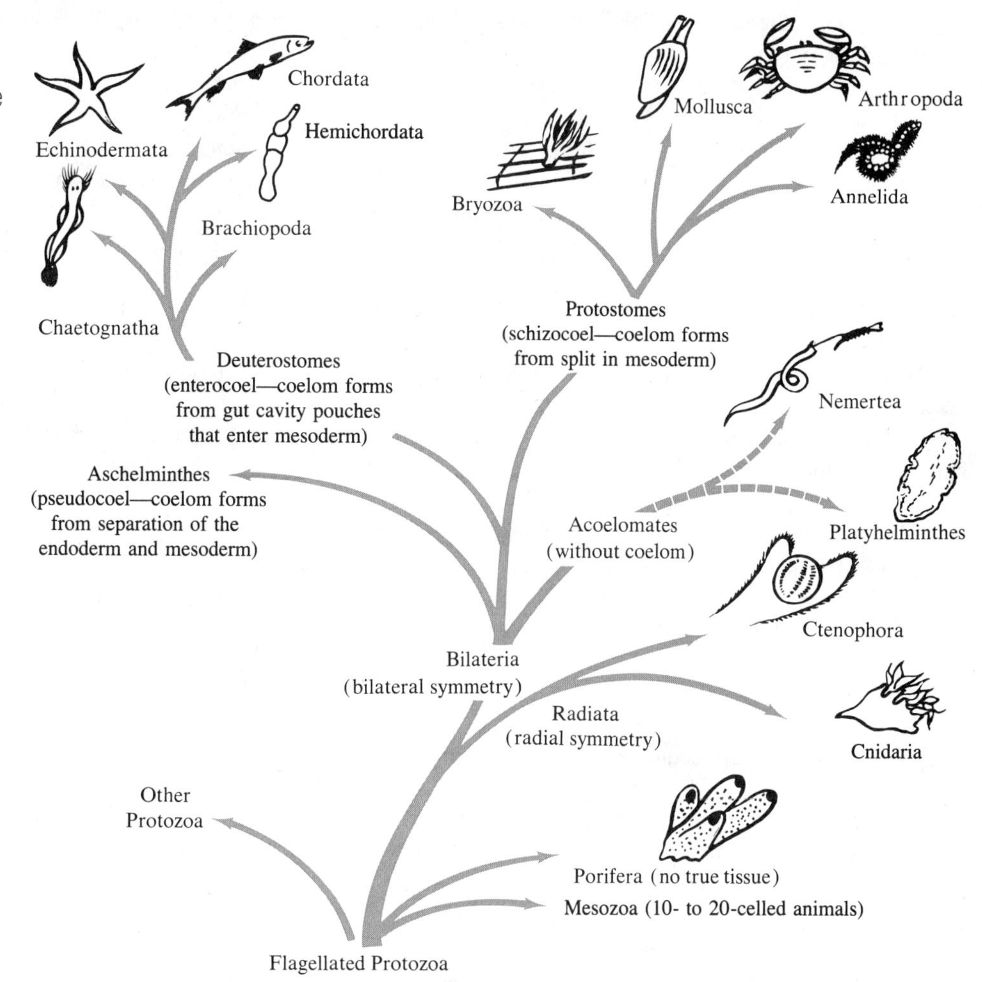

ary tree (Fig. 3.8) come the phyla with radial symmetry: Cnidaria and Ctenophora. Biologists believe that these two phyla, like the sponges, are offshoots of the main evolutionary trunk and the base stock of metazoan evolution. Even though the phylum Echinodermata (e.g., sea stars, sea urchins) have basically a radial symmetry, they are not regarded as radiate phyla because their radial symmetry is clearly a secondary feature. This is evident from the bilateral symmetry of the larval forms.

From the radiate phyla we move to groups with bilateral symmetry. Most phyla have bilateral symmetry. The flatworms are believed to be the lowest phylum having this characteristic. It is the flatworms' evolutionary ancestors, now extinct, that are the common evolutionary stock of most of the animals.

The lowest, or earliest, bilateral animals had a solid construction. This structural feature is characteristic of two phyla important in the marine environment: the flatworms (Platyhelminthes) and the ribbon worms (Nemertea). Although the flatworms are bilateral, they still lack two important features common in most higher animals: a digestive system with separate mouth and anus, and an internal body cavity. Both these features increase the functional efficiency of animals. A gut with separate intake and elimination permits the separation of digestive functions along its length, and a body cavity provides a space in which organs may develop.

Fluid-filled body cavities may also act as a hydrostatic skeleton. A hydrostatic skeleton keeps the body shape by the contraction of muscles against the fluid space—much like applying pressure at one portion of a balloon, thereby causing a change of shape at another. A sea anemone is an example of an animal with a hydrostatic skeleton.

Body cavities are of two types: coelomate and pseudocoelomate. Pseudocoels are cavities that form during embryonic development as a space between the inner and outer layers of cells. A coelom is a cavity that forms in the center of cellular tissue that is between the inner and outer layers. (This tissue is called mesoderm.) The details of embryonic development are not, in themselves, important for you to learn. They are, however, found almost without exception in all species of different animal groups and are clear indications of evolutionary relationships.

A total of sixteen phyla are coelomate, including the major phyla of invertebrates—annelids, molluscs, echinoderms, and arthropods—and the chordates, which include vertebrates. The pseudocoelomate group encompasses seven phyla and as a group are called the Aschelminthes. Aschelminthes are small, generally tube-shaped worms. None are particularly common in the marine environment except for the roundworms, or Nematoda.

In order to understand the differences between the two types of body cavities, we must examine the early events in the development of animal embryos. All animals go through a blastula stage in embryonic development. The blastula is a sphere of cells. In animals that have a pseudocoel, the center of the blastula remains hollow, and this fluid-filled space persists as a body cavity in the adult. In coelomate animals the process is different. The body cavity develops inside the mesoderm (a cell mass that fills the cavity of the early blastula). Although coelom development differs in various species, it provides the basis for a major division of animal phyla.

The coelomate phyla are divided into two groups: the protostomes and the deuterostomes. A **protostome** is an animal in which the embryonic opening, or blastopore, forms the mouth of the adult. A **deuterostome** is an animal in which the blastopore does not contribute to the formation of the mouth. The first group includes the molluscs, annelids, and arthropods. The protostomes are a major group in the animal kingdom and include many highly specialized species with complex anatomical and physiological systems.

The deuterostomes include the enchinoderms, hemichordates, and chordates. Echinoderms (sea stars and sea urchins), for the most part, are slow and some are entirely sessile. They have secondarily developed radial symmetry. The hemichordates are mostly sessile worms that burrow in soft sediments. In this second line of coelomate animals, only the chordates (the phylum that humans belong to) have evolved the complex animal characteristics that are familiar to us.

Coelomate animals are classified into protostome or deuterostome, primarily on the basis of embryology. The second, third, and fourth cell divisions of the developing embryo can be either spiral and radial. Radial division is shown in Figure 3.9. With spiral cleavage, which is more difficult to diagram, the cells in these early divisions assume a spiral alignment. Cleavage in these early stages is also described as being either determinant or indeterminant. In indeterminant cleavage, all of the cells in early divisions are capable of producing an adult form. The occurrence of identical twins in humans is a demonstration of indeterminant cleavage. The experimentally produced quadruplets from the four-cell stage of sea stars is another example. In determinant cleavage, each cell of the early embryo is required to produce some part of the adult. No individual cell, from the two-cell stage onward, can develop into the adult form. In protostomes, early cell division is characterized by spiral, determinant cleavage; in deuterostomes, by radial, indeterminant cleavage.

This scheme of animal phylogeny helps to group together the diverse phyla (listed in Tab. 3.5) and shows the evolutionary relationships between them. The characteristics important to phylogeny are embryonic, however, and thus they are of little help in dealing with the diversity of adult animals. Therefore, the important structural characteristics of each phylum must be considered separately. This is done in Chapters 5 through 8.

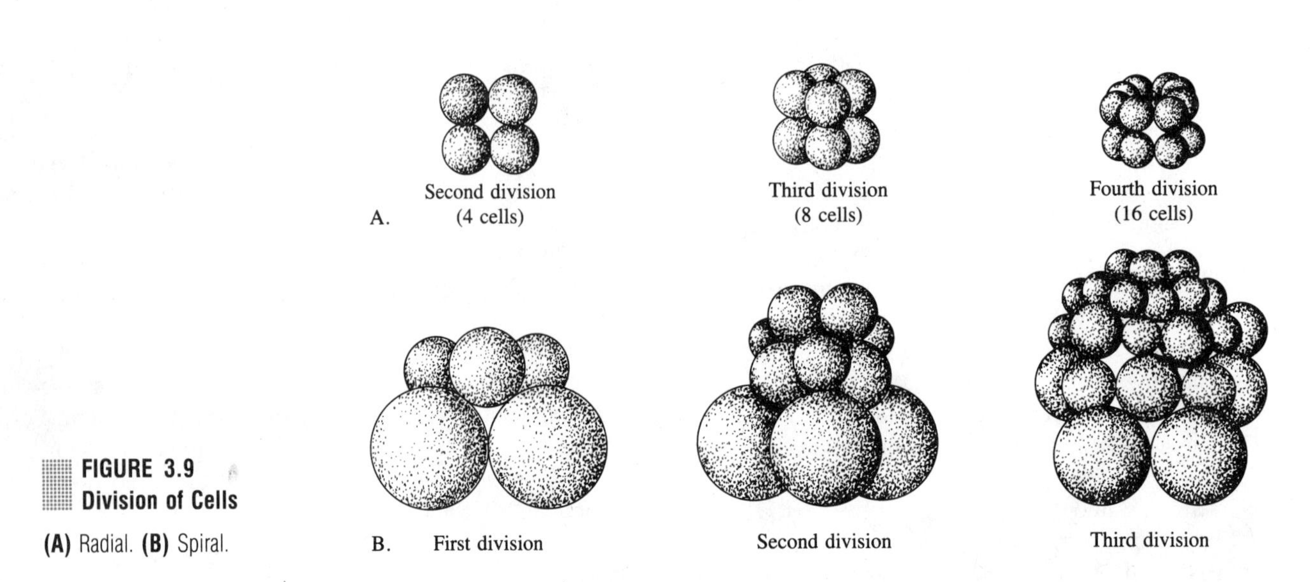

FIGURE 3.9
Division of Cells
(A) Radial. **(B)** Spiral.

KINGDOM PLANTAE

Plants are multicellular, photosynthetic, sexually reproducing eukaryotes. Plants form the dominant vegetation on land but are virtually absent in the ocean. Of the two major phyla (Tab. 3.7), only a few species of flowering plants are able to complete their life cycle submerged in salt water. Another more numerous group is able to tolerate partial submergence in salt water. In general, plants are not particularly important in the marine environment.

The plants, unlike the monerans, protists, and metazoans, did not evolve in the marine environment. Plants' evolutionary affinity with the protists is clear and most biologists agree that they evolved from freshwater green algae. Early fossils are known from 460 million years ago. The first flowering plants, however, do not occur in the fossil record until only 125 million years ago. This means that the presence of plants in the ocean is very recent in geological time.

KINGDOM FUNGI

Fungi are eukaryotes that do not have photosynthesis, that reproduce by the production of spores, and generally are organisms composed of masses of filaments. Essentially, fungi are a multinucleate mass of cytoplasm enclosed in a branching system of tubes. A single tube is a hypha and the hyphae collectively form a mycelium. Fungi are ecologically closely related to plants. Like the plants, fungi are mostly terrestrial, and of the 100,000 species only a few occur in the marine environment.

The evolutionary history of fungi is not well understood, although it is thought that the ancestor is among the protists. The earliest fungus fossil is from 300 million years ago and was in close association with an ancient plant, testimony to the close relationship between these two kingdoms.

TABLE 3.7
Phyla of the Kingdom Plantae

Phylum Bryophyta	Mosses, hornworts, and liverworts (no marine species)
Phylum Tracheophyta	Ferns, gymnosperms, and angiosperms (only the angiosperms have marine species)

SUMMARY

There are more than 500,000 species of marine organisms ranging from bacteria to the blue whale. Taxonomic classification, which ideally reflects the evolutionary history of each species, is used to bring order to this large diversity of life. Of the five kingdoms of life (excluding the viruses), only two—animals and protista—are richly represented in the marine environment. The monerans (bacteria) are moderately represented and plants and fungi have only a few marine species (Tab. 3.8).

Although direct confirmation is not possible, it seems most probable that life evolved in the oceans. The first living things developed rapidly after the earth was formed 4.5 billion years ago. It took 600 million years for the earth's crust and oceans to form. Only 500 million years after that, the first cells (bacteria) had evolved. These first cells were prokaryotic and dominated the earth for 2.4 billion years. At that time, 1 billion years ago, eukaryotic cells evolved. These eukaryotic cells provided the stock for the evolution of the other four kingdoms.

The kingdom Protista has a varied evolutionary past. Dinoflagellates date back to 435 million years ago, although they became common only 200 million years ago. Green and red algae date back between 400 and 500 million years ago. In the protozoa, radiolarians and foraminifera date back at least 400 million years, whereas the tintinnids evolved around 140 million years ago. The diatoms are relatively recent and evolved around 140 million years ago.

By 500 million years ago, most animal phyla were well developed. The animals are believed to have evolved from one or more ancient protists, probably the ancestors of modern ciliated and flagellated protozoa. Little fossil evidence remains of the evolutionary story of animals beyond 500 million years ago. There is, however, strong evidence of the evolutionary linkages between phyla, much of it based on characteristics of early embryology.

Sponges probably evolved first, but their morphology is so different from other animal groups that they are not assumed to be involved in the ancestry of other animal phyla. The simple worm forms were likely the evolutionary stock of animal evolution.

TABLE 3.8
Marine Species As a Percentage of World Total

Kingdom	Phylum	Total Number of Species	Percent of Marine Species
Monera	Schizomycophyta (bacteria)		
	Cyanophyta (blue-green bacteria)	5000	50
Protista	Pyrrophyta (dinoflagellates)	1200	93
	Chrysophyta (diatoms, coccolithophores, silicoflagellates)	12,600	50
	Protozoa		
	Sarcodina (foraminiferans, radiolarians)	30,000	50
	Ciliophora (tintinnids)		
	Chlorophyta (green algae)	7000	13
	Phaeophyta (brown algae)	1500	99
	Rhodophyta (red algae)	4000	98
Mycota (fungi)	Amastigomycota (filamentous fungi)	500,000	0.001
	Gymnomycota (slime molds, yeasts)		
Metaphyta (plants)	Bryophyta (mosses, liverworts)	25,000	0
	Spenophyta (horsetails)	15	0
	Lycophyta (club mosses)	1,000	0
	Pterophyta (ferns)	12,000	0
	Gymnospermophyta (cone plants)	722	0
	Angiospermophyta (flowering plants)	235,000	0.08
Metazoa (animals)	(29 phyla)	1,167,000	32
Total		2,002,037	21 (424,171)

QUESTIONS AND EXERCISES

1. Prepare a table that summarizes the criteria for dividing living things into five different kingdoms.
2. What was the first energy crisis that life on earth faced?
3. What was the major global pollution event associated with the evolution of bacteria?
4. Using the example of a molecule of sugar, how is aerobic respiration more efficient than anaerobic respiration?
5. What are the major strategies of monerans for obtaining energy to support metabolism?
6. Compare and contrast the characteristics of prokaryotic and eukaryotic cells.
7. Why are diatoms and dinoflagellates not plants?
8. Using the following terms, construct a table showing the evolutionary relationship of the animal phyla: radial symmetry, bilateral symmetry, mesoderm, coelomate, pseudocoelomate, protostome, deuterostome, spiral cleavage, radial cleavage, determinate cleavage, indeterminate cleavage.

REFERENCES

Barnes, R.D. 1980. *Invertebrate zoology.* Philadelphia: Saunders.

Crane, J.M. 1973. *Introduction to marine biology: A laboratory text.* Columbus, OH: Merrill.

Haeckel, E. 1974. *Art forms in nature.* New York: Dover.

Margulis, L. 1982. *Early life.* Boston: Science Books International.

Margulis, L. and **Schwartz, K.** 1982. *Five kingdoms: An illustrated guide to the phyla of life on earth.* New York: Freeman.

SUGGESTED READING

BIOSCIENCE

Krogmann, D.W. 1981. Cyanobacteria (blue-green algae): Their evolution and relation to other photosynthetic organisms. 31(2):121–124.

NATURAL HISTORY

Ahmadjian, V. 1982 The nature of lichens. 91(3):30–37.

SCIENCE

Whittaker, R.H. 1969. New concepts of kingdoms of organisms. 163:150–160.

SCIENTIFIC AMERICAN

Cairns-Smith, A.G. 1985. The first organisms. 252(6):90–101.

McMenamin, M.A. 1987. The emergence of animals. 256(4):94–101.

DID LIFE BEGIN ON THE OCEAN FLOOR?

In our discussion of the origin of life on earth, we described a process by which it may have occurred in the surface waters of the ocean. Critics of this hypothesis say the concentration of organic matter would have been too dilute to support the heterotrophic cells postulated. Moreover, the process by which such primitive cells developed the complex cell membrane needed by heterotrophic forms to transport organic compounds is not clearly understood.

John B. Corliss and colleagues have proposed an alternative hypothesis, that life on earth began in the hydrothermal (hot water) vents of the ocean floor. They believe the intense meteorite bombardment of the inner solar system from about 4.2 to 3.9 B.P. (billion years before the present) would have provided up to one-fifth of the earth's mass and produced major volcanic activity. The earth's surface, made molten by this energy release, cooled as the bombardment diminished. Crustal plates formed and were carried across the earth's surface by the convecting magma beneath, initiating the process of global plate tectonics 3.9 B.P. Soon the oceans filled the low-lying basins. Where crustal plates were pulled apart by mantle convection, magma rose to form new ocean floor. Newly formed crust cracked as it was quickly cooled by contact with ocean water, and as the stressed plates were pulled apart, linear fissures developed. As soon as the ocean water migrated down through the broken crust to be heated by the underlying magma and rose again to the ocean floor as a hydrothermal spring, the stage was set for the origin of life.

According to Corliss's hypothesis, the first protocells on earth appeared almost instantaneously after the formation of the first hydrothermal spring, as in the following steps:

1. Heat and carbon dioxide (CO_2) released by the magma interacted with the seawater near the magma contact.
2. The gases released by the crystallization of the magma—hydrogen (H_2), methane (CH_4), ammonia (NH_3), hydrogen sulfide (H_2S), and others—interacted with the heated water.
3. Heated to temperatures as high as 900+°C(1,900°F) at the magma interface, the rising water contained organic compounds of low molecular weight synthesized at high temperature.
4. The water continued to rise, cooling rapidly, through fractures coated with a catalyst, saponite (a magnesium clay formed by the alteration of basalt). In the presence of the catalyst, low-molecular-weight organic molecules polymerized into complex organic molecules that combined to form protocells.

5. Exiting vent water carried out the protocells at temperatures usually below 100°C (212°F), although some existing vents are known to spew 350°C (740°F) water into the 2°C (36°F) ambient water at the ocean floor.

Fossils found in rocks 3.5 to 3.8 B.P. lend support to the deposition of protocells on the ocean floor around hydrothermal vents. The three ancient rock locations where such evidence is available are the Isua in southwest Greenland (3.8 B.P.), Onverwacht in South Africa (3.5 B.P.), and Warrawoona in western Australia (3.5 B.P.). There is a remarkable resemblance between filamentous 'organiclike fossils' in the Onverwacht and Warrawoona and organic structures in scanning electronmicrographs of hydrothermal deposits recovered at 21°N in the East Pacific Rise in 1979 (Fig. 1). All the ancient fossil locations contain rocks interpreted by some investigators to have been deposited in a hydrothermal vent environment.

The chemosynthetic bacteria isolated from 21°N and Galapagos vents may help answer the question of whether or not the first life forms were heterotrophs. Maybe

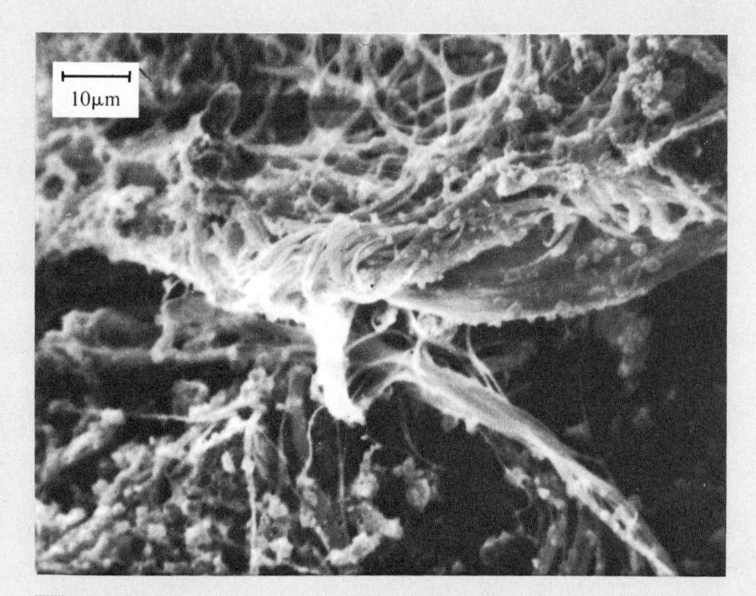

FIGURE 1

Scanning electronmicrograph of thin strands of possibly organic material frequently found in hot chimney rocks at the 21° N site off the tip of Baja, California. These structures resemble fossil structures observed in ancient rocks containing the earliest evidence of life on earth. (Photo courtesy of John Baross, Oregon State University.)

they weren't. If life did originate at the hydrothermal vents as anaerobic chemosynthetic forms, it could have been sustained by the constant effluent of hydrothermal gases—CO_2, H_2S, NH_3, H_2, and possibly H_2SO_4. From this beginning, evolution could have produced the more complex macromolecules that were heterotrophic and reproduced by binary fission.

These early forms, *archaeobacteria,* may have reproduced by budding off chains of new individuals as is observed in protocells synthesized in the laboratory. They may have been *methanogens* that reduced CO_2 and SO_4 (sulphate) to CH_4 by using H_2. Present-day prokaryotes, *eubacteria,* may have evolved from such methane-producing archaeobacteria.

The occurrence of petroleum condensate at the seabed surrounding sediment-covered hydrothermal vents in the Guaymas Basin of the Gulf of California indicates that the plume water must contain large amounts of methane and higher-molecular-weight organic matter. Further investigation may show that such protocells as have been described here are still forming in the hydrothermal vents of today's oceans.

4

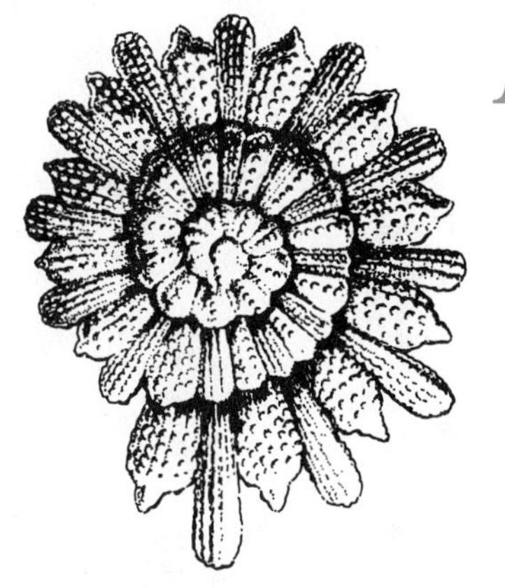

MONERA, PROTISTA, FUNGI, AND PLANTAE

Four kingdoms in one chapter! Clearly the bias in an introductory marine biology course is with the animals (see Chs. 5–8). For the most part this is because the animals are large, conspicuous, and familiar to the beginning student. However, the remaining four kingdoms (Tab. 4.1) are important to marine biology, as we will demonstrate in this chapter.

The Monera are the bacteria. Microscopic bacteria are widespread in the marine environment. These bacteria include types that are capable of **photosynthesis** and **chemosynthesis,** as well as the more commonly known decomposers. This last type are collectively capable of breaking down virtually all types of organic material. The cyanobacteria are photosynthetic and form the base of many food chains. They are found in the plankton and on mud flats.

The kingdom Protista is important in the marine environment because of the large number of groups that use the sun's energy to create energy-rich organic molecules through photosynthesis. The most common primary producers in the marine environment are the microscopic single-celled diatoms, dinoflagellates, and coccolithophores. The Protista also include the large algae or seaweeds that are found only in shallow water close to the shoreline. Also included in this kingdom is a group of single-celled animals called **protozoa.**

The remaining two kingdoms (plants and fungi) are not as common in the marine environment as they are on land. Only a few species of plants are able to live their total life cycle submerged beneath the water. The so-called sea grasses number less than 50 species. Other marine plants are amphibious. For example, plants of salt marshes and mangrove swamps are able to 'get their feet wet' but are unable to live fully submerged beneath the surface. The fungi are not as important as bacteria in the marine environment in the decomposition of organic material.

THE MONERA

The characteristics of the bacteria that warrant their classification into a separate kingdom are discussed in Chapter 3. Here we will describe their distribution and importance in the marine environment.

Bacteria are found everywhere in the marine environment, from the bottom of ice flows in the Arctic and Antarctic to the floor of the deepest ocean trench (Fig. 4.1). Bacteria grow best when they have a substrate to adhere to, and they are much more common in bottom sediments than in the water column. Although bacteria are found in the sediments of the deepest water, they are more common in shallow waters.

The taxonomy of bacteria in the marine environment is poorly known. It is not clear if marine bacteria are the counterparts of those found in soil or fresh water, or whether they are unique in themselves. Marine forms are morphologically variable (Fig. 4.2). Under some conditions they assume rod shapes; under others, they are spherical. The number of marine bacteria that have been described is relatively small. They generally

TABLE 4.1

Major groups of organisms in the kingdoms Monera, Protista, Fungi, and Plantae that are important to marine biology.

Kingdom Monera
- Group I: Saprobs
  - Aerobic Heterotrophic Bacteria
    - Pseudomonads
    - Nitrogen-fixing
  - Anaerobic Fermenting Heterotrophic Bacteria
    - Methane producers
    - Lactic acid bacteria
    - Clostridia-type
    - Spirochaete-type
- Group II: Autotrophic Bacteria
  - Cyanobacteria (formerly called blue-green algae; are photosynthetic)
  - Green and purple sulfur (photosynthetic)
  - Chemoautotrophic
- Group III: Parasitic/Pathogenic Bacteria

Kingdom Protista
- Phylum Pyrrophyta (Dinoflagellates)
- Phylum Chrysophyta (Diatoms, Coccolithophores, Silicoflagellates)
- Phylum Sarcodina (Protozoa: Foraminiferans, Radiolarians)
- Phylum Ciliophora (Tintinnids)
- Phylum Chlorophyta (Green Algae)
- Phylum Phaeophyta (Brown Algae)
- Phylum Rhodophyta (Red Algae)

Kingdom Fungi
- Phylum Ascomycota (yeasts, molds)
- Phylum Basidiomyccota (rusts, mushrooms)

Kingdom Plantae
- Phylum Tracheophyta (flowering plants)

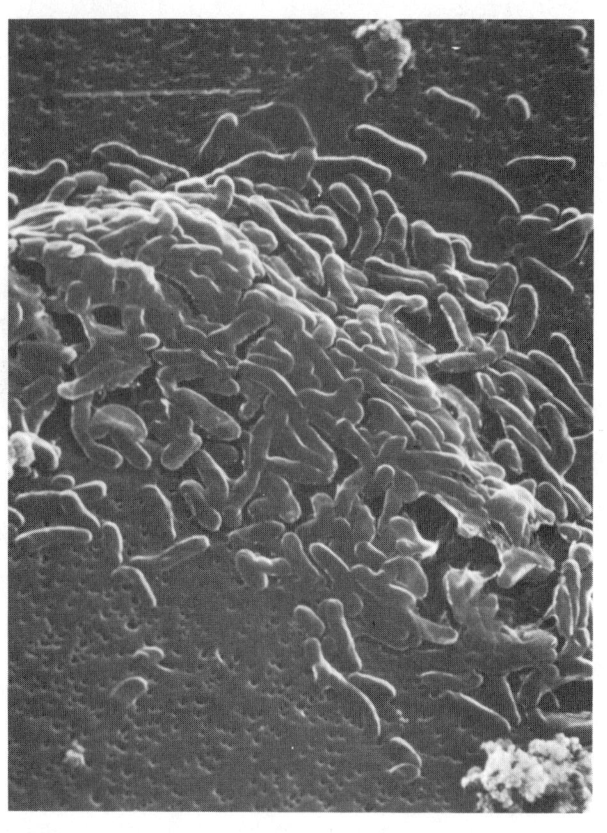

FIGURE 4.1

Bacteria removed from an amphipod trapped at a depth of 10,476 m on the floor of the Marianas Trench. Light bar at upper left is 5 μm long. (Photo courtesy of A.A. Yayanos, Scripps Institution of Oceanography, University of California, San Diego.)

fall into one of three functional groups: saprobic (both aerobic and anaerobic), autotrophic (chemosynthetic or photosynthetic), and pathogenic/parasitic (Tab. 4.1).

Saprobic bacteria are important in the decomposition of organic matter. They reduce dead bodies, feces, and other remains of marine organisms to the inorganic compounds of CO_2, nitrate, and phosphate that are returned to the ecosystem. These compounds become available to autotrophs for the production of organic molecules through photosynthesis. Saprobic bacteria are characterized by the substrate they utilize. Some degrade cellulose, others chitin, while others use proteins, urea, and lipids. As decomposers, saprobic bacteria are the key link in **detritus**-based food webs, which are so important to benthic marine organisms. They can utilize most organic molecules as a source of energy and can even slowly degrade petroleum products released during oil spills.

Autotrophic: Chemosynthetic bacteria are important because they can produce complex organic molecules where photosynthesis is not possible. In the geothermal vents of the ocean floor (see Ch. 11), chemosynthetic bacteria use hydrogen sulphide gas dissolved in water as a source of energy to produce organic molecules.

Autotrophic: Blue-green bacteria (cyanobacteria) are photosynthetic monerans that occur in single cells or in colonies (Fig. 4.3). They are characterized by the presence of pigments, including phycocyanin which causes the common blue-green color and sometimes a reddish color. As with other bacteria, the shape of cyanobacteria cells may differ with environmental conditions or the stages of the life cycle. The taxonomy of this group is not well understood.

The cyanobacteria *Oscillatoria* and *Trichodesmium* are at certain times the most abundant photosynthetic organisms in the phytoplankton. Blooms of *Trichodesmium* are red in color. (The Red Sea gets its name from them.) Some cyanobacteria are common on sediments

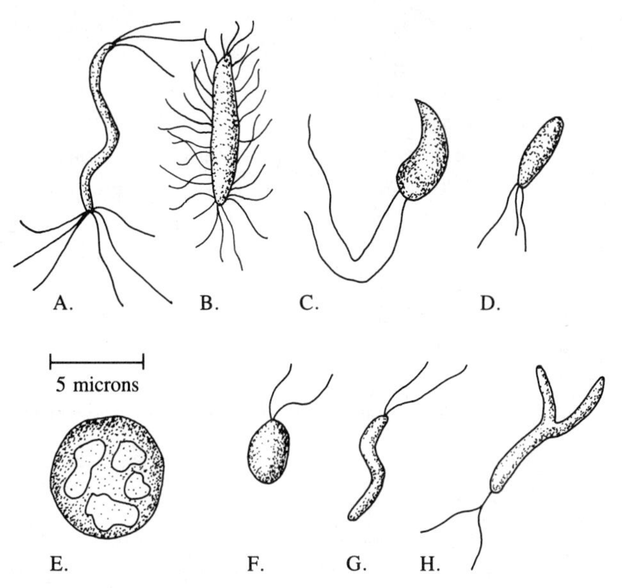

FIGURE 4.2
Shapes of Marine Bacteria

(A) *Spirillum.* **(B)** *Bacillus.* **(C)** *Vibrio.* Bacteria often have a varied morphology depending on environmental conditions. **D–H** show possible morphological shapes for *Pseudomonas.*

as well. Benthic species often form colonies that appear as mats at low tide on tropical mud flats.

Many cyanobacteria are capable of converting nitrogen gas dissolved in water to the essential nutrients nitrate and nitrite. This nitrogen-fixing ability probably explains the importance of cyanobacteria to tropical phytoplankton photosynthesis, where the frequent presence of a permanent thermocline prevents the replenishing of surface waters with nutrients dissolved in deeper water. Cyanobacteria also are important to nutrient enrichment of coral reef water.

Pathogenic/parasitic bacteria in the marine environment are not nearly as well known as the terrestrial pathogenic bacteria studied by medical microbiologists. However, some diseases found in marine mammals, fish, and invertebrates are caused by bacterial infections.

PROTISTA

The kingdom Protista includes organisms that zoologists have historically considered animals (protozoans, nonphotosynthetic dinoflagellates) and that botanists have considered as plants (kelp, diatoms, photosynthetic dinoflagellates). Protists are organisms consisting of single cells or groups of cells that do not specialize into tissues.

Protists are of basic importance in the oceans. While true plants are the dominant photosynthetic organisms on land, it is the protists that are the dominant photosynthetic organisms in the marine environment. On the shoreline and in shallow water, macroscopic algae are the important photosynthesizers. In the water column away from shore the single-celled diatoms and dinoflagellates predominate.

Dinoflagellates

Dinoflagellates (phylum Pyrrophyta) are typically microscopic, single-celled, motile photosynthetic organisms (Fig. 4.4). The phylum is mostly marine, with less than 10 percent being fresh-water species. Marine dinoflagellates have an interesting biology: They make an important contribution to total photosynthesis, are symbiotic with a number of other forms, can occur in great numbers called blooms (one kind that is responsible for red tide), and often are the organisms that cause the bioluminescence observed in surface waters during summer months. Most dinoflagellates have two grooves on the body surface, the *girdle* and *sulcus*. Each groove has a *flagellum* (a tail-like structure used for swimming). Most, but not all, dinoflagellates are photosynthetic. A few heterotrophic species lack chlorophyll; other species have chlorophyll and are photosynthetic but also ingest food. There are two major groups of dinoflagellates in the oceans: free-living species that are part of the phytoplankton, and symbiotic species that live inside the tissues of other organisms, particularly corals and some clams.

The typical free-living dinoflagellate cell has a large nucleus, chloroplasts with chlorophylls *a* and *c*, and numerous other small pigment bodies. Dinoflagel-

FIGURE 4.3
Cyanobacteria

Cyanobacteria are often found in colonies where individual cells are stuck together by mucus. **(A)** *Trichodesmium.* **(B)** *Oscillatoria.* **(C)** *Lyngbya.*

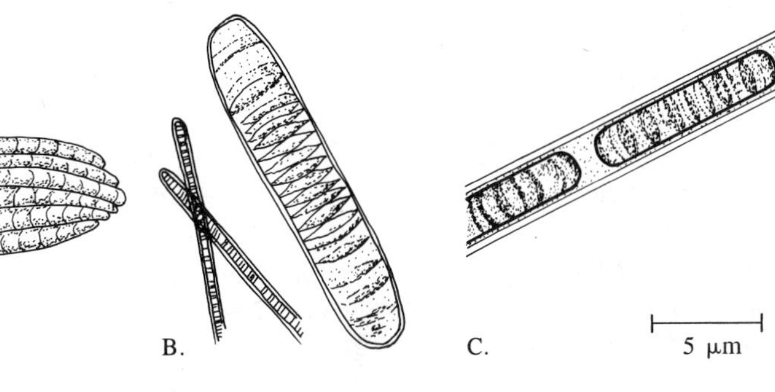

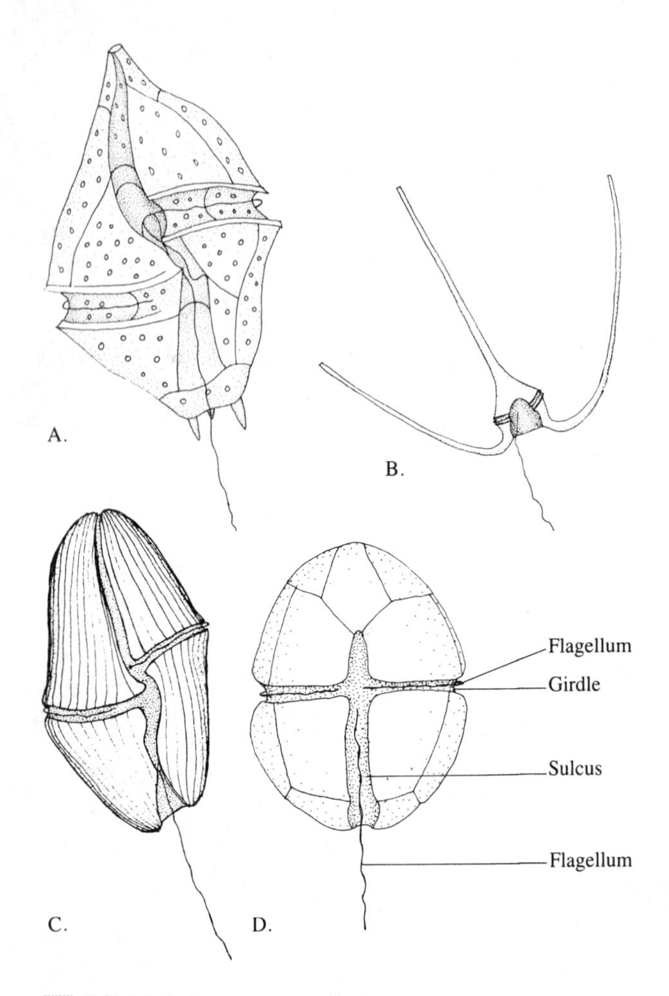

FIGURE 4.4
Marine Dinoflagellates

(A) *Gonyaulax.* **(B)** *Ceratium.* **(C)** *Gymnodinium.* **(D)** *Peridinium.* Note the groove or sulcus found on most species. Generally one of the two flagella lie in the sulcus. (From Crane, J.M. 1973. *Introduction to marine biology: A laboratory text.* Columbus, OH: Merrill.)

lates store photosynthetic energy in food reserves as either oil or starch. Usually, their cell membranes are covered with a thin layer of cellulose. In addition, many species have an external skeleton made of cellulose plates. On the basis of the presence or absence of these cellulose plates, dinoflagellates are classified in two groups: the gymnodiniales (unarmored or naked) and the peridiniales (armored).

Reproduction in dinoflagellates is mostly asexual. The body divides by longitudinal fission and the two new cells are of equal size. Under favorable conditions, dinoflagellates can divide once a day, producing dense blooms.

Dinoflagellates are important phytoplankton in all oceans. Some species are cosmopolitan and are found in coastal and oceanic conditions over a wide range of latitudes. Dinoflagellates are more abundant, however, in tropical and semitropical waters.

A number of dinoflagellates including *Gonyaulax* and *Gymnodinium* (now called *Ptychodiscus*) produce toxins during growth. When environmental conditions are right, blooms of these dinoflagellates may reach concentrations up to 100 million organisms per liter. At these concentrations the water may be discolored red—hence the name red tide. Red tides can kill fish directly: The dinoflagellate cells break and release a neurotoxin (called saxotonin) as they pass through the gills of the fish. Or, shellfish (which are immune to the toxin) may accumulate high enough levels of poison to cause paralytic shellfish poisoning in humans and other vertebrates that eat them.

Often, dinoflagellates are found symbiotically inside the tissues of other organisms. The large tropical clam *Tridacna* derives much of its energy by farming dinoflagellates inside the cells of the outer surface of its mantle. The cells of the host clam contain numerous small spherical dinoflagellate cells called **zooxanthellae.** Photosynthetic products from the zooxanthellae are released to the circulatory system of the clam. The zooxanthellae also contribute oxygen to the clam. In return, the clam provides CO_2 and nitrogenous compounds (the waste products of its metabolism) which are used by the zooxanthellae for photosynthesis. A similar symbiotic relationship occurs between dinoflagellates and hard corals of coral reefs, and some sea anemones.

Phylum Chrysophyta

Chrysophytes are a varied group of mostly microscopic photosynthetic marine organisms. They include three groups of particular importance in the marine environment: the diatoms, silicoflagellates, and the coccolithophores. They all are photosynthetic and contain chlorophyll *a* and *c*. Their color—yellow or golden brown—comes from zanthophyll, an accessory photosynthetic pigment. Skeletons of silica or calcium carbonate are present, and some groups have flagella. Chrysophytes are found in both the benthos and plankton.

Diatoms are the single most important group of photosynthetic organisms in the oceans. They may float free in the plankton or dwell in the benthos. Most are single-celled, but many are found hooked together in chains (Fig. 4.5). There are some 10,000 species, of which 50 percent are marine. Diatoms have no flagella or cilia, and the planktonic species, at least, are not capable of independent movement. They are divided into two groups, based on the shape of their skeleton: **Centric diatoms** (centrales) are circular, dome-shaped, cy-

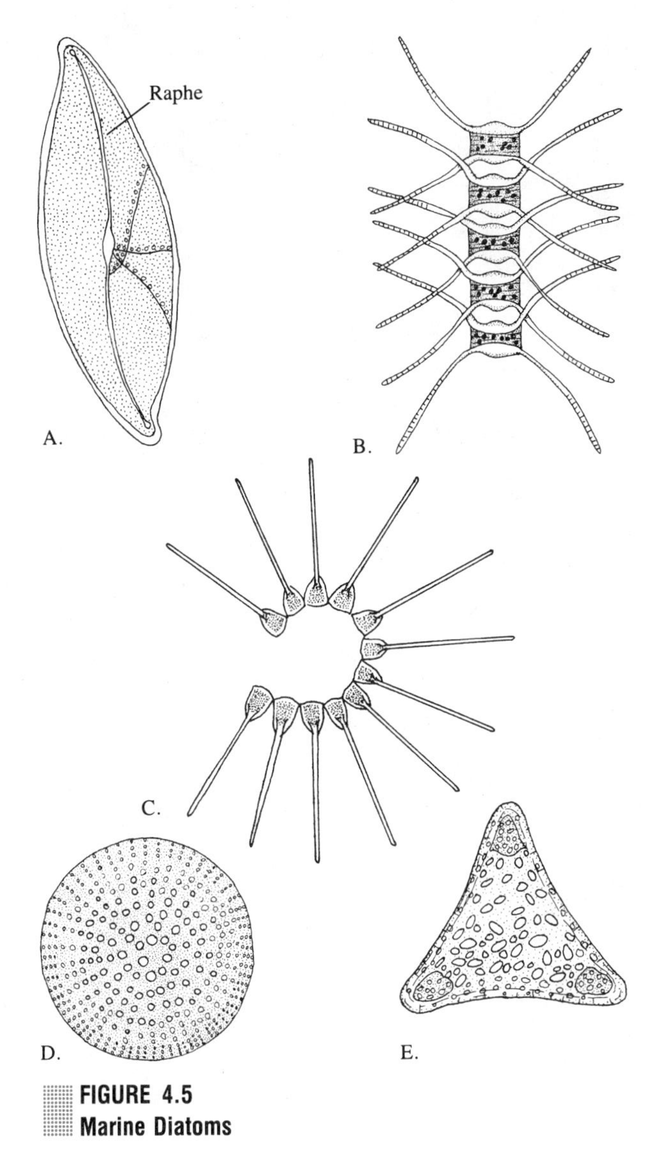

FIGURE 4.5
Marine Diatoms

(A) *Pleurosigma.* **(B)** *Chaetoceras.* **(C)** *Asterionella.* **(D)** *Coscinodiscus.* **(E)** *Triceratium.* (From Crane, 1973.)

lindrical, or triangular in shape. They are mostly planktonic. **Pennate diatoms,** conversely, have oblong, elongate cells often in the shape of a boat. Most pennate diatoms are benthic.

The cell wall of a diatom forms a skeleton. The wall, made of pectin rather than cellulose, is impregnated with silica, forming a rigid, glasslike skeleton called a **frustule.** This frustule is made of two interlocking valves, the epitheca and hypotheca, that fit together rather like a petri dish. The diatom skeleton is different for each species and is the basis of taxonomic classification. The wall of the frustule is ornamented with thin, minute cavities that may form a regular geometric pattern. These thin areas allow the exchange of gases, nutrients, and waste products across the cell wall. In pennate diatoms, which are mostly benthic, the cell wall has a slit, or raphe (Fig. 4.5), that may extend to both valves. This raphe may be quite complex. Cytoplasm extends through the raphe, making contact with the substrate. By extending the cytoplasm, the diatom is able to slowly creep along.

The internal cell structure of diatoms is regular. The cell has one or more vacuoles containing oil droplets which help them maintain buoyancy. Numerous chloroplasts, containing chlorophylls *a* and *c,* are arranged in the cytoplasm beneath the cell wall (Fig. 4.6). Diatoms also have several accessory photosynthetic pigments. Diatomin, the predominant accessory pigment, is golden brownish in color and masks the green of the chlorophyll. At times in geological history massive numbers of diatoms have died and sunk to the bottom, forming diatomaceous earth. Oil droplets in the cells are believed to be one of the important sources of crude oil.

Most reproduction in diatoms is by asexual cell division (see Fig. 4.7). The two valves of the shell separate, each with the products of cell division. New shell valves are produced with edges inside the original. A consequence of this method of reproduction is a continuing tendency toward smaller size with succeeding generations. Thus a population of diatoms of a given species is variable in size, and the largest may be up to 30 times the size of the smallest. When small diatoms have reached the minimum size for the species (about one-third the original size), the cell regains a larger size by forming an auxospore. The diatom discards both the epitheca and hypotheca and grows to the maximum size for the species. New valves are then produced.

Under favorable environmental conditions (adequate light intensity and nutrient availability), diatoms can reproduce at a rate of about one division per day. This rate allows diatoms to bloom into dense growths that color the water a golden yellow. Blooms may contain up to 1 million diatoms per liter.

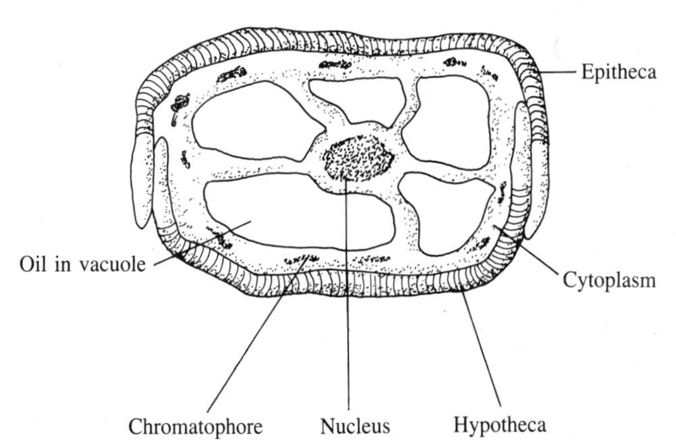

FIGURE 4.6
Diatom Cell

FIGURE 4.7
Diatom Reproduction

(A) During cell division, the epitheca and hypotheca of a diatom separate to become the epitheca of each new cell. The new frustule is completed by the generation of a new hypotheca by each new diatom.
(B) As new generations of diatoms are produced, some cells are crowded into ever smaller frustules. When the size of the frustule becomes critically small, an auxospore forms to allow growth of a larger cell.

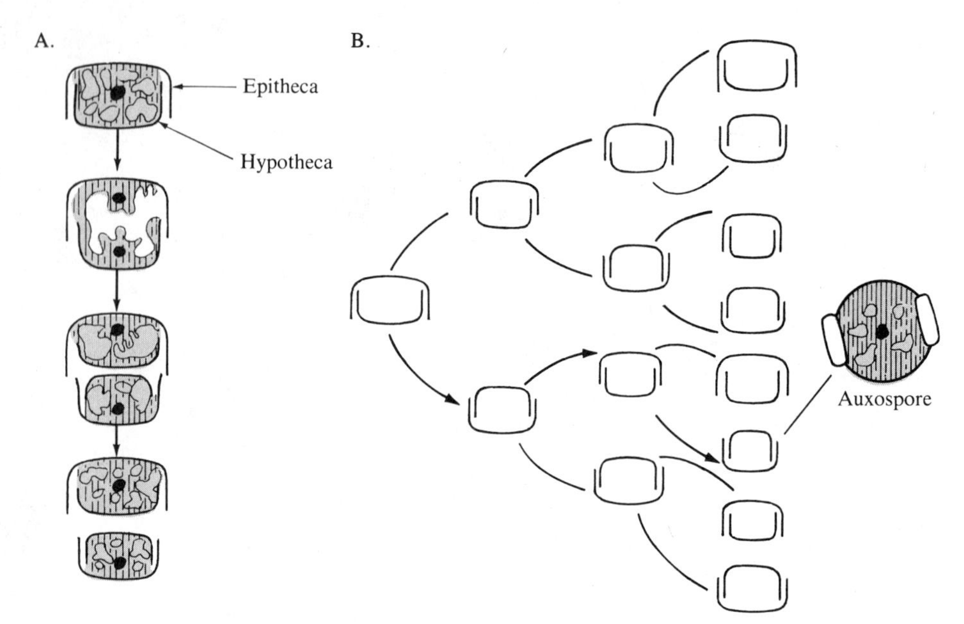

Silicoflagellates are small, single-celled, flagellated organisms. Less than 10 μm in size (called nanoplankton), they pass through the finest mesh of a phytoplankton net (Fig. 4.8). Because of their small size, not much is known about them. They have an external silica skeleton, like diatoms. All species observed so far are marine. In oceanic waters at temperate latitudes, silicoflagellates may outnumber all other phytoplankton. Only recently, marine biologists have recognized the role of these very small organisms in the ocean, finding that they may contribute up to 50 percent of the total photosynthesis. They are important and much more needs to be learned about them.

Coccolithophores are also a group consisting of small, single-celled photosynthetic organisms. Their size is approximately 20 micrometers, making them part of the nanoplankton (Fig. 4.9). Most coccolithophores are planktonic, although some benthic forms are known. The cell walls of coccolithophores are cellulose. The most obvious feature of these organisms, however, is the skeleton, a layer of small calcium carbonate discs (called coccoliths) that are fixed to the outside of the cell wall.

Most of the species are oceanic in tropical waters. Coccolithophores can be important in photosynthesis. Between 75–90 percent of phytoplankton in the Mediterranean Sea may be coccolithophores. Coccolithophores are responsible for the bulk of photosynthesis in the Sargasso Sea. Where coccolithophores are important in the phytoplankton, their skeletons may accumulate in the sediments as calcareous ooze.

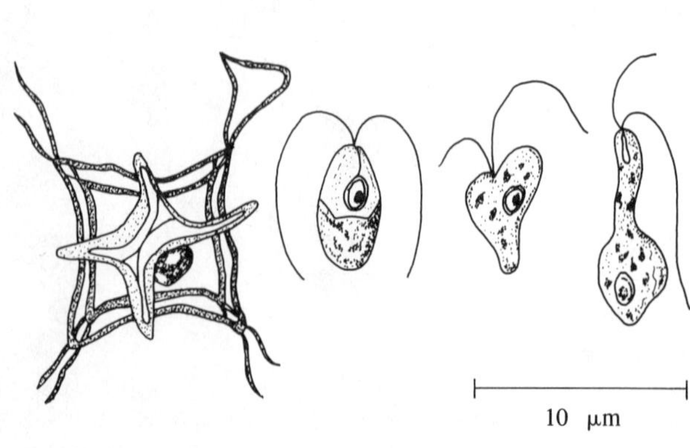

FIGURE 4.8
Silicoflagellates

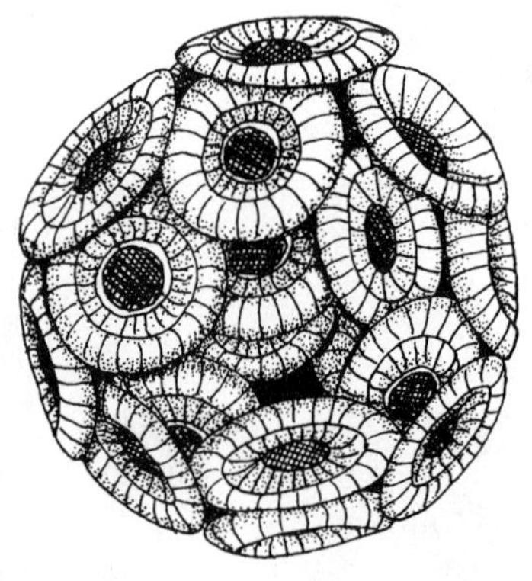

FIGURE 4.9
Coccolithophore

The discs covering the outer surface of the shell are called coccoliths.

Phyla Sarcodina and Ciliophora

This is the phylum of the protozoa. Three groups of protozoans are important to marine biology: the foraminifera, the radiolaria, and the tintinnids. Each of these three groups is widely distributed in the plankton. Foraminiferans also are common in the benthos. Foraminiferans and radiolarians have no flagella or cilia. Instead, they use **pseudopodia,** extensions of the protoplasm, for feeding and limited movement. Both foraminiferans and radiolarians have complex skeletal structures.

Foraminiferans have shells of calcium carbonate. Pores or apertures in the shells allow the pseudopodia to extend. They are primarily benthic, although some planktonic forms are common. Size is generally less than 1 mm (see Fig. 4.10). The shell may have more than one compartment, and the protozoan may abandon one shell and secrete a second. Benthic foraminifera are capable of limited pseudopodial movement of 1 to 6 cm per hour. They eat diatoms and other microalgae.

Planktonic species are not as common as the benthic, but some are important. The genus *Globigerina* is common in tropical waters (Fig 4.10). At death, the shells sink and are major contributors to sediments called **globigerinous ooze.** (Note: Globigerinous oozes also may contain a significant percentage of coccolithophore shells.) Through geological time, globigerinous oozes may compress and form limestone or chalk, as in the White Cliffs of Dover in Great Britain. Some planktonic foraminifera farm dinoflagellates (Fig. 4.11) in a symbiotic relationship similar to that in corals.

FIGURE 4.11
Planktonic Foraminiferan

Electron micrograph of a planktonic foraminiferan *Orbulina.* Note the symbiotic dinoflagellates held in the pseudopods. Magnification 125 ×. (Photo courtesy of O.R. Anderson, Lamont-Doherty Geological Observatory, Columbia University. From Anderson, O.R. 1976. Cytoplasmic fine-study structure of two Radiolarian. *Marine Micropaleontology* 1:84.)

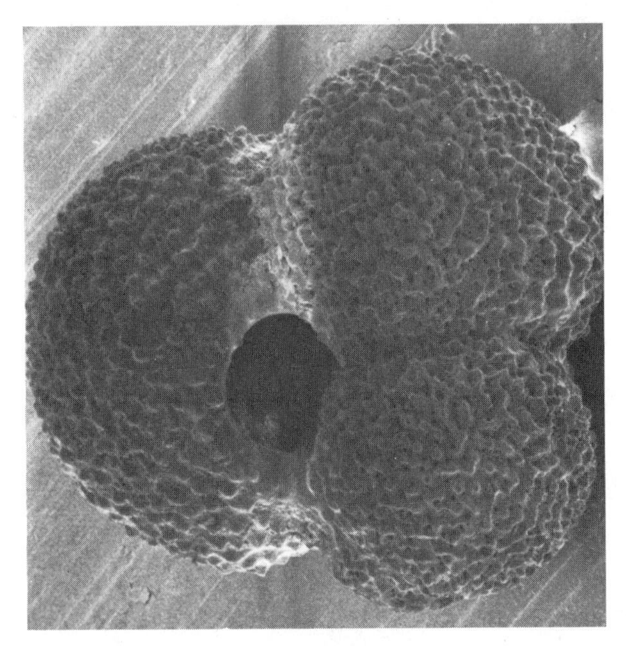

FIGURE 4.10

A scanning electron micrograph of the planktonic foraminiferan *Globigerina.* Actual size is 0.5 mm. (Photo courtesy of J.H. Lipps, University of California, Davis.)

Radiolaria (including the Heliozoa) are pseudopodial protozoa generally having a silica skeleton of intricate latticework extending from the body in spheres or spines (Fig. 4.12). All radiolarians are planktonic, and body size ranges from 0.05 mm to a maximum of 5 mm. Radiolarians feed on diatoms and microzooplankton by catching them with their pseudopodia.

In tropical waters, radiolarians may be abundant enough for their skeletons to accumulate in bottom sediments. Such sediments are called **radiolarian oozes.** The shells of radiolarians often have a beautiful geometry (see Fig. 4.13).

Tintinnids are small ciliated protozoa less than 50 μm long with a shell called a lorica (Fig. 4.14). They are mostly planktonic and at some times of the year may be quite numerous. Food habits of tintinnids are poorly known, but they probably feed on microphytoplankton. Food is ingested through a mouth (cytostome) and is digested in a food vacuole.

The lorica has an opening through which the tintinnid can fully retract. The shell may be made simply of chitinous material secreted by the protozoan, or it may be adorned with particles of sand or coccoliths.

FIGURE 4.12
Radiolarians

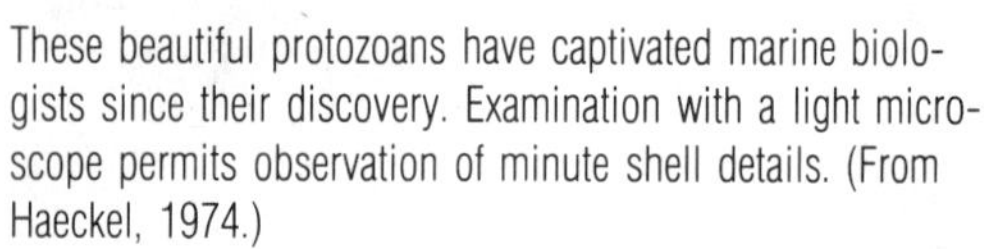

These beautiful protozoans have captivated marine biologists since their discovery. Examination with a light microscope permits observation of minute shell details. (From Haeckel, 1974.)

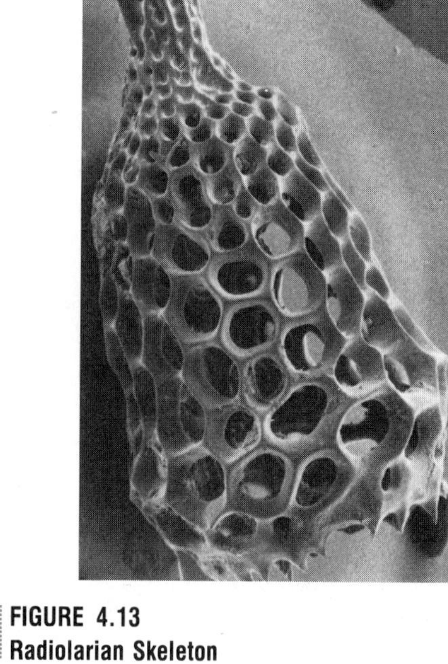

FIGURE 4.13
Radiolarian Skeleton

Scanning electron micrograph of a radiolarian skeleton. Actual size is 0.3 mm. (Photo courtesy of J.H. Lipps, University of California, Davis.)

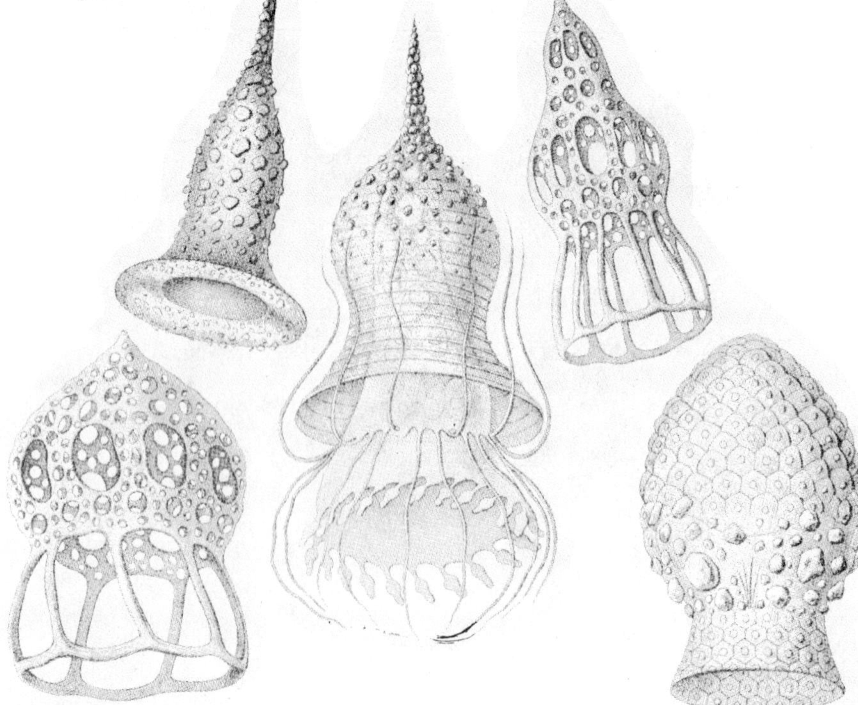

FIGURE 4.14
Tintinnids

Tintinnids are ciliated protozoans. They construct an external shell called a lorica. Of the five loricas shown, only the center one has the tintinnid. (From Haeckel, 1974.)

Phylum Chlorophyta

This phylum is composed of the green algae. Green algae are commonly macroscopic, and appear in the shape of filament sheets or tubules. Of the 6,000–7,000 species of green algae, most are found in fresh water. Only about 15 percent are marine species. The most important are benthic forms found in temperate intertidal areas *(Ulva, Monostroma, Enteromorpha)* and in tropical coral reefs (*Halimeda;* see Fig. 4.15). The primary characteristic of green algae is the clear green color resulting from chlorophylls *a* and *b*. Accessory photosynthetic pigments—common in brown and red algae—are rare in green algae.

Although growth in green algae is usually vegetative, there is a well-defined sexual reproductive cycle. This sexual reproduction, as with red and brown algae, is characterized by a complex alternation of generations. Details are species specific. The green alga *Ulva* will serve as an example.

Ulva has a *gametophyte* and *sporophyte* stage. The two stages differ only in number of chromosomes in each cell. The sporophyte stage is **diploid** (having a double set of chromosomes—2n), and the gametophyte stage is **haploid** (having a single set of chromosomes—1n). Except for this difference, the sporophyte and gametophyte stages of *Ulva* are identical and cannot be separated on the basis of external morphology. During sexual reproduction, cells at the margin of the blade become flagellated (Fig. 4.16). If the plant is in the gametophyte stage, the cells have two flagella; these are the gametes, or 'male' and 'female' cells. A male and female gamete fuse and produce a 2n *zygote* that develops into the sporophyte stage. At maturity, the sporophyte stage develops flagellated cells that have four flagella. These cells are the product of **meiosis** and have the 1n number of chromosomes. Each of these 1n flagellated cells settles to the bottom and grows into a gametophyte stage. Alternation of generations is even more complex in red and brown algae.

Phylum Phaeophyta

This is the phylum of the brown algae. Brown algae are among the largest and most complex of the algae or seaweeds (Fig. 4.17). Kelp (i.e., *Macrocystis, Nereocystis,* and *Eklonia*) may reach lengths of more than 50 m (164 ft). Brown algae are named for the distinct presence of accessory photosynthetic pigments that mask the color of chlorophylls *a* and *c*. The golden-brown pigment fucoxanthin is the most common.

The photosynthetic storage product of brown algae is not a starch but a complex polysaccharide called laminarin. Instead of cellulose in the cell wall, brown algae have alginic acid, a complex carbohydrate. In brown algae, this alginic acid is commonly called algin. In southern California, the kelp *Macrocystis* is harvested for the algin, which is used as an additive in the food industry.

Brown algae are almost exclusively marine (less than 1% of the species is found in fresh water). There are no unicellular planktonic brown algae. The floating populations of the Sargasso seaweed *(Sargassum)* found in the Mid-Atlantic Ocean originated from individuals that were attached to a solid substrate and in some way broke loose. The Sargasso seaweed has numerous floats, or **pneumatocysts,** that allow it to float in the plankton. Although they are found in all oceans, brown algae flourish in the colder water of temperate and higher latitudes. Most species are benthic on rocky shores. They are common in the intertidal zone and are found to depths of approximately 30 m (98 ft).

Brown algae have greater morphological specialization than other algae. Most species can be divided

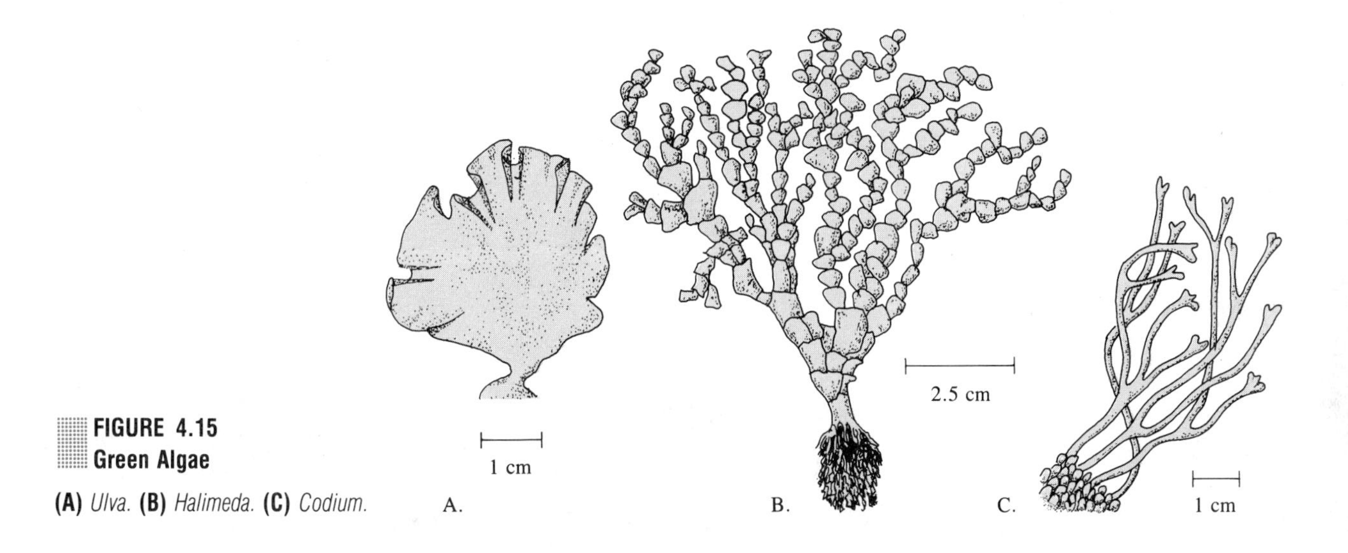

FIGURE 4.15
Green Algae

(A) *Ulva.* **(B)** *Halimeda.* **(C)** *Codium.*

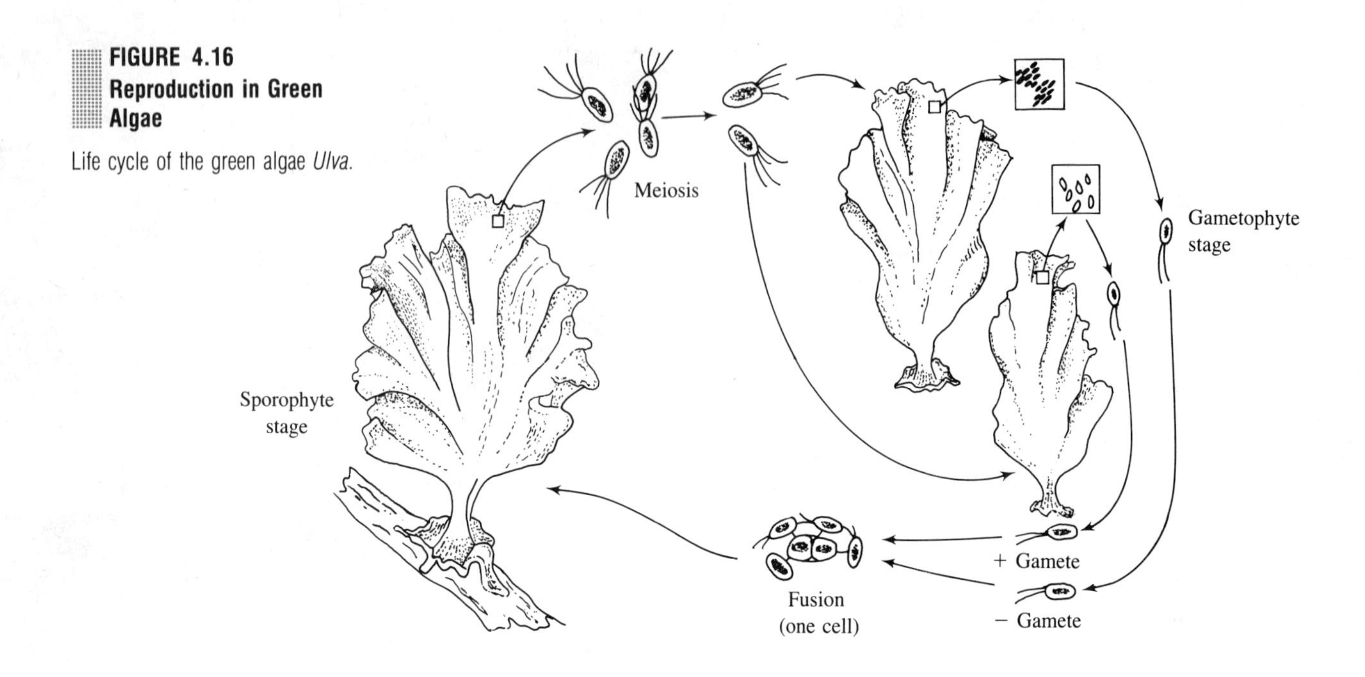

FIGURE 4.16
Reproduction in Green Algae

Life cycle of the green algae *Ulva*.

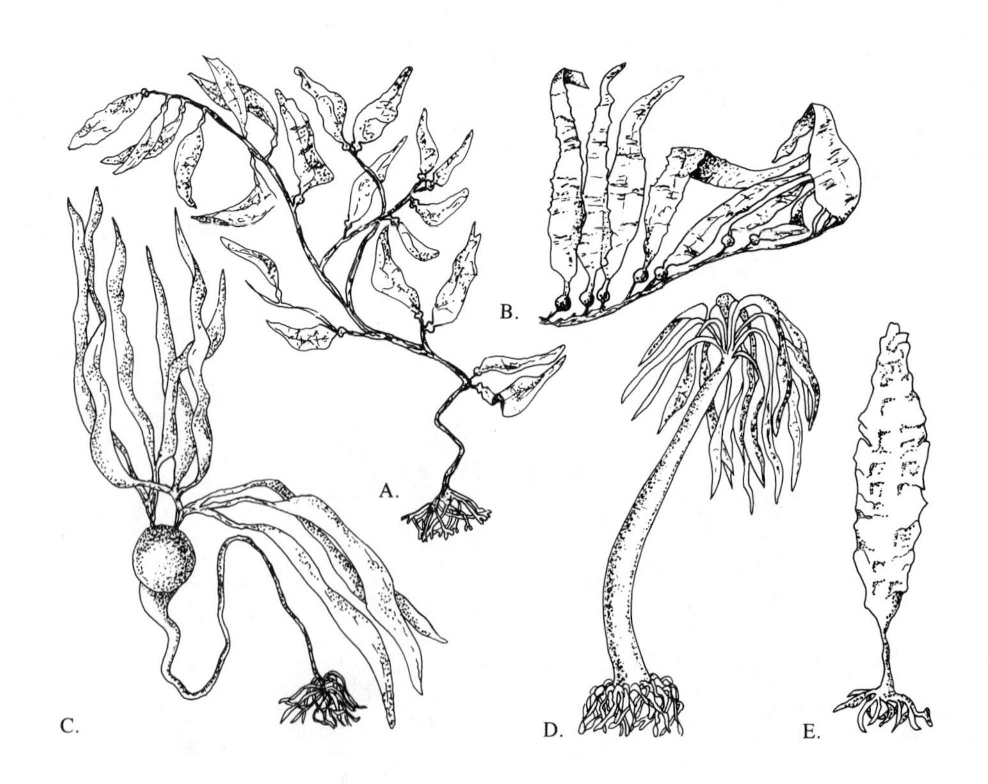

FIGURE 4.17
Brown Algae

(A) and **(B)** *Macrocystis*. **(C)** *Nereocystis*. **(D)** *Postelsia* (sea palm). **(E)** *Laminaria*.

into three parts: holdfast, stipe, and frond (Fig. 4.18). The **holdfast** attaches the algae to the substrate. Holdfasts are made up of numerous branched, rootlike outgrowths that closely adhere to rock surfaces. The **stipe** is an often hollow, stemlike part of the algae. The stipe generally cannot support the weight of the algae fronds out of water. For those species that have holdfasts in deeper water, the stipe allows the fronds to reach shallower water with higher light intensities. **Fronds** are leaflike structures adapted to maximize photosynthesis.

Although the holdfast, stipe, and fronds are described as rootlike, stemlike, and leaflike, the similarity is superficial. Algae have no cellular specialization creating vascular tissue that transports water and nutrients from holdfast to fronds, as plants do from roots to leaves. Another difference is that algae get the nutrients necessary for photosynthesis from the seawater that bathes the fronds. Algae fronds, unlike leaves on terrestrial plants, are the same on both surfaces, and photosynthesis occurs on both the stipe and holdfast. Some brown algae can move organic molecules from the fronds to the stipe and holdfast. *Macrocystis* and *Nereocystis* have cells that function in a way similar to the *phloem* of plants, and can transfer the organic products of photosynthesis from fronds to the holdfast.

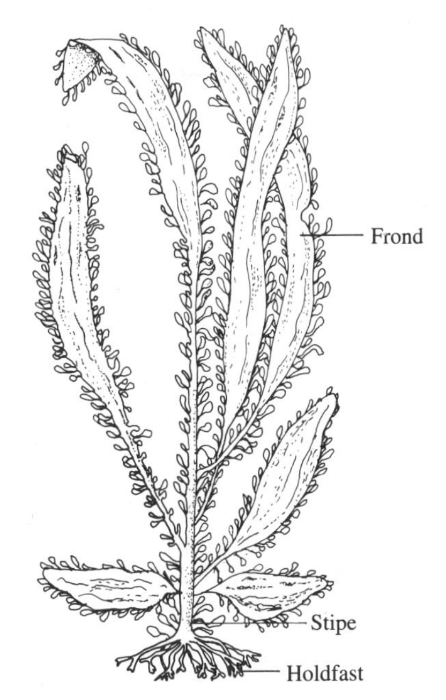

FIGURE 4.18
Morphology of Brown Algae

Most brown algae show morphological specialization into holdfast, stipe, and fronds.

Many brown algae have floats—pneumatocysts—located either on the fronds or stipe. Pneumatocysts are filled with gases and help the algae float. Commonly the gases are those dissolved in seawater—N_2, O_2, and CO_2—but some pneumatocysts contain relatively high concentrations (up to 2%) of carbon monoxide. This carbon monoxide is probably a product of the algae's metabolism.

Reproduction in brown algae is characterized by alternation of generations. In some species—for example, *Fucus* (Fig. 4.19)—the haploid, or gametophytic, generation is a single cell; only the sporophyte is multicellular. In others—for example, *Nereocystis* and *Macrocystis*—the gametophyte stage is small but still a multicellular form, and the sporophyte is the dominant form. In these two algae, specialized cells on the frond, called **sporangia,** produce flagellated microscopic zoospores. These zoospores, haploid (1n) in chromosome number, swim to the bottom and grow into the small gametophytic stage. Gametophytes are either male or female. The male gametophyte produces motile

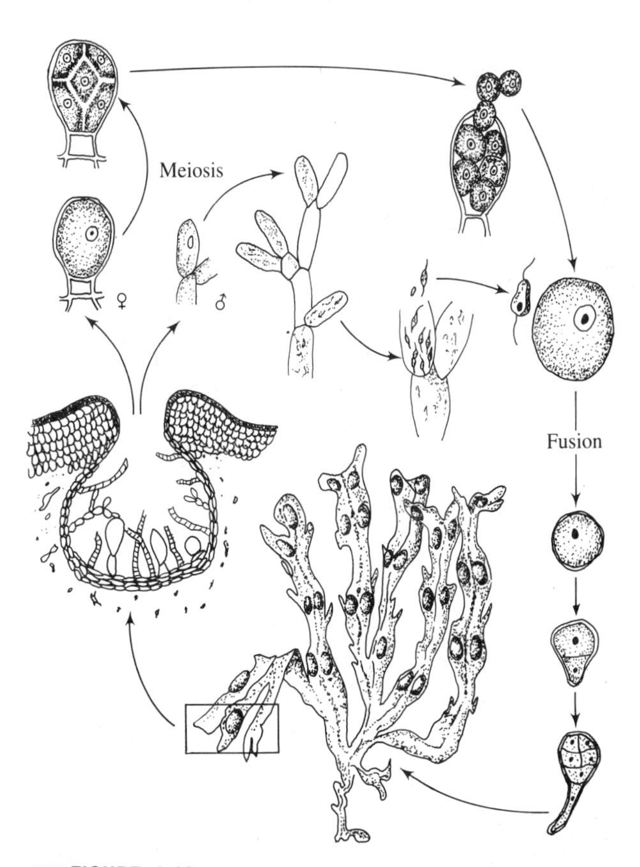

FIGURE 4.19
Reproduction in Brown Algae

Life cycle of the rock weed *Fucus*.

sperm cells that swim to the female (which produces nonmotile eggs). After fertilization, the diploid (2n) zygote attaches to the bottom and grows into the larger sporophyte stage.

Phylum Rhodophyta

The red algae compose this phylum. Red algae are mostly filamentous or sheetlike macroalgae (Fig. 4.20). Their name comes from the presence of red accessory photosynthetic pigments that generally mask the green of the chlorophyll. Most red algae (more than 95% of the species) are marine and are found in the benthos attached by a holdfast. There are no motile red algae; even the reproductive stages cannot swim.

Whereas brown algae reach the largest size of any algae, red algae have a greater number of species than either the browns or the greens. Red algae are most widely distributed in tropical waters. The coralline algae, which secrete calcium carbonate, are of particular importance in the formation of coral reefs.

Red algae grow better than brown or green algae under relatively low light intensities. They have been found deeper than other algae groups (down to 260 m in the tropics). In temperate waters, where brown algae predominate, red algae are often found in the shade of the larger brown algae.

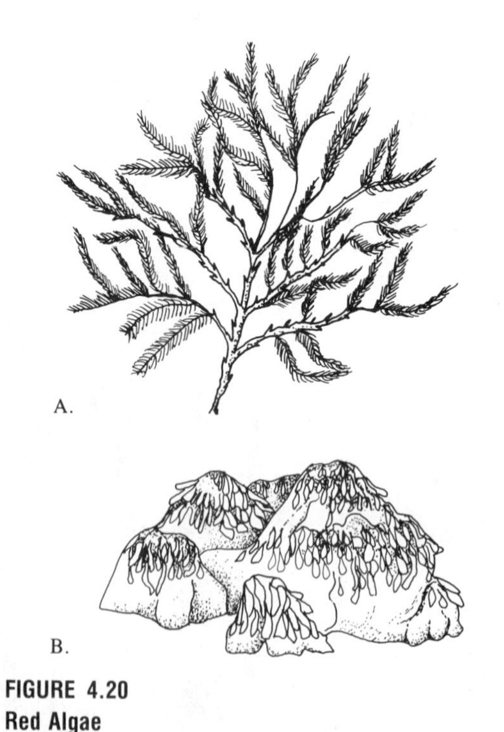

FIGURE 4.20
Red Algae

(A) The foliose *Polysiphonia.* **(B)** The encrusting coralline algae *Lithothamnion.*

The color of red algae comes from the photosynthetic pigment **phycoerythrin.** Many species also have a blue pigment, **phycocyanin.** Only one type of chlorophyll (chlorophyll *a*) is commonly present. The cell walls of red algae are cellulose and have an outer mucilagenous covering made of carrageenan and/or agar. Because these two molecules have great commercial value to the food industry as gelling agents and stabilizers, the harvest of red algae in many locations, such as northeast North America, South America, and Great Britain, is important.

Sexual reproduction in red algae involves the alternation of generations in patterns that often are very complicated. The life cycle of *Porphyra,* an edible seaweed called nori in Japan, is representative (Fig. 4.21). The large edible phase of *Porphyra* is the gametophyte stage with a haploid (1n) number of chromosomes. Mitotic divisions produce small sperm cells that are carried by water currents to the female algae, where the egg is fertilized. The zygote sinks to the bottom and grows into a small sporophyte stage that has a diploid (2n) chromosome number. This sporophyte, called the conchocelis stage, was for many years considered a separate species.

The life cycle of other red algae species may be even more complicated than *Porphyra.* In addition to the haploid gametophyte stage and the diploid sporophyte stage, there may be a third multicellular form in the life cycle called the carposporphyte. This complexity of life cycles has led to many difficult problems in developing the taxonomy of red algae.

FUNGI

The kingdom Fungi has limited distribution in the marine environment, compared to its distribution on land. Of the 50,000 or so species described to date, only 1 percent are found in the oceans. Marine forms, shown in Figure 4.22, are relatively small, seldom exceeding a size of 2 mm.

Fungi are divided into several phyla with one, the Ascomycota, containing most of the marine species. The ascomycetes, which includes the yeasts, may be found in virtually all marine habitats, although they are most common in intertidal and shallow subtidal benthos. Some of the largest marine forms, with fruiting bodies of 2–3 mm, have been found in the deep ocean. Most of the fungi are saprobes and, like bacteria, break down organic matter. In salt marsh and mangrove communities, fungi are a major decomposing component. The role of fungi compared to bacteria in the decomposition of organic material in other marine habitats is not well understood. Bacteria, however, are

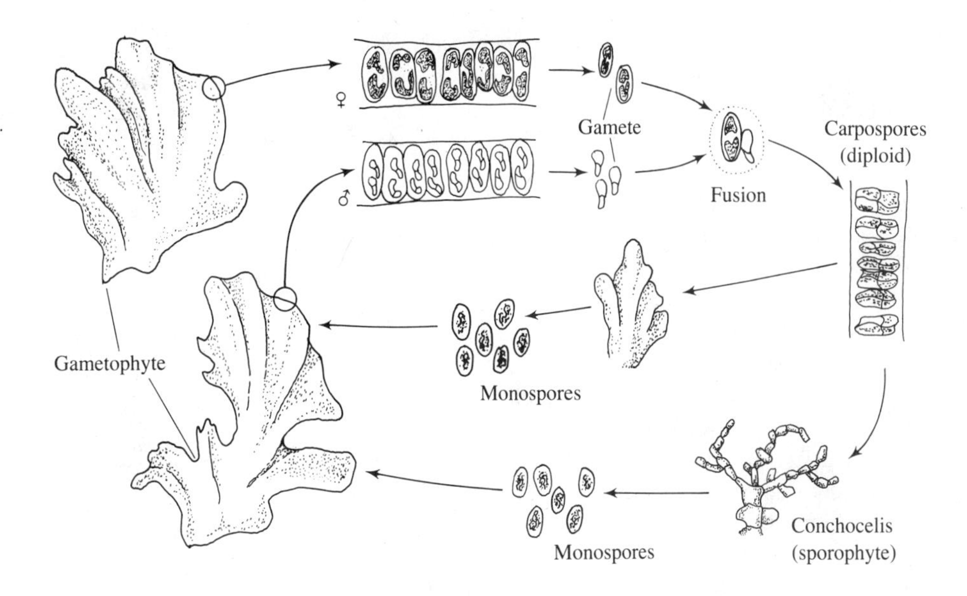

FIGURE 4.21
Reproduction in Red Algae

Life cycle of the red algae *Porphyra.*

thought to be more important than fungi in the breakdown of organic material in most areas of the marine environment.

Fungi are found in symbiotic relationships with other marine groups. Some fungi are parasitic on algae and salt marsh vegetation; others are mutualistic. Parasitic fungi in salt marshes and mangrove swamps have been found infecting algae, diatoms, plant tissue, and invertebrate animals. Parasitic fungi are a problem in mariculture of the red algae *Porphyra*. They also are believed to be a major contributor to, if not the total cause of, the wasting disease that destroyed the seagrass beds of eastern North America in the years 1930–1932.

Lichens are the mutualistic combination of a fungus and green alga that forms a new morphological shape. Most marine lichens are found in the high exposed areas of rocky intertidal zones. The fungal component of the lichen provides a protective covering that retains moisture during the long periods of tidal exposure, and the alga provides the fungus with a source of food.

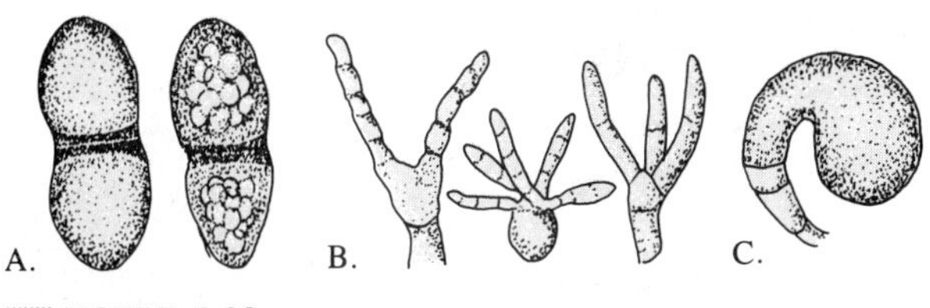

FIGURE 4.22
Marine Fungi

(A) Typical ascomycete. **(B)** Typical basidiomycete. **(C)** Imperfecti fungi. Sizes are less than 2 mm.

PLANTAE

There are seven phyla, or divisions, in the kingdom Plantae. Of these, six phyla (including mosses, ferns, and cone-bearing plants) do not occur in the marine environment. The seventh, the Angiospermophyta, or flowering plants, has a few species that tolerate partial or complete submergence in seawater. The number of such species is small, however. Of the 235,000 species of flowering plants, fewer than 200 live in the sea. Many of them can tolerate only partial immersion; only 50 species live fully submerged in seawater.

The relatively small number of flowering plants in the marine environment is a major difference between the biology of the land and the sea. Flowering plants dominate the terrestrial environment. Not only are they the primary source of food energy (provided through photosynthesis); they provide the structure for most biological communities. In the oceans these tasks, for the most part, fall to organisms of other kingdoms.

Why do flowering plants have such a limited distribution in the sea? The explanation lies in the functioning of the vascular system. Plants rely on the evaporation of water of the leaf surface to provide the force that draws fluids from the roots and through the stems to the leaves. For those plants that are only periodically submerged in seawater, such as mangroves and salt marsh vegetation, this evapotranspiration system can still function, although these plants must devote a considerable amount of energy to desalinating the salty water coming into the root system.

Marine angiosperms include trees, shrubs, and grasses that belong to a wide variety of families. Plants of the marine environment are traditionally studied according to the biological communities they form. Three such communities are: salt marshes, mangroves, and sea grasses. These communities are discussed in more detail in Chapters 12 and 13. Here we will comment on the systematic diversity found in marine plants and mention the important adaptations for living in the marine environment.

More than 18 angiosperm families make up salt marsh vegetation. Most species are monocotolydon grasses *(Spartina* and *Juncus),* although dicotolydons *(Salicornia)* are also represented (Fig. 4.23). Salt marsh plants can tolerate only periodic submergence in seawater and are found in the upper part of the intertidal zone. The adaptations of salt marsh plants to the marine environment are extensive. Most species have rhizomes (underground stems) from which both new roots and stems develop. Roots are of two types. Some are coarse and cork covered, penetrating deeply into the sediments to anchor the plant. Since deeper sediments are mostly anaerobic, salt marsh species have a second, shallow root with many branches and root hairs that absorb water and nutrients. The leaf structure of salt marsh plants is also modified to minimize water loss. Salt marsh species tolerate seawater but do not require it for growth. In fact, salt marsh species grow better in estuarine conditions, where salt water is diluted with fresh. Under experimental conditions, salt marsh species grow best in fresh water.

Salt marsh plants are found mostly in temperate and higher latitudes. In the tropics, the upper intertidal zone of protected areas is likely to have mangrove vegetation. Mangroves are a group of taxonomically unrelated trees that can tolerate having their roots submerged in salt water for at least part of a tidal cycle. Some 100 species belonging to 30 families are included in the mangrove group. In the Atlantic Ocean there are only ten species, while in the Indo-Pacific there are some 90 species. The most widespread genus in both tropical Atlantic and the Pacific is *Rhizophora* (Fig. 4.24). This genus and some other mangrove species have adaptations to minimize water demand by the roots, and their thick, waxy leaves reduce evaporative water loss. Mangrove swamps are typically found in humid conditions that help to reduce water loss. Reproduction also is adapted to the marine environment. Seeds germinate while still attached to the parent tree and develop there into seedlings (Fig. 4.25). Seedlings either root themselves after dropping, or float on tidal currents to a new, suitable substrate.

Both salt marsh and mangrove plants are in a sense amphibious because they live only partially submerged in seawater. A few species of plants—the sea grasses—are able to live and reproduce while completely submerged in seawater. There are some 40 species of sea grasses that belong to two families of angiosperms. Although sea grasses are monocots, they are not true grasses, and are most closely related to the lily family.

Sea grasses are found only in shallow waters (generally less than 10 m). They have no accessory photosynthetic pigments and are therefore not adapted to the low light intensities of deeper waters. Most species are found in shallow, protected waters where they often form extensive meadows. Important genera that form sea grass meadows include *Zostera* (eel grass), which is found in quiet temperate waters (Fig. 4.26); *Thalassia* (turtle grass), which is found in quiet tropical and subtropical waters (Fig. 4.27); and *Syringodium* (manatee grass), which is common in the Caribbean. Although

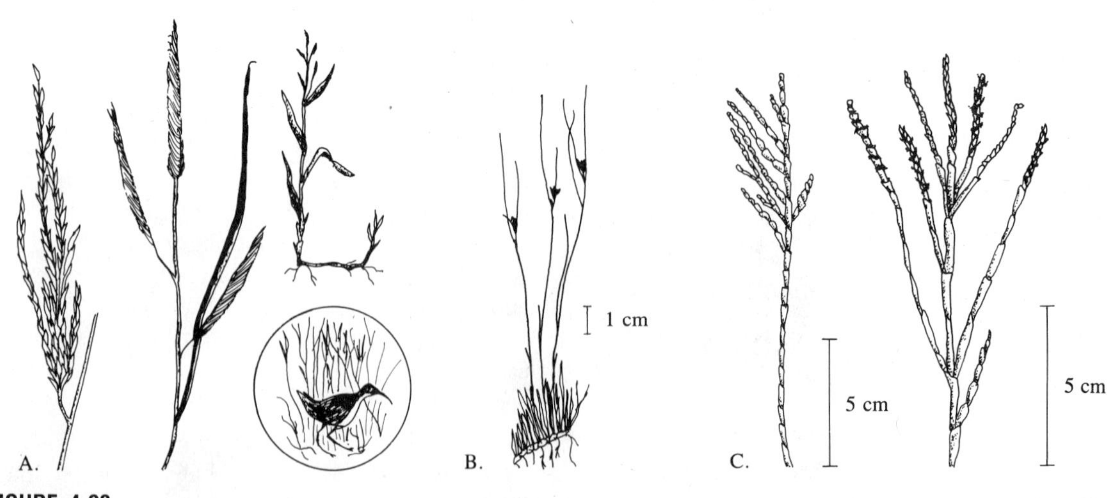

FIGURE 4.23
Salt Marsh Plants

(A) *Spartina.* **(B)** *Juncus.* **(C)** *Salicornia. Spartina* and *Juncus* are monocot grasses, while *Salicornia* is a dicot.

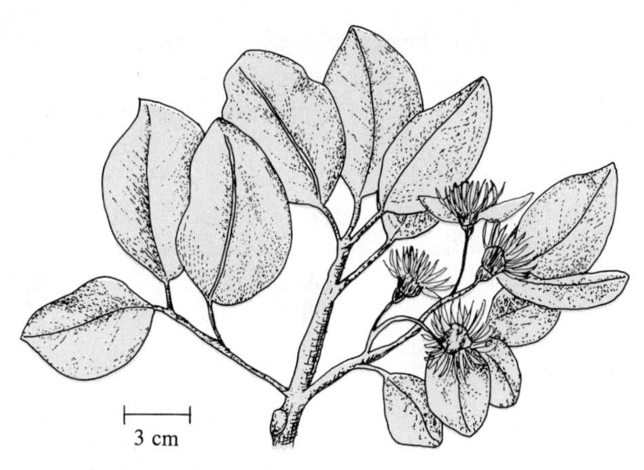

FIGURE 4.24
Rhizophora

Leaf and flower of the common mangrove tree *Rhizophora.*

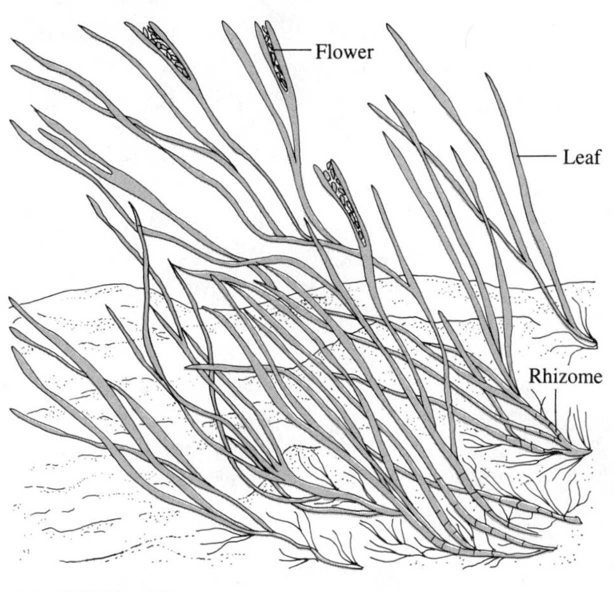

FIGURE 4.26
Zostera

The temperate sea grass *Zostera,* also called eel grass, is common in protected waters of North America.

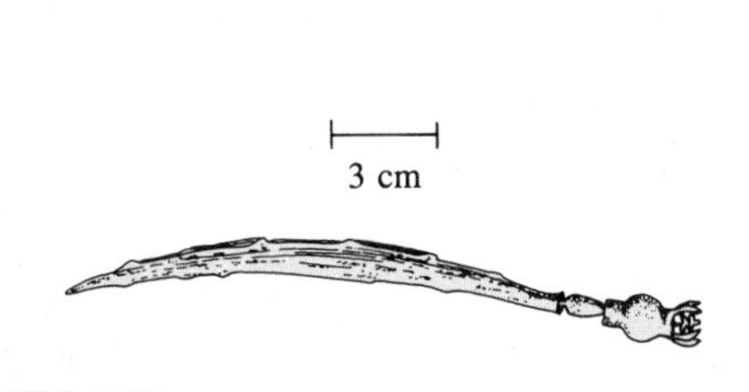

FIGURE 4.25
Rhizophora Seedling

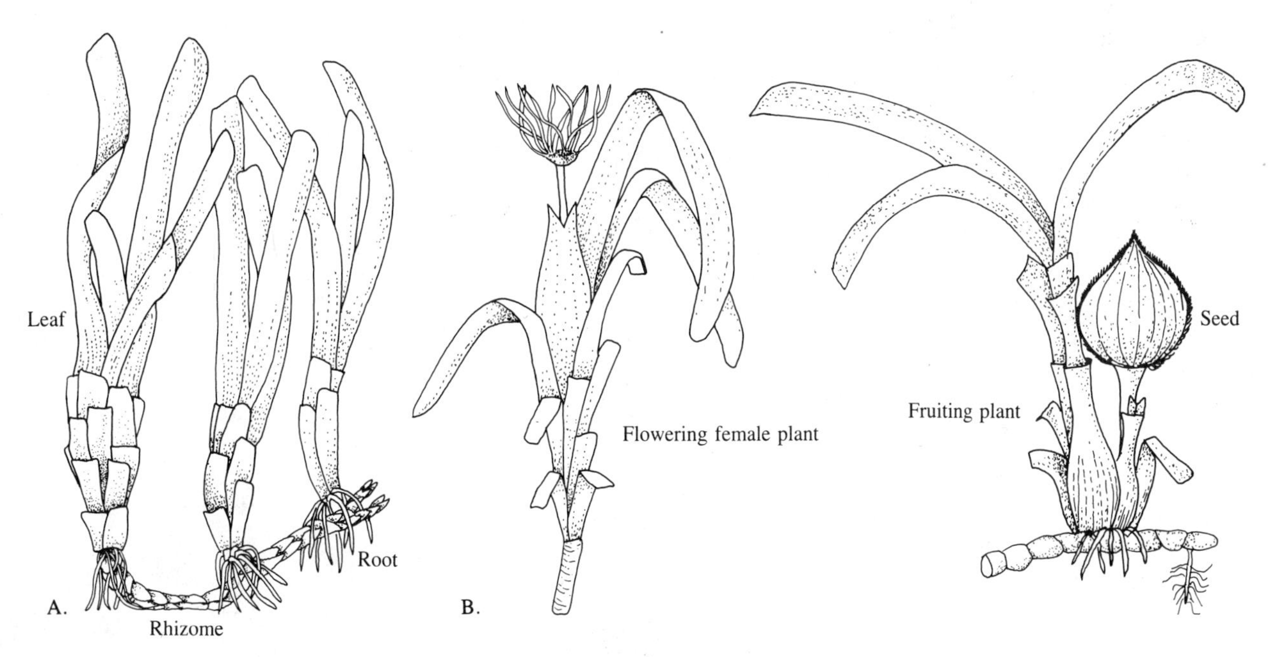

FIGURE 4.27
Thalassia

The sea grass *Thalassia* is also called turtle grass. **(A)** Rhizome, roots, and leaves. **(B)** Flowering parts.

most sea grasses are found in sheltered waters, the genus *Phyllospadix,* or surf grass, lives on rocky shores in the western United States that are exposed to high wave action.

Sea grasses are perennial plants. In temperate latitudes most leaves break off the plants at the end of the growing season, but the rhizomes persist. Growth is primarily vegetative, by the development of new rhizomes. Sexual reproduction with the production of seeds is also common. The pollen is surrounded by a mucilaginous sheath and is carried to the flower by water currents. Sea grass plants are able to obtain nutrients necessary for photosynthesis (NO_4 and PO_4) either directly across the leaf surface from the seawater, or from the sediments through the roots and vascular tissue.

Although the flowering plants of the marine environment are few in number, they are very important in the biology of shallow-water marine areas. The ecology of these plant communities is discussed in Chapter 13.

SUMMARY

Of the five kingdoms of life (excluding the viruses), only two—Metazoa (animals) and Protista (protozoa, algae, diatoms)—are richly represented in the marine environment. The Monera has moderate distribution, and true plants and fungi have only a few marine representatives.

The kingdom Protista consists of a group of phyla of divergent evolutionary history. In general, protists are single-celled eukaryotic cells that function as integrated organisms. Although red, green, and brown algae phyla are mostly large multicellular marcroscopic forms, these algae show very little cell differentiation or development of tissue.

All the phytoplankton, which are responsible for over 90 percent of photosynthesis in the marine environment, are protists. The Chrysophyta (diatoms and coccolithophores) and the Pyrrophyta (dinoflagellates) are particularly important in the photoplankton. Diatoms have no means of swimming, have a sturdy silicon dioxide exoskeleton, and maintain neutral buoyancy by controlling the ion content of a large vacuole. Diatoms are favored in areas rich in nutrients and with strong upwelling currents or turbulence. In temperate and polar latitudes, diatoms tend to dominate over dinoflagellates.

Dinoflagellates swim by means of a pair of flagella and many have an exoskeleton made of cellulose plates. They are favored in water with high nutrient concentrations and stratified circulation with little water movement. Dinoflagellates tend to dominate in semitropical and tropical latitudes. Some dinoflagellates live symbiotically in the tissues of animals such as corals and giant clams.

QUESTIONS AND EXERCISES

1. Suggest possible reasons why there are no trees in the ocean.
2. Discuss reasons for the observation that plants dominate the land and are poorly distributed in the ocean, while algae dominate the shallow water ocean yet are of little importance on land.
3. What are the similarities of the various systematic groups that make up the phytoplankton? What are the differences?
4. Why is the two-kingdom scheme that divides organisms into animals and plants inadequate for classifying the diversity of life in the marine environment?
5. How are the adaptations to life floating in the water similar in the different systematic groups of phytoplankton? How are they different?
6. What are the basic structural differences of the various systematic groups of protozoa?
7. What are the characteristics of the various groups of algae that are used to place them each in separate phyla?

REFERENCES

Austin, B. 1988. *Marine microbiology.* New Rochelle, NY: Cambridge University Press.

Hamlebon, C., Spindler, M., and **Anderson, O.R.** 1989. *Modern planktonic foraminifera.* New York: Springer-Verlag.

Harris, G.P. 1986. *Phytoplankton ecology.* New York: Chapman and Hall.

Laybourn-Parry, J. 1984. *A functional biology of free-living protozoa.* Berkeley, CA: University of California Press.

Round, F.E. 1981. *Ecology of algae.* New York: Cambridge University Press.

Sleigh, M.A. 1987. *Microbes of the sea.* New York: J Wiley.

Woelkerling, W.J. 1988. *The coralline red algae.* New York: Oxford University Press.

SUGGESTED READING

OCEANS

Mays, B. 1976 When the seas run red. (Dinoflagellates) 9(6):52–56.

OCEANUS

Dacey, J.H. 1981. How aquatic plants ventilate. 24(2):43–51.

Jannasch, H.W. 1988. Lessons from the *Alvin* launch. 31(4):34–40.

Moore, R.E., Helfrich, M.P., and **Patterson, M.L.** 1982. The deadly seaweed of Hana. 25(2):54–63.

Takashashi, K. 1982. Minute marine organisms found in tropical waters. 25(2):40–41.

SEA FRONTIERS

Coleman, B.A., Duetsch, R.N., and **Sjoblad, R.D.** 1986. Redtide: A recurrent marine phenomenon. 32(3):184–191.

Johansen, H.W. 1987. Coralline algae: Pink plants of the seafloor. 33(6):438–443.

McFadden, G. 1987. Not-so-naked ancestors. (Phytoplankton structure) 33(1):46–51.

5

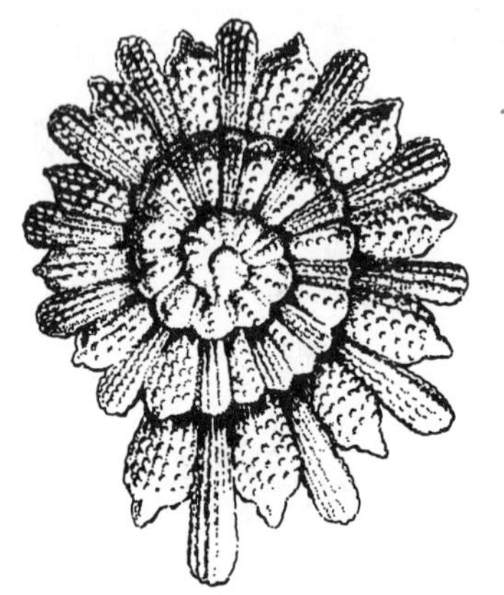

KINGDOM METAZOA: THE ANIMALS

For the beginning marine biology student, the study of life is dominated by the animals because many are familiar, conspicuous, and large. There are at least 2 million described species of animals classified into over 30 phyla (Tab. 5.1). What is important to notice in a survey of the phyla is the number of groups that are almost or entirely exclusive to the marine environment. This is an indication that most groups evolved in the marine environment, and that colonization of fresh water and land was secondary; many phyla did not make this step.

In this chapter we discuss the important marine phyla. Importance is measured by one or more of these criteria: the number of species in the phylum having (a) some unique features of their biology, or (b) having some particular ecological importance. As you work through the phyla, note the order in which they are presented. In general, the order reflects evolutionary relationships, with the groups that are thought to have evolved earliest considered first. However, remember from the discussion in Chapter 3 that the early evolution of animals is not clearly understood. It is clear that animals have their origins in the protists, but the specific groups for the most part are not well known. It is possible that different animal groups evolved from more than one group of protists.

PHYLUM PORIFERA

This is the phylum of the sponges. Sponges (Latin: *porus*-pore; *ferre*-to bear) are very simple animals that appeared early in the evolution of animal life. They are not thought to be on the evolutionary path leading to other animal groups. They reached their greatest abundance and diversity some 100 million years ago.

Given our earlier definition of what an animal is (see Ch. 3), sponges barely qualify. They have no organs or well-defined tissues. Their cells have a relatively high degree of independence, and many are capable of multiple functions. Although some sponges have a distinct and predictable shape, most do not (see Fig. 5.1). Instead, the body forms an irregular mass. Because they are made up of individual cells that grow independently, it is not clear whether the sponges should be called individuals or colonies.

The shape of a sponge is greatly influenced by its environmental factors such as the shape of the surface the sponge is attached to, and the direction and velocity of its water currents. Sponges have no nervous system, and all are sessile. The only clear evidence that sponges are animals is that they reproduce in a way typical of animals: An adult develops from a zygote, or fertilized egg.

TABLE 5.1

Approximate number of species for the animal phyla and the percent of the species that occur in the marine environment.

Phylum	Common Name	Species	Marine (%)
Porifera	Sponges	10,000	98
Cnidaria	Coelenterates	9,000	99
Ctenophora	Sea gooseberry	50	100
Platyhelminthes	Flat worms	13,000	
Nemertea	Ribbon worm	650	98
Nematoda	Round worm	10,000	≈50
Rotifera		1,800	1
Gastrotricha		400	≈50
Acanthocephala		500	0
Nematomorpha	Horsehair worm	230	1
Gnathostomulida		80	100
Kinorhyncha		100	100
Mollusca	Snails, clams	110,000	49
Annelida	Segmented worms	8,700	≈75
Onychophora		80	0
Arthropoda		956,300	
Insects		925,000	<1
Crustaceans		31,300	50
Sipuncula	Peanut worm	300	100
Priapulida		9	100
Entoprocta		60	99
Echiuria		100	100
Tardigrada	Water bear	400	<5
Phoronida		10	100
Bryozoa		4,000	100
Brachiopoda		280	100
Echinodermata	Sea stars	6,000	100
Pogonophora		100	100
Hemichordata		90	100
Chordata		39,000	
Tunicates	Sea squirts	1,300	100
Vertebrates		37,700	>50
Chaetognatha	Arrow worm	50	100

The morphology of sponges differs from all other animals in that it is organized around an internal system of water canals and chambers through which water is actively pumped (Fig. 5.2). When looking at a living sponge you can see one or more openings, or **oscula** (singular: osculum), which allow water to exit from the body. Water (carrying food and oxygen) enters through numerous microscopic pores called **ostia** (singular: ostium). The internal chambers are lined with flagellated cells that create the water current that carries in food through the ostia.

One of the unique features of sponges is the role that individual specialized cells play. The pore size of the ostia is regulated by cells called porocytes. The size of the osculum is regulated by myocytes. The skeletal elements (spicules) are secreted by spongocytes, and amoebocytes function in the movement of food through the body. The most important cell type in sponges is the **choanocyte** (Fig. 5.2B). These cells pump water through the body, capture food particles, and transfer sperm from the water to the egg. Choanocytes are flagellated. The combined beating of the flagella creates the

FIGURE 5.1

The demospongia *Polymastia,* commonly called dead man's fingers. (From Flora and Fairbanks, 1982.)

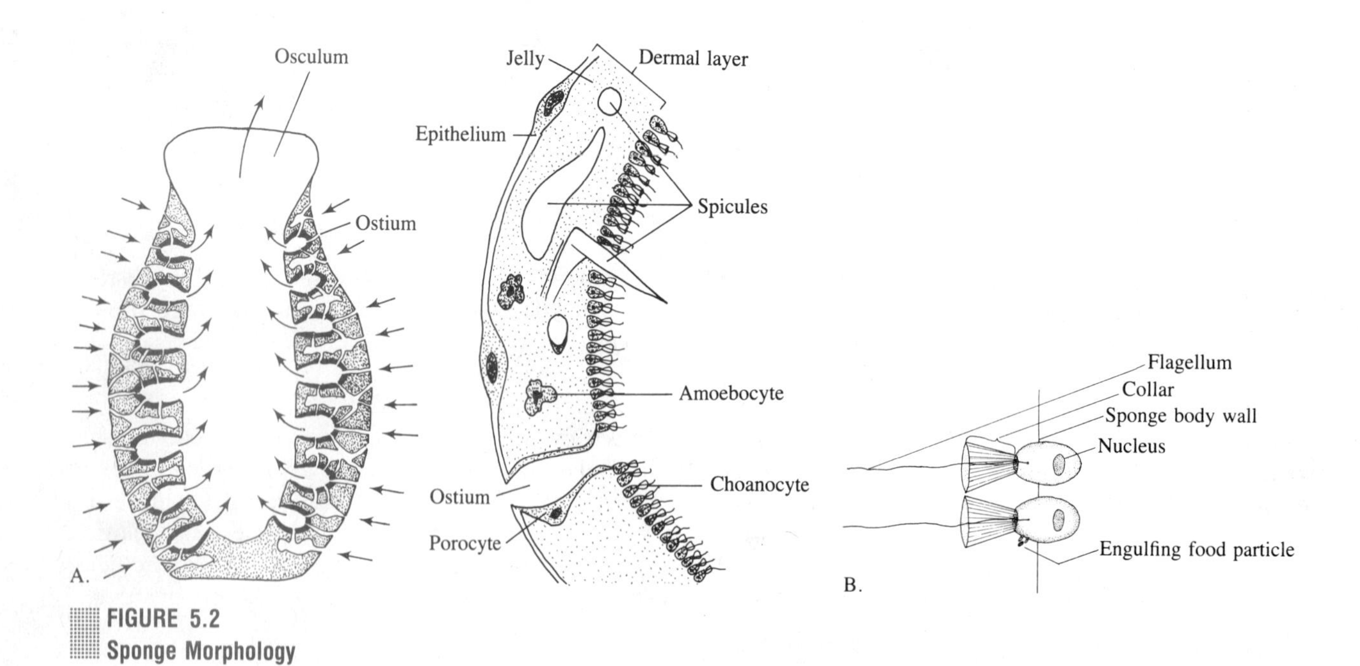

FIGURE 5.2
Sponge Morphology

(A) Water flow through a sponge and various cell types. **(B)** Choanocytes.

water current through the sponge body. Choanocytes also function in feeding. Sponges allow only small particles (less than 50 μm in size) to enter the water system. Many of these particles are engulfed by choanocytes. Digestion is intracellular and may occur in the choanocyte, or the food particle may be transferred to an amoebocyte and digested there.

Choanocytes have yet another function. Sperm, released into the water during reproduction, are carried by water currents into the water system of a prospective mate. In a flagellated chamber, the sperm enters a choanocyte. Both cells lose their flagella and the choanocyte with the sperm inside travels to a nearby egg by amoeboid movement. The choanocyte then fuses with the egg, and the sperm enters the egg cell where the nucleus of the egg and sperm fuse.

The 10,000 or so marine sponges are divided into four main classes: Demospongiae, Hexactinellida (the glass sponges), Clacarea (the calcareous sponges), and Sclerospongiae. Over 95 percent of all sponges belong to the Demospongiae. The main feature used to classify sponges is the nature of the skeleton. Sponges in the class Demospongiae have spicules of silica, spongin, or both (Fig. 5.3). The commercially important bath sponges, which have skeletons of only spongin, belong to this group. Hexactinellids have glass spicules—always with six points; calcareous sponges have a skeleton of calcium carbonate; and sponges belonging to the Sclerospongiae have complex skeletons consisting of spongin, glass spicules, and calcium carbonate.

Although second in terms of number, the glass sponges (class Hexactinellida) are seldom seen since they are deep-water species found generally below 500 m. The body form of glass sponges is often a symmetrical vase or cup shape, and their silica spicules have six points. The height of glass sponges is generally between 10–30 cm (4–12 in). One of the best-known glass sponges is the venus flower basket, which has a pair of shrimp living in its large body cavity. A male and female shrimp enter when they are small and soon grow too large to pass through the osculum. The symbolism of this living arrangement has made the skeleton of the venus flower basket a traditional Japanese wedding gift (Fig. 5.4).

Only about 1–2 percent of sponges are calcareous. Sponges in the class Calcarea have spicules of calcium carbonate rather than silica. Calcareous sponges are found mostly in shallow water.

The class Sclerospongiae has only a few species. This class is characterized by an outer covering of calcium carbonate and an internal skeleton of silica spicules and spongin. Sponges in this class are found only in the cavities and tunnels of coral reefs.

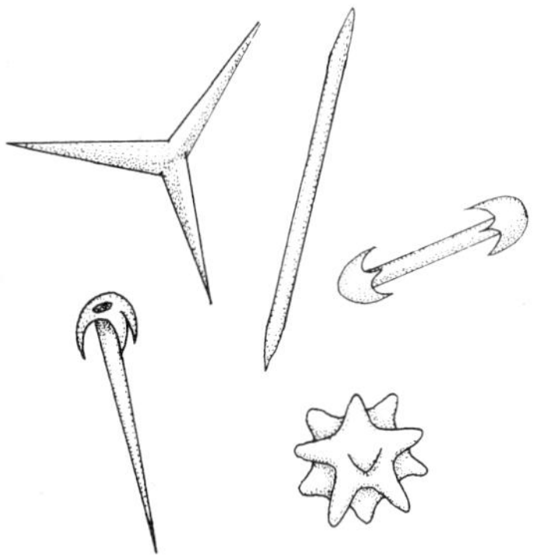

FIGURE 5.3
Sponge Spicules

The basic elements of the sponge skeleton are the spicules. They have many shapes and are valuable tools in identifying individual species.

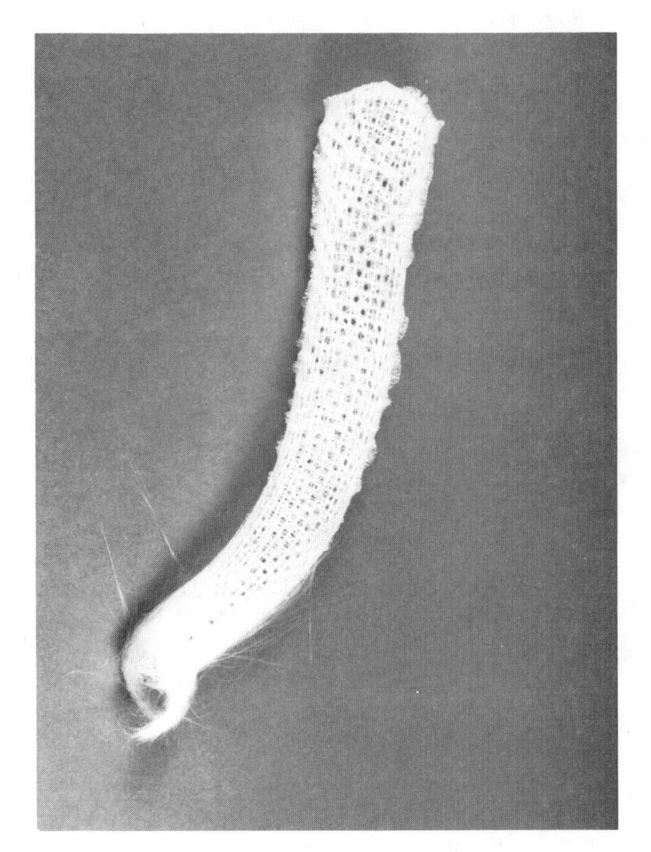

FIGURE 5.4
The Venus Flower Basket, *Euplectella*

(Photo by Alex Kerstitch.)

All sponges are benthic and sessile. Most sponges are found in shallow waters on rocky or solid substrate. Some, however, are found in muddy substrate and extend into the deep-sea benthos. Sponges are tolerant of exposed, turbulent, and quiet water. Only a few species are tolerant of fresh water, and they have a limited distribution in estuaries.

PHYLUM CNIDARIA

The phylum Cnidaria (Greek: *knide*-nettle) includes animals that have radial symmetry, a body with two layers of tissue, and a digestive cavity with a single opening. It is an important phylum, with over 9,000 species widely distributed in the marine environment. Medusae (jellyfish), sea anemones, corals, and sea pens are cnidarians.

As well as radial symmetry, Cnidarians have tentacles bearing stinging cells called **cnidocytes.** These two characteristics provide the formal definition for the phylum.

Cnidarians have a number of important structural features not found in sponges. There is an internal cavity that permits the digestion of large particles of food. Most are carnivores, and although many species feed on small zooplankton, larger medusae can capture and eat fish, and anemones can engulf and digest animals such as snails and crabs.

A second important feature of cnidarians not found in sponges is nerve cells. Although there is no brain, the nerve cells are part of a sensory response system. For example, when food is detected by the tentacles of an anemone, they close up and draw the captured prey into the digestive cavity. Some sea anemones also can respond to chemical stimulation from certain sea stars by detaching from the substrate and awkwardly swimming away.

Although cnidarians have some characteristics of animals not found in sponges, they lack a number of other systems characteristic of higher animals. There is no specialized system for gas exchange, such as gills. The large surface area in contact with the water is adequate for passing oxygen and carbon dioxide to and from the tissues. Moreover, there is no system for waste disposal. The undigested remains of food are expelled from the digestive cavity and the waste products of metabolism (ammonia or urea) diffuse across the body surface.

Cnidarians are relatively simple in construction. A cross section of the body wall shows two layers: an outer ectoderm and an inner endoderm (Fig. 5.5A). The ectodermal layer contains the stinging cells (cnidocytes) characteristic of the phylum. Between the outer and inner layer is a substance called mesoglea. **Mesoglea** is a jellylike material that may occur as a thin layer, as in anemones, or as a thick layer that dominates the shape of the animal, as in medusae.

A.

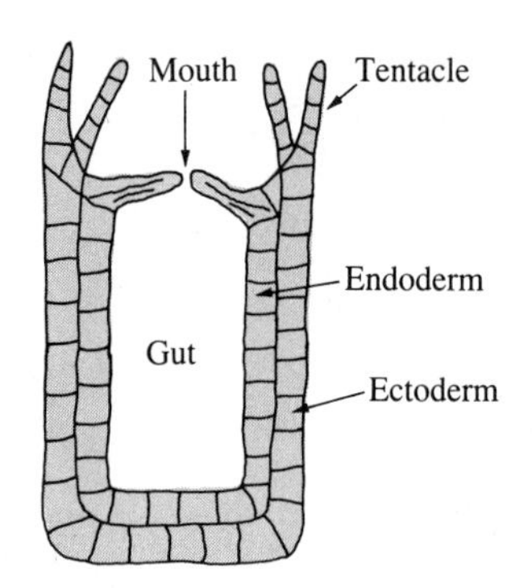

B.

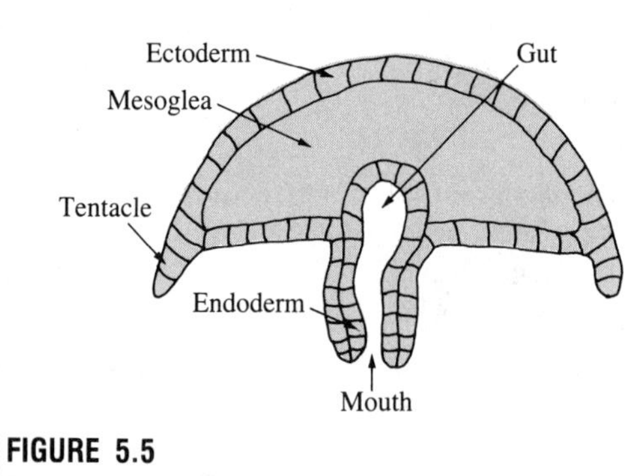

FIGURE 5.5
Body Shapes of Cnidaria

(A) Polyp. **(B)** Medusa.

Cnidarian body shapes are of two basic types: polyp or medusoid (Fig. 5.5). The polyp is sessile, with the mouth end up. The medusa, essentially an upside-down polyp, is planktonic, and the mouth and tentacles hang down. In some cnidarian species there is an alternation of generations. For example, in some classes (hydroids and scyphozoans), a sessile polyp stage is followed by a planktonic medusa stage. Other cnidarians have only one form throughout their lives. Sea anemones and corals, for example, are polyps throughout their entire life.

Although many species of cnidarians are individual animals, an almost equal number are colonial. Colonial forms are found in both benthic and pelagic species. Corals and hydroids are benthic colonial species, while the Portuguese man-of-war (Fig. 5.6) is a pelagic colonial species. Individuals in a colony may show morphological and functional specialization. In reef coral, all individuals are more or less identical in form and function. In the Portuguese man-of-war, however, there is extensive specialization. At first appearance, the man-of-war looks like an individual, and all man-of-wars look similar. Closer examination, however, reveals numerous individuals specialized for various tasks. The central float is an individual specialized for flotation. Suspended from the float are numerous other individuals that specialize in food capture and digestion. Still others have the function of protection or reproduction.

A distinguishing characteristic of the phylum Cnidaria is the presence of stinging cells (Fig. 5.7). These

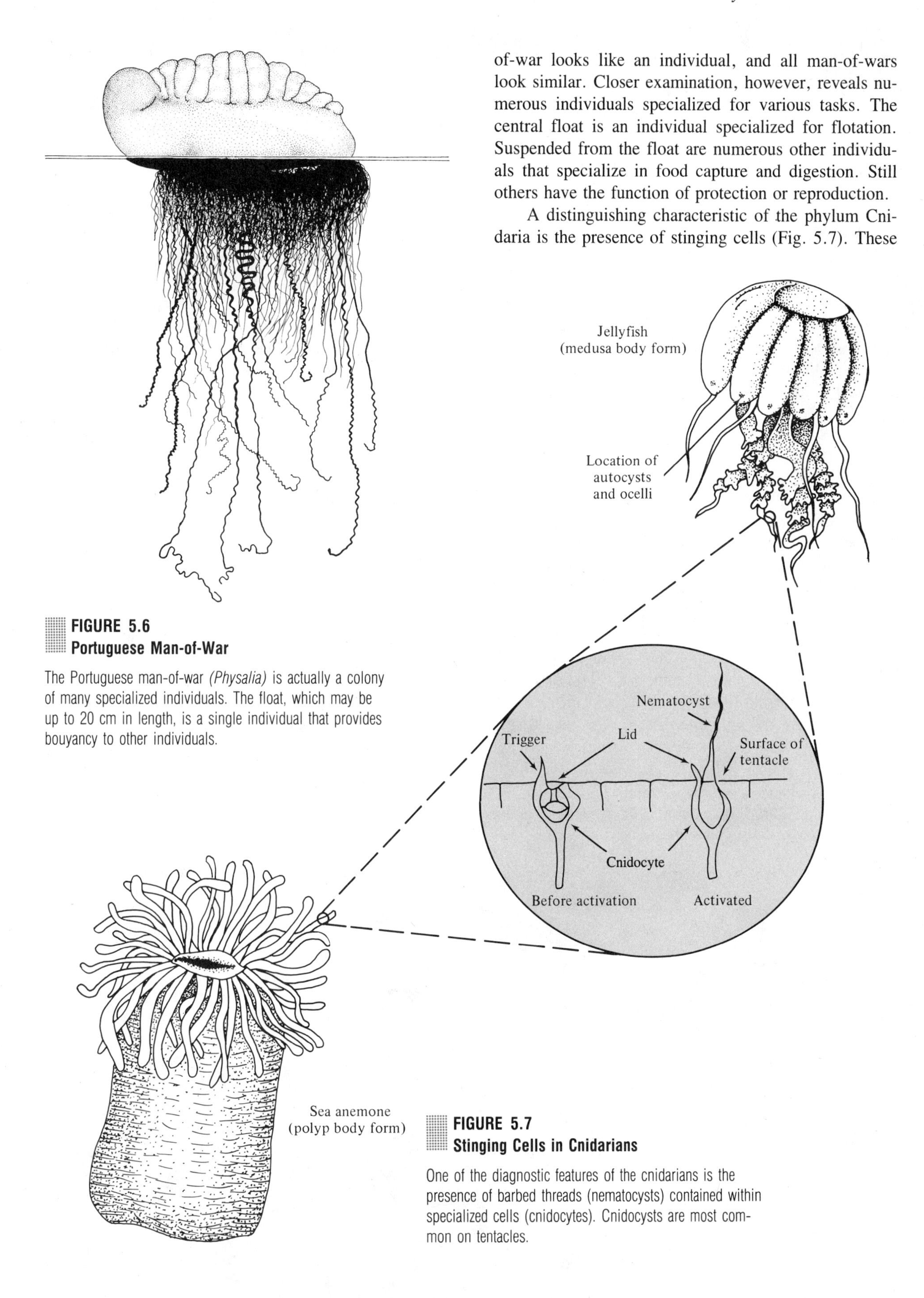

FIGURE 5.6
Portuguese Man-of-War

The Portuguese man-of-war *(Physalia)* is actually a colony of many specialized individuals. The float, which may be up to 20 cm in length, is a single individual that provides bouyancy to other individuals.

FIGURE 5.7
Stinging Cells in Cnidarians

One of the diagnostic features of the cnidarians is the presence of barbed threads (nematocysts) contained within specialized cells (cnidocytes). Cnidocysts are most common on tentacles.

cnidocytes contain threads (nematocysts) that can be forcefully ejected from the cell for use in defense and capturing prey. Cnidocytes are found in all cnidarians and are located mainly on the tentacles. The important parts of the stinging cell are the trigger (cnidocil), the lid, and the capsule inside the cell filled with a thread and barb (nematocyst). Both chemical and tactile stimulation can cause the cell to discharge its nematocyst. Nematocysts may inject poisons (as with sea wasps and the Portuguese man-of-war), or they may secrete a sticky adhesive. When you touch the tentacles of a sea anemone, the sticky feeling is from the discharge of nematocysts. After a nematocyst is discharged, the cnidocyte dies and must be replaced by a new one.

Cnidarians are divided into three classes—Anthozoa, Scyphozoa, and Hydrozoa—each of which we now describe.

Class Anthozoa

The class Anthozoa includes the sea anemones, hard corals, and soft corals. It is the largest of the three classes, with about 6,000 species (60% of the total in the phylum). Anthozoans occur only in the polyp form and may be solitary (e.g., anemones) or colonial (e.g., corals, sea fans).

The basic morphology of anthozoans is best seen in an anemone (Figs. 5.7, 5.8). Two features are important. First is the presence of numerous septa, or sheets, that extend into the digestive cavity and increase the surface area of the digestive tissues. The second is a throat leading from the mouth into the digestive cavity.

The best known of the Anthozoa are the sea anemones and the hard corals. Sea anemones, the largest of the anthozoans, are solitary (Fig. 5.8). Anemones either wait patiently for large prey to unknowingly blunder into their tentacles, as happens to fish, or they capture animals, such as snails and crabs, that have been washed free from their substrate by waves. (Although very simple animals, anemones exhibit interesting behavior. When packed in dense groups, some anemones show intraspecific aggression. This aggression is discussed further in the special feature at the end of this chapter.)

Hard corals are best known for their contribution to coral reefs. They have an external calcium carbonate skeleton which may reach massive proportions in coral reefs (see Ch. 13). Not all hard corals are colonial. Those found in temperate waters or in the deeper sea are generally solitary. In the tropics, however, where coral reefs are common, the hard corals show maximum diversity and abundance. The individual polyps of reef corals are small and are similar in morphology to a small sea anemone (Fig. 5.9). Each individual (or polyp) sits in a calcium carbonate cup called a theca, and is connected to other polyps by tissue. With hard corals, the shape of the individual polyp is not as characteristic as the shape of the colony. Other groups of anthozans include sea pens (Fig. 5.10) and soft corals (Fig. 5.11).

Class Scyphozoa

Scyphozoans (200 or so species) are the largest of the planktonic Cnidaria. The medusa is the dominant phase, but most species also have a sessile polyp stage. The medusae are solitary and do not form colonies. They are relatively fast swimmers for planktonic organisms and a few can reach speeds of up to 6 m/sec (13 m/h) for short bursts. A number of scyphozoans are

A.

B.

FIGURE 5.8
Sea Anemones

(A) *Telia.* **(B)** *Metridium.* (From Flora and Fairbanks, 1982.)

FIGURE 5.9
Hard Corals

Most hard corals are colonial animals. Individual polyps sit in calcium carbonate cups.

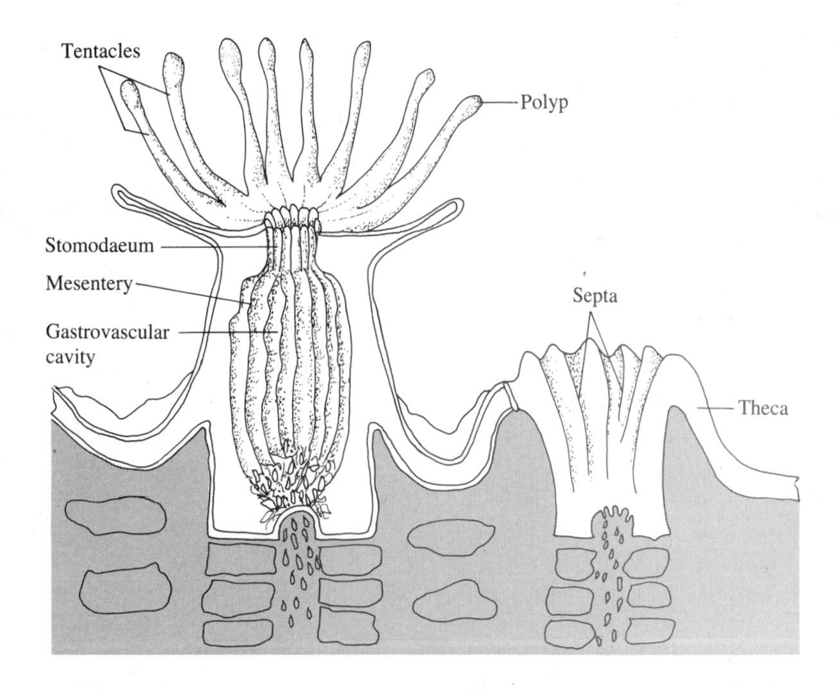

poisonous. The temperate sea blubber or lion's mane *(Cyanea)* has a painful sting, and the tropical sea wasps are among the most poisonous of marine animals. One species of sea wasp has been responsible for over 50 deaths in Australia. A sting from this sea wasp to any part of the body can cause death in less than 30 minutes.

Class Hydrozoa

The hydrozoans, with some 3,000 species, is the second largest group of cnidarians. Hydrozoa generally have a colonial polyp stage (with specialized feeding and reproductive polyps) that alternates with a solitary

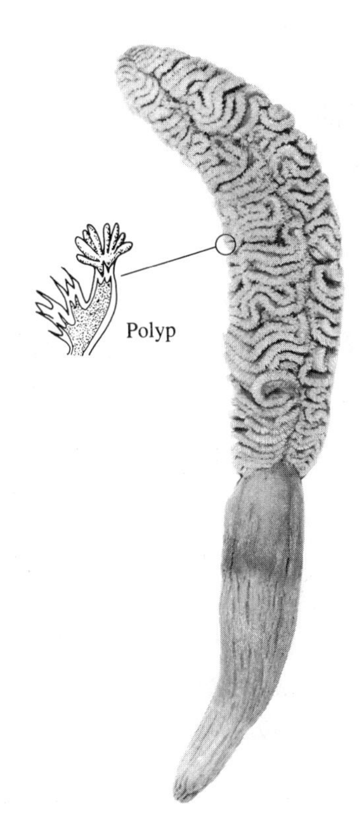

FIGURE 5.10
Sea Pen

Sea pens are colonial anthozoans. Individual polyps are distributed along the leaves of the sea pen's body. Shown is *Ptilosarcus*. (From Flora and Fairbanks, 1982.)

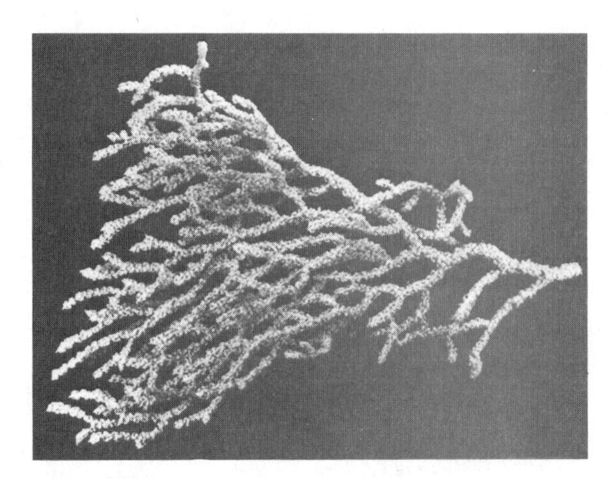

FIGURE 5.11

The soft gorgonian coral *Paragorgia*. (From Flora and Fairbanks, 1982.)

medusa stage (Fig. 5.12). In many cases the medusa stage is not liberated but remains attached to the polyp. The most important group is the hydroid. The polyp stage is dominant in most hydroids and is generally quite small, with colonies reaching 2–3 cm in height. In most species the polyp has an external skeleton called a perisarc that gives support and protection to the colony. Hydroids often are mistaken for turflike algae.

Another important group of Hydrozoa are the siphonophores. These specialized pelagic medusae are widely distributed in the plankton, and occur in deep and shallow water. Perhaps the best known siphonophore is the Portuguese man-of-war (*Physalia;* see Fig. 5.5).

PHYLUM CTENOPHORA

The Ctenophora (Greek: *kteis*-comb; *pherein*-to bear), like the phylum Cnidaria, have radial symmetry. They also share the cnidarian characteristics of a single digestive cavity, tentacles with cells specialized for prey capture and defense, and two body layers with watery mesoglea between. Ctenophorans are a small phylum with only about 90 species, all of which are marine. Most ctenophores are pelagic, and they are an important and common component of the zooplankton in coastal waters. The body shape of most species is spherical (1–2 cm in diameter), giving rise to the common names of sea gooseberry and sea walnut (Fig. 5.13). A number of species, however, have a flattened body shape that in its most extreme form becomes a band-shape organism up to 1 m in length.

Ctenophores differ from the cnidarians in a number of ways. Most important is their locomotion system. Their limited swimming ability comes from a unique and peculiar method of movement. Cilia are systematically grouped into a locomotory structure consisting of combs arranged in plates that overlap each other (Fig. 5.14). There are eight rows of these ciliary combs, starting at the upper end of the body (mouth end) and running most of the way to the bottom (statocyst). The ctenophore swims by beating these ciliary plates in unison. The action of eight rows is coordinated by a balance organ (the statocyst) that, in conjunction with nerve cells, tells the animal which way is up. The comb plates sweep the water in such a way that the direction the animal moves is mouth first.

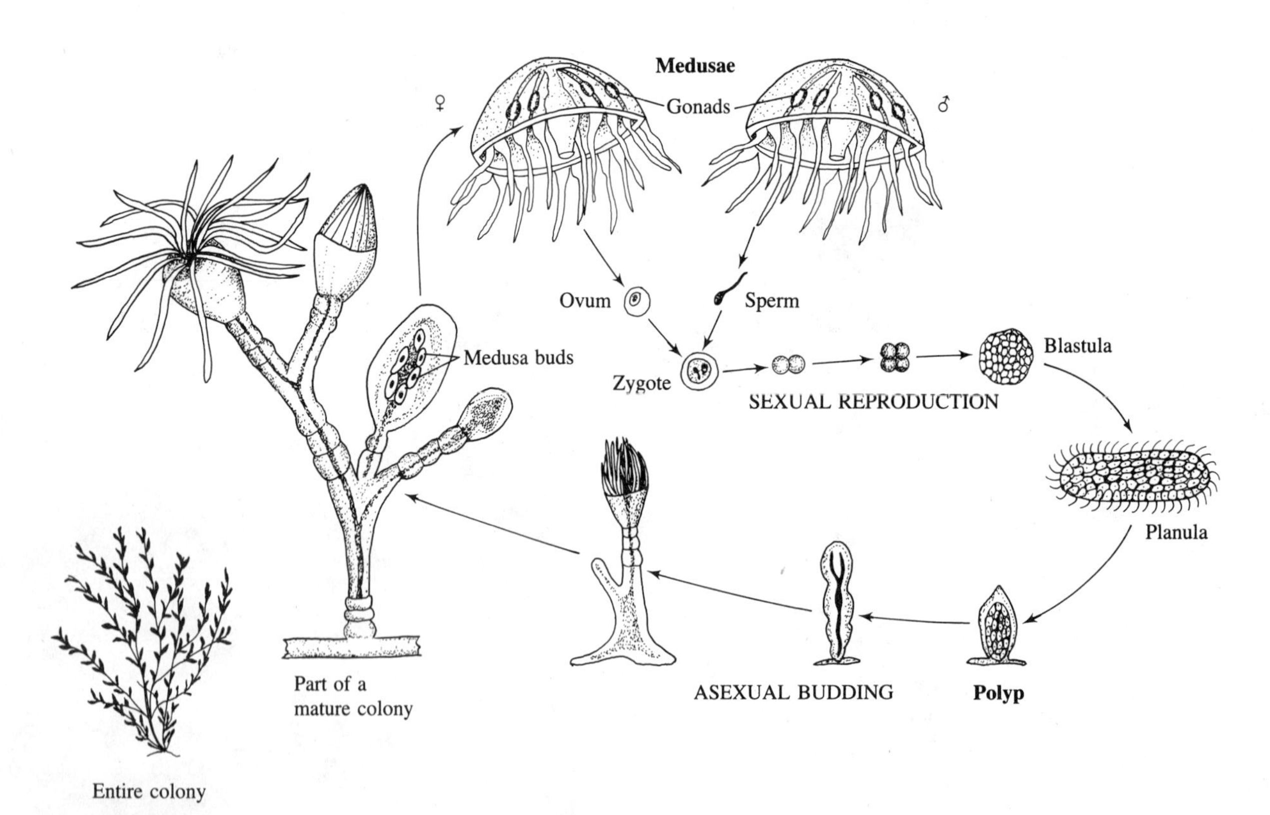

FIGURE 5.12
Hydroid Life Cycle

Many marine hydroids have complex life cycles with a polyp and medusa form. This hydroid *Obelia* has a rather small benthic form, while the medusae are larger (up to 6 cm in diameter).

FIGURE 5.13

The sea gooseberry *Pleurobranchia.* (From Flora and Fairbanks, 1982.)

Ctenophores have on their tentacles some cells similar to the cnidocyte of cnidarians. Like the cnidocyte, the colloblast of ctenophorans contains a coiled thread that is discharged when stimulated, but these threads do not have toxins. Instead, mucus on the end of the thread immobilizes prey (Fig. 5.15). Ctenophores are carnivores and feed mostly on zooplankton. At times some species (e.g., *Pleurobrachia*) occur in tremendous numbers in coastal waters and are voracious predators.

PHYLUM PLATYHELMINTHES

The platyhelminthes (Greek: *platy*-flat; *helmis*-worm) are flat worms. They are bilateral animals, with a head and a tail end. Most flatworms are parasitic. Of the almost 13,000 species, 10,000 are parasites belonging to either the class Cestoda (tapeworms) or the class Trematoda (flukes). The minority of flatworm species (3,000) are free living and belong to the class Turbellaria. These free-living flatworms best represent those features that define the phylum, so we will examine them more closely.

Flatworms have a gut with a single opening and no body cavity (Fig. 5.16). They have a body structure based on three layers of tissues, and the space between

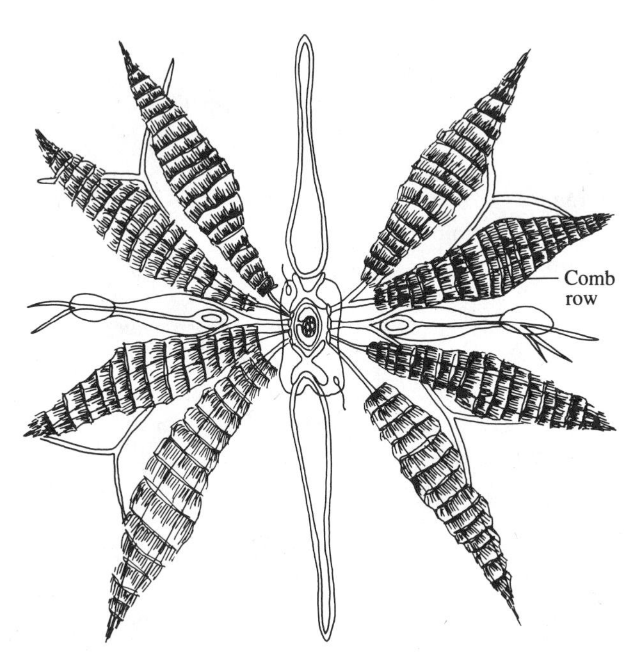

FIGURE 5.14
Comb Rows of Ctenophores

The swimming structure of Ctenophora is made of fused cilia arranged in horizontal rows that run from the upper surface of the body down the sides to the mouth. Shown is top view.

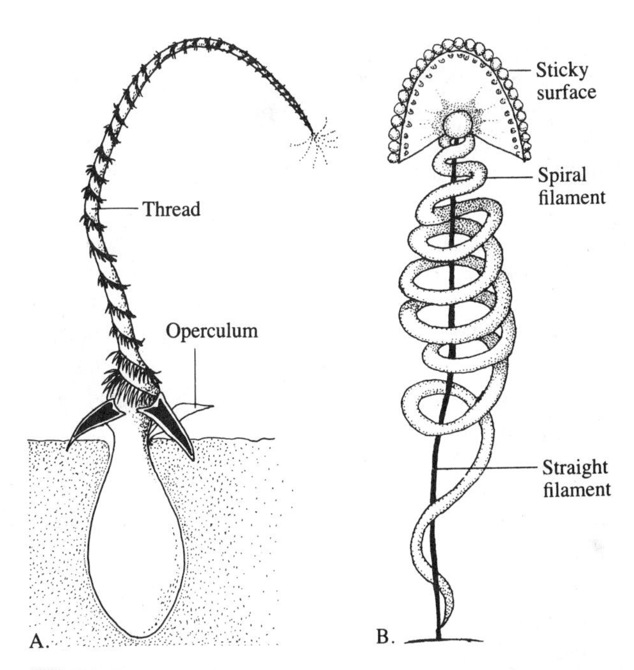

FIGURE 5.15
Nematocysts and Colloblasts

(A) The nematocyst of coelenterates has a barbed tip containing toxins. **(B)** On ctenophorans, the tip of the colloblast is covered with sticky mucous projections.

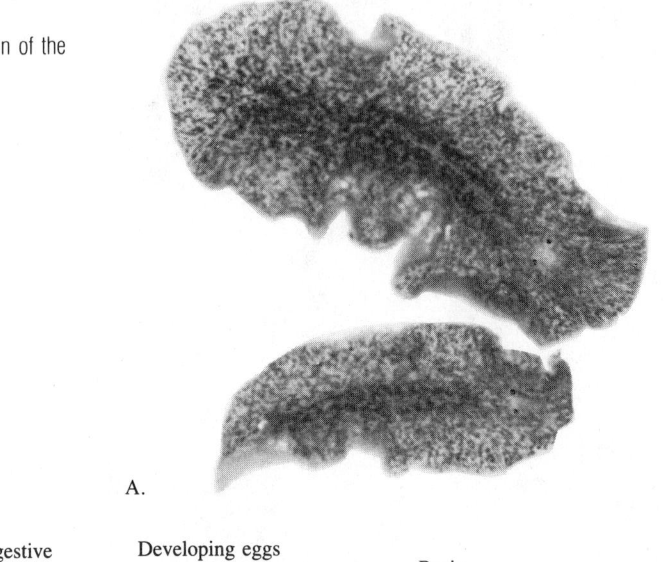

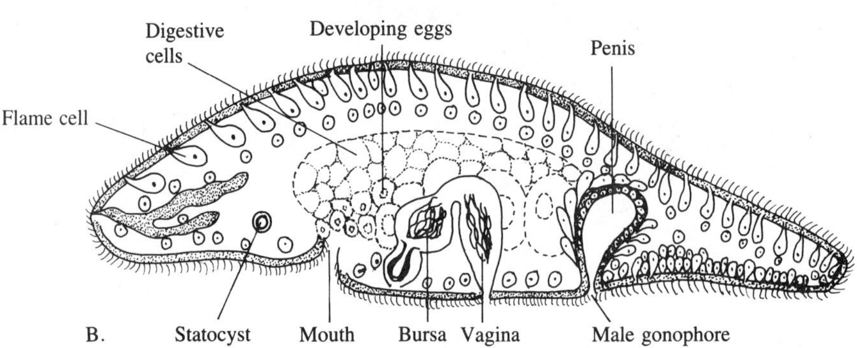

FIGURE 5.16

(A) The turbellarian flatworm *Freemania.* **(B)** Longitudinal section of the turbellarian *Convoluta.* (A from Flora and Fairbanks, 1982.)

the outer layers and internal organs is filled with a soft tissue of cells called **parenchyma.** The free-living flatworms retain many of the primitive characteristics of the phylum. The bilateral shape is an important feature because it gives an anterior and posterior dimension to the body which permits **cephalization,** the concentration of sensory structures and nervous function in the head end. Turbellarians have a ganglion (knot) of nerve cells in the head end called a brain. The head end also has 'eyes.' These eyes are not capable of forming images but are sensitive to changes in light intensity. Turbellarians also have well-developed chemosensory structures that are used for locating food.

Flatworms have two ways of moving. The smaller species use the beating of cilia to creep along a mucus trail that they secrete. Larger species move by muscular contractions of the bottom surface that create a series of undulating waves. This kind of movement is best developed in gastropod snails (see Fig. 5.26).

Turbellarians are predators and scavengers of animal tissue. They capture active prey in the mucus of their feet or browse on sessile organisms such as tunicates and bryozoans. Fewer species are herbivores and feed on benthic diatoms. One unique form of feeding is a flatworm that has a symbiotic relationship with a single-celled green algae. The flatworm *Convoluta* lives on mud flats and comes to the surface in the bright light of low tide, where the symbiotic algae distributed throughout the epidermis of the body actively photosynthesize. Food produced by the algae is used by the flatworm.

Although the majority of turbellarian flatworms are marine, they also are common in fresh water, and a number of species live in damp soil. In the oceans, turbellarians are almost all benthic and are most common in shallow waters. They are found on both soft, muddy substrate and on rocks or shells.

PHYLUM NEMERTEA

Nemertean worms (from the Greek, *nemertes*—unerring—the Greek god *Nemertes* was an archer) are unsegmented worms with a proboscis (generally with jaws) that is used for feeding. The Greek derivative of the phylum name is reference to its feeding method. There are over 650 species of nemertean worms. Nemertean body structure, in some ways, is similar to that of the flatworms. Nemerteans have no body cavity and the space between the digestive tract and the outer epidermis is filled with parenchyma cells. Unlike flatworms, nemerteans have a digestive tract with two openings. Because food can enter one end, be digested, and wastes exit through a second opening, the digestive system is more efficient than that of flatworms or cnidarians. Nemerteans are generally much longer than flatworms (Fig. 5.17).

Small nemerteans move in a fashion similar to small flatworms: by using cilia on a mucus trail. Large nemerteans have well-developed bands of circular and longitudinal muscles. Contraction of the circular muscles extends the body, and contraction of the longitudinal muscles shortens and thickens the body. Some species have an amazing ability to change length. *Cerebratulus* can change its size from 1 m (3.3 ft) in length and 2 cm (0.8 in) in diameter to an extended length of 12 m and diameter of 2 mm. The common name ribbon worm comes from this ability to stretch the body into a ribbon shape.

Another common name for nemerteans is proboscis worm, and the presence of a well-developed proboscis is a distinguishing characteristic of the phylum (Fig. 5.18). The proboscis is a tube housed in a cavity separate from the digestive system. In some nemerteans, it is as long as the body itself and can be rapidly extended by muscular contractions. The proboscis is covered with sticky mucus for holding prey and also may have barbs for piercing and stinging. Nemerteans are active predators and feed primarily on small crustaceans, molluscs, and annelids. Prey is captured by the proboscis and sucked whole into the mouth.

Unlike most marine phyla, which have the highest diversity in tropical latitudes, nemerteans are more abundant in temperate water. They are most common in the benthos of shallow waters, where they burrow in

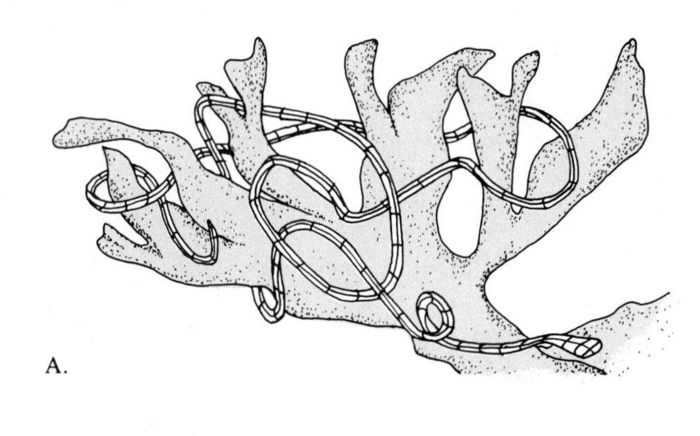

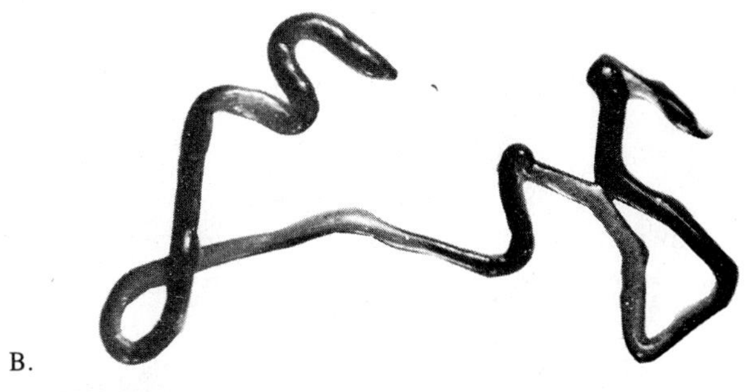

FIGURE 5.17
Nemerteans

(A) *Polia* on a reef coral. **(B)** The intertidal nemertean *Emplectonema,* found in the temperate Eastern Pacific Ocean. (B from Flora and Fairbanks, 1982.)

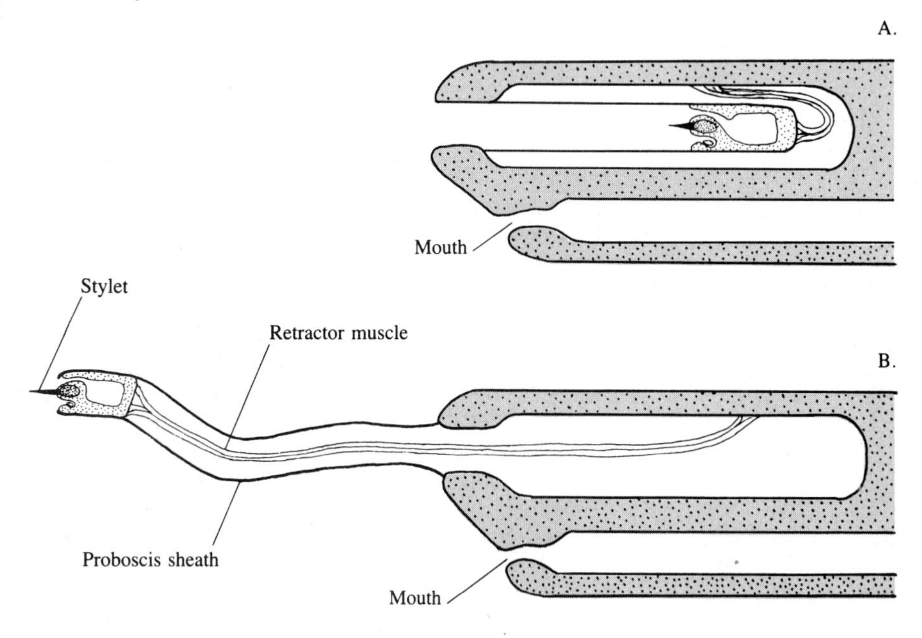

FIGURE 5.18
Proboscis in Nemerteans

(A) Proboscis retracted. **(B)** Proboscis extended. (From Crane, 1973.)

mud or sand, or are found among algae or under rocks and shells. Size is generally up to 20 cm (8 in), but worms up to 10 m (31 ft) are known.

THE ASCHELMINTHES PHYLA

Phylym Nematoda (the round worms; Greek: *nema*-thread)
Phylum Rotifera (Latin *rota*-wheel; *ferre*-to bear)
Phylum Gastrotricha (Greek *gaster*-stomach; *thrix*-hair)
Phylum Acanthocephala (Greek *akanthat*-thorn; *kephale*-head)
Phylum Nematomorpha (Greek *nema*-thread; *morphe*-shape)
Phylum Gnathostomulida (Greek *gnathos*-jaw; *stoma*-mouth)
Phylum Kinorhyncha (Greek *kinein*-to move; *rynchos*-snout)

Generally, members of these phyla are very small, and a microscope is needed to see their features clearly. Most live in sand or mud. Except for the familiar roundworms, the Aschelminthes are known only to specialists; hence the lack of common names. Most Aschelminthes are small; only in the Nematoda, Acanthocephala, and Nematomorpha do some species exceed a few millimeters. These phyla have parasitic species that exceed 10 cm (3.9 in) in length. They also share the characteristic of a circular cross section and elongated, wormlike shape. The remainder of the Aschelminthes are only somewhat elongated and tend to be cylindrical in shape.

Of the seven phyla, the nematodes are most common in the marine environment. This phylum, with over 10,000 species, is widely distributed in the sediments of the oceans. Nematodes are also common in the sediments of rivers and lakes and in soil. Many nematodes are successful parasites, and in the marine environment they parasitize algae as well as animals. There are probably many more species than have been described thus far. Some biologists estimate that there might be as many as 500,000 different nematode species in total. Many nematodes are minute (less than a few millimeters) and have not been well studied. These small species are interstitial and are found among the particles of sea-bottom sediments.

Nematodes are found from the high-tide mark to the deepest parts of the ocean. The food habits of free-living nematodes are diverse. Some are herbivores, feeding on algae and benthic diatoms; others are carnivores, feeding on small animals, including other nematodes; still others are saprobic detritus feeders. Parasitic nematodes feed on seaweeds, invertebrates, fish, and birds. Nematodes rank high on the list of successful animal phyla.

The nematode body, which is circular in cross section (Fig. 5.19), has an external cuticle. Along the body wall are longitudinal bands of muscle which are used in movement. The basic movement of nematodes is side-to-side undulation. When suspended in water, nematodes wriggle frantically back and forth in apparent

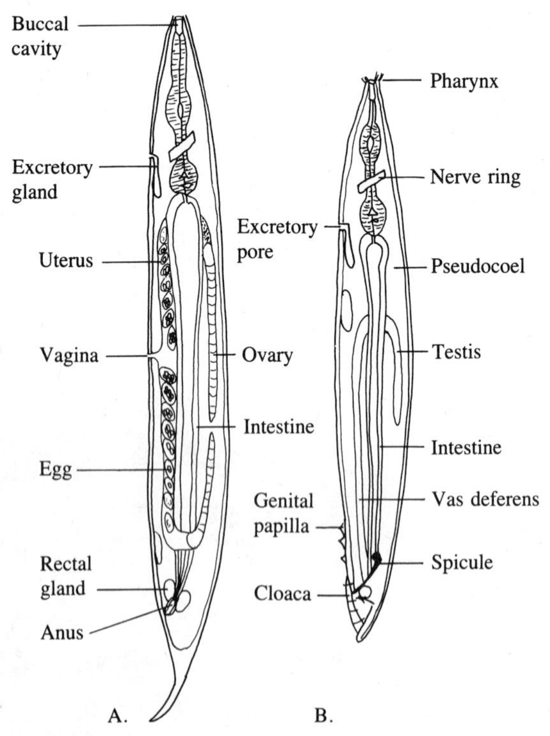

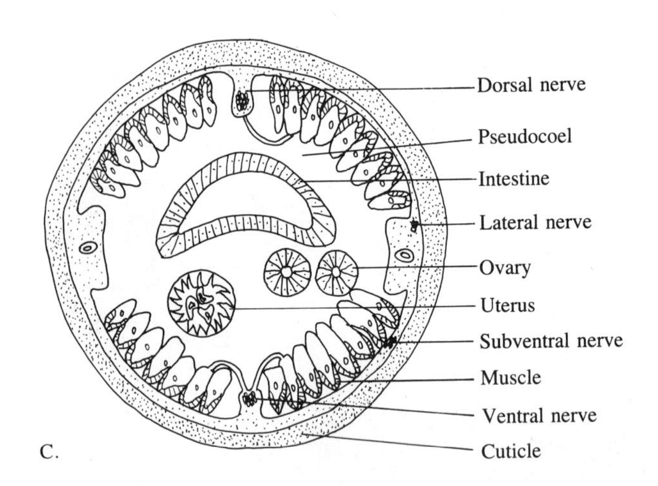

FIGURE 5.19
Nematode Anatomy

(A, B) Longitudinal section of a male and female.
(C) Cross section.

random fashion. They need the small particles of the sediment to use as a pivot for movement and move efficiently and effectively through the interstitial spaces of the sediments.

Compared to the nematodes, the other phyla in the Aschelminthes are of minor importance in the marine environment. The phylum Nematomorpha has 230 species, almost all found in fresh water. The phyla Rotifera is very important to the plankton of fresh water, but only about 50 of the 1,800 species are found in the marine environment. The remaining aschelminth phyla are commonly found only in the interstitial spaces of marine sediments. Most are very small and generally microscopic. The phylum Gnathostomulida is of interest because it was discovered less than 20 years ago (Fig. 5.20). The group now numbers 80 species and more are likely to be found in the interstitial spaces of sediments. The remaining two phyla—the gastrotrichs and kinorhynchs—total some 500 species, most of which are marine (Fig. 5.21). Like other Aschelminthes, they are tiny (under 1 mm), worm-shaped animals found in the interstitial spaces of muds and sands in shallow water.

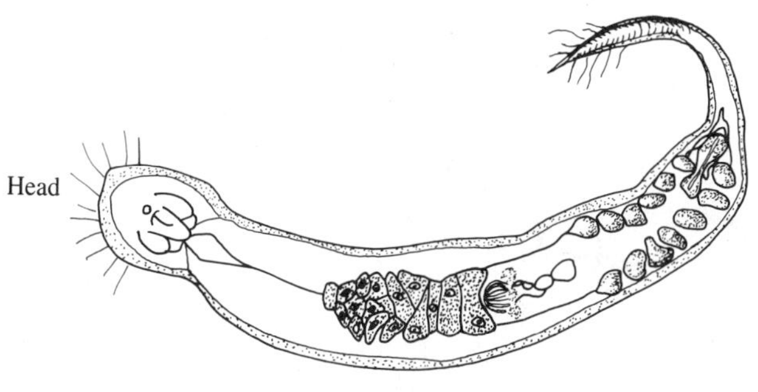

FIGURE 5.20

The gnathustomulid *Gnathostomula.* Natural size is 1 mm.

PHYLUM MOLLUSCA

The Mollusca (Latin: *molluseus*-soft) is a major phylum with a large number of species (many of which you already may be familiar with) including snails, clams, oysters, squid, and octopus. Molluscs have been a dominant group of animals ever since their origin in the Cambrian period over 500 million years ago. Besides the 110,000 existing species, there are over 35,000 fossil molluscs. Molluscs have successfully adapted to fresh and salt water as well as the terrestrial environment. Only half of the mollusc species are marine. The phylum evolved into five distinct classes that show a wide range of body anatomy (Fig. 5.22). At first glance it seems improbable that animals as dissimilar as a squid and an oyster could be related. Because of the extensive evolutionary changes that occurred as molluscs radiated into various habitats, no group has retained all the primitive body features of the ancestral phylum (Fig. 5.23). The species that comes closest is a deep-sea benthic snail *Neopilina. Neopilina,* which was discovered in 1959, shows distinct signs of both internal and external segmentation (Fig. 5.24), which leads some biologists to link it closely to annelids and arthropods.

The general characteristics of this phylum include an external calcium carbonate shell, a prominent foot, a feeding structure called a radula, and a mantle cavity (a tissue-enclosed space between the shell and the foot). The mantle cavity generally contains gills used for obtaining oxygen. No individual species has all of these characteristics, but a general picture of the phylum can be described based on these features.

The shell is secreted by the mantle which, as the name suggests, is a sheet of tissue that drapes the dorsal surface of the body of molluscs. The shell is secreted as a series of layers. The outermost layer of the shell, the

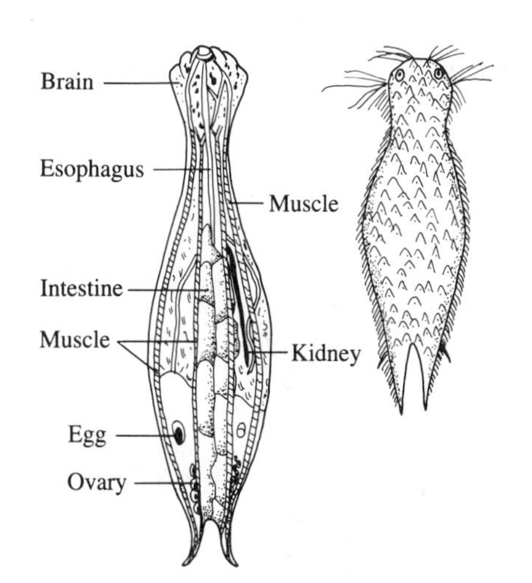

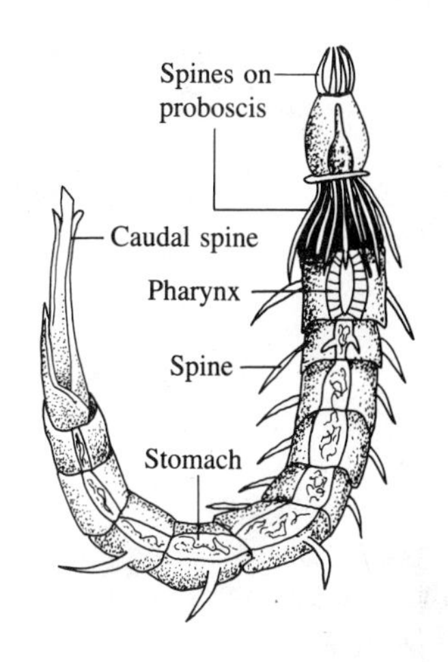

FIGURE 5.21

(A) Gastrotrich. **(B)** Kinorhynch.

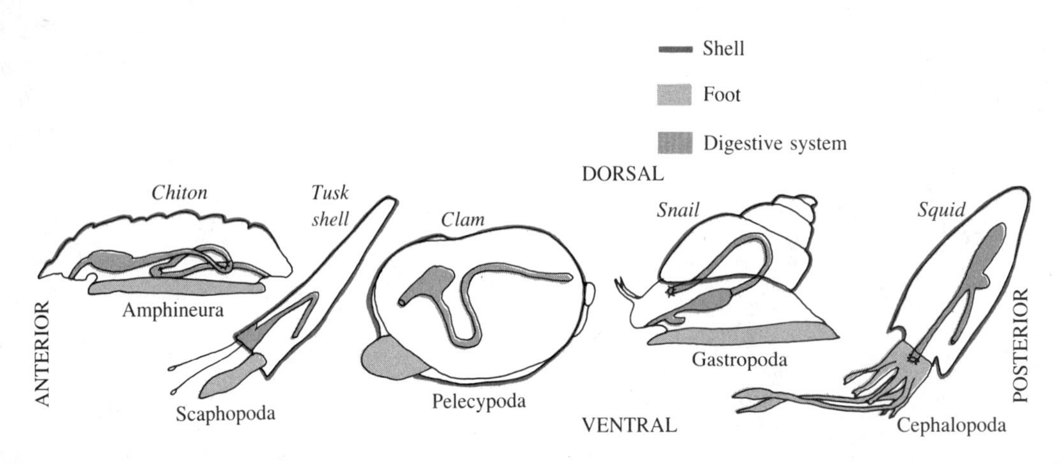

FIGURE 5.22
Molluscs

The five classes of molluscs have quite different basic shapes. (*M*, mouth; *A*, anus)

FIGURE 5.23
Ancestral Molluscan Anatomy

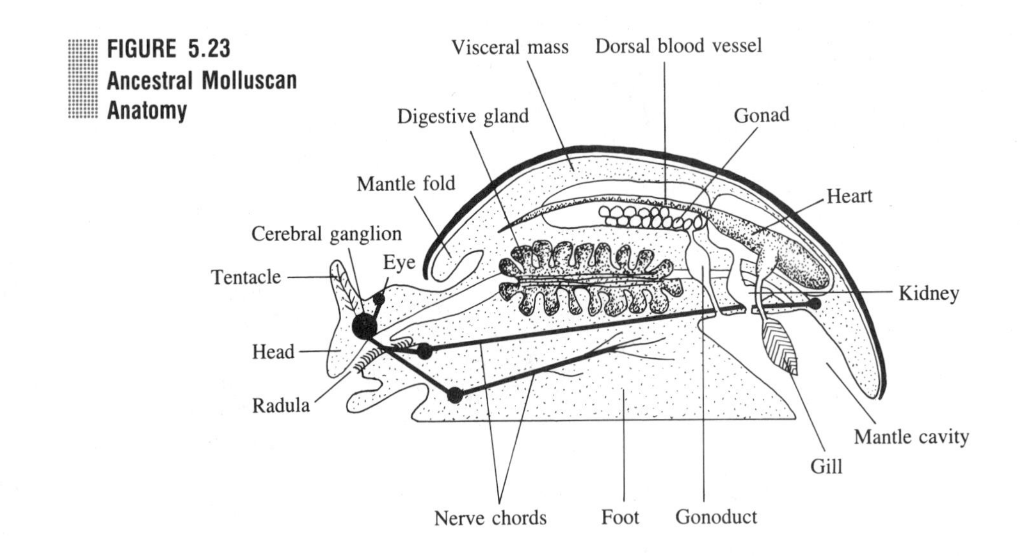

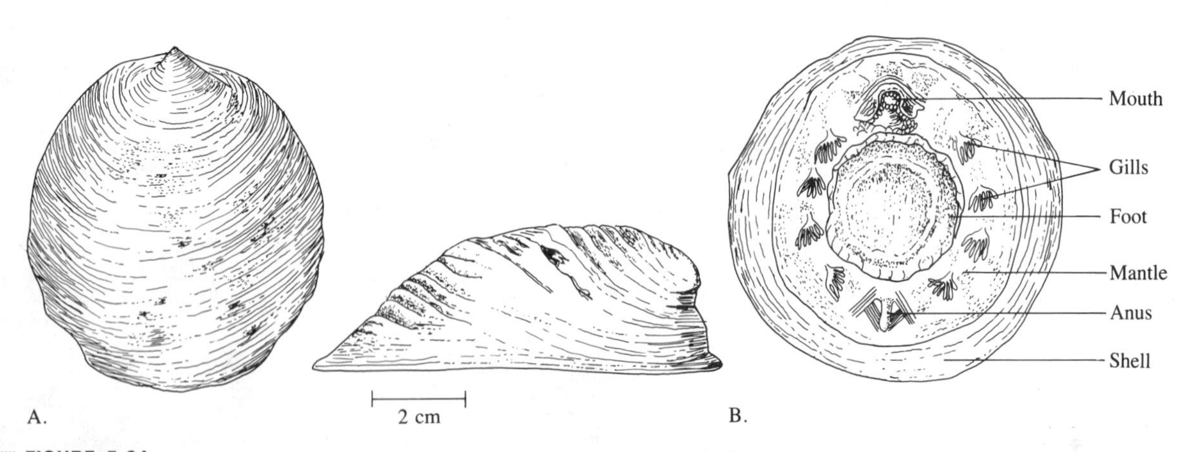

FIGURE 5.24
Primitive Molluscan Anatomy (*Neopilina*, a Monoplacophora)

(A) External views. **(B)** Ventral view.

periostracum, is not calcified but made of a horny proteinaceous material called *conchiolin*. The periostracum can be secreted only by the outermost edge of the mantle (Fig. 5.25), and in many species it becomes worn and eroded and so may not be present in adults.

The central or prismatic part of the shell is secreted in layers by the edge of the mantle. The calcium carbonate is deposited in microscopic prism shapes—hence the name prismatic layer. The innermost layer is called the **nacreous,** or mother-of-pearl, layer and gives the smooth lustrous finish to the inner-shell surface.

The inner-shell layer is secreted by all of the mantle surface and so may grow in thickness. It is this mantle surface that produces pearls. When an irritant comes between the shell and the mantle, the response is to coat it with a layer of nacreous shell. Often the irritant becomes incorporated into the shell. Occasionally, however, it becomes embedded in the mantle tissue and is surrounded by a shell layer perhaps a millimeter in thickness to form a pearl. Some molluscs produce a particularly lustrous mother-of-pearl. The most famous are the pearl oysters found throughout the tropical Pacific. The occasional large, spherical irritant is the source of jewel pearls. Nowadays, foreign objects of the correct shape and size can be purposely implanted into the mantle epithelium of pearl oysters and so-called cultured pearls can be produced easily and routinely in about three years.

The space between the mantle and the body (the mantle cavity) has a number of important functions in molluscs. In the ancestral form (Fig. 5.23) the mantle cavity was the space where the gills were located, where the anus and kidney pores released their wastes, and where eggs and sperm were released. In most specialized modern molluscs, the mantle cavity maintains these functions.

Molluscs have a well-developed foot. In ancestral molluscs the foot was probably a broad flat sole that was ciliated and mucus covered. Movement most likely was by the beating of cilia in a manner similar to that of small, modern flatworms. In modern molluscs many groups retain the broad, mucus-covered, solelike foot, but the main method of propulsion is muscular contraction. The mollusc foot has a complex musculature, including longitudinal, transverse, and dorsoventral muscle bands, and movement depends on an internal hydrostatic skeleton. A hydrostatic skeleton is somewhat like a balloon. Pressure applied to one part of it causes deformation in another. The pressure of blood in the mollusc foot cavity allows muscle contraction to be converted into movement. The mollusc moves as waves of muscle contraction pass across the surface of the foot (Fig. 5.26). Forward movement may originate from the posterior end of the foot (direct wave) or from the anterior end of the foot (retrograde wave).

The solelike mollusc foot also functions as an adhesive gland that works like a suction cup. Using the mucus layer, the mollusc tightly clamps to the substrate. Once disturbed, it is almost impossible to dislodge an abalone or limpet from a rocky surface without breaking the shell.

The final common characteristic of molluscs is the radula. The **radula** is a 'tongue' that can be extended from the mouth to lick and scrape food from the substrate. In the radula are embedded a series of rows of small, chitinous (sometimes mineralized) teeth that are effective scrapers. The radula is constantly forming new teeth at its interior end to replace those that are lost from erosion as they scrape.

The phylum Mollusca is divided into seven classes: Monoplacophora, Polyplacophora, Gastropoda, Cephalopoda, Pelecypoda, Scaphopoda, and Aplacophora. The four most common and familiar are discussed here.

Class Polyplacophora

This is the class of the Chitons. Chitons are a small group of molluscs (some 600 species) found commonly in rocky intertidal areas tropically and on the west coast of North America, and subtidally in New England.

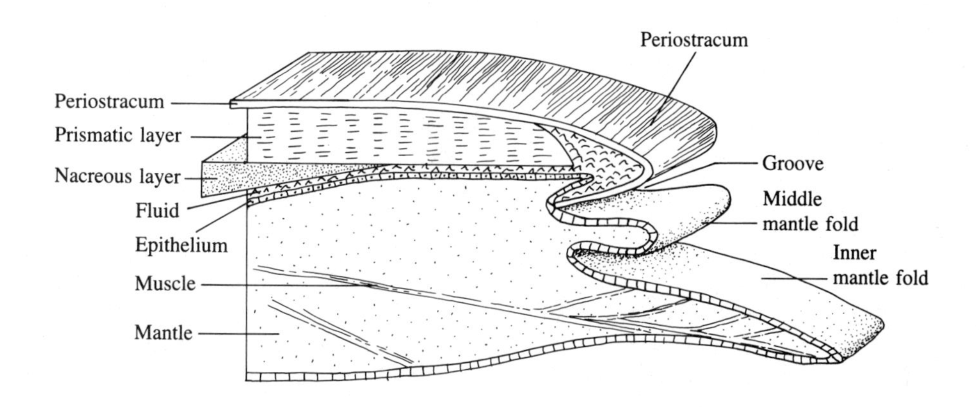

FIGURE 5.25 The Molluscan Shell

Cross section of a bivalve shell.

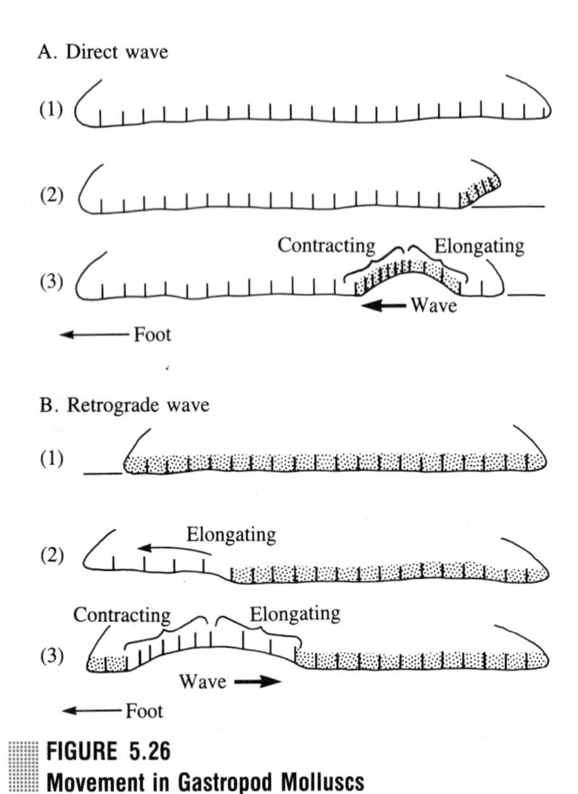

FIGURE 5.26
Movement in Gastropod Molluscs

Gastropods, like many other invertebrates, seem to glide over surfaces. This movement is produced by a series of muscular contractions and elongations that produce a wave. These waves may be direct (moving toward the anterior of the foot) or retrograde (moving in the opposite direction).

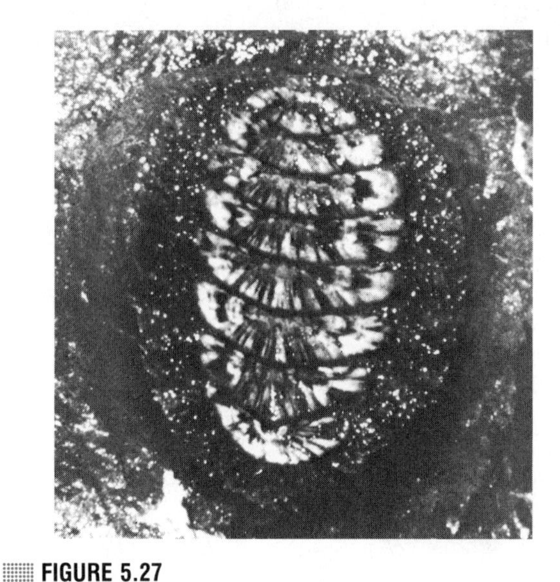

FIGURE 5.27

The chiton *Mopalia*. (From Flora and Fairbanks, 1982.)

They are large animals, ranging from 2–30 cm (0.8–11.8 in). Of all the modern molluscs, chitons retain the greatest number of ancestral characteristics of the phylum (Fig. 5.27). They have a massive flat foot, used to cling securely to a solid substrate. The shell and mantle form a shield over the dorsal surface of the body, and the mantle cavity is a well-developed space between the mantle edge and the foot. This mantle cavity (also called the pallial cavity) is ringed by gills and has the anal and renal pores at the posterior end. The radula of chitons is well developed and the tips of some of the teeth are mineralized with magnetite. There is some evidence that the magnetite becomes magnetized and that chitons can use the magnetic field created by the teeth for navigation. Most chitons are herbivores. They scrape algae with the radula. Some macroalgae feeders scrape benthic diatoms and juvenile forms of macroalgae from the rocks; others eat large green, red, and brown algae.

The shells of chitons are specialized. Instead of a single shield, the chiton shell takes the form of eight overlapping plates. These plates permit the chiton to roll into a ball when dislodged, thus protecting the soft foot. In some species the shell plates are embedded in the mantle tissue and are not external.

Class Gastropoda

The snails and slugs make up the most common class of molluscs. A total of 75 percent of mollusc species (or 75,000) are gastropods. These snails have successfully colonized fresh water and, to a more limited extent, the terrestrial environment. In the marine environment, snails generally are found in the benthos, both on rocky and soft bottoms (Fig. 5.28). They are conspicuous among the intertidal animals but also live at depths of over 5,000 m (16,400 ft). A few have adapted to a pelagic environment and have developed the ability to swim.

Typically the gastropod shell is a pointed tube or cone into which the animal can retract. Probably for reasons of body stability, the tube is coiled to form a shell in the shape of a conical spire (Fig. 5.28). With this seemingly limited body form, gastropods have evolved thousands of different species. Gastropod shells are the favorite of shell collectors; their sculpturing and colors produce dramatic and beautiful specimens. In fact, gastropod shells have become too popular and many areas have depleted populations because of overharvest. Not all gastropods have a coiled shell. Abalone and limpets generally have cone-shape shells, but the body cannot be retracted into the cavity of the shell. The nudibranchs, a major subgroup of gastropods, have lost all traces of a shell.

A.

B.

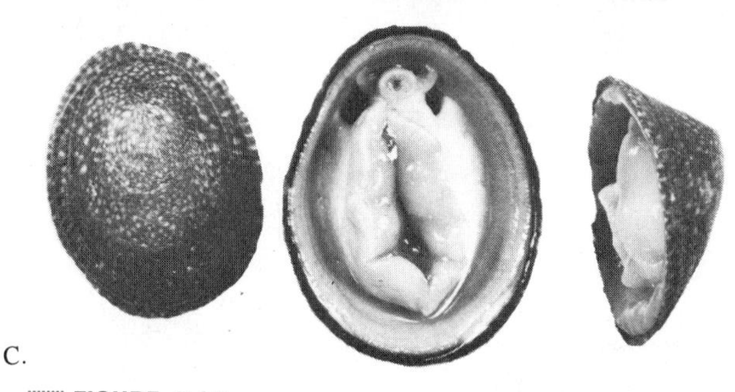
C.

FIGURE 5.28
Diversity of Body Form in the Class Gastropoda

(A) Nudibranch. **(B)** Snail. **(C)** Limpet. (From Flora and Fairbanks, 1982.)

The mantle cavity of gastropods typically houses the gills and the excretory pores of the anus and kidneys. Because of the shell shape, the mantle cavity is relatively small. A series of adaptations insures that water flowing over the gills is not contaminated with excretory products. In many intertidal gastropods, such as limpets, the mantle cavity remains full of water during tidal exposure and helps reduce the threat of drying out.

The radula of gastropods serves a number of functions, although always in some way associated with food gathering. In herbivorous gastropods the radula functions as it does in chitons, scraping material from rocks. In carnivorous gastropods, the radula has fewer teeth and the mouth is generally on the end of a proboscis. Some carnivorous snails use the radula to drill holes through the shells of other animals and insert the proboscis through the hole to eat the prey. In cone shells, radula teeth containing toxin are injected one by one by the proboscis into the prey (polychaete worms and small fish). Paralyzed prey then are swallowed whole. Cone shell venom is toxic to humans, and in tropical areas a number of deaths have been reported.

Class Pelycepoda

The typical bivalve is a sessile mollusc that lives buried in sandy or muddy sediments (Fig. 5.29). Included in the class *Bivalvia* are clams, oysters, mussels, and scallops. Some species (e.g., oysters and mussels), as adults, attach permanently to solid substrates. Many scallops live on the bottom and are capable of limited, awkward swimming. Most bivalves are marine, with a few species found in fresh water.

The shell of a bivalve, as the name suggests, consists of two valves, or plates, that generally completely surround the body. The two valves are hinged at the dorsal (upper) surface of the body and the valves cover each side. The valves are joined together by a hinge ligament (made out of an elastic complex protein) that forces the valves apart. Typically, a pair of adductor muscles closes the valves.

The mantle cavity of bivalves is relatively large. The gills of most species function for food gathering as well as oxygen exchange. To bring water to the mantle cavity, part of the mantle is often modified into a pair of siphons. These siphons bring water to the mantle cavity and carry filtered water and metabolic wastes away. Most bivalves are filter feeders and use the mucus on the gills to trap small particles. Cilia on the gills carry the food particles to the mouth where they are ingested or rejected, depending on size. Food is most often phytoplankton and small zooplankton.

Class Cephalopoda

Cephalopods (meaning 'head', 'foot') are the most advanced class of molluscs, at least in departing from the ancestral mollusc body form (Fig. 5.30). As the name suggests, cephalopods have an emphasis on the development of the head. The highly developed brain has the capacity for learning and memory. Sense organs are likewise highly developed. Eyes, analogous to the vertebrate eye (with lens, retina, and optic tracts), are the dominant sense organ. Balance and chemical receptors are also well developed. With the exception of the *Nautilus,* cephalopods have lost the heavy external calcium carbonate shell. In squid and cuttlefish, an internal skeleton has evolved, whereas in octopods there is no rigid skeleton.

FIGURE 5.29
Bivalves

(A) The scallop *Pecten.* **(B)** The oyster *Ostrea.* **(C)** The clam *Macoma.* **(D)** The clam *Schizothaeurus.* (From Flora and Fairbanks, 1982.)

A.

B.

C.

D.

FIGURE 5.30
Cephalopods

(A) Diversity in the class *Cephalopoda.* **(B)** The squid *Rossia.* **(C)** The octopus *Octopus.* (Photos from Flora and Fairbanks, 1982.)

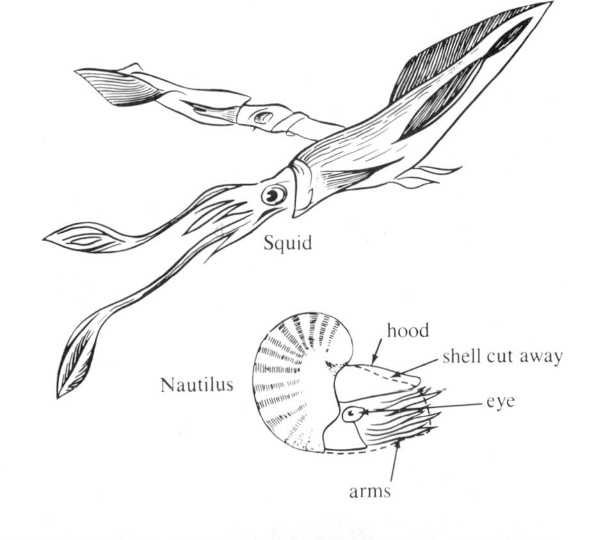

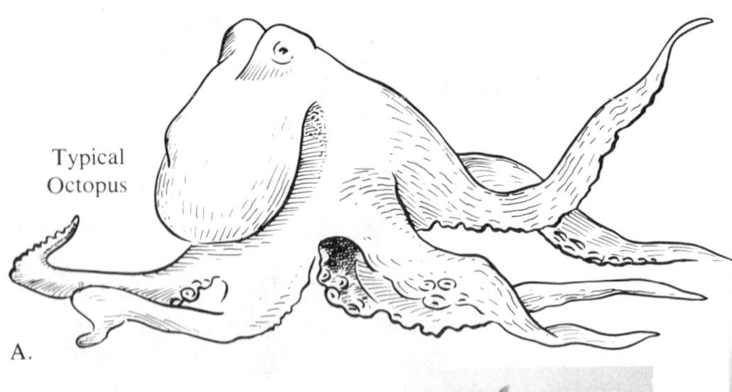

A.

B.

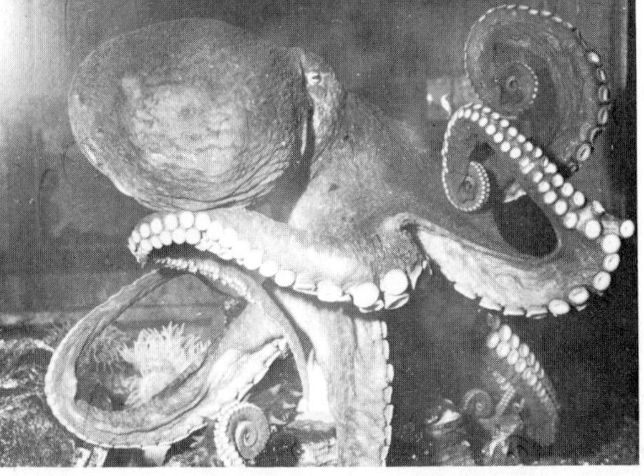

C.

Cephalopods are large, active carnivores. Their average size is the largest of all the invertebrates and compares with the vertebrates. The large squids are up to 12 m (39.4 ft) in length, and some octopods may weigh up to 130 kg (286 lbs). There are some 500 species of cephalopods, most adapted to a swimming existence. Squid are numerous and abundant in the pelagic zone.

The foot of cephalopods is specialized into a radiating set of tentacles surrounding the mouth. These tentacles are used for feeding, reproduction, and locomotion. Although some cephalopods such as the octopus move across the bottom with their tentacles, they are capable of much more rapid movement by jet propulsion. Part of the foot is modified into a siphon. Water is sucked into the muscular mantle cavity and forcefully expelled through the siphon, rapidly propelling the animal. Speeds comparable to those of swimming fish can be attained.

Most cephalopods are carnivores. Food captured by the tentacles is bitten with a pair of jaws shaped like the beak of a parrot. Some species produce toxins that are injected with the bite. The radula, although present, is generally reduced and small, although some octopus have a well-developed drilling radula.

On the basis of numbers of species, squids are the most important cephalopods. There are some 400 species widely distributed throughout the world's ocean (Fig. 5.31). Like the related octopus and cuttlefish, most squid hide during the day and feed actively at night. Squid spend their days at depths with little light and are commonly found in the deep scattering layer.

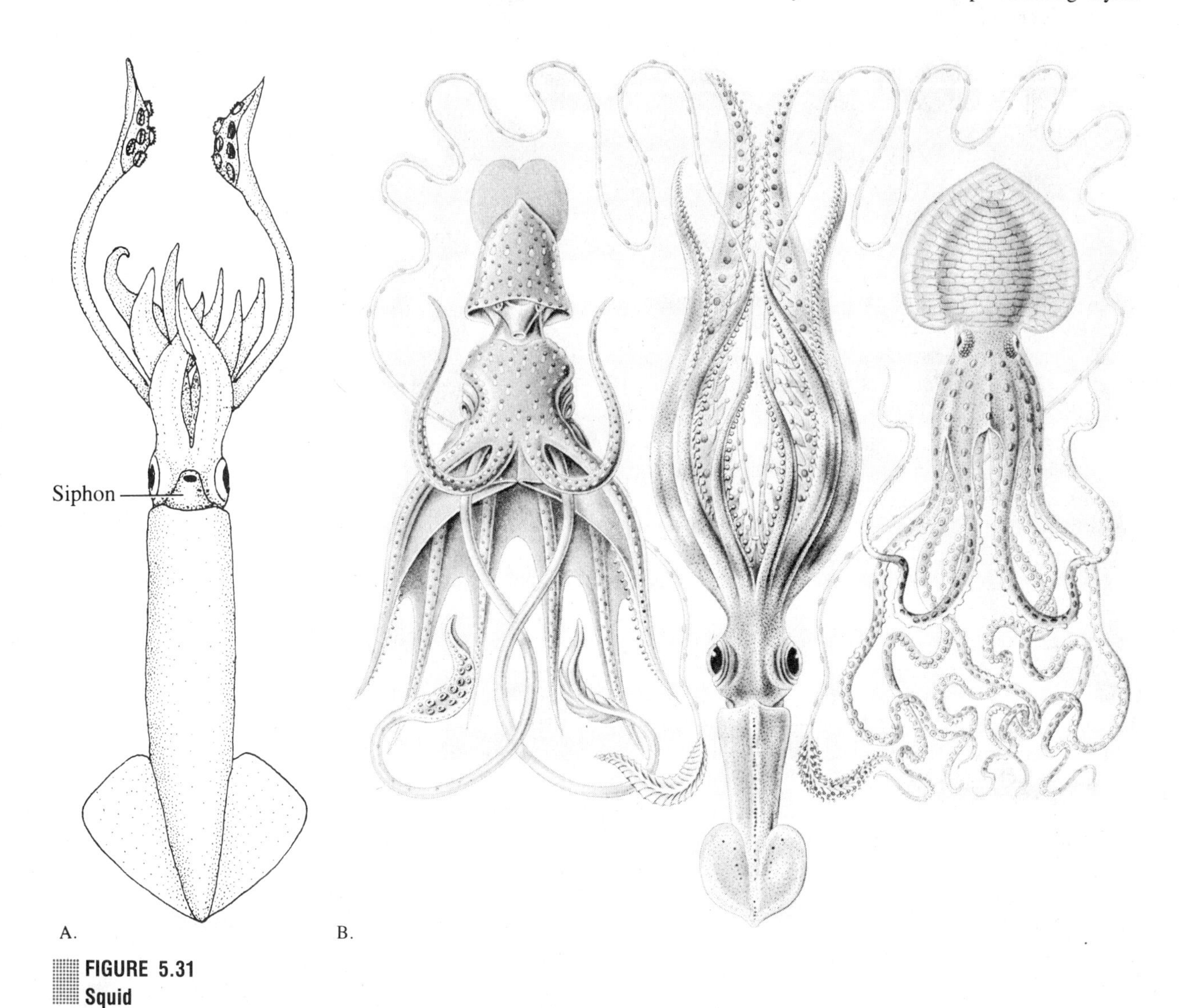

FIGURE 5.31
Squid

(A) *Loligo,* the common squid, is found in shallow waters in many parts of the world. **(B)** Squids from midwaters. (A from Crane, 1973. B from Haeckel, 1974.)

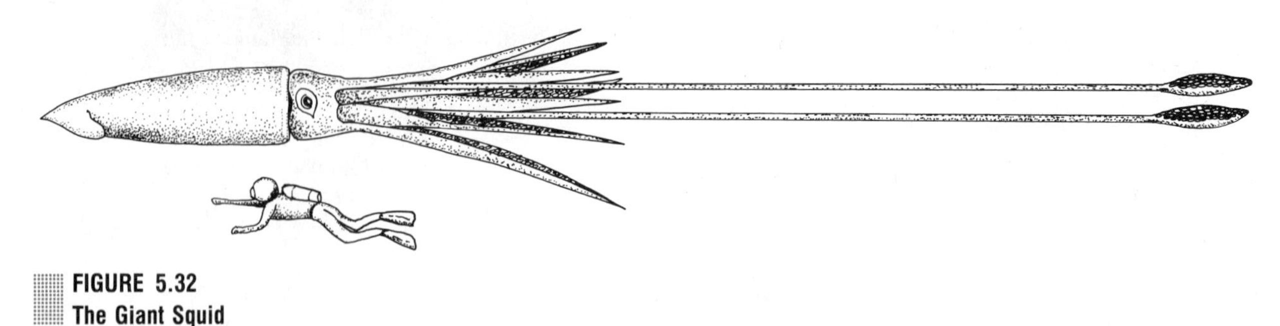

FIGURE 5.32
The Giant Squid

Architeuthis, the largest of the squids, may exceed 18 m in length. The scuba diver indicates relative size.

Some squid spend all of their time in the dim light of the mesopelagic zone.

Squid have a wide range of sizes. The genus *Loligo,* which is common around the world, is approximately 13 cm (5.1 in) in length. Many other larger species range from 0.5–1 m in length. The largest species of squid belong to the genus *Architeuthis* (Fig. 5.32). Specimens may reach a length of 18 m (59.1 ft) and a weight of 4 metric tonnes (4.4 tons). A squid this large has tentacles over 10 m (32.8 ft) long and body circumference of approximately 3.5 m. *Architeuthis* is a squid of the mesopelagic zone and generally is found between 300–600 m depth. It is known to science mostly by corpses that have washed ashore in the North Atlantic. Also known by its common name kraken, *Architeuthis,* along with other squid, is food for sperm whales. Numerous sucker-shaped scars on the jaws and lips of sperm whales are evidence of lively battles between the whale and squid.

Probably the most important characteristic of squid is their abundance in the oceans. Because squid are good swimmers and live mostly in deeper waters, they are difficult to catch with research nets, particularly the larger species. The smaller, shallower squid are more easily caught and the Japanese commercially harvest some 1 million metric tonnes per year.

The best estimates of the size of the squid population in deeper waters comes from the analysis of the stomach contents of sperm whales. Scientists estimate that sperm whales eat up to 110 million metric tonnes (121 tons) of squid per year, and before sperm whale stocks were reduced by commercial whaling, they probably took some 260 million metric tonnes per year. Squid is an important source of food for other animals as well. The southern elephant seal eats some 10–20 million metric tonnes of squid per year. Some dolphins, seals, penguins, albatross, and fish also eat squid. The great abundance of oceanic squid can be better appreciated if we remember that the total world commercial harvest of fish is around 70 million metric tonnes per year.

The nervous system of squid is fairly sophisticated. They have a well-developed cerebral ganglion (brain) that is larger, relative to its body size, than that of fish. The cephalopod cerebral ganglion functions in the coordination of movement, learning and memory, and control of social behavior. Squid also have well-developed eyes, with a cornea, iris, and lens that are functionally very similar to the vertebrate eye. The cephalopod eye can focus on near and far objects and can adapt to a wide range of light intensities. Other senses of cephalopods are probably as important as eyesight; tactile sensation and chemosensation are both well developed on the tentacles.

The tentacles of squid are used mostly for feeding, although they have the additional role of transferring sperm during copulation. The tentacles (ten in squid, eight in octopods) are well supplied with suckers. These suckers, which often have hooks or barbs, are efficient grasping organs. Squid feed mostly on small fish. *Loligo* is a particularly effective predator. This squid swims backward rapidly into a school of small fish, then uses its tentacles to capture a prey and its jaws to bite through the neck to sever the spinal cord. Once the fish is paralyzed, the squid bites it into pieces small enough to swallow.

PHYLUM ANNELIDA

The phylum Annelida (Latin: *anellus*-little ring) is the first of two major phyla (the other is the Arthropoda) that has a body plan based on a series of segments. The basic body form of such phyla, at least in primitive types, is a series of aligned segments called **metameres,** each having similar musculature, internal organs, and body appendages. The evolutionary history of these two phyla is the story of how the basic segmentation has been modified.

Annelids are segmented worms adapted to living in soft sediments. There are some 8,700 species of which almost all are benthic animals that burrow into the sediments or crawl across the surface or under rocks, shells, and mats of algae. Annelids have successfully colonized fresh water, and the familiar earthworm is one terrestrial species. The phylum is divided into three classes: Hirudinea, Oligochaeta, and Polychaeta.

Class Hirudinea

This class is that of the leech. Leeches are primarily freshwater annelids, with only a few terrestrial and marine species. This class is the most specialized of the annelids. Leeches have the fewest number of segments of any annelid, and the body has developed suckers for attachment and blood-sucking (the anterior sucker has the latter function). Approximately 75 percent of leeches are ectoparasitic bloodsuckers.

Class Oligochaeta

Oligochaetes also are basically freshwater annelids but include some terrestrial species such as earthworms. There are some 200 species of marine oligochaetes common in the sediments of intertidal zones, particularly in estuarine sand and mud flats. Most marine oligochaetes are small (less than 2 mm in length) and are part of the interstitial fauna collectively called the meiofauna.

Class Polychaeta

This third annelid class is a widespread group in the marine environment. Although the majority of species are of moderate size (to 5 cm), a few exceed 10 cm (3.9 in). There are over 5,000 species of polychaetes, all marine. Polychaetes are relatively common in the intertidal zone and are familiar to most casual observers.

Polychaetes, for the most part, have the typical annelid segmented body plan (Fig. 5.33). There is little specialization of segments except for the head and tail ends. The first two segments make up the head, and the ultimate segment, which has the anus, is called the pygidium. All other segments are essentially identical and contain similar elements of the organ systems (circulatory, muscular, excretory, nervous, and reproductive).

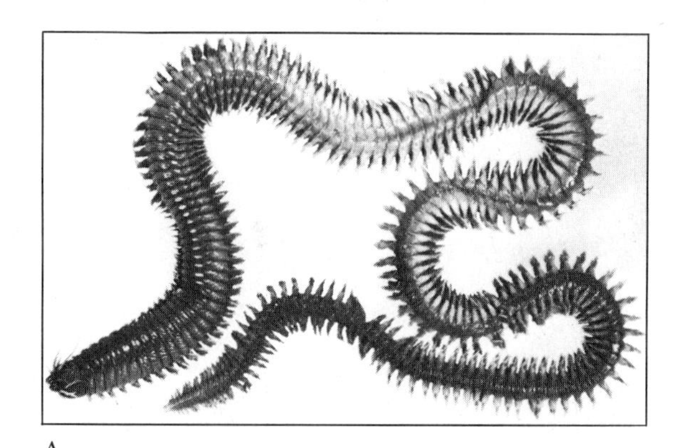

A.

B.

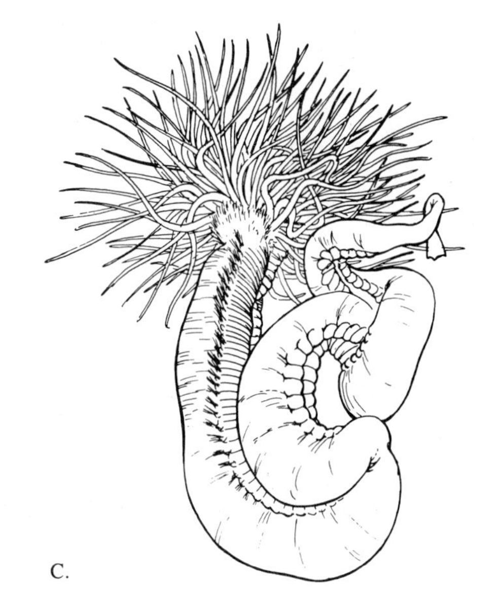

C.

FIGURE 5.33
Polychaetes

(A) *Nereis.* **(B)** *Glycera.* **(C)** *Thelepos.* (A and B from Flora and Fairbanks, 1982. C from Crane, 1973.)

Growth occurs by the adding of new segments at the posterior end of the body, just ahead of the pygidium. In addition to the typical annelid characters, polychaetes have lateral structures on each segment called parapodia. These parapodia are fleshy lobes, usually divided into dorsal and ventral parts, that have chitinous setae, or bristles, embedded in them (Fig. 5.34). Both the segmentation of the polychaete body and the parapodia contribute to the characteristic polychaetic movement.

Polychaetes typically move by means of a hydrostatic skeleton; that is, contraction of either the circular or longitudinal muscles compresses the body cavity fluid. Contraction of the circular muscles extends the body, and contraction of the longitudinal muscles shortens the body. Segmentation divides the body cavity into compartments and makes the hydrostatic skeleton more effective. Polychaetes that live on the surface use parapodia as legs and walk by a movement called parapodial stepping. Bristlelike setae extend from the segments of most polychaetes (and oligochaetes). These bristles are also important in movement, acting as anchors to hold the nonmoving parts of the body in place. Setae are most important in burrowers and tube-dwellers.

The basic annelid anatomy is well suited to burrowing in soft sediments, and polychaetes have a wide range of burrowing strategies. Burrows are generally mucous lined. They may take the form of a complicated series of tunnels with multiple openings to the surface, or they may be simple vertical or U-shaped burrows. Some burrowing polychaetes are carnivorous while others are deposit or filter feeders. Burrow-dwelling polychaetes generally have reduced parapodia, compared with surface crawlers, and rely on setae for anchoring within a burrow.

Surface crawlers live under rocks, in algae mats, among coral rubble, or wherever a measure of protection is provided. They generally are predators and have well-developed parapodia as well as a head end with jaws and sensory structures.

Many polychaetes live in tubes wholly or partially embedded in the sediments (Fig. 5.35). The tube of some polychaetes is little more than sand grains and shell bits stuck into its surface mucus. In others it is a leathery proteinaceous tube that can stand erect from the bottom. The strongest tubes are made of calcium carbonate. Most tube-dwelling polychaetes can withdraw totally into their tubes. Some tube dwellers are active carnivores, capturing prey that comes close to the tube entrance, but most are filter feeders or deposit feeders. Often a cluster of tentacles, used primarily for feeding, is found at the head end. A few species are pelagic. Like other pelagic animals, they are transparent, but otherwise have the regular polychaete worm shape.

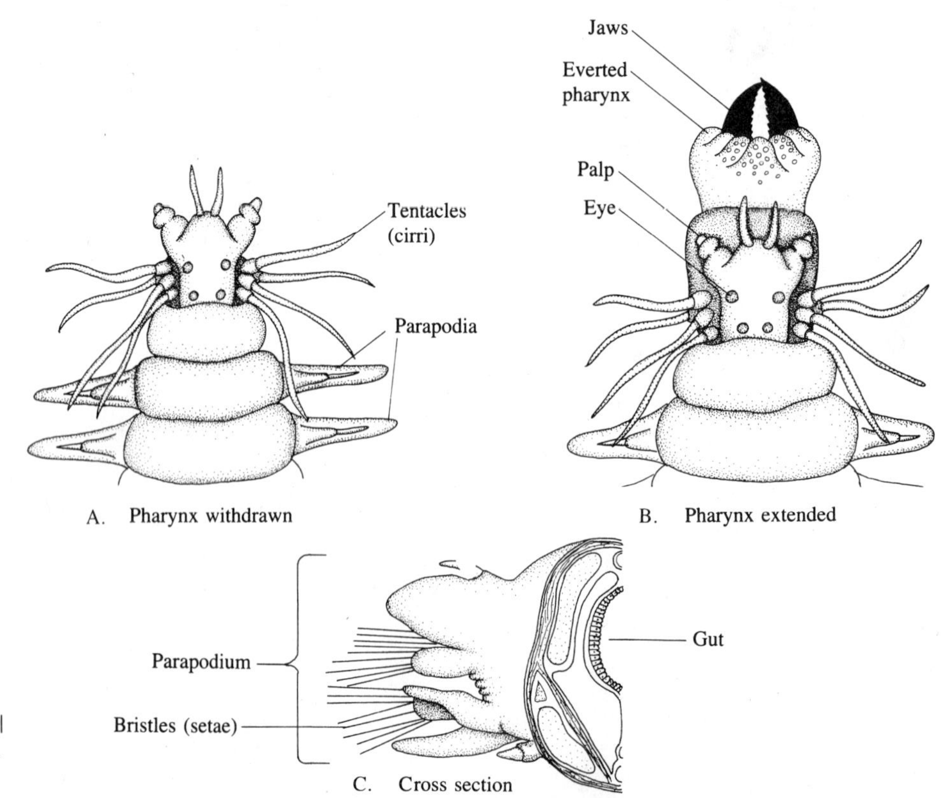

FIGURE 5.34
Anatomy of Polychaetes

The polychaete *Nereis*. Size is 1.2 cm in diameter. A and B are external views of the head. C is a cross section. (From Crane, 1973.)

FIGURE 5.35
Tube-dwelling Polychaetes

A number of polychaetes have evolved to a sedentary existence. From their protective tubes they filter the water or harvest the surface deposits for food. Shown here:the plume worm *Eudestylia.* (From Flora and Fairbanks, 1982.)

PHYLUM ARTHROPODA

Arthropods (Greek: *arthron*-joint; *pous*-foot) include insects and crustaceans. Based on the number of species and wide variety of habitats in which arthropods are found, they rank as the most successful of the animal phyla. There are close to 1,000,000 species of arthropods, more than all the other animal phyla combined. Arthropods are segmented animals, with limbs. The story of arthropod evolution is the story of the adaptation of limbs to various functions such as swimming, burrowing, walking, flying, feeding, reproduction, gas exchange, and sensing. A second defining feature of arthropods is an exoskeltelton made of protein-based chitin. This exoskeleton cannot increase in size. When the animal has outgrown its skeleton, it must discard the old one and secrete a new one in a process called molting, or **ecdysis.**

Another unusual feature of arthropods is an image-forming eye (Fig. 5.36). Many invertebrates have eyes, but only a few are capable of image formation. The eye of cephalopods (molluscs) and arthropods are the two prime examples. The arthropod image-forming eye is called a compound eye because it is made of a large number of light-sensitive units, or **ommatidia,** each of

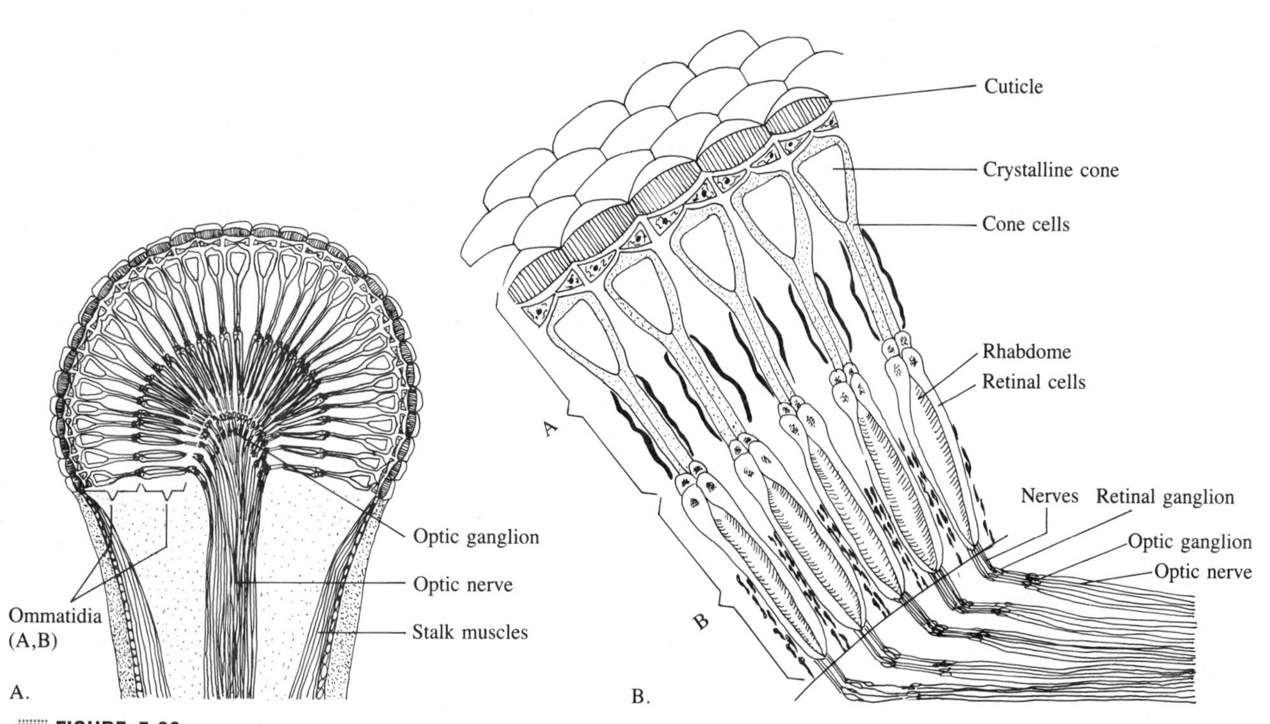

FIGURE 5.36
The Arthropod Compound Eye

(A) Cross section. **(B)** Detail of five ommatidia.

which has a cornea, lens, light-sensitive cells, and neural connectors. The image formed is a composite of those received by each ommatidium.

The Arthropoda is a diverse phylum. In fact, many biologists argue that they should be considered a superphylum and the four subphyla elevated to phyla. The four subphyla are the Trilobitomorpha (trilobites, now extinct); the Chelicerata (horseshoe crabs and spiders); the Uniramia (insects, centipedes, and millipedes); and the Crustacea (crabs, shrimps, and lobsters). A systematic classification of the arthropods is given in Table 5.2. Crustaceans are the most common arthropods in the marine environment.

Subphylum Chelicerata

Chelicerates include the spiders, scorpions, ticks, and mites. They are divided into three classes: Merostomata (the horseshoe crabs, all marine); Pycnogonida (sea spiders, all marine); and Arachnida (terrestrial spiders, ticks, mites). By far the majority of species are found in the terrestrial environment, and there are only a few marine forms. We will consider briefly the horseshoe crabs and sea spiders.

Horseshoe crabs (Fig. 5.37) are found in soft sandy or muddy bottoms. In the water or on the surface of the mud they move awkwardly and slowly, but they are effective burrowers and move easily just below the surface, using their legs to tunnel through the mud or sand. Their food is mostly polychaete worms. Only five species are known in the world, one on the east coast of North America, and four on Asian coasts. During the late spring and summer they are found on the shoreline in large numbers. At that time, females lay eggs in nests at or near high tide during spring tides.

Pycnogonids (Fig. 5.38) also are a small group but they are widely distributed from the tropics to polar environments, and from shallow to deep water. These animals are mostly legs and often are found in hydroid or bryozoan colonies feeding on polyps. Although not familar, pycnogonids are quite common. They have a size range from a few millimeters up to 50 cm (19.7 in), with an average size of 2–3 cm.

Subphylum Uniramia

This, the subphylum of insects, contains the most diverse of all animal groups. Despite the large number of species—over 750,000—only a few are found in the marine environment. A number of insects are found in salt marshes; some beetles are found on sandy beaches; springtails are commonly seen floating on New England

TABLE 5.2
Systematic Classification of the Phylum Arthropoda

The following symbols indicate: (A) An absence of the group in the marine environment; (R) rare in the marine environment; and (C) common in the marine environment.

Phylum Arthropoda

Subphylum Crustacea (C)
- Class Cephalocarida (R)
- Class Branchiopoda (A)
- Class Ostracoda (C)
  - clam shrimp
- Class Copepoda (C)
  - copepods
- Class Mystacocarida (R)
- Class Branchiura (R)
- Class Cirripeda (C)
  - barnacles
- Class Malacostraca (C)
  - Order Euphausiacea (C)
    - krill
  - Order Decapoda (C)
    - crabs, shrimp, lobsters
  - Order Mysidacea (C)
  - Order Cumacea (C)
  - Order Tanaidacea (C)
  - Order Isopoda (C)
  - Order Amphipoda (C)

Subphylum Chelicerata (R)
- Class Merostomata (C)
  - horseshoe crabs
- Class Arachnida (A)
- Class Pycnogonida (C)
  - sea spiders

Subphylum Uniramia (R)
- Class Chilopoda (A)
  - centipedes
- Class Symphula (A)
- Class Diploda (A)
  - millipedes
- Class Pauropoda (A)
- Class Insecta (R)
  - insects

FIGURE 5.37
The horseshoe crab *Limulus*

(**A**) Dorsal view. (**B**) Ventral view.

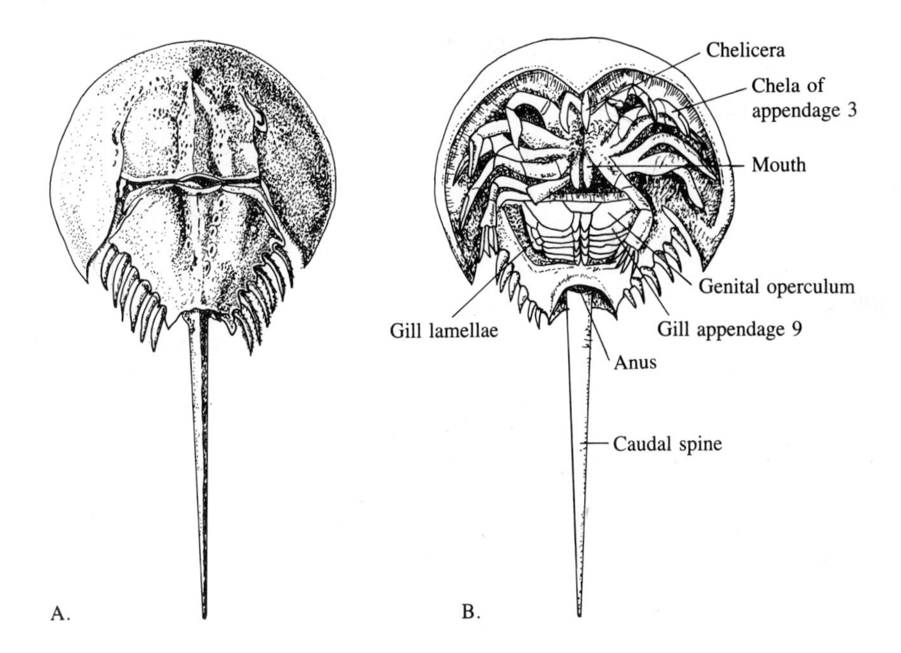

tidepools; and some flies lay their eggs in the intertidal zone. One insect genus—the water strider, *Halobites*—lives out its life cycle on the surface of the ocean. Apart from these examples, insects are animals of the terrestrial and fresh-water environments. Neither insects nor plants, both of which dominate the terrestrial environment, are widely distributed in the marine environment.

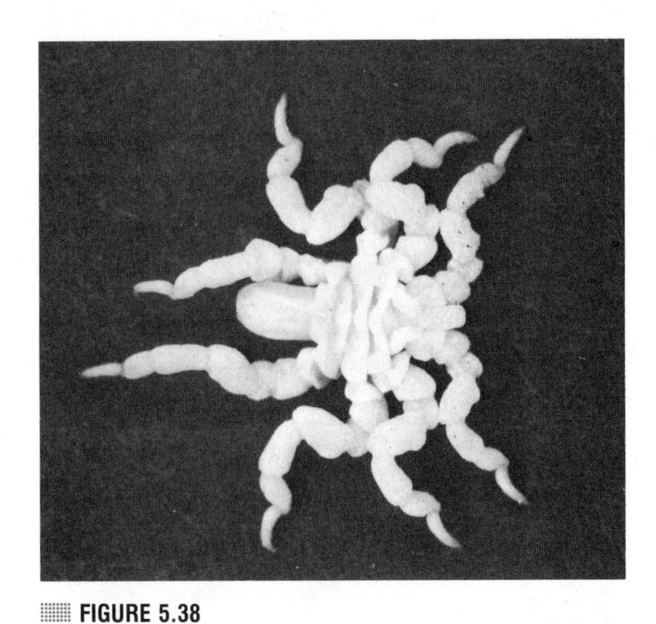

FIGURE 5.38

The pycnogonid *Pycnogonum*. Magnification ×5. (Photo by Gary Cawthon.)

Subphylum Crustacea

Crustaceans (crabs, shrimp, lobsters) are a large and important group of animals to marine biologists, and include many common and commercially important species. Without question, the most numerous animal of the ocean is a planktonic copepod of the genus *Calanus*. These herbivores are a crucial link in the food chains of pelagic waters.

Crustaceans of commercial importance include crabs, shrimp, and lobsters. The 31,000 species of crustaceans are widely distributed, occurring commonly in virtually every marine habitat. Although crustaceans also are numerous in fresh water, only one small group, the isopods or pill bugs, have successfully colonized land.

The diversity of crustaceans is a story of specialization of body segments and appendages. The basic segmented body of arthropods always shows specialization into parts. In crustaceans, the first five segments have fused to form the head. In addition to the concentration of ganglia into a brain, there is generally a pair of compound eyes. The appendages of the head are adapted for sensory function and feeding. Out of the first two segments grow the antennae and antennules, sensory structures that function in tactile and chemosensory ways. The appendages of the third segment are a pair of jaws or mandibles. These jaws, along with feeding appendages on the fourth and fifth segment, surround the mouth. Food can be macerated by the jaws before being passed into the mouth.

The body of crustaceans typically is further specialized into a thorax and abdomen. Appendages of these parts are specialized for walking, feeding, defense, swimming, gas exchange, and reproduction. In many familiar crustaceans, such as shrimp, lobsters, and crabs, the thorax region is covered by a protective dorsal carapace that covers at least the dorsal surface of the body.

Movement in crustaceans is by means of specialized appendages. Some species swim, but most walk using modified thoracic legs. The importance of appendages in forming jaws has already been described. Other appendages are used to capture and hold the food. Filter-feeding species, such as copepods, have branched appendages that form a fine net that filters food particles from the water. Other crustaceans use their appendages to capture larger food. The crab and lobster have one or more anterior appendages equipped with claws for grasping and tearing.

The external skeleton of crustaceans is similar to that of other arthropods. One major difference is that the crustacean exoskeleton is usually calcified into a rigid external covering.

The groups of crustaceans that are important either by virtue of diversity or familiarity to the beginning student are discussed below. The major systematic groups are listed in Table 5.2.

Branchiopoda. Although most members of the class Branchiopoda are found in fresh water, one marine genus is familiar. The brine shrimp *Artemia* occurs in salt lakes and ponds (Fig. 5.39). Brine shrimp are found in great numbers in shallow salt ponds where water is evaporated to reduce commercial salt. Brine shrimp produce desiccation-resistant eggs which are often sold by stores as aquarium fish food. The body anatomy of brine shrimps has characteristics of the primitive crustacean. The body is not divided into the typical head, thorax, and abdomen. Instead, the thorax and abdomen form a single trunk, and the head has no carapace. The trunk has a relatively large number of swimming appendages that also take up oxygen.

Ostracoda. Also known as clam shrimps, ostracods are small pelagic and benthic crustaceans found in both fresh-water and marine environments (Fig. 5.40). They are widely distributed in the oceans. The genus *Gigantocypris* has the largest ostracods, up to 3 cm (1.2 in) in length, and is one of the important members of the deep-sea plankton. The striking feature of ostracod morphology is the development of a calcified carapace into two lateral valves that cover the entire body. At first glance, ostracods look like small bivalves (hence their common name), but a closer examination shows the characteristic crustacean appendages.

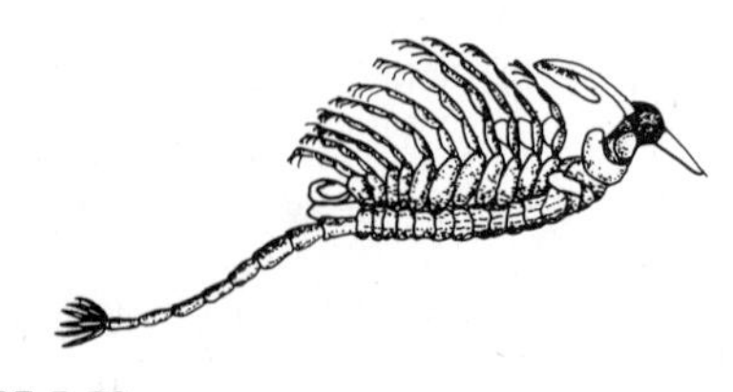

FIGURE 5.39

The brine shrimp *Artemia.* (From Crane, 1973.)

Copepoda. The mostly marine class Copepoda is a large and important group of crustaceans (Fig. 5.41). Planktonic herbivorous copepods of the genus *Calanus* are the most numerous animal in the oceans, both in surface waters and in the deep sea. The pelagic forms are collectively referred to as calanoid copepods. A second major group of copepods are mostly benthic and are collectively called harpacticoid copepods.

Copepods are generally small, in the range of 1–5 mm (0.2 in) in length, and shaped rather like a cylinder. The abdomen is reduced and has no appendages. The thorax has between two and five segments with appendages for swimming or walking. The head is fused to the first thoracic segment forming a cephalothorax, and instead of the typical arthropod compound eye, copepods have a single median (cyclops) eye. The first and second antennae of the head assist pelagic copepods in swimming.

Cirripeda. All the 900 species of the class Cirripeda (the barnacles) are marine. They are permanently attached to some sort of substrate. Both free-living and parasitic species are found. Although barnacles attach to whales, turtles, and other marine animals or to floating logs and other flotsam, most species are benthic, clinging to rocks or pilings.

At first glance, we might be tempted to place barnacles with the molluscs because of their heavy calcareous shell. The larval stages of development and the details of the adult anatomy, however, show the crustacean traits. A barnacle can be thought of as a shrimp-like crustacean standing on its head that uses bristles, or cirri, on the thoracic appendages to sweep through the water, gathering food particles (Fig. 5.42).

There are two kinds of free-living barnacles: acorn barnacles, which are familiar animals of the intertidal zone, and gooseneck barnacles, which occur in some intertidal habitats but are more likely to be found at sea attached to some floating object (Fig. 5.43). Both groups have well-developed calcareous skeletons. This skeleton is not like the typical chitinous arthropod exoskeleton but is, rather, an external box made of calcium carbonate plates. In acorn barnacles, these plates are fused into a cone shape; in gooseneck barnacles, the

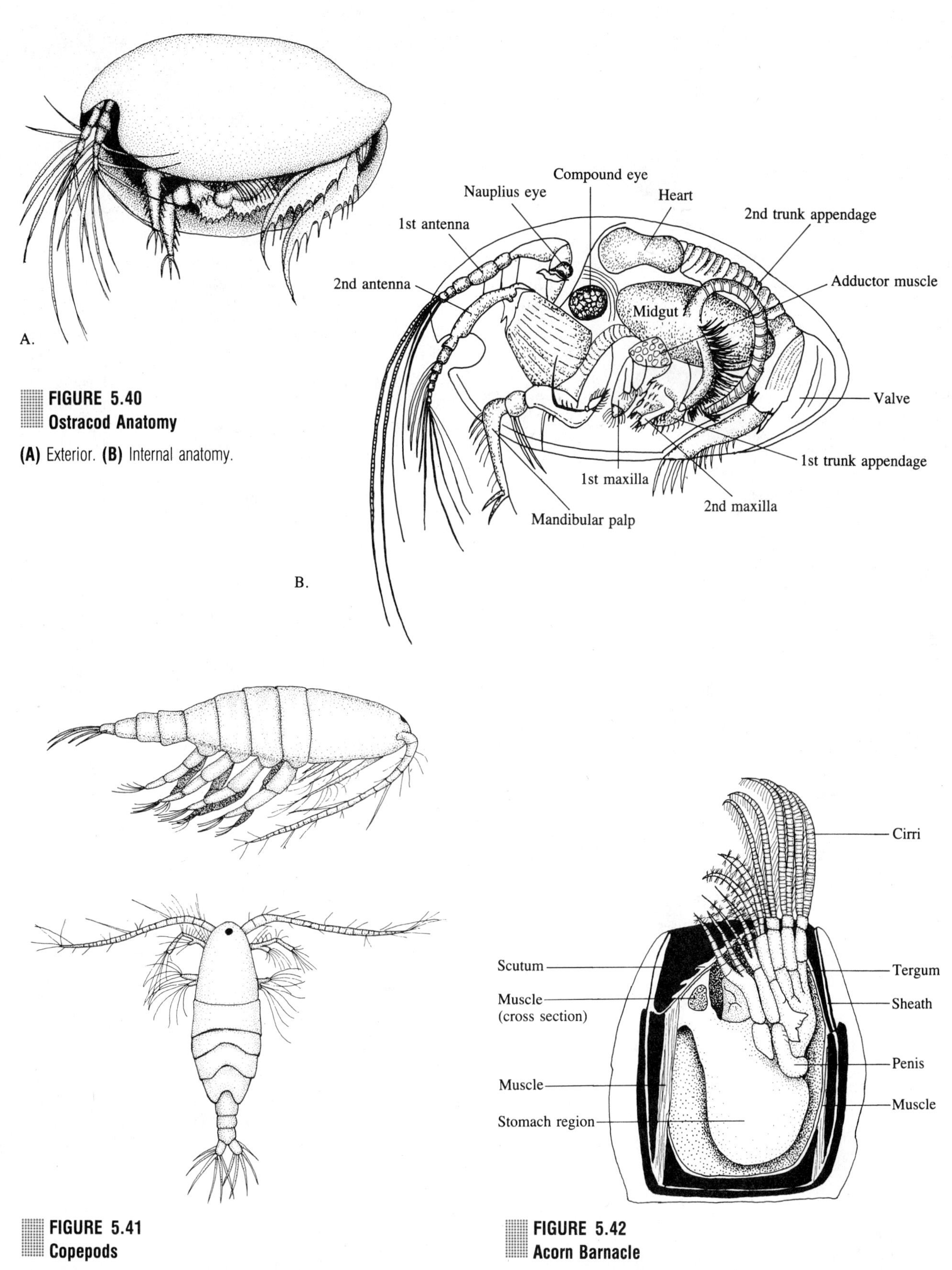

FIGURE 5.40
Ostracod Anatomy

(A) Exterior. **(B)** Internal anatomy.

FIGURE 5.41
Copepods

The common planktonic copepod *Calanus finmarchicus* (lateral and dorsal views). (From Crane, 1973.)

FIGURE 5.42
Acorn Barnacle

Acorn barnacles are shrimplike crustaceans standing on their heads in a calcium carbonate box. (From Crane, 1973.)

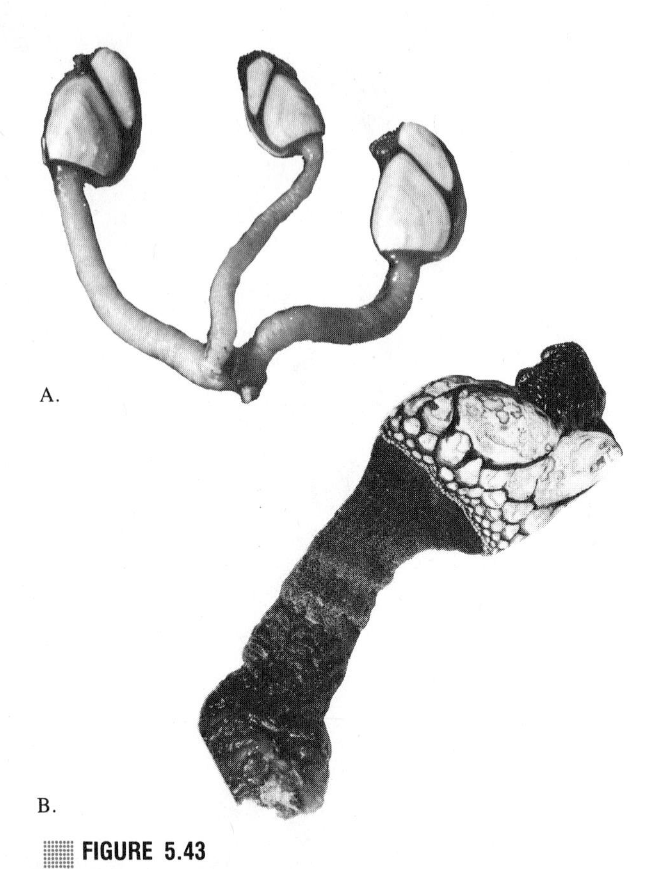

FIGURE 5.43
Gooseneck Barnacles

(A) *Lepas* is typically found on floating objects.
(B) *Mitella* attaches to hard substrates. (From Flora and Fairbanks, 1982.)

plates are generally separate from each other and held together by living tissue.

Euphausiacea. Although the order Euphausiacea has relatively few species (90), they are important and common in the pelagic environment. Euphausiids are shrimplike crustaceans between 2–3 cm (1.2 in) in length. A carapace covers the head and thoracic appendages. They are primarily filter feeders, using the thoracic appendages to create water currents. Many species are herbivors and eat phytoplankton. In polar latitudes, particularly the antarctic, euphausiids called **krill** are the predominant planktonic herbivore and may occur in large schools. Krill, about 5 cm (2 in) long, are an important food source for baleen whales, such as the blue whale. A number of countries are investigating the potential of krill as a fisheries product. Some estimates put the amount of krill that could be harvested at around 70 million tons annually, an amount equal to the current total world harvest of fish.

Decapoda. Decapod crustaceans include shrimps, crabs, and lobsters. With some 8,500 species (mostly marine) it is the largest of the crustacean orders. Decapods make up one-third of all crustaceans. They are relatively large and commonly have a size range of 5–10 cm. Among the largest of the decapods are king crabs, which have a carapace width of over 20 cm (7.9 in) and a total width of 3–4 m (Fig. 5.44). Most decapods are part of the benthos, although a few species are pelagic.

Technically, decapods are crustaceans that have head and thoracic segments covered with a carapace. The carapace extends laterally over the body and forms a cavity in which the gills are found. Each of the last five thoracic segments has a pair of walking or swimming legs (*deca* means 'ten'; hence, *decapod*). The first three thoracic appendages are modified to assist in feeding. In some decapods (shrimps and lobsters), the abdomen is pronounced, while in others (crabs) it is reduced.

Shrimps, also called prawns, have a more delicate body than crabs or lobsters (Fig. 5.45). The thoracic walking appendages are long and slender, and the first two or three pairs often have claws (chelae) at the ends. The carapace of the head generally forms a pointed rostrum (a bill-like extension). Shrimp have a well-developed abdomen that bears appendages (pleopods) adapted for swimming. A number of shrimp are pelagic, commonly occurring in the deeper-water plankton between 200–1,000-m depths. Most shrimp, however, are benthic and use their thoracic appendages to walk over the bottom. For short, rapid movements, the abdomen is quickly contracted and the fan-shaped telson at the tail end provides powerful propulsion.

FIGURE 5.44
Alaskan King Crab

One of the largest crabs is the Alaskan king crab. This specimen has a span of 1.5 m. (From Flora and Fairbanks, 1982.)

FIGURE 5.45
Shrimp

Pandalus (size is about 10 cm). (From Flora and Fairbanks, 1982.)

Crabs are crustaceans with a reduced abdomen tucked back under the thorax (Fig. 5.46). The body is dorsoventrally flattened and often is wider than it is long. This body shape permits the rapid sideways walking characteristic of most crabs. One group of crabs, the portunids (includes the blue crab *Callinectes*), can also swim well. In this group the last thoracic walking leg is modified into a paddle used for swimming. Modification of the fifth pair of legs is also found in other crabs. In hermit crabs, where the fleshy abdomen is unusually large for crabs, the last pair of thoracic appendages is modified to grasp the shell in which the hermit crab lives.

Lobsters (and crayfish, in fresh water), like shrimps, have a well-developed abdomen. Lobsters, however, have very small appendages on the abdomen and cannot swim. Also, they are much more heavily calcified and larger than shrimp. In the American lobster, the first pair of thoracic walking legs is modified into the familiar massive claws.

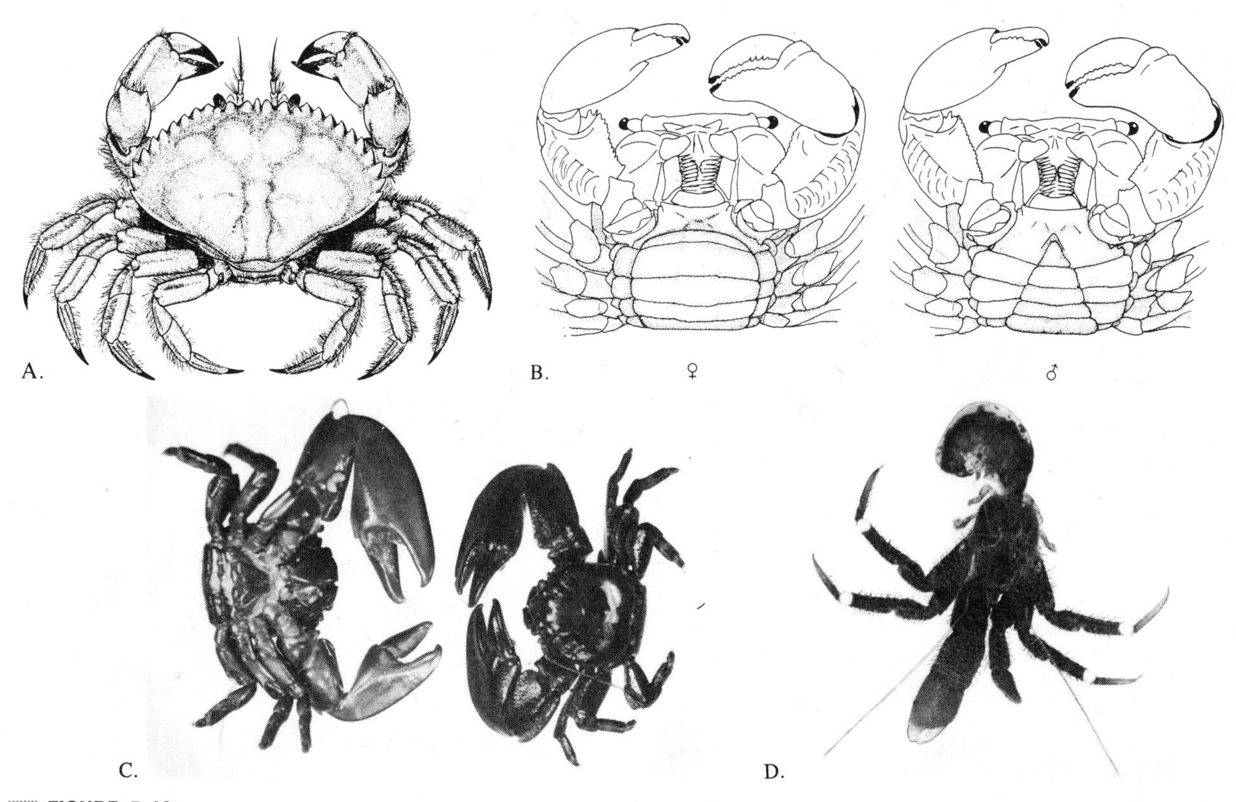

FIGURE 5.46
Various Decapod Crabs

(A) The rock crab *Cancer.* **(B)** Sexual dimorphism in the shore crab *Pachygrapsus.* **(C)** The porcelain crab *Petrolisthes.* **(D)** The hermit crab *Pagurus.* (A and B from Crane, 1973. C and D from Flora and Fairbanks, 1982.)

Isopoda. There are some 4,500 species in the order Isopoda. Most are marine, but there are fresh-water and even some terrestrial species. The familiar pill bugs found in rotting wood and under stones are terrestrial isopods. Most marine species are found crawling in the benthos and are particularly common in deep-sea sediments. The body of isopods is flattened dorsoventrally (Fig. 5.47). There is no carapace and little difference between the thoracic and abdominal segments.

Isopods are mostly scavengers and omnivorous; they do not filter feed. Some marine species are wood eaters. Isopods known as gribbles are an economic nuisance, boring into and destroying pilings and other wood structures. Some gribbles are capable of digesting cellulose directly; others rely on bacteria found in the digestive system.

Amphipoda. Amphipods, or beach hoppers, are shrimplike crustaceans that, like isopods, have no carapace (Fig. 5.48). As with isopods, there is no clear division into head and thorax. They are flattened laterally, however, rather than dorsoventrally. Locomotion is by hopping, crawling, or swimming.

Many amphipods are pelagic but the most important are the bottom dwellers. Amphipods are common in the intertidal and shallow subtidal zones. Some spe-

A.

3 cm

B.

FIGURE 5.47
Isopods

(A) The rock louse *Ligia.* **(B)** The sea slater *Idotea.* (A from Crane, 1973. B from Flora and Fairbanks, 1982.)

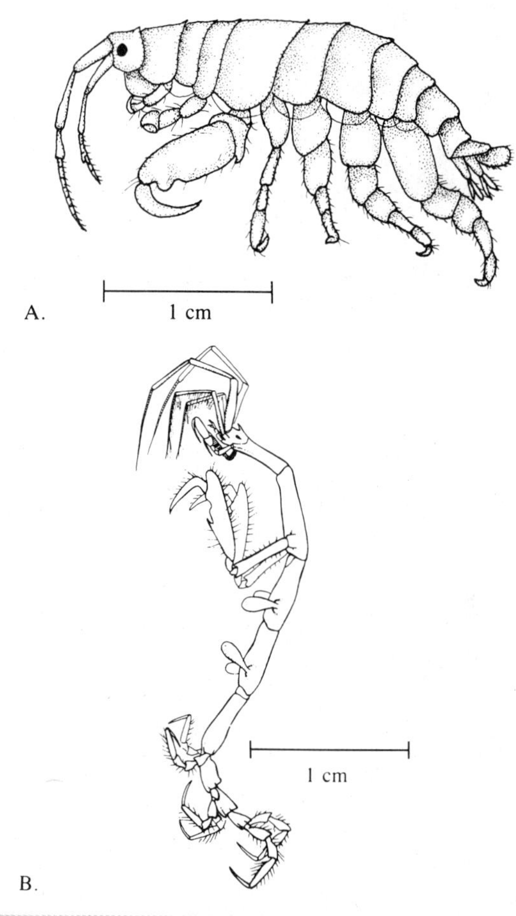

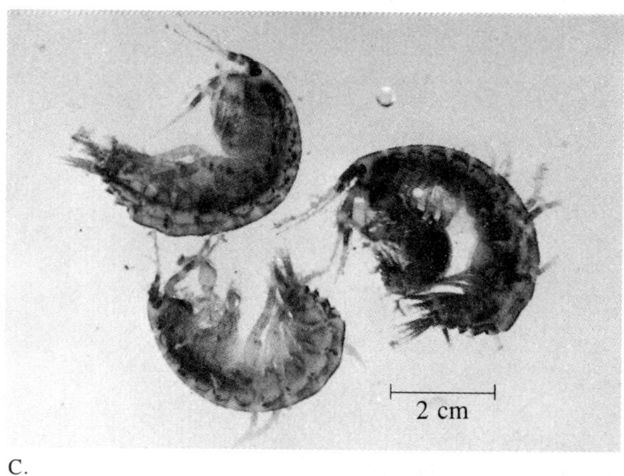

C.

FIGURE 5.48
Amphipods

(A) *Elasmopus.* **(B)** *Caprella.* **(C)** *Orchestia.* (A and B from Crane, 1973. C from Flora and Fairbanks, 1982.)

cies live at or above the high-tide mark and are semiterrestrial. Most amphipods feed on detritus or are scavengers. They can be very numerous and are an important food source to fish, birds, and the gray whale.

The 4,000 or so species are mostly marine, although the relatively few fresh-water species (approximately 100) are important.

MINOR PROTOSTOME PHYLA

Phylum Sipuncula (Latin: *siphunculus*-little pipe)
Phylum Echiura (Greek: *echis*-snake; Latin *ura*-tailed)
Phylum Pogonophora (Greek: *pogon*-beard; *pherein*-to bear)
Phylum Tardigrada (Latin: *tardus*-slow; *gradus*-step)

These phyla are considered minor because they total only about 900 species. They are included as a group because they all share the protostome characteristics of an internal body cavity (the coelom) that is evolutionarily related.

Only the sipunculids (300 species) are familiar enough to have a common name (Fig. 5.49). Also called peanut worms, sipunculids are most common in the shallow water of tropical latitudes, though some species flourish in temperate and even polar latitudes. A few species are found in deep-water sediments; sipunculids have been collected to depths of over 4,800 m.

The phylum Echiura consists of some 100 species of large, unsegmented, sausage-shaped worms (Fig. 5.50). Most live in sand or mud but some are found under rocks and in coral reef crevices. The unique feature of the echiuroid body is a large proboscis extending from the head end of the body. The proboscis cannot be retracted into the body. One species from Japan has a body length of 40 cm (15.8 in) and a proboscis that extends some 150 cm. Like sipunculids, echiuroids are detritus feeders. In some species the detritus sticks directly to mucus on the proboscis. In others, the proboscis secretes a mucus web that filters detritus particles from the water. Periodically the mucus net is eaten and rebuilt.

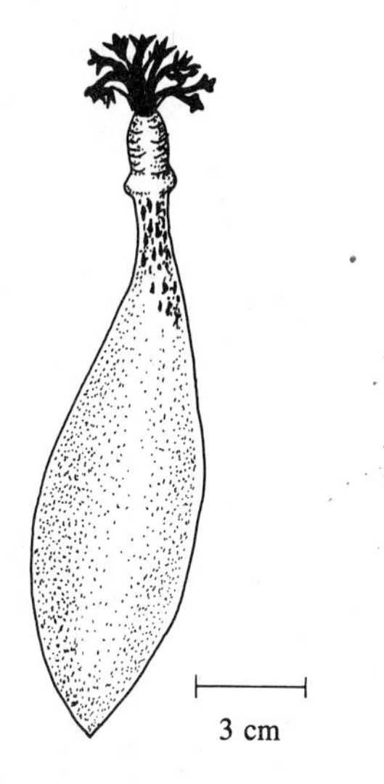

FIGURE 5.49

The sipunculid *Dendrostomum*, with tentacles extended.

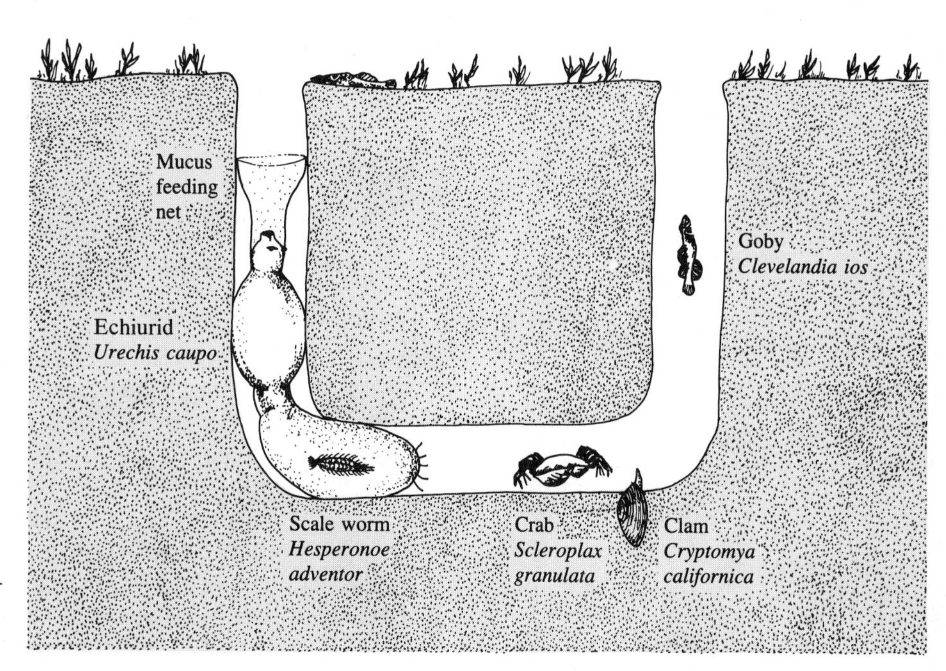

FIGURE 5.50
Echiurid

Urechis caupo burrows in California mud flats. The burrow is often occupied by a number of commensal animals.

The phylum Pogonophora is a facinating group of marine tube-dwelling worms that live in deep water (Fig. 5.51). Pogonophorans were not recognized as a phylum until this century and only 80 species have been described, mostly from muddy sediments of deep water. They are sessile animals that live in chitinous tubes. A large mass of tentacles at the head end is believed to be used in feeding, but its exact role is still a puzzle because all pogonophorans lack a mouth, digestive tract, and anus. There is evidence that some pogonophorans harbor a mass of symbiotic bacteria in the coelomic cavity and feed off the products of the bacterial metabolism. There remain many more interesting unanswered questions about how Pogonophorans feed themselves.

Pogonophorans are widely distributed in the marine environment. Unlike most animal groups, however, they do not occur in shallow water. They are found at depths of 100 m or more and have their greatest diversity in the abyssal ocean. The genus *Riftia* is an important component of the unique biological community recently discovered at geothermal vents on the ocean floor (Ch. 11). Specimens of *Riftia* are the largest of the pogonophorans, reaching lengths of 150 cm (59.1 in) and diameters of 4 cm.

The Tardigrada differ from the other minor protostomes in that they are microscopic and have freshwater as well as marine species (Fig. 5.52). In fact, the majority of tardigrades live on water films of terrestrial

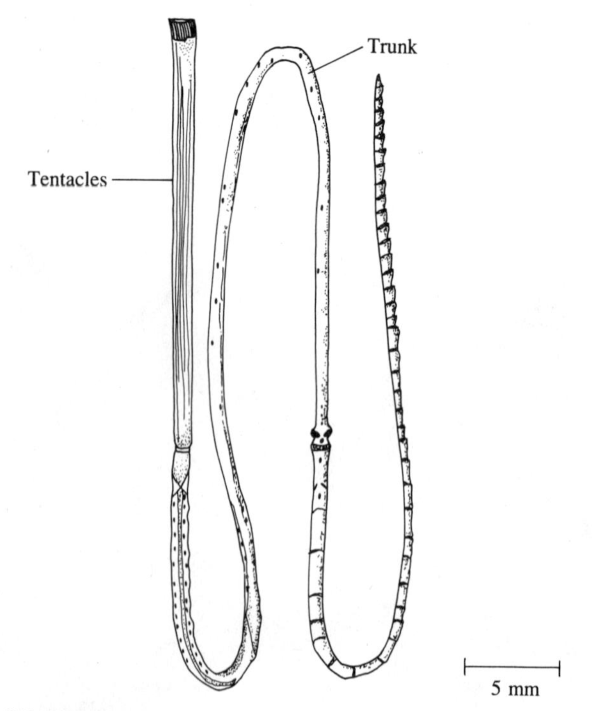

FIGURE 5.51

The pogonophoran *Lamellisabella*.

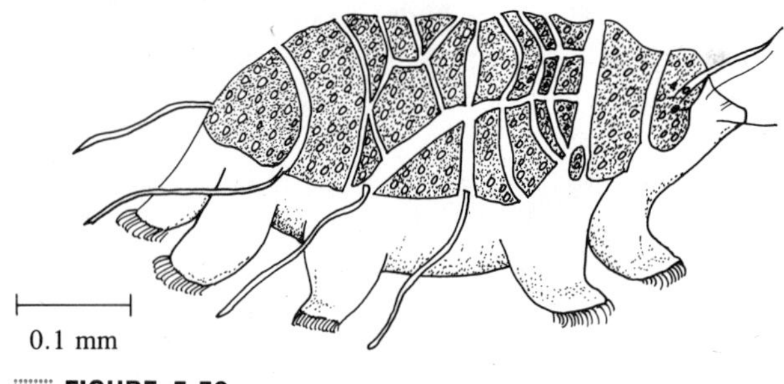

FIGURE 5.52
Tardigrade

Tardigrades, also known as water bears, are typically less than 0.5 mm in length.

lichens and mosses. In the marine environment, tardigrades are found in the interstitial spaces of sands and muds. The tardigrade body shows signs of segmentation having four pairs of legs and an internal nerve cord with a series of ganglia from which lateral nerves spread.

The most striking feature of tardigrade biology is their ability to pass through a stage of suspended animation, technically called a cryptobiotic state. When dried out, tardigrades become shriveled and metabolism drops to a very low rate. In this cryptobiotic condition, tardigrades can survive stresses that are lethal to virtually all other organisms. Individuals exposed to liquid helium (−270°C), absolute alcohol, and salt brines have 'come back to life' when returned to water. The cryptobiotic state has been maintained for longer than seven years in some tardigrades.

THE LOPHOPHORATE PHYLA

Phylum Brachiopoda (Latin: *brachium*-arm; Greek *pous*-foot)
Phylum Bryozoa (also called ectoprocts)
Phylum Phoronida (Greek: *pherein*-to bear; Latin: *nidus*-nest)

The common feature of these three phyla is a complex whorl of tentacles called the lophophore that surrounds the mouth (Fig. 5.53). The primary function of the lophophore is to gather food. Water is pumped by cilia through the lophophore and food particles are trapped on a mucus sheet that covers the tentacles. Cilia on the tentacles move the mucus and trapped food to the mouth. All lophophorate animals are filter feeders, eating primarily phytoplankton.

Although many other invertebrate animals from various phyla have tentacles grouped around the mouth that are used for feeding, the structure of the lophophore is unique and illustrates the close evolutionary affinity of the bryozoans, brachiopods, and phoronids. Of

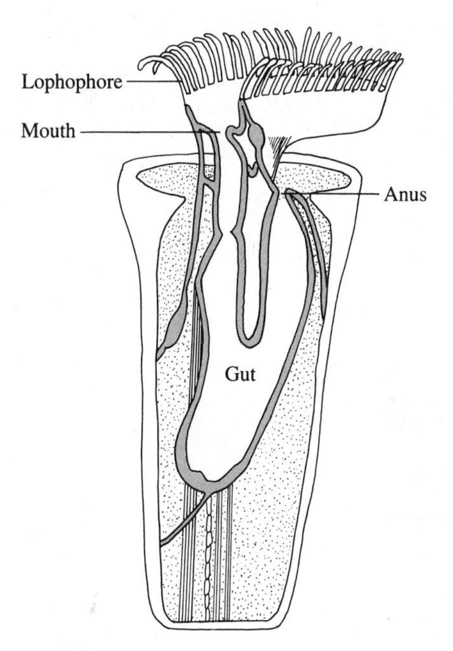

FIGURE 5.53
A Bryozoan, Showing the Lophophore

The lophophore is a *U*-shaped ring of tentacles surrounding the mouth in bryozoans and brachiopods.

these three phyla, only the brachiopods and bryozoans are widely distributed in the marine environment.

The brachiopods have 280 species, all marine. Perhaps the most interesting feature of this phylum is its geological record. Like most invertebrates, brachiopods first appeared some 500–600 million years ago in the Cambrian period. Early on, in the Ordovician period, between 400–500 million years ago, the group flourished and reached its peak of diversity. The fossil record has some 30,000 species. Brachiopods started to decline in diversity over 200 million years ago, and since that time have dwindled to only a few hundred species.

The oldest known genus of animal is a brachiopod: *Lingula*. *Lingula* (Fig. 5.54) has a body form that is virtually identical to a series of fossils that date back to the Ordovician period, 350 million years ago. The species that exist now are a little different from those of millions of years ago, but the morphological differences are minor.

The phylum Bryozoa, which has some 4,000 species, is the most diverse of the lophophorate phyla. Although bryozoans are a fairly large group, they are not quickly recognized. Species tend to be small, and all are sessile and colonial (Fig. 5.55). The individual of the colony, called a zooid, is generally less than 1 mm in length. Bryozoans have an external skeleton which is usually calcium carbonate. The skeletons are box- or tube-shaped, and the individual zooid can retreat entirely within the shell. Many species are even equipped with an **operculum,** or lid, that entirely protects the zooid when it is retracted.

Bryozoans are **polymorphic;** that is, not all zooids are identical. The most numerous zooids, the autozooids, specialize in feeding. The heterozooids are modified zooids. The two most common in bryozoans are avicularia (Fig. 5.56) and vibracula. Avicularia consist almost entirely of a pair of large jaws, yet they have lost the ability to feed. Instead, the jaws are used for

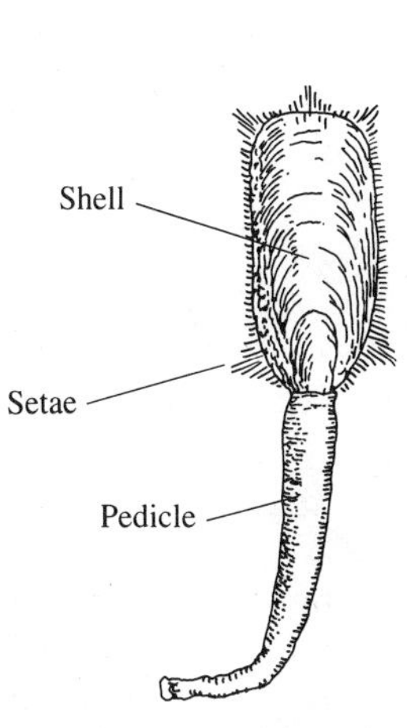
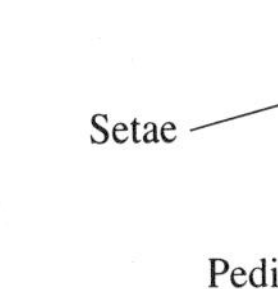

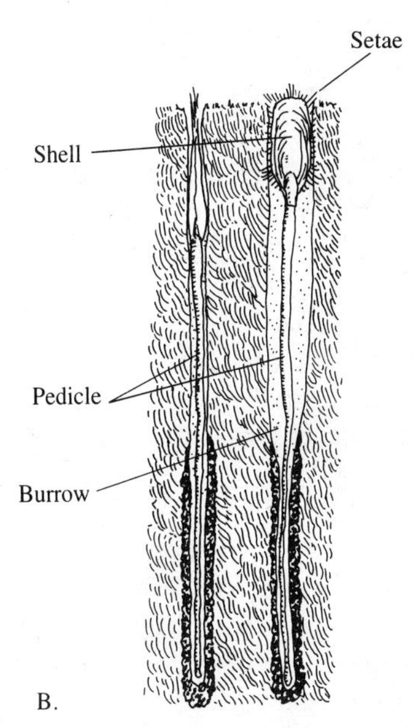

FIGURE 5.54
The Brachiopod *Lingula*

(A) Removed from burrow. **(B)** Cross section in a burrow. The left burrow shows the narrowness of the animal. Size approximately 10 cm. (After Hyman, 1959 and Francois, 1891.)

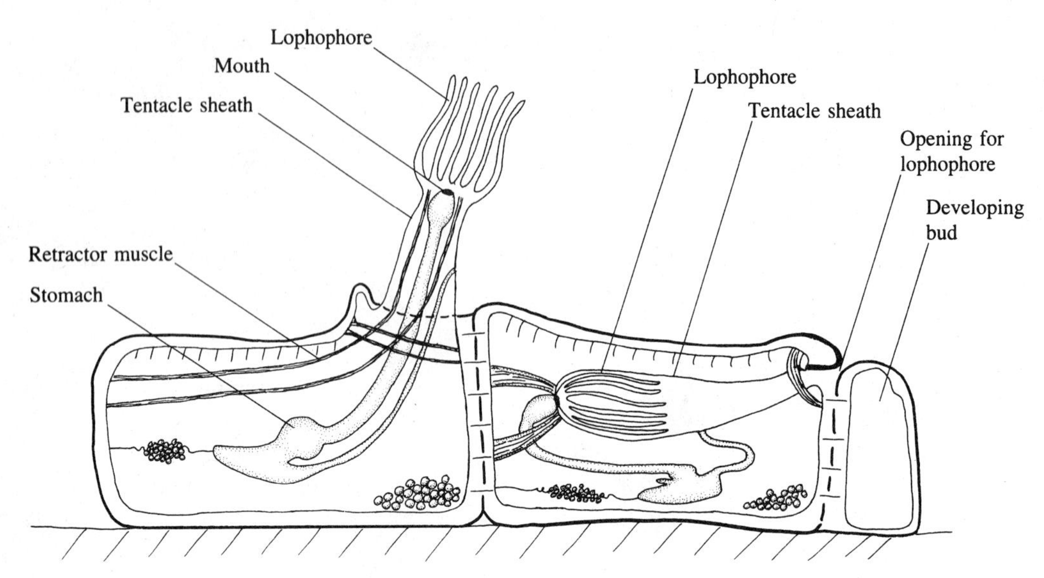

FIGURE 5.55
Bryozoans

Membranipora. (From Crane, 1973.)

defense. Vibracula are long bristles that sweep through the water. Both are used for defense and keeping the colony clean. They remove settling larvae, detritus, and sediments. Bryozoan colonies are generally flat and encrusting, or branched and erect.

Most bryozoans are found attached to hard substrates in shallow waters, although some species live at depths exceeding 8,000 m.

PHYLUM CHAETOGNATHA

Chaetognaths (Greek: *chaite*-hair; *gnathos*-jaw) are planktonic transparent worms about 0.5–1.5 cm in length. The common name, arrow worm, refers to the narrow body and lateral fins (Fig. 5.57) that resemble the feather blades on an arrow. Although there are only 50 species of arrow worms, they are common in the coastal plankton of both temperate and tropical waters. Chaetognaths are able to swim relatively well for plankton, and use their tail for rapid movement. The lateral fins are used for controlling direction. Arrow worms are voracious predators, using the spines on the head as jaws. They feed on a range of planktonic organisms including protozoa and juvenile fish. Their most common food is the planktonic copepod.

Arrow worms have a simple organ system. There are no circulatory, excretory, or respiratory organs. Gas exchange and waste elimination are achieved directly through the body wall. Chaetognaths are hermaphroditic, with adults having both mature eggs and sperm. Generally during copulation two individuals exchange sperm, although self-fertilization has been observed.

PHYLUM ECHINODERMATA

Echinoderms (Greek: *echinos*-spiny hedgehog; *derma*-skin) are animals with radial symmetry, and a water vascular system including tube feet that serve in locomotion, food capture, and gas exchange. Echinoderms are relatively large animals (around 10 cm, or 3.9 in, in diameter), and the 6,000 species are all marine. Included in the phylum are sea stars, sand dollars, sea urchins, and sea cucumbers. The body anatomy is based on five parts located around a central axis. This radial symmetry is clearly a secondary development, for their embryology shows that they evolved from bilateral ani-

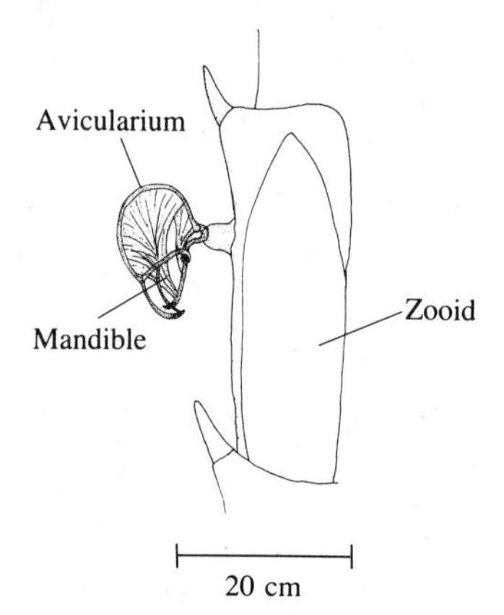

FIGURE 5.56
Specialization in Bryozoans

Bryozoans are polymorphic with individuals specialized for cleaning and, as shown here, for defense.

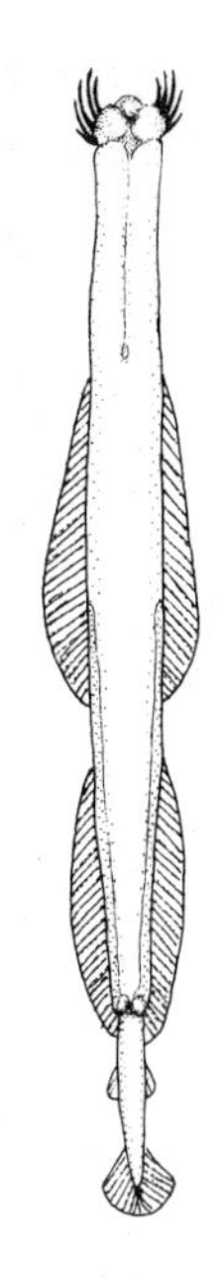

FIGURE 5.57
Arrow Worm

The chaetognath *Sagitta.* (From Crane, 1973.)

mals. Echinoderms are not closely related to the other metazoan phyla that have radial symmetry (cnidarians and ctenophores).

The skeleton of echinoderms is an internal one made of calcium carbonate (Fig. 5.58). The skeletal units, called **ossicles,** may be rod-, plate-, or cross-shaped. In echinoderms such as sea urchins and sand dollars, the ossicles are articulated into a solid structure. In others, such as sea stars, the ossicles fit together in a more irregular way to produce a rough surface. Though the name *echinoderm* means 'spiny skin', the outer surface of all echinoderms is epithelial,

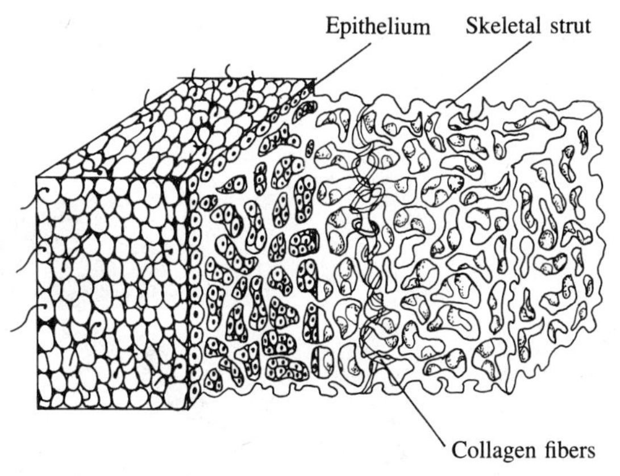

FIGURE 5.58
Echinoderm Skeleton

Echinoderms have an internal skeleton composed of calcium carbonate ossicles.

or soft tissue. This epithelium may have protective structures called pedicellarie that help keep the surface clean of parasites and predators (Fig. 5.59). The nervous system is between the epithelium and the skeleton.

The unique feature of the echinoderm anatomy is a system of internal, fluid-filled canals that connects with the tube feet on the underside of its arms (Fig. 5.60). This water vascular system, also called the **ambulacral** system, has a number of functions in various echinoderms: locomotion, attachment, sensation, feeding, and gas exchange, to name a few. The water vascular system is a hydraulic system that acts as a fluid reservoir for the tube feet. It opens to the outside through a madreporite, which is assumed to be the intake for the water needed to maintain a constant fluid volume.

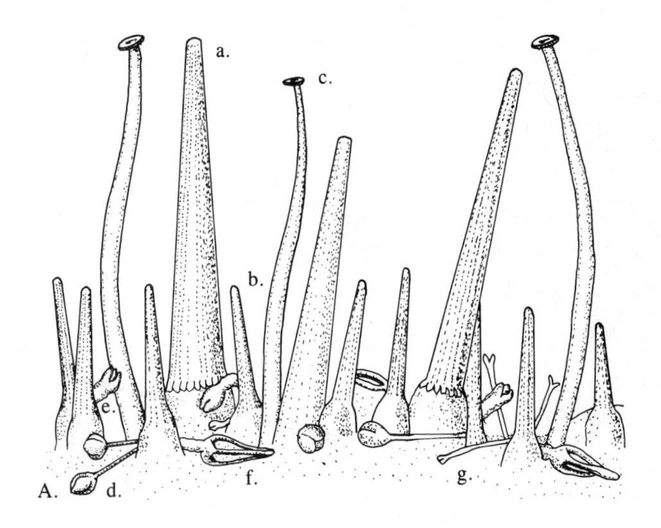

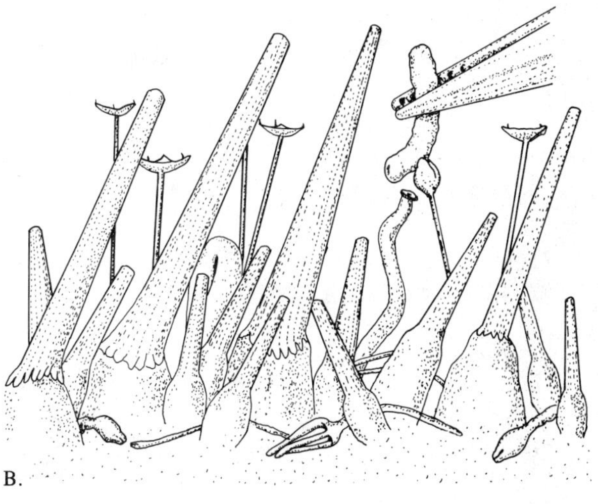

FIGURE 5.59

Echinoderm surfaces have pedicellariae that serve a defensive function. Here we see the surface of a sea urchin. **(A)** a: Spines (around 5 mm in length). b: Smaller spines. c: Tube foot. d–g: Pedicellaria. **(B)** One type of pedicellaria reacts to the tube foot of a sea star.

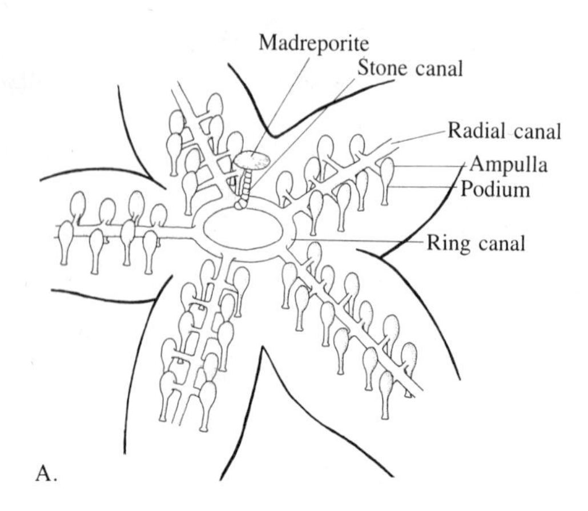

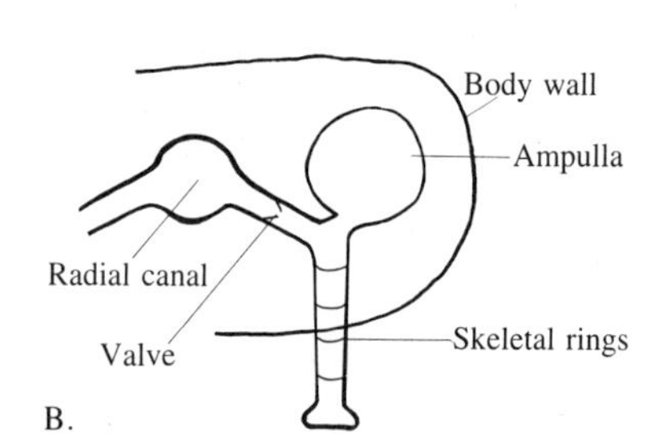

FIGURE 5.60
Water Vascular System in Echinoderms

(A) The ambulacral system in asteroids. **(B)** Tube foot. (From Crane, 1973.)

The function of the tube feet is well known. Each tube foot is supplied with circular and longitudinal muscles and by selective contraction has a wide range of movement. Tube feet typically have adhesive suckers, which are used for attachment as well as locomotion.

The Echinodermata are divided into five classes: Asteroidea (sea stars), Ophiuroidea (brittle stars), Echinoidea (sea urchins), Holothuroidea (sea cucumbers), and Crinoidea (sea lilies).

Class Asteroidea

Sea stars are probably the most familiar of the echinoderms (Fig. 5.61). Slow moving, they are found on rocky, sandy, or muddy substrates. They clearly show radial symmetry and most have five arms, although four or more than five is encountered. The general size is about 5 cm (2 in) in diameter; the largest, *Pycnopodia,* may exceed 100 cm.

Sea stars are active carnivores, feeding on snails, bivalves, polychaetes, and crustaceans. Food is grasped with the tube feet and the stomach extends from the mouth to engulf and digest the prey. This method is particularly effective in capturing bivalves. The sea star stomach can be squeezed through a gap in the two shells of bivalves that is only 0.1 mm wide.

Sea stars have remarkable powers of regeneration. As long as an adequate portion of the central disc is left attached to an arm, the whole body can regenerate. A sea star cut into five pieces can grow into five new individuals.

The water vascular system in asteroids is well developed for locomotion. Sea stars move by using their tube feet. Those species that live on solid substrate have tube feet with adhesive suckers for attachment, while species living on sandy or mud substrate have tube feet with pointed tips for leverage on the soft bottom.

Class Ophiuroidea

Brittle stars are similar in shape to sea stars but are generally smaller (Fig. 5.62). They typically have five arms radiating from a central disc. Brittle star arms are much thinner than those of sea stars. The skeletal elements of brittle star arms are articulated along their length like a series of vertebrae. Muscles attached to

A.

B.

C.

FIGURE 5.61
Sea Stars

(A) *Astropecten.* **(B)** *Mediaster.* **(C)** *Evastarias.* (Photos from Flora and Fairbanks, 1982.)

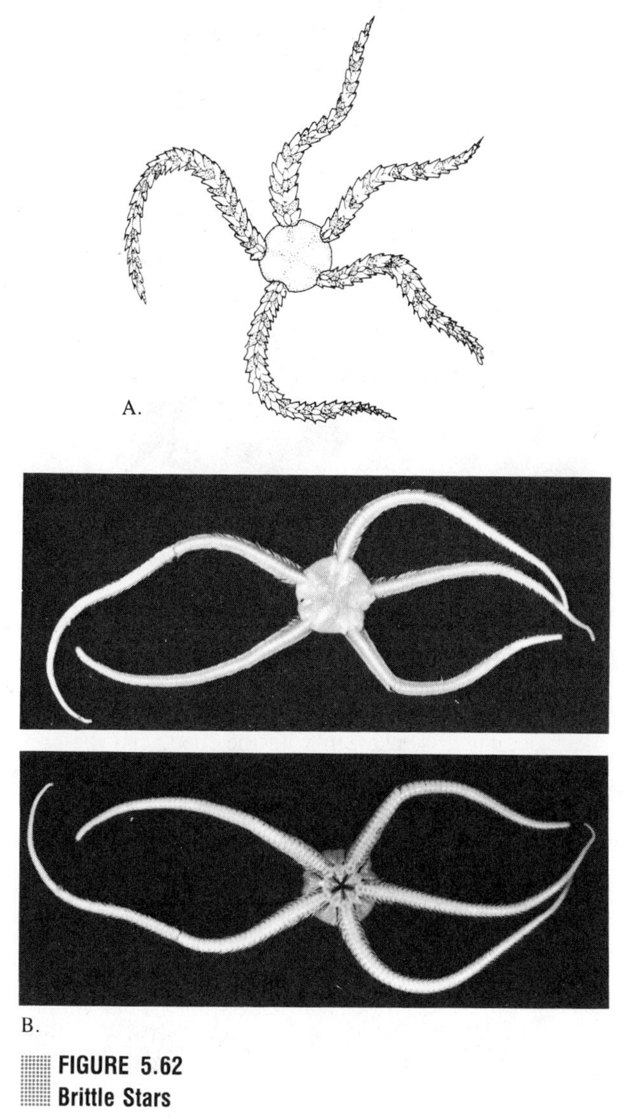

FIGURE 5.62
Brittle Stars

(A) The Serpent star *Ophionereis.* **(B)** The brittle star *Ophiura.* (A from Crane, 1973. B from Flora and Fairbanks, 1982.)

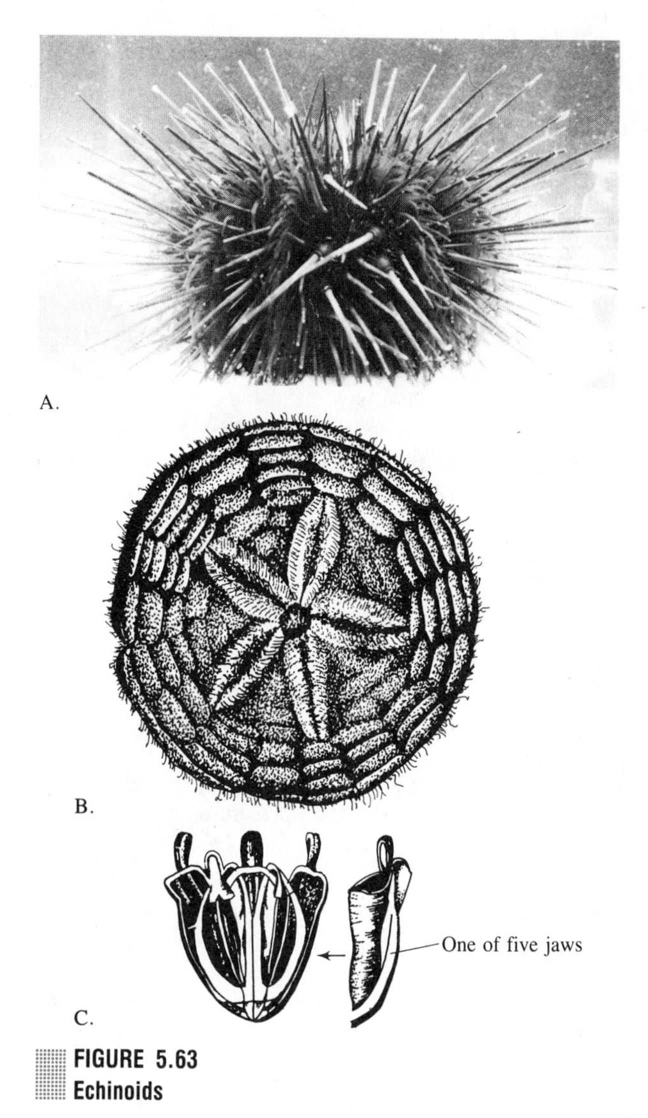

FIGURE 5.63
Echinoids

(A) The sea urchin *Strongylocentrotus.* **(B)** A sand dollar. **(C)** Aristotle's lantern (jaws of a herbivorous sea urchin). (A from Flora and Fairbanks, 1982.)

these ossicles provide for movement. The flexible arms of brittle stars can be moved much more quickly by muscular contraction than the tube feet movement of sea stars. Ophiuroids are the most active of echinoderms. The water vascular system in brittle stars is not well developed.

Food is mostly detritus, but some brittle stars are filter feeders, and others are predators or scavengers. Brittle stars are found on the surface and under rocks and on mud from the intertidal zone to great depths.

Class Echinoidea

Echinoids are echinoderms without arms (Fig. 5.63). The body tends to be spherical (as in sea urchins) or flattened (as in sand dollars). The body surface is covered with spines. In echinoids the skeletal ossicles are plates intricately fused into a shell, called a **test,** with many spines. As in sea stars, the water vascular system in echinoids is well developed. There are five bands of tube feet that pass through holes in the shell. Sea urchins use the tube feet in conjunction with spines for movement. The tube feet are adhesive and permit urchins to attach to the substrate while moving. For rapid movement across the bottom, spines also can be used as levers. Sand dollars depend almost totally on spines for movement.

Echinoids have diverse feeding habits. Some urchins are herbivores, feeding on seaweeds found in the intertidal and shallow subtidal zones. These species have a remarkable set of jaws called Aristotle's Lantern, consisting of five complex calcareous jaws held together by muscles. Other echinoids that are detritus or mud feeders have reduced jaws.

Class Holothuroidea

Holothuroids (sea cucumbers) have a body that is elongated into a cucumber shape with five bands of tube feet running down its length (Fig. 5.64). Sea cucumbers lie on their sides. Generally, their skeleton is reduced to a series of microscopic ossicles, and the body is soft and often flaccid.

The water vascular system in sea cucumbers is well developed and most species have tube feet. Although some species use the tube feet to creep across the bottom, most species burrow in the sand or mud or retreat into cracks and crevices. Most have a large whorl of tentacles around the mouth for feeding. These tentacles are part of the water vascular system. Food particles are trapped in mucus on the tentacle surface. Other cucumbers are deposit feeders and swallow the surface mud.

Class Crinoidea

The crinoids, or sea lilies, are the most ancient of the echinoderms and show ancestral features of the phylum (Fig. 5.65). They have a body form in which the mouth opens up instead of down. In addition, the tube feet of the water vascular system are used to filter food, which is believed to be a primitive function. Crinoids are of two types: sessile (attached to the substrate by a stalk), and free living. There are only 80 species of stalked crinoids, which are found at depths of 100 m or more. The free-living forms are most common in tropical waters and are found from shallow subtidal areas to great depths. Free-living crinoids use their arms for walking and, in some cases, swimming. Their arms are articulated in a manner similar to brittle stars, and movement is by muscular contraction. The water vascular system and the tube feet are used for filter feeding. Food particles adhere to mucus on the tube feet, and cilia move the mucus-trapped food to the mouth.

PHYLUM HEMICHORDATA

Hemichordates, or acorn worms, are unsegmented, elongated, fleshy worms of moderate size (Fig. 5.66). The most conspicuous feature of the body is an anterior proboscis, in the general shape of an acorn, which functions in movement and feeding. Acorn worms that live in burrows use the proboscis to move through the burrow and through crevices in rocks. The proboscis is also used for gathering food. Detritus particles become

FIGURE 5.64

The sea cucumber *Cucumaria.* (From Flora and Fairbanks, 1982.)

stuck to its mucus-covered surface as it moves across sediment. Cilia move the particles to the mouth. Inorganic sediment is taken in with the food, and the burrows of acorn worms generally have a pile of sandy fecal material, called a casting, close by.

Hemichordates have a number of deuterostome traits that show the affiliation of this phylum with both the echinoderms and chordates. In many hemichordates the nerve chord is hollow, as in the chordates, and hemichordates have a dorsal nerve chord rather than a ventral chord as seen in other invertebrate phyla. The stages of larval development in hemichordates show great similarity to echinoderms. Hemichordate embryos develop clefts in the pharynx that persist in the adult as a set of pores of even larger openings. These pores are similar to the gill slits found in the phylum Chordata.

PHYLUM CHORDATA

Chordates (Latin: *chorda*-chord) are a diverse and successful group of animals ranging from sessile, spongelike tunicates to the highly mobile and intelligent mammals. The characteristics shared by all these animals are a dorsal tubular nerve chord, a notochord, and gill slits. The notochord is a flexible, rodlike support that lies above the digestive system but below the nerve chord. The gill slits are the openings in the pharynx area through which water passes to the gills. Clearly, the gill slits and notochord are not found in the adults of many chordates—for example, humans. At some stage of the life cycle of all vertebrates, if only in the embryo or larval stage, the three features can be found.

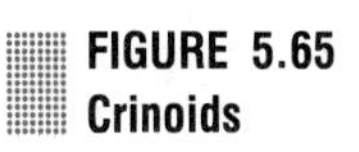

FIGURE 5.65
Crinoids

(From Haeckel, 1974.)

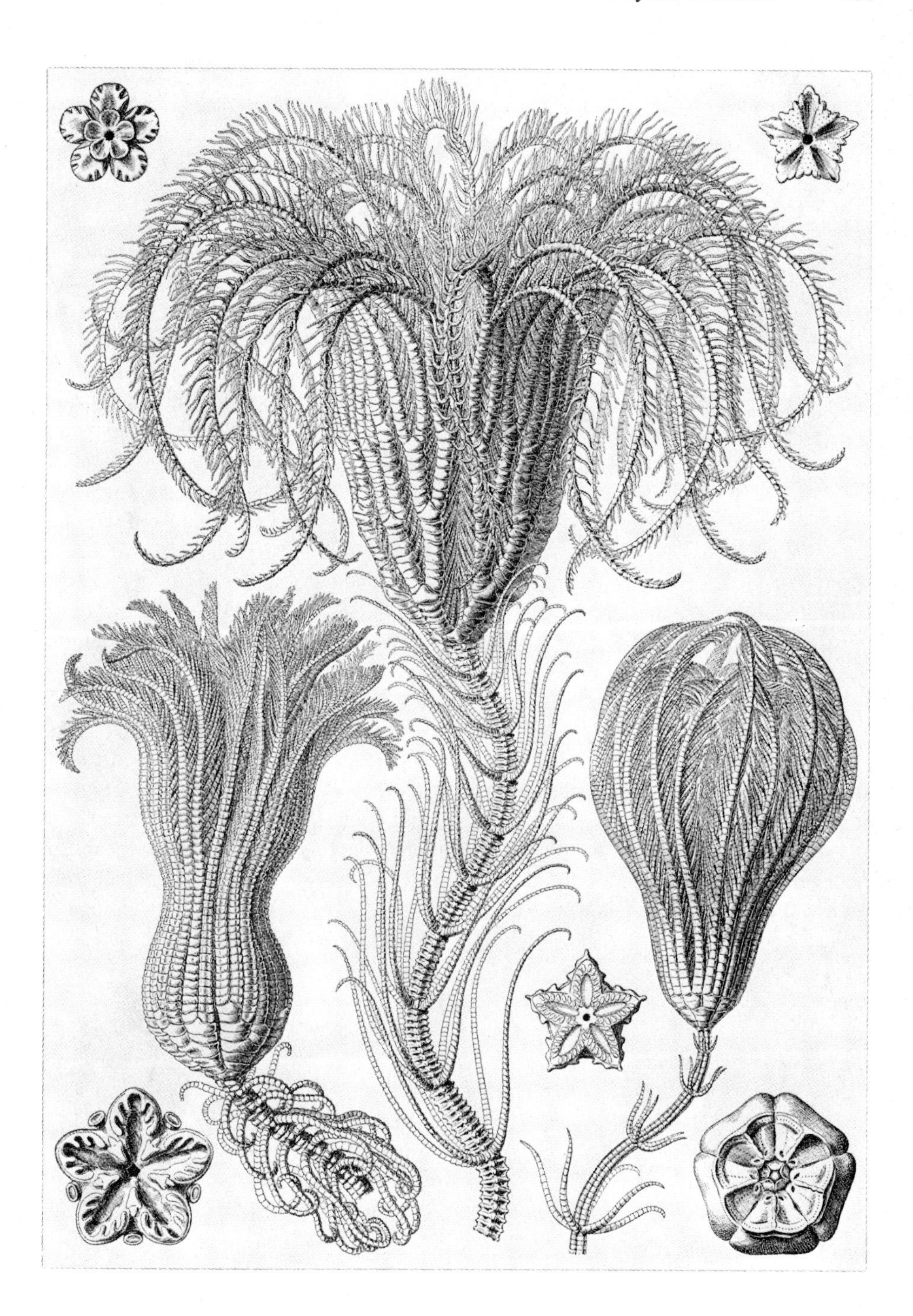

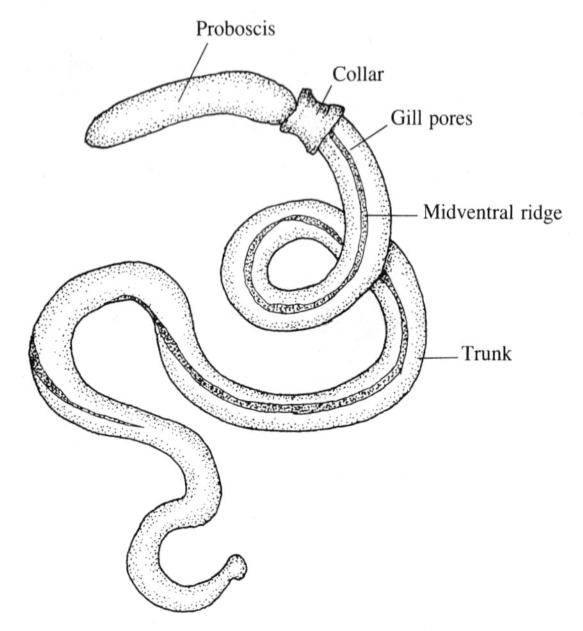

FIGURE 5.66
Acorn Worm

The hemichordate *Balanoglossus*. (From Crane, 1973.)

The two groups of chordates important to the study of marine biology are the tunicates and vertebrates. Tunicates are filter feeders that live as sessile benthic animals or passive floaters in the plankton. They have not evolved the active life style found in the vertebrates. (We will discuss the vertebrates in Chapters 6–8.)

Tunicates (class Ascidiacea) are highly modified chordates, and only show the phylum characteristics in the larval stages. The group, numbering some 1,500 species, is found primarily in the shallow-water benthos (Fig. 5.67), although a few species are highly successful in the plankton.

Sessile tunicates are commonly called sea squirts. They are found mostly on rocky bottoms but also may occur in mud and sand. The basic body form is cylindrical in shape. The major feature of the body is a relatively large pharynx perforated with small slits (Fig. 5.68). This pharynx is used in filter feeding to strain food particles from circulating water. The body has two openings, or siphons. One admits water to the pharynx and the second serves as an exit. The body of the tunicate is covered with an outer layer, or tunic, made of a type of cellulose that is thick and leathery in some species.

Pelagic tunicates, although few in number of species, may be common in the plankton. One pelagic group is the Larvacea, or Appendicularia. These pelagic tunicates retain larval features and are notable because of the gelatinous 'house' they secrete which is actually a fine filter used for feeding.

Tunicates may be solitary or colonial. Colonies may form by asexual reproduction and appear as a cluster of similar-looking individuals. In some species, the individuals remain connected to each other and are embedded in a common tunic. The shape of these colonies

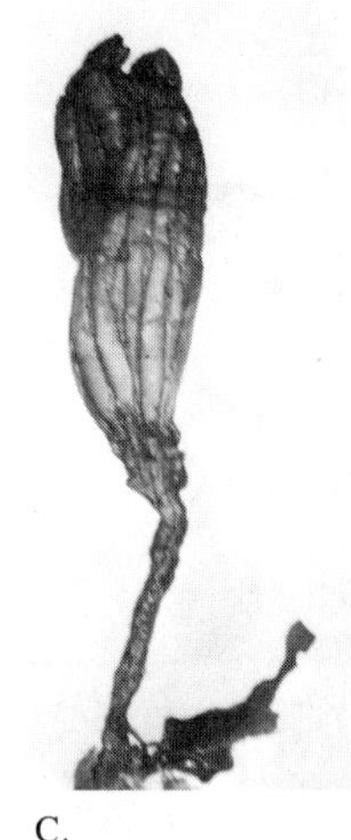

FIGURE 5.67
Benthic Tunicates

(A) Colonial tunicate. **(B)** and **(C)** Solitary tunicates. (From Flora and Fairbanks, 1982.)

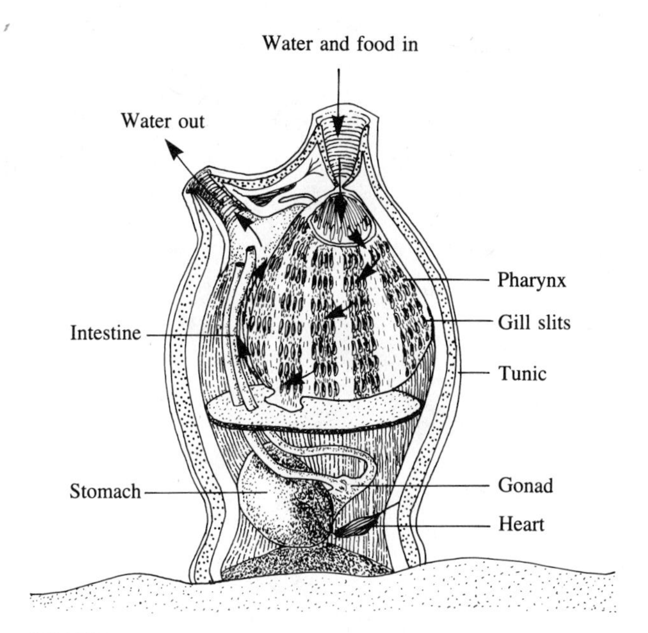

FIGURE 5.68
Internal Anatomy of a Benthic Tunicate

Note the large filtration basket derived from the gill slits.

closely follows the substrate. Such colonies are often difficult to separate, at first glance, from sponges. Often magnification is needed to tell if the organism belongs to the most primitive or most advanced phylum of animals.

SUMMARY

There are over 30 phyla of living animals in the kingdom Metazoa. All these phyla have species in the marine environment. Most of the invertebrates are small. Many are sessile and most have only a limited ability to move.

Invertebrates occur everywhere in the marine environment. Benthic species are found from the highest level of the intertidal zone to the deepest sediments of the sea. Pelagic species are found from the air-water interface to the deepest parts of the sea. The differences between the phyla reflect their evolution over the last 1,600 million years. Little fossil evidence remains of this evolutionary story beyond 500 million years ago. There is, however, strong evidence of evolutionary affinities between the phyla. Much evidence for the early evolution of animals is based on characteristics of their embryology.

Metazoans are believed to have evolved from one or more ancient protists, probably the ancestors of modern ciliated and flagellated protozoa. The sponges probably evolved first, but their morphology (a system of pores and water channels running through the tissues) is so different from other animal groups that they are assumed to not be involved in the ancestry of other Metazoa. Instead, the radiates (Cnidarians and Ctenophorans) were the base stock of metazoan evolution. The radiate phyla have a morpholology based on radial symmetry. One other phylum, the echinoderms, have a type of radial symmetry based on five parts, but the embryology of this group clearly shows that it evolved from a bilateral ancestor.

Many invertebrates are wormlike. Almost all groups of worms are found only in benthic habitats. The worms vary in characteristics of morphological complexity. Flatworms have only one opening in the body cavity; nemerteans have a long proboscis; nematodes have a protective outer cuticle; and annelids have bodies based on segmentation.

The segmentation of annelids is strong evidence of their evolutionary affinity with the phylum Arthropoda (crustaceans). In fact, the evidence suggests that the annelids, arthropods, and molluscs are fairly closely related. This major group of animals includes many dominant benthic and planktonic species.

And what are the evolutionary relationships of the chordates? This is the phylum to which the large, active vertebrates belong. Other phyla closely related to the chordates are the echinoderms and a small phylum of worms, the hemichordates both very distant relatives of the large, actively moving vertebrates.

QUESTIONS AND EXERCISES

1. Discuss the various skeletal systems found in the invertebrates. Your discussion should include exoskeleton, internal skeletons, and hydraulic, or hydrostatic, skeletons.
2. Given the following anatomical characteristics of invertebrate animals, develop a classification system that groups the phyla discussed in this chapter into the fewest number of groupings:
   - radial symmetry
   - nervous system with centralized ganglion or brain
   - bilateral symmetry
   - body of two layers
   - nervous system diffuse with no concentrated ganglion or brain
   - body with three layers
   - single digestive system opening
   - no internal body cavity
   - two digestive system openings
   - coelomic body cavity
   - noncoelomic body cavity
3. What are the five most important invertebrate phyla in the marine environment? Discuss your criteria.
4. Describe the major taxonomic groups of the phylum Arthropoda. Rank each of these groups according to the number of species they have in the marine environment.
5. Define the following terms, then match them with the most appropriate systematic group of animals: choanocyte, nematocyst, colloblast, lophophore, tube feet, parapodia, radula, compound eye, clones (formation of colonies through asexual reproduction).

REFERENCES

Abbott, D., and **Hilgard, G.H.,** eds. 1987. *Observing marine invertebrates:Drawings from the laboratory.* Stanford, CA:Stanford University Press.

Barnes, R.D. 1980. *Invertebrate zoology.* Philadelphia:Saunders.

Buchsbaum, R., Buchsbaum, M., Pearse, J., and **Pearse, V.** 1987. *Animals without backbones, 3rd ed.* Chicago:University of Chicago Press.

Crane, J.M. 1973. *Introduction to marine biology:A laboratory text.* Columbus, OH:Merrill.

Flora, C.J. and **Fairbanks, E.** 1982. *The sound and the sea.* Bellingham, WA:Western Washington University.

Francois, P. 1981. Observations biologiques sur les lingules. *Arch. Zoo. Exp. Gén.* 2(9).

Haeckel E. 1974. *Art forms in nature.* New York:Dover.

Hughes, R.N. 1986. *A functional biology of marine gastropods.* Baltimore, MD:Johns Hopkins University Press.

Hyman, L.H. 1959. *The invertebrates: Smaller coelomate groups.* New York:McGraw-Hill.

Lawrence, J. 1987. *A functional biology of echinoderms.* Baltimore, MD:Johns Hopkins Press.

Morris, R.H., Abbott, D.P., and **Haderlie, E.C.** 1980. *Intertidal invertebrates of California.* Stanford, CA:Stanford University Press.

Simpson, T.L. 1984. *The cell biology of sponges.* New York:Springer-Verlag.

Taylor, H. 1984. *The lobster:Its life cycle.* New York:Pisces Books.

Wells, J.G., ed. 1982. *Encyclopedia of marine invertebrates.* Neptune, NJ:TFH Publications.

SUGGESTED READING

OCEANS

Bavendan, F. 1986. Friends and anemones. 19(6):49–53.

Dillon, J. 1982. Vellela:By-the-wind sailor. 15(6):3–4.

Holt, D. 1985. A plague of urchins. 18(3):14–17.

Johnson, E. 1979. Stinging nudibranchs. 12(4):49.

Kerstitch, A. 1976. Archer of the deep. (Feeding of cone snails) 11(6):50–51.

Kitchell, J. 1978. Buoyancy architecture. (Evolution and buoyancy of the Nautilus) 11(6):44–49.

Mallory, K. 1982. The sea butterfly, *Clione.* 15(3):12–13.

Menard, W. 1976. Mata-Malu anemone. (Dangerous sea anemones) 9(6):25–27.

Pratt-Johnson, B. 1978. Everybody loves an octopus. 11(1):10–15.

Russo, R. 1979. A salute to sea slug. 12(4):42–48.

Trusty, L. 1975. Sponges. 8(2):56.

Ward, D.E. 1978. Unorthodox dining habits of sea stars. 11(5):14–17.

Zann, L.P. 1977. Symbionts of sea fans and sea whips. (Coral reef symbiosis) 10(1):10–15.

OCEANUS

Tamm, S. 1980. Cilia and ctenophores. 23(2):50–59.

SCIENTIFIC AMERICAN

Alldredge, A. 1976. Appendicularians. 235(1):94–105.

Gosline, J.M. and **Dehront, M.E.** 1985. Jet-propelled swimming in squids. 252(1):96–103.

Richardson, J.R. 1986. Brachiopods. 255(3):100–107.

SEA FRONTIERS

Heston, T.B. 1987. Living history:The horseshoe crab. 33(3):195–199.

Messing, C.G. 1988. Sea lilies and feather stars. 34(4):236–241.

Radtke, R.L. 1987. Brine shrimp:Curious crustaceans. 33(2):128–133.

Rodhouse, P.G. 1989. Squid, the southern ocean's unknown quantity. 35(4):206–211.

Sefton, N. 1989. The secret life of sponges. 35(3):159–162.

Wu, N. 1989. The paper nautilus. 35(2):94–96.

AGGRESSION IN SEA ANEMONES

The sea anemone *Anthopleura elegantissima* is a common species of the intertidal zone of the west coast of North America. *A. elegantissima* appears in two forms: (a) a solitary form; (b) an aggregating form that occurs in dense beds composed of individuals that are clonemates. That is, each individual of the bed is genetically identical to the others. Marine biologist Dr. Lisbeth Francis observed that the beds occasionally had bare strips with no anemones (Fig 1). The width of these strips is generally a similar size to the width of the anemone.

Dr. Francis discovered that the strips described the boundaries of different clones, and that individuals on either side of the strips aggressively defended the territory of their respective clones. The aggression is initiated by the contact of tentacles of individuals of different clones. Contact of individuals from the same clone does not elicit aggression.

The aggression of *A. elegantissima* utilizes a special organ called the **acrorhagus** (Fig. 2). The acrorhagus is a blunt, white-tipped protrubance of the body wall, and contains large concentrations of a particular type of nematocyst called the **atrich nematocyst.** The acrorhagus and the atrich nematocyst are used only for

FIGURE 1

Massive clonal aggregations of *Anthopleura elegantissima* on large boulders, exposed during low tide in the intertidal zone at the Scripps Institution of Oceanography, La Jolla, CA. The contracted anemones are covered in a reflective layer of adhering pebbles and shells, making the borders between clones obvious as dark bands of unoccupied rock. (Photo by Dr. Lisbeth Francis.)

FIGURE 2

Underwater attack at a clonal border. An expanded warrior individual (center) has inflated three acrorhagi, which it directs toward contracted members of an adjacent clone. These anemones have very few adhering pebbles, so the columns of contracted individuals appear dark, rather than light colored. (Photo by Dr. Lisbeth Francis.)

the aggressive behavior. Anemones on the edge of the clones have larger and more numerous acrorhagi than do individuals living in the middle of the aggregation.

When individuals from two clones come in contact, the acrorhagi swell and become turgid. During contact, pieces of acrorhagi are attached by the aggressive anemones to the ones that are under attack. The pieces of acrorhagi slowly fire the atrich nematocysts into the body of the attacked anemone. If the attacked anemone cannot move away, it eventually will be killed by the aggressor.

The aggressive behavior between two genetically different anemones follows a series of stages including: (a) contact of the tentacles of the two anemones; (b) inflation of the acrorhagi so that they become turgid and pronounced; (c) movement of the body of the anemone so that the acrorhagi come in contact with the other anemone; (d) a sticking of the ectoderm from the attacking anemone onto the attacked animal; and (e) a cessation of attack, where the anemone returns to its original position. The aggressive sequence is repeated if contact of tentacles occurs again.

This aggressive behavior also has been observed in anemone species in New Zealand and in Tropical Australia.

6

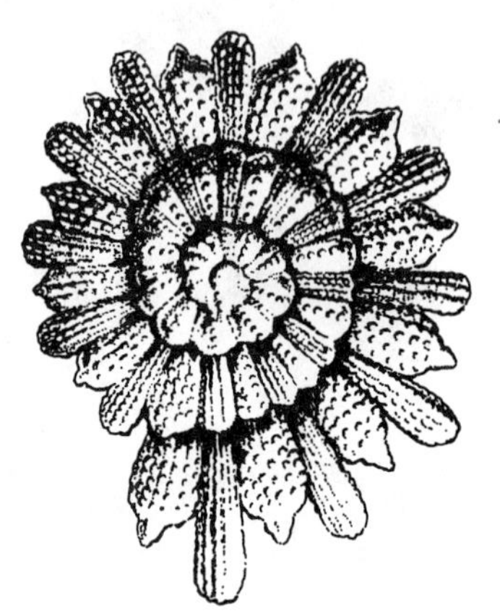

MARINE FISH

There are more species of fish in the marine environment than of any other vertebrate. Fish are found in all regions of the oceans, from the surface to the deepest waters. They are most common in the shallow waters bordering continental land masses, but are also found in oceanic regions of the high seas. The largest of the fish are generally found in the top 100 m (328 ft), the epipelagic zone. The larger fish—such as tuna, swordfish, and halibut—exceed 2 m in length and can weigh over hundreds of kilograms. Deeper-water species typically are much smaller. Therefore, when the full range of species is considered, the average length is only 15 cm, (5.9 in) and the average weight is less than a kilogram. Considered as a group separate from the rest of the fish, the sharks are on average 2 m long.

Fish are aquatic vertebrates that use gills to obtain oxygen from the water. They have fins with a variable number of skeletal elements called **fin rays.** There are three vertebrate classes that—using this definition—have species we call fish: the hagfish and lampreys (Cyclostomata); the sharks, rays, and chimeras (Chondrichthyes); and the bony fish (Osteichthyes) (Fig. 6.1). Only two of these groups—the sharks and rays, and the bony fish—are widely distributed in the marine environment. With at least 20,000 species in all, the sharks, rays, and bony fish are easily the most successful group of the phylum Chordata. Of these 20,000 species, 58 percent are found in the marine environment (Fig. 6.2). They are most common in the warm and temperate waters of the continental shelves (some 8,000 species). In the cold polar waters of the continental shelves there are some 1,100 species. In the oceanic pelagic environment, well away from the effects of land, only approximately 255 species are found. Surprisingly, in the deeper mesopelagic zone of the pelagic environment (between 100–1,000 m depth), the number of species of fish increases. There are some 1,000 species of so-called midwater fish. In the following discussion we will use *fish* to refer only to the bony fish, and will consider sharks and rays separately.

SYSTEMATICS

Class Chondrichthyes

There are 250 species of sharks and 300 species of rays in this class. Sharks and rays have a number of unique characteristics that separate them from bony fish. Sharks and rays are generally larger than bony fish. Unlike bony fish, they have no swim bladder. Their skeletons are cartilaginous, while those of bony fish are generally calcified. Finally, the two groups differ in reproduction. In sharks, eggs are incubated inside the body of the female, and the young are generally born live. Rays deposit their eggs in a capsule in which the young develop. Bony fish, conversely, almost always discharge their eggs into the sea.

Sharks show great variation in body shape (Fig. 6.3). The whale shark and basking shark, both filter feeders, are the largest of all sharks. The family Lam-

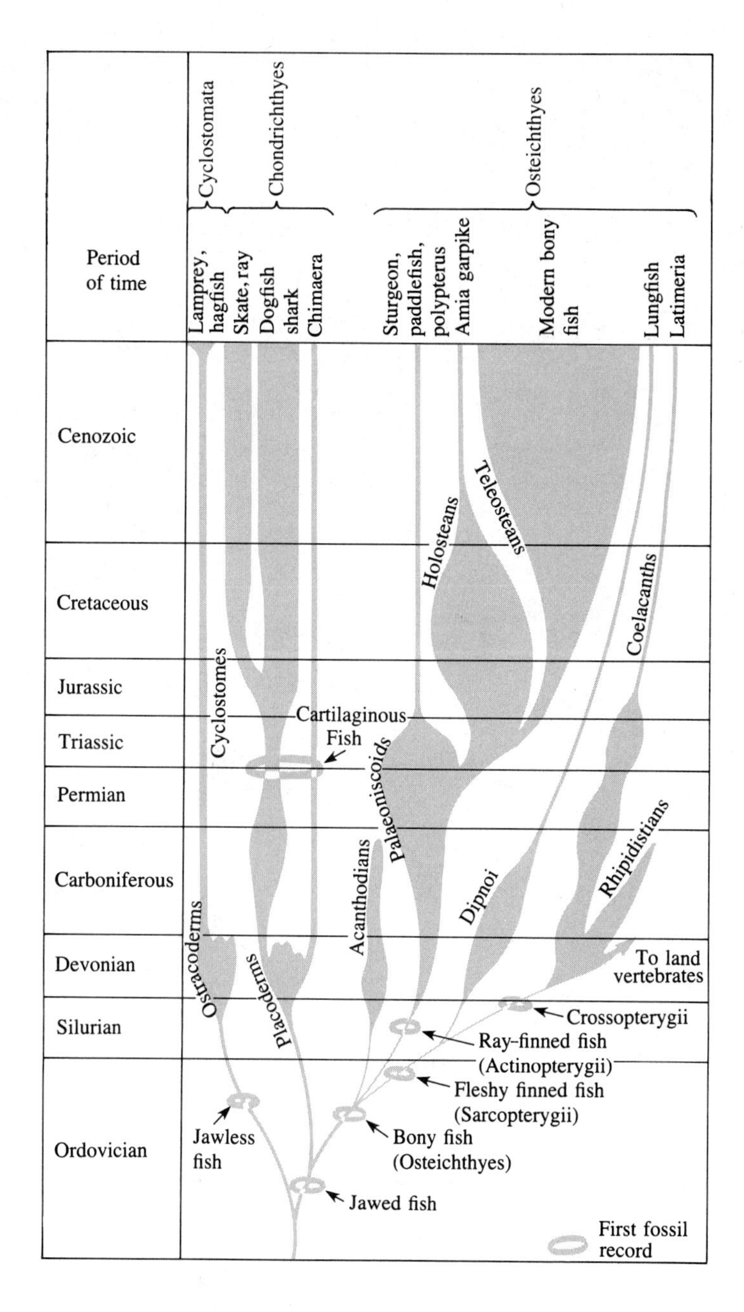

FIGURE 6.1

Fish are a complex group of vertebrates. The two most important groups are the sharks and rays, and the teleost fish. (After Marshall, N.B. 1971. *Explorations in the life of fishes*. Cambridge, MA:Harvard University Press.)

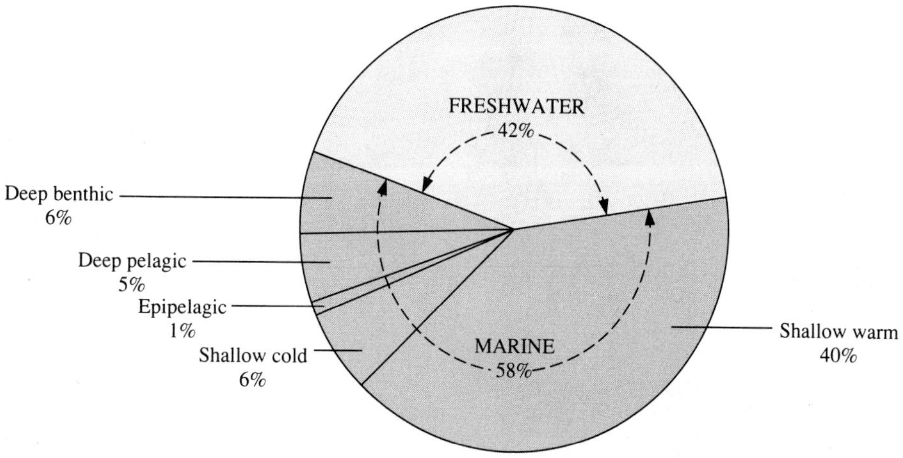

FIGURE 6.2
Numbers of Fresh Water and Marine Fish

Of the some 20,000 species of fish, the majority (58%) are marine. In the ocean, the greatest richness of species is in warm shallow water.

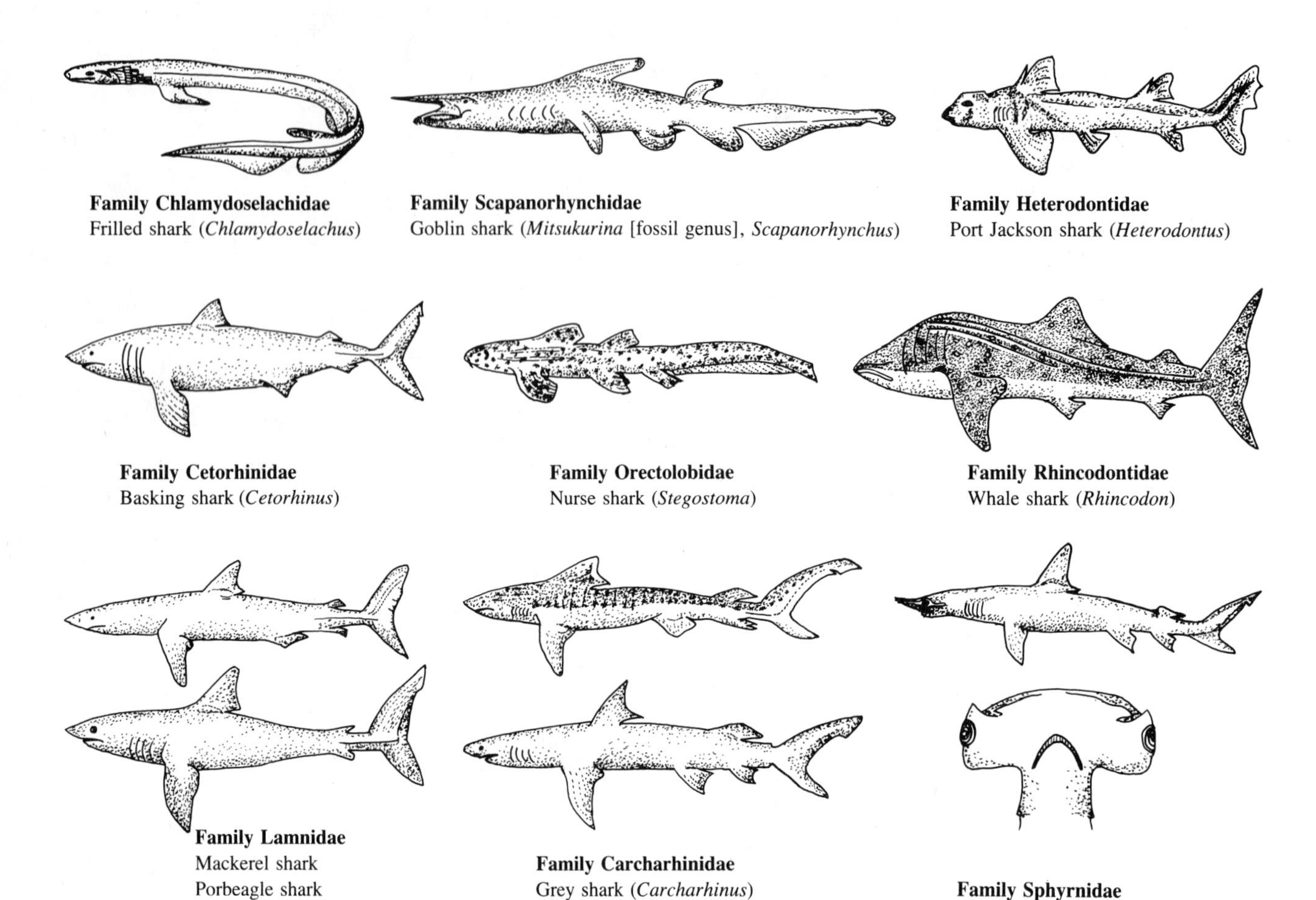

FIGURE 6.3
Sharks

There are some 20 families of sharks. The family Lamnidae (mackerel shark) is the most streamlined and has the best swimmers.

nidae is important because it includes the sharks that roam freely in the marine environment. Among these powerful swimmers are mako sharks, mackerel sharks, and the famous great white shark (Fig. 6.4).

FIGURE 6.4
Great White Shark

The great white shark *(Carcharodon)* is the best known of the large fast-swimming sharks. Although this species has been known to attack and kill humans, the number of such deaths is small.

While sharks usually are found in the pelagic environment, skates and rays (Fig. 6.5) are more frequent in the benthic environment. The pectoral fins of skates and rays are greatly enlarged and are attached to the head, giving them their triangular shape.

Class Cyclostomata

The hagfish and lamprey (Fig. 6.6) that compose this class are distinguished by a lack of jaws. This is thought to be a primitive feature and serves to set hagfish and lamprey apart from the true fish. The mouth is a circular sucking disk that is used to attach and/or burrow into other fish. Both groups are primarily parasitic, although the hagfish is also a scavenger. Lamprey are **anadromous,** that is, they spend their adult life at sea, but reproduce in fresh water. When the St. Lawrence Seaway was completed and removed the barriers to migration, the lamprey spread into the Great Lakes and contributed to the decimation of the trout population.

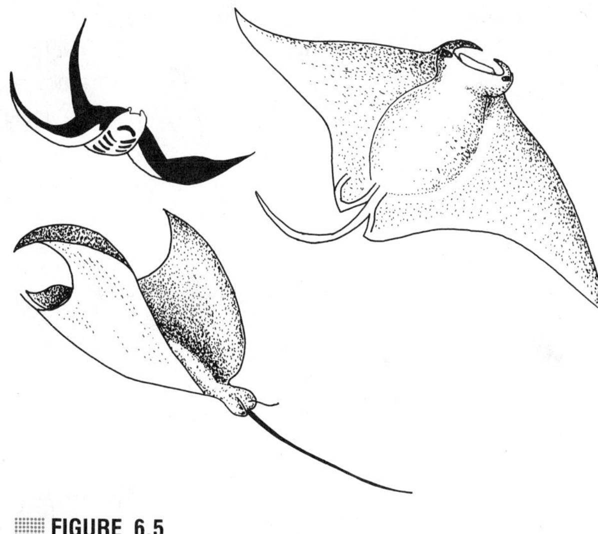

FIGURE 6.5
Skates and Rays

The shape of skates and rays is due to largely expanded pectoral fins which they use for swimming just above the sea floor.

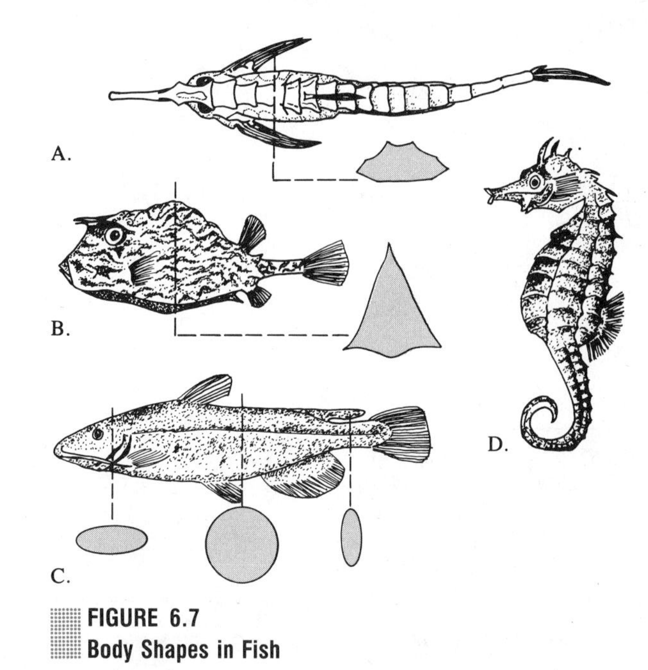

FIGURE 6.7
Body Shapes in Fish

(A) Sea moth. **(B)** Cowfish. **(C)** Bullhead. **(D)** Sea horse.

Class Osteichthyes

There are many more species of bony fish than sharks and rays (20,000 vs. 500). The bony fish owe their success to their skeleton. The heavily armored skeleton of the bony fish's evolutionary ancestors was lost in favor of a light, strong, internal skeleton that provides the support for active swimming and the effective functioning of the jaw. Body shapes are highly varied (Fig. 6.7). Bony fish almost exclusively use a **swim bladder** to maintain neutral buoyancy. This neutral buoyancy has allowed the tail fin to become symmetrical with equal lobes (**homocercal**) since it does not have to provide a lifting force during swimming. Bony fish fins can be moved independently of the rest of the body, and are used for directional swimming as well as for communication, aggressive behavior, and reproduction. Bony fish also communicate by sound, and a majority of species, at times at least, cluster together in schools.

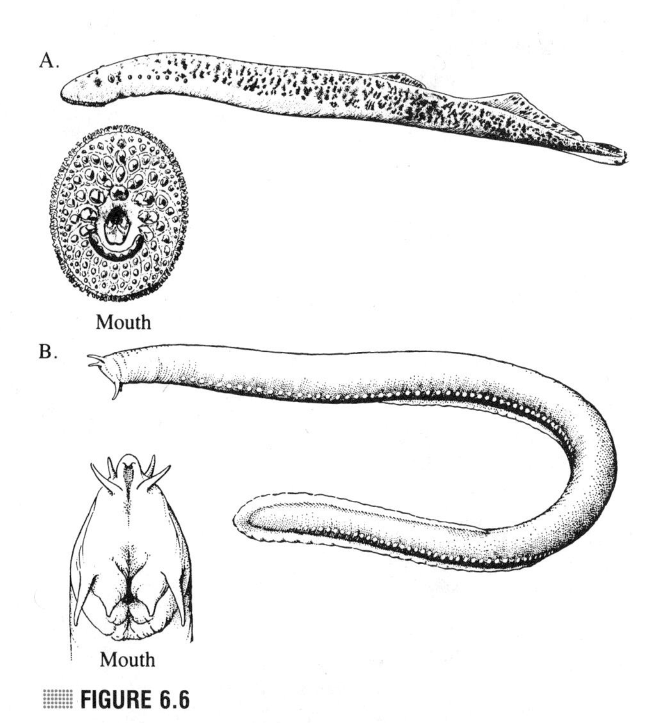

FIGURE 6.6

The lamprey and hagfish show their primitive position in vertebrate evolution by their lack of jaws. **(A)** Lamprey. **(B)** Hagfish.

At least 12 orders of bony fish have marine species that are important in one way or another. There are two orders of spiny-rayed, or **percoform,** fish (Fig. 6.8) that make up the majority of the bony fish. The Percomorphi are 5,000 marine species that include tunas, dolphin fish, most coral reef fish, and many of the intertidal and shallow sublittoral species in their order. Much of the success of the percoform fish is attributed to the anterior migration of the pelvic fins, which allows for much greater flexibility and versatility in the function of these fins. As numerous as percoform fish are, they do not include several common groups such as the flatfish, eels, codfish, herringlike fish, and salmon, all of which belong to different orders. (The greatly modified skeletal structure of the flatfish is discussed in the special feature at the end of this chapter.)

Epipelagic and mesopelagic fish species are quite different from each other. Epipelagic fish, which num-

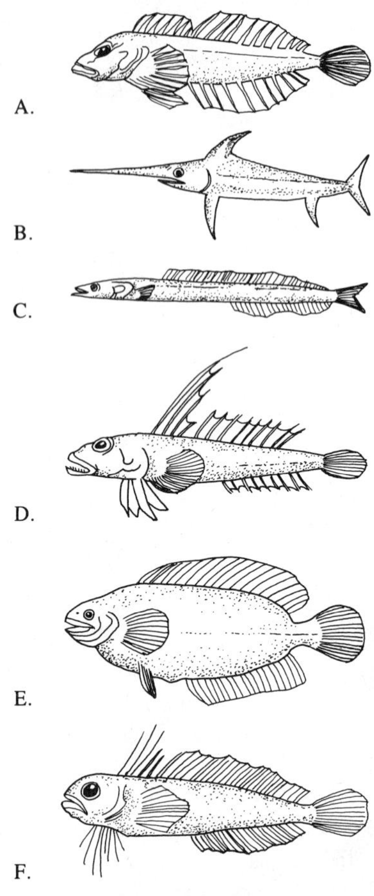

FIGURE 6.8
Percoform Fish

Percoform fish are the most dominant and diverse group. **(A)** Sculpin. **(B)** Swordfish. **(C)** Sand lance. **(D)** Dragonet. **(E)** Ragfish. **(F)** Goby.

ber under 300 species, include flying fish, tuna, and mackerel. The majority of the 1,000 or so mesopelagic species are lantern fish, lancet fish, and the so-called light cyclothone fish.

Epipelagic species, such as the tuna or swordfish, tend to be large (more than 1 m), active, and carnivorous. These species are mostly tropical but range widely and may regularly move to temperate waters during the summer to take advantage of a more abundant food supply. Mesopelagic species, on the contrary, are generally small fish (around 15 cm long) that eat plankton. The lantern fish and cyclothone fishes are examples. The major movements of mesopelagic fish are their daily vertical migrations.

In general, the distribution of marine fish is governed by the availability of food. Epipelagic species are mostly associated with the primary productivity of upwelling or the seasonal productivity of phytoplankton in temperate waters. Typically, the greatest diversity of fish species is found in the shallow, more productive waters of the continental shelf, particularly in upwelling areas. Coastal upwelling is of particular importance to the herringlike fish, which rank first among marine fishes in abundance and biomass.

SWIMMING

As a group, both sharks and fish are good swimmers. Their shape is well adapted to moving through the water efficiently and effectively, and fish are among the fastest-swimming animals of the sea. Some tuna, which cruise at 9 km/hr, can swim for short bursts of speeds up to 70 km/hr (42 mph). Most fish have more moderate swimming speeds (Tab. 6.1).

The major problem with swimming effectively is overcoming the density of water. As a fish (or any other object) moves through the water, it tends to create turbulence. This turbulence causes frictional drag and increases the amount of energy needed to move through the water. The degree of turbulence is affected by the shape of the body and the increasing rate of speed.

The most streamlined body shape for movement through water (or air) is a fusiform shape (Fig. 6.9). Fusiform shapes are circular in cross section, and the point of maximum thickness occurs one-third of the way back from the anterior end. Fish and sharks (as well as many marine mammals) that are active swimmers tend to have a fusiform shape. The major departure from the theoretical fusiform shape in actively swimming fish is the elliptical cross section of the body caused by the shape of the lateral muscles, which provide the powerful thrust for swimming.

The basic swimming movement of fish is initiated by the contraction of lateral muscles, which creates a series of waves that pass down the length of the body, increasing in amplitude towards the posterior end (Fig. 6.10). These waves pass down the body at a speed that is faster than the fish is moving. The actual thrust that pushes the fish forward is caused by the back-and-forth movement of the body. In most species of fish, little or no thrust comes from the tail. Instead, the tail, or **caudal fin,** reduces side-to-side oscillations during swimming and controls direction. Eels, which have no caudal fins, are effective swimmers but have a high degree of sinuosity in the path they follow because they lack this fin.

There are two notable modifications of this basic pattern of swimming. One is a series of morphological and physiological adaptations that allow sustained high-speed swimming, as in tunas, mackerel, and some sharks. The other is the use of the pectoral, pelvic, and dorsal fins for swimming in slow-moving fish such as the sea horse.

TABLE 6.1

Swimming Speeds of Fish

Most fish swim at speeds less than 5 mph. The fastest of the fish is the barracuda, which has been measured at over 27 mph. Values for representative marine mammals are included for comparison. (2.5 cm = 1 in; 1 km/h = 0.6 mph)

Common Name	Scientific Name	Length (cm)	Speed (km/h)
Sea lamprey	*Petromyzon marinus*	37.5	4.3
Lemon shark	*Negaprion brevirostris*	184.2	8.7
Great barracuda	*Sphyraena barracuda*	129.5	44.4
Blenny	*Zoarces viviparus*	6.4	0.8
Eel	*Anguilla vulgaris*	60.0	4.2
Flounder	*Pleuronectes flesus*	27.5	3.9
Lemon sole	*Pleuronectes microcephalus*	8.0	0.4
Plaice	*Pleuronectes platessa*	8.0	1.1
Stickleback	*Gasterosteus spinachia*	10.0	2.6
Cod	*Gadus callarius*	56.0	7.7
Haddock	*Gadus aeglefinus*	42.0	6.6
Whiting	*Gadus merlangus*	20.0	5.8
Herring	*Clupea harengus*	25.0	6.3
Mackerel	*Scomber scombrus*	38.0	10.9
Brown trout	*Salmo trutta*	24.0	8.5
Sea trout	*Salmo trutta*	38.0	11.8
Goldfish	*Carassius auratus*	12.5	5.8
Rainbow trout	*Salmo irideus*	20.0	6.1
Carp	*Cyprinus carpio*	13.5	6.1
Perch	*Perca fluviatilis*	24.0	4.7
Common dolphin	*Delphinus delphis*	200	36.7
Fin whale	*Balaenoptera physalus*	1,300	25.9
Blue whale	*Balaenoptera musculus*	3,000	37.0

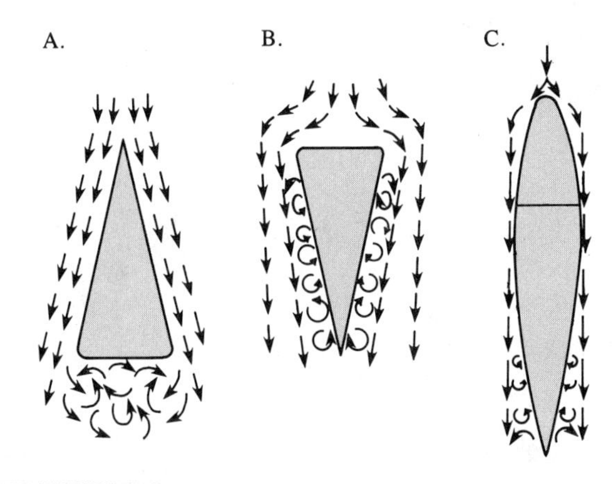

FIGURE 6.9
The Fusiform Shape

The fusiform shape has less drag than a cone in a stream of water. Maximum drag results when the point of the cone is in the stream **(A)**. Although drag is reduced when the blunt end of the cone is in the stream **(B)**, it still is greater than that of the fusiform shape **(C)**.

The tuna and tunalike fish have a number of adaptations that make them very fast and effective swimmers. First, unlike most fish, the primary thrust for swimming comes from the tail fin. Tendons connect the tail to the massive lateral musculature. The relatively small **caudal peduncle,** the region where the tail fin joins the body, is fitted with keels. The tendons pass over these keels, increasing the mechanical advantage of the muscle pulling on the tail. The body shape of tunalike fish, which includes a nearly circular cross-sectional shape, is more fusiform than any other fish.

These fast-swimming fish have a number of adaptations to help decrease friction in water. The pectoral, pelvic, and dorsal fins have depressions into which they can be retracted during times of rapid acceleration. Even the eyes have fairings to reduce friction. There also is a series of finlets on the caudal peduncle that smooth the flow of water over this area.

As a rule, fish and sharks are cold blooded. That is, their internal body temperature is the same as that of the surrounding water. Tunas (and the great white shark) are an exception. They can maintain tempera-

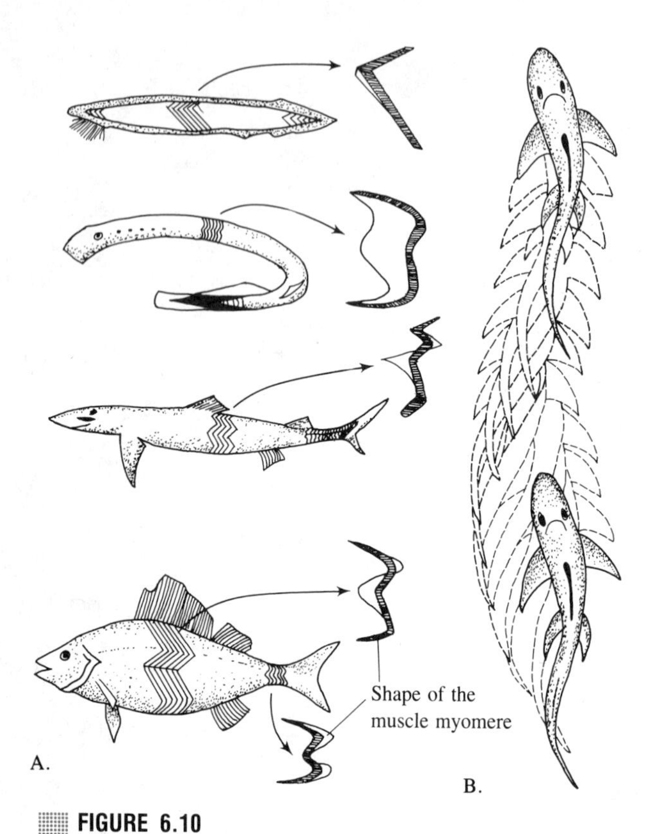

FIGURE 6.10
Swimming in Fish

(A) Swimming in fish is generally initiated by contraction of bands of lateral muscles called myomeres. **(B)** Sequential myomere contraction results in a sinusoidal movement.

tures inside the body that are warmer than the surrounding water. Warmer temperatures permit higher metabolism which, in turn, allows for more rapid swimming. In general, a decrease in water temperature in fish reduces metabolism and slows swimming. The active muscular contractions that provide the energy for swimming produce large quantities of heat. In most fish, when the blood flows through the gills, the heat is quickly lost to the water flowing over the gill surfaces. In tuna, a countercurrent mechanism consisting of a network of blood vessels, the **rete mirabile,** conserves the heat in the internal body mass. The rete mirable is a bundle of fine blood vessels with the blood in adjacent vessels going in different directions. Blood from the warm visceral mass passes through a blood vessel next to one carrying cold blood from gills. The cold blood entering is warmed by the blood leaving. In species that live continuously in warm tropical waters, the muscles that produce the heat—the **red muscle mass**—are located on the lateral surface of the fish. This heat exchanger may be used to dissipate heat to help cool the body. In those species that make seasonal trips to cooler temperate waters, the red muscles used for cruising are deeper in the body, and heat retention is more efficient. As well as conserving heat in the muscles used primarily for cruising, this mass provides warm blood to the viscera of the abdominal cavity and keeps the digestive system warm.

One tuna species is particularly good at regulating internal body temperature over a wide range of water temperatures (Fig. 6.11). The great Atlantic bluefin tuna ranges from the tropics to the edge of the Arctic. These tuna move quickly from one location to another (up to 130 km per day) and normally pass through large changes in water temperature. In waters of 7°C (45°F), the internal body temperature may be some 20° C, and over a range of water temperatures from 5–25°C the internal body temperature of the bluefin tuna can be predicted by the formula shown in Figure 6.12.

These remarkable morphological and physiological adaptations that permit sustained, rapid swimming in tunalike fish are also found in oceanic sharks, particularly in mackerel shark and the great white shark. Both

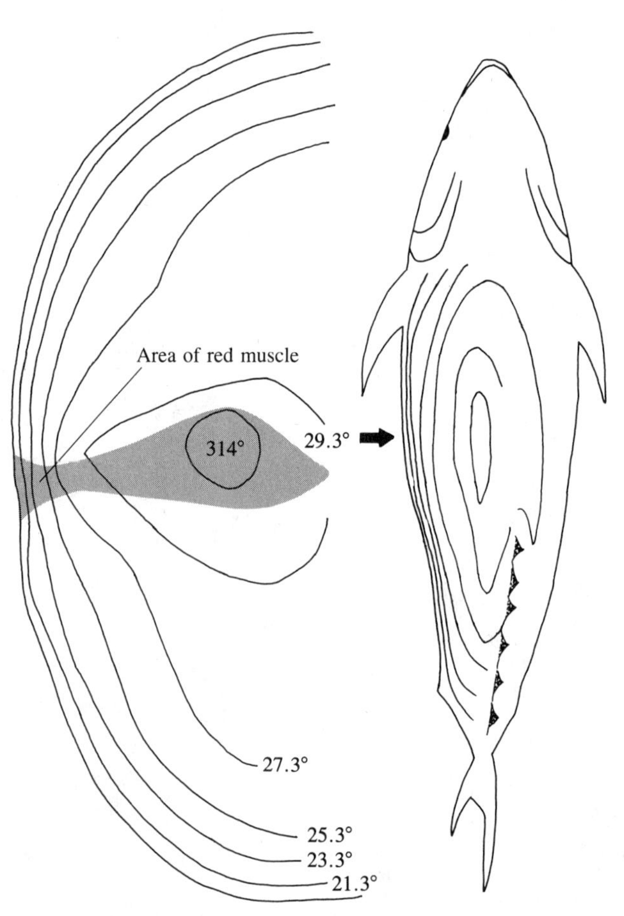

FIGURE 6.11
Heat Retention in Tuna

Temperature distribution in the body of a bluefin tuna. Heat produced by swimming raises the body temperature.

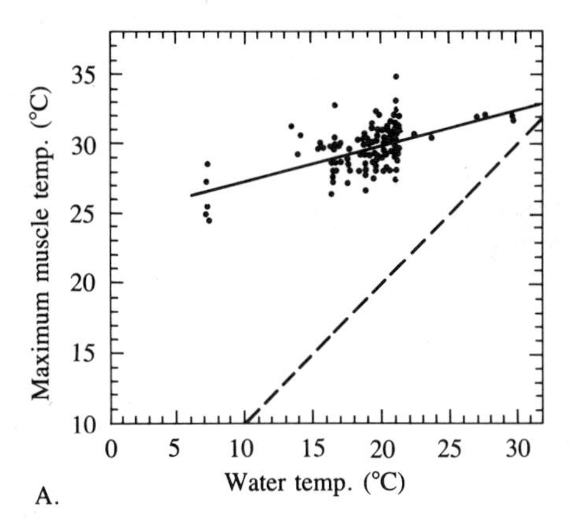

B. $T_m = 0.25T_w + 25°C$

FIGURE 6.12

(A) In the bluefin tuna, muscle temperature varies only 5°C over a 20°C range of water temperatures. **(B)** The relationship of muscle temperature (T_m) to water temperature (T_w) can be expressed by a formula. The dotted line indicates where body temperature equals water temperature. (5°C = 41°F)

are members of the lamnid sharks, a group found in the warm pelagic environment. These sharks have a fusiform shape that is nearly circular in cross section, as well as a narrow caudal peduncle and stiff crescent-shaped tail that gives them thrust in swimming. The physiological mechanisms for maintaining a temperature differential are also present in the sharks. Countercurrent heat exchangers result in the lateral musculature of sharks being up to 7°C higher than the water temperature, and warmed blood is also directed to the intestines.

Fins are important to the swimming of both bony fish and sharks. Sharks have a **heterocercal caudal** fin; that is, the upper part is larger than the lower (Fig. 6.13). The result is that tail movements produce a lift as well as a forward thrust. The pectoral fins of sharks also are used primarily for lift. The combined lift of the pectoral fins and tail counteract the force of gravity. Sharks are negatively buoyant, and pelagic species must continually swim to obtain the lift needed to stay in surface waters. The dorsal and anal fins in sharks are used to control direction. In general, the fins of sharks are not as functional as those of true fish.

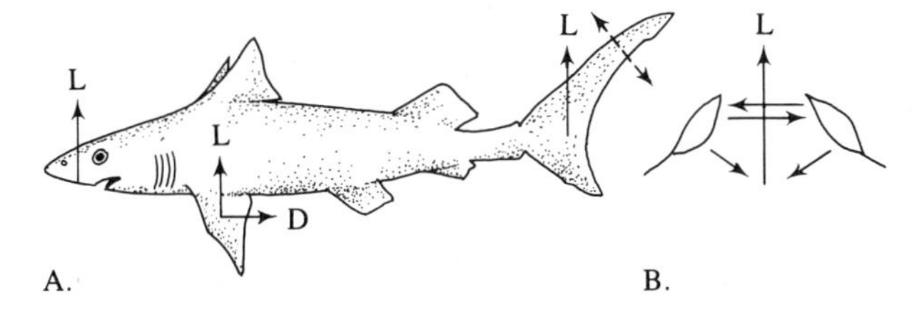

FIGURE 6.13
Heterocercal Caudal Fin

Instead of a swim bladder, sharks have other features that help to maintain neutral buoyancy as they swim. **(A)** Lift (L) is provided by the tail, pectoral fins, and snout. **(B)** Tail lift is produced by a side-to-side movement.

The fins of bony fish are more flexible and play a more active role in swimming (Fig. 6.14). First, because most fish are naturally buoyant (the function of the swim bladder), the fins do not have to provide lift, as they do in sharks. The pectoral fins function in a number of different ways. They can be used for controlling direction in sharp turns, and in some species they are used as paddles. Pectoral fins also can be used for braking and to maintain a constant position against the pumping action of water flowing over the gills. The dorsal and anal fins, located on the centerline of the body, are used mostly to control direction, but in some species also may be used for swimming.

One of the most interesting adaptations of the pectoral and pelvic fins is found in flying fish. When chased by a predator, the flying fish can leave the water and glide through the air for a considerable distance (Fig. 6.15). Flying fish are 15–45 cm (5.9–17.7 in) in length and have large pectoral fins that are over one-

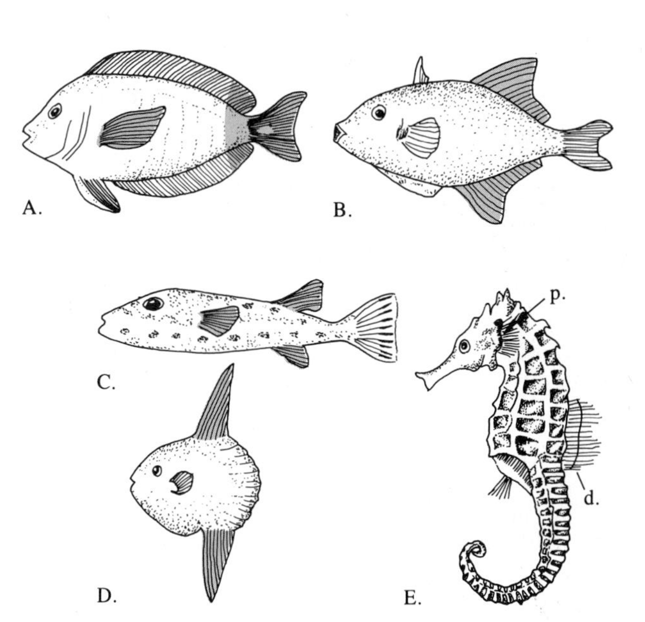

FIGURE 6.14
Use of Fins in Swimming

In many fish, propulsion is achieved by movement of fins (the fins shaded in gray are used in swimming). **(A)** Surgeon fish. **(B)** Triggerfish. **(C)** Puffer fish. **(D)** Sunfish. **(E)** Sea horse, which uses only the pectoral (p) and dorsal (d) fins (not to scale).

FIGURE 6.15
Flying Fish

The flying fish *Cypsilunus*.

half the length of the body. The pelvic fins are also relatively large, and the tail fin, which has an extended lower lobe, is also important in gliding. The fish glides in two phases. Swimming rapidly, they break the surface at a speed of 25–32 km/h (15–19 mph) and open the pectoral fins wide—but the fish is not yet airborne. The second phase involves rapid beating of the tail fin (only the lower lobe is submerged) to increase speed. As the pelvic fins open, the fish leaves the water and glides for one to ten seconds over a distance of 20–100 m (66–328 ft). Flying fish can glide in a straight or curved path. Although not the general pattern, a flying fish may choose not to re-enter the water at the end of a glide but instead will retract the pelvic fins and build up speed again by rapidly beating the tail. Up to five repeated glides have been observed. Flying fish are occasionally found on the decks of ocean-going boats.

Fins may be used for swimming. A number of bony fish (and skates and rays) use the fins as the major or sole means of movement. Skates and rays use primarily the large, lateral, winglike pectoral fins to swim. Many bony fish also use their pectoral fins for swimming. Coral reef species, such as butterfly fish, surgeon fish, parrot fish, and wrasses, often use the pectoral fins as paddles to swim. In other species, such as the puffer and porcupine fish, swimming is by movement of the dorsal and anal fins as well as the pectoral fins. Perhaps the most peculiar method of movement is that of the pipefish and sea horse. These fish live in algae, seagrass meadows, and coral reefs and do not travel large distances; mostly they hover. In sea horses (see Fig. 6.6), the body is rigid and they swim by using the relatively small dorsal and pectoral fins. Many other specialized swimming patterns that depend on the fins are known among the diverse bony fish.

BUOYANCY

One of the important differences between sharks and rays and true fish is the widespread occurrence of a swim bladder in fish, and its total absence in sharks and rays. Even though sharks and rays have relatively large livers (up to 25% of body weight) that contain high concentrations of buoyant lipids, their bodies are still heavier than water and will sink. Sharks and rays must swim or they will sink to the bottom. For oceanic sharks of the epipelagic zone, which travel almost constantly, negative buoyancy appears not to be a great disadvantage. In fact, tuna, which follow a similar life style, have reduced swim bladders or none at all. Constant swimming has a continual energy cost, however, and most species of fish have a swim bladder that gives them neutral buoyancy while both resting and swimming.

There are two types of swim bladders in fish: open and closed. Open swim bladders, which are found in the more primitive bony fish such as salmon, herring, and eels, have a connecting tube from the swim bladder to the esophagus (Fig. 6.16). Most fish, however, have a closed swim bladder, with no connection to the gut. In most marine fish the swim bladder takes up about 5 percent of the body volume, enough to provide neutral buoyancy.

Species with an open swim bladder fill it by swallowing air at the surface, but this apparently simple method has difficulties. Since water pressure increases by one atmosphere every 10 m in depth, a fish that fills its swim bladder at the surface would have it compressed by one-half at only a depth of 10 m. Since most marine fish regularly travel over a much greater depth range, the utility of an open swim bladder is limited.

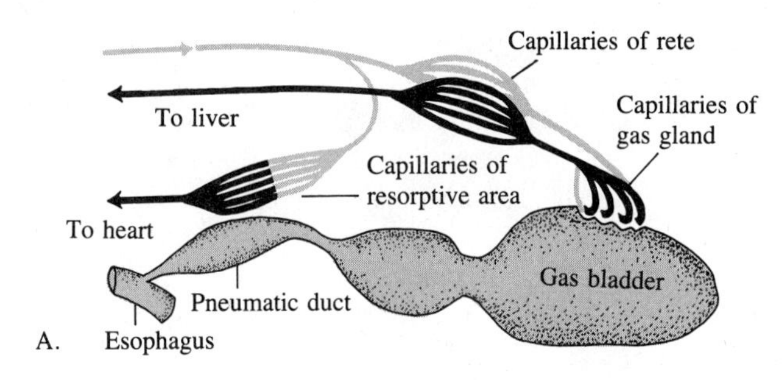

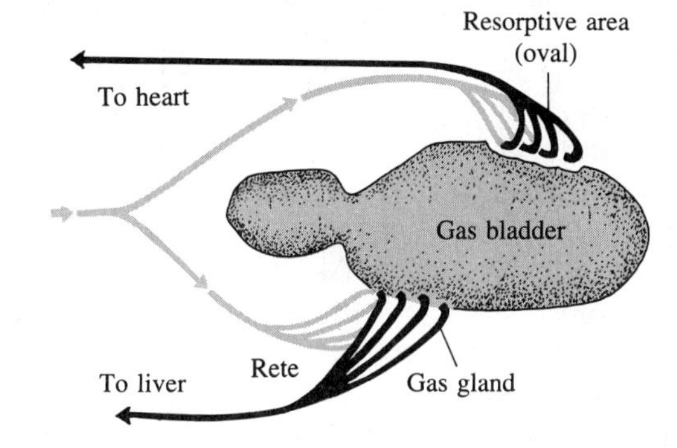

FIGURE 6.16
Swim Bladders

(A) Open swim bladder (connected to the esophagus). **(B)** Closed swim bladder (not connected to the gut). Both types may have gas glands to control the swim bladder volume. Secretion and absorption of gases is made possible by bundles of capillaries called rete.

The gas pressure and volume of closed swim bladders is regulated by a gas gland. The gas gland, made up of a bunch of capillaries, is part of the circulatory system (Fig. 6.17). Its structure is based on a countercurrent system, and gases are secreted and absorbed by a rete mirabile. The countercurrent system is similar to the mechanism tuna use to maintain high internal body temperature (see discussion above on swimming and body temperature). In many fish, the functions of absorption and secretion of gases are found in different parts of the swim bladder. Secretion and absorption by the gas gland is controlled by the nervous system. Nerves regulate the volume of the swim bladder rather than the pressure, and gas is absorbed and secreted to maintain a given volume.

Although the swim bladder is primarily an organ for buoyancy control, it may function in the reception and production of sound. The swim bladder is a more effective sound receiver than the bony otolith of the ear. Swim bladders can discriminate a wider range of frequencies, can better determine pitch, and are more sensitive to sound than the otolith. A majority of fish have some kind of connection between the inner ear and the swim bladder.

Swim bladders also may be used to produce sounds. Many fish, particularly those of coastal waters, have muscles that insert into the swim bladder. When the muscles are contracted, the swim bladder vibrates, producing hollow sounds that have been described as grunts, growls, thuds, and barks. In some species, such as croakers, sounds are made at feeding times; in others, sounds are produced during reproductive activities.

FOOD AND FEEDING

Virtually all marine fish are carnivorous. In some coastal upwelling areas, phytoplankton are an important part of the diet of herringlike fish (e.g., anchovy, sardines), and a few species are primarily scavengers (e.g., mullet, intertidal gobies, blennies). The large majority of fish, however, seek out other live animals. Food capture is by filtration or by active predation of individual prey. Since copepods are the most common and abundant animal of the sea, fish that use copepods as food are likewise abundant. The copious herring and

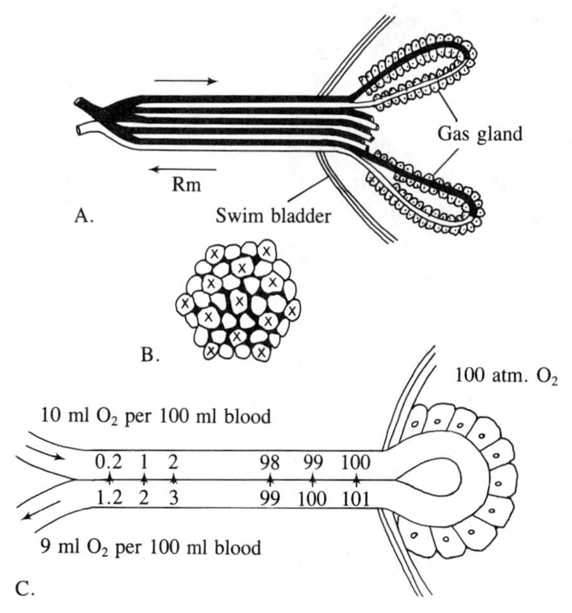

FIGURE 6.17
Rete Mirabile

(A) Rete mirabile (Rm) are formed by bundles of capillaries. Arteries are shown in black, veins in white. **(B)** Arrangement of arteries and veins in cross section. Venous capillaries are marked with an *x*. **(C)** Gradients of gas concentration in the capillaries show diffusion from high to low concentrations away from the swim bladder.

herringlike fish are the most successful plankton feeders. Their jaws open to form a large filtration basket. The **gill rakers** (forward extensions of the gills) form the filtration surfaces that retain the zooplankton as water passes through the mouth, over the gills, and out the opercular flaps. The number of copepods taken can be immense. Young herring (about 18 mm long) eat around 4,000 copepods a day and adults may eat up to 100,000 per day. The herring and related species, in turn, serve as a source of food for a number of larger predators including tuna, swordfish, salmon, haddock, and cod.

Other groups are also notable for their filter feeding. Some species of salmon in the North Pacific consume euphausiids by filter feeding. The basking shark and whale shark, the largest of the fishlike animals, are both filter feeders on zooplankton. Basking sharks can filter up to 200,000 tons of water per hour and the stomach can hold up to 150 gallons of copepods. Basking sharks filter feed at swimming speeds of between 1.4–1.8 km/hr (0.8–1.1 mph), and other filter feeders move at similar slow rates. Although filter feeders must spend more time feeding than active predators, the swimming speeds are slow and energy efficient. An additional advantage of filter feeding is that the young of a species can quickly adapt to this way of feeding.

Active predation is a more complicated method of feeding. Prey usually must be located and captured as they are trying to escape. Most fish use a burst of rapid movement to either capture or escape capture. These swimming bursts are of short duration (10–20 sec). Predator-prey movements are limited by visibility, which is much lower in water (compared to air) because of the rapid absorption of light by water. The maximum swimming speed of fish is approximately 36 km/hr (22 mph). This speed can be reached from cruising speed within 0.5 sec. The typical attack range is ten times the length of the fish. For larger fish, this distance may exceed the visibility of the water, which may be one reason that large predatory fish are found in clear oceanic water. Prey usually are one-fourth the length of the predator. This means that as an individual predator grows it will have different prey species.

SENSES

The senses in fish and sharks are used for other functions in addition to feeding. In general, these free-moving animals are endowed with a full range of excellent sensory receptors, two of which are unique: the lateral line in fish, and the ampullae of Lorenzini in sharks.

Most fish have a conspicuous canal and row of pores—the **lateral line**—extending down the lateral surface of the body, usually on or above the midline. In some species, such as greenlings, these lines may be multiple; in others, such as herring, the lateral line is reduced. The lateral line is an extension of a network of sensory pores on the head that are connected to the inner ear. In fish there is no middle or outer ear. While the ear functions in balance and sound reception, the lateral line apparently is sensitive to water displacement and pressure, and may be used to assist in the location of prey. It also assists schooling fish to maintain position. The lateral line canal is generally mucus-filled. The sensory structures, or **neuromast organs,** are deformed by changes in water pressure when the lateral line nerve is excited.

Sharks have a system of pores on the snout called the **ampullae of Lorenzini.** These sense organs detect small electrical charges. All animals in seawater produce a small electrical field, and sharks can detect this charge at short distances (Fig. 6.18). Sharks are capable of using a range of senses to locate prey. Sound is used for long-range detection (about 1,000 m), while smell is used up to 100 m (328 ft) distance. The lateral line is sensitive to a distance of 30 m, vision to 18 m, and the electrical sense of the ampullae of Lorenzini is used in the centimeter range. Taste is used on contact.

MIGRATIONS AND SCHOOLING

A **migration** is a movement between two places of predictable location at a predictable time. Although it often is difficult to separate foraging movements of oceanic species from migratory movements, many fish clearly have extensive migrations. The North Atlantic bluefin tuna spawns off southern Florida and may migrate to feeding grounds in Norway—some 12,200 m (7,300 miles) away—in only three months. Most migratory movements are related to reproduction and feeding, with spawning in one location and adult feeding in another. Migration distances may be as little as 25 km, but are generally greater. Two of the most famous fish migrations are the salmon of the North Pacific and the eels of the North Atlantic.

Salmon are anadromous; that is, they spawn in fresh water in the late summer and fall and spend the two to six years of their adult life at sea. The feeding grounds of the adult are up to 3,200 km (1,920 miles) from their spawning grounds. Salmon return to the river of their birth. Their migratory movements are so carefully timed that adult fish arrive at the river within days of the time their parents did in their cycles (Fig. 6.19).

North Atlantic eels are **catadromous.** They spend their adult years in fresh water and migrate to the ocean to spawn. The spawning area is in the Sargasso Sea at

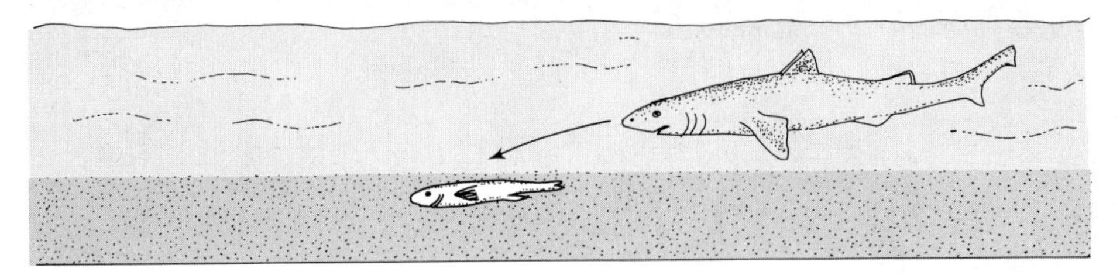

Flounder under the sand elicits the shark's accurate, directed attack.

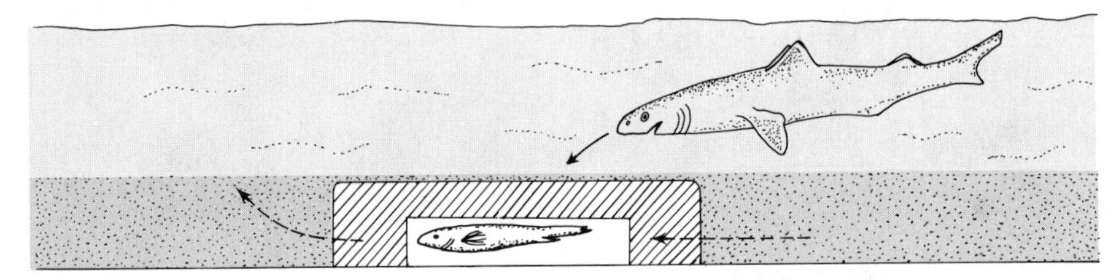

The shark also attacks accurately when the flounder is in an agar chamber, "transparent" to electric current, with a resistance about equal to that of the medium.

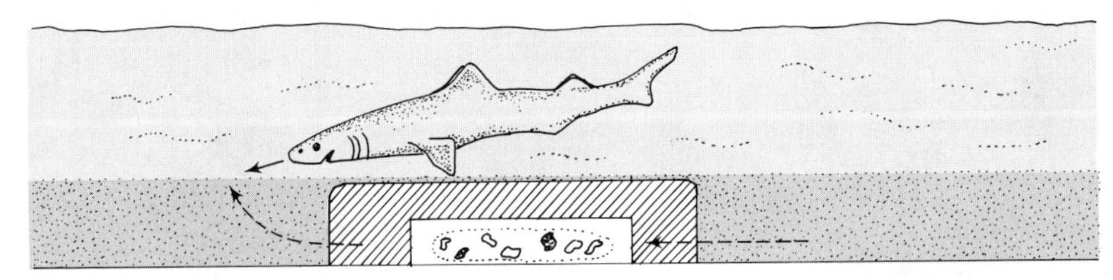

Only if the flounder has been chopped up and frozen for a long time, exaggerating the olfactory stimulus and fragmenting the electric field, will the shark diffusely search in the downstream area.

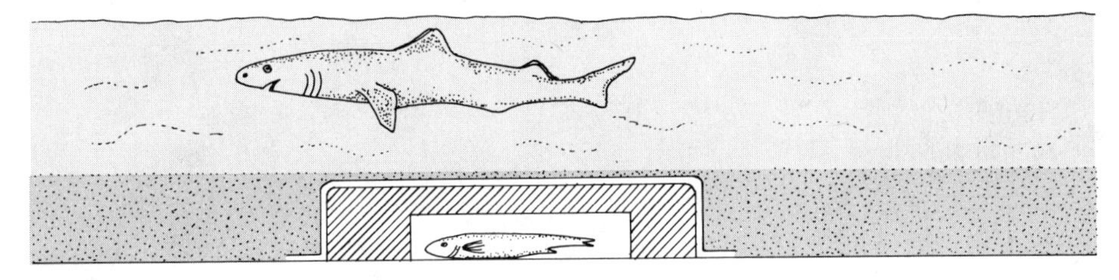

If the agar is covered by an insulating plastic film, the response is lost.

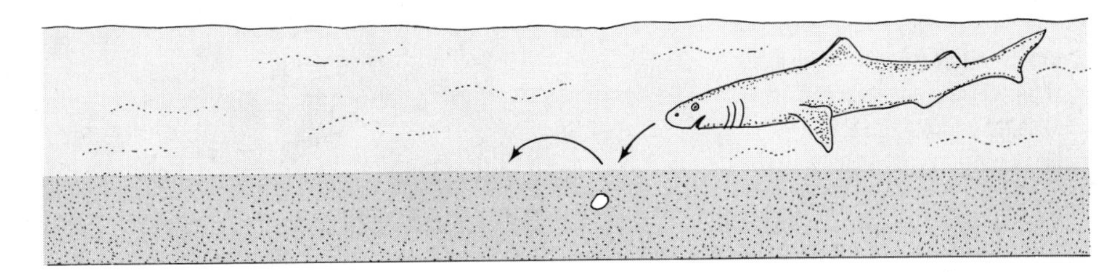

Electrodes producing an electric field like that of the flounder elicit the response.

FIGURE 6.18
Ampullae of Lorenzini

The ampullae of Lorenzini in sharks are sensitive to weak electrical currents. The function of the ampullae was demonstrated through the series of experiments shown here.

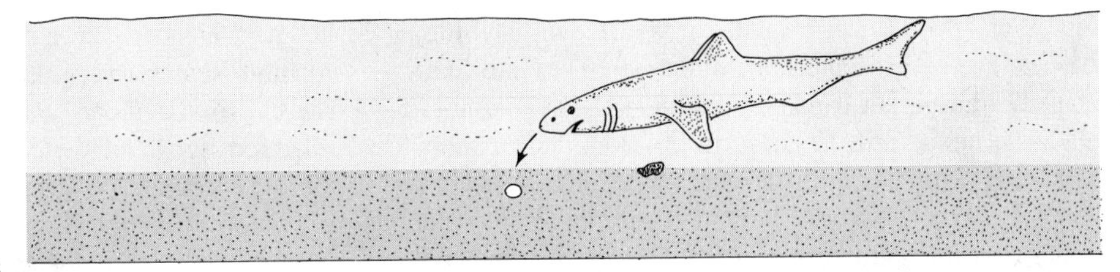

Even in the presence of a piece of food the response is still directed toward the electric field.

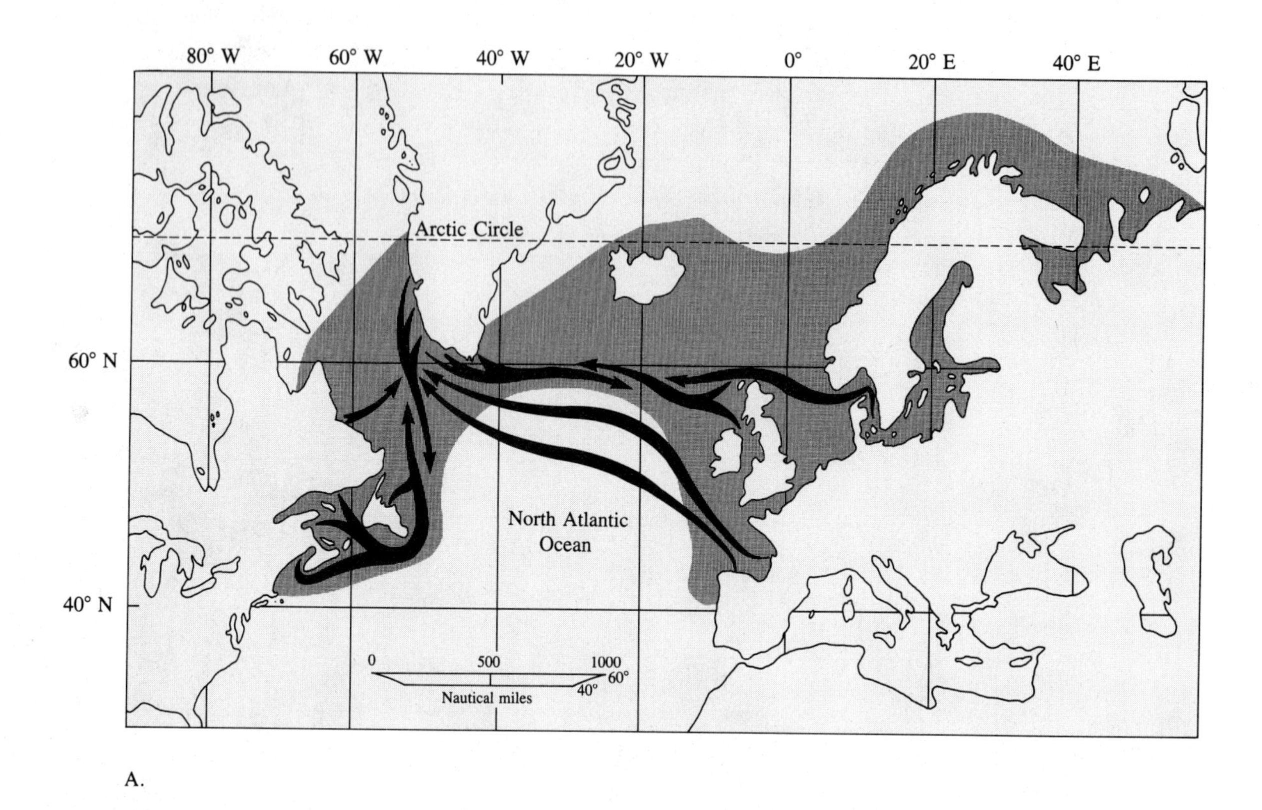

A.

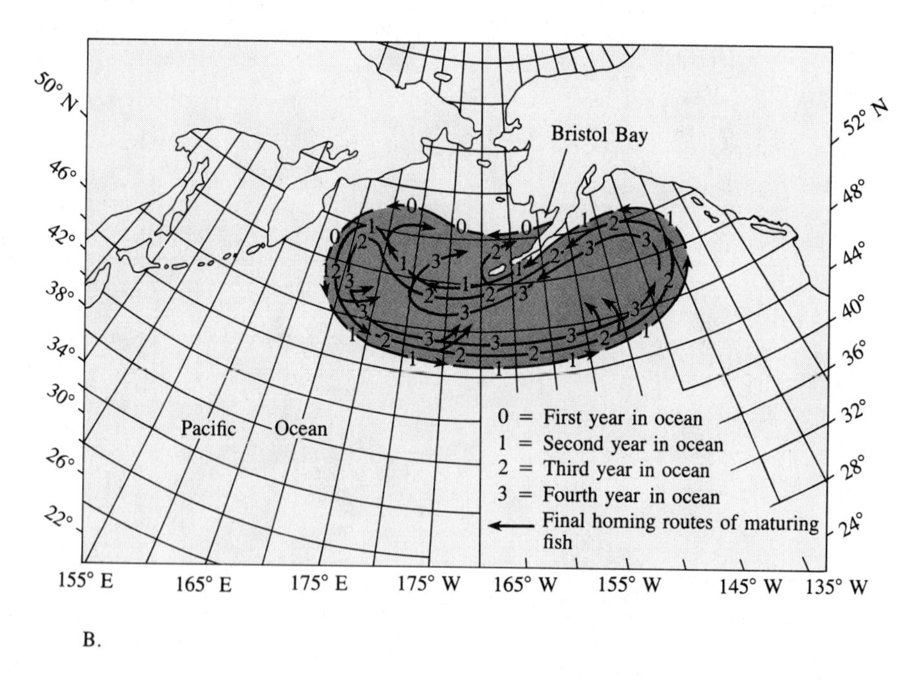

B.

FIGURE 6.19 Salmon Migrations

Salmon are anadromous, spawning in fresh water. Their adult life is spent in the high seas after a migration of thousands of miles. The life cycle of salmon varies from two to six years, depending on the particular species. Shown are the migration paths and general range of the Atlantic salmon **(A)** and the North Pacific sockeye or red salmon **(B)**. There are six species of salmon in the North Pacific. (After Netboy, A. 1974. *The salmon:Their fight for survival.* London:Andre Deutsch.)

depths of 400–700 m (1,300–2,300 ft). On hatching, the eels are called leptocephalus larvae. They are leaf-shaped, transparent, and morphologically quite unlike eels. The larvae drift in the surface waters of the Atlantic for up to three years. On reaching the coastal waters off America or Europe, they metamorphose into the adult eel shape and return to fresh water, where they remain for four to seven years (Fig. 6.20).

And how do fish navigate during their migrations? For high seas travel, salmon use the sun as a compass. There also is evidence that they can orient to a plane of polarized light and the earth's magnetic field. As they approach a river, they are guided by the odor of the water. By continually choosing their path by smell, the salmon are able to locate the exact tributary of their spawning.

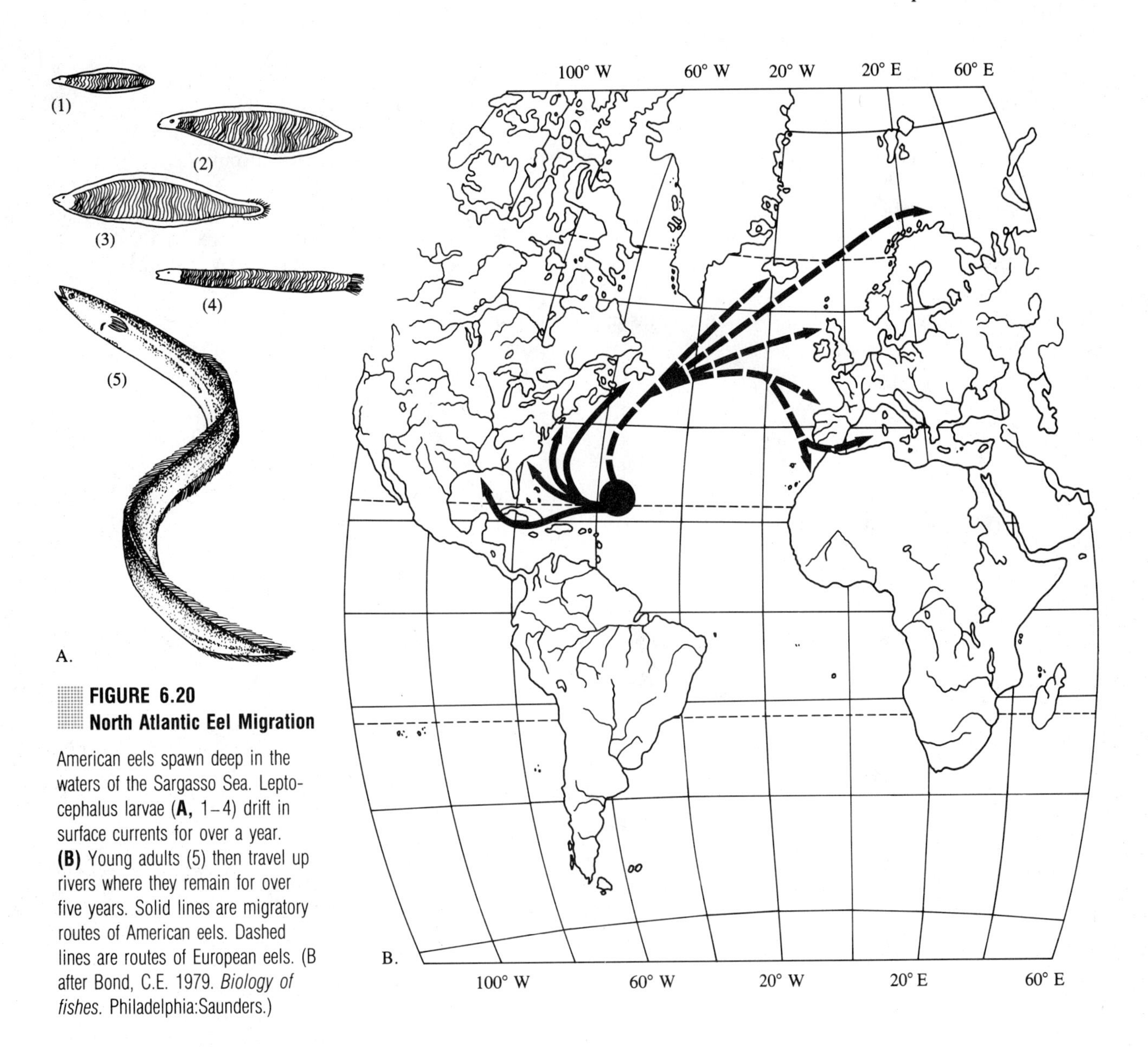

FIGURE 6.20
North Atlantic Eel Migration

American eels spawn deep in the waters of the Sargasso Sea. Leptocephalus larvae (**A,** 1–4) drift in surface currents for over a year. **(B)** Young adults (5) then travel up rivers where they remain for over five years. Solid lines are migratory routes of American eels. Dashed lines are routes of European eels. (B after Bond, C.E. 1979. *Biology of fishes.* Philadelphia:Saunders.)

Patchiness in the distribution of marine organisms is the general condition, and the distribution of most species of fish is no exception. At least 25 percent of all fish species, at some time in their life cycle, gather together in dense schools, or shoals. Most schooling fish group together all through their life cycle; some examples are herring, anchovies, sardines, mackerel, tuna, salmon, cods, and flying fish. Schools contain individuals of similar size, and the different stages of the life cycle of a species are segregated. There is apparently no leadership in fish schools. When attacked or threatened, schooling fish tend to pack more closely together. Schooling reduces the chance of being captured by a predator. Since most predators travel singly or in small groups, schooling fish are less likely to be found by roving predators. Moreover, once the school is detected, predators can only consume a small fraction of the school and then are full. In this way, the total mortality of the schooling species is reduced.

REPRODUCTION

Most marine fish lay eggs that are fertilized externally. Eggs are generally small (approximately 1 mm diameter), and fecundity is high (up to 1 million eggs per female). Eggs are laid at midwater depths and develop into a planktonic larval form that is transparent and must feed soon after hatching. Many fish larvae begin feeding on copepod larvae (the nauplius stage).

As might be expected, there are many exceptions to this general pattern. Some species, such as herring, lay eggs on the bottom. Some, such as shiner perch, have internal fertilization, and the young develop internally, too.

Sharks and rays typically exhibit more complex reproductive behavior than do fish. Males have copulatory appendages called claspers, which are a modification of the pelvic fins. As a rule, fertilization is internal. Fertilized eggs may be laid in an egg case (as in skates), or the young may complete development in the oviduct using only the food energy of the yolk (as in dogfish sharks and certain other sharks).

SUMMARY

The term *fish* is commonly used to describe animals that are quite distinct in their evolutionary history. There are two main groups of fish: the cartilaginous sharks and the bony fish. The evolutionary history of these groups of fish is unclear. Nevertheless, there is considerable evidence that fish evolved in fresh water some 400 million years ago and only secondarily invaded the marine environment.

Fish are the most common of the vertebrate classes, with over 11,000 marine species. In general, they are the smallest of the vertebrates, with an average species length of 15 cm (6 in). Sharks are actually a subgroup of the fish class. Although there are fewer species of sharks (about 250 species), they are on the average larger than fish, with an average length of 2 m.

QUESTIONS AND EXERCISES

1. How do fish and sharks swim? Are there any differences between the two groups?
2. Discuss the relationship between fish size and the type of food consumed.
3. Discuss the role of fins in the swimming of sharks and fish.
4. Discuss the general patterns of distribution of fish in the marine environment. Consider latitude, water depth, and proximity to land.
5. How do fish maintain buoyancy in the water column? How do sharks differ?
6. Discuss the advantages of schooling behavior in fish.
7. How can some fish and sharks maintain an internal body temperature higher than that of water?
8. What is the function of the lateral line in fish? Is there any relation of the lateral line to the ampullae of Lorenzini of sharks? If so, describe this relationship.

REFERENCES

Bond, C.E. 1979. *Biology of fishes.* Philadelphia:Saunders.

Budker, D. 1971. *The life of sharks.* New York:Columbia University Press.

Crane, J.M. 1973. *Introduction to marine biology:A laboratory text.* Columbus, OH:Merrill.

Gold, J.P. 1989. *Sharks in question: The Smithsonian answer book.* Washington, DC:Smithsonian Institute Press.

Joseph, J., Klawe, W., and **Murphy, P.** 1988. *Tuna and billfish:Fish without a country.* La Jolla, CA:Inter-America Tropical Tuna Commission.

Marshall, N.B. 1970. *The life of fishes.* New York:Universe Books.

Marshall, N.B. 1971. *Explorations in the life of fishes.* Cambridge, MA:Harvard University Press.

May, R.M., et al. 1979. Management of multispecies fisheries. *Science* 205(4403):267–277.

Netboy, A. 1974. *The salmon: Their fight for survival.* London:Andre Deutsch.

Pitcher, T.J. 1986. *Behavior of teleost fishes.* Baltimore, MD:Johns Hopkins University Press.

Policansky, D. 1982. The asymmetry of flounders. *Sci. Amer.* 246(5):116–123.

Stafford-Deitsch, J. 1988. *Shark, a photographer's story.* San Francisco:Sierra Club Books.

Stevens, J. 1987. *Sharks.* New York:Facts on File.

SUGGESTED READING

OCEANS

Baldridge, H.D. 1982. Sharks don't swim—they fly. 15(2):24–29.

Branson, B.A. 1976. Sockeye salmon. 9(4):18.

Brash, T. 1976. Stonefish. 9(6):22–24.

Compton, G. 1977. Manta rays and sting rays. 10(1):4–9.

Compton, G. 1982. Dangerous marine animals. 15(2):4–36.

Golanty, E. 1977. Fish survival in subfreezing temperatures. 10(5):52–53.

Gordon, D.G. 1986. The spiny dogfish. 19(1):45–52.

Johnson, P. 1982. Salmon ranching. 15(1):38–54.

McCullough, B. 1976. Beware of pufferfish. 9(6):20–21.

Sharks (special issue). 1977. 10(6):12–47.

Slayter, E.M. 1981. Unraveling the eel. (Atlantic eel migration) 14(1):26–31.

Straus, K. 1981. Grunion run. 14(6):25–27.

Thompson, P.C. 1985. The uncommon mola (ocean sunfish). 18(4):43–52.

OCEANUS

Kalmijin, A.J. 1977. The electric and magnetic sense of sharks, skates and rays. 20(3):45–52.

SCIENCE 81

McCosker, J.E. 1981. Great white shark. 2(6):42–53.

SCIENTIFIC AMERICAN

Eastman, J.T., and **DeVries, A.L.** 1986. Antarctic fishes. 255(5):106–113.

Horn, M.H., and **Gibson, R.N.** 1988. Intertidal fishes. 258(1):64–71.

Webb, P.W. 1984. Form and function in fish swimming. 251(1):72–83.

SEA FRONTIERS

Arrigon, W. 1989. Seahorses. 35(6):342–347.

Martin, R. 1989. Why the hammerhead. (Sharks) 35(3):142–145.

Rodgers, D.L., and **Muench, C.** 1986. Ciguatera, scourge of seafood lovers. 23(5):338–346.

Siezen, R.J. 1986. Antifreeze in polar fishes. 32(5):360–366.

Winkler, M., Sitko, S.E., and **Sund, P.N.** 1983. Tunas—nomads of the sea. 29(1):51–56.

FLATFISHES—WHY ARE SOME LEFT-EYED AND OTHERS RIGHT-EYED?

One of the most typical characteristics of vertebrates is their bilateral symmetry: The left side of their body is a pretty good mirror image of their right side. Flatfishes, however, are a striking exception to this rule. Although they hatch as typical fish small fry, with a body not unlike that of other fishes, this condition soon changes. First, one eye migrates to the other side of the head, so that both eyes are on the same side. Next, they begin to lie on their side on the sea bottom—the side with both eyes, naturally, in the upper position. Too, the bottom side becomes pale or white, while the upper surface develops the ability to change color to match the sediment in which the flatfishes partially bury themselves. From their camouflaged positions on the ocean floor, they lunge upward to capture unsuspecting prey that swim into their range.

It is common to refer to those flatfish that lie with the right side up as belonging to the flounder tribe, and those that lie with the left side up as belonging to the turbot tribe. It is not that simple. All species that are considered to be members of the flounder tribe will have at least a small percentage of individuals that have their eyes on the left side. The same can be said for the turbots. For example, the starry flounder, *Platichthys stellatus,* found in Pacific coastal waters from California to

Figure 1

This member of the family Pleuronectidae was found in the Northeast Pacific Ocean. (Photo by Charles Seaborn.)

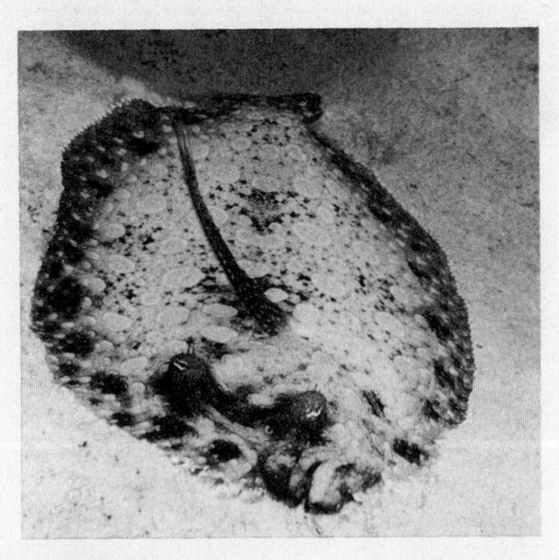

FIGURE 2
Peacock Flounder

(Photo by Marty Snyderman.)

Japan, is closely related to many right-eyed species. However, it is only about 50 percent right-eyed off the coast of California. In Alaskan waters, about 30 percent are right-eyed, and off Japan less than 1 percent are right-eyed!

This phenomenon is very interesting to biologists because such a pattern of change is usually the result of natural selection achieved by a changing environment. Yet it is certainly difficult to visualize how such a condition as having the eyes on the right or left side of the head could be a response to environmental factors. Research (Policansky, 1982) shows the passing of this trait from parents to offspring is clearly genetic in the starry flounder. If this is true, is there some survival benefit that would lead to adaptive selection of right- or left-eyedness? So far, none has been demonstrated. So we are left with no answer as to whether the environment is in any way responsible for the existence of left- or right-eyed genes in flatfishes. This is another of the multitude of questions about life in the oceans that, for an answer, we must wait a bit longer.

7

REPTILES AND BIRDS

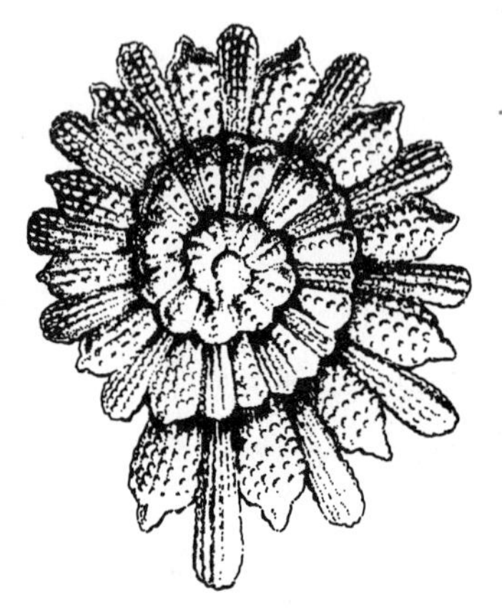

Both marine birds and reptiles are amphibious in that (a) they must get their oxygen from the air rather than the water, and (b) all birds and most reptiles must come onto land in order to reproduce. Marine reptiles are restricted in their distribution to tropical and warmer temperate waters. Birds, however, range widely over the world's oceans and are common in most areas.

Sea turtles and sea snakes are widely distributed in tropical marine waters. Turtles are amphibious in that they must return to land to breed, whereas snakes can give birth to their young at sea. Both turtles and snakes are most common in tropical shallow waters close to shore. The largest turtle (the leatherback) ranges up to 2 m (6.6 ft) in length and 550 kg (1,210 lbs) in weight. Sea snakes are approximately 1 m long.

Sea birds show their amphibious nature in having to reproduce on land. Even so, birds are found in all parts of the ocean, and many species spend their nonreproductive periods far from land. Birds range in size from the stormy petrel, approximately 15 cm (6 in) in length, to the emperor penguin, which is 1 m long and weighs up to 45 kg (100 lbs).

REPTILES

There are approximately 50 different marine snakes (Fig. 7.1). All are highly venomous (they are close relatives of cobras), and most are found in coastal tropical waters of the Indian and Pacific Oceans. Their length

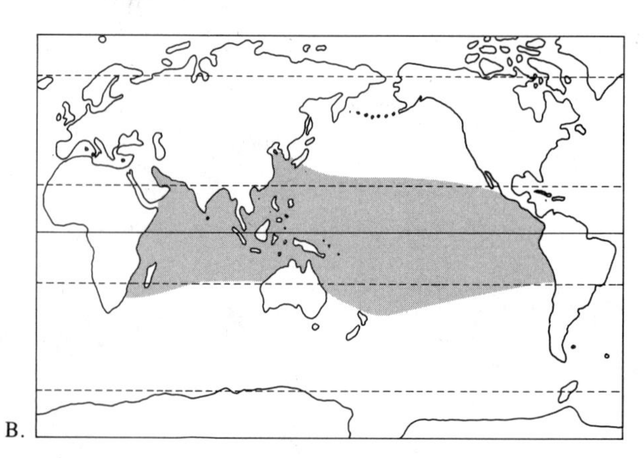

FIGURE 7.1
Sea Snake

(A) Yellow-bellied sea snake. **(B)** Distribution of sea snakes. (After Webb, J.E., et al. 1978. *Guide to living reptiles.* London:Macmillan.)

generally is around 1 m, but specimens exceeding 3 m have been taken. Sea snakes are colorful—dark above and light below, with cross bands of black, purple, brown, gray, green, or yellow. Most sea snakes are **viviparous;** that is, they bear their young live at sea. Sea snakes are active predators and feed mainly on small fish.

Five species of turtles live in the oceans: the hawksbill, the green, loggerhead, the Ridley (Kemp's and Olive), and the leatherback (Figs. 7.2, 7.3). Species are distinguished one from another by morphological features of the head and shell. The heads of the five species differ in the number of scales between the eyes (the prefrontal scales), and the shape of the beak. There are species differences between the shells of marine turtles. The dorsal shell (the carapace) of the leatherback turtle differs from all the rest in having a leathery back without scales. The other four species differ in the number and shape of scales on the carapace.

All but one species are circumglobal in the tropical regions of the world's oceans. Most sea turtles live in the shallow coastal waters of the continental shelves. Only one—the leatherback turtle—travels extensively in the oceanic environment. Sea turtles are generally much larger than their terrestrial counterparts. The largest—the leatherback turtle—may weigh up to 550 kg

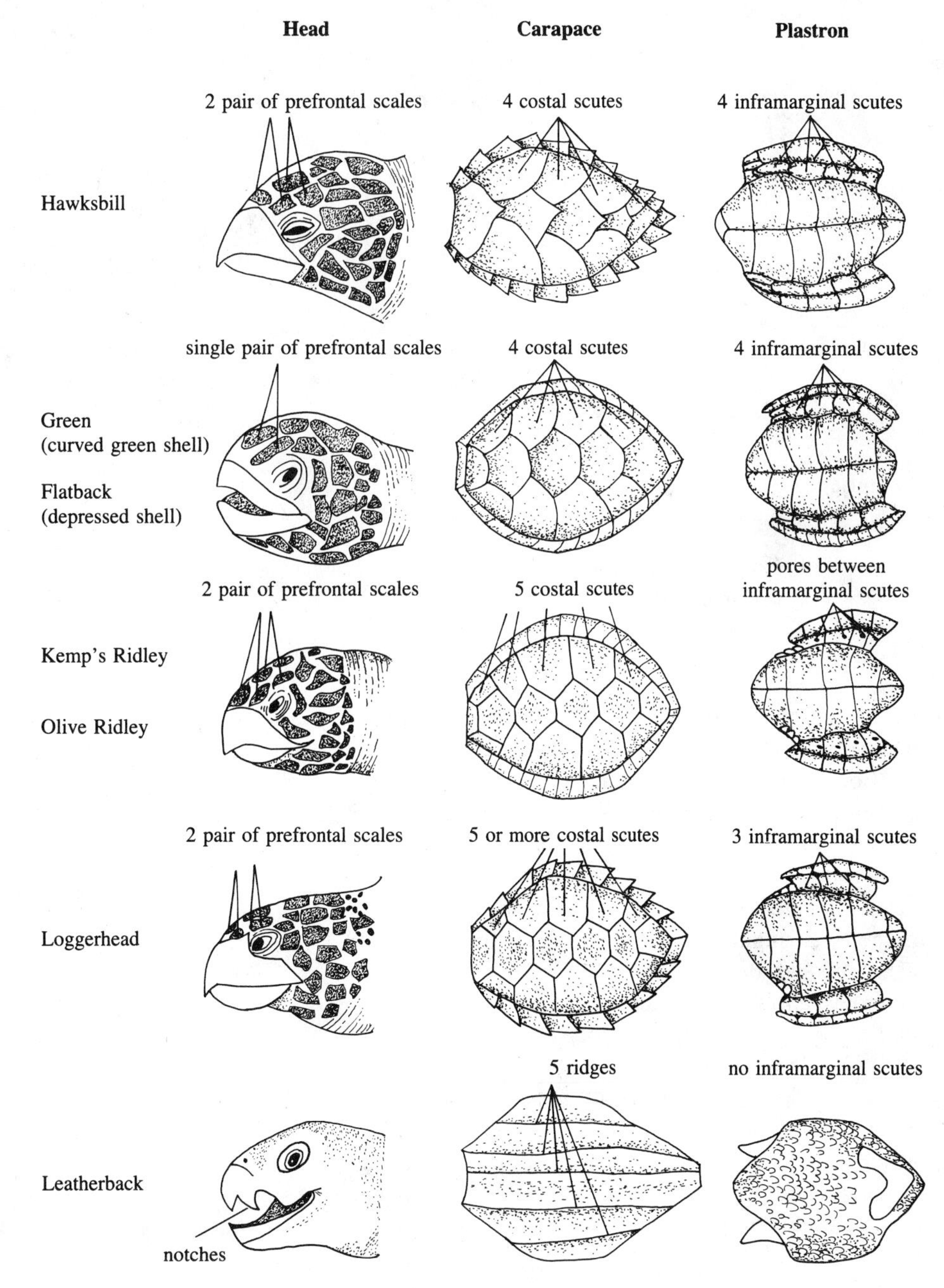

FIGURE 7.2 Classification of Sea Turtles

Sea turtles differ in structure of the head, dorsal surface (carapace), and ventral surface (plastron).

FIGURE 7.3
Major Species of Sea Turtles

(A) Leatherback *(Dermochelys).* **(B)** Loggerhead *(Caretta).* **(C)** Hawksbill *(Eretmochelys).* **(D)** Green *(Chelonia).* (Not to scale.)

(1,210 lbs), and other sea turtles can weigh at least 90 kg at maturity.

Compared to other vertebrates, the metabolism of sea turtles is relatively low. Internal body temperature is similar to that of the surrounding water, and the relatively inefficient reptilian circulatory system slows the metabolic rate. Turtles are capable of only short bursts of speed, as might be needed to escape a predator's attack. Although most of the body is protected by the armorlike shell, the head and flippers cannot be fully retracted into the shell. Sea turtles live to be 15–30 years old.

Sea turtles have special salt glands that assist the kidneys in ridding the body of excess salt taken when turtles swallow seawater. In sea birds, the salt gland is found around the eye, and opens into the nasal passage. In sea turtles, the salt gland opens into the tear duct of the eye. The salt gland excretes a salty brine that the turtle eliminates as tears.

Sea turtles are effective swimmers. The forelegs are modified into sturdy flippers that work like paddles propelling the turtle through the water. Although sea turtles are negatively buoyant when dead, they have no trouble maintaining neutral buoyancy when alive. Some species lie on the surface for hours directly absorbing the heat of the sun, staying afloat by means of the air in their lungs.

All sea turtles lay eggs in the sand above the high-tide line. Sea turtles are awkward out of water. The flippers that serve so well for swimming cannot lift the body high enough so they can walk. Instead, they use the fore and hind flippers to drag themselves across the sand to a suitable nesting site. The eggs are laid in a large nest that they dig in the sand (Fig. 7.4). The nest

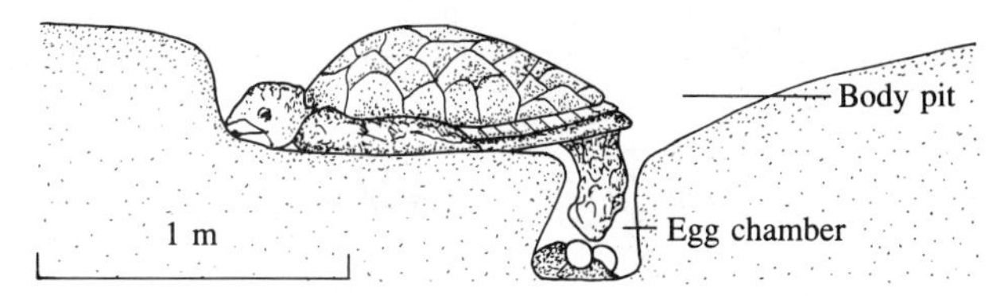

FIGURE 7.4
Sea Turtle Laying Eggs

All sea turtles lay their eggs on land. Shown is a green turtle with an almost completed egg chamber.

is in two parts: a body pit and an egg chamber. The body pit is dug with the front flippers and has a depth about equal to the thickness of the body. Once the body pit is dug, the hind flippers are used to carefully excavate the egg chamber in the bottom of the body pit. About 100 eggs are deposited in the egg chamber in a period of 20 minutes to 3 hours. The general locations of major nesting sites are shown in Figure 7.5.

The food of sea turtles is varied. The green turtle feeds mostly on the leaves of sea grasses, although it also eats jellyfish and sponges when available. Other sea turtles are omnivorous, favoring animals when available. The loggerhead and hawksbill eat a wide range of epibenthic animals, including crabs, shellfish, and tunicates. The largest of the sea turtles—the leatherback—feeds mostly on jellyfish. This food preference makes the leatherback susceptible to death by swallowing plastic bags that can get stuck on their throat spines. This species is the most oceanic and regularly ranges into temperate waters.

Sea turtles are eaten by sharks and killer whales. These predators, however, are not nearly as devastating as overexploitation by humans. Turtle eggs are coveted as food by many people around the world, and the green hawksbill and leatherback are also sought after for their meat. Most species of sea turtles have seriously depleted numbers and are endangered, although conservation programs have been established at many locations.

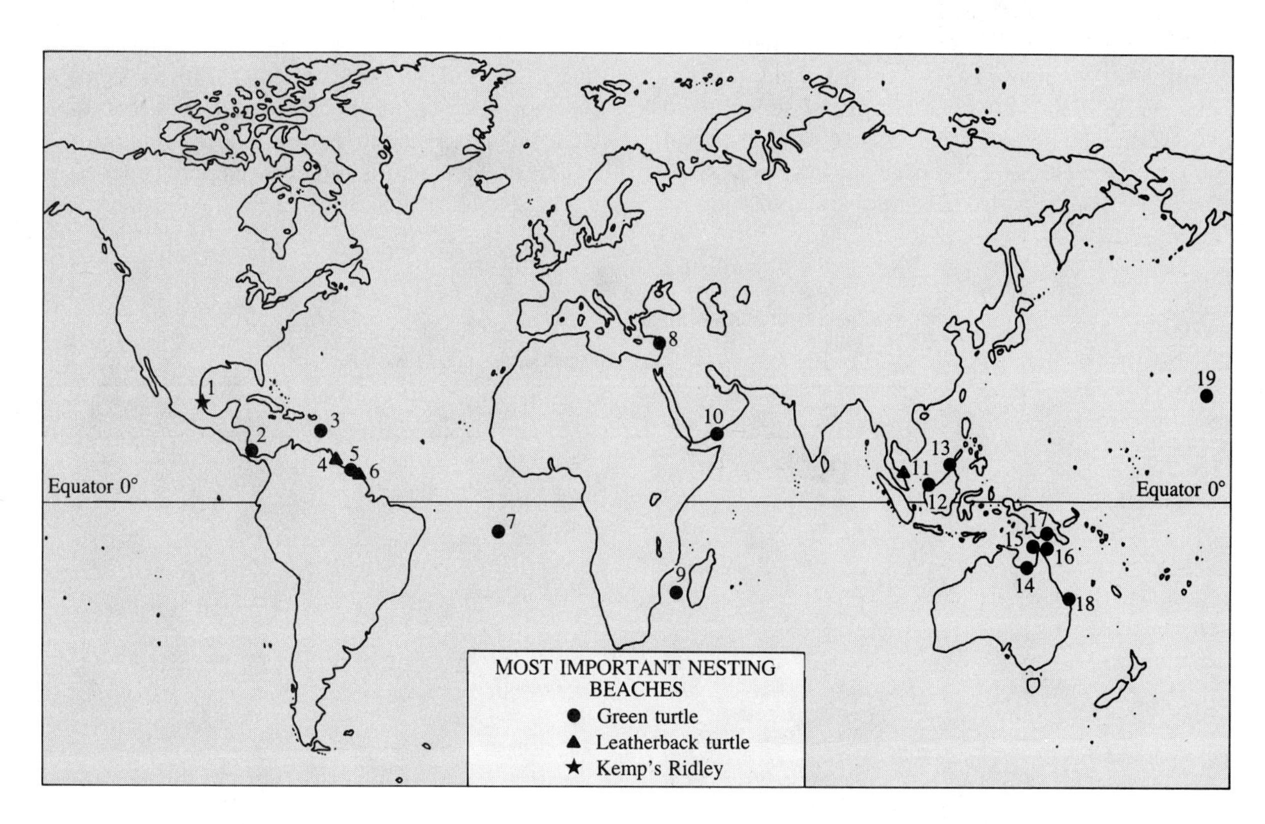

1. Tamaulipas, Mexico
2. Costa Rica
3. Aves Island
4. Guyana
5. Surinam
6. French Guiana
7. Ascension Island
8. Turkey
9. Europa Island
10. Sharma, S. Yemen
11. Trengannu
12. Sarawak Islands
13. Sabah
14. Bountiful Island
15. Crab Island
16. Raine Island
17. Bramble Cay
18. Capricorn-Bunker Cays
19. Western Hawaiian Islands

FIGURE 7.5
Major Nesting Sites of Sea Turtles

THE SEA BIRDS

Many species of birds use the surface waters of the oceans for feeding. Some, like the shore birds, wade in the water to find their food. Other species are only seasonal visitors to the oceans. Many species of ducks, loons, and grebes reproduce and spend the summers inland on fresh water and then spend the winters feeding on the ocean, usually close to shore. A third group of species obtain virtually all their food from the sea, either by skimming the surface, swimming underwater, or diving. These species are considered by marine biologists to be the true sea birds. Four major groups of sea birds are discussed in the following sections: the tubenoses, the pelicans and related species, the gull-like birds, and the penguins.

During their evolution, sea birds have developed a number of adaptations to surviving marine conditions. Morphological adaptations include the ability to swim under the water, often by use of the wings. Sea birds are excellent fliers and are capable of soaring, sailplaning, and diving. One important physiological adaptation of sea birds is the presence of a salt gland. Salt in the swallowed seawater and food (mostly fish high in salt content) must be removed to prevent the blood from becoming excessively salty. The kidneys of sea birds do not serve this purpose. Instead, salt or nasal glands in or around the orbit of the eye or in shallow depressions of the skull excrete a strong brine solution that lowers the salt content of the blood. The brine drips from the salt gland into the nasal cavity. Many sea birds have ridges along the edge of the beak that direct the brine to the tip of the beak, where there is little chance of it being swallowed.

Of all known species of birds (8,600), some 3 percent, or between 260–285 species, are sea birds (Tab. 7.1). Sea birds are more diverse in the Southern than in the Northern Hemisphere (Tab. 7.2). They are most abundant in areas with adequate food near the surface during daylight hours. The neritic zone and areas of upwelling are of particular importance to sea birds. Fewer species are found in oceanic conditions. One sea bird—the albatross—can stay in the air for weeks or months on end, using the strong winds of temperate latitudes to soar above the surface.

Tubenoses

The 100 or so species of tubenose birds are so named because of their well-developed external nostrils used for detecting odors and measuring the speed of passing air (Fig. 7.6). They are the most common of the groups of sea birds, and the total number of tubenoses is three times the number of the other groups combined. Tubenoses are the most oceanic of the sea birds and include albatrosses, petrels, and shearwaters. Both the largest of the sea birds—the Wandering albatross, with a wingspan of almost 4 m (13 ft)—and the smallest—

TABLE 7.1
Species of Marine Birds

(After Lockley, R.M. 1974. *Ocean wanderers.* London:David and Charles.)

Order	Common Name	Number of Species
Procellariiformes	Albatrosses	13
	Fulmars, petrels, shearwaters	60
	Stormy petrels	18
	Diving petrels	5
Pelecaniformes	Tropic birds	3
	Pelicans	1
	Gannets, boobies	9
	Cormorants	35
	Darters	(1)
	Frigate birds	5
Charadriiformes	Sheathbills	2
	Skuas	4
	Gulls	42
	Terns, noddies	39
	Skimmers	3
	Auks and alcids	22
	Phalaropes	2
Sphenisciformes	Penguins	17

TABLE 7.2
Numbers of Sea Bird Species in the Northern and Southern Hemispheres

Ocean	Number of Species
North Pacific	107
North Atlantic	74
Mediterranean	24
Arctic	31
Total	236
South Pacific	128
South Atlantic	73
Indian Ocean	73
Antarctic	44
Total	318

the stormy petrel, which is the size of a robin—are tubenoses.

A majority of the 13 species of albatross are found in the Southern Hemisphere. These birds can stay in the air continuously for weeks and use the strong winds between 30–60° latitude for lift, gliding for hours with little or no expenditure of energy. Albatross travel extensively, and the Wandering albatross is known to circle the globe in a year's travel. (See the special feature at the end of this chapter.)

FIGURE 7.6
Tubenose Birds
(A) Albatross. **(B)** Shearwater. **(C)** Petrel.

Shearwaters are probably the most numerous of sea birds. They nest in an area that ranges from the sub-antarctic through the tropics to the boreal latitudes in the Northern Hemisphere. Shearwaters are so named because of their habit of flying side to side, close to the water's surface, giving the impression of 'shearing' into the water.

Stormy petrels are one of the smallest sea birds. They are common in all oceans in both hemispheres. Millions congregate on the antarctic ice flows during the austral summer. Their small size permits them to fly during storms, and they feed on plankton churned to the surface by waves. They have been observed feeding in winds of up to force 10 (55 mph).

Pelicans and Related Types

The second large group of sea birds includes the pelicans (75 species), the sulids (or gannets) and boobies (75 species), cormorants (29 species), frigate birds (5 species), and tropic birds (3 species) (Fig. 7.7). These birds are globally distributed, although they most commonly occur in tropical and warm temperate latitudes. They usually are large birds. The pelican (around 11 kg) is surpassed in weight only by two species of penguins and the large albatross, and the frigate bird has a wingspan of almost 2.5 m. In general, the pelican group is not as oceanic as the tubenoses, and a number of species are restricted to close to shore. All species in the group are fish-eaters.

Pelicans are found close to shore in warm latitudes. They are large birds without the soaring capabilities of the tubenoses. Most pelicans capture food by plunge diving (see the following section on food and feeding). Gannets and boobies have a well-developed **gular pouch** and look like pelicans, but have a much narrower wingspan. They also feed by plunge diving. Gannets and boobies are more oceanic than pelicans. Most are highly colored during their reproductive period, and are colonial nesters. Cormorants range from the Arctic to the Antarctic, although they are most abundant in tropical and temperate latitudes. They are strong underwater swimmers and use their feet as paddles. Cormorants are birds of coastal waters, and some species will even range to inland fresh waters. They effectively capture fish for food, and some species are tamed and trained to catch fish for humans.

The final two types of sea birds in the pelican group are the frigate birds and tropic birds. The frigate bird, also called the man-of-war, has a large wingspan exceeded only by the albatross. Frigate birds are oceanic birds found in tropical latitudes. They are adept at soaring. The wingshape, which is negatively dihedral

FIGURE 7.7
Pelicanlike Birds

(A) Frigate Bird. **(B)** Blue-footed Booby. **(C)** Peruvian guano bird. **(D)** Booby. **(E)** Frigate bird. **(F)** Comparison of the regular wing shape with that of a dihedral shape.

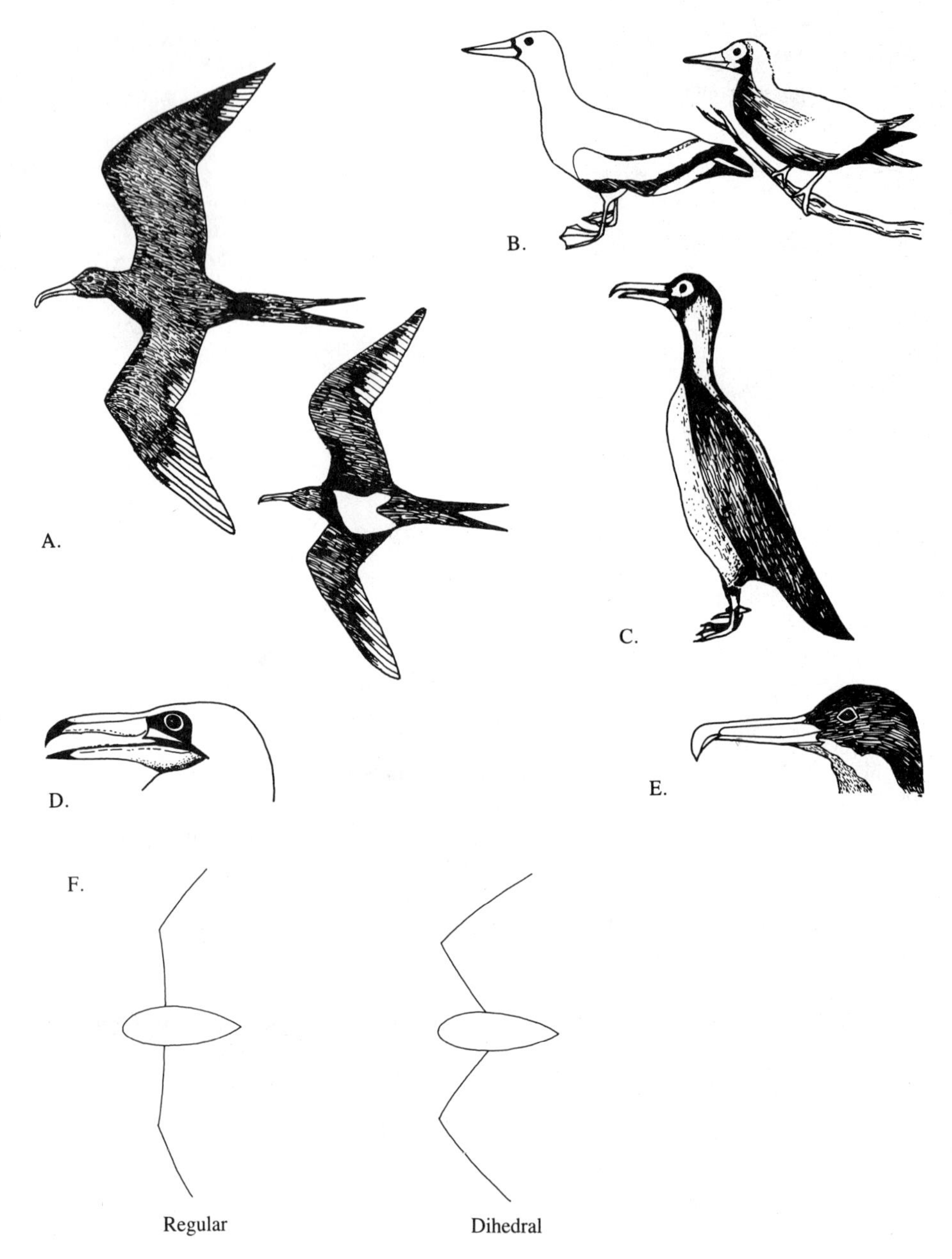

(Fig. 7.7), coupled with their light weight relative to wingspan, permits amazing trick flying. Capable of gliding, soaring, hovering, twisting, and diving, they can outmaneuver any other bird. Frigate birds neither walk nor swim; they must directly perch upon landing, and can take off from the surface of the water only with great difficulty. Frigate birds feed on flying fish that they catch in midair, and on squid and fish that they scoop up from the surface. They also frequently pirate the food of other sea birds by harassing the bird in flight until it regurgitates or drops its catch, or by stealing the eggs and young from gannet and booby nests.

Tropic birds, also called marlinspike birds, are found in tropical waters. The three species are distributed in the tropical Indian, Pacific, and Atlantic oceans. They are large birds with mostly white plumage. Tropic birds are the most oceanic of the pelican group, and travel long distances over the ocean in search of food, roosting at night on the ocean surface. They are good fliers and regularly travel in areas of light winds. Tropic birds are one of the few sea birds that are found in mid-oceanic gyres, like that of the Sargasso Sea, and in the tropical doldrums. They dive for their food, which consists of small fish and squid.

The Gull Group

The third group of sea birds (gulls, auks, terns, and skuas) is the largest in terms of the number of species (Fig. 7.8). As a group they have a global distribution, with most species found close to land masses. The gulls include some 45 species of moderately sized gray and white birds found mostly in the Northern Hemisphere, although some species breed from the Arctic to the Antarctic. Few gulls are oceanic, and a number of species are found far inland. Gulls are graceful fliers and good swimmers, using their feet as paddles. They are buoyant birds and float high in the water; consequently, they seldom dive beneath the surface. Although a few species—such as the kittiwakes—are plankton feeders, most gulls are omnivorous. They can catch fish but are more likely to be found scavenging dead animals. The scavenging habits of gulls have attracted them to areas of human development, where garbage dumps and sewage outfalls support large numbers.

Terns (about 40 species) are more marine than gulls, and may travel over the ocean for prolonged periods in search of food. They also are more tropical in their distribution than gulls. Although some terns breed in the Arctic and some in the Antarctic, most species are found in warmer waters. Like gulls, terns are mostly gray and white, and may breed in inland fresh waters. Terns feed on fish or larger zooplankton, catching them by their characteristic plunge dive. Some terns have extensive migratory movements (see Fig. 7.9). The arctic tern has the longest migration of any marine animal. It breeds as far north as land goes in the Arctic, and winters on the antarctic ice during the austral summer. Many other terns do not make long migrations. Tropical species, in particular, travel little from their breeding areas.

A. B. C.
D. E.
F.

FIGURE 7.8
The Gull Group

(A) Tern. **(B)** Auk. **(C)** Gull. **(D)** Puffin. **(E)** Guillemot. **(F)** Tern heads.

Auks (22 species) include the puffins, murres, and guillemots. They are cold-water birds found only in temperate and high latitudes of the Northern Hemisphere. Moderate to small in size, auks are excellent underwater swimmers, using their wings as paddles. They feed on plankton, small fish, and shallow-water benthos. Auks are relatively poor fliers. Their short, stubby wings must beat very quickly to lift the bird. They are nearshore birds and do not travel beyond the edge of the continental shelf. Some species of the auk group have inconspicuous colors (i.e., dark above and light below). Others, such as the puffin, are adorned with colored feathers that accent their large, bright orange beak.

The final gull group is made up of the skuas (4–6 species). Skuas are sometimes called sea hawks because of their predatory feeding habits. They are large birds, generally brown in color. They have the typical gull shape and are excellent fliers. They breed in the Antarctic and Arctic and spend nonbreeding periods in all oceans, from tropical to arctic waters. Like frigate birds, skuas pirate food from other birds. They harass other smaller feeding birds until they regurgitate their meal. The skuas then catch the food before it hits the water. Sometimes skuas even upend a bird by grabbing its tail feathers. Skuas also rob breeding colonies of eggs and young birds.

Penguins

The final large group of sea birds is that of the penguins (Fig. 7.10). There are about 18 species that make up the millions of penguins found in cold antarctic waters. Penguins are limited to Antarctica, the southern tips of Africa and South America, southern Australia, and the Galapagos Islands (Fig. 7.10). They are not found in the Northern Hemisphere. They are excellent swimmers, using their modified wings to fly through the water. Emperor penguins can swim for sustained periods at speeds between 8–16 km/hr (5–10 mph). Penguins are also efficient divers. They have low buoyancy and special physiological adaptations that allow them to re-

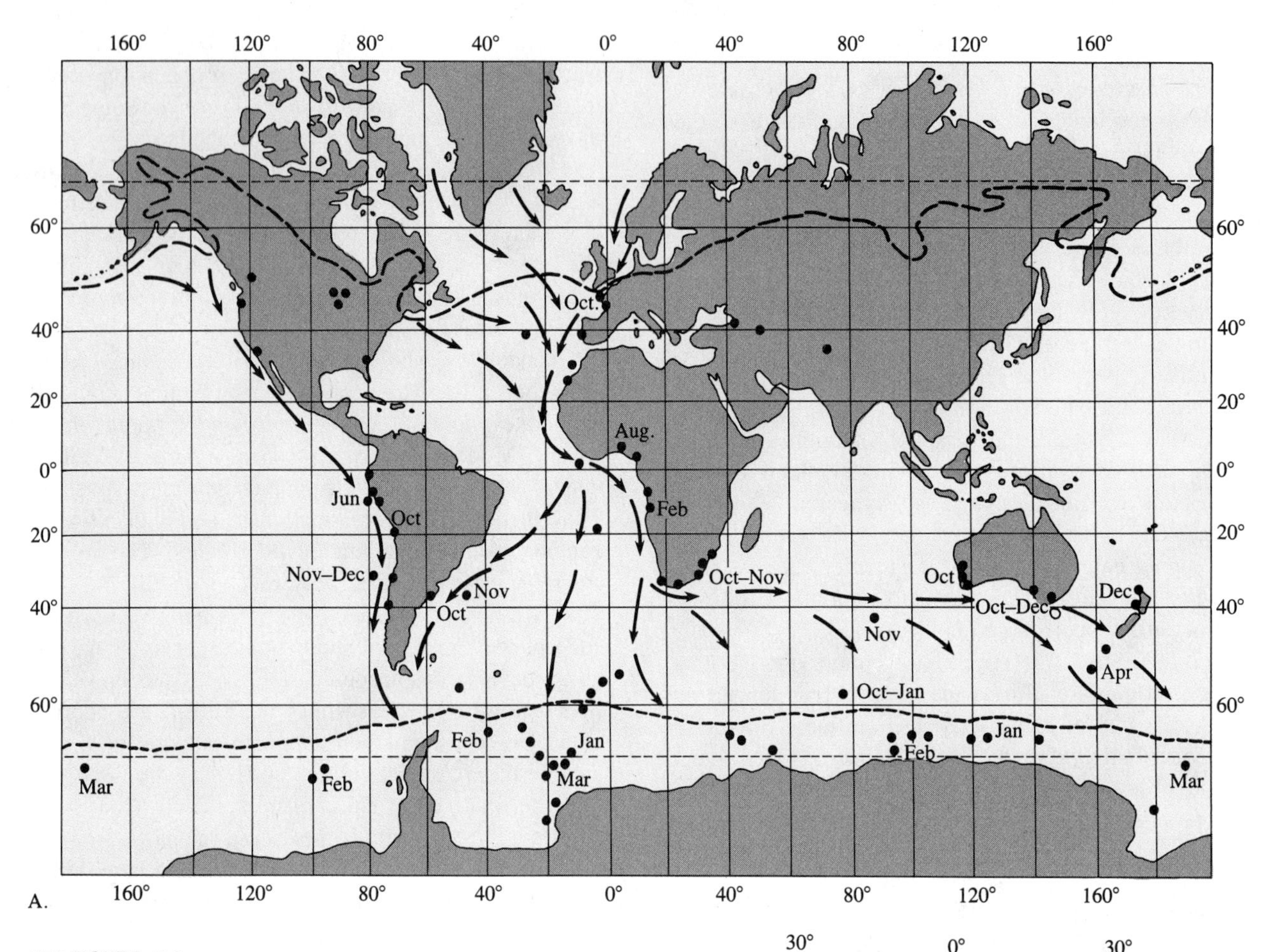

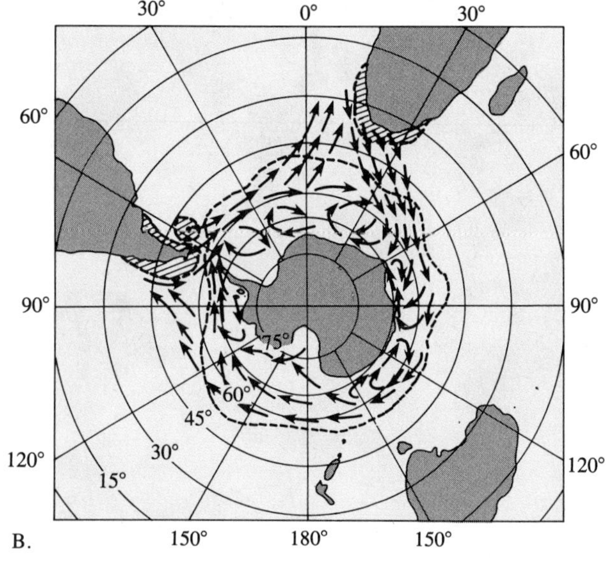

FIGURE 7.9
Migration of the Arctic Tern

(A) The arctic tern breeds in the arctic summer and winters in the antarctic summer. **(B)** As well as traveling from pole to pole, the arctic tern circumnavigates the antarctic continent during the antarctic summer. (After Lockley, R.M. 1974. *Ocean wanderers.* London:David and Charles.)

main underwater. Emperor penguins, for example, regularly stay submerged for up to 10 min, and dives to a depth of 265 m (869 ft) have been recorded. Penguins feed on fish, crustaceans (particularly krill), and squid, all of which are common in the shallow surface waters of the Antarctic.

The penguins' lack of ability to fly is related to their large size. The maximum size of a bird that can use its wings for both flying and swimming, such as the auk, is about 1 kg. Penguins are generally heavier; the largest (the emperor penguin) weighs about 45 kg (100 lbs).

FIGURE 7.10
Penguins and Their Breeding Areas

(A) Penguins. (1) Little penguin. (2) Emperor penguin. (3) Peruvian penguin. (4) Crested penguin. **(B)** Breeding areas of penguins. (After Nelson, 1979.)

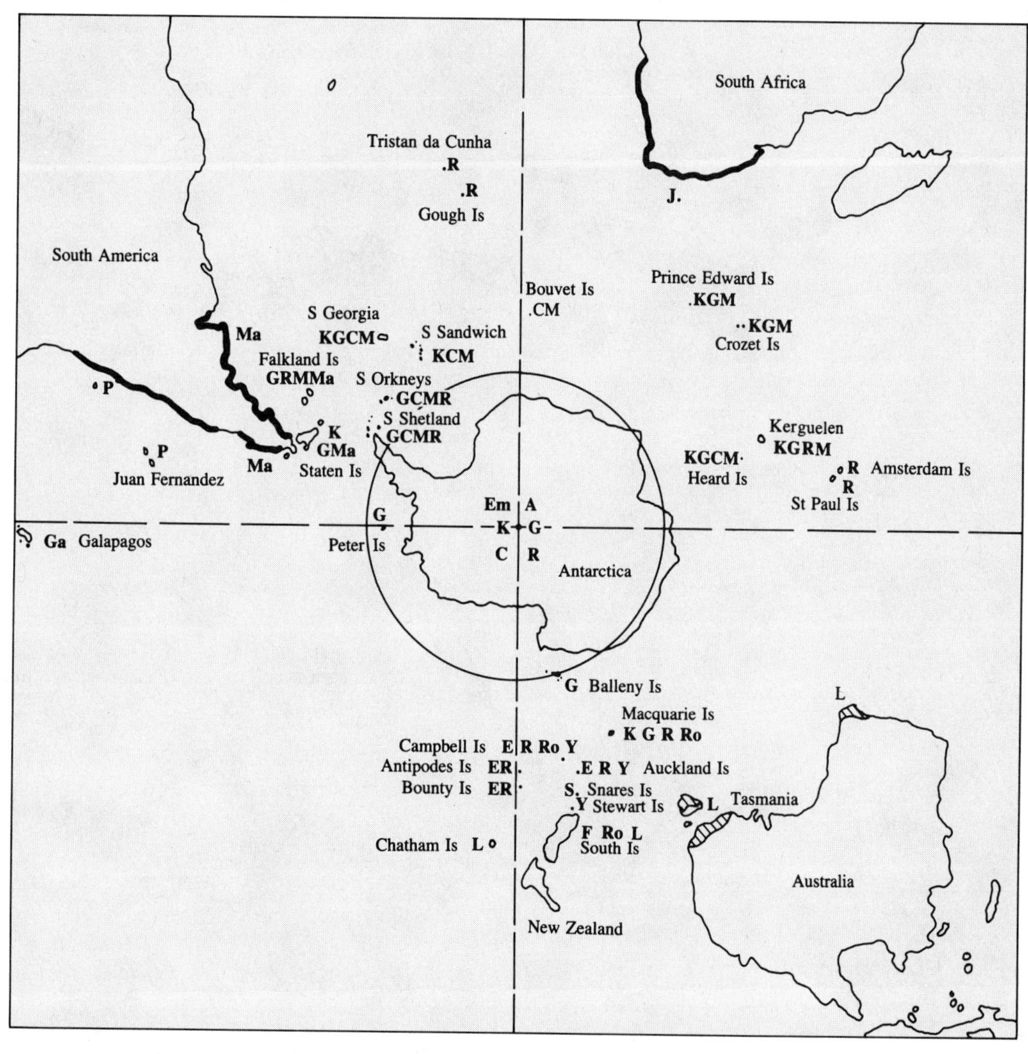

Em Emperor
K King
R Rockhopper
E Erect-crested
M Macaroni
Ro Royal
S Snares Island
F Fiordland Crested
Y Yellow-eyed
G Gentoo
C Chinstrap
A Adelie
J Jackass
P Peruvian
Ma Magellanic
Ga Galapagos
L Little

Food and Feeding

A majority of sea birds are fish-eaters. Some 51 percent of all species eat fish as the principal food of the diet, while 74 percent have fish as an important part of their food. Except for the penguins and auks, which use their wings to swim underwater (generally 5–20 m, or 16–66 ft deep), most birds fish at the surface. A number of methods of capture have developed (Fig. 7.11). The most common method is pursuit diving, in which the bird dives from the surface of the water and pursues the prey by swimming, using either the feet or the wings. Cormorants, penguins, auks, and diving petrels use this method of fishing. Second to pursuit diving is capture while floating or swimming on the surface. This surface dipping method is commonly used by skuas, many gulls, terns, petrels, pelicans, and albatross. Pelicans, when they dive, generally stop at the surface and extend the large bill into the water. They then actively expand the gular pouch (with a capacity of up to 10 l), which sucks in any fish. The water is forced out, and the fish swallowed.

The third general method of fishing is dipping while flying. This method is important to many gulls. The fourth category, although not widely used, is

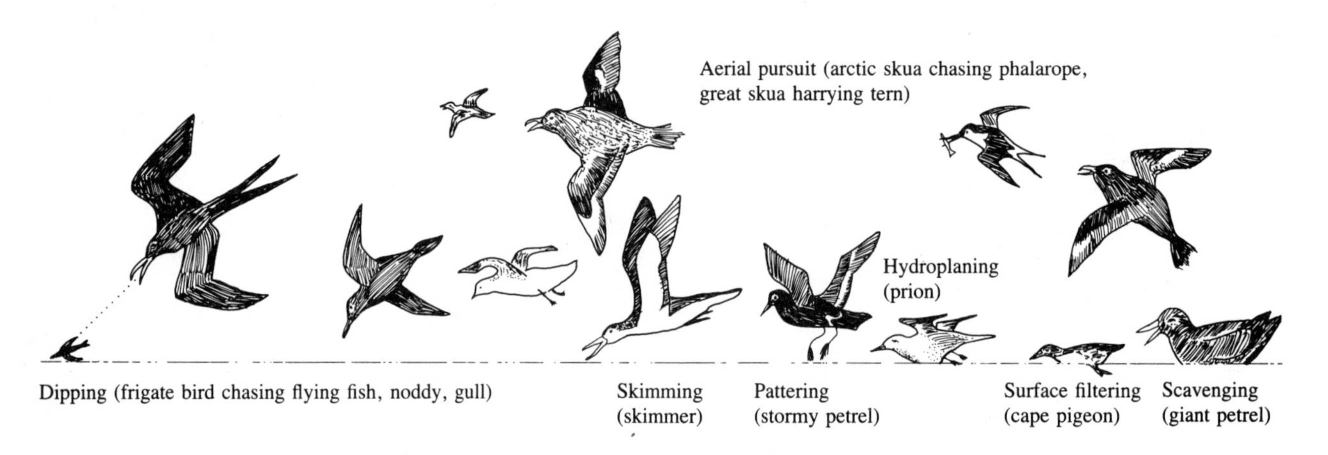

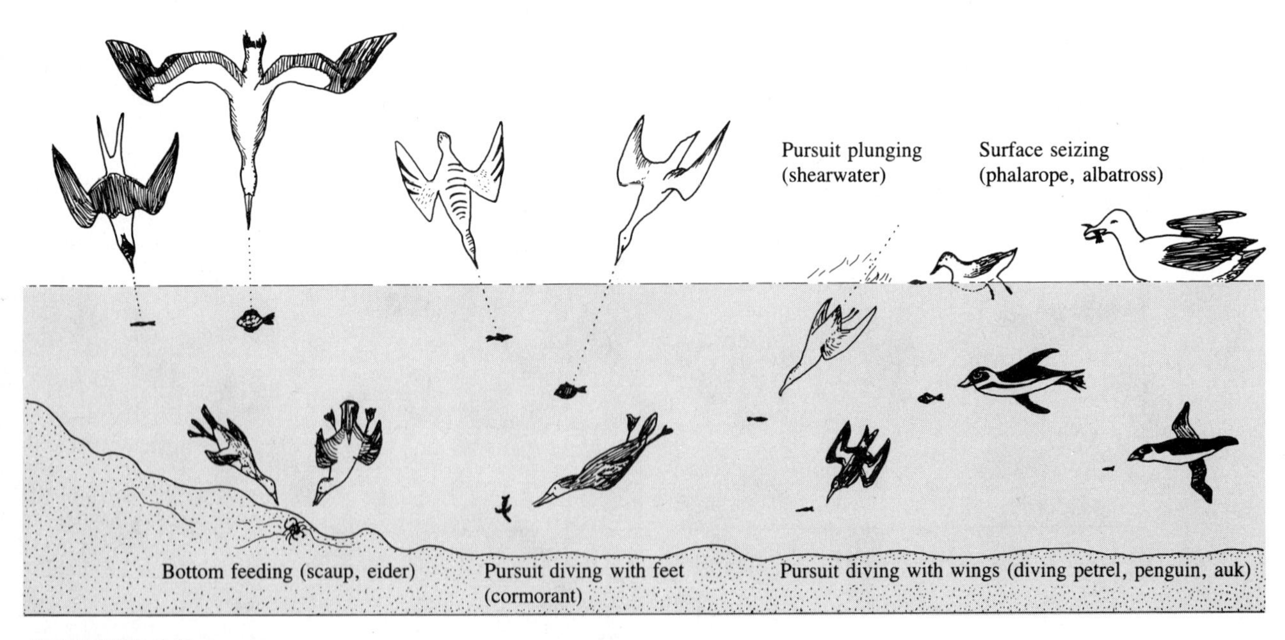

FIGURE 7.11
Feeding Methods of Sea Birds

(After Aschmole, N.D. 1971. *"Sea bird ecology and the marine environment."* In Farna, D.S., et al. *Avian biology, vol 1.* New York:Academic Press.)

plunge diving. In plunge diving, a bird dives while on the wing and captures the prey below the surface. The hover and dive of terns is a well-known example. Some 70 percent of terns plunge dive, as do most of the gannets and boobies (sulids). Other birds, including skuas, gulls, pelicans, and albatross, also plunge dive. Sulids dive from heights of 10–100 m and may go as deep as 10 m. Most plunge divers catch their fish in the top 1 m of the water.

Zooplankton and squid, in addition to fish, are important sources of food for sea birds. About 15 percent of the species of sea birds use zooplankton as their principal source of food, and 66 percent use it for food occasionally (Fig 7.12). Zooplankton is captured by birds feeding from the air, by surface feeders, and by diving and pursuit. Stormy petrels are perhaps best known for their ability to take zooplankton while hovering or flying. Stormy petrels feed on copepods and other zooplankton both day and night, and spend considerably more time collecting their food than do the fisheaters. A few birds take plankton by surface feeding. The cape pigeon (a tubenose), which is common in the Antarctic, paddles with its feet to bring euphausiids to the surface. Gulls and petrels also feed on surface plankton. Pursuit diving is also used to catch plankton. The Adelie penguins feed this way, almost exclusively on krill.

Squid are an important source of food for a number of large sea birds. Some 6 percent of sea birds feed exclusively on squid, and 28 percent of sea birds eat squid at least occasionally. Frigate birds, when not obtaining food by piracy, fly over the surface and capture squid by extending their beaks just below the surface. The albatross feeds on squid at night, waiting for the squid to come close enough to the surface to capture. Some penguins, such as the emperor penguin, dive and pursue squid, and some boobies and gannets take squid by plunge diving.

Sea birds are opportunistic both in the type of food taken and the method of capture, and individuals may quickly switch from one to another. No sea bird gets its primary food by preying on other sea bird species, although the frigate bird and skua may take chicks from breeding colonies and scavenge sick and disabled adults.

Migrations and Foraging

Sea birds are very migratory. Most species travel extensively at some point during their life cycle. Although most use flying for travel, the flightless penguins travel by swimming. There are three main types of long-distance travel: (a) feeding trips during the reproductive period; (b) directed migrations from one specific location to another; and (c) nomadic wandering through the

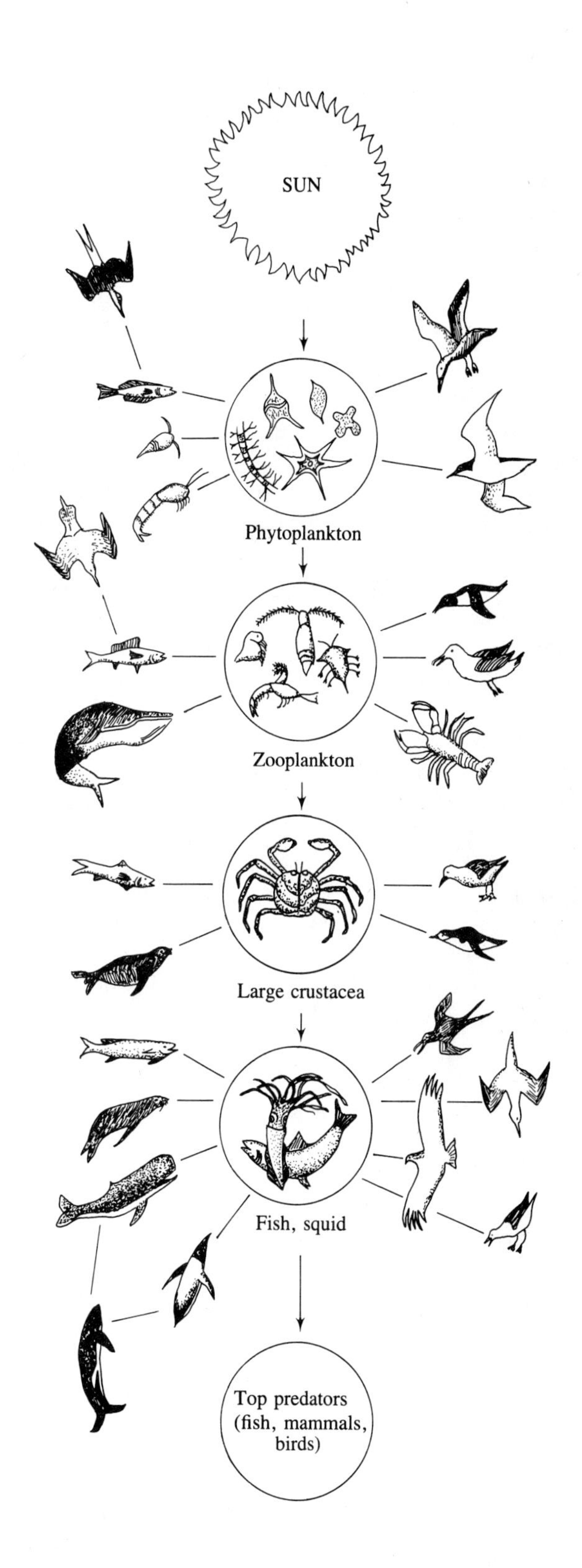

FIGURE 7.12
Food Sources of Sea Birds

Sea birds are part of the food web from the first level (phytoplankton eaters) to upper-level carnivores.

range of the species. Most of these movements are associated with locating food supplies.

When not breeding, most sea birds distance themselves from each other, congregating only in areas of high food concentration. Wandering and migration differ in degree of predictability. Wandering may cover large distances, but there is no specific destination or route. Migrations, however, are wholly or principally guided by genetic-based instincts. The time, location, and routes can be predicted. One of the most fascinating sea bird migrations is that of the arctic tern. This bird nests in the high Arctic during the summer, and 'winters' in the austral summer as far south as the antarctic ice flows (Fig. 7.9). This 13,000-km (7,800 mi) trip is the longest migration between summer nesting and winter feeding grounds of any bird. Arctic terns see little darkness in their lives because nesting grounds are in the continual light of the arctic summer, and the wintering grounds are in the continual light of the antarctic summer. Arctic terns seldom rest or feed during their migration, and the trip is made in only a few weeks. While in the Antarctic, arctic terns circumnavigate the continent, following the prevailing winds, and by the end of the austral summer and the time for the northerly migration, the arctic tern has circled the Antarctic continent (Fig. 7.9). The arctic tern spends four months breeding in the arctic summer and the remaining eight months in the Antarctic traveling a route of some 38,000 km (22,900 mi), or an average of 160 km per day.

Another long migration is that of the short-tailed shearwater, which breeds in Southwest Australia and winters in the North Pacific (Fig. 7.13). On their figure-eight migration, shearwaters rapidly pass through the tropical regions. Other sea birds with extensive migrations include albatross, shearwaters, petrels, and a few terns.

Many other seabirds wander extensively within the boundaries of their species range. Tropical sea birds, in particular, disperse over large areas without any strong directional movements. Sulids (frigate birds and boobies) travel up to 6,000 km on such wanderings, although they do not go farther north or south than 23° latitude.

Sea birds of temperate latitudes also show extensive wanderings. Auks, petrels, gulls, and some of the pelican group wander through the species range between breeding seasons. Some albatross may wander up to 9,600 km (5,760 mi), using the temperate latitude winds to completely circle the globe.

Sea birds often must fly long distances during the breeding period. Since sea birds require land that is safe from predators and disturbance, breeding space is limited and most sea birds form colonies during the reproductive period. The size of the colonies can be staggering. Often hundreds of thousands of birds will locate in a small area. The high density of sea birds means that individuals are required to make extensive trips to find adequate food for themselves and the developing juveniles. Foraging trips may be up to 900 km (fulmar), or 1,000 km (shearwater), although most trips are shorter.

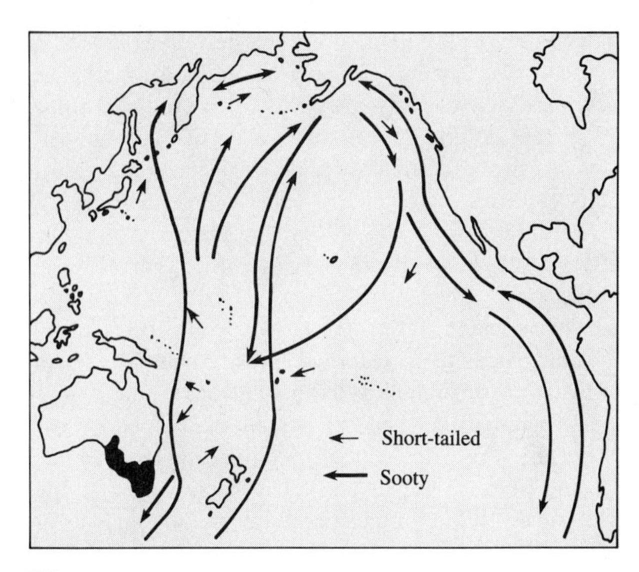

FIGURE 7.13
Migration of the Short-tailed and Sooty Shearwater

The breeding area of the short-tailed shearwater in southern Australia is shown in black. (After Nelson, B. 1979. *Sea birds*. Reading, MA:Addison-Wesley.)

In all these types of movements, sea birds apparently use similar navigation skills. They have an accurate sense of time, both the passing of the hours and the seasons, and use light from the sun, moon, and other stars and planets to set their chronometers. Sea birds also use the position of the stars and planets to determine their latitude and longitude. Sea bird migration is not possible in mist, fog, and clouds, which obscure the heavenly bodies. Migratory routes are genetically imprinted in seabirds, and isolated young can make the migratory travels on their own. They can learn the details of a migratory route, however, and are more efficient in successive migratory trips.

Reproduction

Much of the biology of reproduction in sea birds revolves around obtaining an adequate food supply and a solid place to lay an egg and raise a chick to the point it

A. The parent albatross squirts a mixture of oil and stomach contents forcibly into the trough of its lower mandible. The chick takes it mainly as it emerges from the adult's throat.

B. The herring gull chick pecks at the red spot on the adult's lower mandible and takes the food either as it emerges or after it has been regurgitated onto the ground.

C. The parent tern presents small fishes directly to the chick.

D. All sulids feed their young directly from the adult's throat.

E. Cassin's auklet chick sips from the adult's bill (they are plankton feeders).

F. Tropic birds place their beak into the chick's gape.

FIGURE 7.14

Sea birds generally must find food far from breeding areas and carry it to the young. The young are fed by regurgitation. **(A)** Albatross. **(B)** Herring gull. **(C)** Tern. **(D)** Sulid. **(E)** Cassin's auklet. **(F)** Tropic bird. (After Nelson, 1979.)

can fly. Most sea birds are colonial breeders. Courtship often is accompanied by elaborate behavior. Usually the male and the female share in the incubation and raising of the chick. This allows one partner to tend the nest and the other to make the often extensive foraging trips. Sea birds must actively feed the helpless young. Parents swallow food caught at sea and regurgitate it to the young. One of the most interesting adaptations to carrying food long distances to the young is the stomach oil of the tubenose group. Food is partially digested and stored in an oil form to be regurgitated to the young (Fig. 7.14). The weight of food the adult has to carry is thereby reduced, and longer-distance foraging is possible. When an albatross feeds its young, it literally shoots a stream of oil into the chick's mouth. Penguins secrete oil from a specialized part of the esophagus. The oil is fed to the young and regurgitation is not necessary.

SUMMARY

Birds and reptiles are amphibious in that they obtain their oxygen from the air, and (most) reptiles and (all) birds require land for reproduction.

The common reptiles of the marine environment are snakes and turtles. There are more species of snakes than turtles (50 vs. 5), but the turtles are much larger. Marine snakes are often venomous, and are for the most part ignored by humans. Sea turtles, conversely, are an important source of food to some groups of people, and several species are familiar to most people. All turtles are threatened by overexploitation and pollution.

Marine birds range widely over the ocean's surface, and yearly migrations of thousands of miles are not uncommon. Although not all species make long trips, marine birds best fit the description of wanderers. The arctic tern travels from the northern Arctic to the southern Antarctic and returns every 12 months. Albatrosses may circle the globe once or twice a year without resting on land. Although birds are amphibious by nature, they are very important in marine ecosystems. Sea birds use a variety of pelagic and benthic animals as food. They are top predators in marine food webs.

QUESTIONS AND EXERCISES

1. Discuss the advantages and disadvantages of a large body size in marine birds.
2. In what ways are reptiles and birds only partially adapted to life in the oceans?
3. Discuss the relationship of bird size to type of food consumed.
4. Discuss the adaptations of turtles for swimming.
5. Discuss the general patterns of distribution of reptiles and birds in the marine environment.
6. What is the mechanism used by marine reptiles and birds to remove the excess salt from sea water? (Describe it.)
7. What is the difference between the migratory movements of marine birds and their wandering movements in search of food?
8. Construct a table that describes the various types of flying done by marine birds.
9. Construct a table that describes the different ways that various groups of marine birds catch fish.

REFERENCES

Alderton, D. 1988. *Turtles and tortoises of the world.* New York:Facts on File.

Aschmole, N.D. 1971. Sea bird ecology and the marine environment. In Farner, D.S., et al. *Avian biology, vol 1.* New York:Academic Press.

Carr, A. 1984. *So excellent a fishe: A natural history of sea turtles.* New York:Scribner.

Crane, J.M. 1979. *Introduction to marine biology: A laboratory text.* Columbus, OH:Merrill

Croxall, J.P. 1987. *Seabirds: Feeding ecology and role in marine ecosystems.* New Rochelle, NY: Cambridge University Press.

Furness, R.W. 1987. *Seabird ecology.* New York:Methuen.

Haley, D., ed. 1984. *Seabirds of the eastern north Pacific and arctic waters.* Seattle, WA: Pacific Search Press.

Jouventin, P. and **Weimerskirch, H.** 1990. Satellite tracking of Wandering albatrosses. *Nature* 343(6260):746–748.

Lockley, R.M. 1974. *Ocean wanderers.* London:David and Charles.

Lockley, R.M. 1983. *Sea birds of the world.* New York:Facts on File.

Nelson, B. 1979. *Sea birds.* New York:Addison-Wesley.

Webb, J.E., et al. 1978. *Guide to living reptiles.* London:Macmillan

SUGGESTED READING

NATURAL HISTORY

Strange, I.J. 1989. Albatross alley. July, 1989:26–33.

Trivelpiece, S.G., and **Trivelpiece, W.Z.** 1989. Antarctica's well-bred penguins. December, 1989:28–37.

OCEANS

Andersen, G. 1987. The Kemp's Ridley puzzle. 20(3):42–49.

de Roy, T. 1984. Three ways to be a boobie. 17(3):9–13.

Dillon, J. 1983. Shorebirds. 16(3):3–9.

Easter, D. 1983. Leatherback turtles. 16(1):3–7.

Minton, S.S., and **Heartwole, H.** 1978. Snakes and the sea. 11(2):27–33.

Sea birds. (Includes adaptations to salt water and navigation) 1978. 11(4):18–54.

Stone, L.M. 1981. The loggerhead and his friends. 12(5):37–41.

Whittow, G. 1985. Panting and fluttering. (Heat control in birds) 18(3):32–40.

OCEANUS

Brown, R.G. 1981. Seabirds at sea. 24(2):31–38.

Considine, J.L., and **Winverry, J.J.** 1978. The green sea turtle of the Cayman Islands. 21(3):50–55.

SEA FRONTIERS

Dawson, S. 1985. Deadly but not dangerous. (Sea snakes) 31(5):282–285.

Gregg, S.S. 1988. Of soup and survival: The plight of the sea turtle. 34(5):54–61.

Kenyon, J. 1989. Northern hemisphere albatrosses. 35(6):342–347.

Kirkpatrick, S.J. 1989. Gulls: A sea guide. 35(3):146–153.

Ritchie, T. 1989. Marine crocodiles. 35(4):212–221.

Sefton, H. 1986. Boobies and frigates: Strange bedfellows. 32(4):290–296.

THE AMAZING FLIGHTS OF THE WANDERING ALBATROSSES

The Wandering albatrosses *(Diomedea exulans)* shown in Figure 1 always have been assumed to be the sea bird with the greatest range of foraging flight. Ranging throughout the Southern Ocean between 30–60° south latitude, large males may weigh up to 12 kg (26.5 lb) and have a wing span of 3.2 m (10.5 ft). One egg is produced between November and February during the Austral spring or early summer by a breeding pair. While one parent incubates the egg, the other undertakes long foraging flights during which they feed and may increase their weight by 15 percent. The partners left on the nest can fast for long periods of time during the mate's foraging flight. Females have been known to stay on the nest for 50 days, and males up to 55 days, before abandoning it in search of food. Within three months, the egg hatches, and the chick must be fed regurgitated squid or cuttlefish frequently, and foraging flights seldom exceed four days. This chick brooding may continue through the fall into July.

The first direct evidence of the great length of foraging flights during egg incubation and chick brooding were obtained in early 1989 (Jouventin and Weimerskirch, 1990). French scientists used transmitters (attached to the birds) and two satellites to track five male albatrosses during incubation forages, and one male

FIGURE 1
The Wandering Albatross, *Diomeda exulans*

The wing span of this albatross measures 3 m (10 ft). (Photo by R.L. Pitman.)

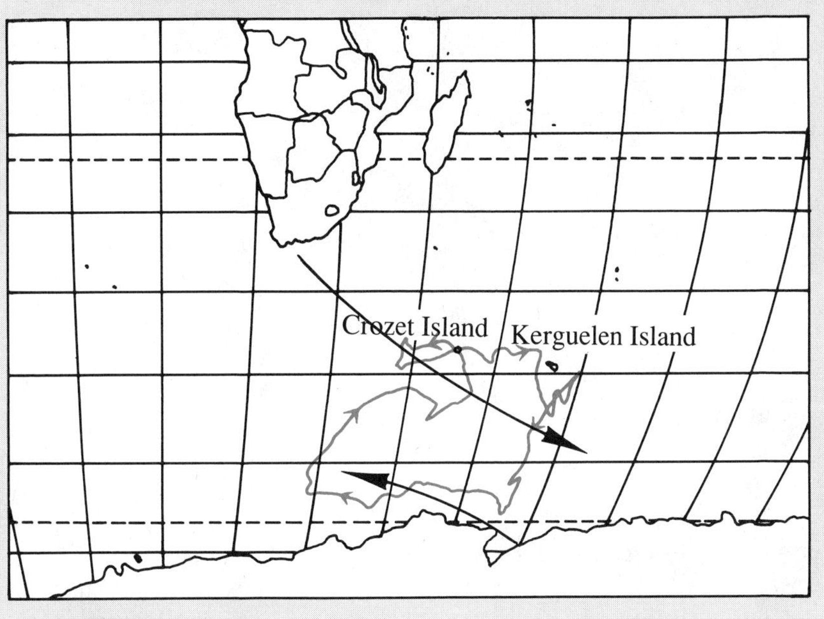

FIGURE 2
Wandering Albatross Foraging Flight

The longest foraging flight of a male Wandering albatross monitored by satellite between February 1 and March 5, 1990. The flight began and ended on Possession Island of the Crozet archipelago. Note that the albatross took full advantage of the strong and consistent prevailing winds that exist throughout much of the Southern Ocean. The two arrows indicate the general direction for the westerly winds and polar easterlies.

on two brooding forages from Possession Island in the Crozet archipelago in the southwestern Indian Ocean (Fig. 2).

Of these flights, the longest was undertaken during incubation and lasted 33 days. This flight covered a total distance of 15,200 km (9,440 mi) from February 1 to March 5. Approximately 96 percent of the distance was covered during daytime, with a maximum speed of flight of 81.2 km/h (50.4 mi/hr). Over one stretch of 808 km (502 mi), the bird averaged 56.1 km/h (34 mi/hr). The birds flew continuously during daylight, with only short stops to feed. They rested at night, but stops rarely exceeded one hour.

The high rate of speed and great distances travelled by these sea birds result from (a) the constant prevailing winds of the Southern Ocean, which blow over this great expanse of uninterrupted ocean, and (b) the exceptional design of the long, slender wings, which permit soaring with a minimal expenditure of energy. The birds usually soared with the prevailing winds on the outward journey, and tacked across the wind on their return. The Wandering albatross clearly evolved in harmony with this unique expanse of open ocean, with its strong and steady winds.

8

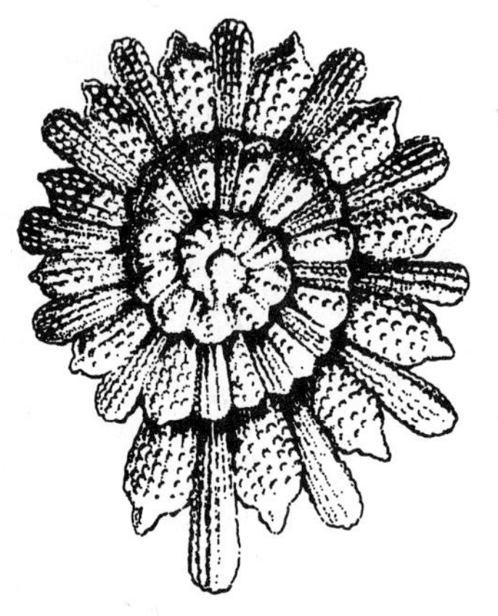

MARINE MAMMALS

In the last two decades, international attention has focused on the mammals that live in the oceans. The increased understanding of their complex behavior from both field and aquarium studies, as well as the relentless commercial exploitation that has decimated the populations of many species, has brought marine mammals into the spotlight. Not only is the question of conservation of resources being debated, but also whether marine mammals deserve special protection because of their affinities with humans in terms of physiological and social behaviors.

Marine mammals are made up of a number of different and evolutionarily unrelated groups (Fig. 8.1). The cetaceans (whales, dolphins, and porpoises) and the pinnipeds (seals and sea lions) are the most common in the oceans. Other marine mammals include the sirens, or sea cows, and the sea otters. Some biologists argue that polar bears also should be included in the marine mammals.

Among marine mammals, whales, dolphins, and porpoises are the most completely adapted to the marine environment. The cetaceans are entirely free of land and bear their young at sea. Hair and fur are missing, and insulation is provided by a thick layer of blubber or fat beneath the skin. The tail is modified into a powerful fluke or fin; the rear legs have disappeared; and the pelvis is reduced to a small, vestigial bone. The front legs persist and form a pair of fins. The nostrils have migrated from near the mouth to well back on the dorsal surface, forming the familiar blow hole. Although marine mammals have to rely on air to get their oxygen, they are capable of deep, sustained diving.

The pinnipeds (seals, sea lions, and walruses) also are successful in the marine environment. Pinnipeds are excellent swimmers and undergo long migrations. They are good divers and can stay below the surface for long periods. Although many pinnipeds have hair, they also have a layer of blubber to assist in insulation. More amphibious than cetaceans, pinnipeds often haul out on land to rest. Too, they must come ashore to deliver their young. They have retained their hind feet, which are formed into flippers used mostly for swimming.

The third group of marine mammals—the sirens—consists only of a few species: the sea cows, dugongs (Fig. 8.2), and manatees. Sirens are the only herbivorous marine mammals. They are mostly tropical and feed on shallow coastal vegetation. They are not numerous anywhere, and their slow ways have made them vulnerable to overexploitation. Two siren species—the Steller's sea cow of the Bering Sea, and the Caribbean sea cow—have become extinct because of overharvest.

Yet another group of marine mammals are the sea otters (Fig. 8.3). Relative newcomers to the marine environment, sea otters evolved only 1.5 million years ago from terrestrial, weasel-like animals. Lacking a layer of blubber, they rely on a thick coat of fur for insulation. Sea otters have the thickest fur of any mammal, and water never penetrates to the skin. This fur is highly prized by humans. Before humans started killing them, sea otters ranged the coasts of the North Pacific Ocean from northern Japan in Asia, through the arctic coasts, and down the west coast of North America to Baja Mexico. Overharvest has caused extinction in most areas.

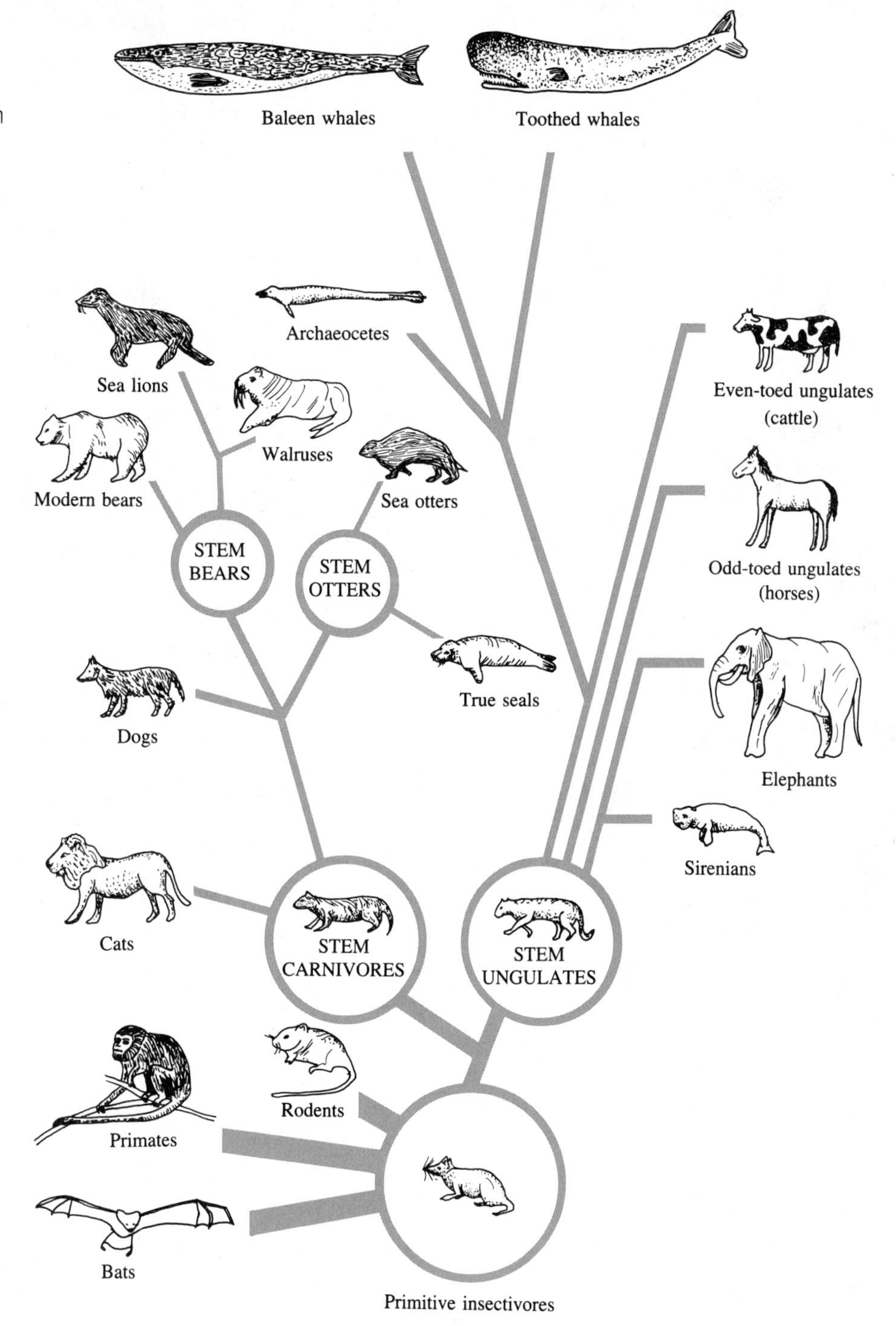

FIGURE 8.1 Evolution of Marine Mammals

Marine mammals have evolved from two basic stocks: the ungulates, which gave rise to the cetaceans; and the carnivores, which gave rise to the pinnipeds.

As mentioned above, one other mammal is sometimes considered to be marine—the polar bear of high arctic latitudes. Polar bears *(Ursus maritimus)* spend long periods of time feeding on the floating ice pack. They hunt for seals along the broken edge of the ice, and often have to swim. Polar bears can easily swim for up to 30 km (18 miles). They use their forelimbs to paddle and trail the hind limbs behind.

In this chapter we focus our discussion on the composition and behaviors of two mammal groups: the cetaceans and the pinnipeds. The majority of mammalian species found in the marine environment belong to either of these groups.

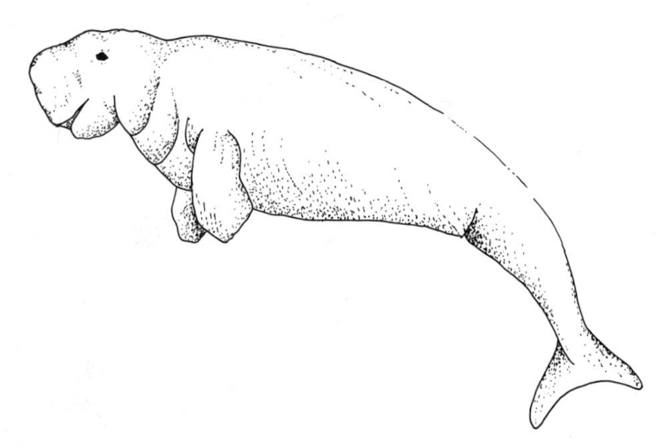

FIGURE 8.2
Dugong

Dugongs are members of the order Sirenia. They are herbivorous and mostly tropical.

FIGURE 8.3
Sea Otter

The sea otter is considered by many marine biologists to be a marine mammal. Insulation comes from its thick fur, rather than blubber.

CETACEANS

Of the groups of marine mammals, the cetaceans (whales, dolphins, and porpoises) and pinnipeds are by far the most important in number of species and extent of adaptation to the marine environment. The cetaceans include around 80 species of marine mammals distributed throughout most of the world's oceans and seas (Fig. 8.4). A few species are even found in fresh water and in lakes. Cetacean species have a wide size range, from the small harbor porpoise (1.5 m, 50 kg) to the largest living animal on earth, the blue whale (typically, up to 30 m in length; weight, 150 metric tonnes). Present-day numbers of cetaceans have become seriously reduced because of overharvest (Tab. 8.1).

Biologists recognize two main groups of cetaceans, the suborder Mysticeti and the suborder Odontoceti. The Mysticeti (*mysti* means 'moustached') are the baleen whales, which capture their food by filtering water through a sieve made of baleen (whalebone). There are 10 species of Mysticeti, the smallest being the pygmy right whale, and the largest, the blue whale.

The 70 species of Odontoceti (tooth cetaceans) have conical teeth that are used mainly to grasp food. The food of toothed cetaceans is mostly fish or squid, which they actively chase. Toothed cetaceans include some whales, but are mostly dolphins and porpoises. In popular usage, the term whale refers mainly to the size of the cetacean. On the basis of the size alone, cetaceans can be classified into three groups. The largest of the cetaceans are the great whales. This group of whales, generally larger than 9 m (29 ft), includes all

TABLE 8.1
Whale Populations

Present whale populations are much depressed from original populations because of overharvest. (Table adapted from Friends of the Earth, 1978; current data do not vary considerably.)

Species	Estimated Original Population	Present Population Estimates
Whale Populations in the Southern Hemisphere		
Fin	375,000–425,000	82,000–97,000
Sei	ca. 150,000	82,000
Minke	220,000–299,000	150,000–298,000
Blue	ca. 200,000	5,000–6,000
Pygmy blue	ca. 10,000	ca. 5,000–6,000
Humpback	30,000+	ca. 3,000
Right	?	900–3,000
Sperm: male	ca. 257,000	ca. 128,000
female	ca. 330,000	259,000–295,000
Whale Populations in the North Pacific		
Fin	42,000–45,000	14,000–19,000
Sei	58,000–62,000	34,000–38,000
Blue	4,700–5,000	1,400–1,900
Humpback	?	1,200–4,000
Right	?	100–1,000
Gray	?	7,500–13,000
Sperm: male	167,000–195,000	ca. 69,000–91,000
female	124,000–152,000	ca. 102,000–125,000

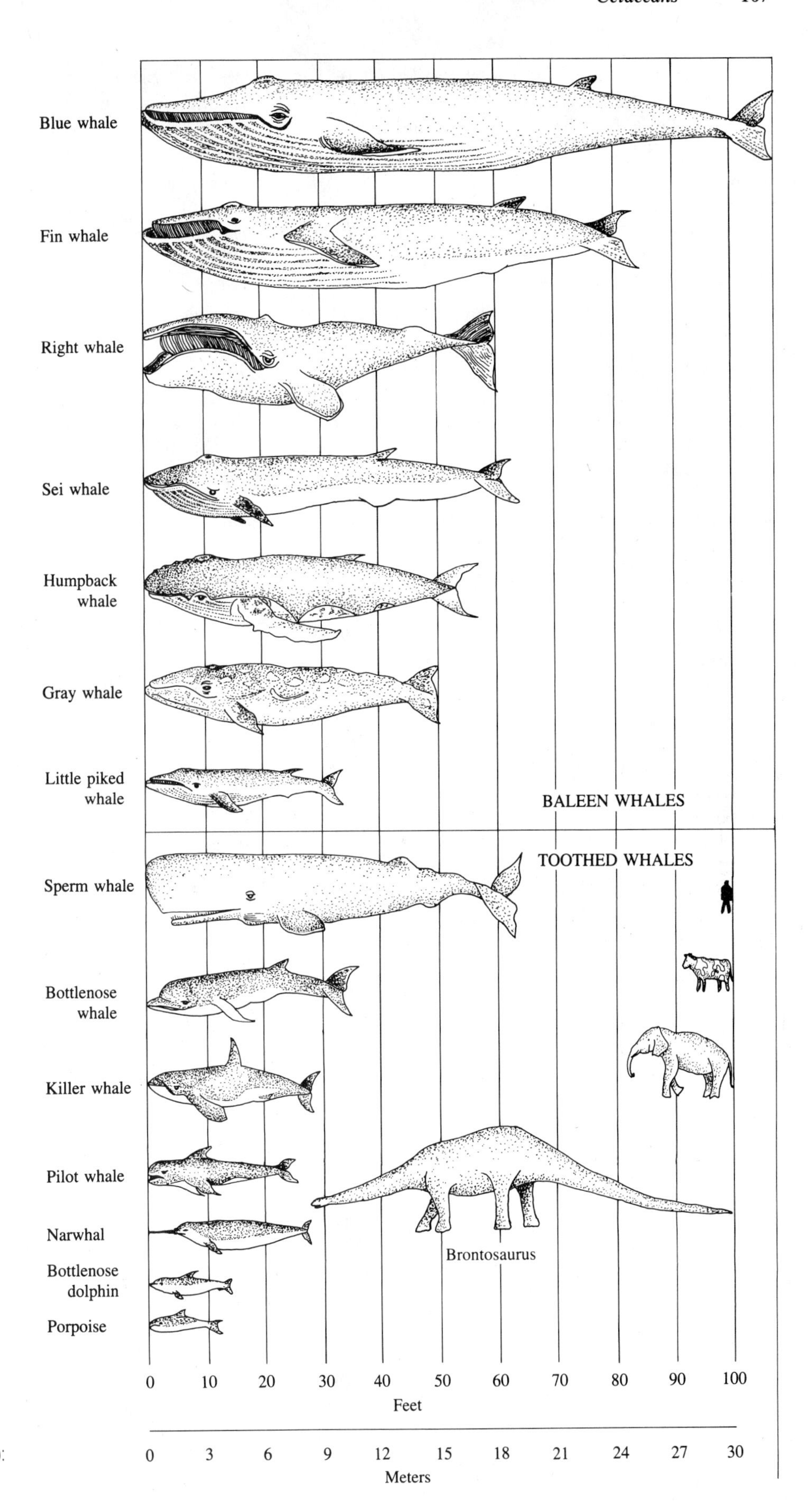

FIGURE 8.4
The Major Cetaceans

(After Friends of the Earth. 1978. *The whale manual.* San Francisco: Friends of the Earth.)

the baleen whales (Mysticeti), and the largest of the Odontoceti (the sperm whale). A second group of smaller whales ranges from 4–9 m in length, and includes killer whales, pilot whales, beluga or white whales, the narwhal, and the beaked whales. The smallest of the cetaceans (1.5–4 m) are the dolphins and porpoises. In common usage, these two names are often used interchangeably. In addition, the mammal dolphin is sometimes confused with the dolphin fish. Dolphin fish, however, usually are less than a meter in length and have a blunt snout. A cursory examination reveals the basic differences between the mammal and the fish.

Marine biologists differentiate between dolphins and porpoises by a number of characteristics. Dolphins have conical-shaped teeth, while those of porpoises are spade-shaped. Porpoises are generally smaller and have a blunt snout, while dolphins generally have a pronounced beak. When present, the dorsal fin in porpoises is low and more triangular; in dolphins, it curves to the posterior. There are six species of porpoises, all belonging to the family Phocaenidae. There are 32 species of dolphins, the majority in the family Delphinidae.

Swimming and Diving

All cetaceans have a fusiform shape, so they can move efficiently and quickly through the water. This streamlining results in a body shape that is a major departure from the basic mammalian form. The neck is absent, and the head molds smoothly with the body. All body projections have been lost or are covered by a fold of skin. External ears are missing, except for small, wax-filled holes. Mammary nipples and sex organs are withdrawn into slits in the body. The posterior limbs, including their skeleton, have been lost during evolutionary history, and the forelimbs, which are formed into stiff flippers, have lost their ability to move—except at the shoulder (Fig. 8.5). The cetacean flipper has all the bones of the mammalian foreleg, including those homologous to the human hand, although most cetaceans have the bones of only four fingers.

An outer layer of blubber smooths the contours, and the skin surface itself is smooth, further enhancements of the streamlined body shape. Both features reduce turbulence and friction. The skin seems to be capable of localized changes in shape that stop eddies from developing as soon as they form.

The swimming power of cetaceans comes from the tail, which forms a pair of paddlelike flukes. Unlike the vertical tail of fish and sharks, the fluke of cetaceans is horizontal. Muscles connect the cetacean tail flukes to the posterior third to half of the body. Swimming speeds of cetaceans are similar to that of the fast fish, such as tuna, salmon, and swordfish, and the power developed in swimming is so great that they can jump free of the water in a movement called breaching.

Another adaptation to swimming is the migration of the nostrils from the normal mammal position at the tip of the snout to well back on the top of the head (Fig. 8.5). This blowhole permits cetaceans to exchange air by rolling in an arc at the surface, and exposing only the dorsal surface of the body. Air exchange is rapid and efficient. One rapid exhalation removes over 90

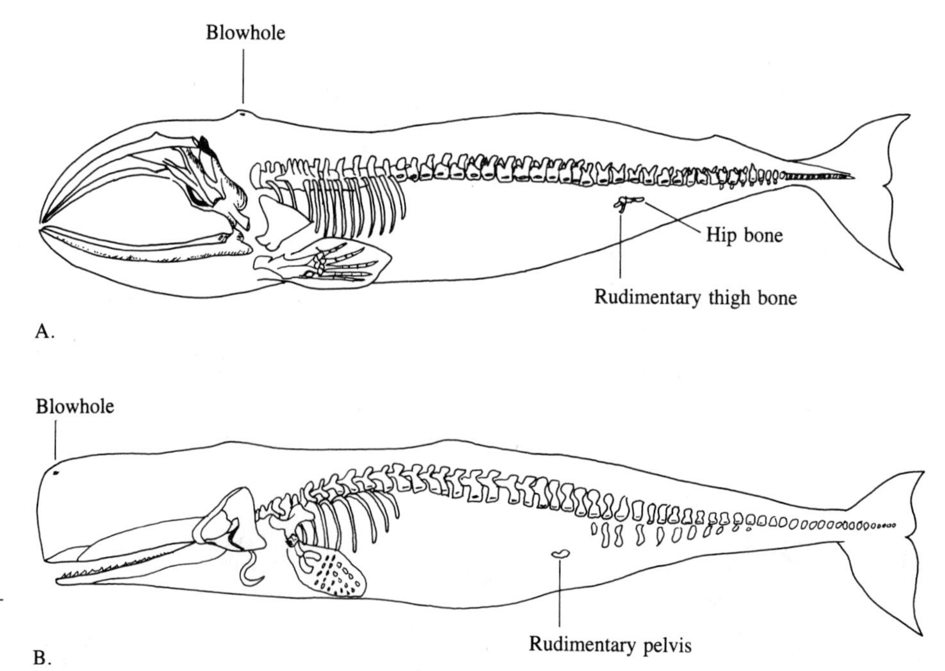

FIGURE 8.5
Skeleton of Cetaceans

In cetaceans, the whole pelvic girdle and rear limbs are virtually lost. **(A)** Greenland right whale *(Balaena mysticetus)*. **(B)** Sperm whale *(Physeter catodon)*.

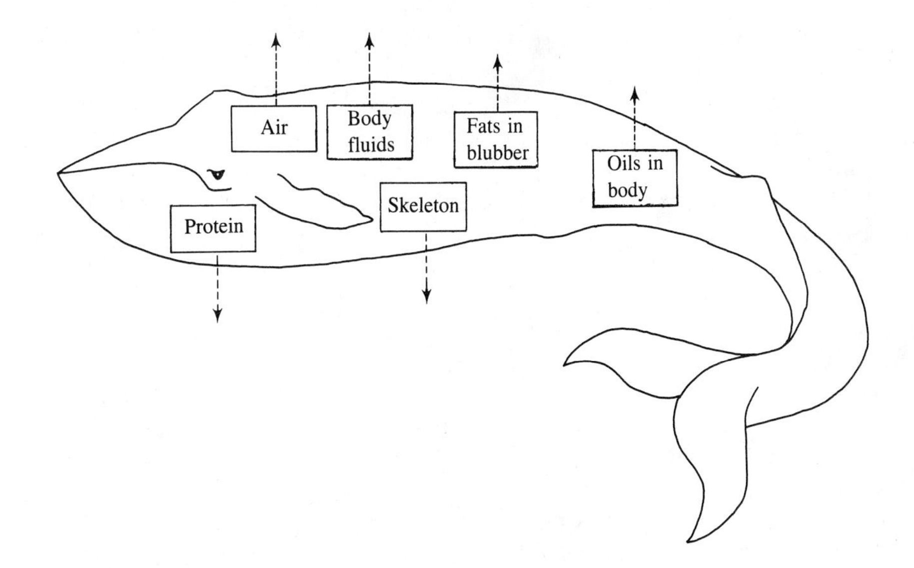

FIGURE 8.6
Buoyancy in Whales

In the blue whale, a combination of oil, fat, air, and body fluids less salty than sea water contribute to neutral buoyancy.

percent of the air in the lungs in less than a second. Exhalation often leaves a cloud of condensed water (up to 6m high in blue whales). Whales can be identified by the particular shape and size of their exhalation. Inhalation is equally rapid, and the full breathing cycle can be completed without breaking the swimming pace. Cetaceans, especially the large whales, use the oxygen in the air of the lungs more efficiently than land mammals. The greater number of capillaries in cetacean lungs allows 50 percent of the oxygen to be removed in a breath, compared to 20 percent in terrestrial mammals.

Most cetaceans are negatively buoyant, and sink when dead. The details of buoyancy control are not well known. In some way, the combined buoyancy of the blubber and the air in the lungs provides neutral buoyancy (Fig. 8.6). Water contributes greatly to the buoyancy of cetaceans. Large whales—for example, the blue whale—cannot support their body mass without the aid of the high-density water. Out of water, the skeleton and internal organs are crushed under their own body weight.

Cetaceans are able to stay under water at moderate depths for prolonged periods. Most cetaceans stay in the top 50 m (164 ft) of the water column, where their food is concentrated. Deeper dives are frequent, however. Porpoises have been trained to retrieve objects at 330 m, and baleen whales can dive over 350 m. The sperm whale regularly feeds down to depths of 1,000 m (3,280 ft), and has been recorded feeding to a maximum depth of 2,200 m. While diving, sperm whales have become entangled in undersea cables and have drowned. These unfortunate incidents have provided the best information on the maximum dives of sperm whales.

The length of dives varies. Generally, the baleen whales in their relatively shallow dives remain submerged for 5–15 min, but can stay down for up to 50 min. The sperm whale can stay submerged for 75–90 min in its deep-sea dives (Tab. 8.2). One of the problems humans have in air-assisted diving is that the pressurized air causes a buildup of nitrogen gas in the blood. If a diver returns to the surface too quickly, without adequate decompression, the nitrogen dissolved in the blood forms bubbles that cause a condition known as the bends. Since cetaceans do not have any

TABLE 8.2
Cetacean Diving

Cetaceans are capable of extended dives. Sperm whales dive longest and deepest. (From Harrison, R.J., and Kooyman, G.L. 1971. *Diving in marine mammals.* New York: Oxford University Press.)

	Duration (min)	Depth (m)
Porpoise *(Phocoena phocoena)*	12	20
Dolphin *(Tursiops truncatus)*	12	170–300
Pilot whale *(Globicephala scammoni)*	15	366
Sperm whale *(Physeter catodon)*	75	900–1,134
Fin whale *(Balaenoptera physalus)*	20	355–500 500
Blue whale *(Balaenoptera musculus)*	49	100

air in their lungs while diving, they do not suffer from the bends. Cetaceans have adaptations that seal off the blow hole while diving. In porpoises (Fig. 8.7), there is a strong sphincter muscle that helps keep the trachea closed.

The physiology of cetaceans is also adjusted to diving. The blood has a higher concentration of red blood cells than that of terrestrial mammals, and therefore can carry more oxygen. When a cetacean dives, the heart rate slows and circulation to the muscle mass of the body is decreased. For example, some seals have a heart rate of 100–150/min on the surface. This drops to around 10/min while diving. While diving, blood is directed primarily to the brain, and only a small amount goes to other vital organs. The muscles, however, are not without oxygen. Cetaceans have up to nine times more myoglobin (similar to hemoglobin) in their muscle compared to land mammals. This myoglobin provides oxygen to muscles during diving. Finally, when the oxygen of the muscles is depleted, cetaceans rely on anaerobic metabolism to provide a continual energy source. Anaerobic metabolism is less efficient than aerobic (see the special feature at the end of this chapter), and cetaceans have a higher tolerance for its by-product, lactic acid, than do land mammals. Marine mammals also are more tolerant of high carbon dioxide concentrations. In land mammals, elevated carbon dioxide concentrations automatically cause breathing; cetaceans do not have this reflex.

Food and Feeding

All cetaceans are carnivorous and capture live animals for food. The basic division of cetaceans into the baleen and toothed species reflects the very different methods of food capture: The baleen whales are filter feeders, and toothed cetaceans actively pursue their prey. Baleen whales feed mostly on small zooplankton, while the toothed cetaceans feed mostly on larger fish and squid.

The filtering mechanism of baleen whales is composed of a series of plates that hang down from the gum region of the upper jaw (Fig. 8.8). These plates are not teeth but, actually, outgrowths of the palate. The plates are made of keratin, a structural protein that in other mammals forms hair, fingernails, and claws. The baleen plates number between 150–400 on each side of the mouth. Their outer edges are smooth, and they resemble the slats of a venetian blind. Their inner edges are frayed into threads that intertwine to produce the filter.

Although all the Mysticeti whales use baleen to filter their food, there are some basic differences in feeding methods. Perhaps the most straightforward feeding method is that of the bowhead and right whales. These whales have the longest baleen (in bowheads the plates are up to 3 m [9.7 ft] long). When the mouth is closed, the baleen folds inward; when the mouth opens, the baleen plates spring forward. Bowhead and right whales swim slowly through the water, sieving small zooplankton from the water. They feed mostly on copepods (called brit by early whalers), but may also take other zooplankton such as euphausiids, sea butterflies, mysids, and amphipods.

A large part of the body of the bowhead and right whale is taken up by the baleen filter, and the head is up to one-third of the total body length. Because of the massive head, the body is not well streamlined, and right and bowhead whales are the slowest swimmers of the great whales. This slow swimming suits their feeding needs, however. The most effective swimming speed for filtering is around two knots. The slow swimming speeds of these whales made them the favorite, or 'right' whale, for early whalers to capture.

The rorqual whales (blue, sei, minke, humpback, and Bryde's) use another method to filter feed. With their smaller heads and more streamlined bodies, these whales are faster swimmers than the right or bowhead whales. The rorqual method of feeding is a form of gulping. A series of pleated folds runs ventrally from the snout to almost the navel, surrounding a cavity that is connected to the throat. This cavity can expand to immense proportions. In the blue whale, the cavity has a capacity six times the volume of the animal. Rorquals feed by locating a dense patch of zooplankton. By inflating the cavity, they suck in a large quantity of water and plankton (up to 100 tonnes). When the mouth closes, the 'inhaled' water is sieved through the baleen and the zooplankton are trapped. Rorquals also eat small schooling fish such as anchovies, herring, and sardines.

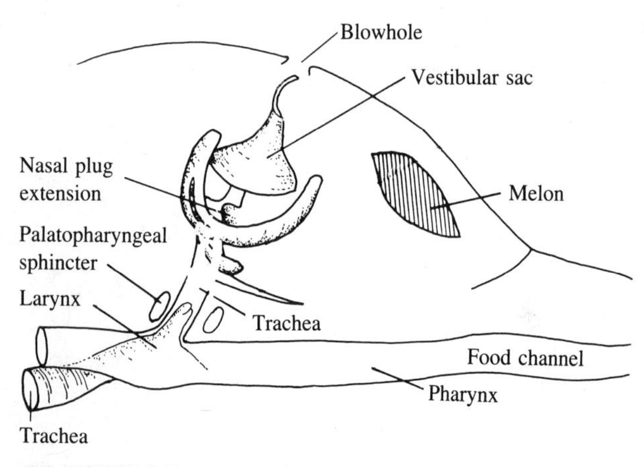

FIGURE 8.7

To prevent water from entering the lungs while submerged, cetaceans have a well-developed cartilaginous larynx that is held firm by a sphincter muscle.

FIGURE 8.8
Filter Feeding in Whales

The simplest method of filter feeding by whales is that of the right whale. Rorqual whales (i.e., humpback and blue whales) feed a little differently (see text). **(A)** To filter small organisms from the water, the baleen whale has a highly specialized mouth. The horny baleen plates hanging down from the roof in a single row on each side act as a sieve when the animal feeds. The cavernous, expandable lower portion of the mouth acts somewhat like a scoop. **(B)** As the whale swims along, it opens its mouth and takes in huge quantities of water filled with small organisms. The whale then closes its mouth and raises its massive tongue. This forces the water out between the baleen plates while small invertebrates and fish are trapped inside, next to the plates.

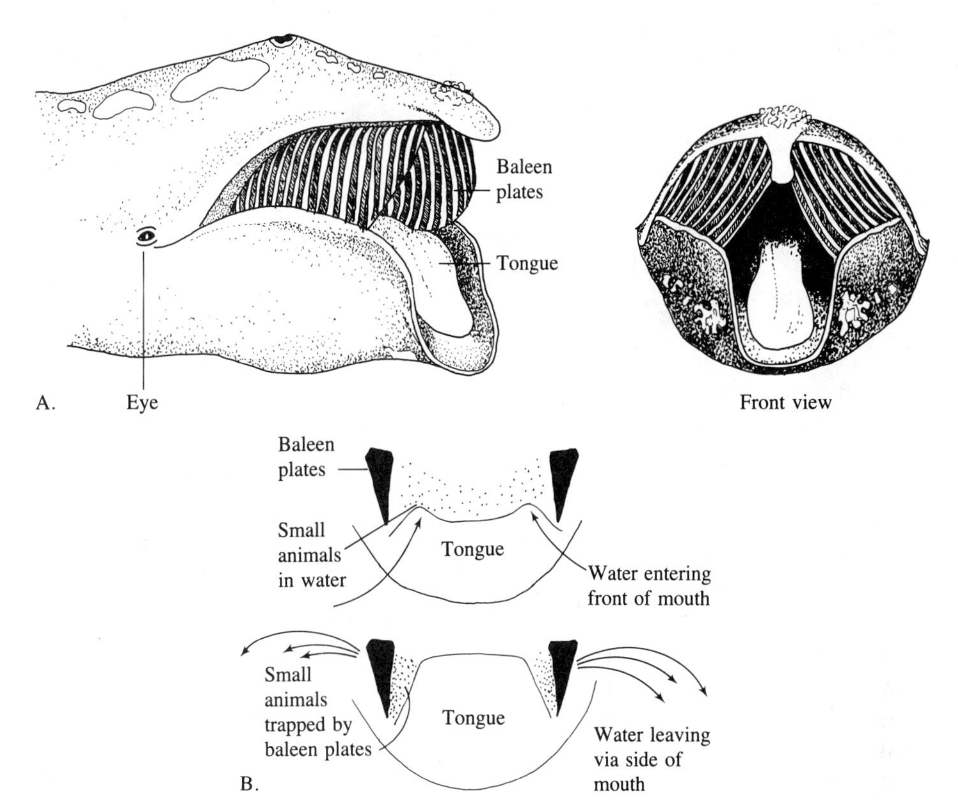

Two species of rorqual whales can concentrate their food by 'herding' it. The finback whale has a large patch of white on the right side of the head and jaw. The whale uses this color to startle small fish into tight schools. When the food has been concentrated, the whale feeds. The humpback whale may concentrate its food by bubble-net feeding. When feeding this way, humpbacks come up from below on the food source (zooplankton or small fish) and gradually encircle it, sending a steady stream of bubbles up from the blow hole. The bubbles concentrate the food, which is then gulped in the regular way of rorquals. Humpback whales bubble-net feed either individually or in pairs.

The North Pacific gray whale uses a third method of feeding. The gray whale has relatively short baleen plates. Its main food source is benthic invertebrates. Although amphipods are the most common food, other benthos—polychaetes, holothurians, tunicates, etc.—are also eaten. The gray whale feeds by swimming on its side across the surface of the bottom, and scoops or sucks up a mouthful of the surface sediment. The sediment is then sieved through the baleen and the animals retained. Gray whales usually feed on their right side, and typically the baleen on this side is more worn. In areas of active feeding, the bottom sediments are marked with shallow trenches from the feeding. Gray whales also feed on small surface fish. They swim quickly up through the school, gulping as many as possible. Since the fish are close to the surface and the whales approach rapidly from below, this type of feeding is accompanied by repeated breaching of the whale. A summary of feeding types is shown in Figure 8.9.

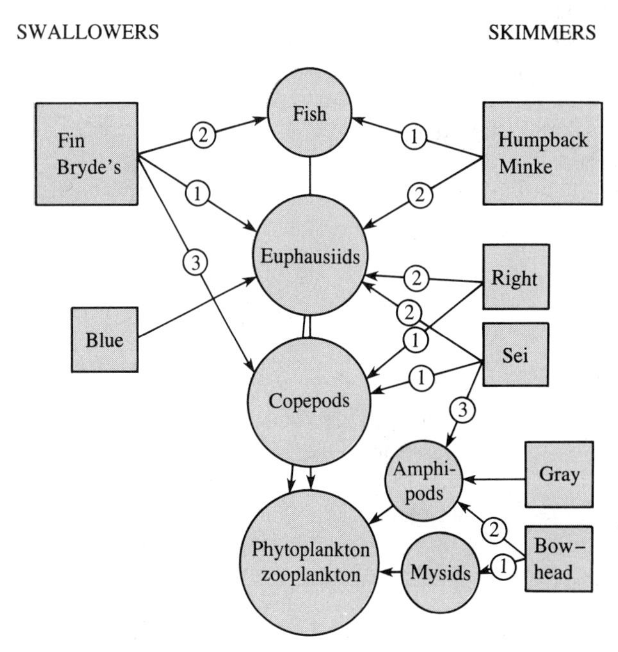

FIGURE 8.9
Feeding Preferences of Baleen Whales

Different whales favor different food. (Circled numbers indicate order of preference.)

All Odontoceti (toothed) cetaceans actively pursue prey and generally take individual animals. Their main foods are fish and squid, but the killer whale may also feed on other marine mammals. The sperm whale, the largest of the Odontoceti, eats mostly squid and ranges widely through the depths of the epipelagic and mesopelagic zones in search of its prey (Fig. 8.10). Like other Odontoceti, the sperm whale swallows these prey whole. Sperm whale teeth, which occur only on the lower jaw, are not used for biting. Some biologists believe that the teeth are used primarily to attract squid. While hunting in the depths of the mesopelagic zone, sperm whales swim with their mouths open. The bright reflectiveness of the 18–25 pairs of white teeth is thought to act as a lure to the squid. Although most of the squid that the sperm whale feeds on are about 1 m long, there are many examples of sperm whales attacking giant squid of the genus *Architeuthis*. These squid, up to 20 m (66 ft) in length and weighing some 4,000 kg (8,800 lbs), can put up a strenuous fight, as evidenced by the numerous scars on sperm whales killed in commercial whaling. Sperm whales also feed close to the bottom, eating benthic fish, octopods, and invertebrate epifauna.

The limited use of the teeth for grasping in most Odontoceti is clearly shown in the beaked whales. These whales feed almost exclusively on squid and have only a few pair of teeth in their jaws. Dolphins and porpoises are mostly fish-eaters, although they take squid when available. They prey on a wide range of fish that they catch with their teeth and swallow whole.

One species of Odontoceti can use the teeth to bite chunks of flesh. The killer whale, with 10–14 pairs of teeth in both jaws, has a wide range of food types, including fish and other marine mammals. One study showed a consumption of 20 seals by one killer whale in a single week. The specific food depends on where killer whales feed. In the Antarctic, killer whales take mostly seals and penguins close to the pack ice. Farther from the ice, the preferred food is the common minke whale. Still farther from Antarctica, killer whales feed mostly on dolphins and fish. In the Puget Sound region of the Pacific Northwest, one group of killer whales living in the Sound eats only fish and coexists peacefully with other marine mammals. Another group living offshore in the open oceans actively pursues marine mammals as well as fish. (See further discussion in the special feature at the end of this chapter, "Competitive Release: Whales, Krill, and Humans.")

There are numerous reports of killer whales attacking large baleen whales. Only the sick, injured, pregnant, or young are usually attacked. When attacked, gray whales may form a circle to prevent the killer whales from reaching any one individual. There are no documented cases of killer whales attacking or killing humans.

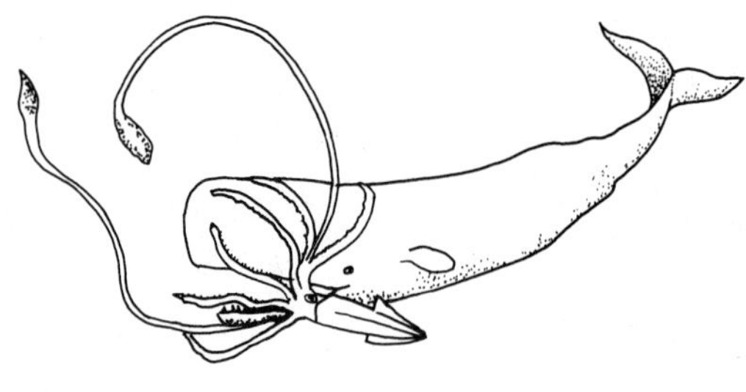

FIGURE 8.10
Sperm Whale

Sperm whales feed mostly on squid.

Migrations

Most cetaceans have a large home range through which individuals continually move. Often the movements are seasonal migrations. In general, the smaller dolphins and porpoises, including the killer whale, do not have set migration routes, whereas the larger whales, including most of the baleen whales and the sperm whale, have well-defined migrations. Migratory movements are related to patterns of feeding and reproduction. Cetaceans generally prefer to have their young in warm tropical waters. Such waters, however, tend to be impoverished in food supplies, so feeding journeys are made to the more productive temperate and high latitudes. The longest migration of any cetacean is that of the North Pacific gray whale, which travels along the west coast of North America (Fig. 8.11). The favored feeding grounds of the gray whale are the Bering Sea and Arctic Ocean. The wintering grounds, where gray whales mate and give birth to young, are on the coast of Mexico. The annual migration may be as long as 16,000 km (9,600 miles). While migrating, they swim at a steady 8–9 km/hr (5 mph) and make the trip in two or three months. Gray whales travel singly or in small groups of up to 12 animals, swimming close to shore, often within a mile of land. They feed little, if at all, while enroute or at the wintering grounds. The fat accumulated in six months of sustained feeding during the summer allows the whales to fast for the rest of the year.

Other active migratory whales include the blue and sperm whales. The huge blue whales feed almost exclusively on euphausiids (krill) during the four to six months they spend on the antarctic feeding grounds. During the summer feeding they eat up to four tonnes of krill each day, around 40 million euphausiids.

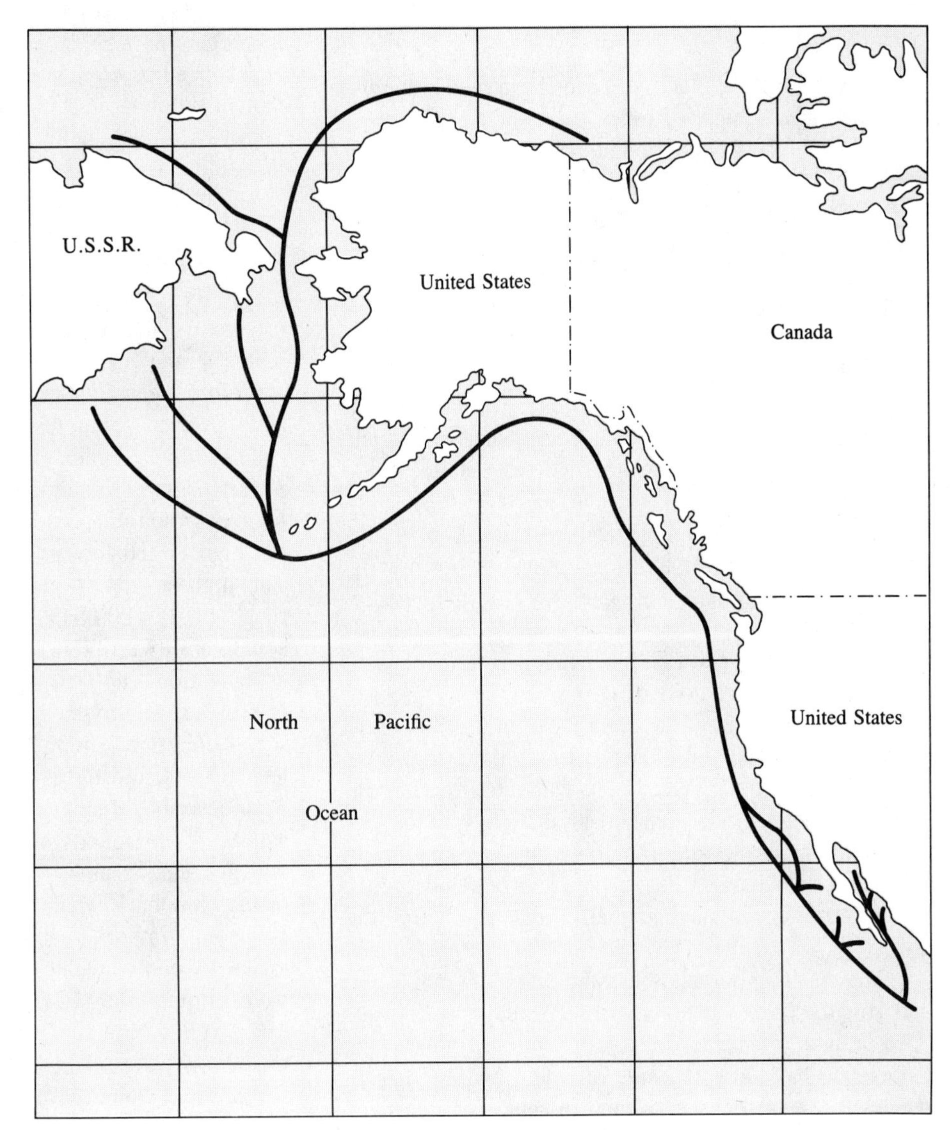

FIGURE 8.11
Gray Whale Migration

The longest cetacean migration is by the California gray whale. Whales move northerly in the spring, and southerly in the fall.

Sperm whales have more complex migratory movements. The females stay within the tropics or travel no farther than 40° north or south. Males, however, make longer migrations, and may travel to higher latitudes for summer feeding.

Not all whales have extensive migrations between the tropics and higher latitudes. Right whales stay the entire year in temperate latitudes, and mostly avoid both the tropics and high latitudes. Finback whales are found off the coast of Mexico at all times of the year.

Reproduction

All cetaceans give birth at sea. In most whales, reproduction is seasonal and an integral part of their annual migratory movement. Dolphins and porpoises may have distinct, seasonal reproductive periods, or be reproductively active all year. Birth is often with the assistance of other mature females, who insure that the newborn is brought quickly to the surface to breathe. Cetacean babies do not suckle, however. Instead, the young animal grabs the female's teat and the mother injects a dose of milk into its throat. The feeding process is completed in a few seconds.

Among the great whales, the reproductive behavior of the baleen whales differs from that of the sperm whales. Baleen whales are monogamous for at least one reproductive period. Females generally give birth every other year, and nurse the young for six to 18 months. Sperm whales, however, are polygamous, which is probably the reason for the much larger size of the males compared to females. The largest of the male sperm whales form harems. Mothers nurse their young for two years and breed only every four or five years.

Senses and Intelligence

Whether whales and other cetaceans have humanlike intelligence is a subject of much discussion and controversy. There is no doubt that cetaceans, like other mammals, have a rich sensory system and exhibit complex behavior. Cetaceans have large brains and, like humans, have a well-developed, folded cerebral cortex (Fig. 8.12). From work with bottlenose dolphins, it is clear that only part of the capacity of the cerebral cortex is needed for integrating sensory input and coordinating motor behavior. Although the general pattern of function of the various areas of the cortex is similar to other mammals, some areas are unique to cetaceans. Much more remains to be learned, however. Echolocation, which is so important to cetaceans, has not yet been associated with any morphological specialization of the brain. A knowledge of brain anatomy has not greatly increased our understanding of the specific intelligence of cetaceans.

Brain size itself can be misleading. Brain size generally increases with an increase in overall body size, so the larger size of the whale brain is not in itself an indicator of intelligence. Within cetaceans, there are some interesting variations. The sperm whale has the largest brain size of all cetaceans, even though it is by no means the largest whale. The relatively large brain size of the sperm whale, killer whale, and bottlenose dolphin is probably related to their active predatory life styles.

The most important senses in cetaceans are sight and sound. Smell, taste, and touch are poorly developed in most cetaceans, and totally absent in some. The Odontoceti have lost the sense of smell, and the baleen whales the sense of taste. Touch in cetaceans is generally restricted to the face and mouth. Touch-sensitive tubercles, many with bristles, are common on the lower and upper lips of many whales. Vision is more important in dolphins and porpoises than the larger whales. Whales, because of the lateral placement of the eyes, have blind spots and do not have binocular vision.

Sound is the single most important sense in cetaceans. This is not immediately apparent because their ears are reduced and they have no sound-producing vocal cords. Cetaceans have evolved, instead, a sound production and reception system as specialized as any in the animal world (Fig. 8.13). Sound travels farther and more clearly underwater than in air. Cetaceans use it for social communication and for navigation and echolocation. Sounds used in social communication are mostly in the audible range of humans, and sounds used for navigation and echolocation are, for the most part, too high or too low in pitch to be audible to humans. The highest sounds perceptible to humans are in the range of 15–20 kHz, whereas cetaceans can detect sounds up to 150 kHz. Bats, which also use sound for echolocation, can hear sounds up to 175 kHz.

Baleen and toothed whales use different sounds for social communication. The low-pitched sounds of baleen whales, mostly in the range of 20–30 kHz, have been described as moans, groans, rumbles, grunts, and squeaks. The sounds of the Odontoceti are more like whistles and screams. Among baleen whales, the 'song' of the humpback whale is perhaps the most interesting. The song, important in courting behavior, is used by males to communicate with each other. Adult males have complex phrases of sounds between 150–180 kHz, well within the human range of hearing. These melodious songs, which may be repeated for between seven to 36 min, have fascinated many people. The songs are specific to individual whales and gradually

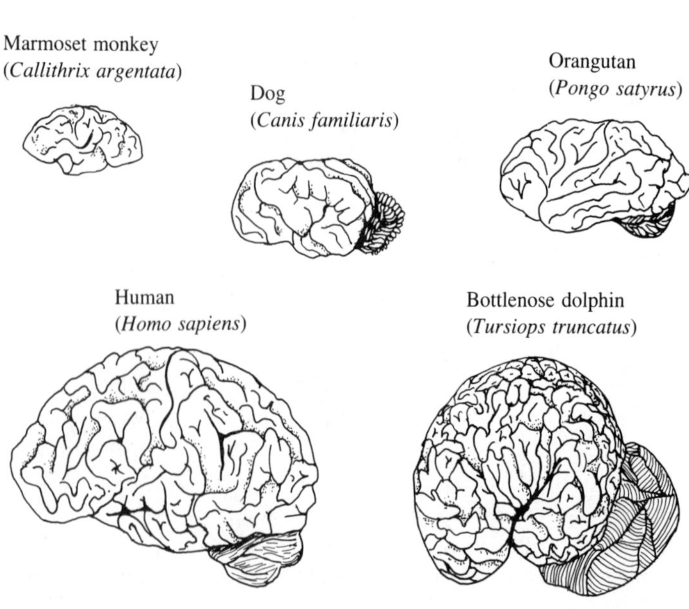

FIGURE 8.12
Cetacean Brain Size

On a per-body weight basis, the size of a dolphin brain (representative of cetaceans) is only slightly smaller than the human brain.

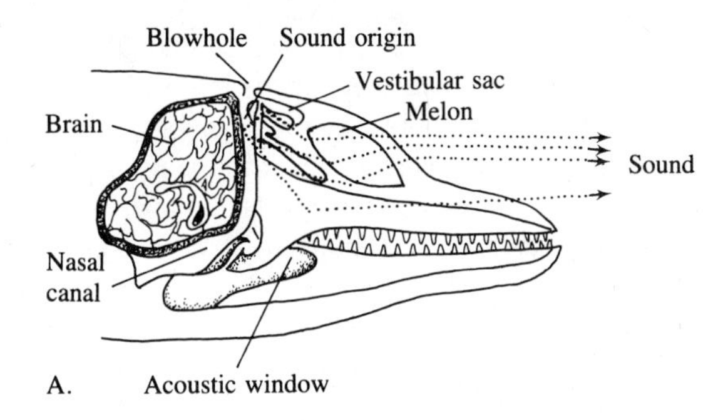

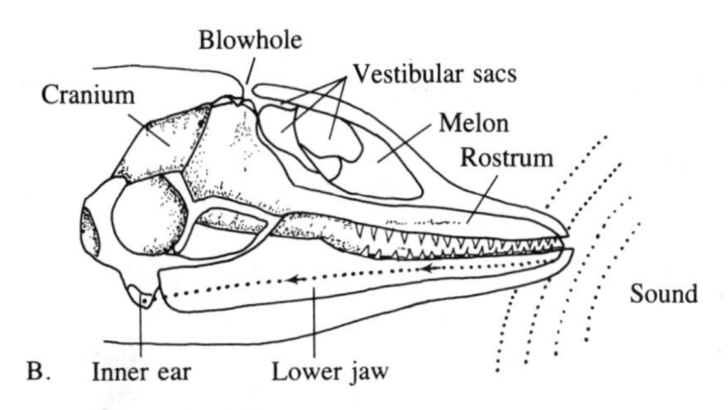

FIGURE 8.13
The Sense of Sound in Cetaceans

(A) Cetaceans produce sounds not only from vocal cords but also from a series of air sacs below the blowhole. **(B)** Sounds are received through the lower jaw and then transmitted to the inner ear.

change during the year. Although each individual sings the same basic song each year, there are small individual differences. Each song is unique.

Other cetaceans also have songs specific to individuals. These sound signatures have been found in a number of baleen whales, as well as sperm whales, killer whales, and dolphins. The signature sounds of individuals allow scientists to identify the positions of various whales in the social order. Whales can recognize the sounds of other species. The California gray whale will turn and flee from the screaming sound of the killer whale, and beluga whales will avoid areas where killer whale sounds are present.

A number of the Odontoceti use a distress whistle in times of danger and attack. In some species, such as the sperm whale and the bottlenose dolphin, the distress whistle brings help, but in others, such as the common dolphin, the distress whistle is avoided by other individuals.

There is some speculation that whales use low-frequency sound for long-distance communication. Sound travels about five times faster in water than air, and decays in intensity much more slowly. Some biologists think that sound sent out at the depth of the SOFAR channel travels hundreds or even thousands of kilometers. Most of the work on this question is classified, presumably for military reasons, and our understanding of this particular use of sound is limited.

Cetaceans use echolocation like radar to navigate and to locate objects. High-pitched sounds, mostly above the range of human hearing, are sent out and the echoes that bounce off solid objects are used to estimate the distance and identify the object. The high-pitched sounds characteristic of echolocation are produced by all cetaceans, including baleen whales. The use of echolocation by baleen whales, however, has not been verified, and the evidence for sperm whales is only circumstantial. Our detailed knowledge of how echolocation is used in cetaceans comes from experiments with dolphins and porpoises. The results are impressive. Dolphins can locate a single lead shot, 1 mm in diameter, thrown into a pool. They also can discriminate between aluminum and copper discs in the water. Their amazing discrimination is not limited to strong sound reflectors. Dolphins can locate a 1 cm-long gelatin capsule or a small piece of fish from a distance of 6–7 m. Dolphins also can focus their echolocation. For longer distances, loud low-frequency sounds that give poor resolution are used. At shorter distances, high-pitched short sounds are used for resolution of fine detail.

The sounds used for echolocation are received by external ears consisting of tubes connecting the inner ear to the exterior, which are often plugged with wax. In dolphins, at least, the jawbone that abuts the ear is apparently used for long-wavelength sound reception (Fig. 8.13). The melon, a lump of soft tissue on the head, is also related in some way to sound reception or production.

There is a great deal that remains to be learned about the role of sound in social communication and echolocation in cetaceans. We know enough, however, to be sure that sound is very important in the sensory world of cetaceans.

Can cetaceans use their sense of sound to communicate with humans? Many scientists passionately hope so and pursue research to that end. What, then, of our original question about the intelligence of cetaceans? Some cetacean scientists believe that whales have an intelligence that is more like that of humans than of any other animal, with the possible exception of other primates. Most, however, caution that until we know more about social communication and the function of the nervous system in whales, dolphins, and porpoises, we should reserve judgment about whale intelligence. Since a clear definition of intelligence in humans remains elusive, intelligence in cetaceans, as with other mammals, can only be used as a general term to describe complex behavior and nervous systems. Specific comparisons between cetacean and human intelligence is premature.

PINNIPEDS

The mammalian suborder Pinnipeda (seals, sea lions, and walruses) consists of some 31 species of mammals (Fig. 8.14). Pinnipeds are amphibious in that they must return to land (or ice flows) to mate and bear their young. Many pinnipeds, such as the California sea lion and harbor seals, stay close to land throughout the whole of their lives, and spend much of their time on land. Other species, like the northern fur seal, spend all their time at sea except for a four-month breeding period. Pinnipeds are found mostly in temperate and polar waters. Warm-water species, such as the Hawaiian monk seal and the Mediterranean seal, are not abundant, and the Caribbean monk seal is now extinct.

There are three families of pinnipeds: the Otaridae (eared seals), the Phocidae (true, or earless, seals), and the Odobenidae (the walrus). Otarids, which have short, external ears, include the northern fur seal and sea lions. The phocids, or true seals, have no external ears and include bearded, ringed, harbor, and elephant seals. There are important differences in locomotion between otarids and phocids. Otarids are walkers, and the phocids are wrigglers. The hind flipper of otarid pinnipeds can be turned under the body, allowing them to walk on all fours (Fig. 8.15). The phocid seals, conversely, cannot turn their flippers forward, so when moving across the land the body must be dragged by the front flippers. This difference affects swimming as well. Otarids use the front flippers as oars and the back flippers as a rudder. Phocid seals paddle with their rear flippers and steer with their foreflippers.

In general, pinnipeds are much smaller than cetaceans. The smallest of the pinnipeds is the ringed seal (90 kg); the largest, the southern elephant seal (629 kg). Although not as completely adapted to the marine environment as the cetaceans, pinnipeds share many characteristics with them. The pinniped body is generally streamlined, with reduced neck and ears and covered genitalia. A thick layer of blubber, which makes up 25 percent of the body weight, provides insulation, buoyancy, a food reserve, and streamlining.

Swimming and Diving

Pinnipeds are swift swimmers. The northern fur seal can sustain a speed of 25 km/hr and the California sea lion 40 km/hr (24 mph). Pinnipeds also are effective divers. The antarctic Weddell seal can dive to 600 m (1,970 ft) for more than an hour. The physiological adaptations of diving in pinnipeds are similar to most of those of cetaceans and include reduction in heart rate, constriction of peripheral blood vessels, and tolerance to high levels of carbon dioxide and lactic acid in the blood.

Reproduction

Pinnipeds are generally gregarious, particularly on the rookery, or breeding grounds. Because pinnipeds require isolated shores protected from predators and breeding sites, the amount of space available for reproduction is relatively small. Most otarids and the ele-

FIGURE 8.14
Pinnipeds

(A) Harbor seal. **(B)** Stellar sea lion. **(C)** Harp seal. **(D)** Walrus.

FIGURE 8.15 Comparative Anatomy of Phocids and Otarids

The difference between the true or earless seals (Phocidae, A) and the fur seals (Otaridae, B) is clearly seen in the skeletons. The otarids can use the hind legs in walking; the phocids cannot.

Phocidae (earless seals)
No external cartilaginous ears
Small foreflippers (used mainly for guidance)
Hind flippers cannot be turned under body (slow and clumsy on land)
Breeding pattern varies
Furred flippers

Otaridae (fur seals)
Visible cartilaginous ears
Large foreflippers (used for standing, swimming, and steering)
Long hind flippers can be turned under body to walk on land
Well-defined rookeries and breeding schedule
Partly furred flippers

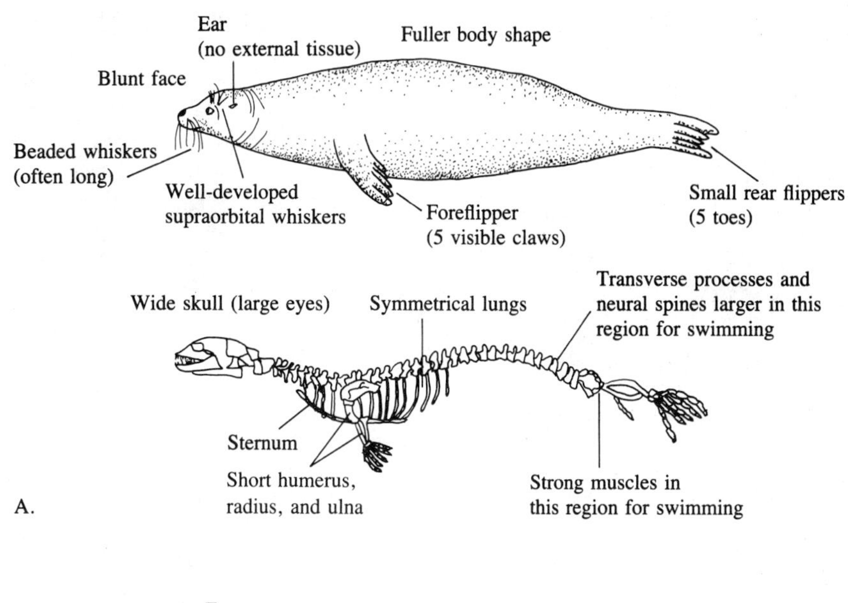

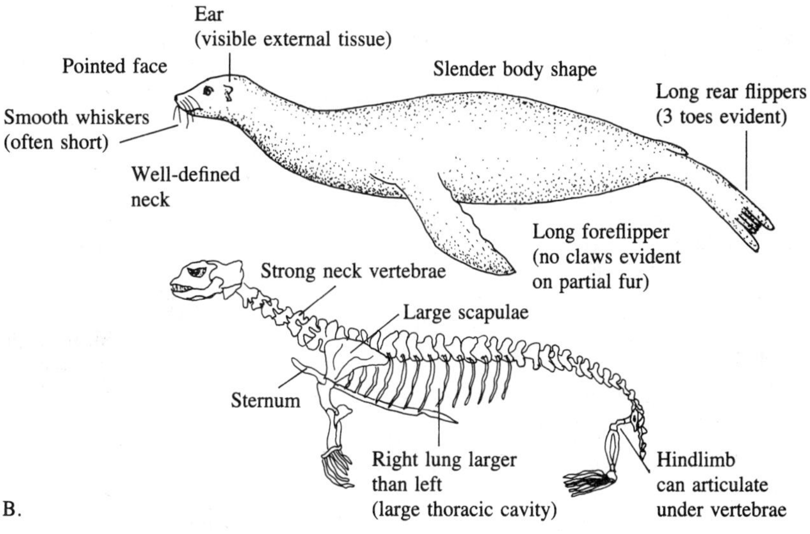

phant seal have a harem structure in which only the large, aggressive, dominant males are able to mate with the females. The dominant males arrive on the breeding grounds first and establish territories to which they try to attract females. These harem bulls do not leave the breeding ground for the approximately two months of the breeding season. They spend considerable time defending their territories from challenging males, often fighting ferociously. A harem bull can maintain his dominance for only two or three years, after which he is deposed by a younger, stronger, more aggressive male.

Walruses also may have harems and be polygamous. The behavior of the harem bull, however, is not as ritualized as that of sea lions and elephant seals. Most phocids, or true seals, do not form harems. Many species of seals are promiscuous but in a few the male may form a monogamous relationship with a female for the length of the breeding season. Other than copulation, male pinnipeds are not involved in family life.

In many pinnipeds, females mate soon after giving birth, but the fertilized egg does not develop immediately. It may lie in a quiet state in the uterus for a number of months. This delayed implantation, which occurs in all pinnipeds except the walrus, insures the correct timing of the birth.

Food and Feeding

Pinnipeds have a wide range of food habits, although all are carnivores. The crab-eater seal of the Antarctic feeds entirely on krill. Diets are highly varied, the most important consideration being the size of the prey. Pinnipeds prefer to swallow their prey whole, and their teeth function mostly for grasping. They can, however, tear large prey to pieces, which are swallowed without chewing.

Harbor seals and northern fur seals eat shellfish incidentally. The walrus, however, has adaptations that permit using clams as the primary food. The mouth and tongue create a strong suction that pulls the clam out of its shell. Walruses swim along the bottom using their tusks like sled runners and the bristles on their cheek pads to dig out the clams. Walruses also use their tusks to haul themselves out on ice flows, where they roost and reproduce. (See the special feature—"California Gray Whales and Walruses, Geologic Agents in the Bering Sea"—at the end of this chapter for further discussion of walrus feeding habits.) At least one pinniped species actively seeks out larger prey: The leopard seal of Antarctica feeds mostly on penguins and other seals.

Senses

Sight and sound are the most important pinniped senses, although smell is important in reproduction. Females recognize their young on the basis of smell, and males determine if females are sexually receptive by smell. Vision is good on both land and in the water. Unlike cetacean eyes, pinniped eyes are close together, thereby allowing binocular vision. We do not know if sound is as important in pinnipeds as it is in cetaceans. Many pinnipeds are vocal, and the sounds used for social communication are in the range of human hearing. Some pinnipeds are sensitive to the high-pitched sounds used by cetaceans in echolocation, and California sea lions themselves may echolocate. In the absence of confirming evidence, however, we must conclude that sight is more important to pinnipeds than sound.

SUMMARY

Marine mammals adapted to ocean conditions only recently. The first fossil remains date back to around 45 million years ago. Marine mammals are clearly derived from terrestrial ancestors, and probably reached the oceans through fresh water. The modern groups of whales appear in the fossil record about 25 million years ago. The whales and porpoises (cetaceans) are not closely related to the seals and sea lions (pinnipeds).

Marine mammals range widely over the world's oceans, and yearly migrations of thousands of miles are not uncommon. All groups of marine mammals are good divers. Cetaceans spend most of their time under water. All marine mammals are dependent on the air for a supply of oxygen.

Marine mammals are top predators. The one species that stands above the rest is the killer whale. It eats fish as well as other marine mammals including sea otters, seals, and sea lions, and even attacks large whales.

The largest of the marine mammals are the great whales. Of these, the largest is the blue whale, which may reach 30 m in length and 150 metric tonnes in weight. All the great whales are predators, but only the sperm whale actually pursues individual prey. Sperm whales dive to depths over 1,000 m (3,280 ft) to capture squid and other animals. The other great whales, collectively called baleen whales, are predators on zooplankton. Their method of feeding is to filter zooplankton through the sheets of baleen in their mouths.

QUESTIONS AND EXERCISES

1. Discuss the advantages and disadvantages of a large body size in marine mammals.
2. Which of the marine mammals are amphibious, or do not spend all their lives in water? What are the advantages of this amphibious existence, if any?
3. Discuss the relationship between marine mammal size and the type of food consumed.
4. Discuss the general patterns of distribution of marine mammals in the marine environment. Consider latitude, water depth, and proximity to land.
5. Discuss the adaptations of marine mammals to deep diving. Draw a graph that show the depths to which various marine mammals dive.
6. What can absolute and relative brain size tell us about intelligence in marine mammals?
7. What do marine biologists mean when they say that marine mammals use sound in the way that terrestrial animals use sight? Does this apply to all marine mammals?
8. How does the echolocation system of dolphins work?

REFERENCES

Bare, C.S. 1986. *Sea lions.* New York:Dodd Mead.

Crane, J.M. 1973. *Introduction to marine biology: A laboratory text.* Columbus, OH:Merrill.

Doak, W. 1989. *Encounters with whales and dolphins.* New York:Sheridan House.

Ellis, R. 1981. *The book of whales.* New York:Knopf.

Ellis, R. 1982. *Dolphins and porpoises.* New York:Knopf.

Friends of the Earth. 1978. *Diving in marine mammals.* New York:Oxford University Press.

Haley, D. 1986. *Marine mammals.* Seattle, WA:Pacific Search Press.

Harrison, R., and **Bryden, M.M.**, eds. 1988. *Whales, dolphins, and porpoises.* New York:Facts on File.

Jones, M.L., **Swartz, S.L.**, and **Leatherwood, S.**, eds. 1984. *The gray whale,* Eschrichtius robustus. Orlando, FL:Academic Press.

King, J.E. 1983. *Seals of the world.* Ithaca, NY:Cornell University Press.

Minasian, S.M., and **Balcomb, K.C., III.** 1986. *World's whales.* New York:Norton.

Nakamura, T. 1988. At sea with the humpback whale. San Francisco, CA:Chronicle Books.

Nelson, H. and Johnson, K.R. 1987. Whales and walruses as tillers of the sea floor. *Sci. Amer.* 256(2):112–118.

Nickerson, R. 1984. *Sea otters: A natural history.* San Francisco:Chronicle Books.

Unterbrink, M. 1984. *Manatees: Gentle giants in peril.* St. Petersburg, FL:Outdoor Publishers.

SUGGESTED READING

OCEANS

Allan, D. 1984. Diving with crabeaters. (Antarctic crabeater seal) 17(6):3–7.

Barr, L. 1975. Steller sea lion. 8(4):18.

Butcher, A. 1981. Navy's underwater allies. (Navy training program for dolphins) 14(6):3–5.

Curtis, P. 1987. Contact with dolphins. 20(1):18–23.

Edelbrock, J. 1975. Manatees: Sirens of the sea. 8(6):66.

Ellis, R. 1987. Why do whales strand? 20(3):24–29.

Gordon, J. 1982. One whale less. 15(1):30–31.

Haley, D. 1980. The great northern sea cow. (The extinct Steller's sea cow) 13(5):7–11.

Hoyt, E. 1977. *Orcinus orca.* (The killer whale) 10(4):22–26.

Leatherwood, S., and **Cornell, L.** 1985. Birth of a dolphin. 18(1):46–49.

Lilly, J.C. 1977. The cetacean brain. 10(4):4–7.

Marine mammals: Biology and ecology. 1981.

McNally, R. 1977. Echolocation. (Cetaceans' sixth sense) 10(4):27–33.

Nichols, G. 1975. *Eschrichtius robustus.* (The gray whale) 8(3):60.

Pich, W.C. 1985. The whale's pearl. (Ambergris) 18(3):23–25.

Pinnipeds, seals, sea lions, walrus. 1980. 13(3):11–57.

Reynolds, J.E., III. 1977. Precarious survival of the Florida manatee. 1975. 10(5):50–53.

Sea mammals. 1975. 8(3):32–65.

Storro-Patterson, R. 1980. Larr Foster whale photographs. 13(1):14–20.

von Ziegeson, O. 1983. At ease in the environment. (Pygmy killer whale) 16(6):36–45.

Whales. 1984. 18(4).

OCEANUS

Marine mammals. 1978. 21(2).

Watkins, W.A. 1977. Acoustic behavior of sperm whales. 20(2):50–58.

Whales and dolphins. 1989. 32(2).

SCIENCE 80

Ellis, R. 1980. The rarest large animals. 1(6):82–87.

SCIENTIFIC AMERICAN

Johnson, R. 1987. Whales and walruses as tillers of the sea. 256(2):112–118.

Kooyman, G.L. 1969. The Weddell seal. 221(2):100–107.

Rudd, J.T. 1956. The blue whale. 195(6):46–65.

Whitehead, H. 1985. Why whales leap. 252(3):84–93.

Wursig, B. 1979. Dolphins. 240(3):136–148.

Wursig, B. 1988. The behavior of baleen whales. 258(4):102–107.

Zapol, W.M. 1987. Diving adaptations of the Weddell seal. 256(6):100–107.

SEA FRONTIERS

Aube, L.J. 1989. *Orca.* 35(6):320–325.

Forcier-Beringer, A.G. 1986. Talking with dolphins. 32(8):84–92.

McHugh, J.L. 1986. Whales have riders too. (Whale ectoparasites) 32(4):252–259.

Paulson, A.C., and **Lang, J.S.** 1988. The Pacific walrus. 34(3):152–159.

Sobey, E.J. 1986. Listening to gray whales. 32(1):10–19.

Volger, G., and **Volger, S.** 1989. Northern elephant seals. 35(6):342–347.

COMPETITIVE RELEASE: WHALES, KRILL, AND HUMANS

Although the principles of fishery management seem straightforward enough, humans have had no success in using these techniques to maintain healthy fisheries. The history of the average fishery goes something like this:

1. When a virgin fishery stock is identified, it is larger than it ever will be after fishing begins, and it contains a larger percentage of mature individuals than it ever will again.
2. Since fishing reduces life expectancy, the percentage of large mature fish decreases over time. Therefore, the average growth rate increases as the percentage of fast-growing younger fish increases.
3. As the fishing effort increases, a **maximum sustainable yield** (MSY) is eventually reached. This is the largest catch that the fishery can support on a year-to-year basis.
4. Continued increase in the fishing effort causes the MSY to be exceeded and the cost per unit of catch to increase significantly.
5. The catch finally declines to the point where it is not economically feasible to conduct the fishery. It has collapsed.

Although the fishery population is not near extinction at this point, it never recovers its virgin dimensions. The lack of recovery may be due to the growth of competing populations of less desirable animals, which accompanies the decline in the preferred fishery species population.

A study of baleen whale populations in the Antarctic between latitudes 70° and 130° E and their variation since the beginning of modern whaling operations in 1930 provides some insight into the effects that exploitation of one population can have on competing populations (Fig. 1). Because the ecology of the Antarctic is relatively

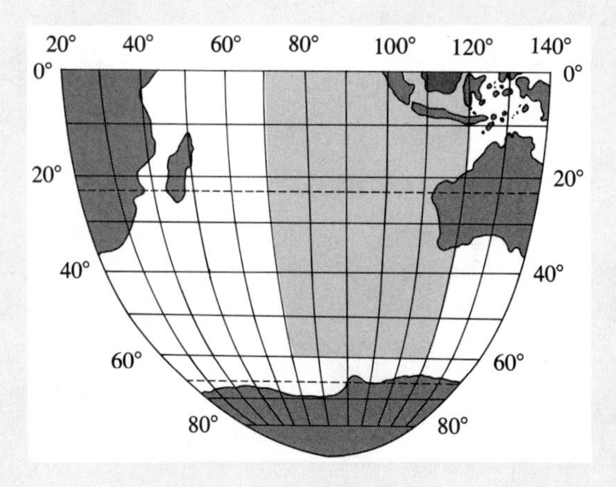

FIGURE 1
Study Area

simple, this study can tell us much about the effects of the decrease in a desirable fishery population on the populations of less desirable competitors.

In the Antarctic, whales, birds, seals, fishes, and squids feed on shrimplike krill, which feed on phytoplankton. *Krill* is a Norwegian term meaning 'tiny fish' that is applied to the 5- to 8-cm-long (2–3 in) euphausiid crustaceans that occur in swarms up to a kilometer across. Such swarms may contain up to 16 kg (35 lb) of krill per cubic meter. Krill-eating mammal populations change slowly, and the size of the total population is always closely related to the number of mature animals.

Targets of the antarctic whale fishery, in order of decreasing preference, were the humpback, blue, fin, sei, and minke. Because humpback whales became a major target as early as 1910, they are not considered in this comparison. As the 1930 whale fishery developed, the blue whale population, with some individuals that grew to lengths exceeding 27 m (89 ft), decreased quickly (Fig. 2). By 1950 it was only one-tenth of the 1930 population. The fin whale population, containing individuals up to 22 m in length, was reduced to about the same level by 1962. The 15-m sei whales were not exploited significantly in the early years of the fishery because of their smaller size. About 1940, when the blue and fin whale populations had been severely reduced, the sei whale population began a gradual increase. It reached a population in 1960 that was 20 percent larger than it had been in 1940. Increased harvesting that began in the mid-sixties, however, decreased the sei population to a third of its 1930 level by the mid-seventies. The minke whale population, composed

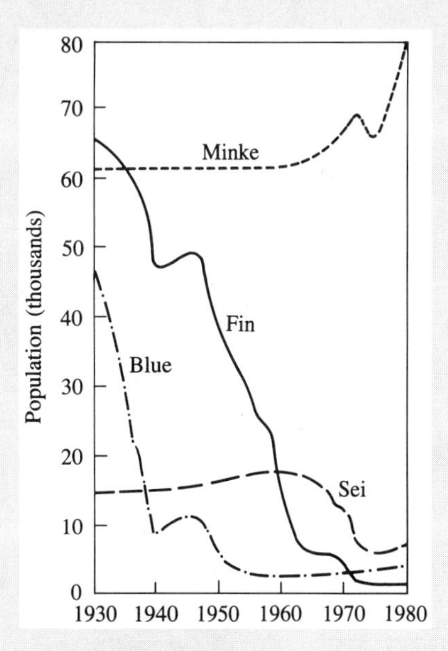

FIGURE 2
Population Changes in the 1930 Antarctica Whale Fishery, 1930–1980

of individuals that rarely exceeded 8 m in length, remained stable until 1960, when it began to increase. By 1980 it had reached a level 30 percent above that of the virgin population, even though it was being harvested. The total biomass of Antarctic baleen whales in 1980 was about one-sixth of its estimated pristine values.

In addition to the minke whales, other krill-eating mammal populations have increased. Where they were in competition with baleen whales, southern, Adelie, and gentoo penguins, as well as other seabird populations, have increased in concert with the decrease in baleen whales. These increases are possibly the result of what ecologists call **competitive release,** the expansion of certain populations as a result of the decrease in the populations of their competitors.

As a result of the decreases in the antarctic baleen whale population, a surplus of some 150 million metric tonnes (165 million tons) of krill has developed in antarctic waters. This surplus has aroused interest in developing a krill fishery. Should the krill fishery continue to grow, the potential level of baleen whale population recovery would surely be reduced. Considering the ineffectiveness of past fishery management, the result of humans exploiting the competitive release made possible by the decrease in the biomass of antarctic baleen whales may well be negative. Some investigators believe the expansion of the krill fishery will surely result in a reduction of the baleen whale population below present levels.

CALIFORNIA GRAY WHALES AND WALRUSES—GEOLOGIC AGENTS IN THE BERING SEA

It has been known since the late 1970s that pits and furrows that could not be explained by physical processes were present on the floor of the northeastern Bering Sea. These odd depressions are found in the sediments of the shallow Chirikov Basin and Norton Sound regions, where water depths are usually less than 50 m (164 ft; see Fig. 1).

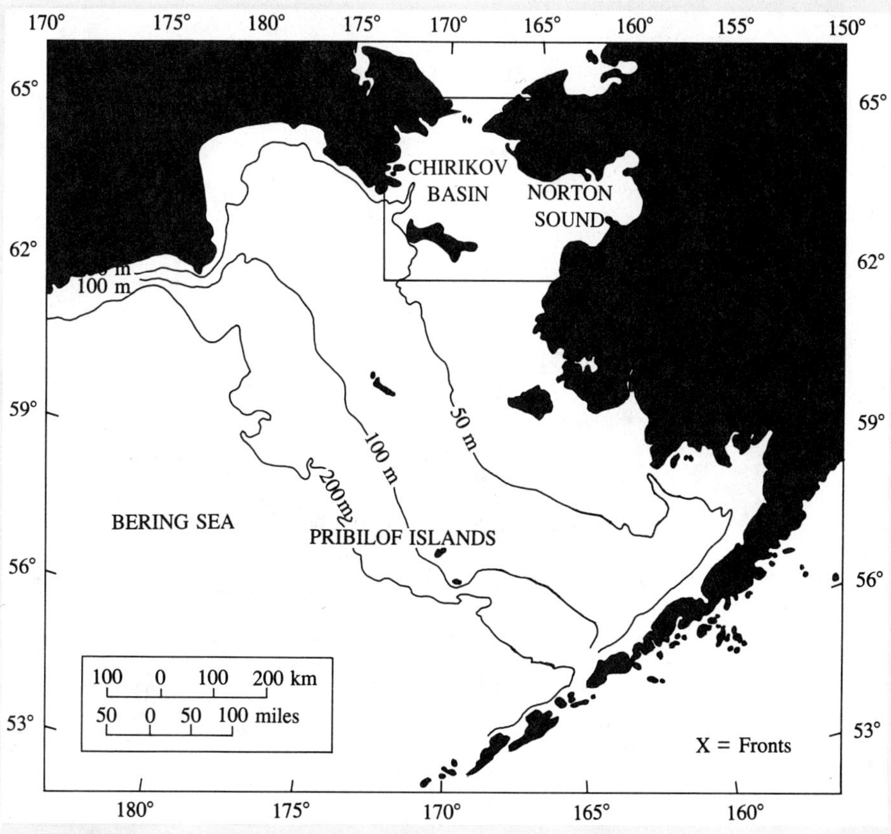

FIGURE 1
Northeastern Bering Sea

The Bering Sea lies between the Aleutian Island Arc to the south and the Bering Strait to the north. The shallow Chirikov Basin and Norton Sound in the northeastern Bering Sea are the favorite feeding grounds of the California gray whale and walruses, respectively. The whales feed primarily in the central Chirikov Basin, where the sediments are sandy. Walruses prefer the muddy bottoms of Norton Sound and the perimeter of the Chirikov Basin.

Subsequent studies (Nelson and Johnson, 1987) have shown that the pits are created by the feeding activities of California gray whales. The whales feed on bottom-dwelling amphipods and other invertebrates that live in the thick sand deposits of the Chirikov Basin (Fig. 2). The whales apparently swim with the side of their head parallel to the sea floor, and suck up the sediment and any organisms contained therein. Each whale may create a series of pits that average 2.5 m (8 ft) in length before surfacing for air. The pits typically are 1.5 m (5 ft) in length and 10 cm (4 in) deep. Larger pits (e.g., 8 m—26 ft—in length, and 4 m—13 ft—in width) are also found. These larger pits appear to be enlargements of the smaller feeding pits created by bottom currents removing loose sand (Fig. 3).

The furrows that average 47 m (154 ft) in length and 0.4 m (3 ft) in width are believed to be created when walruses feed on their favorite prey—clams and other bottom-dwelling invertebrates. These feeding tracks are more common on the mud bottoms that encircle the Cririkov Basin, and are the preferred environment for clams. Walruses apparently make the furrows with their snouts while swimming head down in search of clams. They excavate the clams by blowing jets of water into the sediment. The clams are then sucked from their shells, which are deposited alongside the furrows.

FIGURE 2
Amphipod Mat

This mat of mucous-lined burrows is created by ampeliscid amphipods, the preferred prey of California gray whales. These mats hold the loose sand grains together so the sediment is not eroded by the strong bottom currents. When the whales destroy the mats while feeding, bottom currents then can remove the sand and enlarge the feeding pits. (Photo courtesy of Kirk R. Johnson, Yale University.)

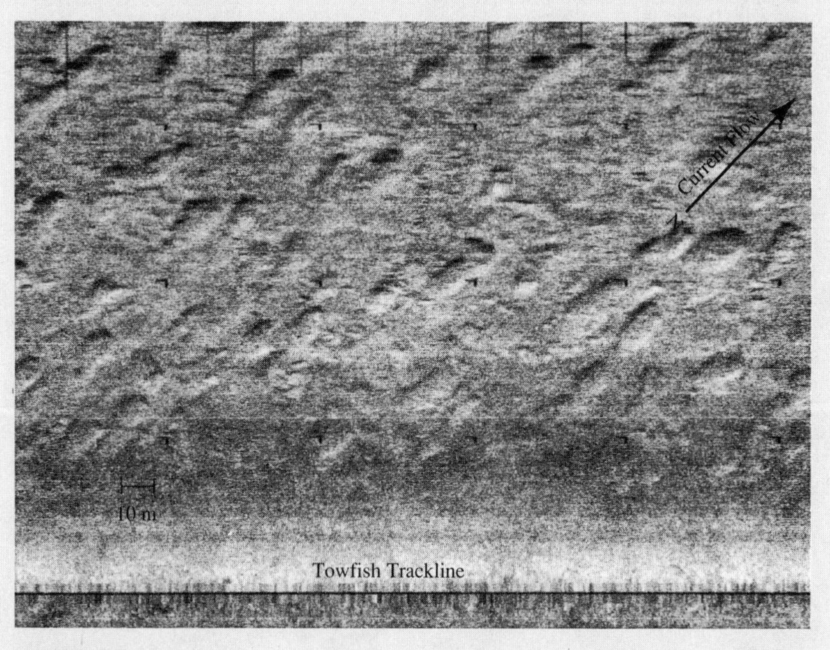

FIGURE 3
Side-Scan Sonograph

This sonograph shows pits on the floor of the Chirikov Basin created by California gray whales feeding on bottom-dwelling amphipods and other invertebrates. The smaller 2–3 meter- (6.5–10 ft-) long pits produced by the initial feeding do not show up well at this scale. All the readily visible pits have been enlarged by currents that flow in a northerly direction. The pits show this north–south alignment. (Sonograph courtesy of William K. Sacco, Yale University.)

Whale feeding resuspends three times as much sediment as the Yukon River discharges into the Bering Sea. Walruses resuspend approximately 60 percent as much sediment as the gray whales. Not only are these mammals significant geologic agents—the whales appear to prevent the sand that amphipods require from being covered with a layer of mud. During the May–November feeding season, California gray whales suspend the clays that would form mud deposits, thereby freeing them to be carried away to the north by currents. The whales also aid the benthic ecology by suspending the nutrients needed by the phytoplankton that support the benthic community.

These examples are but two of many that demonstrate the way biological and physical processes are closely interrelated to the benefit of marine ecology.

PART THREE

MARINE HABITATS

Now that we have provided you an overview of the relationships among marine organisms and the physical ocean environment, a more in-depth discussion of these physical factors is in order. Chapter 9, Ecological Principles, focuses on the effects of the availability of solar radiation and inorganic nutrients on photosynthetic productivity. The factors that relate to the cycling of mass within ecosystems and the flow of energy through them are considered. This overview of ecological principles serves as preparation for the consideration of how organisms make a living in the various marine habitats discussed in Chapters 10–13.

Chapters 10, 11, and 12 begin with discussions of the physical phenomena that are particularly relevant to the ecology of each of the habitats covered in the chapters. Biological adaptations to the physical environment and the interactions of populations within the ecosystem are then discussed.

Chapter 10, The Photic Zone, begins with a discussion of ocean currents that provide a significant portion of the physical framework within which photic zone organisms exist. Of course, the effects of solar radiation on these organisms provides the fabric of the chapter.

The geology of the ocean floor, including marine sediments and the nature of sea-floor spreading, are the subject of the physical ocean discussion that opens Chapter 11, The Deep Ocean: Benthic and Pelagic Zones. This seems appropriate, given the fact that the ocean floor within this habitat represents about 80 percent of the benthic environment. Deep-sea benthos are much more abundant and diverse than are the pelagic forms in the overlying deep water. Also, marine organisms contribute significantly to the sediment of the deep ocean, and the most recent and exciting discoveries of biocommunities are associated with deep-sea hydrothermal vents generated by the process of sea-floor spreading.

The periodic wave motions of ocean water, wind-generated waves, and tides have a major influence on life forms discussed in Chapter 12, The Intertidal Zone. To provide an understanding of the forces that provide the great amount of energy released by short-period pulses of breaking waves, and that drive the more leisurely paced alternation of exposure and inundation of this zone by the tides, we first consider the origins of these wave phenomena. Chapter 13, The Coastal Zone, extends the discussion into the subtidal depths of the continental margins, and considers habitats ranging from the soft sediments of the continental shelf and their retiring infauna to the wonderously flamboyant inhabitants of coral reefs.

9

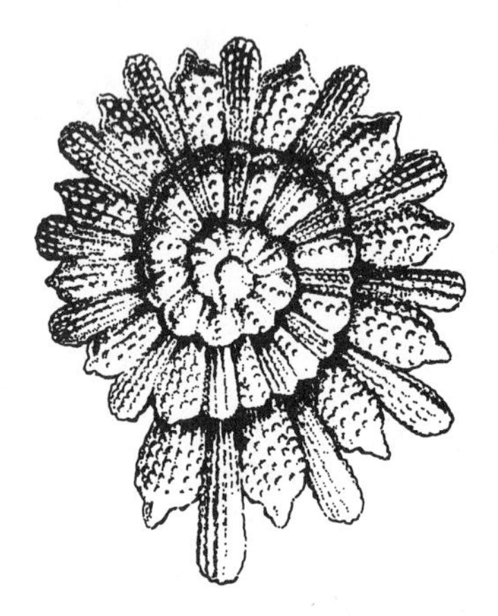

ECOLOGICAL PRINCIPLES

In this chapter we will discuss the principles of ecology that affect marine life. **Ecology** is defined as the study of organisms in relation to one another and the environment in which they interact. There are clearly ecological relationships that tie all marine organisms together in the oceanwide view of marine ecology, and we will address some of the fundamental relationships of marine ecology in this chapter. The relationships we will consider are related to the introduction of energy into the biotic community, and the flow of energy through and the cycling of mass within the marine ecosystem.

To more specifically focus on the processes and communities that are significant in the marine ecosystem, we will divide it into the photic zone, deep ocean, intertidal zone, and coastal zone. The following four chapters (Chs. 10–13) cover each zone, in depth.

THE MARINE ECOSYSTEM

An **ecosystem** is a self-contained unit that may range in size from a small kelp forest to the ocean as a whole (Fig. 9.1). It includes the **biotic community** and the **abiotic** (nonliving) **environment** with which it interacts. The biotic community is composed of a number of interacting populations, and each **population** is composed of a group of organisms of the same species. Within an ecosystem, there are three categories of organisms: producers, consumers, and decomposers.

Phototrophs and chemotrophs (some bacteria) are autotrophic **producers;** that is, they have the capacity to produce enough food to meet the needs of the whole community through photosynthesis or chemosynthesis. The consumers and decomposers are heterotrophs; they depend on the organic compounds produced by the producers or other heterotrophs for their food supply.

Consumers are divided into three categories: herbivores, carnivores, and omnivores. **Herbivores** feed directly on phototrophs. **Carnivores** feed only on heterotrophs; **omnivores** feed on both. With the more restricted use of the words plant and animal which, in the past, may have been used in place of phototroph and heterotroph in the preceding sentences, some other terms are coming into common use to refer to more specific groups of consumers. For instance, **planktivores** feed on plankton, and **bacteriovores** feed on bacteria.

Decomposers, such as bacteria and fungi, break down the organic compounds in dead organic matter while converting the energy represented by some of these compounds to meet their own energy requirements. The oxidation of organic molecules represented by the decomposition process results in the release of simple inorganic salts into the environment, where they are available for use as nutrients by the producers.

Within an ecosystem, species interact in many ways. In addition to the feeding relationships described above, there are more subtle symbiotic relationships. **Symbiosis** is an intimate and prolonged relationship between two or more species in which at least one of the species benefits. In the simplest terms, these relationships may be classified as commensalism, mutualism, and parasitism. In **commensalism,** a smaller or less-dominant participant, or **symbiont,** benefits without

FIGURE 9.1
Ecosystem Interactions

Energy enters an ecosystem supported by phototrophs as radiant energy from the sun. It is transferred as chemical energy to the consumers by feeding, and to the decomposers as dead organic matter. The phototrophs consume carbon dioxide and produce oxygen while the consumers and decomposers generate carbon dioxide and deplete environmental oxygen.

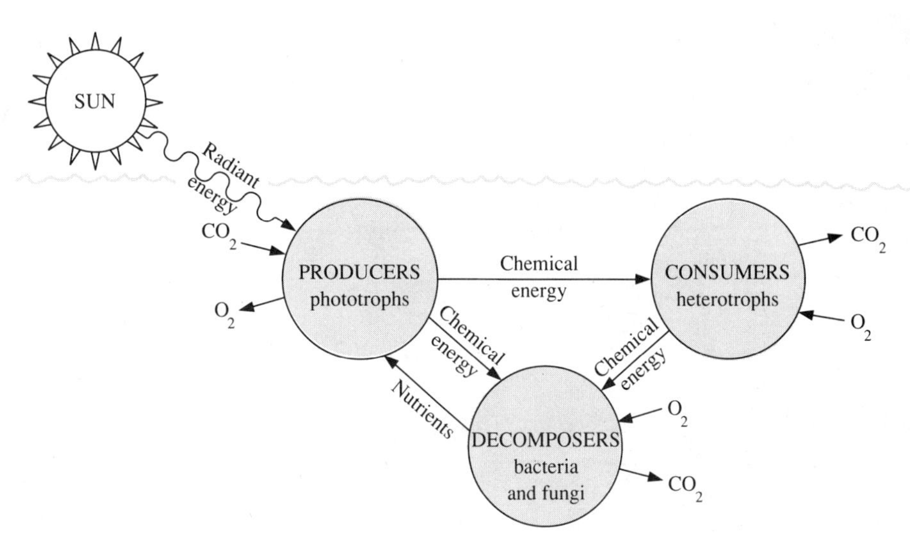

harming the **host,** the participant that affords subsistence or lodgment to the other. Examples are the remoras *(Remora* and *Echeneis),* fishes that attach to sharks, whales, and turtles. The remoras obtain food in the form of small fragments of the host's prey. The anterior dorsal fin of the remoras is flattened into a suction disc that is used to attach to the host and obtain effortless transportation. In **mutualism,** both participants benefit. Such a relationship exists between the large reef fish and the cleaner fish that keep them free of parasites. Although the exact nature of the relationship is controversial, that of the clown fish, *Amphiprion,* and the sea anemones, *Actinia* and *Stoichactis,* appears to be mutualistic (Fig. 9.2). The clown fish brings food to the anemones, while the stinging tentacles of the sea anemones serve as a deterrent to predators seeking out the clown fish. **Parasitism** is a relationship where one participant, the **parasite,** benefits at the expense of the host. Many fish are hosts to isopods that attach to the fish and derive their nutrition from the body fluids of the fish, thereby robbing the host of its energy supply.

FIGURE 9.2
Clown fish and Sea Anemone

PRIMARY PRODUCTION

Primary production is defined as the amount of carbon fixed by autotrophic organisms through the synthesis of organic matter from the inorganic compounds CO_2 and H_2O, using energy derived from solar radiation or chemical reactions. The major process through which primary production occurs is photosynthesis. Although new knowledge of the role of chemosynthesis in supporting hydrothermal vent communities on oceanic spreading centers has emerged, chemosynthesis is much less significant in overall marine primary production than is photosynthesis. Because of the far greater importance of photosynthesis in the overall productivity of the oceans and our greater knowledge of the factors that affect marine photosynthesis, the following considerations will relate primarily to this process.

Photosynthetic Productivity

The total amount of organic matter produced by photosynthesis per unit of time represents the gross primary production of the oceans. Phototrophs will use some of this energy for their own maintenance through respiration. That which remains is the net primary production, which is manifested as growth and reproduction products. It is the net primary production that supports the

the heterotrophic marine populations—animals, protozoans, and bacteria.

Methods of Analysis. In the 1920s, the **Gran Method** (Fig. 9.3) of estimating primary productivity was developed. This method is based on the fact that oxygen is produced by photosynthesis in proportion to the amount of carbohydrates synthesized. The method involves putting equal quantities of phytoplankton into a series of bottles, all of which contain a measured amount of dissolved oxygen. The bottles are then arranged in pairs, one being transparent and the other totally opaque. These pairs are suspended on a line through the euphotic zone, where they are left for a specific period of time. After the bottles are brought to the surface, the oxygen concentration is determined for each bottle.

Photosynthesis, which is confined to the transparent bottles, will add oxygen to the water, whereas respiration will reduce the oxygen content in both the transparent and opaque bottles. Increased oxygen concentration in the transparent bottles is proportional to the amount of photosynthesis that has occurred, minus the oxygen consumed by plant respiration. This value represents a relative measure of the net photosynthesis, or net increase, in biomass of the phytoplankton within the transparent bottle. Decreased oxygen content within the opaque bottles corresponds to the respiration rate. For any depth, an assessment of gross photosynthesis can be made by adding the oxygen gain or loss in the clear bottles to the oxygen loss in the opaque bottles.

The depth at which the oxygen production and the oxygen consumption are equal is called the **oxygen compensation depth.** It represents the light intensity below which plants do not survive. Since respiration goes on at all times, during the daylight hours phytoplankton must produce, through photosynthesis, biomass in excess of that which is consumed by respiration in any 24-hr period, if the total biomass of the community is to be maintained or to increase (Fig. 9.3). An analogy can be made with the common paycheck:

Gross net photosynthesis (Gross pay earned) =
Oxygen change in clear bottle (Take-home pay) +
Oxygen loss in dark bottle (Income tax withheld).

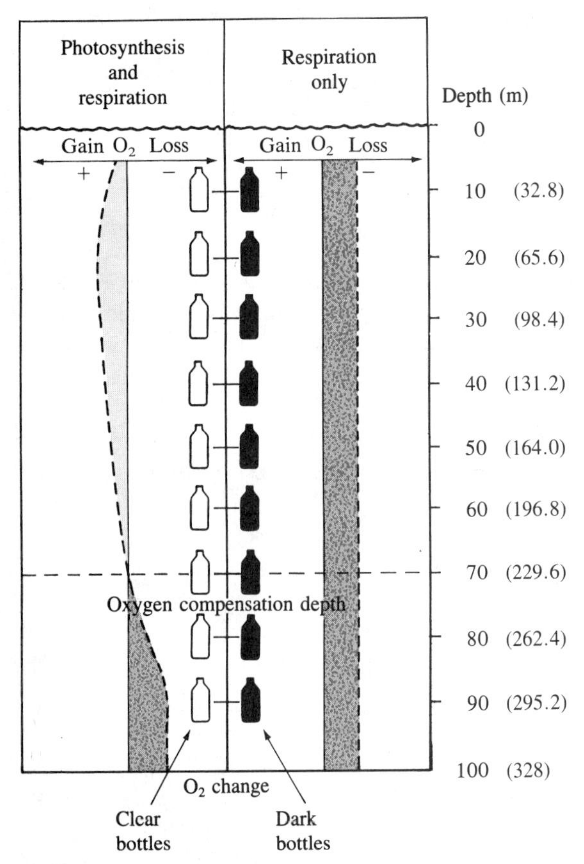

FIGURE 9.3
Gran Method—Compensation Depth of Oxygen

Biological assay of phytoplankton production and consumption of oxygen (O_2) shows to what depth light penetrates with sufficient intensity to sustain photosynthesis at a level sufficient to increase phytoplankton mass. This occurs only above the oxygen compensation depth, the depth at which oxygen production by photosynthesis is equal to consumption through respiration. The change in oxygen content in the clear bottles represents a net photosynthesis resulting from gross photosynthesis less the consumption of oxygen by respiration. The loss of oxygen in the dark bottles gives a value for the amount of respiration that occurred in the clear bottles. This amount, added to the net photosynthesis that occurred in the clear bottles, gives a value for the gross photosynthesis in the clear bottles.

During the 1950s, a method involving the use of radioactive carbon (^{14}C) was developed. It has been refined and is currently the most often used method for determining marine primary productivity. The procedure is similar to that described above, since it involves the use of a series of paired clear and opaque bottles. Each bottle contains identical phytoplankton samples, equal amounts of CO_2 containing ^{14}C, and equal amounts of CO_2 containing stable carbon. The phytoplankton sample is filtered from each bottle after the system has been suspended in the ocean for a sufficient period of time. The amount of radiation emitted by each sample is measured by a radiation-counting device, and the rate of assimilation of ^{14}C is computed from these measurements. High levels of radioactive emission indicate high levels of productivity.

A third method used in studies of marine primary productivity is the measurement of chlorophyll *a* content of living phytoplankton samples taken from ocean surface waters. Although not as precise a method as the measurement of net primary productivity using ^{14}C, there is a direct relationship between the amount of chlorophyll *a* in a phytoplankton sample from a given volume of water and gross primary productivity.

Geographic Variation in Primary Production

The production of carbohydrates by phytoplankton through the process of photosynthesis will vary considerably throughout the areal extent of the ocean, as well as over time. Photosynthetic productivity in the oceans depends on: (a) the availability of solar radiation, and (b) the availability of nutrients. Both variables are related to the seasonal patterns that affect all the world's oceans to some degree.

Photosynthetic productivity of the oceans varies from about 0.1 g of carbon per square meter per day ($gC/m^2/d$) in the open ocean to over $10gC/m^2/d$) in highly productive coastal areas. This variability is primarily the result of the uneven distribution of nutrients throughout the photosynthetic zone. Some of the causes of this variability are discussed in the following section.

Temperature Stratification and Nutrient Supply.

Solar radiation penetrates only the surface layer of the ocean. Infrared wavelengths that can be converted directly to heat are essentially 100 percent absorbed in the upper 2 m (6.5 ft) of the ocean. It is not surprising that the surface waters are warmed. This condition produces a thermocline that separates the surface waters, which are relatively warmer, from the deep water masses that are colder and more dense. Thus a well-developed thermocline produces a density separation between the upper and deeper water masses. In this case, the density change is a direct result of change in temperature and increasing pressure.

Thermoclines are relatively permanent in tropical latitude areas, while a seasonal thermocline may be present during the summer months in the midlatitudes. They are generally nonexistent in polar latitudes. In the following discussion of productivity, we will see that the presence or absence of the thermocline will have an important effect on nutrient availability.

In considering the availability of solar radiation and nutrients, we divide the oceans into polar, temperate, and tropical regions. We will concern ourselves primarily with the nature of the open ocean in all three regions. The polar region will be discussed first, as the productivity considerations are simplest there.

Polar Region.

As an example of productivity on a seasonal basis in a polar sea, we will consider the Barents Sea off the northern coast of Europe. Peak diatom productivity there is during the month of May, and tapers off through July (Fig. 9.4A). This brings about a peak period of zooplankton development consisting primarily of small crustaceans of the genus *Calanus* (Fig. 9.4B). The zooplankton biomass reaches a peak in June and continues at a relatively high level until winter darkness begins in October. In this region, which is above 70° N latitude, there is continuous darkness for about three months of winter, and continuous illumination for a period of three months during summer.

In the Antarctic region, particularly at the southern end of the Atlantic Ocean, productivity patterns are similar to those found in the Barents Sea, except that the seasons are reversed, and productivity is somewhat greater. The most likely explanation for this greater productivity in Antarctic waters is the continual upwelling of water that has sunk in both the North Atlantic and the Antarctic. Due to the lack of a thermocline development in polar regions, during each winter, atmospheric cooling lowers the temperature of the surface water to the degree that it achieves a density great enough to sink into the deep ocean. This sinking maintains the oxygen supply needed to maintain deep-sea life. The water that sinks in the North Atlantic, the North Atlantic Deep Water, moves south through the deep ocean. When it encounters the Antarctic Bottom Water, which is even more dense, it surfaces hundreds of years later, carrying high concentrations of nutrients (Fig. 9.4D).

To illustrate the very great productivity that occurs during the short summer season in polar oceans, we can consider the growth rate of baby whales. The largest of all whales, the blue whale, migrates through the temperate and polar oceans at the times of maximum zooplankton productivity. As a result of this timing, they are able to develop and support calves that during a gestation of 11 months reach lengths in excess of 7 m (23 ft).

The mother suckles the calf for six months with a teat that actually pumps the youngster full of rich milk. By the time the calf is weaned, it is over 16 m in length, and in a period of two years will reach a length approaching 23 m (75 ft). In three and a half years, a 60-ton blue whale will have developed. This phenomenal growth rate gives some indication of the enormous biomass of copepods (Fig. 9.4B) and krill that these large mammals feed upon.

In polar waters, there is little density stratification to prevent the mixing of deeper water with shallow water. Polar waters are relatively isothermal (Fig. 9.4C). In most polar areas, the surface waters freely mix with

FIGURE 9.4 Productivity in Polar Oceans

(A) Diatom mass increases rapidly in the spring, when the sun rises high enough to cause deep penetration. Once the diatom food supply develops, the zooplankton begin feeding on it. The zooplankton biomass reaches its peak by early summer. **(B)** The copepod *Calanus*. (Photo courtesy of Scripps Institution of Oceanography, University of California–San Diego.) **(C)** There is little temperature stratification in polar water; nearly isothermal conditions are common, from surface to bottom. **(D)** The continual upwelling of the North Atlantic Deep Water insures that antarctic waters will remain nutrient rich.

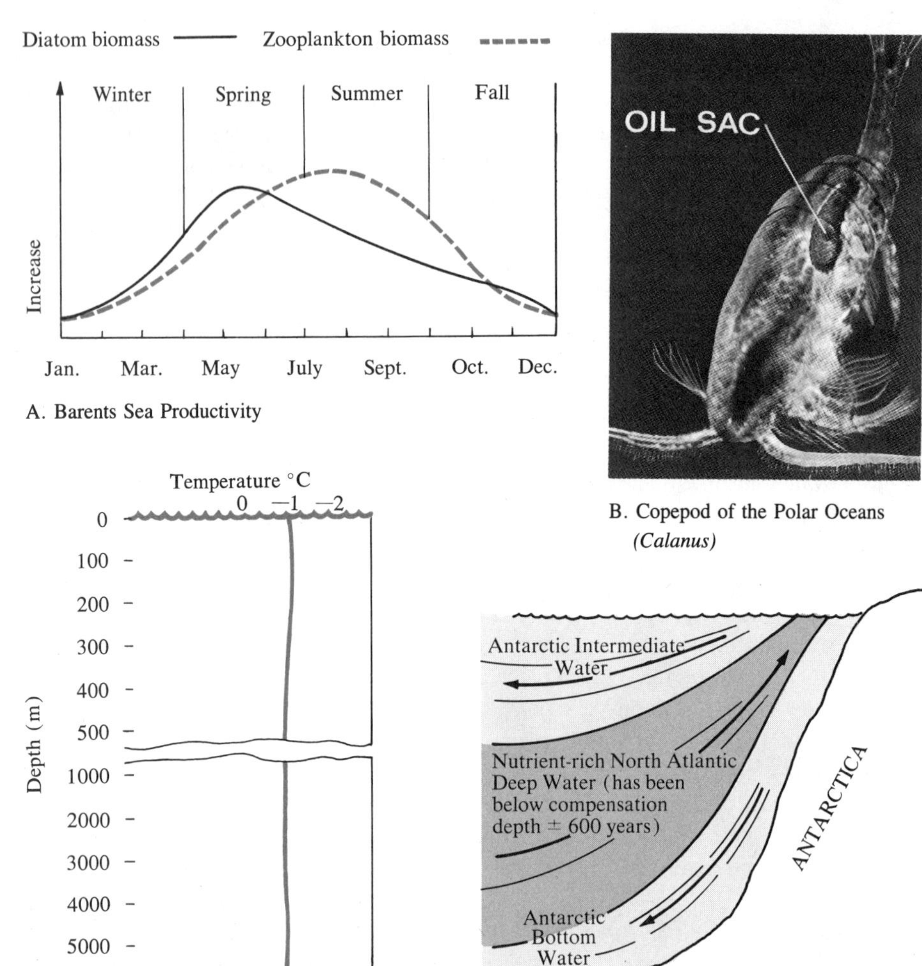

the deeper, nutrient-rich water. There can be, however, some density stratification of water masses due to the summer melting of ice, which lays down a thin low-salinity layer that does not readily mix with the deeper waters.

There are usually high concentrations of phosphates and nitrates in the surface waters. Thus, photosynthetic productivity in the high latitudes is more commonly limited by the availability of solar energy than by the availability of nutrients. The productive season in these waters will be relatively short but will be characterized by an outstandingly high rate of production.

Tropical Region. In direct contrast with high productivity associated with the summer season in the polar seas, low productivity is the rule in the tropical regions of the open ocean. Light penetrates much deeper into the open tropical ocean than into the temperate and polar waters. This produces a very deep oxygen compensation depth. In the tropical ocean, however, a permanent thermocline produces a relatively permanent stratification of water masses and prevents mixing between the surface waters and the nutrient-rich deeper waters (Fig. 9.5A).

At about 20° latitude we typically will find concentrations of phosphate and nitrate in surface waters to be less than 1/100 of the concentrations of these nutrients in temperate oceans during winter. Nutrient-rich waters within the tropics lie for the most part below 150 m (492 ft), and the highest concentration of nutrients occurs between 500–1,000 m (1,640–3,280 ft) depth.

Generally, there is a steady, low rate of primary productivity in tropical oceans. Although the rate of tropical productivity is low, when we compare the total annual productivity of tropical oceans with that of the more productive temperate oceans, we find that tropical productivity usually is at least half of that found in the temperate regions.

Within tropical regions there are three environments where productivity is unusually high—regions of equatorial upwelling, coastal upwelling (Fig. 9.5B,C), and coral reefs. In areas where trade winds drive west-

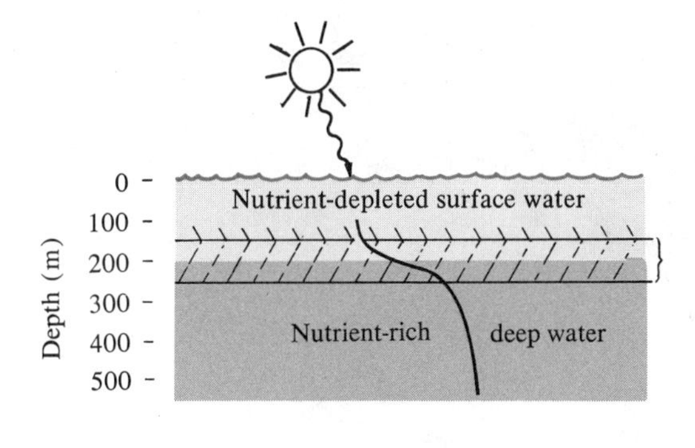

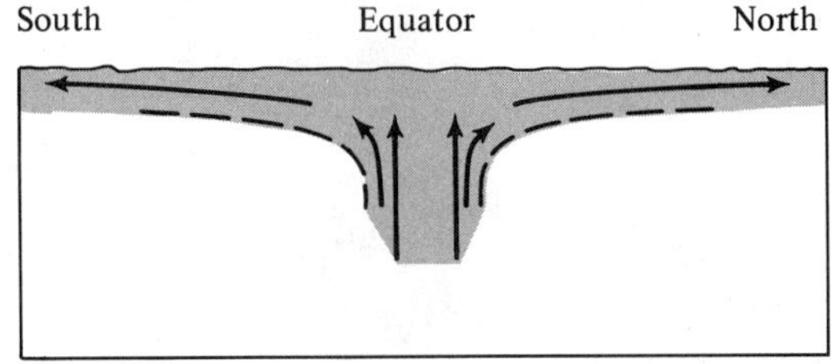

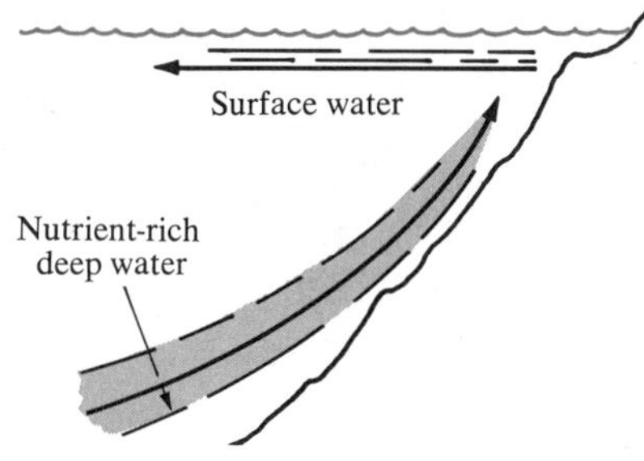

FIGURE 9.5
Productivity in Tropical Oceans

(A) The primary productivity of tropical oceans is generally low because the permanent thermocline prevents mixing of surface waters with nutrient-rich deeper waters. Near the equator **(B)** and along the coasts **(C),** however, upwelling makes for unusually high productivity.

erly equatorial currents on either side of the equator, surface water moves off toward higher latitudes and is replaced by nutrient-rich water that surfaces from depths of up to 100 m (330 ft). (The mechanics of upwelling are covered in the discussion of ocean currents in Chapter 10.) The condition of equatorial upwelling is probably best developed in the eastern Pacific Ocean. Where the prevailing winds blow toward the equator and along the western margin of continents, the surface waters are driven away from the coast. They are replaced by nutrient-rich waters from depths of 200–900 m (660–2,950 ft). As a result of this coastal upwelling of nutrient-rich water, there is a high rate of primary productivity in these areas, which supports large fisheries. Such conditions exist along the southern coast of California and the southwest coast of Peru in the Pacific Ocean. Upwelling also occurs along the coast of Morocco and the southwest coast of Africa in the Atlantic Ocean.

The relatively high productivity of coral reef environments is not related to upwelling. In fact, coral reefs can prosper only in nutrient-poor waters. This seemingly enigmatic condition is discussed in the special feature, "Coral Reefs and the Nutrient Level Paradox", at the end of this chapter.

Temperate Region. We have discussed the general productivity picture in the polar regions, where productivity is limited primarily by the availability of solar radiation, and in the tropical low-latitude areas, where the limiting factor is the availability of nutrients. We next will consider the temperate regions, where an alternation of these factors controls productivity in a pattern that is somewhat more complex.

Productivity in temperate oceans is at a very low level during the winter months, although high concentrations of nutrients are available in the surface layers. In fact, the nutrient concentration is higher during the winter season than at any time throughout the year. The limiting factor on productivity during the winter season in the temperate ocean is the availability of solar energy. Since the sun is at its lowest elevation above the horizon during this season, a higher percentage of solar energy is reflected and a smaller percentage absorbed into the surface waters than in summer. The oxygen compensation depth for basic producers such as diatoms

FIGURE 9.6
Productivity in Temperate Oceans

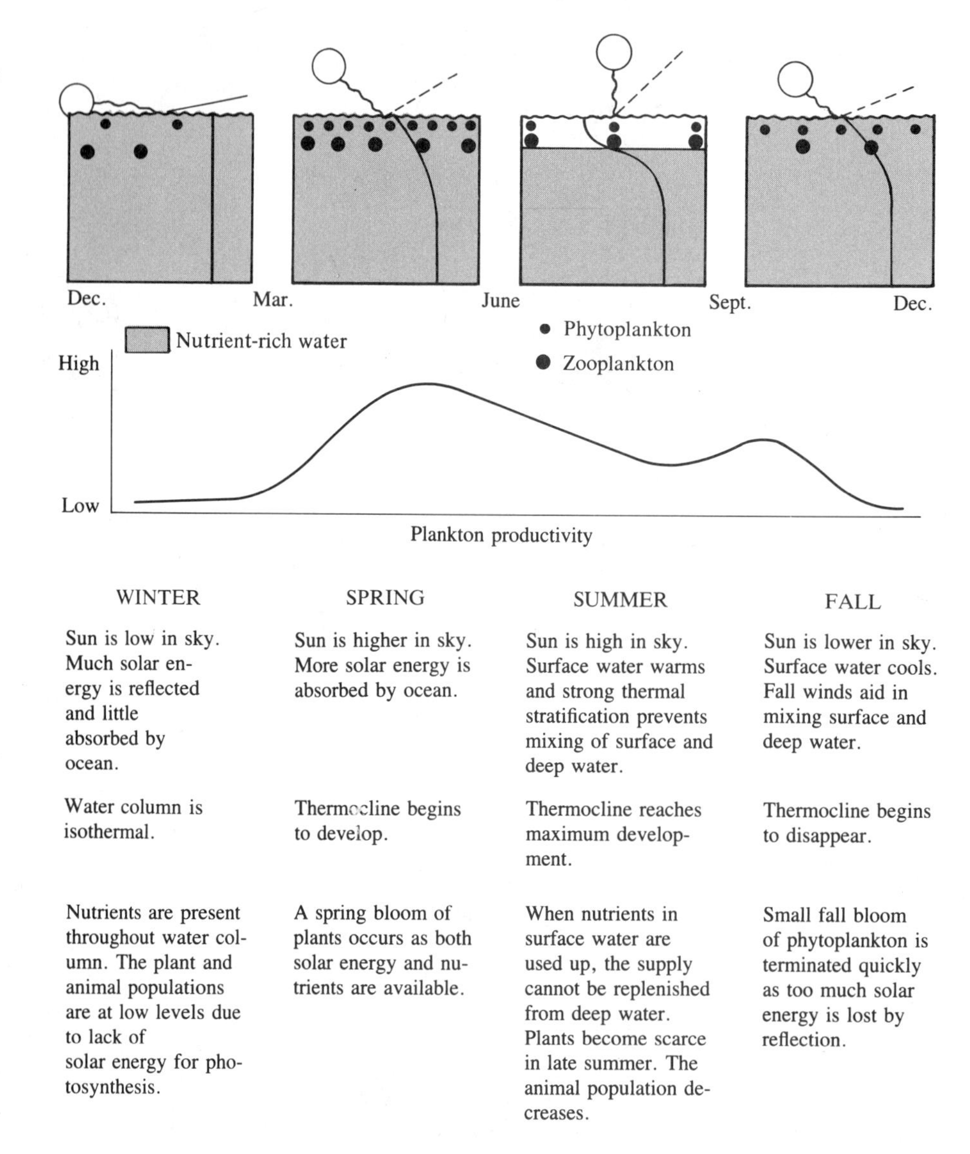

is so shallow that it does not allow growth of the diatom population (Fig. 9.6).

As the sun rises higher in the sky during the spring season, the compensation depth deepens as the amount of solar radiation being absorbed by the surface water increases. Eventually, there is sufficient water volume included within the water column above the compensation depth to allow for the **spring bloom,** a rapid growth of the diatom population. This expanding population puts a tremendous demand on the nutrient supply in the euphotic zone. In most Northern Hemisphere areas, decreases in the diatom population will occur by May as a result of depleted nutrient supply.

As the sun rises higher in the sky during the summer months, the surface waters in the temperate parts of the ocean are warmed, and the water becomes separated from the deeper water masses by a seasonal thermocline that may develop at depths of approximately 15 m (50 ft). As a result of this thermocline development, there is little or no exchange of water across this discontinuity, and the nutrients that are depleted from surface waters cannot be replaced by those available in the deep waters. Throughout the summer months, the phytoplankton population will remain at a relatively low level but will again increase in some temperate areas during the autumn months.

The **autumn bloom** of phytoplankton is much less spectacular than that of the spring because of the decreasing availability of solar radiation resulting from the sun dropping lower in the sky. This causes a decrease in surface temperature, and the breakdown of the summer thermocline. A return of nutrients to the surface layer occurs as increased wind strength mixes it with the deeper water mass in which the nutrients have been trapped throughout the summer months. This bloom is very short-lived because of the decreased availability of solar radiation. The phytoplankton population begins to decrease rapidly. The limiting factor in this case is the opposite of that which reduced the population of the spring phytoplankton. In the case of the spring bloom, solar radiation was readily available, and the decrease in nutrient supply was the limiting factor.

Additional Considerations. Coastal waters with high nutrient levels are highly productive, or **eutrophic.** Most of the open ocean has a low level of nutrients and productivity. This low productivity is referred to as an **oligotrophic** condition. Figure 9.7 shows the general patterns of ocean productivity based on photosynthetically fixed carbon.

Measurements of photosynthetic productivity of oligotrophic waters in the North Atlantic and North Pacific Oceans indicate they may be from two to seven (or even more) times as productive as ^{14}C data indicate. Instead of using the small bottles used in the ^{14}C measurements and suspending them for a short time in the ocean, some physical oceanographers have used a much larger 'bottle'. In their studies of the circulation within the subtropical oceans, physical oceanographers use 'bottles' that are 'capped' by the pycnoclines produced by the thermoclines beneath the warm layers of surface water within the subtropical oceans. Depending on the design of the study, these bottles contain from tens to thousands of km^3 of water that record the average results of photosynthetic activity over periods of from a few months to decades.

Studies that analyze the effect of photosynthesis on the oxygen concentrations in (a) the euphotic zone and (b) immediately beneath the euphotic zone have provided interesting results. In the North Pacific Ocean, oxygen saturations in **subsurface oxygen maximums (SOM)** at depths between 50–100 m (165–330 ft) were from 110–120 percent at latitudes of from 30–40° N during summer months, based on data obtained from 1962 through 1979. The excess oxygen (anything over 100%) within the euphotic zone may be due to trapped photosynthetically produced oxygen.

A North Atlantic study determined the rate at which oxygen is used up beneath the euphotic zone (100 m) by decomposing organic matter falling toward

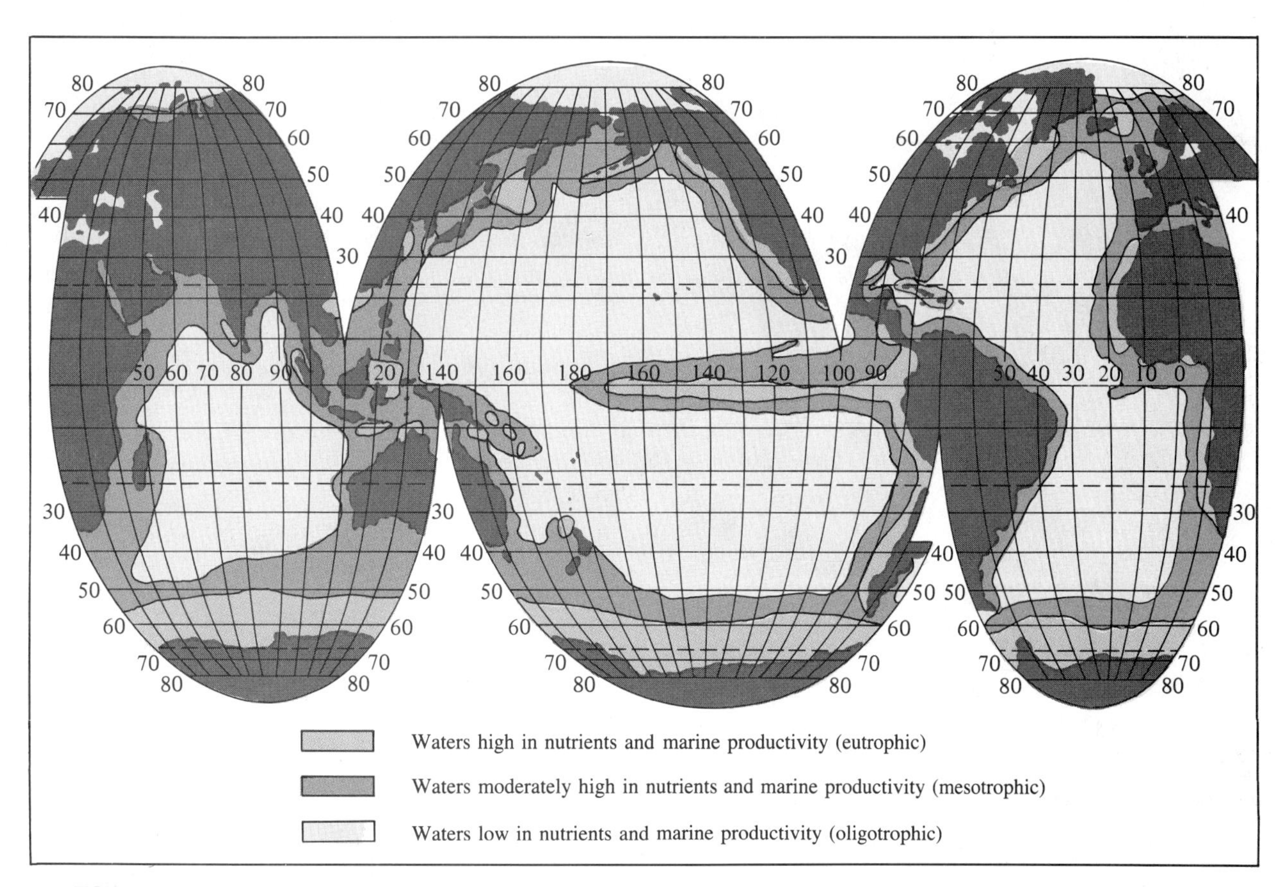

FIGURE 9.7
Plant Productivity in the World Ocean

(Base map courtesy of National Ocean Survey.)

the ocean floor. **Oxygen utilization rates (OUR)** of twice the accepted mean oligotrophic rate indicate there may be much more organic matter synthesized in the oligotrophic euphotic zone than is indicated by ^{14}C data.

The problem with the ^{14}C method may be that by using such a small sample and exposing it for such a short time, periods of unusually high productivity may be missed and not averaged into the results. Some interesting findings may help explain how some of the biological productivity of oligotrophic water may have been missed.

Mats of diatoms belonging to the genus *Rhizosolenia* have been found floating in the North Pacific and the Sargasso Sea. Averaging 7 cm (2.8 in) in length, these mats possess symbiotic bacteria that fix molecular nitrogen (N_2) into nitrate (NO_3), usable as a nutrient by the diatoms (Fig. 9.8). Also, bundles of a nitrogen-fixing cyanobacteria, *Oscillatoria spp.*, have been found in the Sargasso Sea. Such aggregations are readily broken up by net tows and missed by standard bottle casts.

Recent studies in the Pacific and Atlantic Oceans indicate that a large percentage (60% in one study) of oligotrophic photosynthesis is achieved by picoplankton (0.2–2.0 mm; 0.00008–0.0008 in) that pass through many filters used in productivity studies. Continued investigation of this problem may indicate the overall productivity of the oligotrophic ocean waters may have been greatly underestimated.

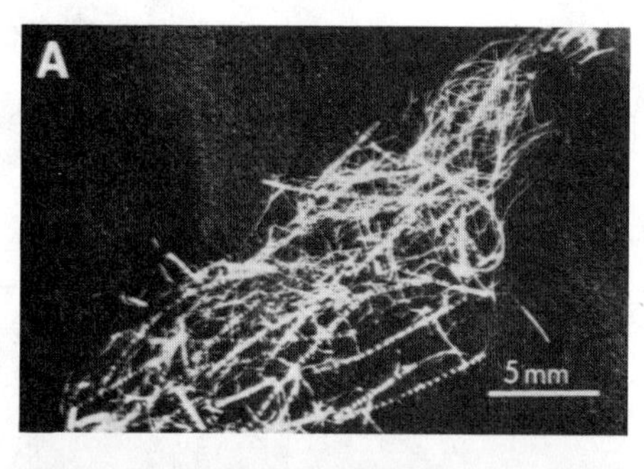

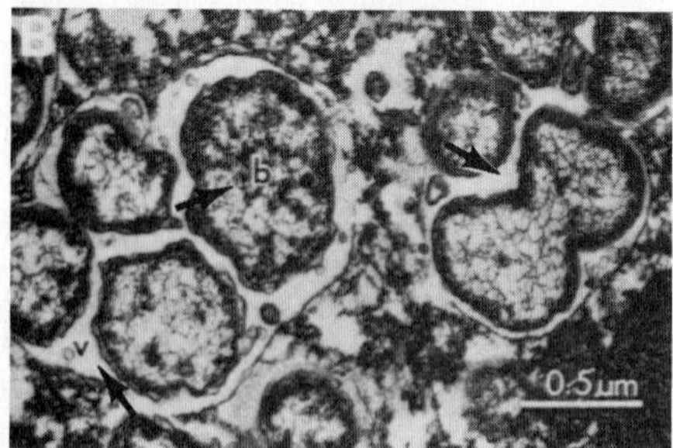

FIGURE 9.8
Symbiotic Bacterial Cells Fix Nitrogen for Diatoms Living in the Oligotrophic Ocean Waters

(A) Typical mat of *Rhizosolenia* about 5 cm long. It is composed of intertwining chains of *R. castracanei* (wider cells), and *R. imbricata* (narrower cells.) (Photo by James M. King.) **(B)** Transmission electron micrograph (TEM) showing cross section of *R. imbricata* containing intracellular bacteria (b) within a vacuole (v). Note the dividing bacterial cells (arrow). (Photo by Mary W. Silver.)

Chemosynthetic Productivity

Another source of significant biological productivity in the oceans occurs in the rift valleys of the oceanic spreading centers, at depths of over 2,500 meters (8,200 ft), where there is no light for photosynthesis. In regions where ocean water seeps down fractures in the ocean crust to depths where it is heated by underlying magma chambers, hydrothermal springs exist. Once the water is heated, it rises to the ocean floor, dissolving minerals from the crustal rocks as it does so. Deposits of minerals are associated with biotic communities. A significant component of these communities is the vestimentiferan worm population, a sample of which can be seen in Figure 9.9. These 1 m (3.28 ft)-length tube worms and other large benthos (e.g., clams, mussels) reach unusually large sizes, in association with autotrophic bacteria that derive energy to produce their own food from the hydrogen sulfide gas (H_2S) dissolved in the water of the hydrothermal springs. By oxidizing the gas to form free sulfur (S) and sulfate (SO_4), the bacteria produce chemical energy that they use to carry on synthesis of carbohydrates. Similar communities have been discovered near cold water seeps at the base of the Florida Escarpment in the Gulf of Mexico, and are discussed further in Chapter 11. Since the bacterial conversion of inorganic nutrients into carbohydrates depends on the release of chemical energy, it is called chemosynthesis. The true significance of bacterial productivity on the deep ocean floor will not be fully understood until much more research on the phenomenon is conducted.

Recently, biochemists have discovered that bacteria can obtain the chemical energy needed for the synthesis of organic molecules through the oxidation of a large variety of compounds containing the metals iron, manganese, copper, nickel, and cobalt. These microorganisms may well be an important factor in the deposition of ore-quality deposits of the oxides of these metals on the ocean floor, in the form of manganese nodules.

ENERGY TRANSFER AND NUTRIENT CYCLING

We have been discussing the general considerations related to availability of nutrients. Our attention now will be turned to the flow of energy through the ecosystem,

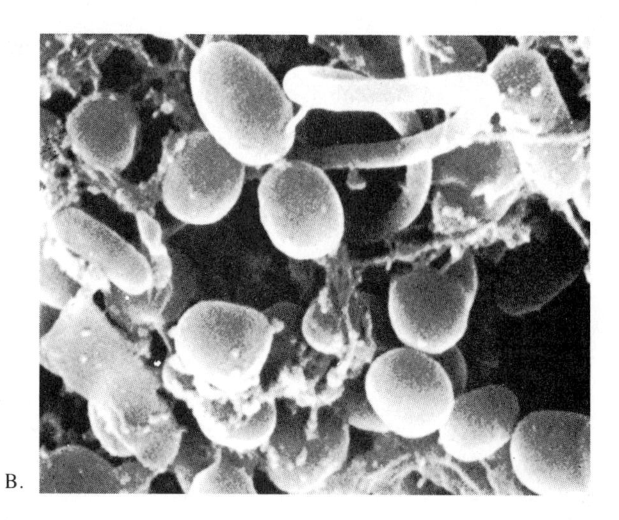

A. B.

FIGURE 9.9
Organisms of the Hydrothermal Vents

(A) Large vestimentiferan worms have been found in the hydrothermal biotic community of the Galapagos Rift and at other deep-ocean hydrothermal vents. **(B)** The worms possess no gut and have embedded in their tissue sulfur-oxidizing bacteria that chemosynthetically produce food from inorganic nutrients dissolved in the deep-ocean water. Magnification 20,000×. (Photos courtesy of Woods Hole Oceanographic Institution.)

and the cycling of specific important classes of nutrients.

Energy Flow

The flow of energy through a marine biotic community may be initiated by the grazing of herbivores on phytoplankton or benthic algae. The net primary productivity less the loss from grazing and sinking below the photosynthetic zone is the **standing crop** or **biomass** of the phototrophs at a given time. The standing crop of phytoplankton can be determined by filtering the phytoplankton from a known quantity of ocean water. Acetone then is used to dissolve the chlorophyll from the sample. The amount of light absorbed by the solution of pigmented acetone can be used as an indirect measure of the amount of phytoplankton in the ocean at the time the sample was taken. Standing crop measurements can be made for any component of the biological community, but they usually are more accurate for plankton. Commonly, biomass is expressed as dry weight of carbon per cubic meter of ocean water. It is not uncommon, especially in subtropical and tropical seas, for the grazing of herbivores to be so significant that the standing crop of phytoplankton is less than 5 percent of the net primary productivity measured in an area. Therefore, the standing crop cannot be used as a direct measure of productivity.

Before considering the **biogeochemical cycles** that involve the cycling of matter in inorganic and organic stages, we will consider the flow of energy in general. Energy is put into most biotic communities through phototrophs. It then follows a unidirectional path that leads to a continual transformation of energy, culminating in an increased entropy. **Entropy** is a thermodynamic property that reflects the degree of randomness in a system. As can be seen in Figure 9.10, which depicts the flow of energy through a biotic community sup-

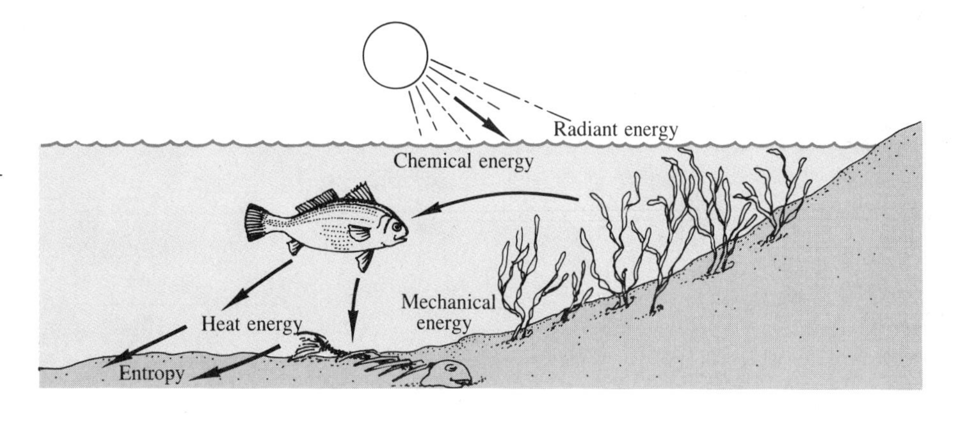

FIGURE 9.10
Energy Flow

Energy enters the biological community as solar radiation, a high grade of energy. The energy in the system follows a path of degradation to lower forms of energy until it achieves a high degree of entropy and is no longer available to do work.

ported by photosynthesis, energy enters the system as radiant energy. Photosynthesis converts it to chemical energy that is used for plant respiration. It then is transferred in this form through the ecosystem via feeding, and along the way is converted to kinetic energy and heat energy.

Trophic Levels. The transfer of chemical energy stored in the mass of the ocean's phototrophic population to the heterotrophic community occurs through the feeding process. Zooplankton feed as herbivores on the diatoms and other phytoplankton, while larger animals feed on the macroscopic algae and grasses found growing, attached to the ocean bottom near shore. The herbivores are fed upon by larger animals (carnivores) who, in turn, will be fed upon by another population of larger carnivores, and so on. Each of these feeding levels is a **trophic level.**

In general, the individual members of a feeding population will be larger, but not too much larger, than the individual member of the population on which they feed. Although we can consider this to be generally true, there are outstanding exceptions to this condition. The blue whale, which is the largest animal known to have existed on earth, feeds upon the krill that grow to maximum lengths of 6 cm (2.4 in).

We should recall that all consideration of energy transfer must be approached with the understanding that the transfer of energy from one population to another represents a continuous flow of energy. All energy that enters the organic community is inevitably lost to increased entropy in the end.

Transfer Efficiency. In the transfer of energy between trophic levels, we are greatly concerned with efficiency. There is a relatively high degree of variability in the efficiency by which various plant species convert and transfer the radiant energy they receive to biomass under various laboratory conditions. As an average, we consider the percentage of light energy absorbed by plants and ultimately synthesized into organic substances made available to herbivores to be about 2 percent.

The **ecological efficiency** at any trophic level is the ratio of energy passed on to the next higher trophic level, divided by the energy received from the trophic level below. When we examine the transfer of energy from feeding population to feeding population within the ocean, we can readily see that of the biomass representing the food intake of a given population, only a portion will be passed on to the next feeding level.

Figure 9.11 shows that some of the energy taken in as food by herbivores is passed by the animal as feces, and the rest is assimilated by the animal. Of that portion of the energy that is assimilated, much is quickly converted—through respiration—to kinetic energy for maintaining life, while what remains is available for growth and reproduction. Only a portion of the mass of food ingested is available to be passed on to the next trophic level through feeding.

Many investigations have been conducted into the efficiency of energy transfer between trophic levels. There are many variables, such as the age of the animals involved, to be considered. Young animals display a higher growth efficiency than older animals. The availability of food also can alter efficiency. When food is plentiful, animals are observed to expend more energy in digestion and assimilation than when food is not readily available. Most ecological efficiencies that have been determined range between 6–20 percent. Al-

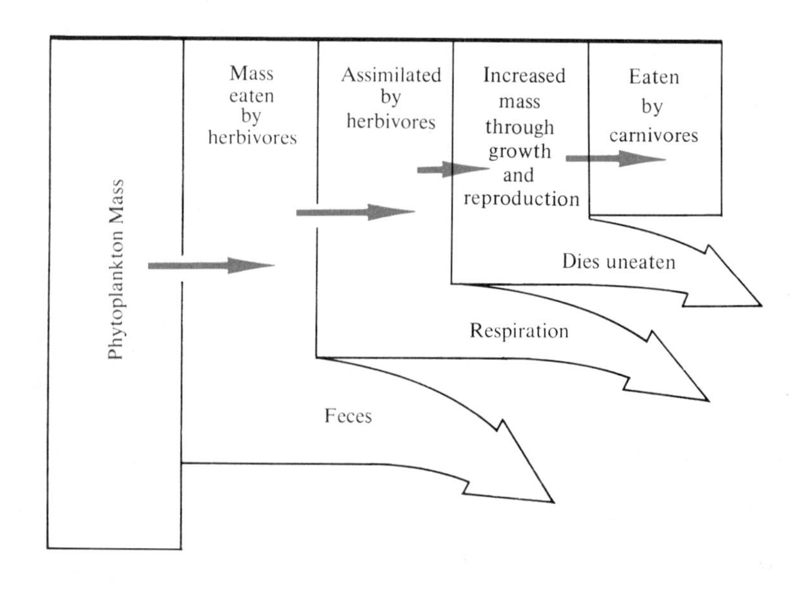

FIGURE 9.11
Passage of Energy Through a Trophic Level

The ecological efficiency of a population at a given trophic level is equal to the energy passed on to its predator population as biomass, divided by the energy represented by the mass of food it ingested. This example of the means by which energy is lost within a trophic level would serve well for any feeding population. Some of the ingested mass is not digested and excreted as feces. Of that which is assimilated, a significant amount is consumed by respiration, which releases energy required for maintenance of life. Some of that which represents increased mass of the population through growth and reproduction dies uneaten.

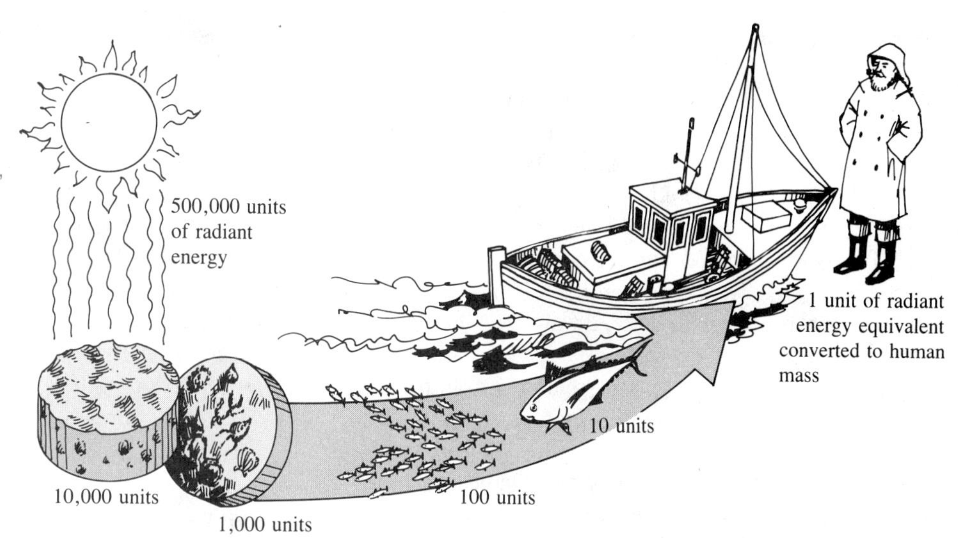

FIGURE 9.12
Ecosystem Energy Flow—Efficiency

Based on probable efficiencies of energy transfer within the ecosystem, one unit of mass equivalent is made available to the fifth trophic level (the tertiary carnivores) for every 500,000 units of radiant energy available to the producers (plants). This value is based on a 2 percent efficiency of transfer by plants, and 10 percent efficiency at all other levels.

though it is far from being documented, it is common to see ecological efficiencies in natural ecosystems presented as averaging 10 percent. Out of convenience, we do the same in our illustrations. However, there is some evidence that in populations important to our present fisheries, this efficiency may run as high as 20 percent. The true value of this efficiency is of practical importance to us because it determines the fish harvest we can anticipate from the oceans. Figure 9.12 represents diagrammatically the passage of energy between trophic levels through an entire ecosystem, from the solar energy assimilated by plants to the mass of the ultimate carnivore.

Food Chains. **Food chains** represent a sequence of organisms through which energy is transferred from the primary producers through the herbivore and carnivore feeding populations, until one feeding population does not have any predators, which marks the end of the chain. In nature, it is rare to find food chains comprising more than five trophic levels. Based on our previous considerations of the efficiency of energy transfer between trophic levels, we see that for a population within a food chain, it may be beneficial to feed as close to the primary producing population as possible. This arrangement increases the biomass available for food, and the number of individuals in the population to be fed upon.

An example of an animal population that is an important fishery, and usually represents the third trophic level in a food chain, is the herring off the coast of Newfoundland. Although some herring populations are involved in longer food chains, the Newfoundland herring feed primarily on a population of small crustaceans and copepods that, in turn, feed upon diatoms (Fig. 9.13).

Food Webs. It is, however, uncommon to see simple feeding relationships of the type described above in nature. More commonly, the animals representing the last step in a linear food chain feed on a number of animals that have simple or complex feeding relationships, constituting a **food web.** The overall importance of food webs is not well understood, but one consequence for animals that feed through a web rather than a linear chain is their greater likelihood of survival if population extinctions or sharp decreases occur within the web at or below their feeding level. Those animals involved in

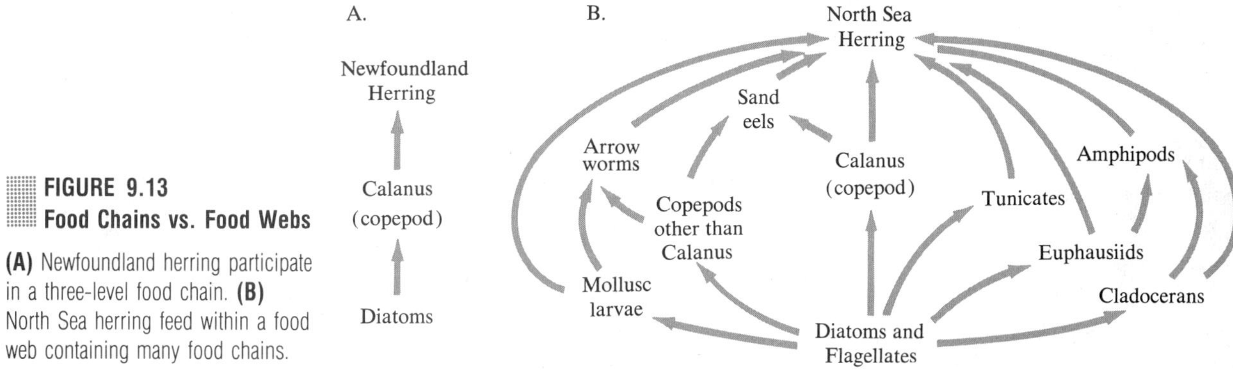

FIGURE 9.13
Food Chains vs. Food Webs

(A) Newfoundland herring participate in a three-level food chain. **(B)** North Sea herring feed within a food web containing many food chains.

food webs—such as the North Sea herring in Figure 9.13—are less likely to suffer from the extinction of one of the populations upon which they depend for food, than are the Newfoundland herring, which feed only on copepods. The extinction of the copepods in the latter case certainly would be expected to have a catastrophic effect on the herring population.

The Newfoundland herring population does, however, have an advantage over its relatives who feed through the broader-based food web. The Newfoundland herring more likely have a larger biomass to feed on since they are only two steps removed from the producers, while the North Sea herring represent the fourth level in some of the food chains within its web.

A study of 55 marine ecosystems indicated an average of 4.45 trophic levels. Only two trophic levels were encountered at rocky shores in New England and Washington, and at a Georgia salt marsh. Seven links were found in the longest food chains in Antarctic seas, while two tropical communities in the Pacific Ocean were observed to have seven and ten trophic levels. Tropical epipelagic seas and Pacific Ocean upwelling zones had eight trophic levels. The study found that the amount of primary productivity and the degree of variability of the environment have little influence on the length of food chains within an ecosystem. The only variable that showed a clear relationship to the length of food chains was whether or not the environment was two- or three-dimensional. Three-dimensional (pelagic) marine ecosystems averaged 5.6 trophic levels, compared to an average of 3.5 for two-dimensional (benthic) ecosystems.

Biomass Pyramid. It is obvious from considering the energy losses that occur in each feeding population that there is some limit to the number of levels in a food chain, and that the population at each level must necessarily contain much less biomass than the population upon which it feeds. It was mentioned previously that the individual members of a feeding population tend to be larger than their prey, but it seems they should also be less numerous. Each feeding population must contain a smaller biomass than the population upon which it feeds. This concept of a decreasing mass in successively higher trophic levels is referred to as an Eltonian pyramid, or **biomass pyramid** (Fig. 9.14).

Biogeochemical Cycles

Composition of Organic Matter. We have discussed the noncyclic nature of flow of energy through the biotic community, and saw that it is a unidirectional flow. Now let us consider the biogeochemical cycles involving matter that is not lost to the biotic community, but is cycled by being converted from one chemical form to another by the various members of the community.

In view of the fact that living organisms are composed of a variety of organic compounds, let us first concern ourselves with the elements that go into the make-up of these substances. About 20 elements in these basic organic compounds are considered to be essential to their production:

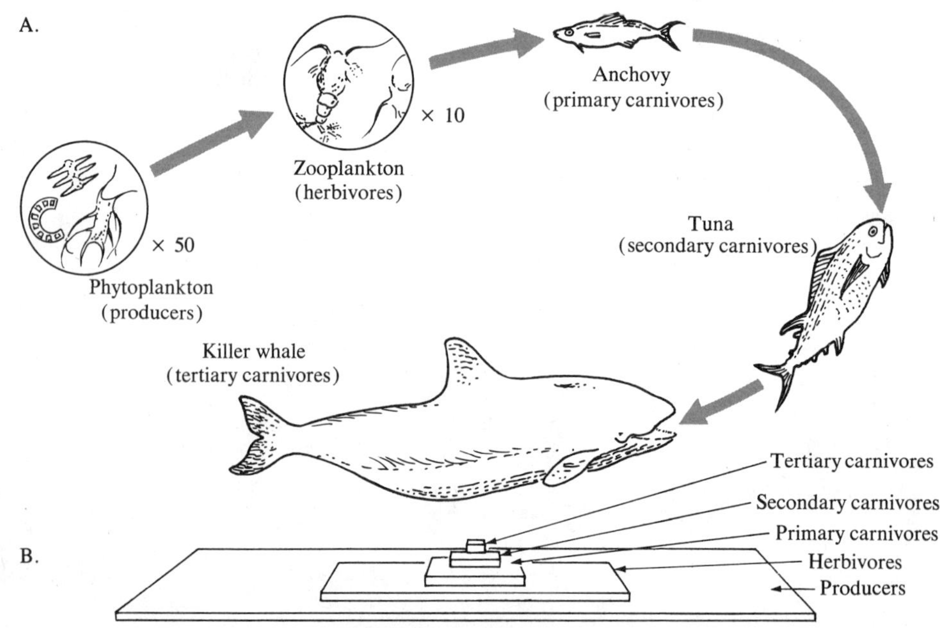

FIGURE 9.14
Biomass Pyramid

(A) The higher on the food chain an organism is found, the larger it is as an individual. **(B)** The total biomass represented by a population high on the food chain is less than that for a population at a lower level.

1. The *major constituents* of organic materials are those that individually make up at least 1 percent (each) of organic material on a dry weight basis. They are, in the order of their relative abundance, carbon, oxygen, hydrogen, nitrogen, and phosphorus.
2. *Secondary constituents* occur in concentrations of from 500 ppm to 10,000 ppm by weight. They are the elements sulfur, chlorine, potassium, sodium, calcium, magnesium, iron, and copper.
3. *Tertiary constituents* are a third group occurring in concentrations of less than 500 ppm of dry weight. They are boron, manganese, zinc, silicon, cobalt, iodine, and fluorine.

Considering the availability of these elements as solutes in ocean water, we find that the secondary constituents of organic composition have been considered to be in sufficient concentrations to make it unlikely that members of this group would ever be limiting factors in plant productivity. However, recent evidence indicates low concentrations of iron may be responsible for limiting photosynthesis in antarctic and North Pacific waters, where other nutrients are available in high concentrations. The tertiary constituents, which occur in very low concentrations in organic compounds, are also found to be in very low concentrations in the marine environment. It seems probable that the absence of tertiary constituents might create conditions that would limit productivity.

Returning to a consideration of the major constituents, we find that carbon dioxide and water should certainly, on the basis of their concentrations, provide enough carbon, oxygen, and hydrogen to assure that these elements would never limit productivity. Nitrates and phosphates are the nutrients containing nitrogen and phosphorus that we need to consider in more detail since they do, under many conditions, limit marine productivity.

In biogeochemical cycles, elements follow a pattern where an inorganic form is taken in by an autotrophic organism that incorporates it into organic molecules. The food is passed through a food web that usually ends in decomposition of the organically produced compounds into the inorganic forms. The inorganic forms may again be used in autotrophic production to complete the cycling.

A New Role for Bacteria Although it has been widely accepted that zooplankton are the primary grazers of phytoplankton, new evidence indicates that free-living bacteria may consume up to 50 percent of the production of phytoplankton. The bacteria are thought to consume the dissolved organic matter that is lost from the conventional food web by three processes:

1. Phytoplankton *exudate*. As phytoplankton age, they lose some of their cytoplasm directly into the ocean.
2. Phytoplankton *'munchates'*. As phytoplankton are eaten by zooplankton, cytoplasm is 'spilled' into the ocean.
3. Zooplankton *excretions*. The liquid excretions of zooplankton are dissolved into ocean water.

Free-living heterotrophic bacteria absorb this dissolved organic matter and use it as food. Then, the organic matter re-enters the conventional food web, primarily through the grazing of microscopic flagellate bacteriovores (Figure 9.15). Other larger zooplankton may be important in this process locally.

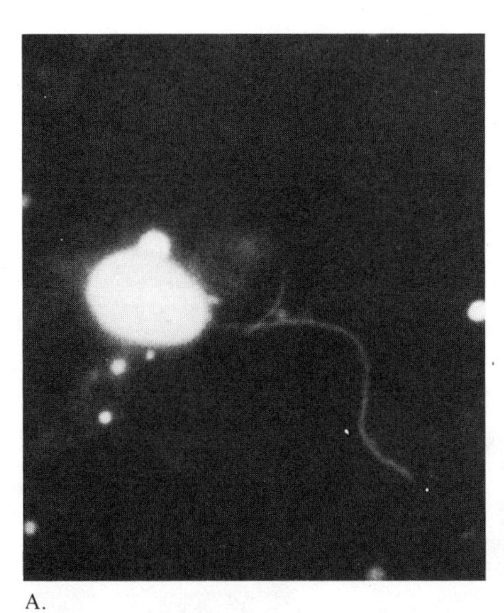

A.

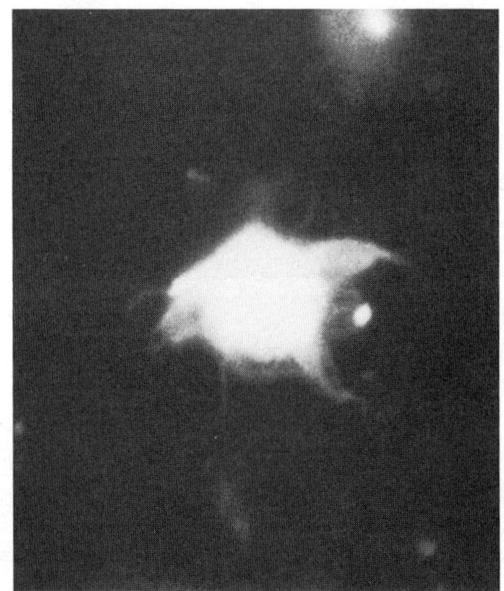

B.

FIGURE 9.15
Nanoplanktonic Flagellate and Ciliate That Feed on Bacteria

If bacteria are important absorbers of dissolved organic matter in the oceans, they could enter the conventional food web as a result of being consumed by nanoplankton such as the 7 μm-long flagellate **(A)** and the 7 μm (0.003 in) ciliate **(B),** which were recovered from an estuary in Georgia where they are known to consume bacteria. These nanoplankton are known to be grazed by copepods that are a major link in the marine food web. (Photos courtesy of Evelyn and Barry Sherr, University of Georgia Marine Institute, Sapelo Island.)

The details of the process by which free-living bacteria participate in the marine food web are still under investigation, but microbial ecologists studying the problem are convinced the evidence indicates bacteria have an important role in the transfer of energy through the marine ecosystem. Thus, bacteria may have more varied roles in the biogeochemical cycling of matter in the oceans than had previously been known. See the special feature, "Married Couple Investigate Pelagic Food Web", at the end of this chapter for a discussion of some of the problems under investigation in this field.

Carbon, Nitrogen, and Phosphorus. Although there is no concern that carbon concentrations in the marine environment are a limiting factor in marine productivity, we will examine the carbon cycle, since this element is the basic component of organic compounds. To appreciate the excess of carbon dioxide in the world's oceans as it is related to the requirements for photosynthetic productivity, we should know that only about 1 percent of the total carbon present in the oceans is involved in photosynthesis.

Comparatively, the soluble nitrogen compounds incorporated into phytoplankton biomass may be ten times the total nitrogen compound concentration in seawater that can be measured as a yearly average. This level implies that the soluble nitrogen compounds must be cycled completely up to ten times per year. Available phosphates may be turned over up to four times per year.

Comparing the ratio of carbon to nitrogen to phosphorus in dry weights of diatoms, we find that proportions are 41:7:1. This ratio is also observed in the zooplankton that feed on the diatoms, as well as in ocean water samples taken throughout the world. Thus, phytoplankton take up nutrients in the ratio in which they are available in the ocean water, and pass them on to zooplankton in the same ratio. When these phytoplankton and zooplankton die and decompose, carbon, nitrogen, and phosphorus are restored to the water in this ratio.

The Carbon Cycle. This cycle involves the uptake of carbon dioxide by phytoplankton. Carbon dioxide is returned to the ocean water primarily through respiration of phytoplankton, animals, and bacteria, and secondarily by autolytic breakdown of dissolved organic materials. **Autolytic decomposition** results from the action of enzymes present in the cells of organic tissue, and does not require bacterial action (see Fig. 9.16).

Figure 9.17 shows that as carbon dioxide (CO_2) combines with water (H_2O), carbonic acid (H_2CO_3) is formed. Carbonic acid may dissociate and produce a hydrogen ion (H^+) and a bicarbonatc ion ($H_2CO_3^-$). On further dissociation of the bicarbonate ion, an additional hydrogen ion may be released to produce a carbonate ion (CO_3^{2-}). The ocean is essentially saturated with Ca^{2+} ions, so when CO_3^{2-} ions become available, these ions combine to form calcium carbonate ($CaCO_3$). Carbon may be removed from ocean water by precipitation of $CaCO_3$ as sediment on the ocean floor. As Figure 9.17 shows, all the reactions mentioned can be reversed.

Figure 9.18 shows that the presence of carbon dioxide and the carbonate ion in ocean water is almost mutually exclusive. In ocean waters of normal pH levels (8.1), bicarbonate ions are about ten times more abundant than the carbonate ions. The carbonate ions are ten times more abundant than the carbon dioxide molecules, which is about 100 times more abundant than carbonic acid.

A rapid rate of carbon dioxide consumption resulting from a phytoplankton increase will result in an in-

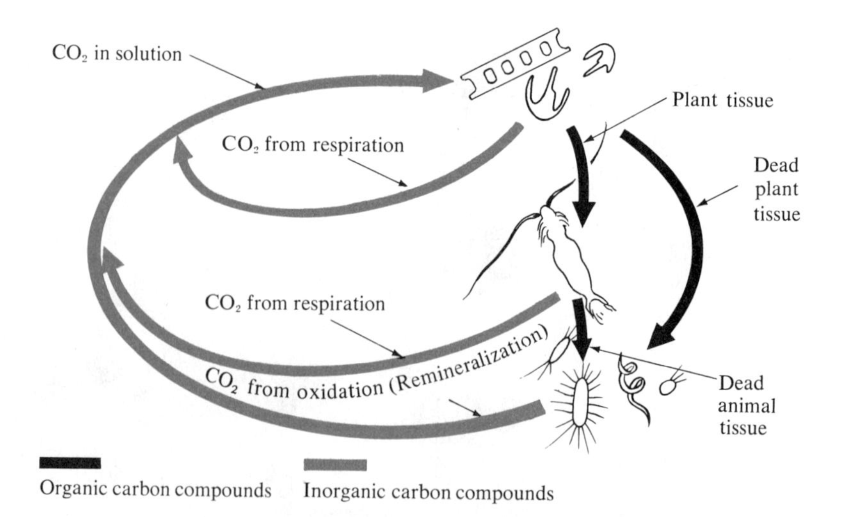

FIGURE 9.16
The Carbon Cycle

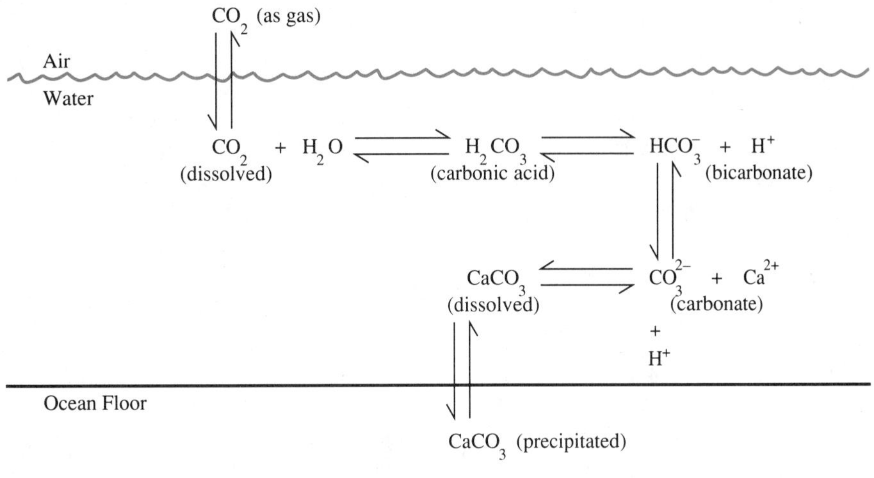

CO_2 (as gas)

Air

Water

CO_2 (dissolved) + H_2O ⇌ H_2CO_3 (carbonic acid) ⇌ HCO_3^- + H^+ (bicarbonate)

$CaCO_3$ (dissolved) ⇌ CO_3^{2-} + Ca^{2+} (carbonate) + H^+

Ocean Floor

$CaCO_3$ (precipitated)

FIGURE 9.17
The Abiotic Carbon Cycle

Atmospheric carbon is dissolved in ocean water and combines chemically with water to produce carbonic acid. Carbonic acid molecules can dissociate to produce a bicarbonate ion and a hydrogen ion. The bicarbonate ion can dissociate to produce a carbonate ion and a hydrogen ion. The carbonate ion, with its double negative charge, combines with a calcium ion that has a double positive charge to produce calcium carbonate. Calcium carbonate then can be precipitated on the ocean floor, which removes the carbon dioxide from the ocean water. These reactions can be reversed. If the pH of the ocean drops too low, the reaction moves toward carbon dioxide and removes hydrogen ions from the water to combat the drop in pH. Should the pH of the ocean water rise, the reaction moves toward the formation of more carbonate and increases the hydrogen ion content of the water. This helps hold down the pH.

crease in the pH of the water. This causes a conversion of bicarbonate to carbonate, and releases hydrogen ions to combat the rise in pH. A high rate of respiration, which produces increasing concentrations of carbon dioxide, will lower the pH. The increase in carbon dioxide will be accompanied by a conversion of carbonate ions to bicarbonate ions. This removes hydrogen ions from the water to combat the drop in its pH. Extreme ranges of pH are rare in the ocean due to this buffering activity of the carbon system that quickly restores near-normal pH values. The pH values of ocean water usually remain within a range of 7.5–8.3.

The Nitrogen Cycle. Nitrogen is necessary for the production of the nucleic acids, DNA and RNA, as well as amino acids, the building blocks of proteins. These organic compounds are consumed by animals and free-living bacteria. They then are passed on to the saprobic (decomposing) bacteria, as dead plant and animal tissue and excrement.

The saprobic bacteria gain energy from breaking down these compounds. This breakdown leads to the liberation of inorganic compounds, such as nitrates, that are the basic nutrient salts used by plants. Most nitrogen in the ocean is found as molecular nitrogen (N_2).

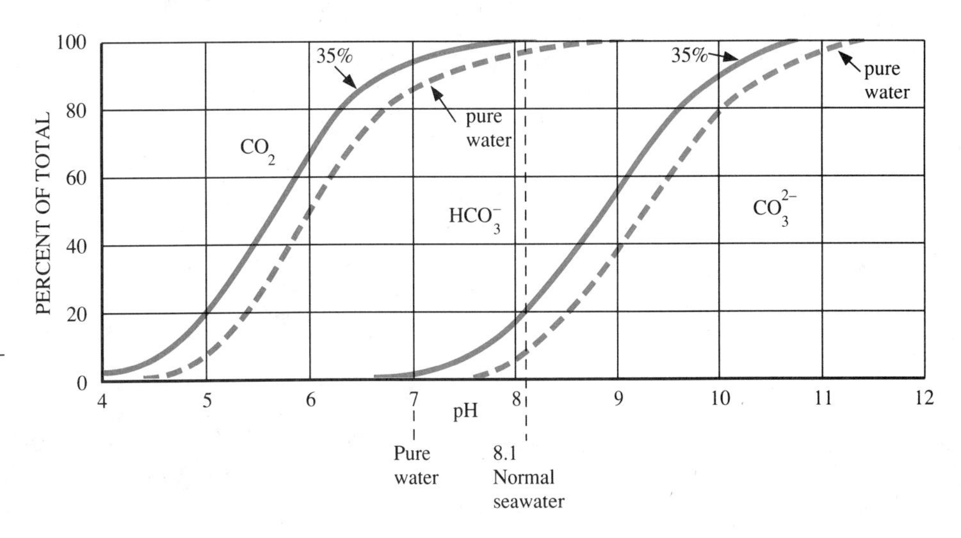

FIGURE 9.18
Distribution of Carbon Species in Water

Pure water has a normal pH of 7, while ocean water has a normal pH of 8.1. The figure shows that as the pH of either is raised, the concentration of carbon dioxide is reduced and the concentration of carbonate increases. The reverse reaction occurs if the pH drops below normal values.

The most common combined forms are nitrite (NO_2^-) and nitrate (NO_3^{2-}). These are oxidized states of nitrogen. The most abundant reduced form is ammonia (NH_3^+).

Different bacteria involved in the nitrogen cycle make it somewhat complicated. Although most of the bacteria play the heterotrophic roles of consuming dissolved organic matter, or converting organic compounds into inorganic salts, there are some that are able to fix molecular nitrogen into combined forms. These are **nitrogen-fixing bacteria. Denitrifying bacteria** make up another special group whose metabolism depends upon the breakdown of nitrates and the liberation of molecular nitrogen. The nitrogen cycle is graphically presented in Figure 9.19.

Autotrophic organisms of various types can use ammonium nitrogen and nitrites. However, the most important nutrient form of inorganic nitrogen—nitrate—can be utilized more efficiently by phytoplankton. Studies of nutrient cycles in various portions of the world's oceans indicate that the availability of nitrogen is clearly a limiting factor in determining the level of productivity during summer months. This condition arises from the fact that the processes involved in converting particulate organic matter to nitrate salts through bacterial action may require up to three months. The conversion begins in the lower portions of the photosynthetic zone as the particulate matter is sinking toward the ocean bottom.

There are three basic stages in the conversion of particulate organic matter to nitrate salts that involve three distinct bacterial types. The first oxidizes the particulate organic nitrogen into ammonium nitrogen, which is then acted upon by the second bacterial community, which converts it to nitrite. A third action is required to convert the nitrite to the prime organic salt species, nitrate. By the time this conversion is completed, the nitrogen compounds usually are below the euphotic zone and thus unavailable for photosynthesis. They cannot readily be returned to the euphotic zone during summer, due to the strong thermostratification that exists throughout much of the ocean surface. If nitrates are to again become available to plants in these regions, their return must wait until the thermostratification disappears, allowing upwelling and mixing during the winter to effect the movement of the nutrients into the surface waters.

The Phosphorus Cycle. The phosphorus cycle is simpler than the nitrogen cycle primarily because there are fewer ionic forms of phosphorus than nitrogen. There is only one stage of bacterial action involved in breaking down the organic phosphorus compounds. This difference can be studied by comparing Figures 9.19 and 9.20. The rate at which the organic compounds can be decomposed into inorganic **orthophosphates** (the phosphorus compounds primarily used by

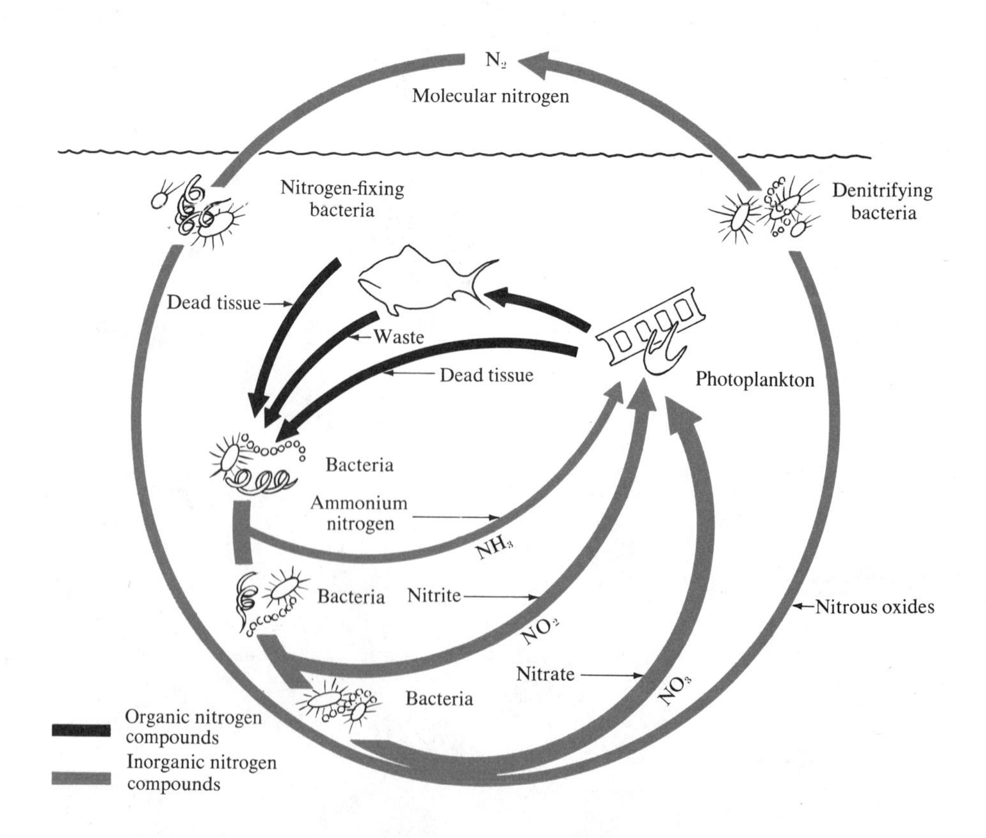

FIGURE 9.19
The Nitrogen Cycle

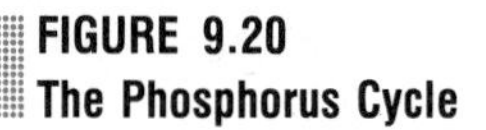

FIGURE 9.20
The Phosphorus Cycle

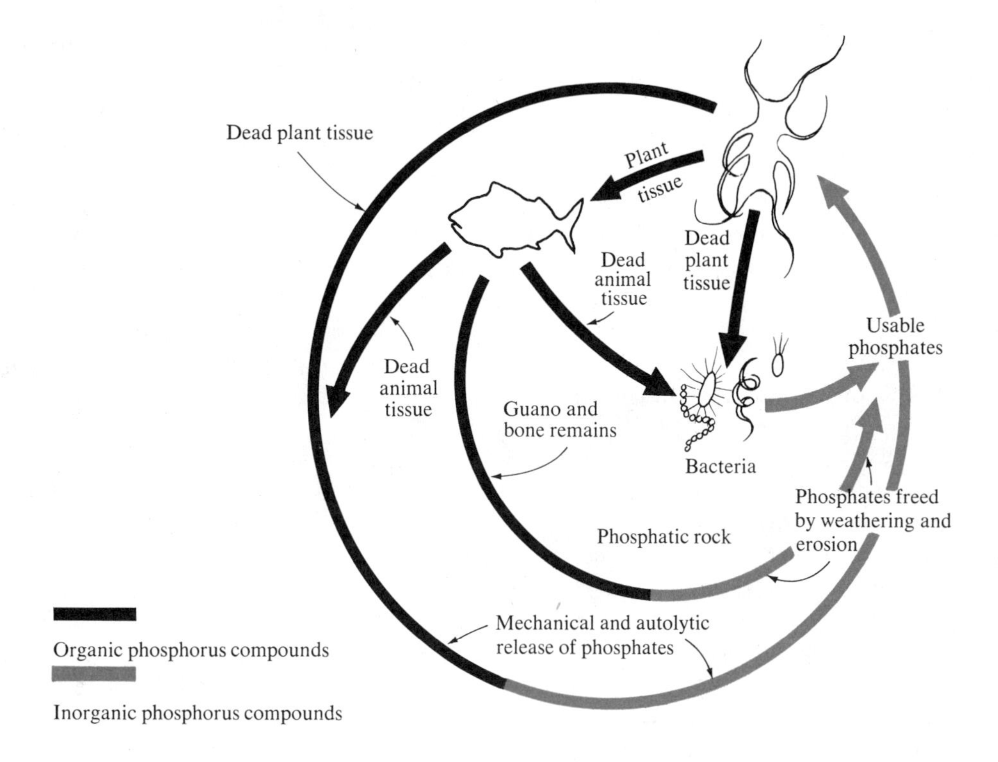

phytoplankton) is much more rapid than that of nitrogen breakdown. Also contributing to the increased rate of converting organic nitrogen to nutrient form is autolytic breakdown of phosphatic organic material by enzymes. As a result of the greater rate of breakdown for organic phosphorus, much of it can be completed above the oxygen compensation depth. It therefore is made available to plants within the photosynthetic zone. Although the concentration of phosphorus in the oceans is only about one-seventh that of the concentration of nitrogen, the fact that the recycling can take place within the photosynthetic zone usually allows adequate quantities of inorganic phosphorus to be available for plant productivity. The lack of phosphorus is rarely a limiting factor of photosynthetic productivity.

The Silicon Cycle. As previously discussed, frustules of diatoms are composed of silica (SiO_2). It is available in the form of dissolved silica. Although the availability of silica can be a limiting factor in the productivity of diatoms, it rarely is a limiting factor of total primary productivity since not all phytoplankton require silica as a protective covering. There is very little likelihood that silicon will ever be a limiting nutrient in total productivity because it is very abundant in the rocks that make up the earth's crust, and the small fraction of the earth's crustal silica that eventually will find its way into the sea is very great in relation to the silicon needs of the diatom population. Silica concentrations range from unmeasurable up to 400 mg/m^3. Fluctuations in concentration roughly coincide with those observed in nitrogen and phosphorus. However, the fluctuation in silica concentrations displays a much greater amplitude than fluctuations for nitrogen and phosphorus. This condition is probably due to the fact that silica does not undergo bacterial decay, and is taken directly into solution after undergoing autolytic breakdown.

SUMMARY

Ecology is the study of organisms in relation to one another, and the environment in which they interact. It is studied in the context of an ecosystem, which contains populations of producers, consumers, and decomposers. Within an ecosystem, two or more species may have an intimate and prolonged relationship that benefits at least one of the species. Such a relationship is termed symbiosis, and it may take the form of commensalism, mutualism, or parasitism.

Photosynthetic productivity of the oceans is limited by the availability of solar radiation and of nutrients. The depth to which sufficient light penetrates to allow plants to produce only that amount of oxygen required for their respiration is the oxygen compensation depth. Phototrophs cannot survive below this depth, which may occur at less than 20 m (65 ft) in turbid coastal waters, or at a probable maximum of 150 m (492 ft) in the open ocean. Nutrients are most abundant in coastal areas due to runoff and upwelling. In high-latitude areas thermoclines are generally absent, so upwelling can readily occur, and productivity is commonly limited more by the availability of solar radiation than by the lack of nutrients. In low-latitude regions, where a strong thermocline may exist year-round, productivity is limited except in areas of upwelling. The lack of nutrients is generally the limiting factor. In temperate regions, where distinct seasonal patterns are developed, productivity peaks in the spring and fall and is limited by lack of solar radiation in the winter and lack of nutrients in the summer.

In addition to the primary productivity through photosynthesis, organic biomass is produced through bacterial chemosynthesis. Chemosynthesis, recently observed on oceanic spreading centers in association with hydrothermal springs, is based on the release of chemical energy by the oxidation of hydrogen sulfide.

Radiant energy captured by phototrophs is converted to chemical energy and passed through the biotic community. It is expended as mechanical and heat energy, and ultimately increases the entropy of the system.

As energy is transferred from phototrophs to the herbivore and various carnivore feeding levels, only about 10 percent of the energy taken in at one feeding level is passed on to the next. The lost energy is represented by the mass of food that is not assimilated and is excreted as feces and the energy expended through respiration. The ultimate effect of this decreased amount of energy that is passed to trophic levels higher in the food chain is a decrease in the number of individuals and total biomass of populations higher in the food chain. This process can be expressed as a biomass pyramid.

There is, however, no loss of mass. The mass used as nutrients by phototrophs is converted to biomass. Upon the death of organisms, the mass is decomposed to an inorganic form ready again for use as nutrients for phototrophs. Of the nutrients required by phototrophs, compounds of nitrogen are most likely to be depleted, and restrict photosynthetic productivity. Since the total decomposition of organic nitrogen compounds to inorganic nutrients requires three stages of bacterial decomposition, these compounds may have sunk beneath the photosynthetic zone before decomposition is complete, and therefore are unavailable for photosynthesis.

QUESTIONS AND EXERCISES

1. Define ecology, and discuss the components of a biological community found within an ecosystem.
2. Define oxygen compensation depth, and explain the use of the dark and transparent bottle technique for its determination. Also discuss how the quantity of oxygen produced by photosynthesis in each clear bottle (gross photosynthesis) is determined.
3. Compare the biological productivity of polar, temperate, and tropical regions of the oceans. Include a discussion of seasonal variables, thermal stratification of the water column, and the availability of nutrients and solar radiation.
4. Discuss chemosynthesis as a method of primary productivity. How does it differ from photosynthesis?
5. Describe the components of the marine ecosystem.
6. Describe the flow of energy through the biotic community, and include the forms to which solar radiation is converted. How does this flow differ from the manner in which mass is moved through the ecosystem?
7. How is the energy taken in by a feeding population lost so that only a small percentage is made available to the next feeding level? What is the average efficiency of energy transfer between trophic levels?
8. If a killer whale is a third-level carnivore, how much phytoplankton mass is required to add each gram of new mass to the whale? Assume 10 percent efficiency of energy transfer between trophic levels. Include a diagram.

9. Describe the probable advantage to the ultimate carnivore of a food web over a single food chain as a feeding strategy.

10. What are the proportions by weight of carbon, nitrogen, and phosphorus in ocean water, phytoplankton, and zooplankton? Suggest how these amounts may support or refute the idea that life originated in the oceans.

11. Explain why nitrogen is much more likely than phosphorus to be a limiting factor in marine productivity.

REFERENCES

Alldredge, A.L., and **Cohen, Y.** 1987. Can microscale chemical patches persist in the sea? *Science* 235(4789):689–691.

Craig, H., and **Hayward, T.** 1987. Oxygen supersaturation in the ocean: Biological versus physical contributions. *Science* 235(4785):199–201

Ducklow, H.W. 1983. Production and fate of bacteria in the oceans. *BioScience* 33(8):494–501.

Grassle, J.F., et al. 1979. Galapagos: Initial findings of a deep-sea biological quest. *Oceanus* 22(2):2–10.

Hallock, P., Hine, A.C., Vargo, G.A., Elrod, J.A., and **Jaap, W.C.** 1988. Platforms of the Nicaraguan Rise: Examples of the sensitivity of carbonate sedimentation to excess trophic resources. *Geol.* 16(2):1104–1107.

Hallock, P., and **Schlager, W.** 1986. Nutrient excess and the demise of coral reefs and carbonate platforms. *Palaios* (1):389–398.

Jenkins, W.J. 1982. Oxygen utilization rates in North Atlantic subtropical gyre and primary production in oligotrophic systems. *Nature* 300:246–248.

Littler, M.M., Littler, D.S., Blair, S.M., and **Norris, J.N.** 1985. Deepest known plant life discovered on an uncharted seamount. *Science* 227(4683):57–59.

Martin, J.H., Cordon, R.M., and **Fitzwater, S.E.** 1990. Iron in Antarctic waters. *Nature* 345(6271):156–158.

Martinez, L., Silver, M., King, J., and **Alldredge, A.** 1983. Nitrogen fixation by floating diatom mats: A source of new nitrogen to oligotrophic ocean waters. *Science* 221(4606):152–154.

Paerl, H.W., and **Bebout, B.M.** 1988. Direct measurement of O_2-depleted microzones in marine *Oscillatoria:* Relation to N_2 fixation. *Science* 241(4864):442–445.

Parsons, T.R., Takahashi, M., and **Hargrave, B.** 1984. *Biological oceanographic processes, 3rd edition.* New York: Pergamon Press.

Platt, T., and **Sathyendranath, S.** 1988. Oceanic primary production: Estimation by remote sensing at local and regional scales. *Science* 241(4873):1613–1619.

Platt, T., Subba Rao, D.V., and **Irwin, B.** 1983. Photosynthesis of picoplankton in the oligotrophic ocean. *Nature* 301(5902):702–704.

Russell-Hunter, W.D. 1970. *Aquatic productivity.* New York:Macmillan.

Sherr, B.F., Sherr, E.B., and **Hopkinson, C.S.** 1988. Trophic interactions within pelagic microbial communities: Indications of feedback regulation of carbon flow. *Hydrobiologia* 159(1):19–26.

Shulenberger, E., and **Reid, J.L.** 1981. The Pacific shallow oxygen maximum, deep chlorophyll maximum, and primary productivity reconsidered. *Deep Sea Research* 28(9):901–919.

SUGGESTED READING

SCIENTIFIC AMERICAN

Benson, A.A. 1975. Role of wax in oceanic food chains. 232(3):76–89.

Childress, J.J., Feldback, H., and **Somero, G.N.** 1987. Symbiosis in the deep sea. 256(5):114–121.

Levine, R.P. 1969. The mechanism of photosynthesis. 221(6):58–71.

Pettitt, J., Ducker, S., and **Knox, B.** 1981. Submarine pollination. 244(3):134–144.

SEA FRONTIERS

Arehart, J.L. 1972. Diatoms and silicon. 18(2):89–94.

Bell, T. 1989. The besieged bays of the world: Brown tide and other novel blooms. 35(4):238–245.

Coleman, B.A., Doetsch, R.N., and **Sjblad, R.D.** 1986. Red tide: A recurrent marine phenomenon. 32(3):184–191.

Ebert, C.H.V. 1978. El Niño: An unwanted visitor. 24(6):347–351.

Idyll, C.P. 1971. The harvest of plankton. 17(5):258–267.

Jensen, A.C. 1973. WARNING—Red tide. 19(3):164–175.

Johnson, S. 1981. Crustacean symbiosis. 27(6):351–360.

McFadden, G. 1987. Not-so-naked ancestors. 33(1):46–51.

Oremland, R.S. 1976. Microorganisms and marine ecology. 22(5):305–310.

Philips, E. 1982. Biological sources of energy from the sea. 28(1):36–46.

CORAL REEFS AND THE NUTRIENT LEVEL PARADOX

At numerous locations around the tropical ocean, coral reef development appears to be retarded despite the fact that the physical factors known to limit the development of reefs are favorable to reef development (Hallock et al., 1988; Hallock and Schlager, 1986). One of these apparent anomalies is found on the Nicaraguan Rise that extends from the coastline of Honduras and Nicaragua across the Caribbean Sea to Jamaica (Fig. 1). Although the water salinity and temperature should allow for coral reef development across much of the feature, the occurrence of coral reefs decreases from being well developed in Jamaican waters to being negligible at the western end of the rise.

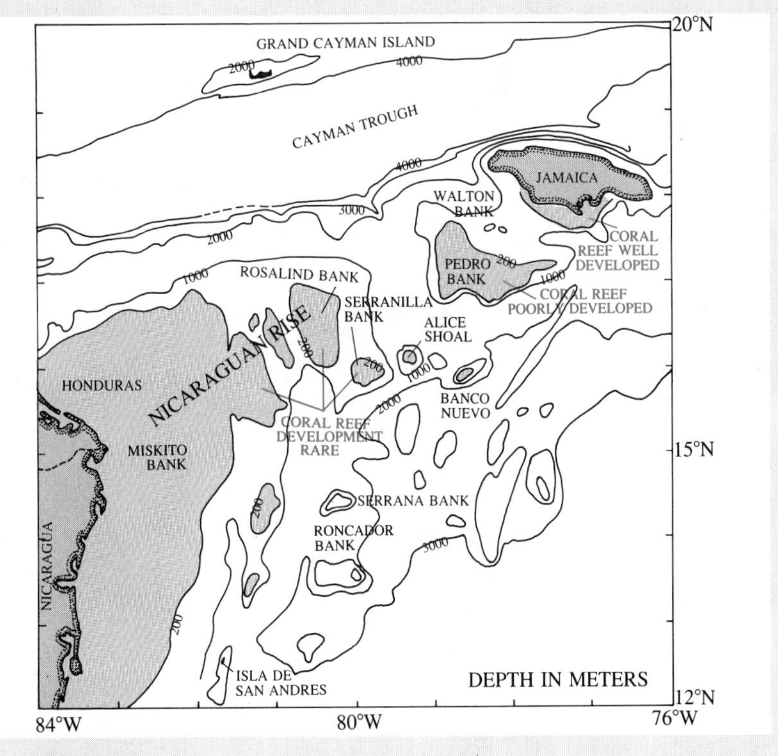

FIGURE 1
Nicaraguan Rise

Extending from the Honduran and Nicaraguan coastline east–northeastward across the Caribbean Sea to Jamaica, the Nicaraguan Rise has numerous shallow banks that rise to near the ocean surface. The quality of coral reef development along the Nicaraguan Rise is best along the southern coast of Jamaica. The reef community on the Pedro Bank is less well developed, and reef-building coral is rarely found on Miskito Bank and the smaller banks clustered around its eastern end. (Map courtesy of Pamela Hallock, University of South Florida.)

Well-developed Jamaican reef structures are replaced by inferior Florida Keys-type development on the southeast flank of Pedro Bank, southwest of Jamaica. Further southwest—on Serranilla Bank, Rosalind Bank, and the eastern end of the Miskito Bank—sponges and frondose brown algae become dominant. Calcareous algae occur on the bank tops, but reef-building corals are rare.

Recent research has shown that there is a correlation between the nutrient availability in surface waters and the type of benthic community that inhabits the shallow tropical ocean floor. As nutrient levels decrease, the dominant benthic community changes from heterotrophic suspension feeders at high nutrient levels, through fleshy benthic plants at moderate levels, to phototrophic animal-plant symbionts such as reef-building corals at low levels. Photosynthetic pigment concentrations (chlorophyll *a* plus phaeophytin; see Fig. 2) were chosen to indicate nutrient levels because they have been shown to be directly related to nutrient supply.

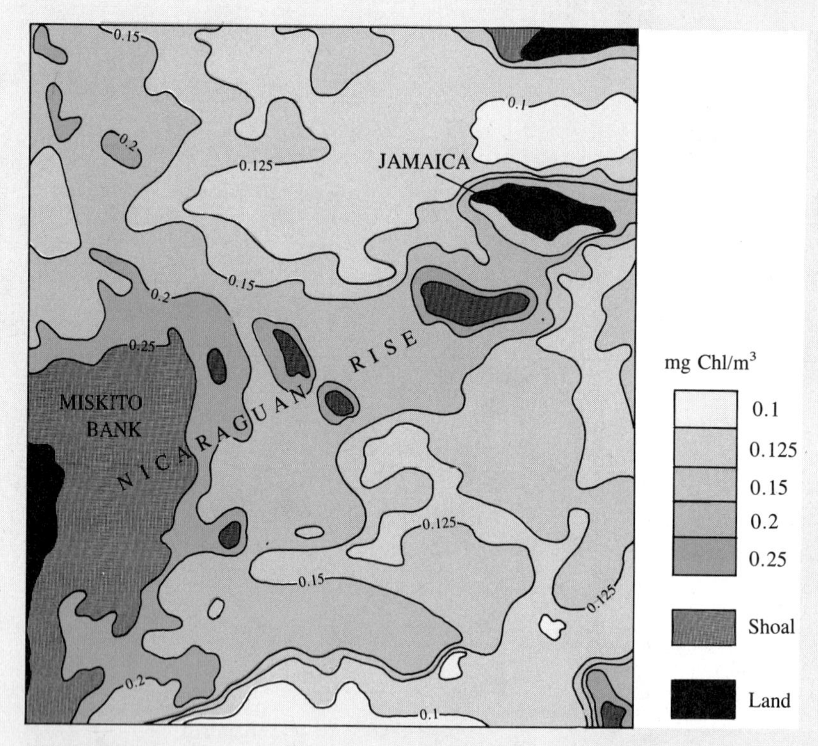

FIGURE 2
Mean Biotic Pigment (Chlorophyll *a* plus Phaeophytin) Concentrations in Surface Waters in the Area of the Nicaraguan Rise

Pigment value data were obtained from the Coastal Zone Color Scanner (CZCS) aboard the *Nimbus 7* satellite. Where water depths are less than 40 m (shoal), pigment values may be elevated due to bottom reflectance. (Map courtesy of Pamela Hallock, University of South Florida.)

Although there appears to be only a modest doubling of the nutrient supply westward along the Nicaraguan Rise—from levels just above 0.125 mg chlorophyll/m^3 in Jamaican waters, to 0.25 mg chl/m^3 along the edge of the Miskito Bank (levels up to 0.36 mg chl/m^3 have been recorded on Miskito Bank)—this variation clearly determines the type of benthic community.

The fact that high levels of nutrients put stress on reef-building corals has not been widely recognized because high levels of gross primary productivity have been reported in studies dating from 1949 to 1957. The significance of that productivity has been, in large part, misunderstood. This high gross primary productivity apparently results from two processes occurring within the reef community. Corals and other reef organisms that possess symbiotic algae in their tissue are very efficient at recycling phosphorus, and nitrogen-fixing algae are common. Corals and other carbonate reef producers with algal symbionts are highly adapted to nutrient-deficient environments.

In areas of high nutrient availability, phytoplankton mass exceeds that of benthic algae, and the most successful benthic animals are those tied to the phytoplankton food web. Increasing densities of plankton reduce the depth of light transmission, which reduces the depth ranges within which reef-building corals and calcareous algae can grow. In nutrient-rich waters, the zooxanthellae algal symbionts of the corals fix more carbon than the coral can utilize, and the corals shed it as a mucous. In the most nutrient-rich waters, corals secrete more mucous than they can shed, and bacteria blooms that develop in the mucous kill the corals.

A high level of competition for space occurs, and coral competitors and predators whose larval forms are supported by the phytoplankton are favored over the larval coral planulae, which live on internal reserves and nutrition provided by algal symbionts. Fast-growing filamentous algae, barnacles, sponges, bryozoans, and ascidians overgrow and crowd out slower-growing corals.

Predation and bioerosion are at high levels. The fast-growing heterotrophic suspension feeders and algae displace the slower-growing corals, and actively destroy the reef structure through bioerosion. Chemical and mechanical bioerosion both occur, with sponges and sea urchins being examples of groups that are very destructive. Ultimately, the reef framework is weakened and collapses under existing physical forces.

It has been predicted that plankton primary productivity of 150–200 mg carbon (C)/day should significantly increase the abundance of bioeroding organisms. In the central and western Pacific, where coral reef development is maximum, plankton primary productivity is about 100 mg C/day, while in the Caribbean it averages about 150 mg C/day. This may well explain the marginal coral reef development throughout much of the Caribbean Sea.

MARRIED COUPLE INVESTIGATE PELAGIC FOOD WEB

Evelyn and Barry Sherr of the University of Georgia Marine Institute on Sapelo Island have been among the leaders in recent research into the roles of bacteria and protozoa in the pelagic food web (Fig. 1). They have developed the primary method for evaluating the rate of bacteria consumption by protozoans—fluorescently labeled bacteria (FLB; see Fig. 2). This method was recently adapted by others to label nanoplankton algae, in order to determine grazing rates of ciliates and microcrustaceans. In this application it is called fluorescently labeled algae (FLA).

The Sherrs have shown that nanoplanktonic ciliates and flagellates may be major trophic links in the pelagic food web. Because of their rapid growth, grazing, and nutrient-regeneration capabilities, these protozoans may play a pivotal role in regulating carbon and energy flow within ecosystems.

The Sherrs' present and future research efforts include *in situ* determination of the feeding preferences of heterotrophic flagellates and ciliates for bacteria and for small algal cells. The Sherrs also intend to investigate the degree to which flagellates

FIGURE 1

Drs. Evelyn and Barry Sherr collect water samples in a tidal embayment adjacent to Sapelo Island off the Georgia coast.

may be able to use high molecular mass dissolved organic compounds as a food source. They also plan studies of the rates of growth of protozoa, and the predator-prey interactions between populations of flagellates and ciliates, as well as between microcrustaceans and metazoan larvae in coastal waters.

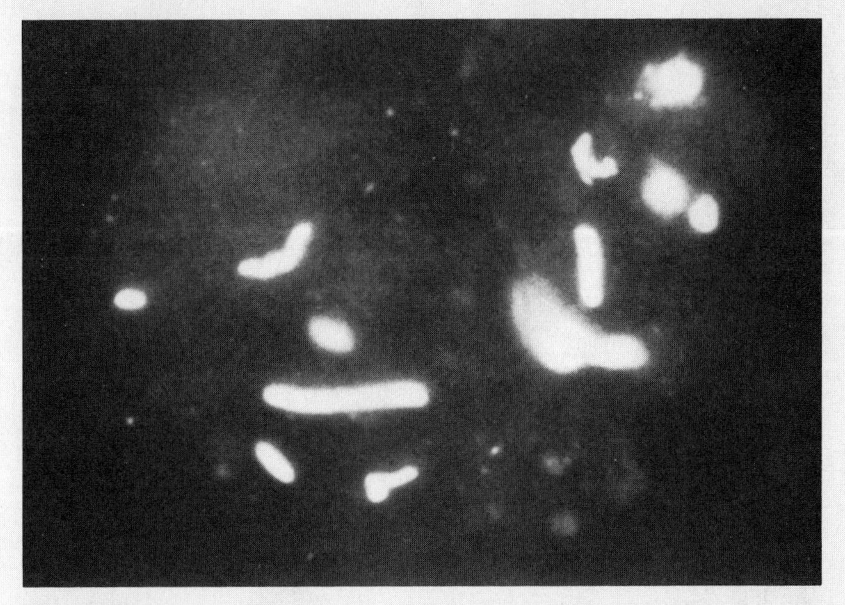

FIGURE 2

Bacteria that have been subjected to the fluorescently labeled bacteria (FLB) process. The longest is 4 μm long and 0.5 μm wide.

10

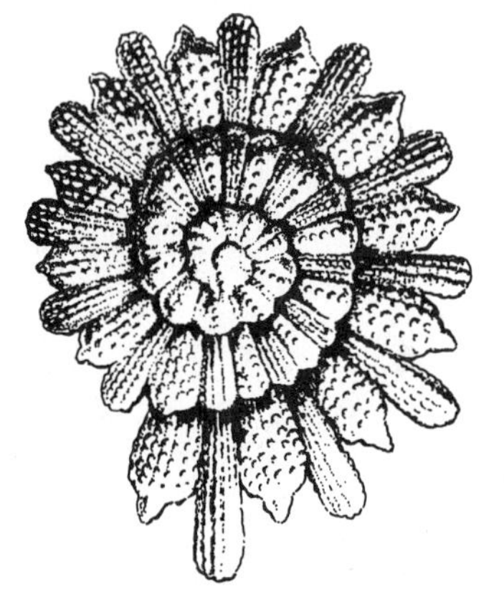

THE PHOTIC ZONE

The **photic zone** is defined as the depth of the sea to which sunlight penetrates. In clear tropical waters, this can be as deep as 1,000 m (3,280 ft). It is less in coastal areas with higher turbidity, but still includes all the water above the continental shelves. The volume of the photic zone is immense. (Remember: Surface ocean waters cover 70% of the earth.) Only the deep-ocean environment exceeds the photic zone in volume.

Marine biologists traditionally have divided the photic zone into two parts: the euphotic (or epipelagic) zone and the disphotic zone (also called the mesopelagic or twilight zone). The euphotic zone has adequate light to support photosynthesis, and the disphotic zone has light intensities adequate for animal responses, but not enough for photosynthesis. The oxygen compensation depth marks the boundary between the two zones. Although there are these important differences between the euphotic and disphotic zones, there are many more similarities that allow the photic zone to be regarded as a single habitat.

CURRENTS

An important physical process that influences life throughout the photic zone is ocean circulation. Ocean current systems are driven by two forces. Wind stress at the ocean surface drives the horizontal currents, which seldom extend to depths greater than 100 m (328 ft). However, **density-driven circulation** often maintains these currents to depths greater than 1 km (0.62 mi), moving deeper water in the same direction as the wind-driven surface flow. Density differences are totally responsible for the vertical circulation that sends cold, dense surface water in high latitudes slowly to the bottom of the ocean, along with the dissolved oxygen required to sustain life beneath the photic zone. This oxygen-rich water slowly travels across the deep-ocean floor, carrying this life-supporting gas the length of ocean basins. The uneven heating of the atmosphere generates the prevailing wind systems that drive the ocean current systems. This uneven heating occurs due to two phenomena: (a) the solar radiation striking the earth is not evenly distributed about its surface, and (b) water vapor, which readily absorbs infrared radiation, is not evenly distributed within the atmosphere (Fig. 10.1). The angle of the earth's rotation affects what areas of the earth's surface receive the most direct solar radiation throughout the year, as illustrated in Figure 10.1. The maximum rate of atmospheric heating occurs at the equator, and the minimum rate at the poles. Because of this distribution pattern, warm, humid, low-density air rises above the equator, and cold, dry, dense air descends over the poles. There is also a belt of high-density air at about 30° latitude descending over the subtropics, and a belt of low-density air rising at about 60° latitude, as shown in Figure 10.2. These convection cells produce the prevailing wind system at the earth's surface. The three systems are named the **polar easterlies, westerlies,** and **trade winds.**

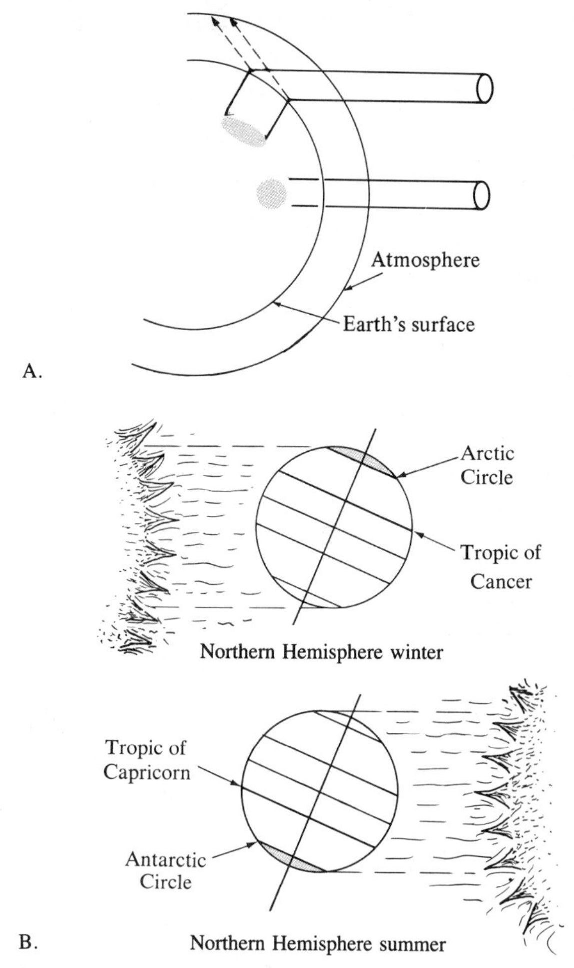

FIGURE 10.1
Solar Radiation

(A) The amount of heat received at higher latitudes is less than that at lower latitudes for three reasons: (1) A ray of solar radiation that strikes the earth at a high latitude is spread over a greater area than an equal ray that strikes the earth's surface perpendicularly at a lower latitude; (2) the high-latitude ray passes through a greater thickness of atmosphere; and (3) more of its energy is reflected, due to the low angle at which it strikes the earth's surface.
(B) During the Northern Hemisphere winter, areas north of the Arctic Circle receive no direct solar radiation, while areas south of the Antarctic Circle receive continuous radiation for months. The reverse is true during the Northern Hemisphere summer. Throughout the year, the sun is directly over some latitude between the tropics.

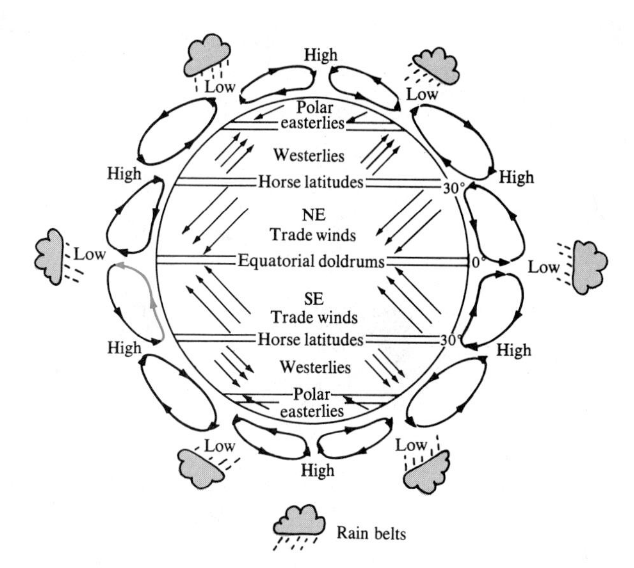

FIGURE 10.2
Prevailing Wind Belts

Low-pressure belts develop along the equator and the 60°-latitude regions due to rising columns of warm air. Descending cool dry air produces high-pressure belts in the 30°-latitude regions. Air movements are primarily vertical in these belts. Strong lateral movements of air between the belts produce the westerlies and trade winds. Cold air masses overlying the polar regions produce additional high-pressure conditions there.

Horizontal Circulation

The air masses that are moving away from the subtropics (30° N and S latitudes) toward the equator and high latitudes are moving basically south and north. For the sake of our discussion, we will consider only the Northern Hemisphere. In the case of the trade winds, they do not appear to blow directly out of the north but out of the northeast, while the westerlies, as their name implies, blow out of the southwest. Why do we observe this east-west component of motion in air masses that are moving north and south? The explanation lies in a phenomenon called the **Coriolis effect.** Any object on the earth's surface (such as an air current or projectile) moving horizontally over a long distance for a relatively long period of time will be observed to veer to the right in the Northern Hemisphere, and to the left in the Southern Hemisphere. The magnitude of this effect increases with the velocity and the latitude of the object. It is zero at the equator, and maximum at the poles. The Coriolis effect works in the following manner.

As the earth rotates on its axis, points at different latitudes rotate at different velocities (Fig. 10.3A). The rotational velocity of a point is proportional to its distance from the axis of rotation, and ranges from 0 km/h at the poles to over 1,600 km/h (994 mi/h) at the equator. An air mass near the subtropical region at 30° N

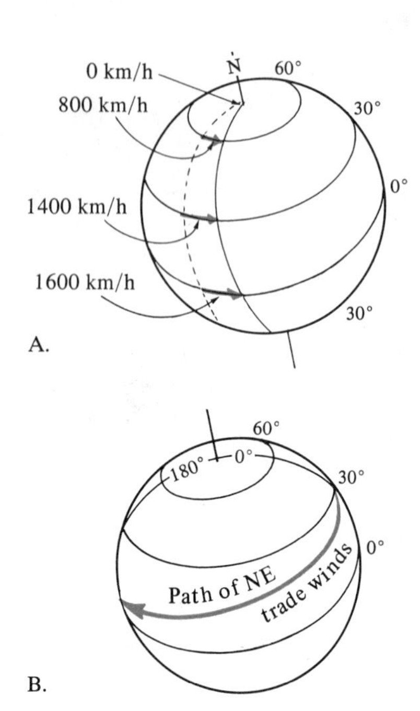

FIGURE 10.3
Coriolis Effect

(A) Points at different latitudes on the earth's surface rotate at different velocities, ranging from 0 km/h at the poles to 1,600 km/h at the equator. **(B)** An air mass that leaves the 30° latitude—where it is rotating with the earth at 1,400 km/h—and heads south toward the equator at 32 km/h will not reach the equator until 100 h later. Since the equator is rotating east at 200 km/h faster than the air mass, it will meet the equator at a point 20,000 km west of the point on the equator that was directly south of the air mass when it left its position on 30° latitude. This air mass could represent the NE trade winds, which follow a similar path.

latitude is moving with the rotating earth in an easterly direction at about 1,400 km/h (870 mi/h). In the case of the trade winds, the air mass is moving toward a point on the equator at the same longitude, which is moving east at 1,600 km/h. The distance that the air mass must cover to reach the equator is about 3,200 km (2,000 mi). If it moves to the south at 32 km/h, it will require 100 h to arrive at the equator.

The air mass is moving south at 32 km/h (20 mi/h) and east at 1,400 km/h. For every hour that the air mass moves in the southerly direction, the point on the equator toward which it started moves east 200 km (124 mi) farther than does the air mass (1,600 km/h − 1,400 km/h). In the period of 100 h it takes the air mass to make the trip, the point on the equator toward which it started will be 20,000 km (100 h × 200 km/h) east of the air mass when it reaches the equator. While the air mass was moving the 3,200 km in a southern direction, it appears as a result of the earth's rotation to have moved 20,000 km (12,420 mi) in a westerly direction. Stated another way, as the air mass covered 30° of latitude from north to south, it appeared to move across 180° of longitude in a westerly direction. Certainly, if we were to encounter this air mass aboard ship at some point between the equator and 30° N latitude, it would appear to be coming out of the east-northeast (Fig. 10.3B).

The effects described for this south-moving air mass are valid for movements in any direction. Objects veer to the right of their intended path in the Northern Hemisphere, and to the left in the Southern Hemisphere.

The prevailing wind belts set the ocean surface in motion as energy is transferred to the surface layer of ocean water through frictional stress. The trade winds blowing out of the southeast in the Southern Hemisphere and northeast in the Northern Hemisphere provide the backbone of a system of ocean surface currents. Due to a phenomenon called the **Ekman spiral,** the surface water set into motion by a wind system moves at an angle of 45° to the right of the wind in the Northern Hemisphere, and 45° to the left in the Southern Hemisphere (Fig. 10.4). Layer by layer, this energy

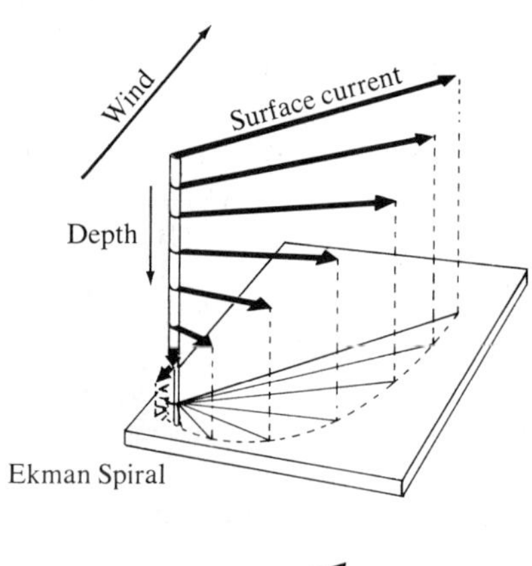

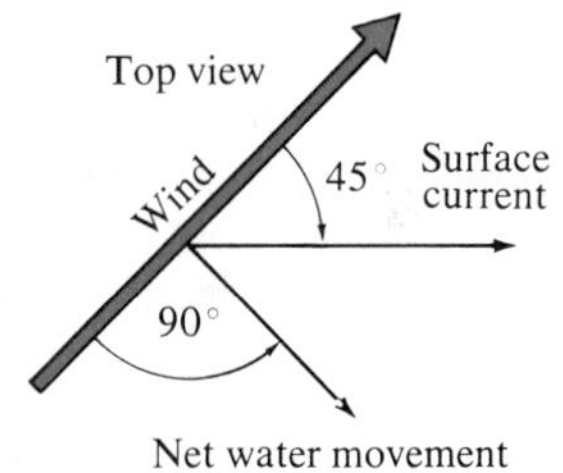

FIGURE 10.4
Ekman Spiral

Wind drives surface water in a direction 45° to the right of its path in the Northern Hemisphere. Deeper water layers continue to deflect to the right, and move at a slower speed with increased depth. The net water movement is at right angles to the wind direction.

is transferred to water beneath the ocean surface. Each layer of water moves with a velocity a little less and in a direction to the right (or left) of the layer that set it in motion. At some depth, usually between 100–200 m (328–656 ft), the wind energy is used up and wind-generated current flow stops. Because each layer of water moves progressively to the right in the Northern Hemisphere and to the left in the Southern Hemisphere, there is a net transport of water 90° to the right or left of the wind direction. This net transport is called **Ekman transport.**

Because of this phenomenon, the trade winds in each hemisphere create westerly flowing equatorial currents, and the westerlies produce easterly flowing currents at higher latitudes. The Coriolis effect causes these systems to merge into clockwise-rotating gyres in the Northern Hemisphere, and counterclockwise-rotating gyres in the Southern Hemisphere (Fig. 10.5). The Ekman transport pushes water toward the center of rotation of each **gyre,** or subtropical convergence, causing a hill of water up to 2 m (6.6 ft) high within the rotation. As a result of a balance reached among forces growing out of the earth's rotation and energy flow through the current system, the top of the hill is near the western margin of the gyre. The velocity of the **geostrophic** (earth-turning) **current** is proportional to the slope of the hill, and is therefore much greater on the west side of ocean basins where the slope of the hill is greater (Fig. 10.6). These **boundary currents,** which rapidly move warm water toward the poles along the western margin of ocean basins and carry cold water more slowly toward the equator on the east side, have a significant effect on the distribution of life in the oceans.

Studies of the Gulf Stream have revealed a phenomenon that is probably common to all boundary current systems, but is best observed there. The volume of water carried by the Gulf Stream near the Tail of the Banks south of Newfoundland is about half its volume off Chesapeake Bay. Just how this loss of water occurs

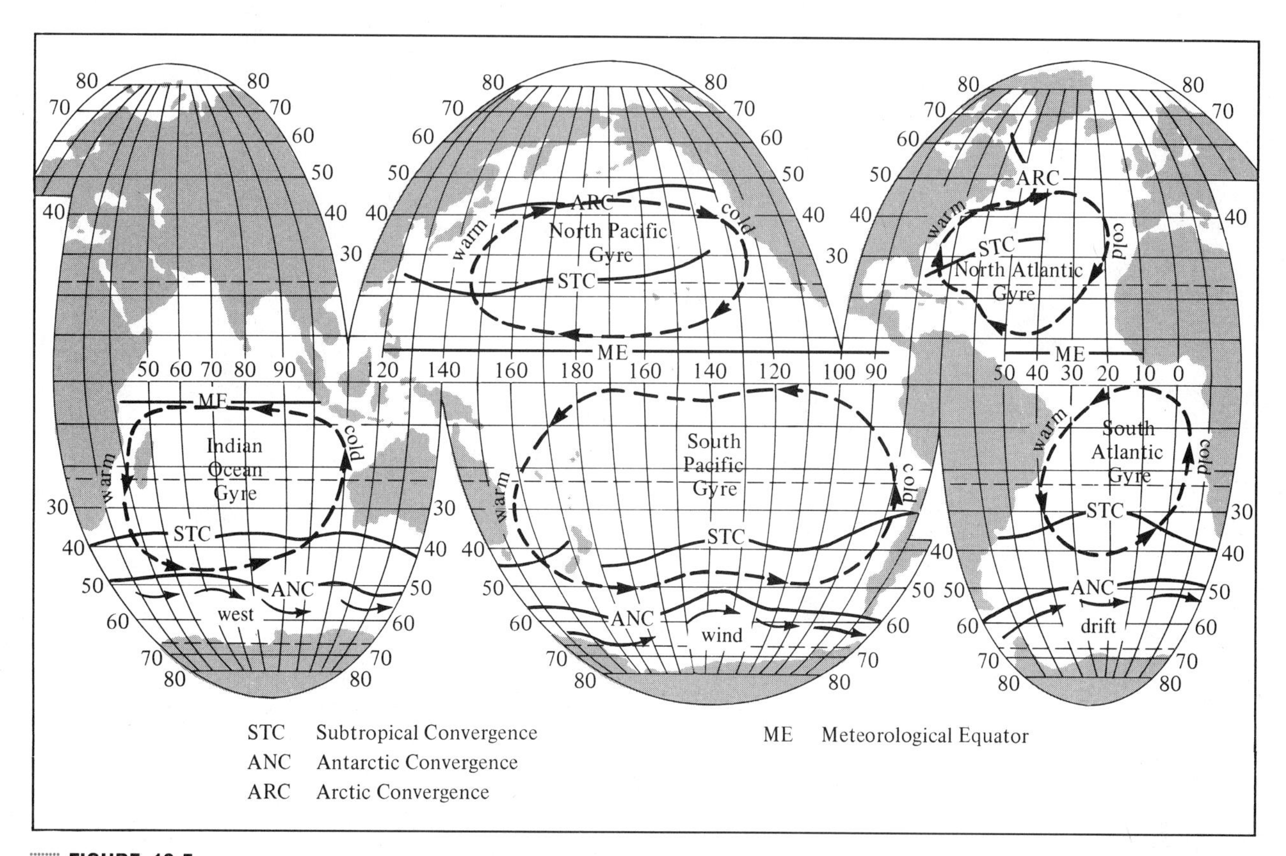

FIGURE 10.5
Wind-Driven Surface Currents in February and March

Note that subtropical gyres centered at 30° N and S latitudes dominate the surface circulation of the world's oceans. They rotate clockwise in the Northern Hemisphere, and counterclockwise in the Southern Hemisphere. Cold currents carry water toward the equator on their eastern margins, and warm currents flow toward higher latitudes along their western margins.

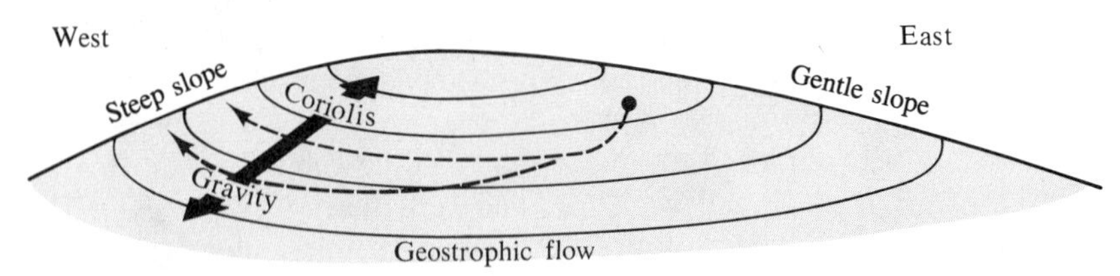

FIGURE 10.6
Geostrophic Current

Water piles up inside the gyre to form a 'hill' with an apex closer to its west margin due to the earth's rotation to the east. The geostrophic current flows nearly parallel to the contour of the hill. Due to the offset of the apex to the west, the current moves more rapidly along the steeper western margin.

is yet to be determined, but much of it may be achieved by meanders (sinuous curves in the course of a current) pinching off to form large eddies. As shown in Figure 10.7, meanders north of the Gulf Stream pinch off and trap warm **Sargasso Sea water** in eddies rotating in a clockwise direction. These eddies move southwest toward Cape Hatteras, where they rejoin the Gulf Stream at speeds of 3–7 km/day (1.9–4.3 mi/day). South of the Gulf Stream and west of 50° W latitude, eddies with a core of cold inshore water rotating counterclockwise move south and west. These large eddies, ranging in diameter from 100–500 km (62–310 mi), and extending to depths of 1 km (0.62 mi) in the warm rings and as deep as 3.5 km (2.2 mi) in the cold rings, could remove large volumes of water from the Gulf Stream. (A discussion of the impact of these eddies on the biology of the North Atlantic is found in the special feature, "Gulf Stream Rings and Marine Life", at the end of this chapter.)

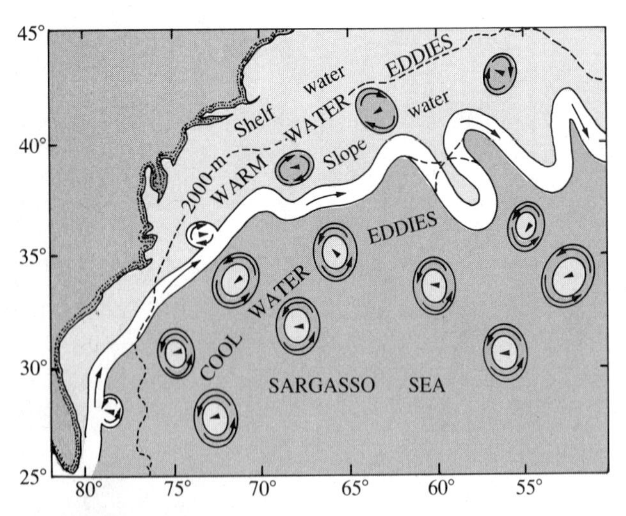

FIGURE 10.7
Gulf Stream Eddies

Large eddies form to the north and south of the Gulf Stream as meanders pinch off. Those to the north contain warm water and rotate clockwise, while those to the south contain cool slope water and rotate counterclockwise as they move southwest to rejoin the mighty current. (From Richardson, 1976.)

Vertical Circulation

Horizontal motions of surface water may bring about vertical movement of ocean water within the upper water mass. Steady winds blowing across the ocean surface create shallow convection cells with alternate right-hand and left-hand circulation (Fig. 10.8A). The axes of these cells are parallel to the wind. Organic debris and microscopic organisms may accumulate in the areas of downwelling in the zones where surface water converges between adjacent cells. This phenomenon, first explained by Irving Langmuir in 1938, is called **Langmuir circulation.** Langmuir circulation rarely extends to depths greater than 6 m, and produces windrows of concentrated organic matter that include plankton and detritus.

Coastal upwelling and **equatorial upwelling** extend to greater depths, and thus nutrient-rich water from beneath the photosynthetic zone comes to thc surfacc (Fig. 10.8B and C). The unusually high biological productivity of coastal areas, such as the area off Peru, results from coastal upwelling, and the deposit of radiolarian ooze (See Fig. 11.6) is the result of equatorial upwelling enriching the equatorial waters of the Pacific Ocean. (See Chapter 14 for a more complete description of the coastal upwelling phenomenon.)

Because there is little thermal stratification in the high-latitude water columns, surface water there can become dense enough to sink to the ocean bottom, carrying with it the oxygen needed to support life in the deep-ocean basins. Only in the North Atlantic Ocean regions off Greenland and in the Norwegian Sea, as well as in the South Atlantic Ocean area of the Weddell Sea, do surface water masses sink to the deep-ocean floor. These masses are called the **North Atlantic Deep Water** (NADW) and the **Antarctic Bottom Water** (AABW). When the southward-flowing NADW meets

Habitats in Color

The Open Ocean

The open ocean, or pelagic zone, is where organisms live without contact with the bottom. While some pelagic species can float, this habitat requires most organisms to swim constantly. The pelagic habitat is home to the largest species of marine life: the great whales.

The color of pelagic life is quite predictable, whether it be a shark, fish, or marine mammal. Species tend to be dark on top, and light on their undersurfaces. This is an adaptation to reduce the chance of being seen in the water. From above, the dark top surface blends with the darkness of the water. From below, the light color of the undersurface blends with the lightness of the sky above.

Whales dwarf the human scale. This Southern right whale is ten times longer than its human companion.

Dolphins are very social animals, and live in groups called pods. The spotted dolphin reaches a size of around 3 m (10 ft). The spots that give this species its name do not change the basic color pattern of dark on top and light below.

The majesty of size of the great whales is best noted when they are on land.

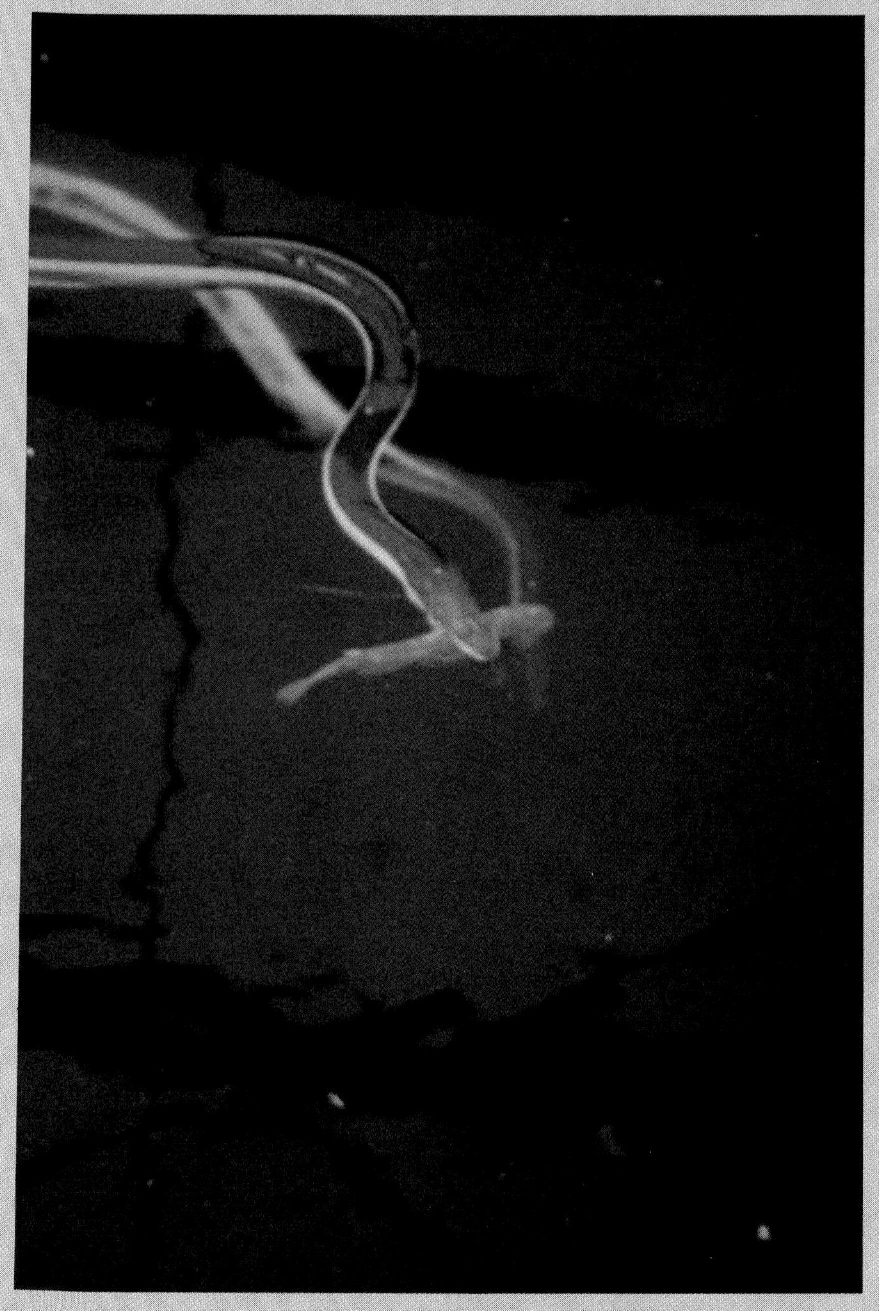

Not all pelagic animals have drab colors. This sea snake, with a fish in its mouth, is brightly colored yellow and black. The coloration is probably a warning to other species that it is venomous.

The Intertidal Zone

The intertidal zone is where the land meets the sea. When the shoreline is bed rock, the features of the shore are dominated by the type and nature of the rock. When the rock particles are not solid and can be moved by waves, the nature of the shoreline is dictated by wave action. These wave-sorted particles constitute what is technically considered a beach.

Color in the intertidal zone is dominated by the nature of the substrate and the overlying plant and animal life. Since light is abundant in the intertidal zone, much of the color is dominated by the greens, browns, and reds of the photosynthetic organisms.

This is a portion of the intertidal zone of the exposed coastline of the Pacific Northwest. The bare black rock is the zone that is exposed too much of the time for any marine life to attach, yet is covered with enough wave spray to prevent its colonization by any terrestrial plants. The zonation of life in the intertidal zone is evident in this picture. The high zone is white, due to barnacle shells (in particular, the acorn barnacle Balanus, *and the gooseneck barnacle* Mitella*). The blue color in the white of the barnacles is due to mussels* (Mytilus)*. The next zone down is dominated by brown algae. Travelling lower yet, the zone is dominated by bright-green surf grass. In the lowest zone, underneath the leafy green algae, encrusting red algae cover the rock.*

In lower areas of the intertidal zone—where stress from exposure is minimal—many animals compete with the macroalgae for space on the rocky substrate. The pink color seen here is an encrusting sponge. It is not clear if this brilliant color serves any function for the sponge. The color of the anemones, however, is related to their biology: The green results from the presence of plantlike cells in the tissues of the anemone. The brightly lighted intertidal zone is the optimal environment for photosynthesis, and symbiotic relationships between animals and diatoms or dinoflagellates are common.

On most rocky shores, the intertidal macroalgae dominate. This rocky shelf is covered with brown and green algae. The green color is from chlorophyll. Brown algae also has chlorophyll, but the green is masked by other pigments. These additional brown pigments assist in photosynthesis.

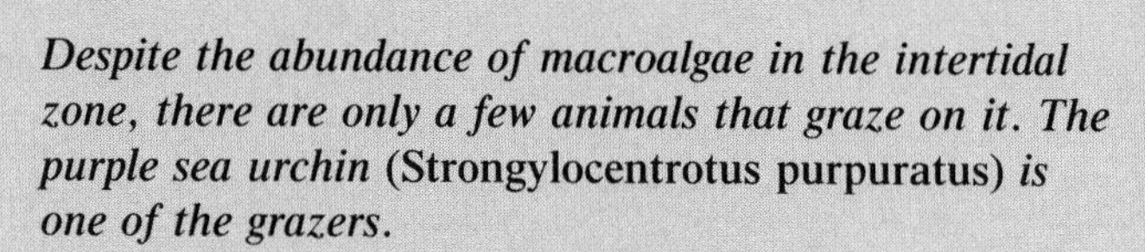

Despite the abundance of macroalgae in the intertidal zone, there are only a few animals that graze on it. The purple sea urchin (Strongylocentrotus purpuratus) *is one of the grazers.*

Where wave action can carry and sort particles of a size between 0.5–1.5 mm (.02–.06 in), a sandy beach is formed. The gray color reflects the color of the mineral particles. If the particles were made up of shell fragments, the beach would be white. In a very few locations, the beach consists of particles of lava basalt—therefore, the beach is black sand. The unstable nature of the substrate prevents benthic animals from attaching, and there is no conspicuous surface life as is found in the rocky intertidal.

Where sediments are protected from direct wave action, fine particles form mud shores. Small particles of organic material (detritus) contribute to the dark brown or black color. The upper portion of muddy shores frequently is covered with salt marsh vegetation. In the spring, the salt marsh grasses are bright green with new growth; as summer comes, the mature vegetation takes on a golden color.

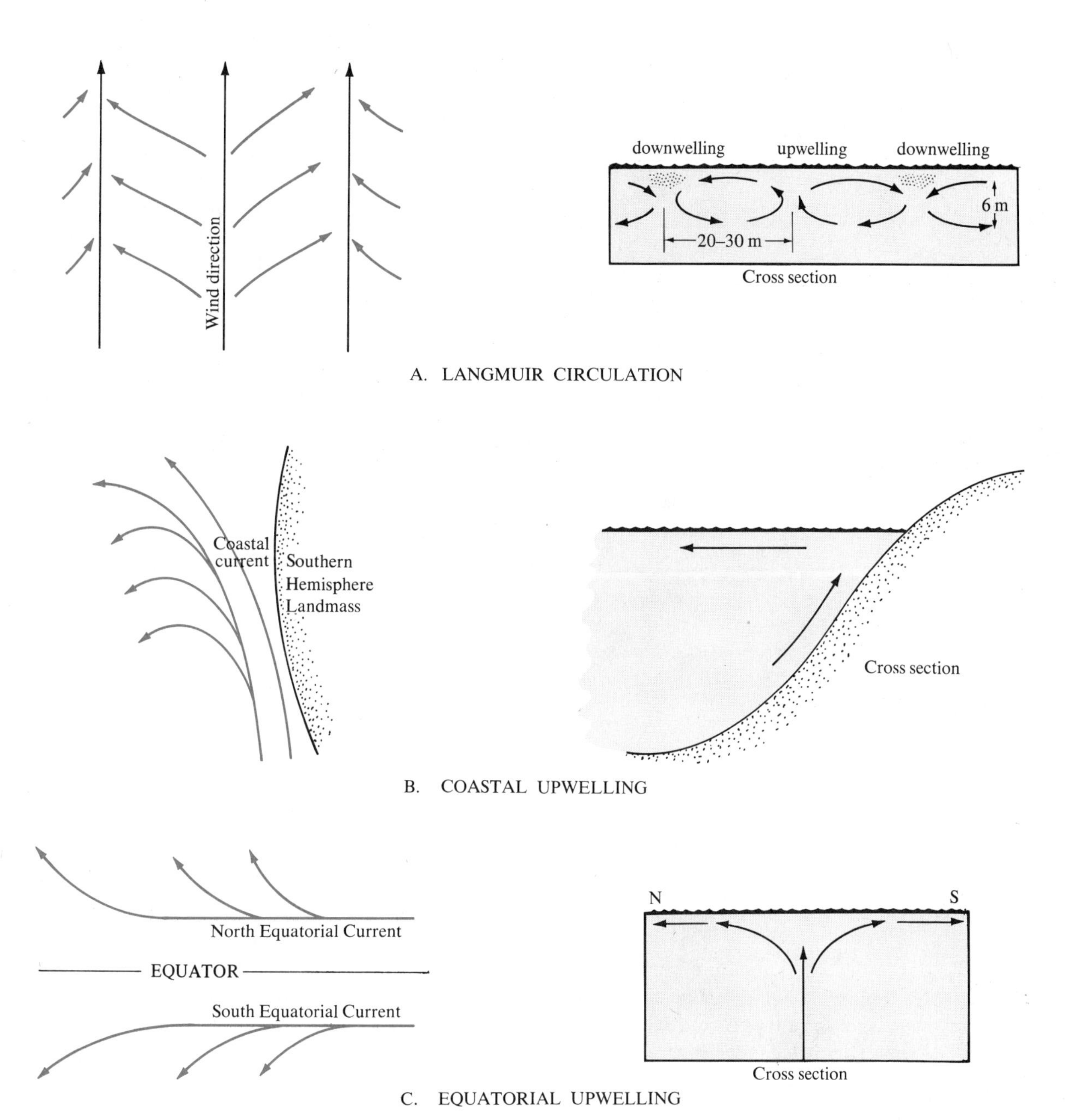

FIGURE 10.8
Wind-Driven Vertical Circulation

(A) Steady winds blowing across the ocean surface create convection cells with alternate right- and left-hand circulation. The axes of these cells run parallel to wind direction. Organic debris accumulates in downwelling zones of convergence and produces windrows that run parallel to the wind direction for great distances. **(B)** In areas where wind-driven coastal currents flow along the western margin of continents and toward the equator, the Ekman transport carries surface water away from the continent. An upwelling of deeper water replaces the surface water that has moved away from the coast. **(C)** The Coriolis effect acting on westward flowing equatorial currents pulls surface water away from the equatorial region. This water is replaced by subsurface water.

the northward-flowing AABW, the NADW moves up off the bottom and over the denser AABW. The nutrient-rich NADW surfaces again off the antarctic continent some 300 years after sinking (Fig. 10.9). These water masses are carried into the other ocean basins by the **Circumpolar Current** that moves from west to east around Antarctica. Since there is no source of oxygen for ocean water except that found in the atmosphere and that produced by photosynthesis in the sunlit layer, life in the deep ocean would be impossible without the for-

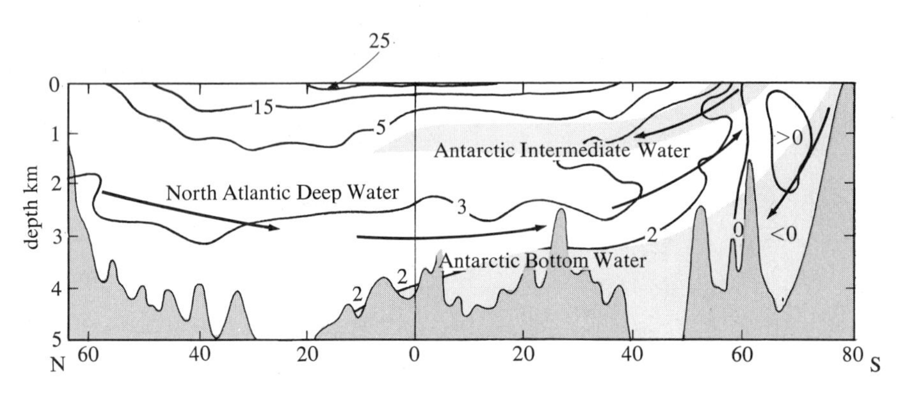

FIGURE 10.9
Thermohaline Circulation

The North Atlantic Deep Water (NADW) sinks toward the ocean floor in the Norwegian Sea, carrying life-sustaining oxygen into the deep ocean. Near the equator, it encounters an even denser water mass, the Antarctic Bottom Water (AABW). This Antarctic Bottom Water forms in the Weddell Sea at the southern end of the Atlantic Ocean. The North Atlantic Deep Water is forced to the surface in antarctic waters, where it provides a huge amount of plant nutrients that support the massive summer biological production characteristic of the Antarctic. (After Sverdrup, H.; Johnson, M.; and Fleming, R. 1942. *The oceans: Their physics, chemistry and general biology*. Englewood Cliffs, NJ:Prentice-Hall.)

mation of these deep-water masses. The constant enrichment of antarctic surface water by the upwelling of North Atlantic Deep Water makes these waters an excellent feeding ground for baleen whales.

DIVISIONS OF THE PHOTIC ZONE

The immense photic zone has a number of subdivisions. We are already familiar with the traditional division into an upper epipelagic and a lower mesopelagic zone. The phytoplankton are restricted to the epipelagic or euphotic zone, and only zooplankton and nekton are found in the dimly lit mesopelagic zones. A number of other well-defined geographic subareas of the photic zone are recognized: the air-water interface, the subtropical gyres, upwelling areas, the neritic zone, and mid- and high-latitudes. In all these areas, *differences* in environmental conditions and life styles are not as important as the *similarities* that organisms show in adapting to a three-dimensional water space with changing intensities of sunlight.

The Air-Water Interface

The organisms that live at the air-sea interface, or at least in the top few millimeters of water, are known collectively as the **neuston.** The neuston are a highly productive part of the ocean and may support a biomass of life many times greater than the biomass just a few centimeters below the surface. The surface film has a higher concentration of salts and dissolved organic matter, including high concentrations of lipids, fatty acids, polysaccharides, and proteins, than the water below. This organic matter supports bacterial populations with 10–1,000 times the density of the water below the neuston. These bacteria provide food for a relatively high concentration of Protozoa, including tintinnids, radiolarians, and foraminiferans. Other larger zooplankton are also present in the neuston, among them certain species of copepods and larval forms of decapod crustaceans. Floating fish eggs are often present.

Larger animals are attracted to the neuston. Large zooplankton such as *Physalia* (the Portuguese man-of-war) and *Vellela* (the purple sail) float among the neuston, as do the pelagic gastropods *(Glaucus* and *Ianthina)*. One animal that lives on top of the water surface is the water strider *Halobates,* one of the few insects that is truly a marine animal (Fig. 10.10). *Halobates* feeds on the small zooplankton of the neuston. Other visitors to the neuston from the air side include a number of oceanic birds that skim the neuston for food.

A striking characteristic of neuston animals is their often intense blue color. This blue color may afford protection from predation or absorb the harmful ultraviolet radiation found in the surface layer.

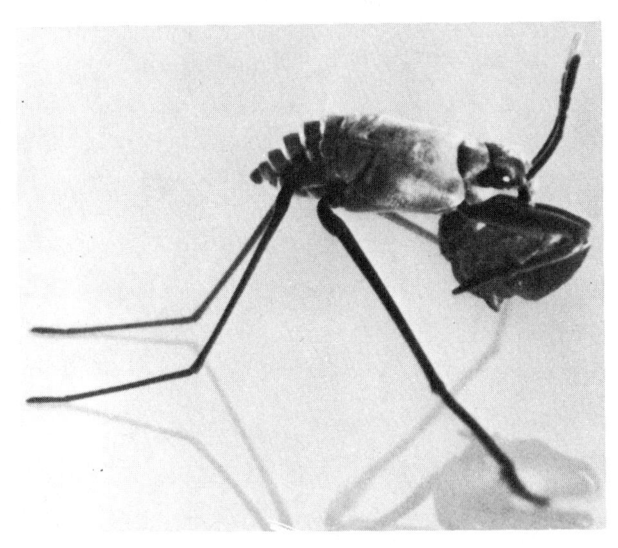

FIGURE 10.10

The oceanic insect *Halobates* strides across the ocean surface in search of food. (Photo courtesy of Scripps Institute of Oceanography, University of California, San Diego.)

Subtropical Gyres

The largest subareas of the photic zone are the large rotating gyres found in the subtropical regions of both hemispheres of the Atlantic Ocean and Pacific Ocean, and the southern hemisphere of the Indian Ocean. These gyres are surrounded by the major surface currents of the oceans. The five gyres cover some 50 percent of the world's oceans (Fig. 10.5). Although they often are regarded as uniform features, their structure is quite complex. The boundaries of the gyres tend to meander and some may at times be split into two parts by currents. In addition, numerous ring eddies pinch off the currents and meander into the gyres.

The gyres are warm, clear, and convergent. The upper mixed layer is between 150–200 m (492–656 ft) thick. Below the mixed layer is a permanent thermocline of some 100–200 m thickness. The waters of the upper 500 m (1,640 ft) of the gyres have high salinity, relatively high dissolved oxygen concentration, and low nutrient content. There are only slight seasonal changes and no opportunity for nutrients to reach the mixed layer from below the thermocline.

Biologically, gyres are the poorest parts of the ocean. The standing stock of phytoplankton is uniformly low throughout the year. Phytoplankton species are small, the majority of the primary producers being small flagellates. Zooplankton biomass is also lower than in any other part of the ocean. While there may be 10,000 copepods per cubic meter in coastal waters, there are only 10–100/m^3 (35.3 ft^3) in gyre areas. Zooplankton species tend to be relatively small. About 25 percent of the biomass of net-caught zooplankton is small microzooplankton. Deeper down, the deep scattering layer is poorly developed in gyre regions. The number of benthos of the sediment under gyres is less than in any other area of the ocean floor.

Although density and biomass are minimal, the species diversity of zooplankton is higher, for unknown reasons, in the gyres than in other ocean locations. Gyres also may have species found in no other locations. For example, the euphausiid *Euphausia brevis* is found in all five oceanic gyres but in virtually no other location.

Upwelling Areas

In a number of relatively restricted areas of the ocean (less than 0.1% of the ocean surface), wind and current conditions are such that subsurface waters are brought to the surface. These nutrient-rich waters generally promote phytoplankton growth which in turn supports relatively rich zooplankton and nekton populations. The importance of upwelling areas is underscored by the fact that 50 percent of the world's commercial fisheries catch comes from the 0.1 percent of the ocean's surface where there is active upwelling.

Upwelling is of two types, coastal and equatorial. Upwelled water does not immediately support rich phytoplankton growth. Patches of upwelled water have been followed off the coast of Peru for up to one week, with no increase in phytoplankton (Fig. 10.11). In the equatorial upwelling systems, up to 50–150 days are required to fully develop the phytoplankton community. By this time, the patches of upwelled water may have traveled some 2,000 km (1,242 mi). In these equatorial upwellings, nekton such as small fish and cephalopods may take four months to develop.

Phytoplankton production rates are much higher in upwelled water than in surrounding water. Equatorial upwelling may have ten times the productivity of adjacent waters. The phytoplankton of upwelled water has a high percentage of relatively large diatoms. Where the phytoplankton has high diatom concentrations, herbivorous copepods predominate. In equatorial upwelling, 50 percent of the zooplankton may be herbivorous calanoid copepods. In the coastal upwelling of Peru, diatoms may be large enough to serve as a direct source of food to the generally dominant anchovy fish. This is the only example of a major world fishery in which a significant proportion of the food of the catch (anchovy) is phytoplankton. In virtually all other commercial fisheries, there are a number of trophic layers before a harvest-

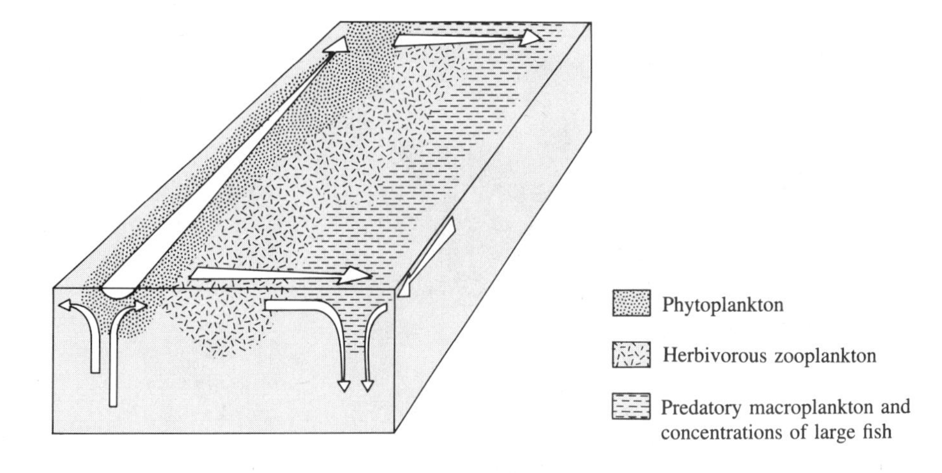

FIGURE 10.11
Biomass Composition in Upwelling Areas

Since upwelled water is part of surface currents, the biological events that depend on nutrient enrichment may occur at some distance from the point of upwelling. Arrows indicate water movement.

able size is reached. Euphausiids are also important herbivores in upwelling areas. Again, the relatively large size of the diatoms may permit these larger zooplankters to feed directly on the phytoplankton. In general, upwelling areas are characterized by the rapid movement of food energy to large animals.

As nutrients in the upwelled water are depleted, smaller dinoflagellates and coccolithophores become the dominant phytoplankton, and the majority of the copepods found are carnivorous species.

The Neritic Zone

The waters that cover the continental shelves are often referred to as the neritic zone. Approximately 10 percent of the ocean's surface is within the neritic zone. In general, the neritic zones are similar to adjacent oceanic areas, and since continental shelves span large differences in latitude, there is much variation in their biology. As a whole, the neritic zones are more productive than adjacent oceanic waters. Approximately 99 percent of the world's fishery catch comes from the waters of the continental shelf (this includes upwelling areas). The primary productivity of neritic areas is some ten times higher than adjacent oceanic waters.

The high productivity of these waters is promoted by shallow water and strong mixing forces. Besides wind mixing, neritic waters are subjected to tidal forces that help to mix the water column. The higher mixing insures a continual supply of nutrients from deeper waters and results in greater growth of phytoplankton.

Neritic plankton occurs in definite assemblages. Both phytoplankton and zooplankton have groups of species that are characteristic of the neritic zone, and different from adjacent oceanic areas. The neritic zooplankton is also characterized by a greater number of meroplankton (larval forms). The many benthic species of the continental shelf contribute a large component to the zooplankton during reproductive periods.

ADAPTATIONS TO THE PELAGIC ENVIRONMENT

The three-dimensional liquid space of the pelagic environment has no counterpart in the terrestrial environment. Organisms that live in the photic zone must contend with the relentless pull of gravity and the changing light intensities at different depths. To resist sinking, zooplankters either swim or develop some mechanism to enhance buoyancy. Changing light intensities affect the availability of food and vulnerability to predation. Specialized coloring, transparency, and bioluminescence allow zooplankters to foil predators, and vertical migrations help to insure ample food supplies.

Even with salt water's greater buoyancy, the specific gravity of a pelagic organism's tissues is generally greater than that of the surrounding water. Pelagic organisms must have flotation or neutral buoyancy mechanisms or else swim; otherwise, they sink.

An important characteristic of the pelagic environment is the absence of anywhere to hide. Even though vision under water is restricted, compared to vision in air, prey cannot conceal themselves in the pelagic environment. Therefore color, or the lack of it, becomes an important adaptation to life in the buoyant medium.

Buoyancy

The rate of sinking of an organism depends on three main factors: size, the difference in density between the organism and the surrounding water, and the viscosity of the water (Tab. 10.1). For small organisms, the relationship between these three factors is expressed by Stokes law. For physical particles up to a size of 60 μm (0.02 in), Stokes law predicts that the smaller the particle, the slower the rate of sinking. For planktonic organisms, Stokes law can be extended to organisms up to 500 μm (0.197 in) in size; that is, planktonic organisms of 500 μm or less sink more slowly if they are smaller.

TABLE 10.1
Sinking Rates of Pelagic Organisms

(From Parsons, Takahasi, and Hargrave, 1977.)

Group	Sinking Rate (m/day^{-1})
Phytoplankton	
Living	0–30
Dead, intact	<1–510
Protozoans	
Foraminifera	30–4800
Radiolarians	~350
Zooplankton	
Amphipoda	~875
Chaetognatha	~435
Cladocera	~120–160
Copepoda	36–720
Ostracoda	400
Pteropoda	760–2270
Salpa	165–253
Siphonophora	240
Fecal pellets	36–376
Fish eggs	215–400

Planktonic organisms can further slow their rate of sinking by developing the elaborate spines that extend from the body. Spines and long appendages increase the frictional drag of the body and decrease the rate of sinking. Spines have an additional function in diatoms. The spines of some diatoms cause the body to rotate during slow sinking, insuring that all parts receive uniform illumination. Yet another advantage of spines may be to reduce the pressure of predation. Larger organisms are usually less susceptible to predation. One function of spines and long appendages may be to make the body appear larger so as to effectively reduce the number of predators that can prey on it.

A mechanism of reducing density is the accumulation of oils and fats. These lipids have a specific gravity of less than one. By accumulating a relatively high concentration of lipid, total density is decreased. A wide range of plankton from diatoms to copepods accumulate lipids. For most of these species, lipid accumulation is primarily for energy storage and the buoyant effect is secondary. Some planktonic forms use lipids specifically for flotation. Fish eggs that develop in the plankton have a small oil droplet that gives the eggs neutral buoyancy.

Finally, density in plankton may be reduced by air or other gas bubbles. The permanent gas float of some siphonophores, such as the Portuguese man-of-war, and the gas bladder of fish are examples. The blue-green algae *Tricodesmium* has intracellular gas bubbles that aid in flotation. The pelagic snail *Ianthina* produces bubbles that stick to its foot. Generally, plankton that use gas bubbles for flotation are part of the surface plankton (neuston). Some siphonophores and fish, however, are able to secrete gas against the great hydrostatic pressures found in the deep sea. Siphonophores and fish that use gas for natural buoyancy can be found in the deepest parts of the ocean.

Swimming

Only a few organisms rely solely on small size or some passive mechanism to maintain neutral buoyancy. Bacteria, most radiolarian Protozoa, photosynthetic diatoms, and coccolithophores use size or physiological adjustments to maintain neutral buoyancy. Most other plankton groups use swimming to help in maintaining their position against the force of gravity. Their swimming abilities may be feeble—dinoflagellates beat a flagellum, and tintinnid Protozoa beat their cilia—but these efforts are enough to combat gravity or move the plankter to a new nutrient or food zone.

In general, the slow swimming of planktonic organisms is relatively efficient. The energy required for slow swimming is only a tenth of that required for an animal to walk across the land. The limited swimming of the zooplankton, at speeds of 20 m (66 ft) per hour or less, is adequate to maintain position in the water column, to filter feed, and to migrate vertically.

Color

The zooplankton of the epipelagic zone are for the most part transparent. Except for eye spots and the gut contents of herbivores, most zooplankton have no color. This transparency reduces the visibility of the animal and reduces the chance of being discovered by a predator.

Zooplankton (and small fish) in the mesopelagic part of the photic zone are generally red or silver in color. The silver color of an animal at this depth produces a mirrorlike reflection. In conditions of low uniform light, a mirrored surface becomes invisible. Red wavelengths, conversely, are not present in deeper water, and the color as seen by available light would appear black. The red color, however, is probably an adaptation to the light of bioluminescence. This light is mostly blue-green in color. The red of the animals of the mesopelagic zone is a strong absorber of the blue-green light and thus makes it difficult for a predator to form a sharp image in bioluminescent light.

Vertical Migrations

For over 150 years marine biologists have known that surface waters, virtually devoid of zooplankton during daylight hours, suddenly teem with plankton and small fish during the night hours. Many species of zooplankton spend the day in deeper water and return to the surface in the evening. This daily, or diel, vertical migration is the most easily observable of a number of movements through the water column that zooplankton species regularly make (Fig. 10.12). Diel vertical migrations are most common in the top 100 m of the water (epipelagic zone), but may extend as deep as 1,000 m (3,280 ft). Diel migrations are known for all oceans and all latitudes from tropical to polar waters. Although most species have diel migrations, not all of them do. Some studies have found that up to 50 percent of the zooplankters in an area do not have daily migrations. Sometimes a species will migrate in one area and not in another. Even with these exceptions, diel vertical migration is a widespread and common phenomenon that can be expected of most zooplankton in the photic zone.

Vertical migrations are a unique and important characteristic of life in the pelagic zone. Although vertical migration allows an organism to stay in areas of uniform light intensity at all times, it subjects the migrating organism to changing conditions of temperature, pressure, and food supply. Organisms that migrate vertically alternate between surface waters that are warm, well lighted, and rich in food and deeper waters that are dark, colder (by 10–20°C or 18–36°F), and have a low food supply.

Daily vertical migrations are for the most part characteristic of the zooplankton and smaller nekton. Phytoplankton generally do not show vertical migrations, although the dinoflagellate *Noctiluca* can migrate some 15 m (49 ft) in depth each day.

All zooplankton phyla from the smallest (tintinnid Protozoa) to the largest (squid) exhibit vertical migrations. Zooplankton have species-specific depths at which they are found during daylight hours. These depths reduce the visibility of the zooplankton and afford relief from predation. In addition, the colder temperature slows metabolism and reduces the energy

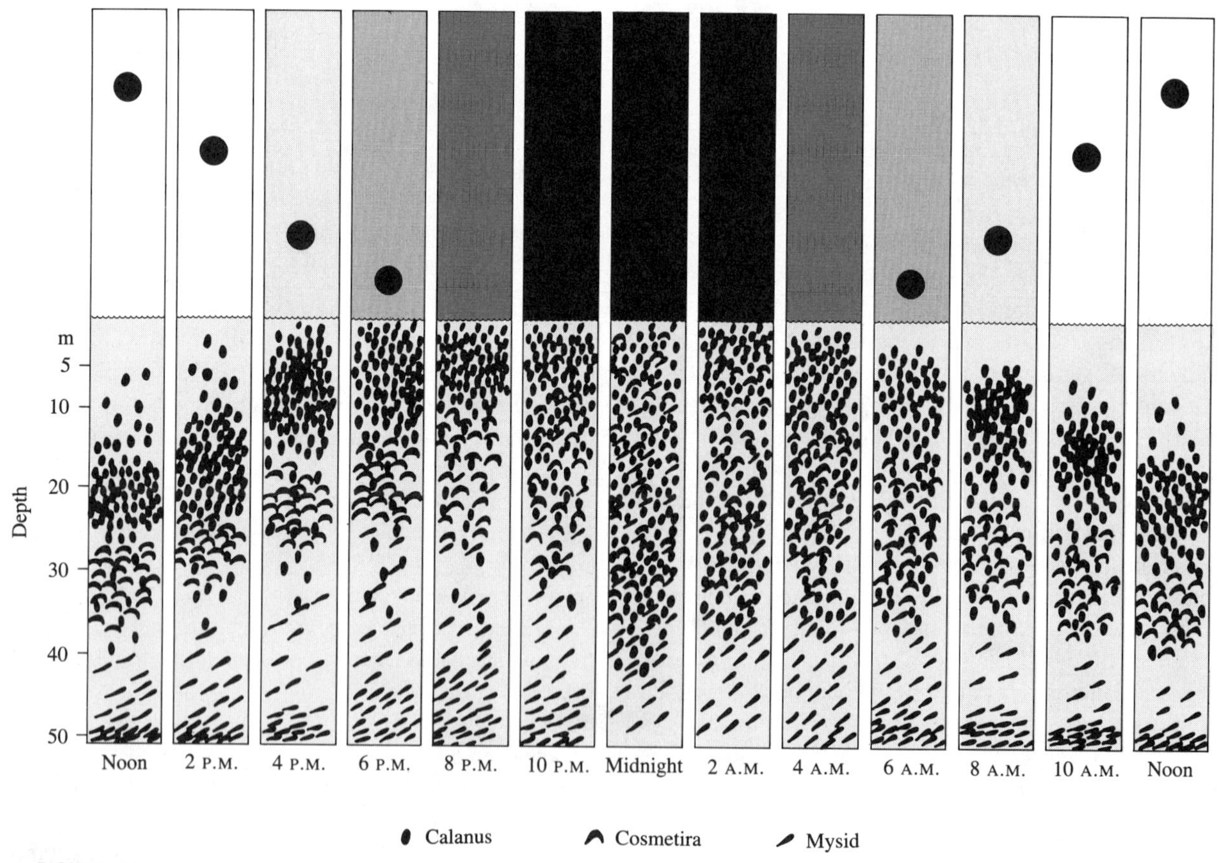

FIGURE 10.12
Daily Vertical Migration of Zooplankton

Many planktonic species in the top 1,000 m of water spend the daylight hours in deeper water and move toward the surface during night hours. Shown are the vertical migrations for the shallow-water copepod *Calanus*, the medusa *Cosmetria*, and the mysid *Leptomysis*.

required while resting at lower depths. Because phytoplankton remain in surface waters, herbivorous zooplankton must rise to the surface to feed. With the herbivorous zooplankton come the carnivorous zooplankton and the small fish. In some cases, species will feed for a few hours and return to the depths, but most species remain at the surface until the increasing light of dawn triggers a return to the depths. Larger zooplankton have larger daily migrations than smaller species. Smaller species are generally found during the day in the top 100 m (328 ft). Larger species, particularly those of the deep scattering layer, are found during the day at depths between 100–400 m (328–1,312 ft).

Migrating species have different rates and times of migration. This means that a given species in any 24-hour period migrates through constantly changing populations of other plankters. Not all species come to the surface. The deeper species, in particular, may spend the night hours at intermediate depths.

There is no question that light intensity is the major environmental factor that controls vertical migration of zooplankton and nekton. It appears that each species is sensitive to a given ambient light regime that is preferred for daylight. As this particular light intensity decreases, the upward migration begins. With the onset of dawn the return migration begins.

Although light intensity is without question the most important environmental factor influencing vertical migration, other mechanisms may play a part. Some species of zooplankton have an **endogenous rhythm;** that is, some sort of biological clock cues the zooplankter that it is time for upward and downward migrations. This conclusion is based on observations of migrating zooplankton in laboratory conditions of total darkness. The number of species that have an endogenous rhythm is small, however, and its general importance is open to question.

Scientists have a number of theories about the purpose of vertical migration in the zooplankton. The most obvious explanation is that it helps the migrator to avoid predation. By melting into the shadows during the daytime, the chance of being detected is decreased. Although this generalization is popular, it has some flaws. Many species that migrate vertically spend their daylight hours at a depth where there is sufficient light for location by predators. Other species go deeper in the day than is necessary to avoid predation. Moreover, a number of migrating species are transparent, and little advantage is added by sinking into the shadows. Possibly, some species seek deep, cold water to take advantage of a slower, more efficient rate of metabolism that can be achieved there.

Another advantage of vertical migration for the zooplankton is that it permits horizontal movement beneath the surface. Currents at lower depths are of a different velocity and often a different direction than they are at the surface. By spending the day at lower depths, zooplankters are transported horizontally and rise to the surface at a different location in the evening. Horizontal movements of 16 km (10 mi) with each diel migratory cycle are common. This horizontal movement may be used by herbivorous zooplankton to both search out and avoid phytoplankton. Herbivores are specific in feeding, with regard to both the type of phytoplankton and its density. Through vertical migrations, new sources of food can be explored and dense blooms of inedible phytoplankton can be avoided.

Bioluminescence

The light through most of the photic zone is of a fairly uniform blue-green color. The longer wavelengths of visible light are absorbed in shallow water, and the blue-green light decreases with depth. Most of the deeper levels of the photic zone are dimly lighted, although in oceanic areas there is still enough light to cast a shadow. Below these depths there is only enough light to form a silhouette, and bioluminescence becomes an important source of light.

Many marine animals are bioluminescent, including bacteria, phytoplankton, zooplankton, benthic invertebrates, and fish. However, luminescence is most widespread and of particular importance in the photic zone.

The color of bioluminescence in marine animals is generally a blue to blue-green, but the wavelength of luminescent light can vary. Some deep-sea fish produce a pink or yellow light, and one deep-sea squid can produce red and white light. The more common blue-green wavelength has the greatest penetration in water. Bioluminescent light can be quite bright compared to the daylight found in the mesopelagic zone. Theoretically, the light produced by six krill in a jar of water would be just adequate to read a newspaper by.

In the photic zone, bioluminescence is common in both epipelagic and mesopelagic waters. At the surface, bioluminescence is responsible for the phosphorescence of the water at night. This light is often made by dinoflagellates such as *Peridium* and *Gonyaulax,* ctenophora, copepods, euphausiids, and medusae. In deeper water of the mesopelagic zone, bioluminescence is more common in larger animals such as fish and squid. Organisms use luminescence in a number of different ways. Bacteria luminesce constantly, but all other organisms can turn their luminescence on and off.

Some animals excrete their luminescence. Some copepods and the mysid *Gnathophausia* excrete a substance containing luminescent bacteria from the mouth area in a cloud that lasts for up to 60 sec. One of the most brilliant excretions is from a squid that lives in the mesopelagic zone. Since light levels there are very low, the cloud of black ink that squid generally release when attacked is of little use. The midwater squid releases a luminescent cloud that startles and confuses the predator.

In other cases, bioluminescence is intracellular. Protozoa and phytoplankton luminesce from inside the cell. In larger zooplankton and fish, specialized organs may be developed to utilize the luminescence. These complex organs may include a lens to direct the light. Still other animals harbor symbiotic bacteria that are luminescent. Some shallow-water squid and mesopelagic fish maintain luminescent bacteria in specialized photophores. The significance of bioluminescence in the photic zone is not yet well understood. One function may be to startle and blind predators.

Fish of the mesopelagic zone use bioluminescence to reduce their visibility in the daylight. Myctophid fish have a series of photophores on the ventral surface. These fish can adjust the intensity of the light in the photophores to match the intensity of downwelling light. From below they are virtually invisible. Some 60–75 percent of mesopelagic fish, as well as some squid and cuttlefish, use such ventral photophores to luminesce. This **counterlighting** is an important advantage in reducing the chance that a fish will be detected by a predator.

ORGANISMS OF THE PHOTIC ZONE

Despite the large size of the photic zone, it contains relatively few species. Of the 200,000 or so species of organisms found in the pelagic zone, less than 4,000 (2%) are found in the pelagic photic zone. The majority of these are planktonic species.

Most marine phyla (including the bacteria and blue-green algae) have representative species in the plankton. Although all of them share the characteristic of limited mobility, they vary greatly in size. Most plankton are microscopic. Among these are bacteria, Protozoa, phytoplankton, and many zooplankton. A few plankton species, however, reach a large size. The Portuguese man-of-war, for example, may be 0.5 m (1.6 ft) in diameter and have tentacles 30 m (98 ft) long. Marine biologists have a number of schemes that divide pelagic organisms on the basis of size. Large planktonic species may be larger than small nekton species. The key to the definition of plankton is the inability to move against currents, rather than the size.

The photic zone has the largest diversity and number of plankton in the oceans; at least 90 percent of the plankton live there. Although light intensity below 100 m (328 ft) limits photosynthesis, it is sufficient to support other biological activities that have lower minimal light requirements (Fig. 10.13). For example, color vision is possible down to almost 600 m (1,968 ft) in clear ocean water, and around 200 m (656 ft) in coastal waters. Crustaceans can use the available light to form images as deep as 900 m (2,952 ft) in the open ocean, and the vertebrate eye can form an image at the maximum depth of the photic zone.

The Phytoplankton

The phytoplankton of the ocean are often referred to as the 'pastures of the sea'. The role of phytoplankton in the capture of the sun's energy in the marine environment is important. At least 90 percent of the photosynthesis in the marine environment is due to phytoplankton. The remaining 10 percent is from littoral and shallow-water algae, sea grasses, salt marsh, plants, and mangroves. Phytoplankton differ from these other forms primarily in their small size. The maximum size of phytoplankton cells is around 100 μm, and smaller forms are in the size range of 2–5 μm.

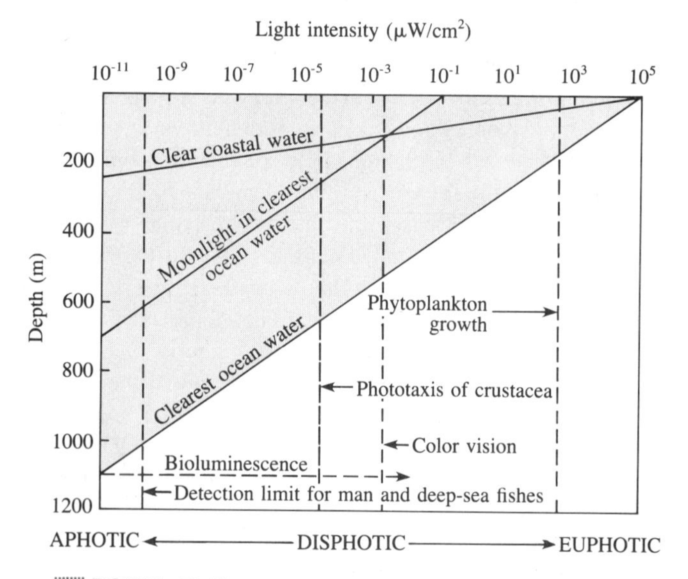

FIGURE 10.13
Light Intensity and Depth

Although the intensity of light below 100 m is not strong enough to support photosynthesis, other important light-dependent biological functions can be carried on. This chart shows the depths to which phytoplankton growth occurs, color vision is possible, crustaceans can see images, and the detection of light is possible in clear coastal water and clearest ocean water. It also shows the intensity of moonlight detected at different depths in clearest ocean water.

Seldom are phytoplankton visible in the ocean, although occasionally dense blooms color the water or leave an oily slick on the surface. Generally, though, the blooms are not dense enough to be visible to the naked eye. Population densities of phytoplankton in a given segment of the water column are generally characterized by a low biomass. By weight there is usually more zooplankton than phytoplankton. At first glance, this seems to be contrary to the ecological concept of the food pyramid. On land, there is generally ten times the plant material for every unit weight of herbivore. In the oceans, the herbivores often eat the phytoplankton as fast as it grows, so the standing crop is less than that of the grazers. Phytoplankton not immediately eaten rapidly age and sink out of the euphotic zone. However, the rate of production for phytoplankton is often much greater than that for zooplankton. For example, it is not uncommon for phytoplankton populations to double their mass in one day, while it may take a zooplankton population a month to double its mass.

There is one exception to the rule that pelagic photosynthetic organisms are microscopic. In the Sargasso Sea there are large quantities of floating macroscopic brown algae, the sargassum weed. Although the sargassum algae has its origin on a solid substrate, the floating pieces are able to reproduce vegetatively and continue to grow for years. Even though there is a relatively large biomass of sargassum weed in the Sargasso Sea, its contribution to the photosynthesis in the photic zone is minor.

The phytoplankton at any given location is generally a mixed population of species belonging to a number of groups. The most important phytoplankton group is the diatoms. The other important groups are the dinoflagellates, coccolithophores, microflagellates, and blue-green algae.

Diatoms are the dominant photosynthetic organisms in areas of upwelling (coastal and equatorial), in the neritic zone where turbulent mixing by winds and tides is high, and in mid- and high latitudes during the spring, when light is adequate and turbulent mixing high. They reproduce rapidly, though, and predominate in locations where nutrients are abundant.

Dinoflagellates also are common in neritic and midlatitude waters. They predominate in conditions of high nutrients and low turbulent mixing. Dinoflagellate blooms or swarms are often colored and may form red tides. In oceanic areas of both the tropical and temperate latitudes, coccolithophores and microflagellates are generally the most frequently encountered phytoplankton. Cyanobacteria are characteristic of oceanic areas of warmer waters.

Species diversity of phytoplankton differs with latitude and in neritic and oceanic waters. At low latitudes, there tends to be a greater number of species, fewer numbers of each species, and lower total biomass. Near shore and in temperate and high latitudes there is a greater number of individuals and biomass, but a lower number of species. A similar pattern is noticeable on the vertical scale in the euphotic zone. There tends to be more individuals but fewer species toward the surface, and a greater number of species but fewer numbers with depth.

Phytoplankton are shade-tolerant, and their optimal light intensity for photosynthesis is found at some moderate depth beneath the surface (Fig. 10.14). Full sunlight has an intensity of 8,000–10,000 footcandles at the surface. The light intensities at which phytoplankton are saturated is much lower, however. For phytoplanktonic green algae, saturation levels are between 500–700 footcandles; for diatoms, between 1,000–2,000 fc; and for dinoflagellates, between 2,500–3,000 fc.

Accessory photosynthetic pigments are important in the photosynthesis of phytoplankton. Chlorophyll absorbs wavelengths of 600 nanometers (0.24 in) or longer, while the accessory pigments selectively absorb shorter wavelengths. Since light at wavelengths of 600 nm or longer is quickly absorbed by seawater, the accessory pigments take on a particular importance.

When phytoplankton encounter light intensities higher than the optimal, two effects are common. The rate of photosynthesis is inhibited, and the cells become 'leaky.' Organic molecules produced during active photosynthesis escape through the cell membrane into the surrounding water. Up to 24 percent of the products of phytoplankton (mannitol and various polysaccharides) may be lost to the surrounding seawater. Presumably, this *exudate* is used as a source of food by bacteria.

The Zooplankton

The most common animals of the photic zone are small. Since their source of food is also small, this is not surprising. Although zooplankton are the drifting animals of the sea, most are capable of limited swimming. This movement, however, is far exceeded by the distance the zooplankton is carried by currents in the top 100 m (328 ft) of the water. The zooplankton fall into two categories: *holoplankton* (lifelong members of the plankton) and *meroplankton* (parttime members of the plankton).

The Holoplankton.

Three phyla of Protozoa from the kingdom Protista and seven from the animal kingdom contain species that are holoplankton. The relatively small proportion of animal phyla with holoplanktonic species (about 25%) is testimony to the demanding conditions of life in the three-dimensional water space.

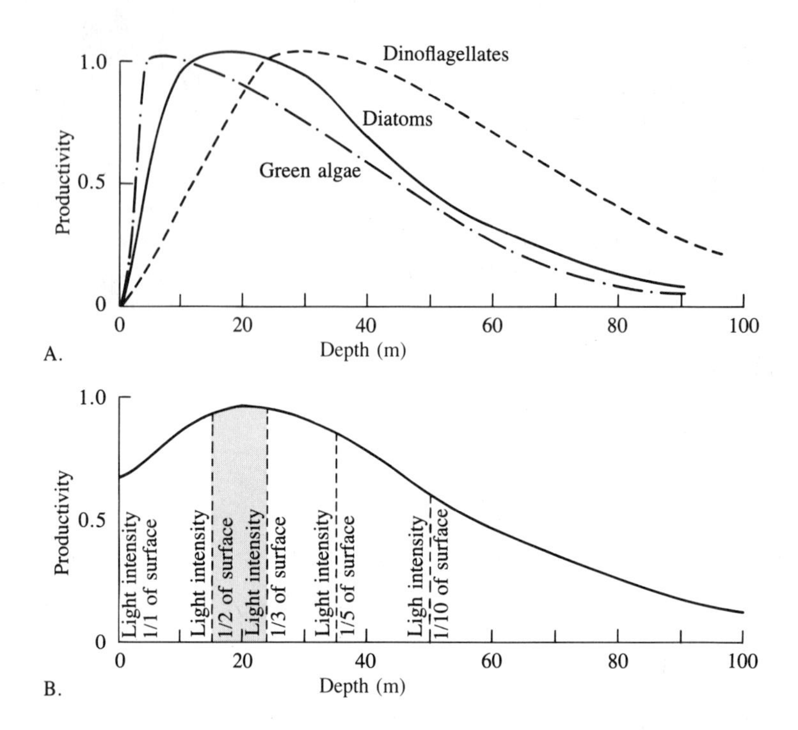

FIGURE 10.14 Light Intensity and Productivity

Phytoplankton are generally shade-tolerant organisms, although the presence of accessory photosynthetic pigments causes different optimal light intensity for different groups. **(A)** Optimal light intensities for green algae, diatoms, and dinoflagellates differ. **(B)** On average, optimal intensity occurs at a moderate depth beneath the surface. Maximum productivity is at some depth below the surface, where the light intensity is one-third to one-half of surface intensity. In clear water, this may be at a depth of about 20 m (66 ft).

Among the protozoans, the foraminiferans, radiolarians, and tintinnids (and other ciliates) are important in the zooplankton. All are generally microscopic. The Foraminifera are mostly benthic; less than 2 percent are found in the plankton. Most planktonic foraminiferans are in the family Globigerinidae. Foraminifera are carnivorous and use pseudopodia to capture small plankton. Radiolarians also use pseudopodia to capture small copepods, diatoms, and other small plankton, and are common in the zooplankton at certain times.

The third group of planktonic protozoa are the ciliates. Of particular importance are the tintinnids (see Chap. 4). These protozoa use cilia to swim. Tintinnids and other ciliates are relatively small, generally less than 10 μm (0.004 in). They often are abundant in the zooplankton. If a water sample is analyzed for all its planktonic components, generally ciliates make up at least one half, if not more, of the number of organisms. As a group, however, ciliates are not well known enough for us to be able to assess their importance in the sea. They may be important links in food chains, particularly in areas where phytoplankton cells tend to be small, such as the stratified waters of the tropics.

Of the invertebrates, two phyla (Ctenophora and Chaetognatha) are almost exclusively planktonic. Chaetognaths (arrow worms) total less than 60 species. They are common and often are the dominant members of the zooplankton, however. Chaetognaths are predators and feed mostly on copepods. The tail flukes and lateral fins of the body are used for active swimming and attack. Chaetognaths are found at all latitudes and extend throughout the depths of the photic zone.

The Ctenophora is another phylum that has a small number of species, but is generally common in the plankton. Ctenophores are widespread in surface waters and are also found throughout the depths of the photic zone.

The phylum Cnidaria has a moderate number of planktonic species, some of them ranking among the largest of the zooplankters. There are two groups of cnidarians that are important in the plankton: the Scyphozoa and the Siphonophora. The Scyphozoa, or jellyfish, are feeble swimmers. Their body, which is mostly mesoglea (jelly), can reach a relatively large size and still have virtually no biomass beyond its water content. Scyphozoa are carnivores that use tentacles with nematocysts to capture prey. Many species of Scyphozoa have a sessile polyp stage as part of the life cycle, and technically are part of the meroplankton. Many other jellyfish are totally planktonic, however. Generally, medusae are in the size range of a few centimeters.

Some of the largest of the zooplankton, such as the Portuguese man-of-war, are siphonophores. Siphonophores found at the surface have the familiar blue color of the neuston. In some areas, siphonophores produce a sharp echo as part of the deep scattering layer. These siphonophores found at 200–400 m (656–1,312 ft) in the photic zone are often so numerous as to form a continuous net of tentacles that effectively traps particles sinking through the water column.

The phylum Annelida has only a few permanent members in the zooplankton. They are not of particular importance and are not dominant in any situation.

In the phylum Mollusca, the only holoplankton species are in the class Gastropoda. At least three different groups of snails have evolved planktonic forms. The so-called violet snails (of the genus *Ianthina*) are found mostly in tropical and subtropical waters, although at times they may be carried by the Gulf Stream as far north as Cape Cod. Violet snails have not lost their shells in adapting to planktonic life. Instead, they maintain their buoyancy by secreting a raft of bubbles around the edge of the foot. Violet snails are often found in large swarms. They are carnivorous and feed almost exclusively on *Physalia,* the Portuguese man-of-war.

The second group of planktonic gastropods is the larger Heteropoda. The mostly transparent heteropods have a reduced shell and large lateral 'wings' which are projections of the foot used for swimming. Like violet snails, heteropods are found mostly in tropical waters. Some are carnivorous and feed on small fish and medusae, while others are filter feeders.

The final group of planktonic gastropods is the Pteropoda, or sea butterflies. Pteropods have only rudimentary shells and, like heteropods, they have the winglike projections off the foot (Fig. 10.15). Pteropods are brightly colored and are found primarily in warmer waters, often in large swarms.

In the phylum Chordata, there are two groups of tunicates that are holoplanktonic. The first consists of the salps, doliolids, and the genus *Pyrosoma*. *Pyrosoma* is a colonial form. All have a highly specialized adult morphology, including barrel-shaped bodies through which they filter water. Salps and doliolids alternate between solitary and colonial stages. They are more common in warmer water but may at times swarm in large numbers in temperate latitudes. Salps, doliolids, and *Pyrosoma* have effective filtration devices that can remove the finest nanoplankton. They are most common in surface water, but have limited distribution deeper in the photic zone.

The second group of planktonic tunicates is the Larvacea. These holoplankton are peculiar because they retain the basic shape of the tunicate tadpole larvae through the whole of their life cycle. Larvacea produce a gelatinous 'house' that is a filter mechanism. The swimming of the Larvacea in its 'house' creates a current from which small plankton are filtered. Once the fine filter becomes clogged, the Appendicularium leaves and builds a new one. This abandoned 'house' becomes an important source of food for animals living in deeper water. Larvacea are common in temperate waters.

By far the most important phylum contributing to the holoplankton is Arthropoda. A number of groups of crustaceans are important, including copepods, amphipods, mysids, euphausiids, and some decapods, but copepods are the most numerous. Excluding the Protozoa, copepods are the largest component of the zooplankton (Fig. 10.16). A sample of zooplankton collected with a net can be expected to contain about 95 percent copepods.

For the most part, copepods are herbivorous filter feeders and serve as the main link between the phytoplankton and animals in pelagic food webs of the photic zone. They are generally quite small, most in the range of 0.5–3 mm (0.02–0.03 in) in length. They are most common in the top 100 m (epipelagic zone), but are also found throughout the photic zone. Deeper species are larger and are more likely to be carnivorous. Copepods are common in all oceans and at all latitudes.

Euphausiids are also important members of the holoplankton. Generally larger than copepods, euphausiids range from 2–5 cm (0.8–2 in) in length. In Antarctica, *Euphausia superba* (krill) is the dominant surface herbivore and filter feeds on diatoms. *E. superba* forms large swarms that are food for a number of species of baleen whales. In mid- and temperate latitudes, euphausiids are important members of the deeper zooplankton where they are mostly carnivorous and feed on copepods.

Although mysids are for the most part shallow-water benthic animals, some species are found in the permanent plankton. *Gnathophausia ingens* is a cosmo-

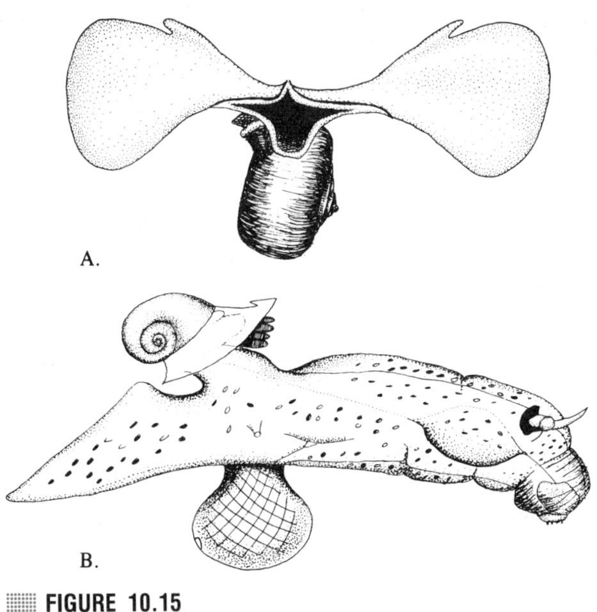

FIGURE 10.15
Planktonic Gastropods

(A) Pteropod. **(B)** Heteropod.

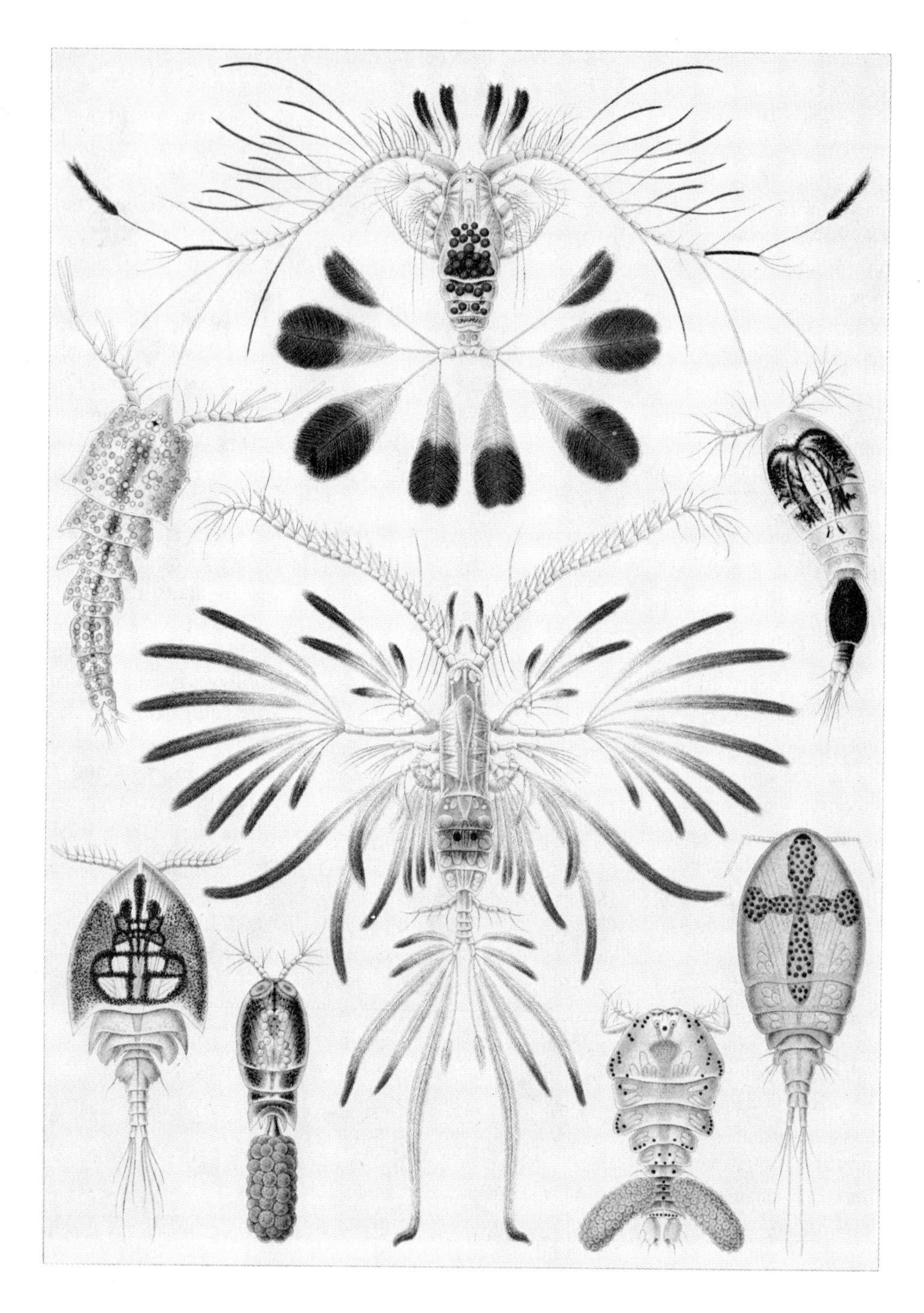

FIGURE 10.16
Copepods

Copepods are the most important and numerous of the planktonic herbivores. (From Haeckel, 1974.)

politan mysid found in the deeper zooplankton (500–1,000 m or 1,640–3,280 ft). *G. ingens* is a large plankter reaching lengths of 30 cm.

Amphipods, like mysids, are primarily benthic but have a few planktonic species (Fig. 10.17). The main planktonic group is the hyperiid amphipods. These zooplankters, 1–2 cm (0.4–0.8 in) in length, are found mostly in deeper water of the photic zone, where they are parasitic on siphonophores and medusae.

Only a few decapod crustaceans are planktonic. *Sergestes,* a pelagic shrimp, is an example (Fig. 10.18). They may occur in large numbers and are important predators on smaller plankton.

The Meroplankton. Meroplankton are species that spend only part of their life cycle in the plankton. Although most meroplankton are larval forms of benthic species, a few adult meroplankton are known. Mysids in shallow waters spend the daytime in the benthos and the nighttime in the plankton. Scyphozoa medusae of the plankton often have an asexual sessile stage in their life cycle.

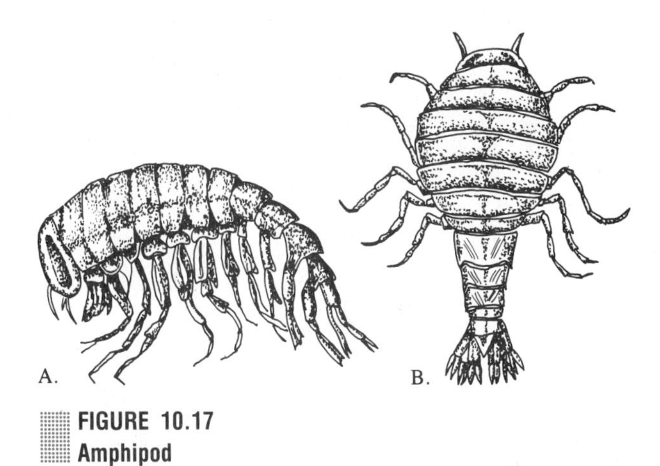

FIGURE 10.17
Amphipod

Only a few amphipods are found in the plankton. Shown is a hyperid amphipod. **(A)** Side. **(B)** Top.

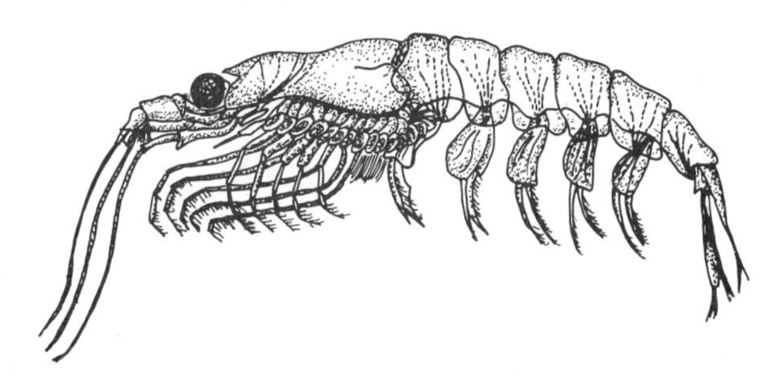

FIGURE 10.18
Decapod Crustacean

Only a few decapod crustaceans are planktonic through all of their life cycle. Shown is a planktonic shrimp.

The larval meroplankton are derived from shallow-water benthos living on or in the sediments of the continental shelf, the littoral zone, and the shallow subtidal zone (coral reefs, seagrass meadows, and kelp and seaweed beds). For the most part, the benthos of the deep sea do not have larval forms that enter the photic zone. Most animal phyla have benthic species with planktonic larvae (Fig. 10.19). Since species-specific traits in larval forms are difficult if not impossible to recognize, larval forms from a number of species are identified as a single type. As a larval form proceeds through metamorphic stages, different stages may be given different names.

A mobile larval stage is a great advantage to sessile benthic animals. Since sessile animals lack the ability to move, the larval form is important for dispersing the species and colonizing new habitat. The larval stage may also be a method of using the relatively rich food supply found in the plankton during part of the year. Larvae are generally filter feeders on the small phytoplankton and ciliate protozoa. For example, the cypris larvae of the acorn barnacle near Scotland are found in the plankton during the seasonal bloom of diatoms.

FEEDING AND FOOD WEBS

Zooplankton are of two basic feeding types: filter feeders and raptorial feeders. Raptorial feeders take selected prey, and filter feeders remove particles of a certain size range from the water. There is not a close correspondence between filter feeding and herbivores, and raptorial feeding and carnivores. Most filter feeders are phytoplankton-eaters. Because of the general small size and dispersed nature of the plankton, however, some carnivores are filter feeders. The baleen whales, for example, use filter feeding with great success. Raptorial feeders, likewise, are not always carnivores. For example, when large diatoms occur in the phytoplankton, copepods, the most widespread of the phytoplankton filter feeders, can seize and eat individual cells.

Filter feeders capture food particles through a ciliary-mucous system, common in a number of phyla, or a setal system, as in crustaceans. The success of the crustacean system of filter feeding is demonstrated by the calanoid copepods. These copepods are the dominant phytoplankton feeders in many regions of the photic zone. A number of the copepod appendages are involved in the filter feeding. The actual net is formed by a pair of appendages close to the mouth, generally the maxillae, which are covered with fine branches of setae (hairy bristles). Depending on their spacing, the setae can retain all but the smallest plankton. The range of filtration effectiveness differs from species to species, but for most copepods is between 5.0–50 μm (0.002–0.02 in), which closely parallels the size of most phytoplankton (Fig. 10.20). As the copepod swims, the abdominal appendages create a current that brings particles toward the mouth. It has been estimated that a single *Calanus finmarchicus* copepod can filter up to 100 cm^3 (6.1 in^3) of water per day.

Copepods are able to discriminate the size as well as the quality of food. Appropriately sized zooplankton are also filtered, and the stomach contents of copepods contain tintinnids and radiolarian Protozoa as well as other small crustacea, including larval copepod stages. Food selectivity can occur. For example, the copepod *Calanus finmarchicus* filters most diatoms, but may avoid or reject certain species.

Some zooplankton filter plankton using ciliary mucus instead of setae. Tintinnid protozoa use cilia to create a current that draws particles toward the animal. The cilia then filter out the food, including detritus, bacteria, flagellates, and diatoms. Gastropod molluscs (heteropods and pteropods) first immobilize their prey using the mucus on the flat foot and the 'wings'; then

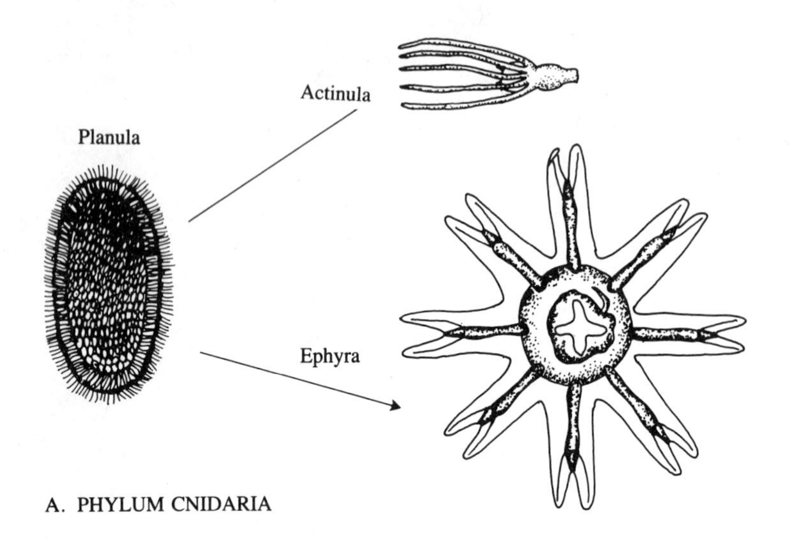

A. PHYLUM CNIDARIA

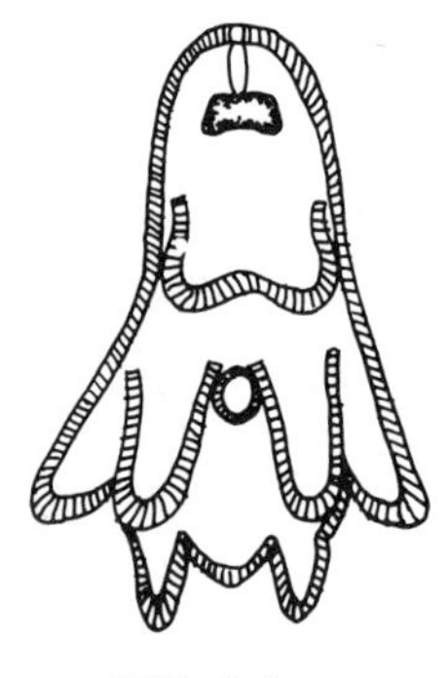

B. PHYLUM PLATYHELMINTHES

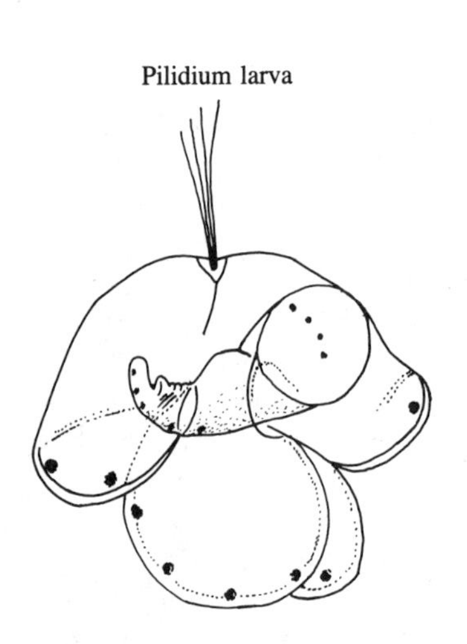

C. PHYLUM NEMERTEA

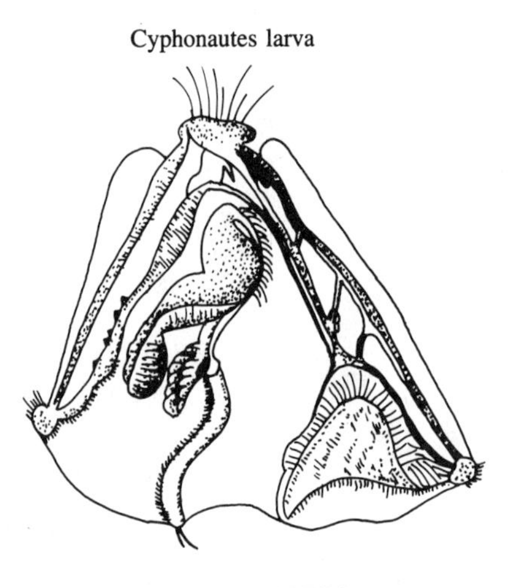

D. PHYLUM BRYOZOA

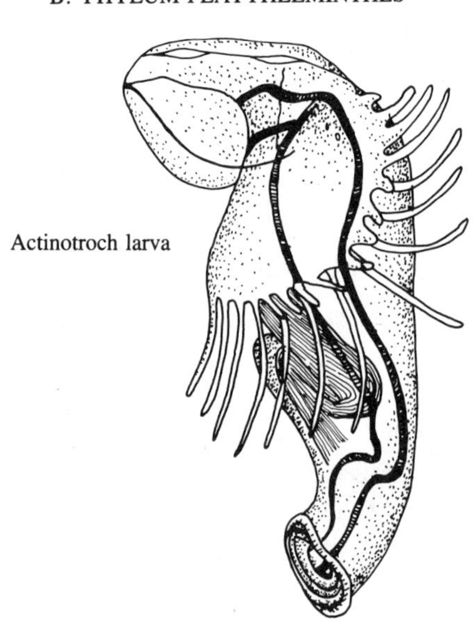

E. PHYLUM PHORONIDA

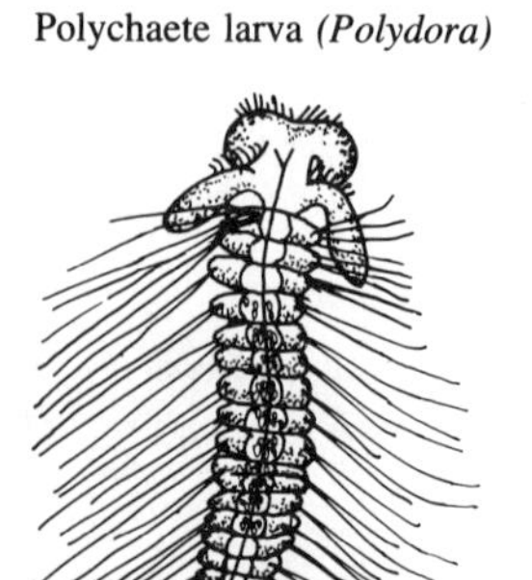

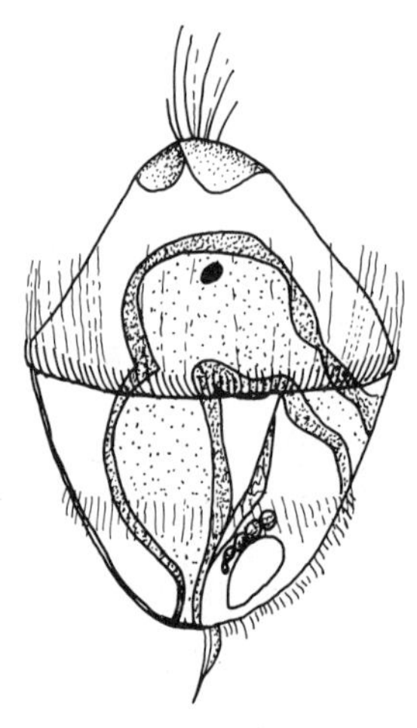

F. PHYLUM ANNELIDA

FIGURE 10.19
Larval Meroplankton of Benthic Species

Many benthic species have larval stages that spend some time in the plankton. A representative sample of larval types is shown. (F, G, and I from Crane, J.M. 1973. *Introduction to marine biology: A laboratory text.* Columbus, OH:Merrill.)

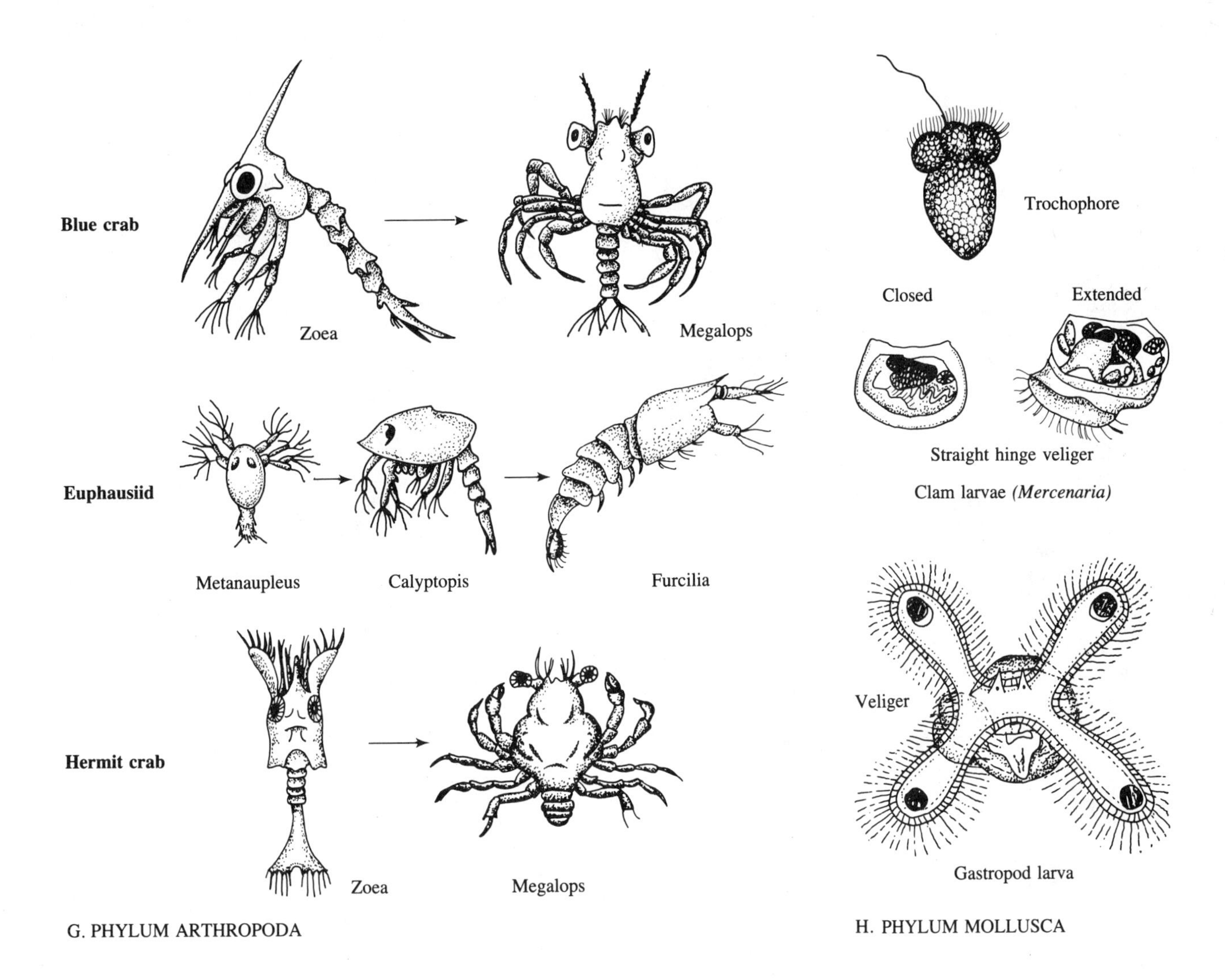

G. PHYLUM ARTHROPODA

H. PHYLUM MOLLUSCA

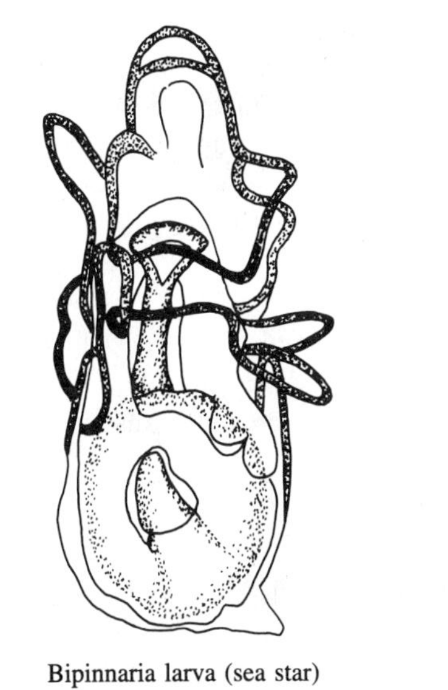

Bipinnaria larva (sea star)

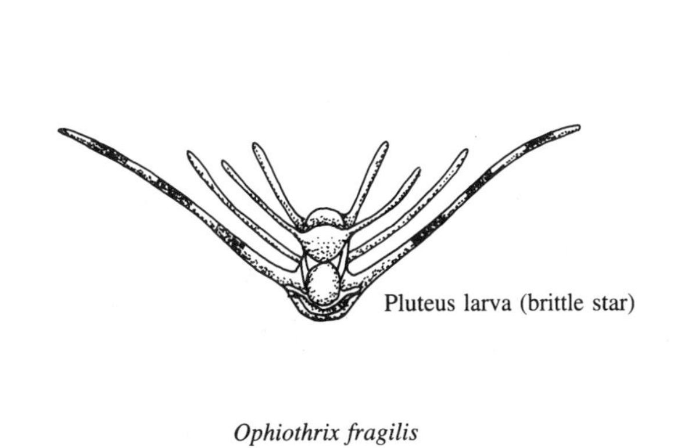

Ophiothrix fragilis

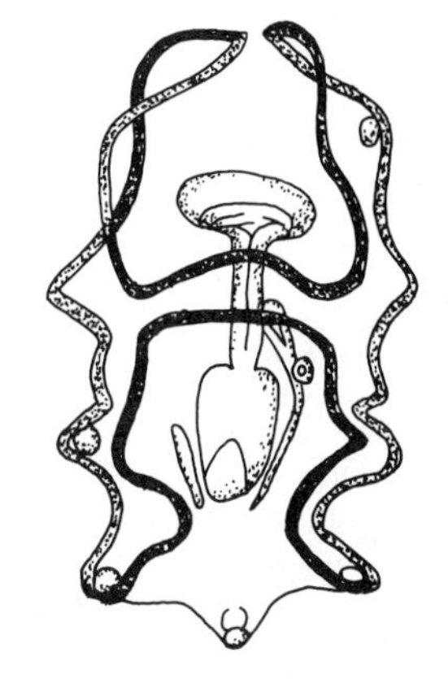

Auricularia larva (sea cucumber)

I. PHYLUM ECHINODERMATA

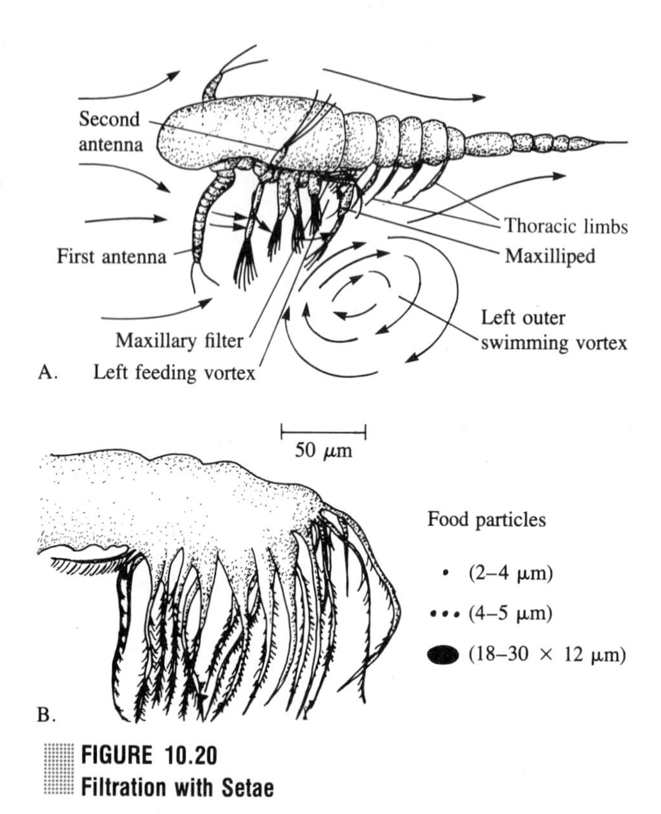

FIGURE 10.20
Filtration with Setae

The success of herbivorous copepods is mostly due to their filtration system. **(A)** Appendages create currents that bring the plankton to the mouth region. **(B)** The filtration appendage, and relative sizes of food particles.

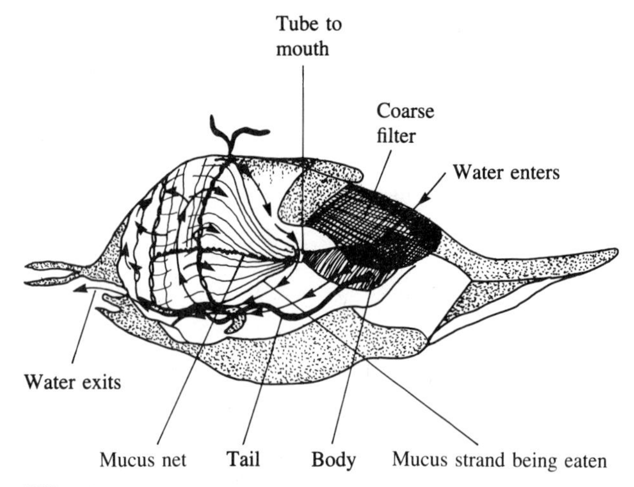

FIGURE 10.21
Filtration with a Mucus Net

Pelagic tunicates such as *Okiopleura* have the finest of filters in the zooplankton. Water enters through the coarse filter, and fine particles of food are trapped in the mucus net.

the mucus and trapped food particles are drawn to the mouth by cilia. The mucous nets of the planktonic chordates are perhaps the most effective filters of all the zooplankton filter feeders (Fig. 10.21). The Larvacean *Okiopleura* can retain particles just 0.1 μm (0.0004 in) in size. In this filtering system, cilia push water through a mucous net where fine particles are trapped. The mucous net ends in a strand that is eaten by the animal.

Raptorial feeding may be either active or passive. In the phyla Cnidaria and Ctenophora, food is captured by passive contact with tentacles. Nematocysts and colloblasts, respectively, are used by these two groups to immobilize and in some cases poison the prey. Passive raptorial feeding is also found among the foraminiferan and radiolarian Protozoa. Both these protozoan groups extend pseudopodia from the main part of the body. Phytoplankton and small zooplankton stick to the cytoplasm and are digested in the pseudopodia.

Other raptorial zooplankton, such as copepods, actively grasp their prey. Many other crustacean zooplankton use the same combination of limbs to grasp and pass to the mouth a variety of prey. Food may be large individual cells, as with antarctic euphausiids and diatoms, or other zooplankton, as with deep-water oceanic euphausiids.

Chaetognaths (arrow worms) capture their prey, mostly copepods, with specialized jaws. Arrow worms are voracious carnivores and actively swim to capture their prey.

PATCHINESS

On land the numbers and kinds of plants and animals found in a particular place is determined mostly by gradients of moisture and temperature. In the photic zone of the marine environment, conditions are more complex. The numbers and kinds of organisms are influenced by gradients of temperature, light, and nutrients. The distribution of life in the photic zone is best visualized as being like clouds in the sky. At times plankton forms layers that cover large areas; at other times it forms smaller clouds which vary in size and have distinct shapes. Regardless of the size of the cloud, distinct layers are found at different 'heights'.

Life in the photic zone, then, is not uniformly distributed. Aggregation, or **patchiness,** as it is called in marine biology, is a characteristic of virtually all kinds of organisms in the photic zone (from phytoplankton to marine mammals). If organisms were not aggregated, they would be too dilute to serve as a source of food for animals. Commercial fishermen rely on the patchiness of fish and mammals to make catches worthwhile. If krill, the primary food of baleen whales in the Antarctic, were not patchy, the whales would have to spend more energy filtering the water than the food would provide. Finally, if phytoplankton were not patchy, the filter-feeding zooplankton herbivores could not obtain enough food energy to sustain the filtration of the water. Patchiness has a horizontal, vertical (Fig. 10.22), and seasonal component.

Horizontal Patchiness

Patchiness is generally used to describe the small-scale distribution of plankton across the surface of the ocean. Phytoplankton patches vary in size from a few meters to hundreds of kilometers in diameter. In the oceanic area, most patches are elliptical in shape and about 50 km (31 mi) across. Closer to shore, in the neritic zone, phytoplankton patches are more likely to be around 1–3 km (0.6–1.8 mi) in size.

Zooplankton patch sizes are similar to those of phytoplankton. Neritic zooplankton have patch sizes as small as 20 m (66 ft) in diameter. In the open ocean, zooplankton patches of around 20 km (12.4 mi) diameter are common. Patches of krill *(Euphasia superba)* in the Antarctic approximate 30 m (98 ft) in diameter. These large patches, also called swarms, are composed exclusively of the single species. Marine biologists from the Scripps Institute of Oceanography in 1981 found one patch that was estimated to contain over 1 million metric tonnes of krill.

Patches of both phytoplankton and zooplankton may be caused by a number of factors, both physical and biological. Small-scale patchiness of plankton may be caused by Langmuir circulation (see above). The alternate zones of downwelling and upwelling create areas of divergence and convergence, concentrating the plankton (Fig. 10.23). Buoyant phytoplankton tend to concentrate in areas of convergence. The response of zooplankton is more complex and depends on their depth in their daily migration.

Water movements can also cause patchiness on a large scale. Turbulent mixing of surface waters by wind may cause patches of high plankton density where these turbulent currents reach the surface. Upwelling is an important cause of plankton patchiness. The upwelling at the equatorial divergence and localized coastal areas produces surface waters relatively rich in nutrients. These nutrient-rich waters, in turn, support relatively dense and patchy populations of phytoplankton.

Patchiness also may be caused by biological factors. According to one theory, grazing by zooplankton may virtually eliminate phytoplankton in some areas. Zooplankton feed in an area of dense phytoplankton until it is gone; then the zooplankton move elsewhere.

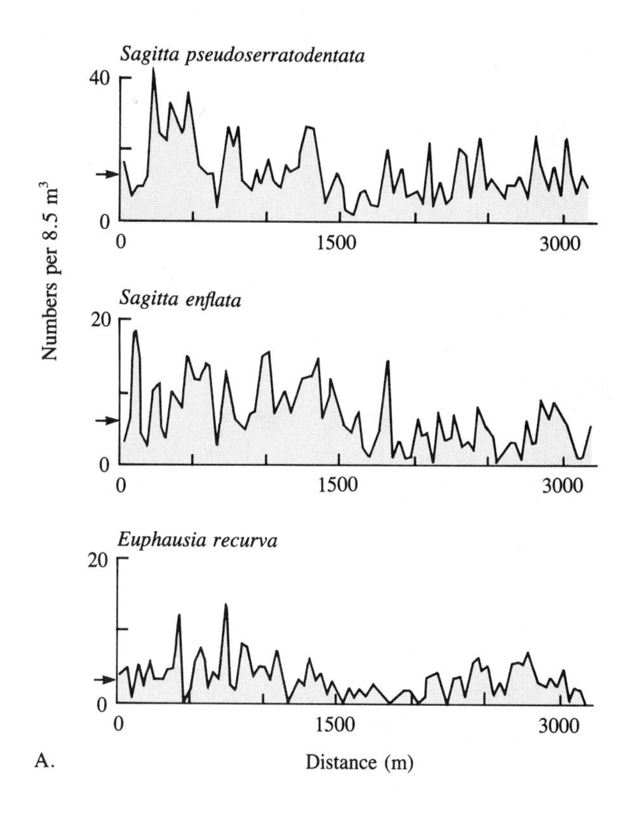

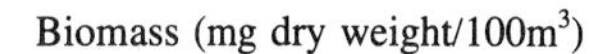

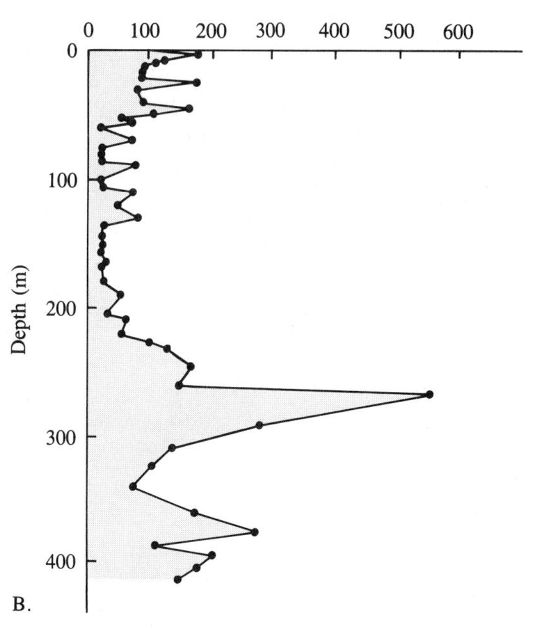

FIGURE 10.22
Patchiness in the Plankton

Patchiness in plankton has both a horizontal and a vertical component. **(A)** Horizontal patchiness of three species of zooplankton. **(B)** Vertical patchiness of all plankton in the top 400 m. (After Parsons, T.R., Takahashi, M., and Hargrave, B. 1977. *Biological oceanographic processes.* New York:Pergamon.)

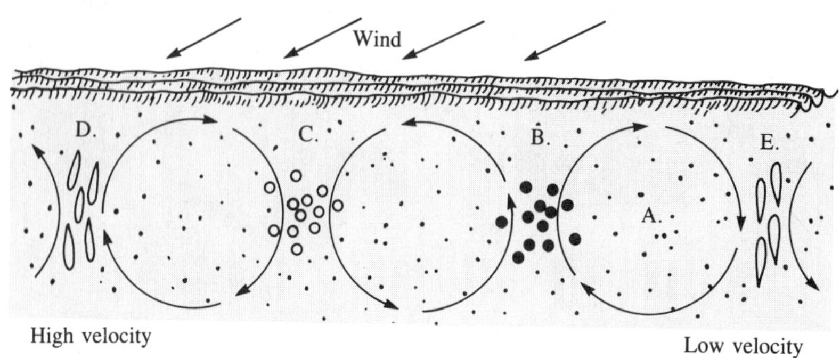

FIGURE 10.23
Langmuir Circulation and Patchiness

Langmuir circulation concentrates plankton on the basis of buoyancy characteristics. **(A)** Neutrally buoyant particles distribute randomly. **(B)** Particles tending to sink aggregate in upwellings. **(C)** Particles tending to float aggregate in downwellings. **(D)** Organisms aggregated in high-velocity upwellings swim down. **(E)** Organisms aggregated in low-velocity downwellings swim up.

Some patches of phytoplankton, however, repel zooplankton. Some dense growths produce metabolites, which the zooplankton find irritating and even toxic. Red tides (blooms of certain dinoflagellates) are avoided by planktonic grazers.

Vertical Patchiness

Both phytoplankton and zooplankton have higher concentrations at certain depths in the photic zone. Phytoplankton seek out depths with optimal light intensities for photosynthesis. Most species of phytoplankton are adapted to shade. Optimal light intensities generally occur below 1 m depth, although photosynthesis occurs down to the so-called *1-percent light level* (the depth at which 1 percent of full daylight is found). The 1-percent light level is commonly considered to be the oxygen compensation point for photosynthesis. The rate of photosynthesis at any deeper level will not provide adequate energy for the phytoplankton to maintain itself.

One of the most striking features of vertical phytoplankton patchiness is the almost universal feature called the **subsurface chlorophyll maximum** (SCM). The concentration of chlorophyll in water is often used as a measure of the total phytoplankton population. Many recent observations have found maximum concentrations of chlorophyll below the accepted 1-percent level. The SCM is widespread in the oceans and is a permanent feature of stratified waters, where surface temperatures have warmed and formed a thermocline between surface and deeper water. Stratified waters are nutrient poor because the thermocline prevents nutrient enrichment from deeper waters. The low levels of nutrients limit the growth of phytoplankton. The SCM is found in the permanently stratified water of the low- to midlatitude ocean current gyres. The SCM is also found at higher latitudes during summer stratification.

The SCM is generally found at the depth of the thermocline, often at 100 m (328 ft), well below the traditional boundary of the euphotic zone. Phytoplankters may concentrate at the thermocline to obtain nutrients from the nutrient-rich waters just below the thermocline.

The SCM is relatively narrow. Measurements have shown it to be only a few meters thick and at the very most 10 m (32.8 ft). The SCM is present at many locations in the Pacific Ocean, where it is a permanent feature from Alaska to New Zealand. From east to west it has been reported from the coast of California to Easter Island. It also is a universal feature of the tropical Atlantic and the Indian Ocean.

The SCM is composed mostly of small flagellated phytoplankton, particularly dinoflagellates and chlorophyte flagellates. The role of the SCM phytoplankton in the overall primary production of an area is not yet known. It is possible that the SCM phytoplankton migrate to higher light intensities after accumulating nutrients. It is also possible that SCM phytoplankters are resting stages. In the estuary of Puget Sound, where nutrients are plentiful, light limits phytoplankton growth the year round. Here an SCM has been observed in the winter season, when light is generally inadequate to support photosynthesis.

Zooplankton are also distributed in a patchy manner throughout the vertical extent of the photic zone. The greatest concentration of zooplankton is generally in the top 200 m (656 ft). A subsurface maximum occurs at the mixing layer. In those areas where the mixed layer extends to deeper levels, the subsurface zooplankton maximum also extends deeper.

In addition to this subsurface maximum, there is often a well-defined daytime layer of zooplankton at 300–600 m (984–1,968 ft). This layer of plankton (and sometimes small nekton) is called the deep scattering layer (see Ch. 2). The layered nature of the zooplankton is most distinct during the day. In general, a zooplankton species stays as deep as possible during the day without having to pay an excessive energy cost to rise toward the surface during the evening to feed. Larger species are found deeper than smaller species, and juvenile stages of a life cycle are more likely to be in shallower water than adults are. The depth at which zooplankton are found during the day is influenced by light intensity, temperature, and the density difference between the plankter and the surrounding water. The optimal depth is species specific.

The depth at which a zooplankton species is found is also influenced by season. Many zooplankton winter in deeper water. One of the best-known examples of this is the distribution of calanoid copepods, particularly *Calanus finmarchicus* in the North Atlantic. In the winter, when food sources are low, the copepods are deep in the water column (between 400–600 m or 1,312–1,968 ft). In the spring, with the start of the diatom bloom, the copepods move toward the surface where they are closer to the abundant food sources.

With few exceptions, marine life is dependent on the energy stored by plants in the photic zone. It will become clear as we discuss the deep ocean in the next chapter that the dynamics of transferring food from the photic zone to the dark abyss has a significant impact on the nature of life in the deep ocean.

SUMMARY

The photic zone is that part of the pelagic zone where sunlight penetrates. It includes two subzones: the epipelagic (or euphotic) zone, which has enough light to actively support photosynthesis, and the mesopelagic (or disphotic) zone, where the light is dim but adequate to see shapes and shadows.

The uneven heating of the atmosphere results in atmospheric convection currents that produce three prevailing wind belts in each hemisphere: the polar easterlies, the westerlies, and the trade winds. Although these air masses move across the earth's surface in basically north-south directions, the Coriolis effect causes them to veer to the right in the Northern Hemisphere and to the left in the Southern Hemisphere. This effect occurs because the earth's surface rotates at different velocities at different latitudes, ranging from 0 km/h at the poles to 1,600 km/h (994 m/h) at the equator.

The major ocean gyres are driven by the trade winds blowing toward the equator, and the westerlies at higher latitudes. Due to the Coriolis effect and Ekman transport, these rotating water masses move in a clockwise direction in the Northern Hemisphere and counterclockwise in the Southern Hemisphere. High-velocity boundary currents carry warm water from the equator into higher latitudes along the western margin of ocean gyres, while low-velocity cold currents move toward the equator on the east side. Coastal and equatorial upwelling significantly increases biological productivity in surface waters, while thermohaline circulation, the sinking of surface water into the deep ocean at high latitudes, supplies the oxygen to make life possible in the deep ocean.

Organisms of the photic zone are all swimmers or floaters. The swimming mechanism may be as simple as the beating of cilia or flagella of a single-celled phytoplankton or protozoan, or it can be as complex as the musculature and body shape of pelagic fish like sharks or tuna. The ability to swim is not restricted to only animals of the photic zone. Many phytoplankton species, such as dinoflagellates, counteract the force of gravity by swimming. Others, such as certain medusae and tunicates, float passively in the water.

Animals in the photic zone use a number of morphological and physiological adaptations to maintain bouyancy. The first is size itself. For organisms less than 500 μm (0.02 in) in size, the smaller they are, the slower they sink. Spines and leafy appendages increase effective surface area and also promote slower rates of sinking. Many organisms, including phytoplankton, zooplankton, and fish, use gases in floats to maintain bouyancy. Other plankton increase bouyancy by actively maintaining lower concentrations of certain ions in their body fluids. Finally, many animals increase bouyancy by concentrating certain lipids or fats that are lighter than water.

Apparently, the waters of the photic zone are not an easy place to live. The photic zone contains one-tenth the number of species found in the underlying benthic areas. Of the 29 invertebrate phyla, only one-fourth have species in the photic zone. Photosynthesis in the photic zone is almost exclusively from single-celled phytoplankton, primarily diatoms, dinoflagellates and coccolithophores. Although the surface waters of the ocean generally have lower productivity than the benthic algae and plants of the coastal zone, the photic zone because of its immense size accounts for 90 percent of the total photosynthesis of the marine environment. The majority of the phytoplankton are part of a grazing food chain; that is, the phytoplankton are eaten by small zooplankton which are eaten by larger zooplankton or small fish. Most animals of the photic zone are small. The most common animal, the copepod *Calanus,* is only a few millimeters in length. The larger predatory animals of the photic zone, such as tuna, swordfish, dolphins, and killer whales, have the longest food chains of any animals on earth.

Plankton are not evenly distributed in the photic zone but occur in patches, or concentrations, which change over time in both horizontal and vertical directions. Phytoplankton are found in patches about 3 km (1.9 mi) in diameter in the coastal zone, and about 50 km (31 mi) in oceanic areas. Zooplankton are found in somewhat smaller patches. Patches may be related to physical factors—concentrations of nutrients, currents, and other oceanographic conditions—or biological factors. Phytoplankton are generally restricted to the euphotic zone, although in many areas the maximum concentration of chlorophyll is found below depths that can support photosynthesis. Most zooplankton stay in dense concentrations at lower depths during the day and rise to the surface and disperse at night. These daily vertical migrations generally are restricted to the top 100 m, although larger plankton and small fish may make diel migrations from depths of 100–900 m (328–2,952 ft), within the deep scattering layer.

QUESTIONS AND EXERCISES

1. Describe the major rotational gyres in ocean basins of the Northern Hemisphere and Southern Hemisphere, as well as the role of the Coriolis effect, Ekman spiral, and Ekman transport in producing them.
2. There are at least two conditions of biological significance related to the sinking of water masses from the surface into the deep ocean. Discuss them.
3. What is the neuston? Where is it found? What organisms are found in the neuston?
4. Discuss the important biological features of upwelling zones.
5. Discuss the advantages and disadvantages of the small size of plankton as it relates to living in a three-dimensional water space.
6. Discuss the adaptations for bouyancy found in the zooplankton.
7. Discuss the advantages and disadvantages of vertical migration in the zooplankton.
8. Discuss the roles of bioluminescence in marine animals.
9. List and briefly describe a few of the more important phytoplankton, as well as holoplanktonic and meroplanktonic zooplankton.
10. What is planktonic patchiness? Where and under what conditions is patchiness encountered?

REFERENCES

Bougis, P. 1976. *Marine plankton ecology.* New York:Elsevier.

Crane, J.M. 1973. *Introduction to marine biology: A laboratory text.* Columbus, OH:Merrill.

Haeckel, E. 1974. *Art forms in nature.* New York:Dover.

Longhurst, A.R., ed. 1981. *Analysis of marine ecosystems.* New York:Academic Press.

Newell, G.E., and **Newell, R.C.** 1977. *Marine plankton.* London:Hutchinson.

Nicol, J.A.C. 1971. *The biology of marine animals.* London:Pitman.

Parsons, T.R., Takahashi, M., and **Hargrave, B.** 1984. *Biological oceanographic processes,* 3rd ed. New York:Pergamon.

Platt, T., Subba Rao, I.V., and **Irwin, B.** 1983. Photosynthesis of picoplankton in the oligotrophic ocean. *Nature* 301(5902):702–704.

Richardson, P. 1976. Gulf Stream rings. *Oceanus* (Spring).

Weibe, P. 1976. The biology of cold-core rings. *Oceanus* 19(3):69–76.

Weibe, P. 1982. Rings of the Gulf Stream. *Sci. Amer.* 246(3):60–70.

Wickstead, J.H. 1976. *Marine zooplankton.* London:Edward Arnold.

SUGGESTED READING

OCEANS

Lohe, J.F. 1977. Krill. 10(3):54–55.

Staller, J. 1986. The next El Niño. 19(5):18–23.

OCEANUS

El Niño. 1984. 27(2):3–62.

Nealson, K., and **Arneson, C.** 1985. Marine bioluminescence: About to see the light. 28(3):13–19.

SCIENTIFIC AMERICAN

Baker, D.J., Jr. 1970. Models of ocean circulation. 222(1):114–121.

Stewart, R.W. 1969. The atmosphere and the ocean. 221(3):76–105.

Ward, P., Greenwald, L., and **Greenwald, O.E.** 1980. The buoyancy of the chambered nautilus. 243(4):190–203.

Webster, P.J. 1981. Monsoons. 245(2):108–118.

SEA FRONTIERS

Korgen, B.J. 1986. Lighted houses in the sea. (Larvacean bioluminescence) 32(1):4–9.

Michel, H.B. 1985. The fuzzy sea butterfly. 31(3):181–182.

Nicol, S. 1984. Krill swarms in the Bay of Fundy. 30(4):216–222.

GULF STREAM RINGS AND MARINE LIFE

The Gulf Stream separates two distinct water masses. To the north lies the Slope Water, with an average temperature of less than 10°C (50°F), and to the south lies the Sargasso Sea Water, with temperatures ranging from 15°C (59°F) to as high as 25°C (77°F). The average width of the meandering stream is perhaps 100 km (62 mi), and periodically the meanders cut themselves off and move as individual rings (eddies) on either side of the Gulf Stream (Fig. 1).

When a meander breaks off and forms a ring (Fig. 2) on the north side of the Gulf Stream, it may be up to 200 km (124 mi) in diameter and have a lens of warm Sargasso Sea Water trapped inside it. The warm water may go to a depth of 1,500 m (4,920 ft). Although of great variety, the concentration of life in this 'warm ring' is meager and has not been studied in detail. Conversely, the 'cold rings' that break off on the south side may be up to 300 km (186 mi) in diameter and contain within them a plug of cold Slope Water that extends to the ocean floor. The depth of the ocean in this region may be in excess of 4,000 m (13,120 ft). This cold water contains a much smaller number of species, but a much greater biomass than the Sargasso Sea Water trapped in the warm rings.

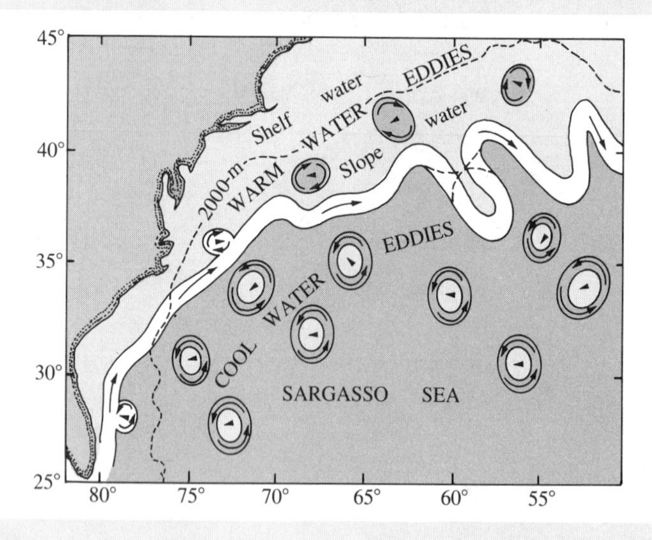

FIGURE 1
Atlantic Ocean Surface Currents

The path of the Gulf Stream separates cold, biologically productive Slope Water on the north from the warm, less-productive Sargasso Sea Water to the south. As the Gulf Stream meanders, some meanders pinch off to form rings—warm rings of Sargasso Sea Water enter the cold Slope Water to the north, and cold rings of Slope Water enter the Sargasso Sea to the south. (From Richardson, 1976.)

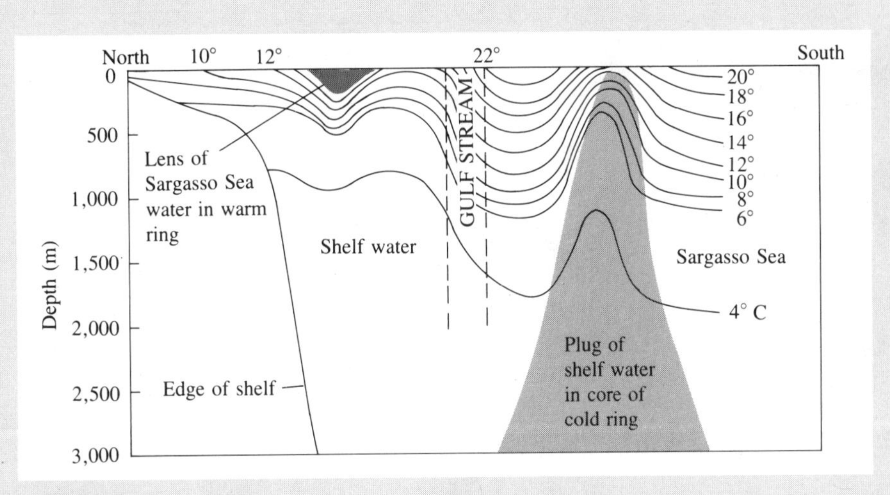

FIGURE 2
Temperature Structure of Warm Ring and Cold Ring

Because of their greater concentration of marine life, the cold rings have been studied extensively by marine biologists (see Weibe, 1976, 1982). The following describes the life history of the cold rings and the responses of the Slope Water biotic community to physical changes within the ring structure.

Although most cold rings have the energy to remain intact for 2–4 years, the average life of a ring is thought to be about 1½ years. Most cold rings break away from the Gulf Stream between 60°–75° W longitude, and move southwest at an average speed of 4 km/day (2.5 mi/day) through the Sargasso Sea. They rejoin the Gulf Stream off the Florida coast north of the Bahamas.

While the rings are detached, the cold core water is modified. It warms up and becomes more saline, and the depth of the oxygen minimum drops from the 150 m (492 ft) typical of Slope Water to around 800 m (2,624 ft), which is characteristic of the Sargasso Sea. A striking feature of this change is the collapse of the dome of cold water extending up into the ring. Figure 3 shows the difference between the temperature structure in a newly formed cold ring and one that has completed its trip across the Sargasso Sea.

As the cold-water environment in the ring decays, the Slope Water life forms in the core become stressed. Most severely affected are the planktonic forms that do not have the ability to swim away to a more favorable environment. The phytoplankton population in the core of newly formed rings is most abundant at a depth of about 30 m (98 ft), as is indicated by maximum chlorophyll concentrations of up to 3.0 mg/l of water. Toward the flanks of the rings, the phytoplankton concentration decreases and is at a maximum at a depth of about 60 m (197 ft). This condition is typical of the Sargasso Sea. Observations in one ring showed a rapid decrease in chlorophyll content. At an age of 7 months, the ring contained only an eighth of the chlorophyll

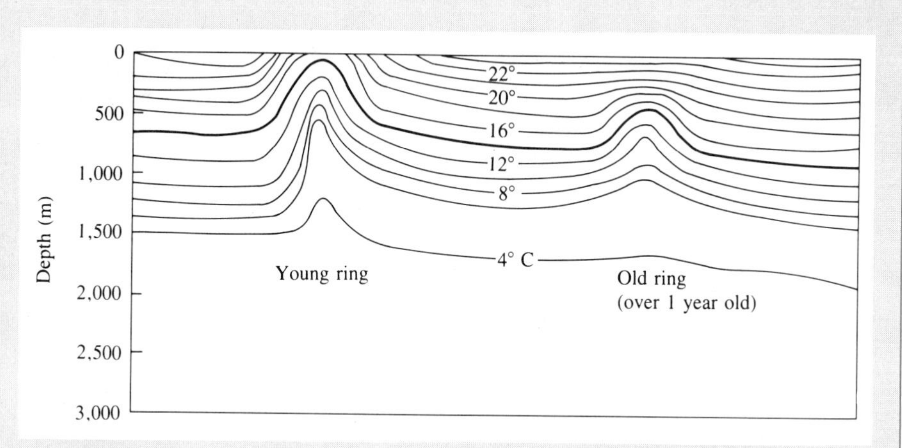

FIGURE 3
Comparison Between Temperature Structure in Young and Old Cold Rings

it had contained four months earlier. This rapid reduction in phytoplankton activity results because nutrient supply cannot be replenished in the trapped Slope Water.

The zooplankton population of a young ring core is quite similar in species abundance and depth range to that of Slope Water. It includes some euphausiids (small omniverous shrimplike crustaceans) and a herbivorous snail that is found only in the Slope Water. The euphausiids (Fig. 4) and other herbivorous zooplankton populations are greatly reduced by the time the rings are six months old because of the rapid decline in phytoplankton.

One of the shrimplike crustaceans, *Nematoscelis megalops,* prefers temperatures ranging between 8–12°C (46–54°F; see Fig 5). This population lives almost entirely above a depth of 300 m (984 ft) in the Slope Water when it is initially trapped in the ring. As the core water warms from the top and the isotherms (lines of equal temperature) sink, the depth range of *N. megalops* also descends. As the population

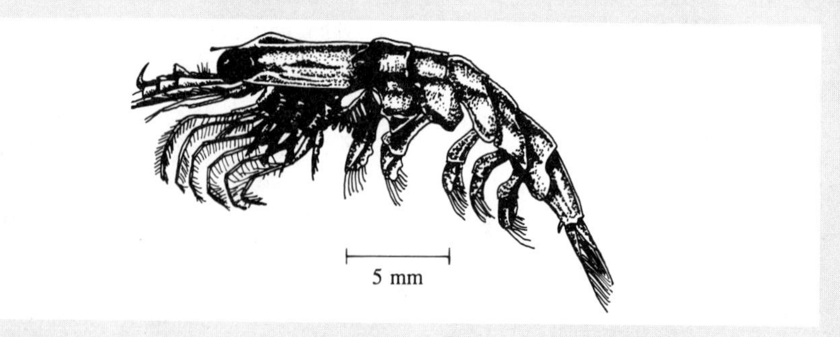

FIGURE 4
Slope-Water Euphausiid *Euphausia krohnii*

(Courtesy of Scripps Institution of Oceanography, University of California, San Diego.)

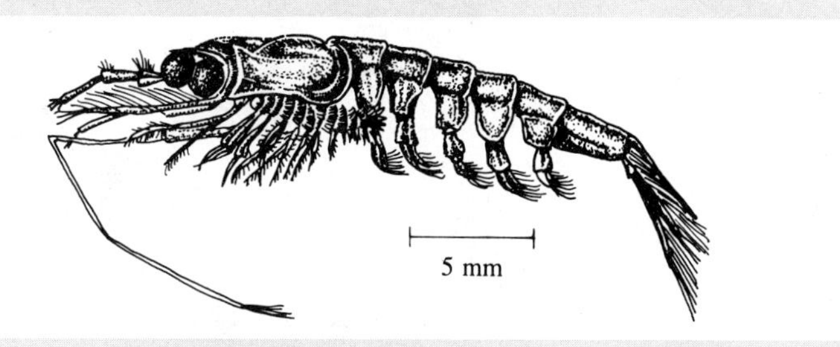

FIGURE 5
Slope-Water Shrimplike Crustacean *Nematoscelis megalops*

(Courtesy of Scripps Institution of Oceanography, University of California, San Diego.)

sinks deeper into the ring, it becomes separated from the higher concentration of food near the surface. By the time the ring is 6 months old, most of the *N. megalops* population will be found between 300–800 m (2,624 ft) depth. They begin to starve. Resulting physiological and biochemical changes include a declining rate of respiration, increased water content, and reduced lipid, carbon, and nitrogen content. In a 17-month-old-ring, no *N. megalops* remained. The same fate may await many of the cold-water species trapped in the ring.

Some of the Slope Water life forms may be left behind as the ring moves across the Sargasso Sea. Since the annulus of Gulf Stream water around the cold core extends only to a depth of 1,000–1,500 m (3,280–4,920 ft), deep-living forms may escape beneath this barrier. It is possible that some *N. megalops,* as well as a number of carnivorous species, may be removed from the core water in this way.

As the cold core decays, Sargasso Sea populations move into the ring structure. They gradually replace the Slope Water populations as conditions become favorable to them.

Rings similar to those formed along the Gulf Stream are formed in association with other major ocean current systems. Continued study of these systems will help increase our understanding of the physical processes of the oceans and the relationships of marine organisms to the processes.

11

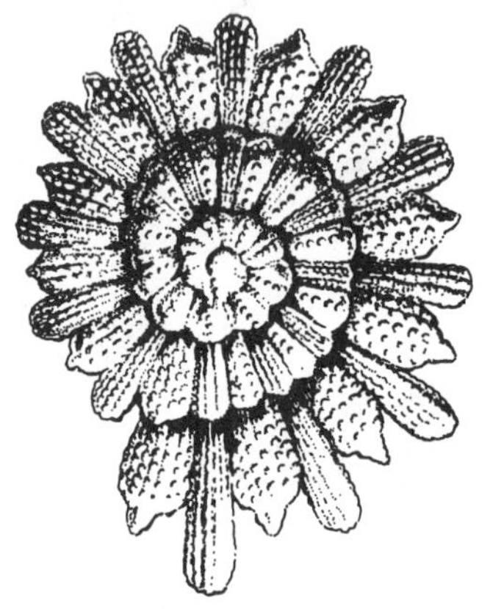

THE DEEP OCEAN: BENTHIC AND PELAGIC ZONES

The deep-sea environment, which starts at a depth of 1,000 m (3,280 ft), is distinctly different from the surface environment. There is no sunlight; water temperatures are cold, salinity constant, and hydrostatic pressures high. Conditions throughout the deep-sea region are more constant than those in any other environment on earth.

The distribution of deep-sea species, for the most part, does not overlap that of shallow water. Certain deep-sea species can be used to measure the boundaries of the deep-sea environment in specific locations. Based primarily on the distribution of the benthos, it is clear that at high latitudes the deep-sea environment comes closer to the surface, and in the tropics it may be deeper than 1,000 m (Fig. 11.1). This tendency of deep-sea species to approach the surface at high latitudes is called **polar emergence** (or equatorial submergence). Marine biologists have long recognized that many species of both pelagic and benthic animals occur at greater depths toward lower latitudes.

The deep sea is the largest marine habitat. Its benthic component covers more than 80 percent of the ocean floor, an area two times the total land mass above sea level (Fig. 11.2). Its pelagic component contains 75 percent of the ocean's water. Despite its large size, the deep sea is impoverished with respect to animal life. Over 80 percent of the biomass of organisms in the oceans is found in the surface 1,000 m of water (Fig. 11.3). The biomass of benthic animals decreases with increasing depth. In the productive shallow seas, there is up to 5,500 g/m^2 (1.13 lb/ft^2) of benthic biomass. Over the continental shelf this value is around 200 g/m^2 (0.04 lb/ft^2). In the deep ocean, values are much lower (Table 11.1). For example, in the abyssal plains of mid-oceanic areas, biomass values are around 0.001 g/m^2 (0.0000033 lb/ft^2). On most of the deep ocean floor the animal population has a density similar to that found in the inhospitable deserts of land.

The environmental conditions of the deep sea are exceedingly uniform. Water temperatures range from a minimum of −1.8° C (28.7° F) near the poles, to 7–8° C (44.6–46.4° F) at a depth of 1,000 m at mid- and low latitudes. There is only one area where deep waters exceed 10° C (50° F). In the Mediterranean Sea, from 1,000–6,000 m (19,680 ft), temperatures are between 10–13° C (55.4° F).

No sunlight penetrates to the deep sea. Some animals are bioluminescent, however, and many species retain eyes to respond to this light. Salinity is uniform through the deep sea and measures around 35‰. Dissolved oxygen values are relatively high in the deep sea, with a few minor exceptions. High hydrostatic pressure is a fact of life for all deep-sea animals. Pressure increases every 10 m (32.8 ft) by an amount equal to the total atmospheric pressure at sea level. At 10,000 m (32,800 ft), depth pressures are 1000 kg/cm^2 (7 tons per sq in). These high pressures affect the physiology and biochemistry of deep-sea animals.

The final important environmental factor in the deep sea is current flow. Many parts of the deep sea, particularly the western boundary areas, have slow but persistent currents caused by movement of intermediate

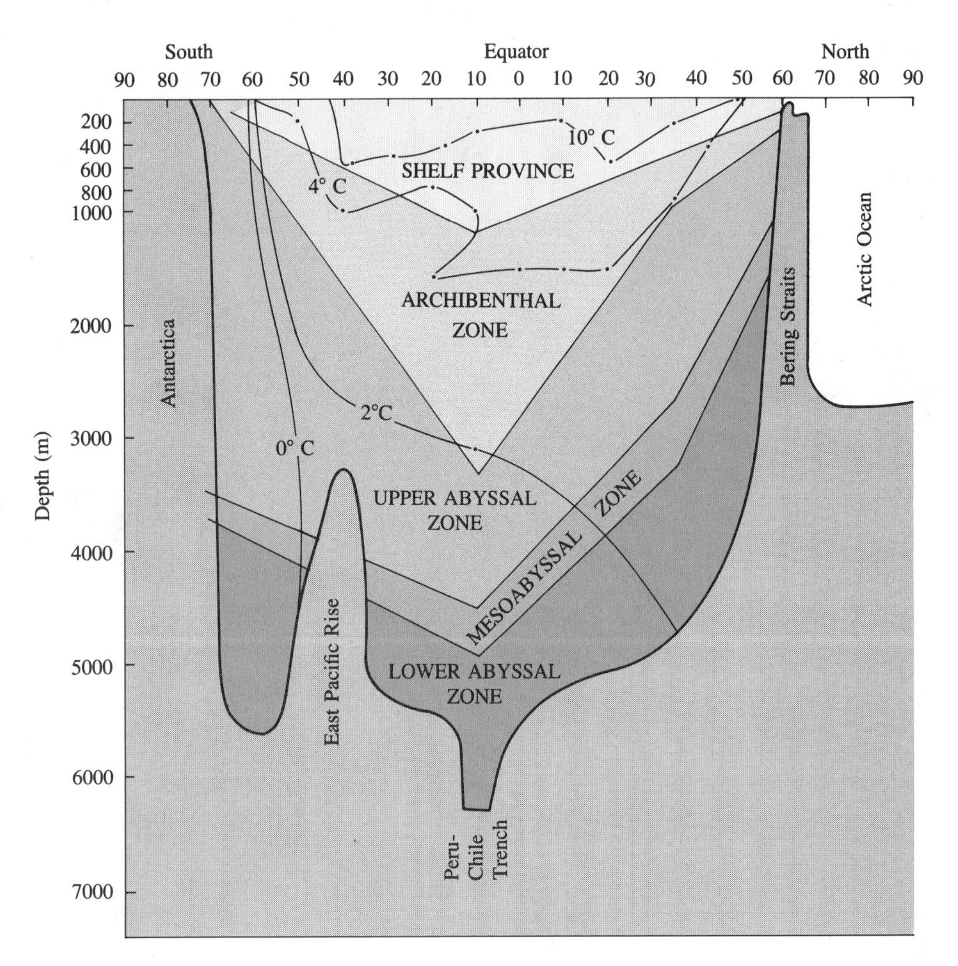

FIGURE 11.1
Boundaries of the Deep-Sea Environment

Although the deep ocean is defined as that part of the marine-environment below 1,000 m (3,280 ft) depth, the biological communities typical of that zone are found in shallower water at higher latitudes. Likewise, the shallower shelf communities may be found below 1,000 m near the equator. (From Menzies, R.J., George, R.Y., and Rowe, G.T. 1973. *Abyssal environment and ecology of the world's oceans.* New York:Wiley. Copyright by John Wiley & Sons, Inc.)

and deep-water masses. These currents are slow compared to surface currents, and move with velocities of around 1 m/sec (slower than a slow walk). Many benthic animals orient toward the currents to filter feed.

The uniformity of environmental conditions over space is intensified by the uniformity of conditions over time. There are no patterns of change from day to night, such as those in the mesopelagic and epipelagic zones, and throughout the year there are virtually no changes in temperature, salinity, and dissolved oxygen. Despite the uniform environmental conditions throughout the deep sea, individual species have distinct patterns of distribution. Some 1,030 species of benthic sponges, cnidarians, barnacles, isopods, decapods, sea spiders, echinoderms, and beardworms (Pogonophora) are included in the benthos of the deep sea. Of these, only 4 percent are found in all three of the major oceans (Pacific, Atlantic, Indian). In fact, only 15 percent of the species are found in more than one of the three oceans. On a latitudinal scale, species diversity of benthos is greater under the warm water of the ocean. An increase in species diversity in both the terrestrial and shallow marine areas of the tropics has long puzzled ecologists. A similar pattern in the deep-sea benthos adds further to the question. See the special feature "Deep-Ocean Benthos" at the end of this chapter for an interview with Dr. Robert R. Hessler, the eminent deep-sea biologist. In this interview, Dr. Hessler discusses questions related to the evolution and diversity of deep-sea benthos.

To fully understand the biology of the deep ocean, it is necessary to have a knowledge of the fundamental aspects of marine geology and physical oceanography that determine the marine environment to which these organisms must adapt. Since we have already discussed the basic concepts related to the physical properties and currents, we now will concentrate our attention on the geological environment.

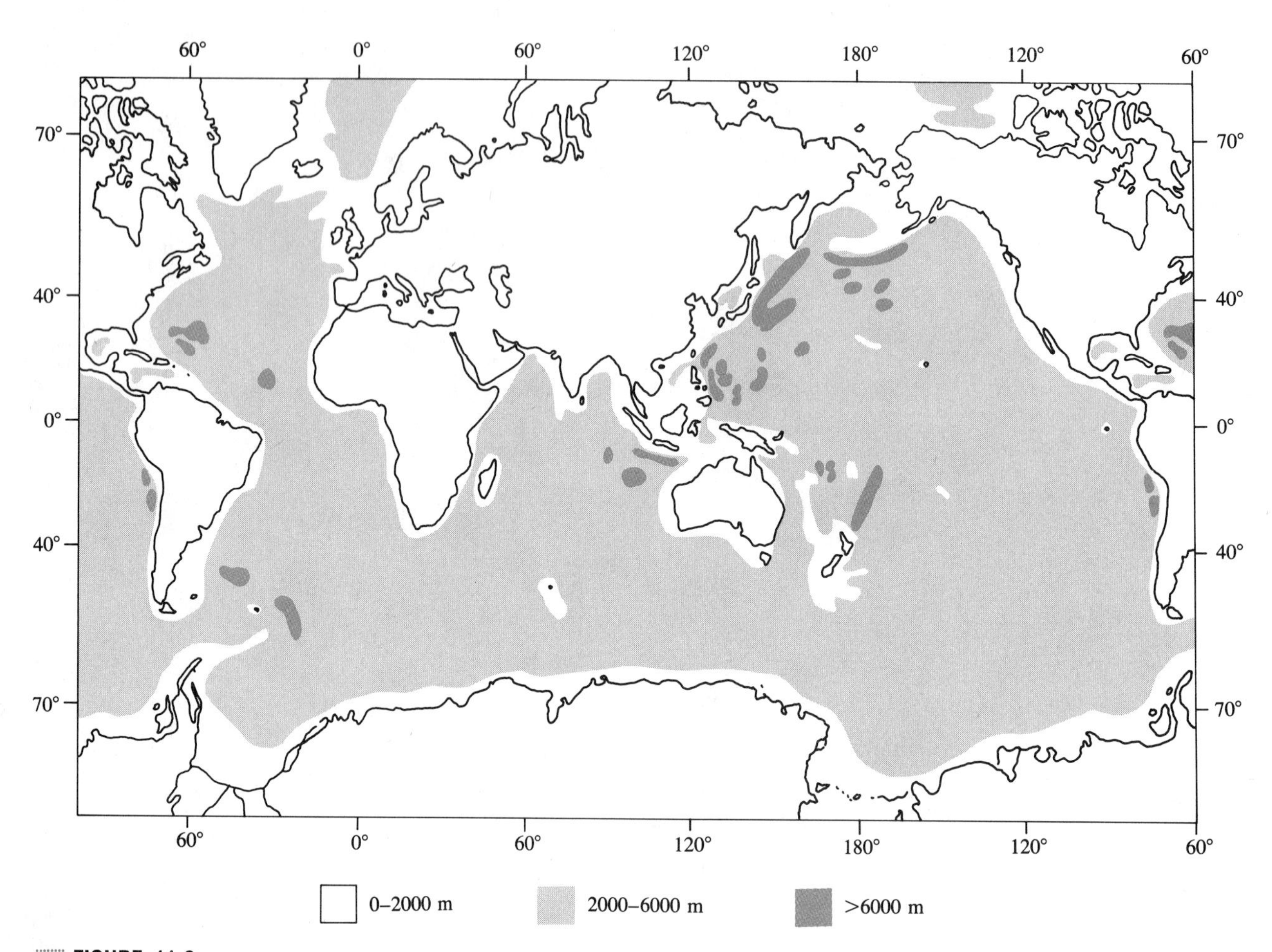

FIGURE 11.2
The Benthic Environment of the Deep Sea

Over 80 percent of the ocean floor is deeper than 1,000 m (3,280 ft) beneath the surface. The deep ocean area is the single largest uniform environment on earth.

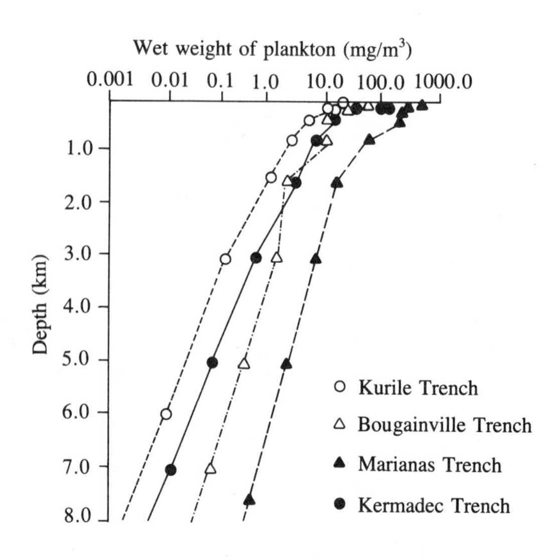

FIGURE 11.3
Biomass of Plankton in the Deep Sea

Values are for total biomass at various depths in the water column over a number of deep-sea trenches. Note that the logarithmic scale of weight tends to mask the large differences with depth. (After George, R.Y. 1981. "Functional adaptation of deep-sea organisms." In Vernburg, F.J., and Vernburg, W.B., eds. *Functional adaptation of marine organisms.* New York:Academic Press.)

TABLE 11.1
Biomass of Benthos in the Deep Sea

The concentration of benthos also decreases with increasing depth. Even though the area of the ocean bottom that is less than 200 m deep is only 7.6 percent of the total, it contains over 82 percent of the biomass.

	Area		Mean Biomass	Total Biomass	
Depth (m)	$(km)^2 \times 10^6$	%	g/m^2 or tn/km^2	$t \times 10^6$	%
0–200	27.5	7.6	200	5500	82.6
200–3000	55.2	15.3	20	1104	16.6
>3000	278.3	77.1	0.2	56	0.8
Whole ocean	361	100	18.5	6660	100

NOTE: t = metric tonnes

SOURCE: Cushing, D.H., and Walsh, J.J. 1976. *The ecology of the seas.* London:Blackwell Scientific.

OCEAN SEDIMENTS

Sediment accumulating in the oceans can be grouped into two basic types of deposits. **Neritic sediments** are those that accumulate rapidly around the margins of continents, and are composed primarily of lithogenous particles. **Oceanic sediments** are found in the deep-ocean basins, and accumulate at a much lower rate than neritic sediments. Depending on conditions, the dominant deposit may be of lithogenous, biogenous, or hydrogenous origin.

Marine organisms make a significant contribution to many oceanic sediment deposits. Oceanic deposits can be divided into two basic types: **abyssal clay,** which contains less than 30 percent biogenous particles and more than 70 percent lithogenous clay, and oozes, which contain more than 30 percent biogenous particles by weight.

Chemically, there are two basic compounds that make up most of the biogenous sediment: **silica** (SiO_2) and **calcium carbonate** ($CaCO_3$). The silica is derived primarily from diatoms and the protozoan *Radiolaria.* Calcium carbonate is derived primarily from the hard parts of the protozoan *Foraminifera,* and from coccolithophores. Pteropods (phylum Mollusca) also contribute to the calcareous deposits (Fig. 11.4).

Three processes are important in determining whether or not an ooze will form on the deep-ocean floor: the *productivity* of the plankton population, the rate at which the hard parts are taken back into *solution* by the ocean water, and the rate of *dilution* by other types of particles. Siliceous biogenous particles are usually thin and delicate, so they are commonly dissolved into ocean water before they accumulate on the ocean floor. Diatom oozes accumulate only around Antarctica and in the North Pacific, where rates of productivity are very high and little lithogenous material reaches the ocean floor. The only significant deposit of radiolarian ooze is in the equatorial Pacific, where upwelling allows a high rate of productivity. Calcareous oozes are the most abundant sediment now accumulating on the deep-ocean floor (Fig. 11.5).

A major feature associated with the formation of calcareous ooze is the **calcium carbonate compensation depth,** the depth at which the ocean is able to dissolve all the calcium carbonate particles falling from above. In shallow surface water, high temperature reduces the ability of the ocean to dissolve gases such as carbon dioxide. Here also phototrophs are using carbon dioxide in photosynthesis, so the water has a low capacity to dissolve calcium carbonate.

Thus, calcium carbonate is rapidly accumulating on shallow ocean bottoms in areas such as the Bahamas. With increasing depth, the water becomes colder and large amounts of carbon dioxide can be held in solution. With increasing depth, the carbon dioxide content of the water gradually increases until it dissolves all the calcium carbonate it encounters. This occurs at the calcium carbonate compensation depth, and no calcium carbonate can accumulate on the bottom below this depth. Beneath the calcium carbonate compensation depth, the dominant oceanic sediment is abyssal clay. Note in Figure 11.6 that nearly all the floor of the North Pacific Ocean is covered with abyssal clay because this is the deepest of the ocean basins. Essentially, all of it lies beneath the calcium carbonate compensation depth, which occurs at an average depth of about 4,000 m (13,120 ft).

A.

FIGURE 11.4
Microscopic Skeletons from the Deep Sea

(A) Magnified 160 times by scanning electron microscopy, the skeletons are foraminifers and radiolarians found in sediment cores taken at Site 55 in the Caroline Ridge area of the west Pacific Ocean by the Deep Sea Drilling Project. They were recovered in water 2,843 m (9,315 ft) deep, with further penetration of 130 m (426 ft) into the ocean bottom. Rounded globose forms are forminifers, whose skeletons are calcareous. Forms with reticulate skeletons are radiolarians and are siliceous. In life, these tiny organisms lived near the ocean surface, but upon death their skeletons fell to the sea bed to become entombed in the sediments. By carefully studying them, scientists can determine how long ago they lived and can learn much about the history of the ocean. (Courtesy of the Deep Sea Drilling Project, Scripps Institution of Oceanography, University of California, San Diego.) **(B)** Coccoliths, magnified about 10,000 times. (Photo courtesy of Deep Sea Drilling Project, Scripps Institute of Oceanography.) **(C)** Foraminifer, *Orbulina universa,* is magnified 300 times. (Photo by Howard J. Spero)

B.

C.

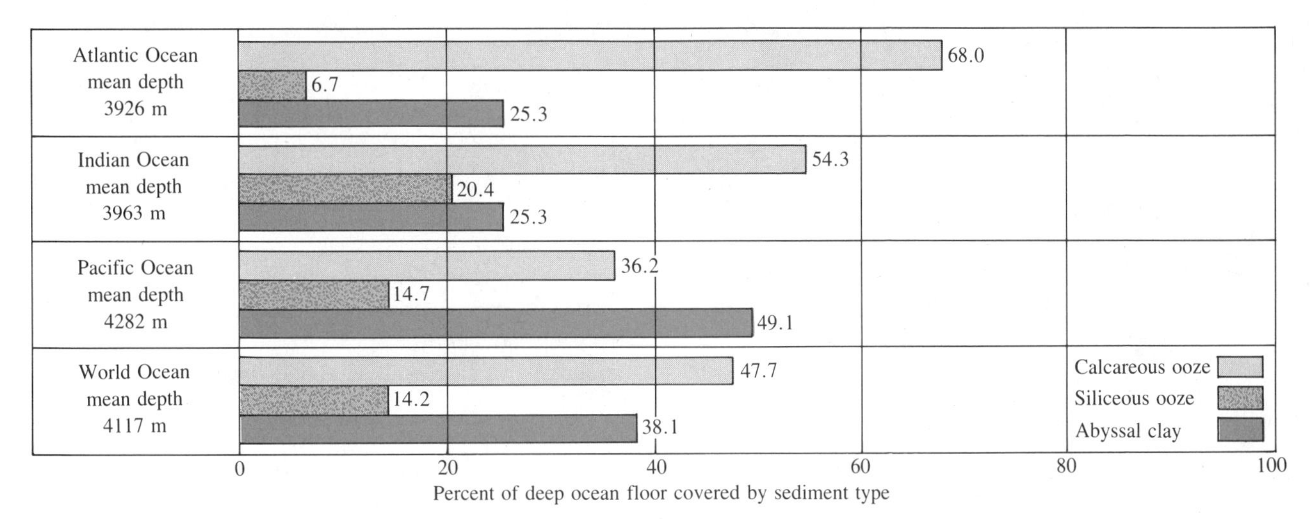

FIGURE 11.5
Distribution of Oceanic Sediment

The percentage of an ocean basin floor covered by calcareous ooze decreases with increasing mean depth of the basin. This is because the deeper an ocean basin is, the greater the percentage of its floor that lies beneath the calcium carbonate compensation depth. The mean depths calculated for this chart exclude the shallow adjacent seas, where little oceanic sediment accumulates. Notice that the dominant oceanic sediment found in the deepest ocean basin, the Pacific Ocean, is abyssal clay, while calcareous ooze is the most widely deposited abyssal sediment in the shallower Atlantic and Indian oceans. (After Sverdrup, H., Johnson, M., and Fleming, R. 1942. *The oceans: Their physics, chemistry and general biology.* Englewood Cliffs, NJ:Prentice-Hall.)

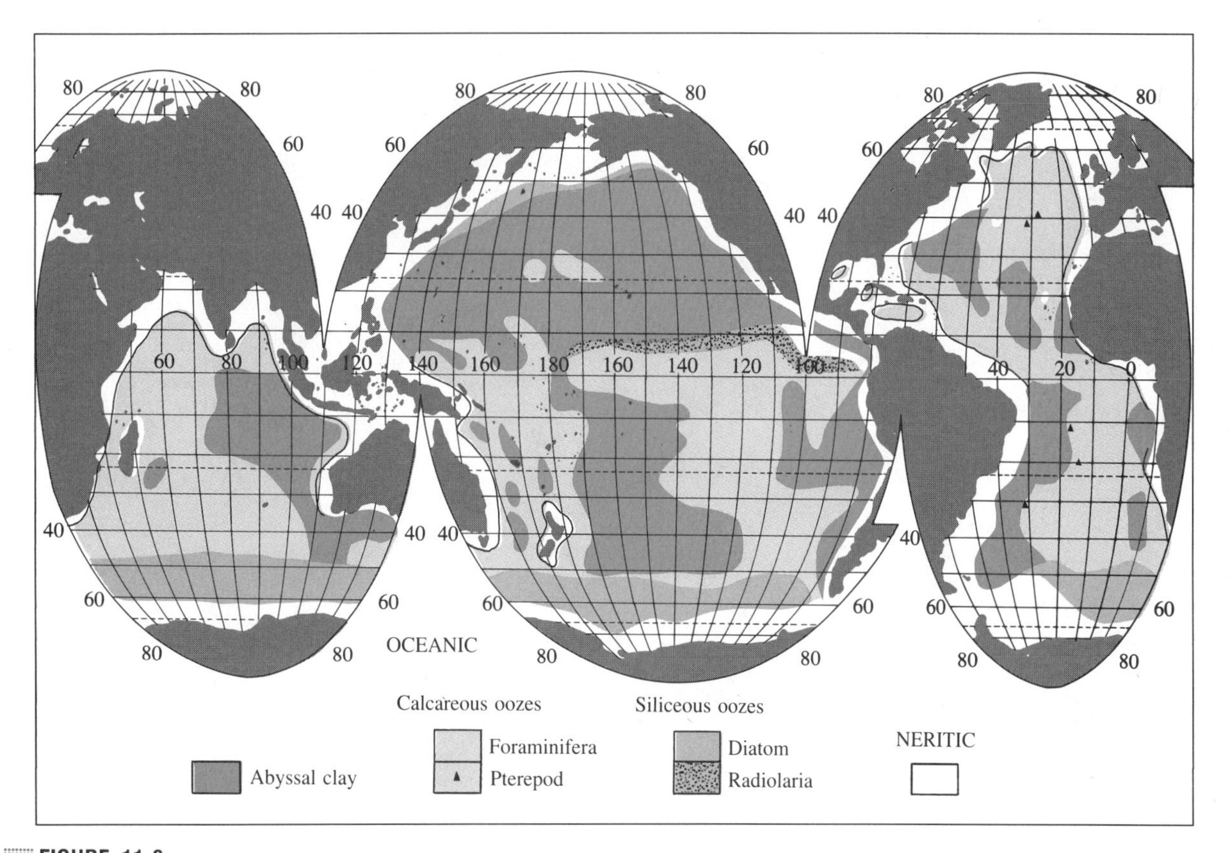

FIGURE 11.6
World Distribution of Neritic and Oceanic Sediments

(Base map courtesy of National Ocean Survey.)

THE DYNAMIC OCEAN FLOOR

In the late 1960s, a revolutionary process was discovered to be occurring at the bottom of the ocean: The ocean floor is moving! The earth's surface, including the ocean floor, is broken into six clearly identifiable large plates and at least six smaller ones (Fig. 11.7). Associated with the motion of these plates are the oceanic ridges, where the plates are supplied with new material by volcanic processes. The plates move away from the oceanic ridges, and by the time they have traveled across the ocean floor for 150–200 million years they descend back into the mantle beneath oceanic island arc-trench systems, or folded continental mountain ranges bordering the margins of continents. The rate of movement is not breath taking—1–10 cm/year (0.4–4 in). Because the process of movement involves the shifting of ocean floor away from the axes of oceanic ridges, it is often referred to as **sea-floor spreading.** It is much more than that, however, and a more appropriate term for the process that has built all the world's mountain ranges while creating earthquakes and volcanism is **global plate tectonics.** It has revolutionized marine geology, and provided marine biologists with an amazing new ecosystem to study—the deep-ocean hydrothermal vent community.

The Spreading Mechanism

Heat released by radioactivity deep within the earth finds its way to the earth's surface through convection. Heat rising beneath the oceanic ridges heats the materials of the upper mantle and crust, causing them to swell into the system of submarine mountain ranges that runs through the ocean basins. A rigid layer of rock including the basaltic ocean crust, the continental granitic crust, and the underlying upper mantle is generated by volcanic and intrusive activity near the axes of these mountainous ridges. It is called the **lithosphere** (rock sphere). After the lithosphere is generated at the oceanic ridges, it moves away at right angles to the ridge axis at rates of from 1–10 cm/yr. As it moves away, it cools, contracts, and thickens, and the ocean becomes deeper (Fig. 11.8). The North Atlantic Ocean is widening at a rate of about 3 cm/yr, which means that North America is moving west at about 1.5 cm/yr. It is possible for the rigid lithosphere to move across the earth's surface because of a plastic zone of mantle material beneath the lithosphere that is near its melting temperature. This **asthenosphere** (weak sphere) is able to flow without breaking if the rate of deformation is slow; its behavior is similar to that of Silly Putty.

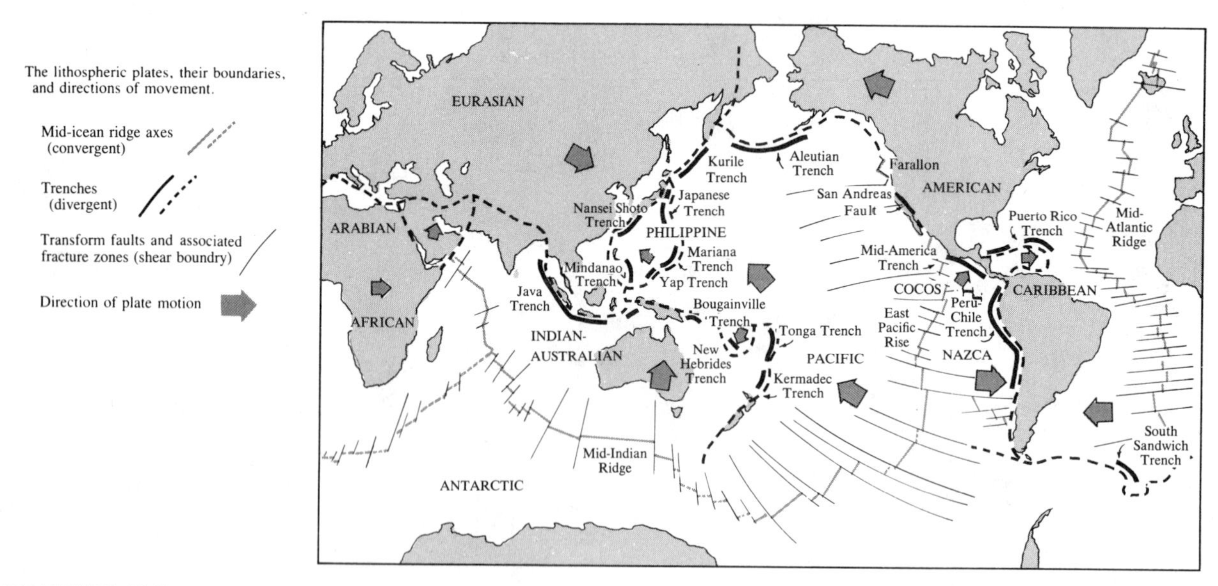

FIGURE 11.7
Lithospheric Plates and Their Direction of Movement

As plates move about on the earth's surface, they come together at convergent boundaries to produce oceanic trenches and associated island arcs. At divergent boundaries—oceanic spreading centers—the plates are moving apart. Plates slide past one another at transform fault boundaries (i.e., shear boundaries), which offset the axes of spreading centers.

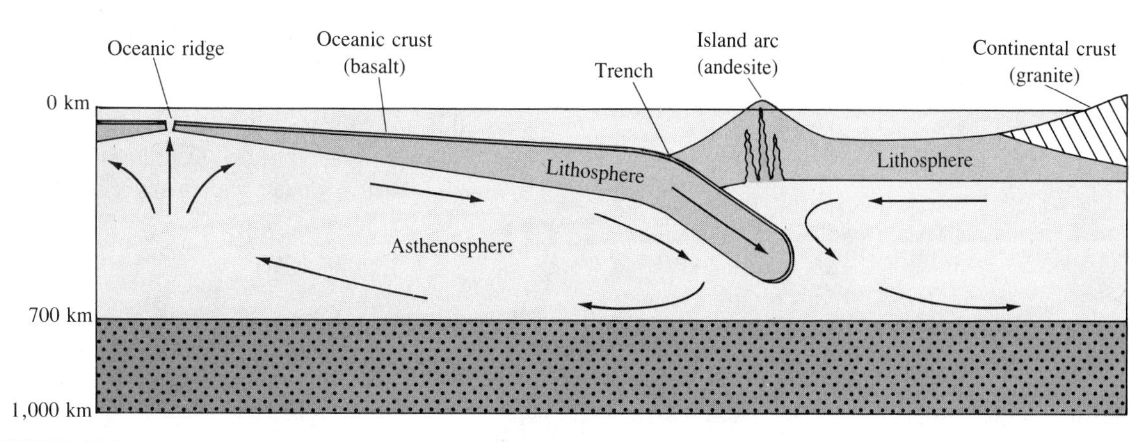

FIGURE 11.8
Movement of Lithospheric Plates

Lateral movement of the lithosphere, the rigid upper portion of the mantle including the crust, is made possible by lateral flowing of the asthenosphere beneath it. The asthenosphere is a low-strength plastic zone within the mantle, where pressure is low enough and temperature sufficiently near the melting point of the material to allow it to flow slowly.

Because the earth is a sphere, lithosphere moving away from one oceanic spreading center must ultimately collide with lithosphere originating from another. These motions set up stresses that break the lithosphere into six major plates: the African, Indian-Australian, Eurasian, Pacific, American, and Antarctic. There are at least six minor plates: the Arabian, Philippine, Juan de Fuca, Cocos, Nazca, and the Caribbean (Fig. 11.7). The boundaries of the **lithospheric plates** are the **oceanic spreading centers** where plates diverge, ocean trenches and folded mountain ranges at the margins of continents where they converge, and **transform faults** where plates slide past one another. At convergent boundaries, such as trenches and folded mountain ranges, one lithospheric plate **subducts** beneath another to descend back into the mantle and be melted.

Oceanic trench systems are usually associated with volcanic island arcs (Fig. 11.9). An example of this type of trench is the Tonga Trench. Others are so close to the margin of a continent that their volcanic arc takes the form of a chain of volcanoes along the margin of a folded continental mountain range. The Peru-Chile Trench is an example. Sometimes, subduction occurs

FIGURE 11.9
Earthquake Activity in Oceanic Trench Systems

Earthquakes occurring in association with a lithospheric plate descending beneath an ocean trench originate near the surface near the trench. Approaching the continent beneath which the subduction is occurring, parallel bands of deeper and deeper earthquake foci can be mapped. Below a depth of about 650 km, the descending plate melts sufficiently so that no earthquakes are produced.

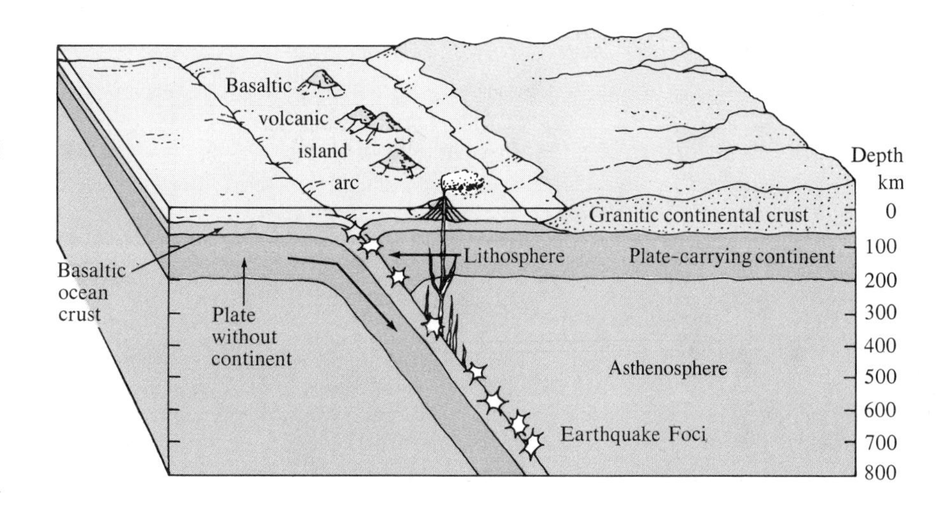

beneath a folded mountain range and no trench develops, as is the case where the Himalaya Mountains have formed at the convergent boundary of the Eurasian and Indian-Australian plates. The oceanic ridges are offset along transform faults. These faults are created by stress from the rotational motion of rigid plates moving across the spherical surface. A well-known example is the San Andreas Fault, which cuts across California from the head of the Gulf of California to Point Arenas, north of San Francisco (Fig. 11.10).

Volcanic and earthquake activity on the earth's surface is closely associated with lithospheric plate boundaries. The Cascade Range, a range of volcanic mountains extending from northern California through Washington, and including such volcanic peaks as Mount Ranier, Mount St. Helens, Mount Hood, and Mount Shasta, is the product of the subduction of the Juan de Fuca Plate beneath the American Plate.

DIVISIONS OF THE DEEP-SEA ENVIRONMENT

The benthic and pelagic components of the deep-sea environment are commonly subdivided into three regions. The bathyl (bathypelagic and bathybenthic) ranges from 1,000–4,000 m (3,280–13,120 ft); the abyssal (abyssal pelagic and abyssal benthic) from 4,000–6,000 m (13,120–19,680 ft); and the hadal (hadal pelagic and hadal benthic) below 6,000 m. In the bathyl and abyssal environments and in the hadal pelagic zone, the distribution of organisms is determined primarily by depth and distance from land masses. Although there are some distinct differences in the pelagic fauna of the bathyl, abyssal, and hadal pelagic subdivisions, these differences are not as important as the overall pattern of distribution of pelagic life. The deeper one travels, the fewer organisms that are found. Both species diversity

A.

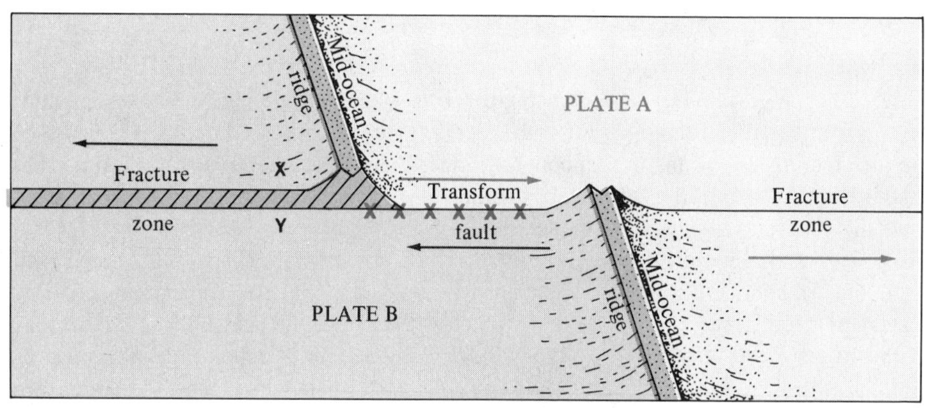

B.

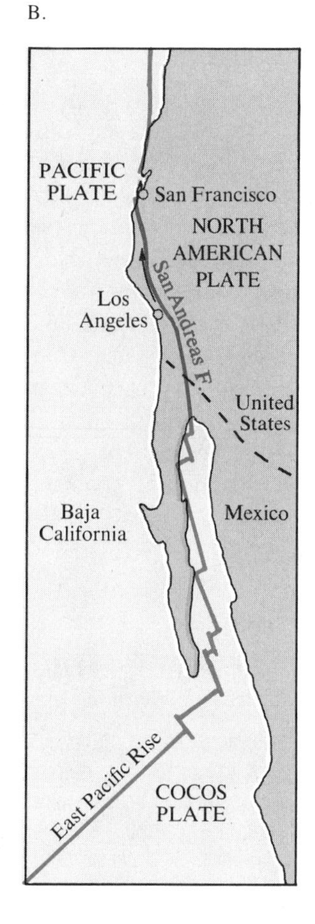

FIGURE 11.10
Transform Faults—Fracture Zones

(A) The axes of oceanic ridges are offset by transform faults due to the stresses resulting from the motion of rigid plates moving across the spherical surface of the earth. Earthquakes are common and of relatively great magnitude along the transform faults. They are called transform faults because they cut across the larger physiographic features of the ocean floor, oceanic ridges, and rises. Extending beyond the offset ridges are fracture zones, where earthquakes are rare because there is no relative movement on the opposite sides of the fracture zone. Fracture zones are scars of old transform fault activity and are maintained as significant topographic features because the ocean floor on one side of the fracture zone is older than the ocean floor immediately opposite it on the other side of the fracture zone. Point *X* represents younger ocean floor than point *Y*. Therefore, point *Y* has subsided more (because of cooling and thermal contraction) than has the ocean floor at point *X*. Thus, there is an escarpment along the fracture zone in Plate B that faces the older ocean floor on the side of the fracture zone where point *Y* is located. There is also an escarpment on Plate A, but it faces the opposite direction. **(B)** The San Andreas Fault is a classic example of a transform fault that offsets the oceanic ridge, or rise, and forms a boundary between plates that are sliding past each other. Along this contact, the Pacific Plate is moving north, relative to the North American Plate.

and abundance decrease steadily with increasing depth because species in shallower water have the first access to the food source, which comes only from the surface. Benthic species diversity is greater than pelagic species diversity in all regions.

In the benthic division of the deep sea, a number of patterns of zonation of benthic organisms have been described. The hadal component, consisting almost entirely of the deep-ocean trenches, includes the unusual **endogenous communities** that originated within and are confined to the trenches. Based on the amount of organic material present in the sediments, areas may be described as eutrophic or oligotrophic (Fig. 11.11). The eutrophic areas (which have sediment organic content of 0.5–1.5%) are found close to the continent, where turbidity currents feed organic material to the deep sea. Eutrophic areas are also found where surface productivity is high, that is, in areas of equatorial upwelling and in temperate and high latitudes.

The differences between eutrophic and oligotrophic sediment of the deep sea are striking. A biomass of benthic animals of up to 3.5 g/m^2 (0.0102 lb/ft^2) can be expected in eutrophic areas, compared with 0.2 g/m^2 (0.000702 lb/ft^2) in oligotrophic areas at the same depth. The oligotrophic areas, which are found primarily under the large oceanic gyres of surface currents, have sediments of a fine red clay and relatively few deposit feeders.

ADAPTATIONS TO THE DEEP-SEA ENVIRONMENT

The deep-sea environment is vast, cold, and dark, and organisms that live there must contend with high hydrostatic pressures and a scarcity of food. Before it was explored, the deep sea was thought to harbor sea monsters. Indeed, at first examination, many of the fish that have been discovered in the deep-ocean environment strike us as misshapen and grotesque by the standards of shallow-water fish. And yet if we look further into the morphological features of deep-sea animals, we find that many animals are not radically different in ap-

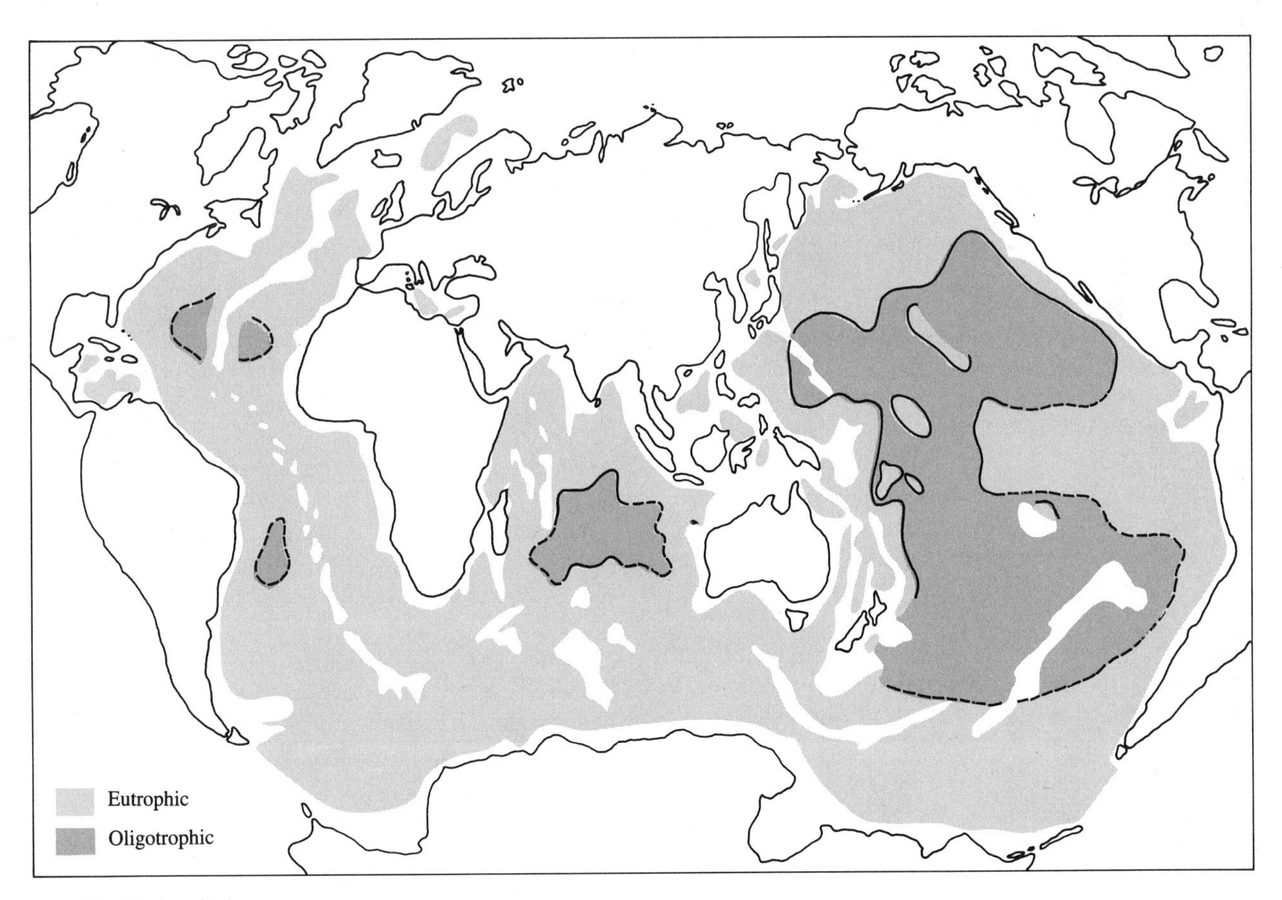

FIGURE 11.11
Food Energy in the Oceans

Based on the supply of food energy from surface waters, the deep ocean can be classified as oligotrophic (little food energy) or eutrophic (greater food energy).

pearance from similar shallow-water groups. A number of biochemical, physiological, and behavioral adaptations, however, are unique to deep-sea organisms.

Morphological Adaptations

Of the three groups of animals living in the deep sea (the benthos, zooplankton, and nekton), only the nekton have distinct morphological adaptations to the deep sea. The most striking feature of pelagic deep-sea fish is their relatively large jaw (Fig. 11.12). In an environment where large food particles are scarce, it is an advantage to take as much food at one time as possible. Deep-sea fish can swallow a food particle that is as large as they are. Some angler fish can take in a particle up to three times their size.

The angler fish, one of the very successful deep-sea fishes, is characterized by strong sexual dimorphism and the use of bioluminescence in food capture. Males are always smaller than females, and in many species he is reduced to being a parasite on the female. The larger female has a fishing lure on the head that attracts prey.

Demersal fish that live on or near the bottom are generally long and eel-like. Their peculiar morphological feature is the angle of the tail to the body, which allows them to swim with the head down, close to the bottom, and the trunk and tail raised above the bottom. One of the most peculiar groups of deep-sea fish is the tripod fish, which stands on its fins to avoid sinking into the mud.

The morphology of the deep-sea benthos is similar to parallel groups from shallow waters. Most phyla with benthic species in shallow water also have species in the deep-water benthos. The deep sea does favor some groups that are rare in shallower water. Crinoid echinoderms, for example, are more important in the deep-sea fauna. In the holothuroids (sea cucumbers), the elapsoids are more important in the deep-sea fauna. And the peculiar mollusc, *Neopilina,* unique because of its apparent segmentation, is found only in the deep ocean.

Some benthic species are relatively large. One polychaete worm *(Onuphis tores)* may be up to 200 cm (79 in) long, much larger than related shallow-water species. In early studies of the deep-sea benthos, reports of large species led to the assumption that *gigantism* is common in the deep sea. Smaller species are the rule, however, and giant species the exception.

Among the zooplankton, the only morphological difference associated with depth is a tendency toward a large body size. Copepods reach their largest size at about 2,000 m (6,560 ft). Euphausiids also get bigger with depth and may reach a length of 10 cm (4 in) in the deep sea. One deep-sea ostracod, *Gigantocypris,* is the largest species known from the group (8 mm, or 0.3 in).

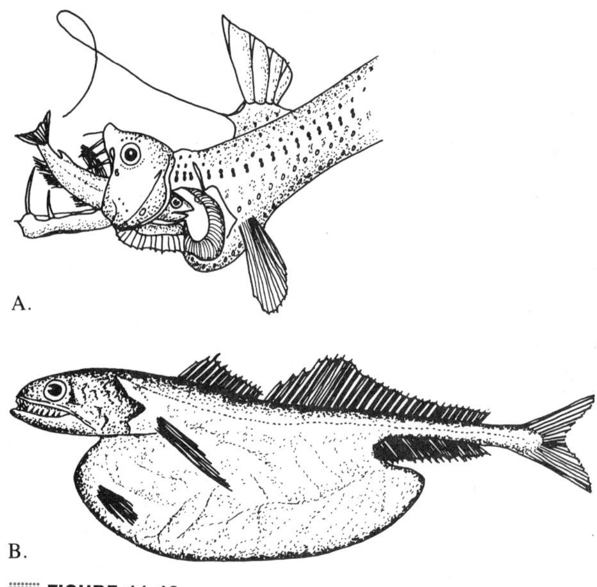

FIGURE 11.12

Deep-sea fish are capable of eating large particles. **(A)** The jaws can hinge in swallowing. **(B)** A prey larger than the predator distends the belly.

The sensory systems of deep-sea animals is different from their shallow-water counterparts. Because of the absence of light, loss of eyes is widespread, as it is in species adapted to cave dwelling. Most of the deep-sea benthic species are blind, whereas over 90 percent of benthic animals from the photic zone have some sort of light receptors.

In the deep-sea plankton and nekton, functional eyes are common, although they are much reduced compared to their striking development in fish of the mesopelagic zone. Although their function is poorly understood, the eyes seem to respond to the light of bioluminescence. The use of elaborate and complex 'fishing lures' by deep-sea angler fish testifies to the importance of bioluminescence in the deep sea: Prey must be able to see it.

The color of deep-sea animals is generally dark brown or black. These colors absorb the wavelengths of light produced by bioluminescence to the maximum, and probably help in avoiding detection and predation.

Bioluminescence

As a general rule, bioluminescence is not as frequent in the deep sea as it is in the photic zone. The back lighting that reduces the visibility of midwater fish is of no advantage in the aphotic deep sea. Besides the 100 or so species of angler fish, few other deep-sea pelagic fish are luminescent. Even fewer of the demersal deep-sea fish are luminescent.

Bioluminescence is frequently found among the deep-sea benthic invertebrates. A number of sea stars, brittle stars, and sea cucumbers are luminescent. Sea pens and anemones (cnidarians) and a large pycnogonid *(Colossendeis colosea)* are also known to produce light. The function of luminescence in the benthos is a matter of conjecture. The sea anemones attract prey by flashing. Other benthos may use light to startle potential predators. We have much to learn about the role of bioluminescence in the deep-sea benthos.

Smell and Sound

The sense of smell is well developed in a number of deep-sea species. Deep-sea eels, for example, use odor to identify and locate mates. In other deep-sea organisms, odor is used to identify and locate food. Scavenger species (fish, amphipods, and shrimp) can detect and quickly locate food. Bait lowered to the bottom at 1,800 m (5,904 ft) attracted some 800 amphipods in less than one hour. Although amphipods detected the bait first, rattail fish and shrimp also soon congregated around it. Congregation of scavenger fish around bait occurs to a maximum depth of 6,000 m (19,680 ft), and scavenger amphipods are attracted to bait down to a depth of 9,600 m (31,488 ft) (Fig. 11.13).

Sound may have a biological function in deep-sea fish. Most deep-sea species have a well-developed lateral line system for sound reception. The intraspecific communication in mesopelagic fish, however, is probably not important in deep-sea fish because few species have sound-sending mechanisms.

Reproduction

The reproduction of deep-sea animals reflects the low food supply available to larval and juvenile stages. The general pattern is for an animal to produce a relatively small number of large yolk-filled eggs. The larval stage is mostly bypassed, and the developing young use the food energy of the egg yolk to develop into small adult-looking juveniles. This pattern is followed by virtually all benthic invertebrates. The planktonic larval stage so common in shallow-water benthos is missing in deep-water species.

Deep-sea zooplankton show similar reproductive adaptations. The eggs of copepods do not hatch until the nauplius stage is passed. With euphausiids, the hatching young immediately adopt the adult feeding pattern of preying on copepods.

Although the deep-sea fish also tend to follow the pattern of a few large yolk-filled eggs and development to a relatively advanced stage, there are some important exceptions. Angler fish have a large number of small eggs that hatch into a larval form. These larvae migrate through the water column to the food-rich epipelagic zone. After about two months, the larvae descend to between 1,000–3,000 m (3,280–9,840 ft) and metamorphoze into the adult stage. Demersal rattail fish also have a larval stage that migrates to the epipelagic zone.

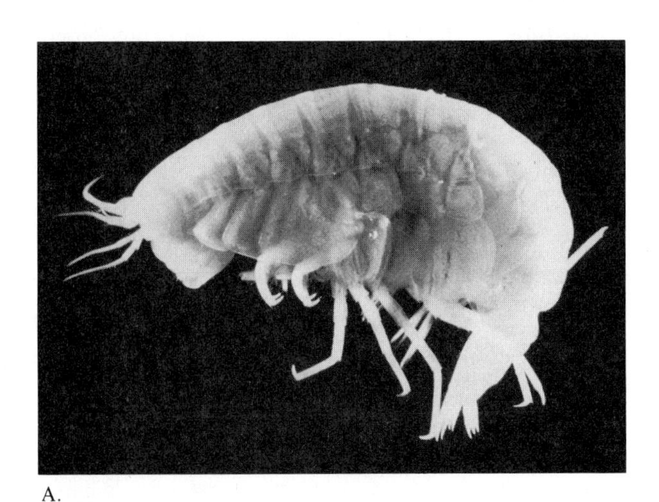

A.

B.

FIGURE 11.13
Deep-Sea Amphipods

(A) A close view of an amphipod recovered in a trap from the floor of the Marianas Trench below 10,400 meters (34,112 ft). **(B)** Amphipods swarm around bait placed on the floor of the Phillipine Trench at a depth of 9,600 meters (31,488 ft). (Photos courtesy of A.A. Yayanos, Scripps Institution of Oceanography, University of California, San Diego.)

Metabolism

One aspect of the physiology of deep-sea animals is dramatically different from shallower species: The metabolic rate of deep-sea animals is relatively low (Tab. 11.2). All deep-sea animals (except those living around hydrothermal vents) have rates of metabolism that are about 100 times lower than shallower species. These low metabolic rates are also found in bacteria. The rate at which bacteria degrade organic material and the rate at which benthic invertebrates and bathypelagic fish take up oxygen is also 100 times slower than in shallow water. In part this is a reflection of the higher content of water and lipids in the tissue of deep-sea animals. This reduces the percentage of body mass that must be maintained by metabolism. Deep-sea fish, in general, and a number of the pelagic species reduce their tissue density with depth. Their body tissues have lower concentrations of protein and skeletal material and higher lipid and water content.

Because of low rates of metabolism, deep-sea organisms are thought to live to great ages. Although few measurements of age have been made, the most celebrated example is the common small benthic bivalve *(Tindaria callistiformis)*. This clam takes 50 years to reach sexual maturity and 100 years to reach its maximum size of only a few millimeters in length.

Hydrostatic Pressure

Despite the high pressures of the deep, many fish still have functional gas-filled swim bladders, and siphonophores maintain neutral buoyancy by means of gas-filled floats. At the biochemical level it appears that deep-sea animals have special adaptations to hydrostatic pressure. Deep-sea animals are almost always dead when brought to the surface. This has been used as evidence that the relatively low pressures at the surface are lethal, but animals brought to the surface from the deep also suffer temperature shock. New methods of collecting animals in the deep sea—so they do not undergo temperature shocks—have been much more successful.

We do know that surface animals do not do well when exposed to deep-sea conditions. A range of shallow-water and surface animals (including zooplankton, fish, and mammals) show a similar response to increased hydrostatic pressures. Hyperactivity occurs at 50–80 atmospheres (500–800 m depth).

Even though death at the surface may not be due to pressure changes, experimental evidence makes it clear that the biochemical enzymes of deep-sea organisms work in a different way from those of surface forms (Fig. 11.14). In one particular pattern, the structure of an enzyme is determined by its number of subunits. These subunits require moderate to high hydrostatic pressure to stabilize the functional shape of the enzyme.

Buoyancy

The mechanisms used by deep-sea pelagic animals to reduce their rate of sinking, or to maintain neutral buoyancy, are on the whole similar to those mechanisms used by shallower species (Fig. 11.15). Deep-sea plankton are too large to gain an advantage, through Stoke's law. In general, all deep-sea pelagic animals have body tissues that reduce the density difference between body and seawater. Reduced protein content, relatively high water content, and relatively high lipid

TABLE 11.2
Oxygen Consumption Among Deep-Sea Animals

Group	Species	Minimum and Maximum Depth of Occurrence (m)	Experimental Temperature (°C)	Average Rate of Respiration (μl/gm/hr)
Copepod	*Bathycalanus bradyi*	900	3.0	3.90
Isopod	*Anuropus bathypelagicus*	200–800	3.0	8.00
Ostracod	*Gigantocypris agassizii*	900–1300	—	1.50
Hagfish	*Eptratretus deani*	1230	3.5	2.20
Macrurid	*Coryphaenoides acrolepis*	1230	3.5	2.40
Polychaete	*Hyalinoecia artifex*	800–1600	4.0	2.00
Solitary coral	*Thecopssammia socialis*	700–1200	4.0	4.62

NOTE: Deep-sea organisms generally have a low metabolism compared to shallow-water species, whose respiration rates range from 10–75 μl/gm/hr.

SOURCE: George, R.V. 1981. "Functional adaptation of deep-sea organisms." In Vernburg, F.J., and Vernburg, W.B., eds. *Functional adaptation of marine organisms.* New York:Academic Press.

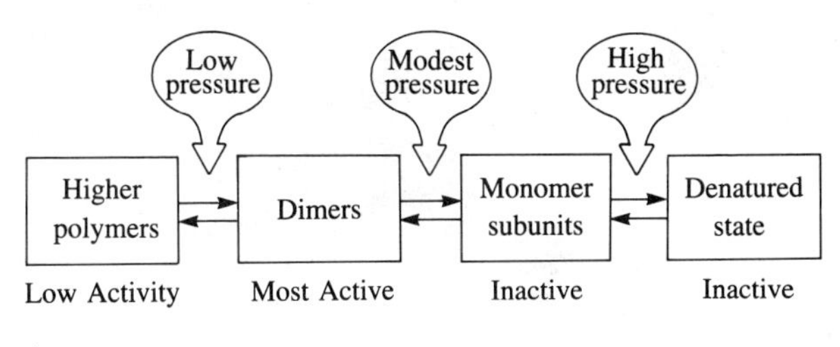

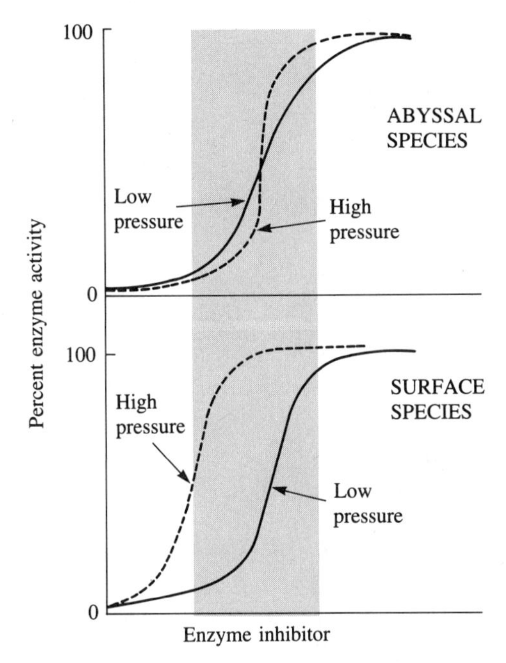

FIGURE 11.14
Hydrostatic Pressure and Enzyme Activity

(A) Hydrostatic pressure can affect the activity of enzymes through the binding properties of monomer subunits. **(B)** Deep-water species seem to be more independent of pressure effects on enzymatic action. (After George, 1981.)

content are common among deep-sea animals. One notable example is the group of deep-sea sharks. These fish spend much of their time near the bottom, but may range widely through the pelagic zone. These sharks have large livers with high concentrations of squalene, an oil with specific gravity of 0.86. These mechanisms well fit the low metabolic rate found in most deep-sea animals.

Of the deep-sea fish, the truly pelagic angler fish and *Cyclothone* species do not have gas-filled swim bladders, but 90 percent of demersal species have swim bladders. These demersal species are found to a depth of 7,000 m (22,960 ft). Swim bladders of deep-sea fish differ from those of many shallow-water species in one important respect: They are all closed. Many surface fish have a connection between the swim bladder and the esophagus that permits the rapid exchange of gases.

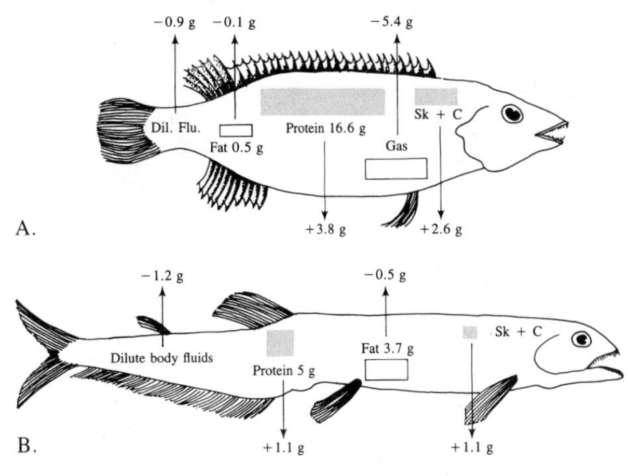

FIGURE 11.15
Buoyancy Mechanisms in Deep-Sea Fish

Deep-sea fish without a swim bladder achieve neutral buoyancy by controlling the density of their body tissues. **(A)** A shallow-water fish with a swim bladder. **(B)** A deep-water benthic fish with no swim bladder. Positive values (+) are for components heavier than seawater. Negative values (−) are for those components lighter than seawater. The values represent the lifting or sinking effect per 100 g of fish. (After Herring, P.J., and Clarke, M.R. 1971. *Deep oceans.* New York:Praeger.)

In deep-sea fish this connection is not found, and gas secretion and absorption is by a tissue called the rete mirabile.

That pelagic deep-sea fish do not have swim bladders while deeper demersal species do is most likely an adaptation to impoverished food supplies. Secreting gases into the swim bladder requires large expenditures of metabolic energy. Maintaining neutral buoyancy through body tissues with high water and fat content, conversely, has a relatively low energy cost.

FEEDING AND FOOD WEBS

Food sources in the deep sea are skimpy, and the availability of food is the principal factor that limits the number of species, as well as abundance and biomass, in the deep sea. Two important sources of energy provide food for the deep sea. The first is sunlight and photosynthesis in the euphotic zone. Gravity brings food particles to the animals of the deep sea. The second source of energy is organic molecules produced by chemosynthesis. Compared to photosynthesis, chemosynthesis is more limited.

In one sense, all the deep-sea animals are directly or indirectly detritus feeders (Fig. 11.16). They rely on the force of gravity to bring dead organisms (or fragments of dead organisms) to the bottom. However,

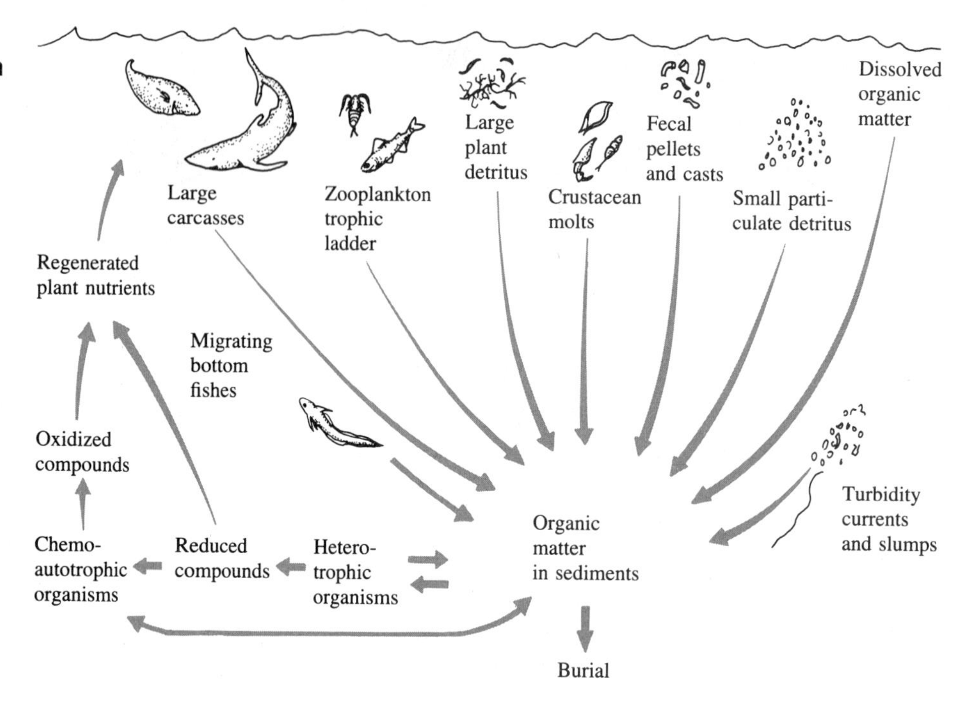

FIGURE 11.16 Food Sources of Deep-Sea Animals

The sources of energy for food of deep-sea animals are limited to shallower waters.

predatory behavior will usually occur if the opportunity presents itself. This detritus may take the form of a carcass (i.e., fish, shark, or whale), fragments of plankton (called 'snow'), or fecal matter from different animals. In addition to this pelagic contribution to the food of the deep sea, there is a contribution from the benthos. On the continental shelves, rates of sedimentation are relatively rapid. In many areas, sediments build up at the edge of the shelf or at submarine canyons. This sediment material is periodically carried down the continental shelf slope and onto the abyssal plains. These sediments generally have high organic concentrations compared to the sediments of the abyssal plains, and are an important source of food for the benthos.

Turbidity currents probably deliver more food to the deep-sea environment, through **slumping,** than does the rain of particulate matter from the surface plankton. The highest concentration of organic carbon in the sediments and the highest density of abyssal benthos is found close to the margin of the continents. The lowest organic content in the sediment and the lowest density of benthos is found far away from the continents under the midoceanic gyres. These areas are characterized by fine red clay sediments, with only 0.3 percent organic matter. These abyssal clay sediments have the lowest levels of biomass (0.01 g/m^2) found in the marine benthos.

The relative productivity of surface waters affects the richness of deep-sea life. Areas of the deep sea under eutrophic surface waters have a greater diversity and biomass of life than areas under the less productive midoceanic gyres. Deep-sea sediments under the areas of equatorial upwelling and in temperate and polar latitudes are enriched compared with those of lower latitudes, particularly beneath midoceanic gyres.

Since the majority of biomass in the photic zone is plankton, we will consider in more detail this contribution to food in the deep sea. Scientists who have directly explored the depths of the mesopelagic zone and the upper areas of the deep sea (in deep submersibles) have observed clouds of flocculent material in the water so thick that it resembled snow. This rain of 'snow' is believed to be dead and fragmented plankton on its way to the bottom. Initial observations gave the impression that this 'snow' contained a great deal of organic material, but it is now apparent that only a small fraction of organic material produced in surface waters reaches the deep sea. The estimates of the recycling of organic matter in the photic zone show that between 87–95 percent of all organic carbon produced by photosynthesis in the euphotic zone is consumed in the top 400 m (1,312 ft) of water. Very little reaches the deep sea. This is not surprising, given the slow rate of fall of small detritus particles, 1–3 meters (3.3–10 ft) per day. At these rates, it takes ten years to reach a depth of 4,000 m (13,120 ft).

Not all detrital particles sink so slowly, however. Copepod fecal pellets sink at a rate of 50–900 m (164–2,952 ft) per day and may reach depths of 4,000 m (13,120 ft) in 18–80 days. In fact, the majority of the detrital particles that do finally reach the sea floor are copepod fecal pellets. Since copepods are by far the most numerous of surface zooplankton animals, they make the largest contribution to the rain of fecal pellets.

Most of the large animals, such as demersal fish and the epifaunal echinoderms, obtain their food by eating the surface layers of the sediments directly. This is the food habit of most groups found in the deep-sea benthos. The food source of these deposit feeders may be the particles of detritus themselves or, more likely, small benthic organisms that feed on the detritus. We know, for example, that bacteria are common in deep-sea sediments, and more numerous in sediments with relatively higher organic content. We know also that meiofauna are more important in deep-sea sediments than in shallow waters. It is likely that the meiofauna are feeding on the bacteria that use the detritus for food energy. The role of foraminiferan Protozoa is thought to be of particular importance in this chain. Finally, we know that the majority of the infaunal organisms are small (less than 10 mm, or 0.4 in) and consist primarily of deposit-eating polychaetes and bivalves. It is likely that the infaunal organisms feed primarily on the meiofauna and on bacteria of the benthos. Most of these feeding transactions occur in the top centimeter or two of the sediments. The deposit-feeding larger epifauna (echinoderms and fish) probably feed, then, on a range of animals, the most common being bivalves, polychaetes, and foraminiferan Protozoa.

Carcasses that fall to the deep-ocean floor are eaten first by scavengers. Shrimp, amphipods, and fish (rattails and brotulids, in particular), using their sense of smell, quickly congregate around these large food sources. It is thought that some scavengers require a rich food meal to complete the life cycle. With scavenging amphipods, at least, most individuals on the deep-ocean floor are one stage of development removed from sexual maturity. The presence of a meal provides adequate energy to complete the production of egg and sperm.

The scavengers are important in exporting food energy to the adjacent benthic infauna. The fecal products of the scavengers are deposited at some distance from the large central food source and serve as an important food source for deep-sea benthos.

Not all deep-sea benthic organisms are deposit feeders. Filter-feeding sponges, barnacles, crinoid echinoderms, and serpulid polychaetes are widespread in the deep sea. They take small zooplankton or suspended detrital particles. Although filter feeders are most likely to be found in areas with relatively strong currents, it is still difficult to understand how they obtain adequate food because pelagic zooplankton is sparse. It may be that there is a special zooplankton community associated with the ocean floor. Samples of copepods have been collected from just above the bottom by the deep submersible *Alvin*. These 'benthic' copepods, which are more abundant than in the pelagic zone away from the bottom, may feed on detritus particles that are resuspended by currents or other factors.

Carnivores are infrequent in the deep-sea macrobenthos. Carnivorous fish and sea stars are generally restricted to the shallower parts of the deep sea (to a maximum depth of 3,000 m, or 9,840 ft). Carnivorous gastropod snails are well represented at all depths. They feed mostly on polychaete worms and seldom reach a size of more than 5 mm (0.2 in).

The bulk of the food supply to copepods in the deep sea is fecal pellets. These fecal pellets, rich in bacteria, are probably the major source of nutrition. Much remains to be learned about the mechanisms by which deep-sea copepods obtain their food.

Copepods, in turn, serve as the principle food source of other deep-sea zooplankton, such as arrow worms (chaetognaths), euphausiids, ostracods, and prawns. The top predators are the angler fish (which eat a range of foodstuffs including zooplankton and fish) and the black bristlemouth *(Cyclothone)*.

ORGANISMS OF THE DEEP OCEAN

Zooplankton

The relatively low number of species of zooplankton found in the deep sea is related to two general observations. First, there are fewer pelagic animals than benthic animals in all of the animal phyla. Second, there is a continual decline in species diversity, density, and biomass of animal life with increasing depth. Among the cnidarians, both scyphozoan medusae and colonial siphonophores are found in the deep-sea plankton. These species of medusae, which may be up to 25 cm (10 in) in diameter, do not have the sessile polyp stage in the life cycle that is characteristic of shallow-water species. Deep-sea medusae are often maroon or purple in color, have firm mesoglea, and are luminescent (Fig. 11.17). They are known to be predators of fish. Siphonophores, which have a number of deep-water species, use gas-filled floats for bouyancy despite the tremendous hydrostatic pressures of the deep.

The comb jellies, or Ctenophora, are poorly represented in the deep-sea plankton, with only two genera reported. Chaetognaths, which are important in all areas of the epipelagic and mesopelagic zones, are also numerous in the deep-sea plankton. They are widespread in occurrence in the deep sea, and have been collected to depths of 6,000 m (19,680 ft).

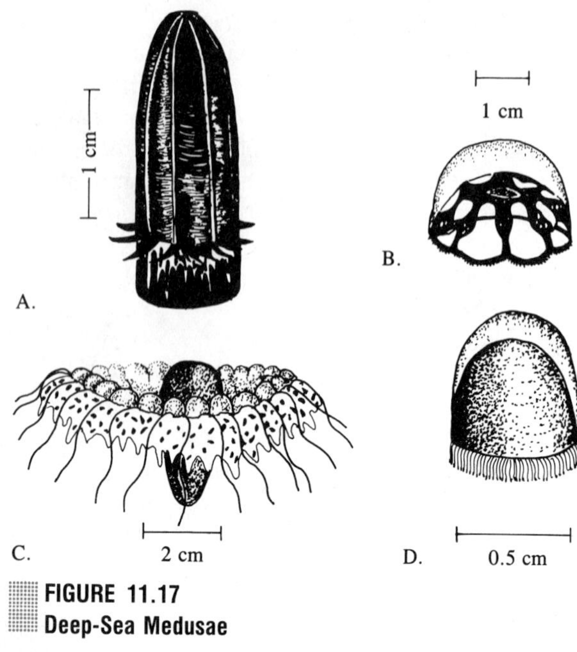

FIGURE 11.17
Deep-Sea Medusae

Deep-sea jellyfish are more colorful and are often larger than shallow-water species. **(A)** *Aglisera* is a flaming red color. **(B)** *Halicreas* has bright red markings. **(C)** *Atolla* is red, cream, and purple. **(D)** *Crossota* is dark brown.

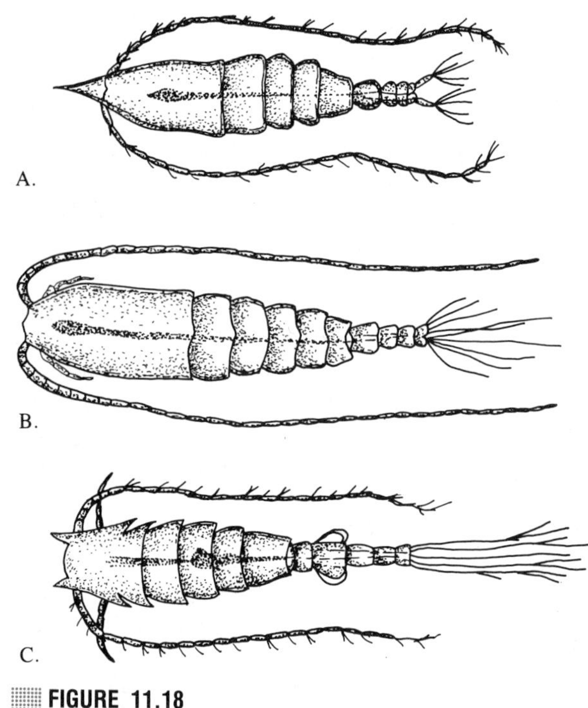

FIGURE 11.18
Deep-Sea Copepods

Most deep-ocean copepods are carnivorous. These three deep-water genera are similar in shape to shallow-water copepods. **(A)** *Haloptilus.* **(B)** *Megacalanus.* **(C)** *Lucicutia.*

Copepods are the dominant group in the deep-sea zooplankton (Fig. 11.18). The largest copepod species are reported from the deep sea. At 2,000 m (6,560 ft) depth, a maximum size of 17 mm (0.7 in) is reached. Below this depth, size decreases. Deep-sea copepods are either predators or detritus feeders.

Although there are mysids and euphausiids in the deep-sea plankton, they are not as important as the ostracods. Ostracods, which are relatively uncommon in the plankton of shallow waters, make up about 10 percent of the deep-sea plankton. Perhaps the most striking of the deep-sea ostracods are of the genera *Macrocypridina* and *Gigantocypris* (Fig. 11.19). These ostracods may be up to 8 mm (0.32 in) in length and, unlike shallow water species, have large eyes.

Pelagic gastropod snails (heteropods and pteropods) and tunicates, which are important and common in the shallow water zooplankton, are not found in the deep sea.

Nekton

The actively swimming animals of the deep sea include crustaceans, prawns, cephalopod squid, octopods, and fish. There are important differences between those that are pelagic and those that are demersal. (Demersal species will be considered later in the chapter.)

In the pelagic zone, cephalopods (squid and octopods) are common in both the epipelagic and mesopelagic zones of the top 1,000 m (3,280 ft). Although not as common in the deep sea, they are still important (Fig. 11.20). The so-called vampire squid has been observed to depths of 5,000 m (16,400 ft), and one species of octopus *(Cirrothauna)* is blind. Deep-sea cephalopods are less muscular and more fragile than those of shallow water. Undoubtedly, the deep-sea cephalopods move more slowly than shallow species. Crustacean shrimp or prawns, which are not particularly common in the nekton

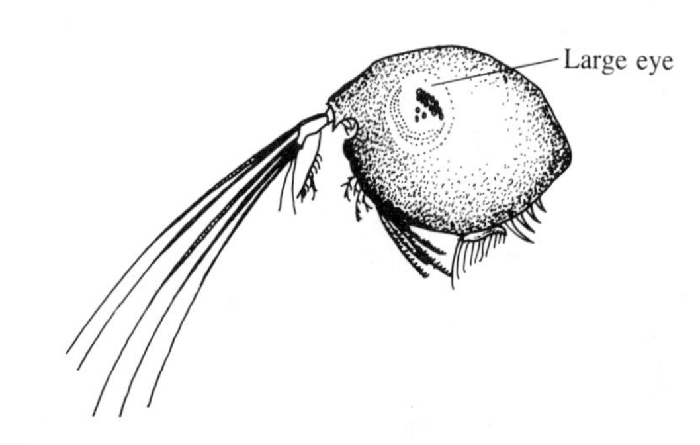

FIGURE 11.19

Ostracods, which are rare in the surface plankton, are relatively common in the deep sea. The species *Macrocypridina* is 6–7 mm in size, much larger than shallow-water ostracods.

FIGURE 11.20
Deep-Sea Cephalopods

Many of the deep-sea cephalopods have a specialized morphology. **(A)** *Amphitretus.* **(B)** The swimming octopus *Cirrothauma,* the only blind octopus discovered, is gelatinous and floats like a jellyfish. **(C)** *Chiroteuthis.*

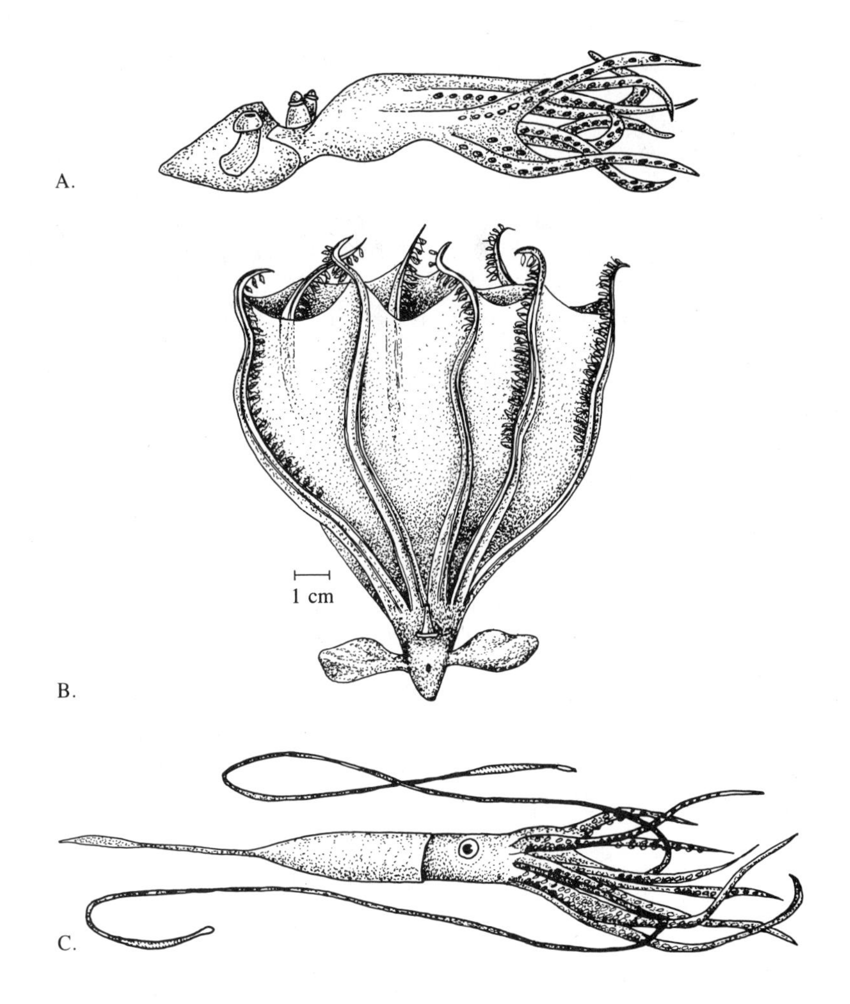

of the photic zone, are also uncommon in the deep sea, although they have been found throughout the deep-sea pelagic zones from depths of 1,000–5,000 m.

The predominant nekton of the deep sea, as in shallow waters, are the fish. As with other groups of marine organisms, diversity is lower in the deep sea than in the photic zone. Some 150 species of fish live in the pelagic zone of the deep sea. This is only 10 percent of the kinds of fish found in the mesopelagic zone.

Of the 150 species of deep-sea fish, 75 percent are angler fish (Fig. 11.21). All angler fish have a stalked lure that is tipped with bioluminescent 'bait'. This lure, which has evolved from the forwardmost ray of the dorsal fin, is used to attract food to the fish. Although the size and shape of the 'rod' and the 'bait' vary a great deal, the anglers do not 'fish' for a particular prey. Angler fish hang motionless with their large mouths open waiting for an animal to be attracted to the lure. A large range of prey, including zooplankton and other fish, are taken. Because of the huge jaw, the angler fish is able to swallow a prey as big or even bigger than itself.

Female angler fish may reach 6–8 kg (13.2–17.6 lb), but males do not exceed a few grams. Males are more streamlined fish with well-developed eyes and olfactory senses in the juvenile stage. Parasitic males attach to the female by burrowing head first through the

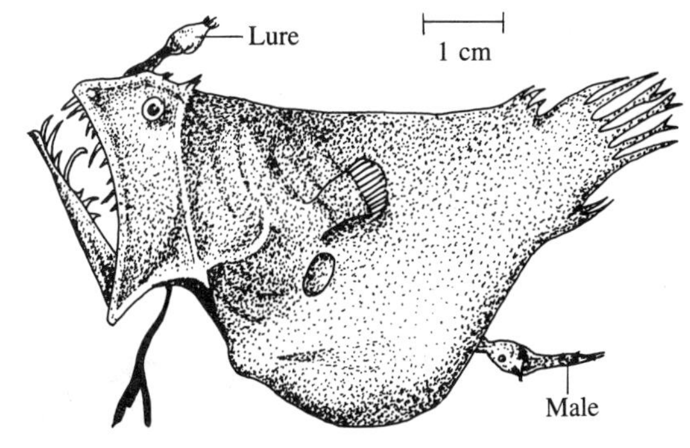

FIGURE 11.21
Angler Fish

The most diverse of the deep-sea pelagic fish is the group known as angler fish. The females are grotesquely shaped. The males are small and free-living or are parasitic on a female. Most female angler fish have a rodlike lure that is luminescent.

skin and then degenerate into little more than a bag of sperm fed by the female. Sperm are released when the female sheds its eggs.

Like other deep-sea pelagic species, angler fish are most common in the shallower regions of the deep sea (1,000–2,000 m). They do not occur in schools. In fact, they space themselves about 30 m (98 ft) apart. They probably use the bioluminescence of the lure and other photophores on the body to maintain their spacing.

Although the greatest species diversity of deep-sea pelagic fish is found in angler fish, they are not the most numerous. The greatest number of individuals belong to the genus *Cyclothone* (Fig. 11.22). Cyclothone species are also common in the mesopelagic zone. In the deep sea they are dark in color, while in the mesopelagic zone, they are mostly silver. Smaller than angler fish, cyclothone have a maximum length of between 5–6 cm (2.0–2.4 in). Black cyclothones of the deep sea have relatively larger heads than shallower species. They feed mostly on zooplankton (chaetognaths and ostracods). The gill rakers are developed into an effective filtering mechanism that can retain prey as small as 1 mm (0.04 in) in diameter. Cyclothone are sexually dimorphic, females being about twice the size of males. Cyclothone do not school; instead, they space themselves apart in the deep sea at a distance of about 3 m (9.8 ft) from their neighbors. Cyclothone are well distributed through the depths of the deep sea and are found between 1,000–7,000 m (3,280–22,960 ft).

Gulper eels, with their large jaws, can take prey ranging from zooplankton to other fish (Fig. 11.23). They are true eels, and some species have the characteristic large leptocephalus larvae that migrate to shallow water and return to the deep at metamorphosis. Other deep-sea pelagic fish are the whale fish (which have jaws with small teeth and feed on zooplankton) and the frog fish. The general features of these deep sea fish are compared to shallower species in Table 11.3.

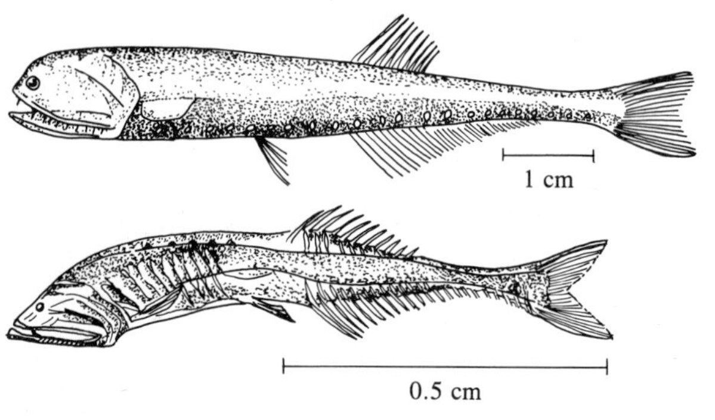

FIGURE 11.22
Cyclothone

Although not as diverse as angler fish, the black bristlemouth of the genus *Cyclothone* are more numerous.

Benthos

The bottom of the deep sea is an immensely large habitat, encompassing some 80 percent of the area of the ocean floor and 60 percent of the earth's surface. Most of the deep-sea bottom is covered with fine mud. A few limited areas—the steep sides of trenches, the current-swept surfaces of sea mounts, and the freshly produced bedrock of oceanic rises—are solid surfaces. At these locations, the benthic animals that require a solid substrate for attachment are found. However, the great majority of deep-sea benthic animals must contend with soft sediments. Animals find an easier existence in the deep-sea benthos than in the pelagic area above, as shown by the greater species diversity, density, and biomass of life in the deep-sea benthos compared to deep-sea pelagic life. Benthic communities are found even in the deepest trenches, although there is a continual decrease in number of species and abundance with increasing depth.

For a profile of an extraordinary deep-sea researcher, see the special feature located at the end of this chapter, "Ruth D. Turner—Investigator of Deep-Sea Benthos".

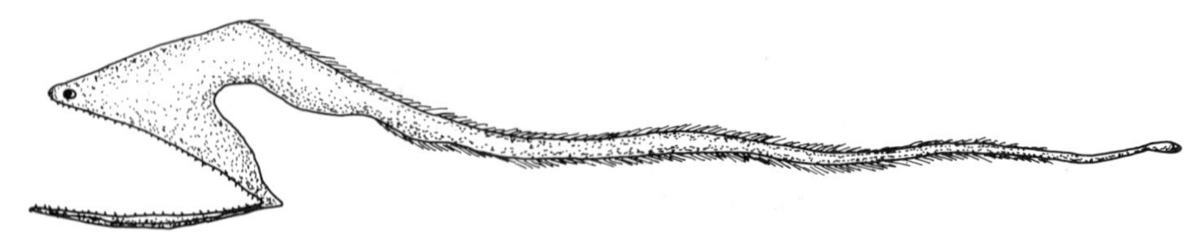

FIGURE 11.23
Gulper Eel

Some deep-sea pelagic fish are misshapen compared to shallow water fish. The gulper eel is the largest, reaching lengths of 60 cm.

TABLE 11.3
Features of Deep-Sea Fish

Features	Mesopelagic, Plankton-consuming Species	Bathypelagic Species
Color	Many with silvery sides	Black
Photophores	Numerous and well developed in most species	Small or regressed in gonostomatids; a single luminous lure on the female of most angler fish
Jaws	Relatively short	Relatively long
Eyes	Fairly large to very large with relatively large dioptric parts and sensitive pure-rod retinae	Small or regressed, except in the males of some angler fishes
Olfactory organs	Moderately developed in both sexes of most species	Regressed in females but large in males of *Cyclothone* species, and angler fish (most species)
Central nervous system	Well developed in all parts	Weakly developed, except for the acoustico-lateralis centers and the forebrain of macrosomatic males
Body muscles	Well developed	Weakly developed
Skeleton	Well ossified, including scales	Weakly ossified; scales usually absent
Swim bladder	Usually present, highly developed	Absent or regressed
Gill system	Gill filaments numerous bearing very many lamellae	Gill filaments relatively few with a reduced surface
Kidneys	Relatively large with numerous tubules	Relatively small, with few tubules
Heart	Large	Small

SOURCE: After Marshall, N.B. 1979. *Deep sea biology*. London:Blanford University Press.

Bacteria. Bacteria are present in sediments at all depths of the deep sea, and are the principal organisms responsible for decomposition. It is still unclear whether the benthic deep-sea bacteria are specifically adapted to deep-sea conditions or whether they are surface types that are simply tolerant of high hydrostatic pressures and low temperatures. Some studies have found deep-sea bacteria to be similar to those of shallow water or even land. These *barophilic* bacteria (tolerant of deep-sea conditions) show the low metabolic rates characteristic of deep-sea life. Other studies have found barophilic species with higher metabolic rates. It may be that barophilic species with higher metabolic rates are suited to deep-sea benthic regions that are richer in organic material. One of the most favorable locations for bacterial growth is the digestive tract of animals. Bacteria isolated from the guts of deep-sea amphipods have the highest metabolic rate of barophilic species.

Protozoa. The benthic Protozoa of the deep sea are similar to shallow-water species. Both ciliate and ameboid protozoa are found. One remarkable group of ameboid Protozoa commonly found in the deep-sea benthos is the Xenophyophoria. These ameboid forms may be up to 25 cm (10 in) in diameter, not including the pseudopodia extending from the body (Fig. 11.24). Radiolarians and tintinnids are not found in the sediment. Foraminifera are especially abundant in the deep-sea benthos and their abundance increases with increasing depth. In abyssal areas, Foraminifera may be the most common benthic organism and have a biomass greater than all other benthic organisms combined.

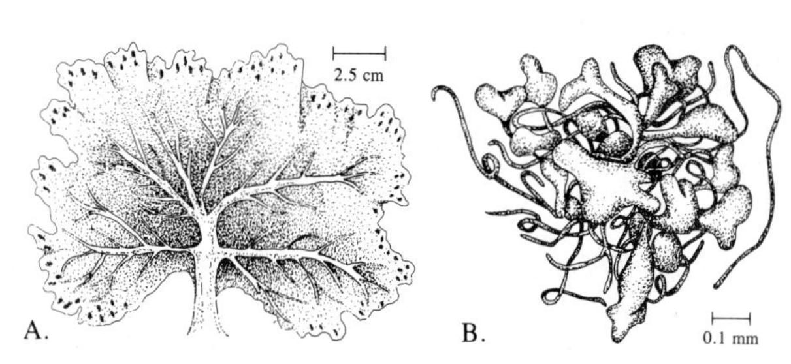

FIGURE 11.24
Deep-Sea Protozoa

Some large peculiar-shaped Protozoa have been discovered in the deep. **(A)** Rhizopod protozoan; **(B)** Foraminiferan. (After Marshall, 1979.)

Meiofauna. Although the number of meiofauna species decreases with depth, the decrease is not as rapid as that of the larger benthic animals. Meiofauna are more abundant in the deep-sea benthos than in shallow waters. Nematodes are generally most common, followed by harpactacoid copepods. The majority of meiofauna live in the top 1 cm (0.4 in) of the sediments, and they do not occur deeper than 5 cm (2 in). Species diversity of the meiofauna increases with depth. Whereas in shallow water benthos there are a large number of a few species, in the deep-sea benthos there are many more species with fewer individuals.

Large Benthos. Most invertebrate phyla have some species in the large benthos of the deep sea, at least to a depth of 6,000 m (19,680 ft). Below this depth, some groups, such as brachiopods and crustaceans, do not occur. As with other groups of animals, species diversity and abundance decrease with increasing depth. In most cases the large benthos of the deep sea are unique to that zone, although they are similar in morphology to species found in shallow water. One basic difference between shallow and deep-sea benthos is that, while 90 percent of shallow-water species of benthos have some kind of photoreceptor or eye, most deep-sea species are blind. Less than 10 percent of deep-sea species can detect light.

Some benthic groups are more common in the deep sea than in shallow water (Fig. 11.25). For example, crinoids and pogonophorans have their centers of distribution below depths of 1,000 m (3,280 ft). One group of sea cucumbers, the elasipoid holothurians, are found only in the deep-sea benthos.

The sponges (phylum Porifera) most common in the deep sea are the glass sponges (Hexactinellids). Glass sponges, which have been collected below 6,000

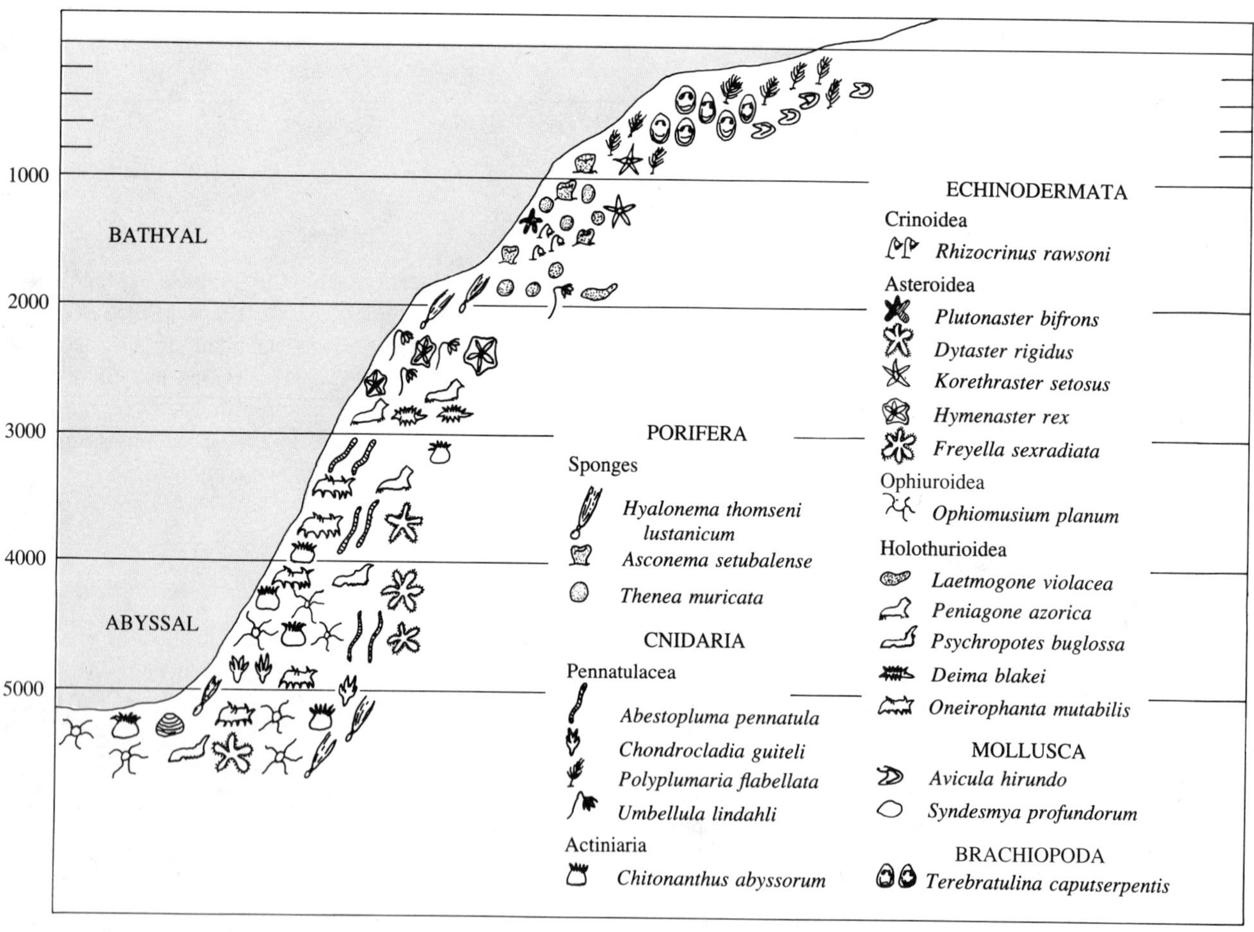

FIGURE 11.25
Distribution of Deep-Sea Invertebrates

Some invertebrate groups are more common in the deep sea than in shallow water. With increasing depth, the echinoderms become more important. (After Heezen, B.C., and Hollister, C.D. 1971. *The face of the deep.* New York:Oxford University Press.)

m (19,680 ft), may be up to 1 m (3.3 ft) in length. All species are epifaunal. They use a stalk to keep themselves out of the mud or are attached to a rocky substrate. Glass sponges are rare in those areas with low surface productivity.

The demospongiae, which occur mostly as encrusting forms in shallow water, are, curiously enough, found in the areas with the poorest food supplies. The more efficient filtration system of the demospongiae may permit these filter feeders to persist in low food conditions. Like shallow-water species, all deep-water sponges are filter feeders.

The most widely observed group of the phylum Cnidaria in the deep sea is the Anthozoa (anemones and corals). Both soft and hard corals have representatives in the deep sea. Soft corals are limited to areas of relatively rapid current movement. Hard corals are found only in solitary form (shallow-water forms are colonial) and are limited to a depth of 5,000 m (16,400 ft) or less. Deeper than this, the calcium carbonate of the skeleton spontaneously dissolves. Both soft and hard corals are suspension-feeding predators on zooplankton.

Sea anemones have been found at the maximum depths of the deep sea (more than 10,000 m). Most species attach to other epibenthic animals (sponges or crinoids) or a rocky surface. Some regular sea anemones are able to live in the mud, and the tube-dwelling types are also found in the deep sea. Little is known about the food habits of sea anemones in the deep sea although some species may be deposit feeders.

The wormlike minor phyla (Sipunculids, Priapulids, Enteropneusts, and Echiurids) have representatives in the deep-sea benthos. All are detritus feeders. The enteropneusts are abundant at depths greater than 4,000 m (13,120 ft) in the South Pacific and at high southern latitudes.

Polychaete worms, abundant in the soft sediment of shallower waters, are equally important in the benthos of the deep sea. Deposit-feeding species, however, are more frequent in the deep sea. Deposit feeders predominate in the organically enriched sediment, while the few species that are suspension feeders (serpulid worms) are more likely to be found in impoverished sediments. Polychaetes have been found in the deepest parts of the ocean bottom.

All classes of the phylum Mollusca are well represented in the deep-sea benthos, both as infauna and epifauna. Most species are small compared to shallow water species (deep-sea bivalves, for example, seldom exceed 5–6 mm in length). These bivalves are primarily deposit feeders while shallow-water species are mostly suspension feeders. Snails have some species that are deposit feeders and some that are carnivores. Of particular interest is the limpetlike genus *Neopilina*. Discovered only some 30 years ago, it is unique among molluscs in showing signs of a segmented body. It has five pairs of gills along the edge of the mantle cavity. Gastropods are found down to the deepest parts of the deep sea, although the shells of species from the trenches (6,000–10,000 m) are more fragile and lightly calcified. This is likely due to the high solubility of calcium carbonate below 5,000 m (16,400 ft).

The phylum Arthropoda, although well represented in the deep sea, is not as well represented as in shallow water. Pycnogonids are found in the deep sea crawling across the mud (Fig. 11.26). Both amphipods and isopods are abundant in the deep-sea benthos (Fig. 11.27). Amphipods are both scavengers and deposit feeders. Scavengers can move rapidly and have been photographed in large clouds around bait set out at great depths. Isopods are found at all depths and are deposit

FIGURE 11.26
Pycnogonid

Deep-sea pycnogonids are similar in form to shallow-water species. Shown is *Colossendeis,* a large deep-water genus.

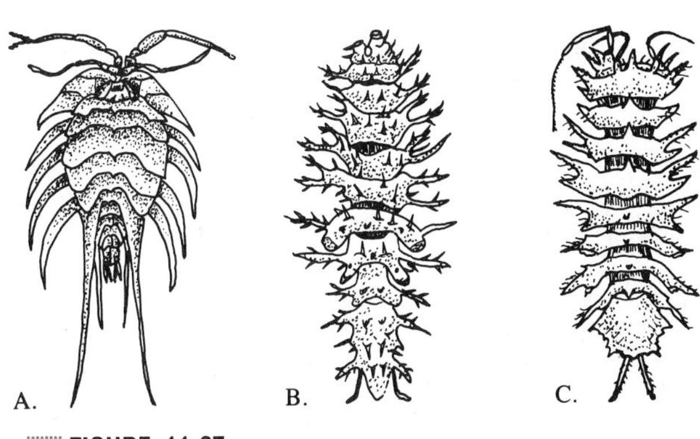

FIGURE 11.27
Deep-Sea Isopods

Isopods, as well as amphipods, are important in the deep-sea benthos. **(A)** *Munnopsis;* **(B)** *Ianirella;* **(C)** *Iolanthe.*

eaters (Fig. 11.27). The color of deep-sea isopods ranges from dull grey to white, whereas shallower species are generally darker (red or black).

Barnacles are important in the deep-sea benthos. There are some 50 species (in two genera) that are distributed to 7,000 m (22,960 ft) depth. Like shallow-water barnacles, deep-sea species are filter feeders. They attach to virtually any hard surface, including other animals such as sponges, coral, tunicates, bryozoans, and pogonophorans.

Decapod crustaceans, so abundant in the shallow-water benthos (crabs, shrimps, and prawns), have a limited distribution in the deep sea. Crabs are not found deeper than 4,300 m (14,100 ft) and shrimps and prawns are found only to 6,000 m (19,680 ft). The limited distribution in the deep sea is thought to be related to the scavenger feeding pattern of decapods. Scavengers have little food at great depths.

The single most important and conspicuous phylum found on the surface of the deep-sea sediments is the Echinodermata. All of the echinoderm masses found in shallow water are also found in the epifaunal deep-sea benthos. One class, the crinoids, has its maximum diversity in the deep sea. Crinoids are filter feeders and keep themselves off the bottom by a stalk that is attached directly to a hard surface or stuck into the mud (Fig. 11.28). Crinoids feed by an intriguing method. The group called the feather stars extend their arms perpendicular to the direction from which food comes. In currents, the arms face into the moving water. In quiet water, the arms are held basketlike to intercept particles falling from above. The arms have the regular echinoderm tube feet. These tube feet are clustered together in small groups. When a food particle touches a tube foot, special glands eject threads of mucus which trap the food particle. Other tube feet direct the strands of mucus to a central food groove on the foot. Cilia carry the mucus and food particles down the food groove toward the mouth.

Brittle stars (ophiuroids) and sea stars (asteroids) both have numerous species in the deep-sea benthos. Both groups are found to depths greater than 6,000 m (19,680 ft). The brittle stars are primarily surface deposit feeders but include some species that can filter particles out of the water. Sea stars are mostly carnivorous, feeding on foraminiferans, polychaetes, and bivalves, although a number of species, particularly in deeper water, are deposit feeders.

The sea cucumbers (holothuroids) are considered to be the most successful of the deep-sea echinoderms, and in areas of relatively rich sediments they are the dominant epifauna. One group, the Elasipoda, is found only in the deep sea (Fig. 11.29). Elasipods are up to 0.5 m (1.6 ft) in length and most have only a few tube feet. Some species have a series of lateral papillae that are used like legs to walk across the mud. Another type of elasipod cucumber has a well-developed 'sole' that leaves footprints as the cucumber walks across the mud. Elasipods readily take to the water and swim. In fact, a few species have become entirely pelagic and do not even rest on the bottom.

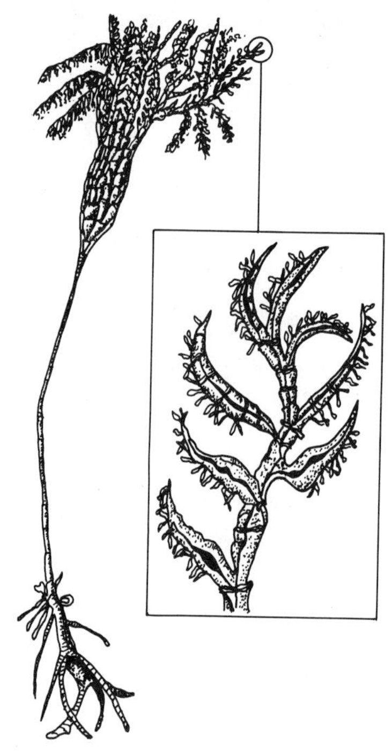

FIGURE 11.28
Crinoids of the Deep Sea

Crinoids have their maximum diversity in the deep sea. The arms bear the characteristic tube feet for filtering food from the water. Shown is *Bathycrinus.* (After Marshall, 1979.)

The phylum Pogonophora (beardworms) is another group of benthos that is more common in the deep sea than in shallow waters. One group of pogonophorans is well represented in some hydrothermal vent communities. Other species of Pogonophora are found to 10,000 m beneath surface waters of high productivity. Since all Pogonophora lack mouth, digestive tract, and anus, their nutrition is of particular interest. They are able to take into their bodies dissolved organic molecules from the surrounding seawater, but their nutrition is probably more complex, as we will see later in this chapter.

The final phylum found in the deep-ocean benthos is the phylum Chordata. Tunicates are found to a depth of 8,000 m (26,240 ft), but only as solitary animals. Colonial species are restricted to shallower waters. They are filter feeders and maintain this method of feeding in the deep sea.

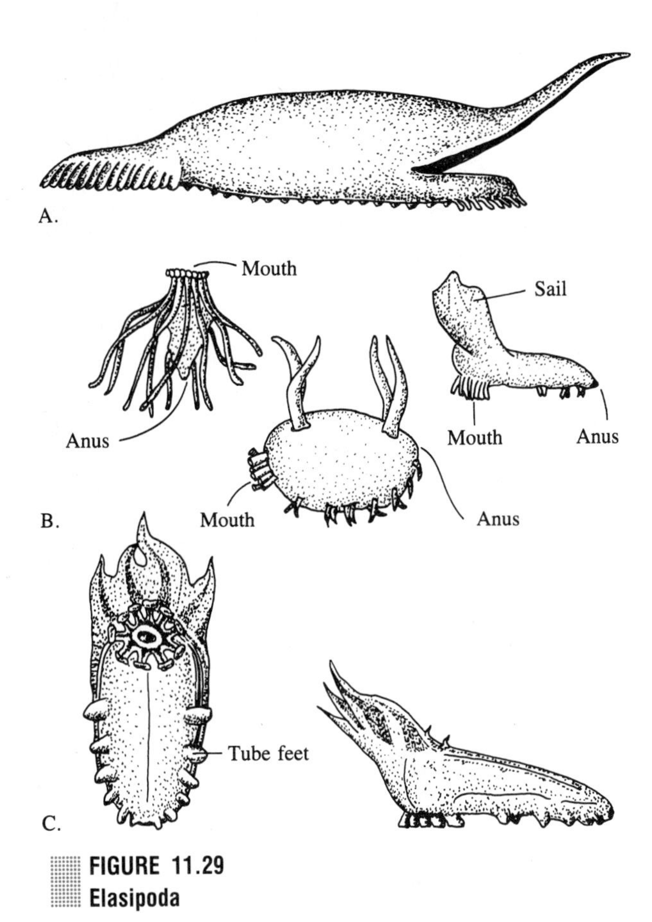

FIGURE 11.29
Elasipoda

Elasipoid sea cucumbers are a group of holothurians found only in the deep-sea benthos. **(A)** Among the largest of deep-sea animals, they may be up to 0.5 m in length. **(B)** Some forms have adopted a pelagic mode of life and are part of the plankton. **(C)** The modified tube feet may be used for walking.

Demersal Fish. The diversity of fish that live on or near the bottom of the deep sea is greater than that of the deep-sea pelagic fish. Some 1,000 species of demersal fish have been found between 1,000–7,000 m (3,280–22,960 ft) depth. Their overall size is also greater than that of the pelagic fish. Demersal species fall mostly in the range of 20–50 cm (8–20 in), the largest reaching lengths of 1.0–1.5 m (3.3–5 ft). As with other animal groups in the deep sea, the diversity and abundance of demersal fish decreases with increasing depths. Again this is related to the decreasing availability of food with increasing depths.

Deep-sea demersal fish have generally elongated bodies and are relatively slow swimmers. Most species can maintain neutral buoyancy by either a functional swim bladder or a reduced muscular and skeletal system. Demersal sharks, like their pelagic counterparts, have relatively large livers with high concentrations of the light oil squalene.

The deep-sea elasmobranchs include members of the skates, rays, and sharks. Deep-sea skates and rays, like shallow-water species, feed on epifaunal and infaunal invertebrates. Skates and rays are found only down to 3,000 m (9,840 ft). The food available below this depth seems to be inadequate.

Some sharks are found in very deep water. The sleeper shark is the giant of the deep-sea fish, reaching lengths of 7 m (23 ft). Carnivorous, it feeds on squid, fish, and crustaceans between 1,000–2,000 m (3,280–6,560 ft). Smaller species of sharks are found at greater depths.

The deep-sea species of demersal bony fish belong to groups that are mostly distinct from shallow water groups. A few true cod species are found below 1,000 m. Eelpouts and snail fish are also found. In fact, a snail fish has been collected at 7,200 m (23,616 ft), the greatest depth reported for any fish. The common deep-sea demersal fish, however, are not found in shallow water. The three most common deep-sea bottom fish are rattails (also called grenadiers), halosaurs, and brotulids. Some 250 species of rattails live in the deep sea, and they are the most common bottom fish at depths greater than 5,000 m (16,400 ft) (Fig. 11.30). All have well-developed swim bladders and feed either on surface depostis or small benthic animals and detritus particles. Others sieve the sediments for worms and bivalves. Rattails swim with their mouth near the bottom and the body at an angle of 30–40° to the bottom. Most rattails are 20–60 cm (8–24 in) in length, with a maximum size of 1.5 m (5 ft).

Halosaurlike fish (including some eels and the notocanths) are all eelshaped fish with a snout that extends over the mouth (Fig. 11.31). Like rattails, they are most common in the shallower waters of the deep sea, but may be found to depths of 3,000 m (9,840 ft). All halosaurs are epifaunal feeders, eating crustaceans, sea anemones, and sea pens. All species produce leptocephalus larvae that spend time in shallower water.

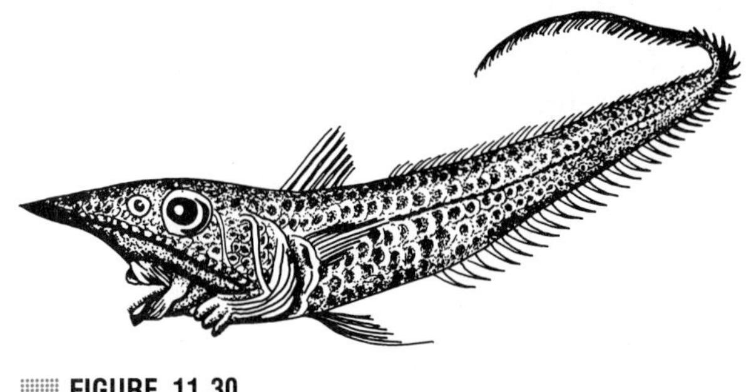

FIGURE 11.30
Rattail

Rattails are common benthic fish of the deep sea. Shown is *Coelorhynchus*, over 50 cm in length.

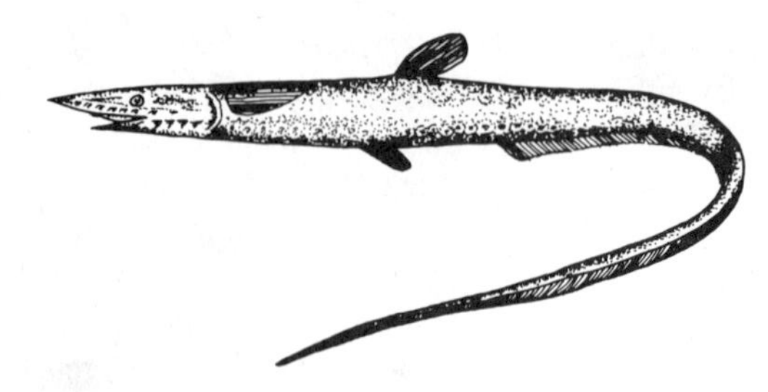

FIGURE 11.31
Halosaur

Shown is the genus *Aldrovandia.*

Brotulid fish have large heads and a tapering tail with continuous dorsal and anal fins (Fig. 11.32). They are found to a depth of 5,000 m (16,400 ft).

One of the most interesting of the deep-sea bottom fish is the tripod fish (Fig. 11.33). There are 18 species in this group, all of which have elongated pelvic and anal fins on which they stand to keep the body above the mud. They face into the slow-moving current and probably feed on plankton. Tripod fish are most common in the deep sea. The shallowest species is found at 500 m, but diversity increases below 1,000 m (3,280 ft). Their maximum depth is 5,800 m (19,024 ft).

Except for the deep-water elasmobranchs, which retain the shallow-water characteristic of bearing their young live (viviparity), the deep-sea fish generally lay eggs. The eggs are large and yolk-filled (lecithotrophic). Development is advanced before the egg hatches, and the young bypass a larval stage. Little is known about food sources of the young.

How do the male and female deep sea benthic fish locate each other? Rattails and brotulids have a drumming structure as part of the swim bladder. Presumably, like species in the mesopelagic zone, sound can be used

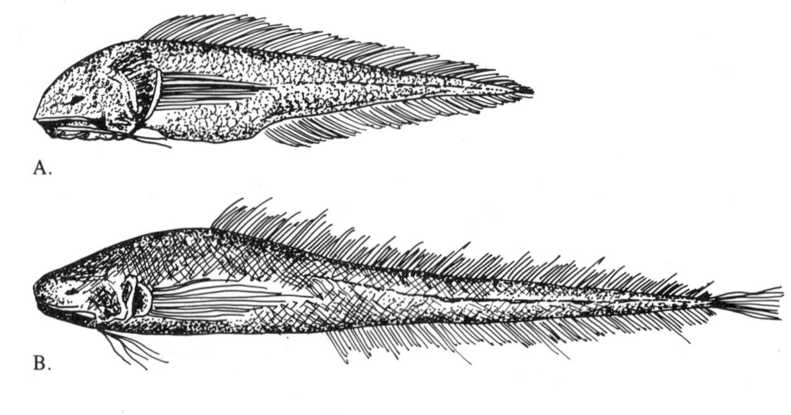

FIGURE 11.32
Brotulids

(A) *Abyssobrotula.* **(B)** *Grimaldichthys.*

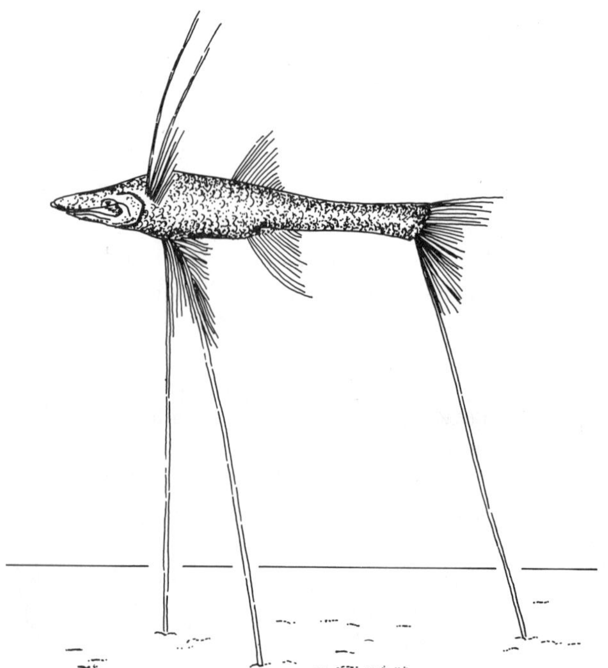

FIGURE 11.33
Tripod Fish

The tripod fish uses extended pectoral and tail fins to stand on the muddy bottom.

for intraspecies communication. Other deep-sea demersal fish, however, do not produce sound. Since very few demersal species have any bioluminescence, it is likely that odor and the sense of smell is used for locating members of the same species.

SPECIAL COMMUNITIES IN THE DEEP OCEAN

Three areas of the deep-ocean environment deserve special recognition: hydrothermal vents, cold water and hydrocarbon seeps, and deep-sea trenches. Biological communities of deep-sea trenches have the deepest living marine life. The biological communities of hydrothermal vents contradict most generalizations about deep-sea life. The communities consist of large, actively metabolizing animals with much greater biomass than anywhere else in the deep ocean. The hydrothermal vent community is unique, too, in that life is not directly or indirectly dependent on the sun for energy. Energy comes, instead, from a geological source—the earth's mantle.

Benthos of Deep-Sea Trenches

The trenches, which cover about 1 percent of the ocean floor, contain the deepest of the ocean's waters (Fig. 11.34). As their name suggests, their sides are relatively steep, often of bare rock. The soft-sediment bottoms are flat. There are some 31 deep-sea trenches around the world ranging from 8,000–10,000 m (26,240–32,800 ft) in depth. Most of them (18) are in the Pacific Ocean. All of the trenches are formed by tectonic movement of the ocean floor.

Benthic communities of the deep-sea trenches tend to have a unique assemblage of benthic animals compared to shallower waters, and the richness of the biological community is affected by their proximity to land. Several characteristics of the trench fauna support the concept of a distinctive hadal zone. The total number of species in the hadal zone (below 6,000 m, or 19,680 ft) decreases and faunal composition changes. Groups such as fishes, tunicates, barnacles, bryozoans, and sponges are remarkably reduced in numbers. Virtually no decapod crustaceans, brachiopods, or turbellarian worms are present. These groups, common and widely distributed in the benthos of shallower waters, are not found in trenches (Tab. 11.4). The reasons for these distributions are not clear but are probably related to food preferences, mode of feeding, and the difficulty of maintaining a calcified skeleton below the calcium compensation depth.

Other benthic groups, including pogonophorans, echiuroid worms, holothuroid echinoderms, and isopod

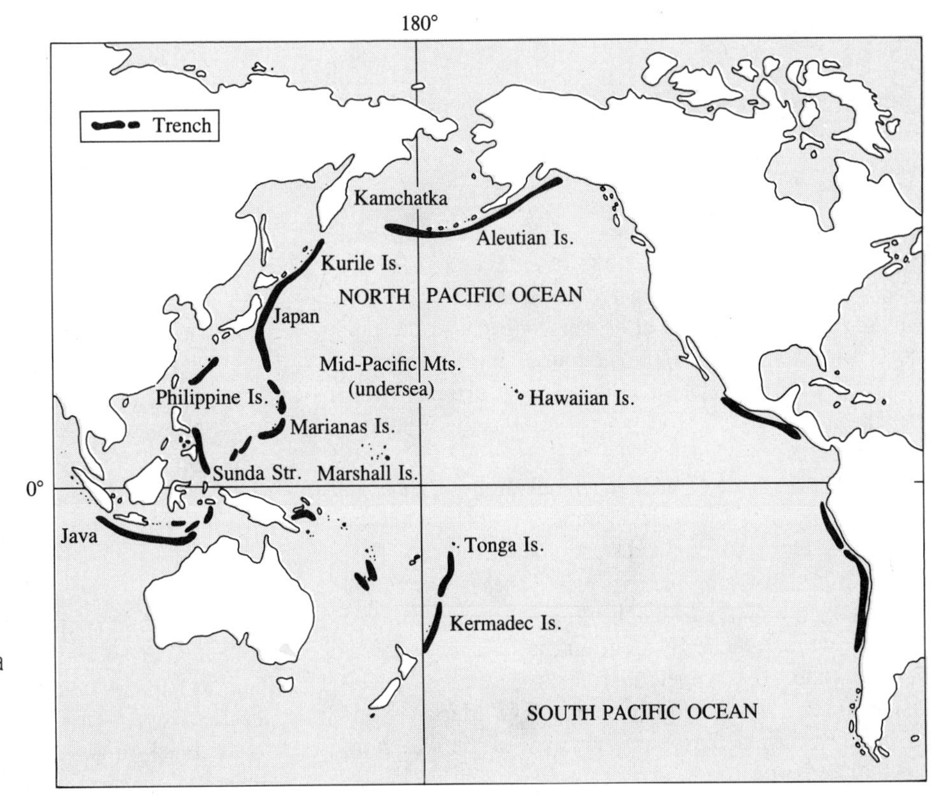

FIGURE 11.34
Deep-Sea Trenches

Most of the trenches of the deep sea are found in the Pacific Ocean. These cuts are close to land and plunge almost 11 km (6.8 mi) straight down.

TABLE 11.4
Benthos of the Trenches

Group	Genus and Species	Depth (m)	Area
Foraminifera	*Sorosphaera abyssorum* Saidova	10,687	Kurile-Kamchatka Trench
Porifera	*Chondrocladia concrescens* (Schmidt)	8,660	
Hydrozoa	*Halisiphonia galatheae* Kramp	8,300	Kermadec Trench
Actiniaria	*Galatheanthemum hadale* Carlgren	10,210	Philippine Trench
Turbellaria	*Turbellariomorpha* sp.	6,200	Milne-Edwards Deep
Polychaeta	*Poecilochaetus vitjazi* Levenstein	10,687	Tonga Trench
Cirripedia	*Scalpellum (Arcoscalpellum) formosum* Hoek	6,860	Kurile-Kamchatka Trench
Cumacea	*Leucon* sp.	7,246	Aleutian Trench
Tanaidacea	*Leptognathia armata* Hansen	8,006	Bougainville Trench
Isopoda	*Macrostylis* sp.	10,710	Mariana Trench
Amphipoda	*Pardaliscoides longicaudatus* Dahl	10,000	Philippine Trench
Decapoda (Anomura)	*Tylaspis anomala* Henderson	4,343	Central South Pacific
Mysidacea	*Amphyops magna* Birstein and Tchindonova	7,230	Kurile-Kamchatka Trench
Pycnogonida	*Nymphon profundum* Hilton	6,860	Kurile-Kamchatka Trench
Gastropoda	*Aclis kermadecensis* Knudsen	8,230	Kermadec Trench
Bivalvia	*Phaseolus* n. sp. Filatova	10,687	Tonga Trench
Crinoidea	*Bathycrinus* sp.	9,735	Idsu-Bonin Trench
Holothuroidea	*Myriotrochus* sp.	10,687	Tonga Trench
Echinoidea	*Pourtalesia aurorae* Koehler	7,290	Banda Trench
Asteroidea	*Hymenaster* sp.	7,657	Mariana Trench
Ophiuroidea	*Ophiosphalma* sp.	7,230	Kurile-Kamchatka Trench
Pogonophora	*Zenkevitchiana longissima* Ivanov	9,500	Kurile-Kamchatka Trench
Pisces	*Careproctus (Pseudoliparis) amblystomopsis* Andriashev	7,587	Japan Trench

SOURCE: Menzies, R.J., George, R.Y., and Rowe, G.T. 1973. *Abyssal environment and ecology of the world's ocean.* New York:Wiley.

crustaceans are well represented in trench benthos. The infauna of the trenches is dominated by polychaete worms and bivalve clams.

The benthic species found in trenches have a higher frequency of endemism than do those in shallow waters, as outlined in Table 11.5. About 70 percent of the benthic species of a trench are **endemic;** that is, they are not found anywhere else. This is because the uniform and constant conditions found in the trenches, coupled with the relative isolation of one trench from another, tend to promote the development of unique communities.

The abundance of benthic animals in trenches is affected primarily by their closeness to land masses. Trenches close to continents receive sediments relatively high in organic matter. Such trenches have a relatively high biomass of benthos (up to 10 g/m^2, or 0.33 oz/ft^2), while trenches away from the influence of continents (and areas under unproductive surface waters) have a very low benthic biomass. The Marianas and Tonga Trenches have a biomass of 0.008 g/m^2 (0.00003 oz/ft^2).

Hydrothermal Vent Communities

As we have seen, life in the deep sea is made up of species of relatively small size. They represent a much smaller biomass per unit of area than is found on the continental shelves. In the deep ocean, the pace of life is slow and metabolic rates are low. The discovery of a new type of deep-sea ecosystem at the Galapagos Rift in 1977 provided a striking exception to the general condition. Hot water emanating from vents along the axis of this equatorial Pacific spreading center supported relatively large animals that rapidly grow to maturity and produce large numbers of larvae. It is believed that this behavior is the result of the fact that hydrothermal vents are short-lived—lasting possibly tens of years. Continued exploration of Pacific and Atlantic spreading centers resulted in the discovery of additional hydrothermal vent biocommunities, shown in Figure 11.35. Although they were not found at the initial discovery, many vent fields have been found to include 'black smokers', spewing 350°C (660°F) black water, rich in metallic sulfides from tall chimneys. The

TABLE 11.5
Endemism in Deep-Sea Organisms

Group	Number of Hadal Species 6,000 m *(A)*	Number of Hadal Species <6,000 m *(B)*	Number of Species Exclusively at Depths 6,000 m *(A–B)*	% Hadal Endemic Species
Foraminifera	128	73	55	33
Cnidaria-Spongia	26	3	23	89
Other Cnidaria	17	4	13	76.6
Polychaeta	42	20	22	52.4
Echiuroidea	8	3	5	62.5
Sipunculoidea	4	4	0	0
Cirripedia	3	2	1	33.4
Cumacea	9	0	9	100
Tanaidacea	19	4	15	79
Isopoda	68	18	50	73.6
Amphipoda	18	3	15	83.5
Solenogaster	3	3	0	0
Gastropoda	16	2	14	87.5
Bivalvia	39	6	33	84.7
Crinoidea	11	1	10	91
Holothuroidea	28	9	19	68
Asteroidea	14	6	8	57.2
Ophiuroidea	6	2	4	66.7
Pogonophora	26	4	22	84.6
Pisces	4	1	3	75
Total species	489	168	321	68

SOURCE: Menzies, R.J., George, R.Y., and Rowe, G.T. 1973. *Abyssal environment and ecology of the world's ocean.* New York:Wiley.

chimneys are also composed of metallic sulfides that have precipitated from the vent waters.

Each vent has its unique suite of species as well as those that are found elsewhere. Associated with the hydrothermal vents are at least 25 new families or subfamilies, more than 50 new genera, and 100 new species. Although many questions remain to be answered, one fact is clear: The food source that supports these rich communities is bacteria that use chemical energy (rather than light energy) to produce organic molecules from inorganic sources (Fig. 11.36).

For years, the discoveries of new hydrothermal vent communities was confined to the Pacific Ocean, where most were characterized by 1-m (3.2-ft) tube worms and 25-cm (10-in) clams. Neither of these vent species possesses a gut. They obtain most, if not all, of their nourishment from chemosynthetic bacteria that live within their tissues.

A comparison of metabolic rates of hydrothermal vent animals to other deep-sea benthos shows the rates for the vent clams to be 50 times greater than other deep-sea clams. The adult clams obtain a much larger adult size, and mature in one-tenth the time required by other deep-sea clams.

The large tube worm, *Riftia pachyptila,* has been found in the Galapagos Rift and at East Pacific Rise vent fields. The body of the worm, which can be fully retracted into the 4 cm (1.6 in) diameter tube, consists of four basic regions: the tentacles (or obturaculum), the vestimentium, the trunk, and the segmented opisthosome (Fig. 11.37). Each worm has some 230,000 tentacles, each one containing paired blood vessels. The trunk is also important to the animal's nutrition. The trunk has a large blood-filled cavity which contains a large mass of tissue, liberally supplied with blood vessels, called the trophosome. The trophosome tissue is actually a tightly packed mass of bacteria. It appears that the bacteria, like those of the vent, use hydrogen sulphide (H_2S) as a source of energy to convert carbon dioxide into organic molecules. *Riftia* probably

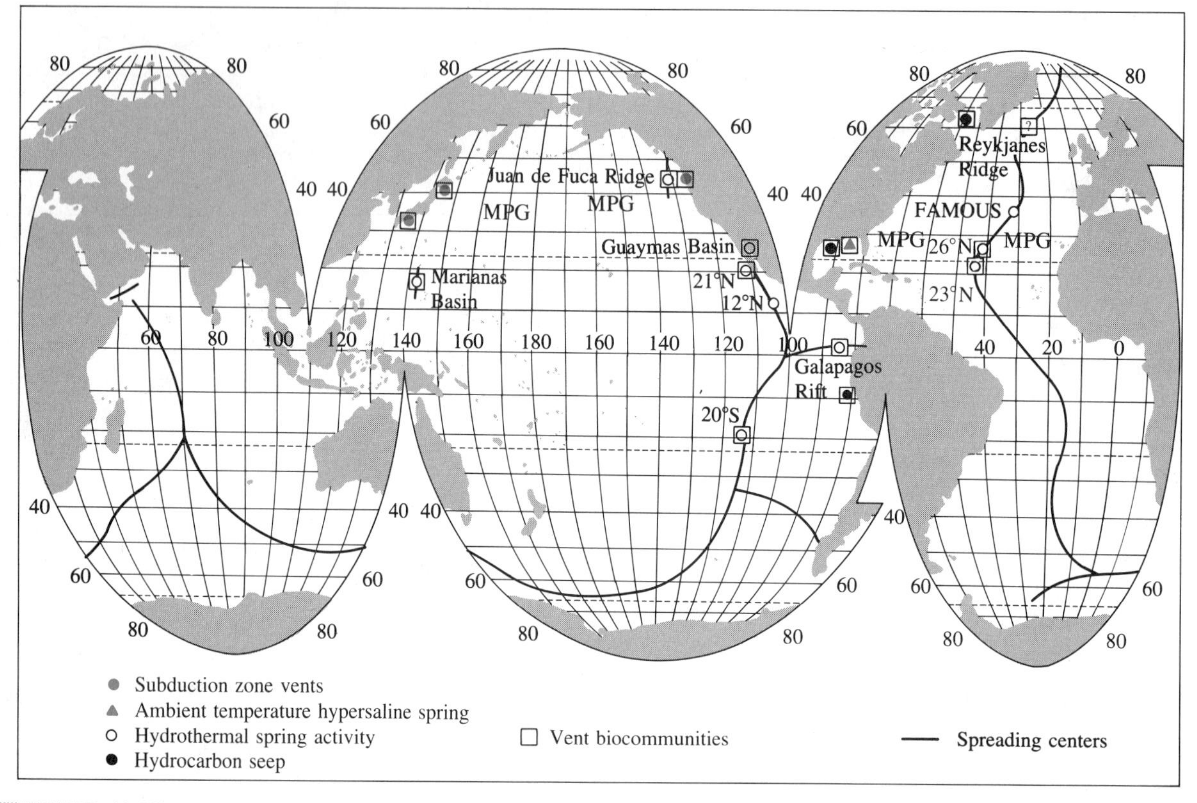

FIGURE 11.35
Submarine Springs and Vent Biocommunities

Locations of submarine ambient and hydrothermal water springs, hydrocarbon seeps, and vent spring and seep biocommunities.

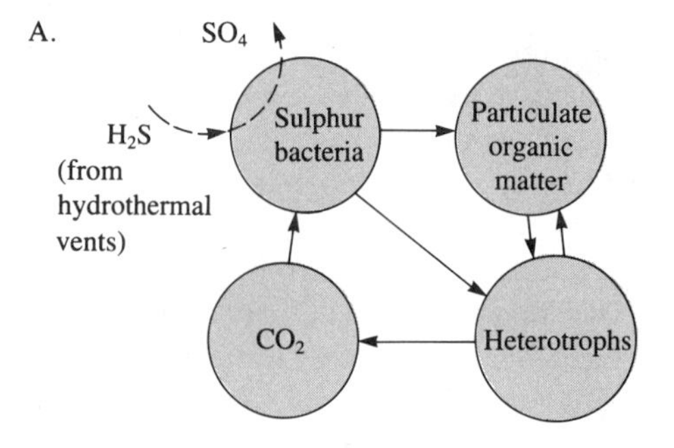

FIGURE 11.36
Chemosynthesis in Hydrothermal Vent Communities

(A) Sulphur bacteria in the vents utilize the energy in hydrogen sulphide gas to build complex organic molecules and grow. The source of carbon is carbon dioxide gas, which is also dissolved in the water. The bacteria are either eaten directly by heterotrophs or die and form particulate organic matter (detritus) that in turn is used as a source of food by heterotrophs. **(B)** Comparison of photosynthesis and chemosynthesis. (A after Longhurst, A.R. 1981. *Analysis of marine ecosystems.* New York:Academic Press.)

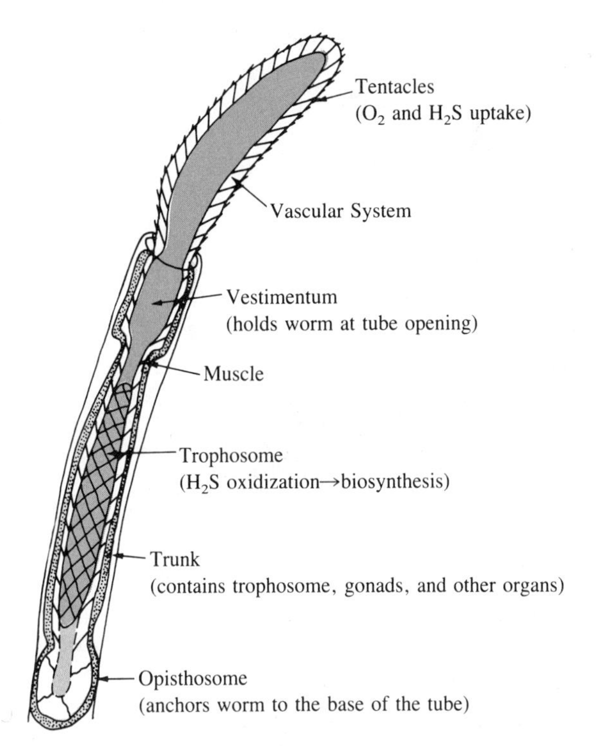

FIGURE 11.37
Physiology of Vestimentiferan Tubeworms

FIGURE 11.38
Mid-Atlantic Hydrothermal Vent Biocommunities

This swarm of shrimp belonging to the genus *Rimicaris* are feeding on chemosynthetic bacteria that live on the surface of the metallic sulfide mounds associated with hydrothermal vents. They have been found at 23° N and 26° N on the Mid-Atlantic Ridge. For further information on these communities, see the color insert, "Habitats in Color". (Photo courtesy of Peter A. Rona, NOAA.)

derive much of their metabolic requirements from this symbiotic relationship. The blood of *Riftia* carries large quantities of O_2 and CO_2 along with H_2S.

It was not until 1985 when hydrothermal vent biocommunities were discovered in the Atlantic Ocean. Two vent fields were located on the Mid-Atlantic Ridge at 23° and 26° N latitude. These were dominated by two species of shrimp of the Genus *Rimicaris* (rift shrimp; see Fig. 11.38). Neither possess the typical shrimp eye stalk or cornea. Interestingly, one of the species, *Rimicaris exoculata,* does have two light-sensing organs, and is only found on 'black smokers'. A biologist at Woods Hole Oceanographic Institution, Cindy Lee Van Dover, wondered if they might use sensors to find the 'black smokers'. She took a picture of a 'black smoker' on the Juan de Fuca Ridge 290 km (180 mi) off Vancouver, Canada, with a device sensitive to very low levels of light. Sure enough, the 350°C water exiting from the chimney produced a distinct glow, which has been named the Van Dover glow. The photograph was taken with an electronic charge-coupled device, which is many times more light sensitive than conventional film. It may well be that these shrimp use the light-sensitive organs to find the 'black smoker' chimneys, since they have not been found on vents where the water temperature is less than 200°C (392°F). They feed by scraping the sulfide from the chimney wall. The rift shrimp ingest the sulfide, digest the chemosynthetic bacteria living on its surface, and excrete the sulfide. There are still many unanswered questions about the vent biota, and the coming years should provide exciting information about the ecology of the hydrothermal vent biocommunities.

Cold Water Seep Communities

Three additional submarine seep environments have also been found to have chemosynthetically supported biocommunities. During the summer of 1984, an investigation of an ambient-temperature, hypersaline (46.2‰) seep at the base of the Florida Escarpment in the Gulf of Mexico (Fig. 11.39) revealed a biocommunity similar in many respects to the hydrothermal vent communities. The seeping water appears to flow from joints at the base of the limestone escarpment and move out across the clay deposits of the abyssal plain at a depth of about 3,200 m (10,496 ft).

The hydrogen sulfide-rich waters support a number of white bacterial mats that carry on chemosynthesis in a fashion similar to that of the bacteria of the hydrothermal vents. These and other chemosynthetic bacteria may provide most of the support for a diverse commu-

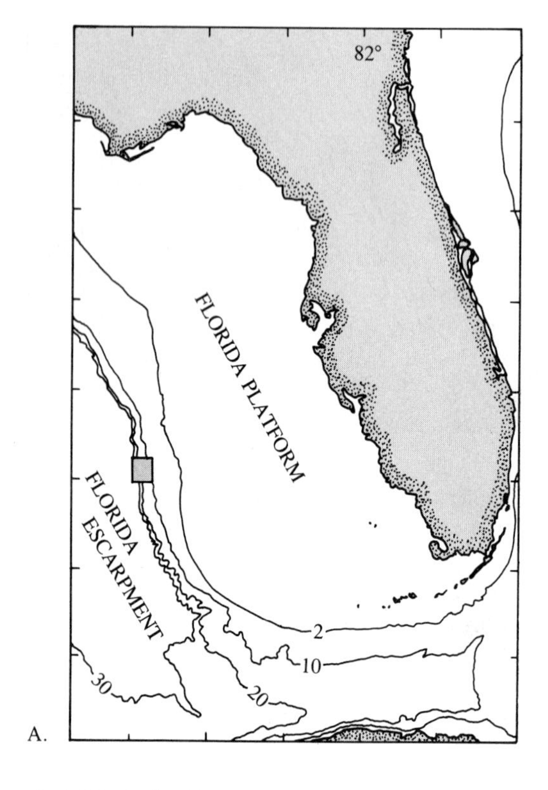

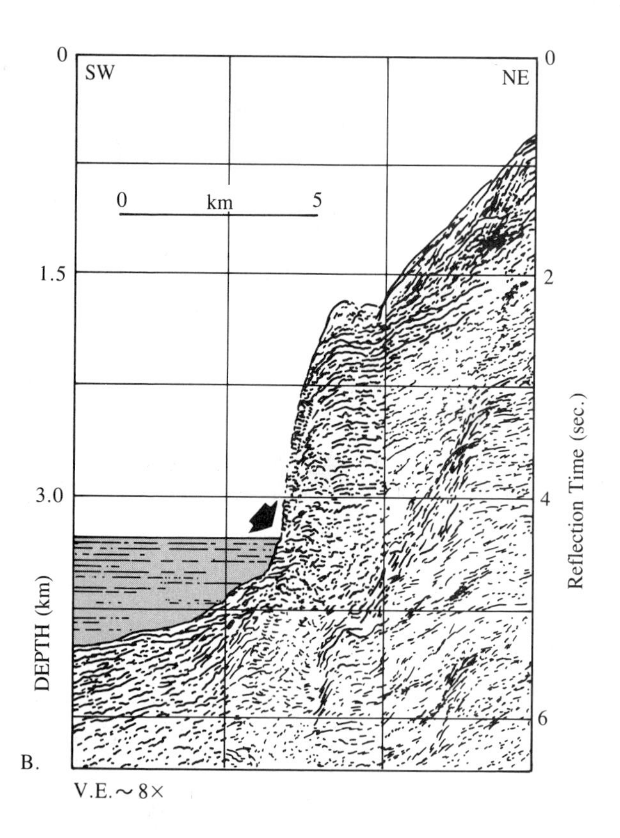

FIGURE 11.39
Florida Escarpment Cold Water Spring Biocommunity

(A) Square at the base of the Florida Escarpment locates the area where the seeps and associated biological communities were found. **(B)** Seismic reflection profile across the Florida Escarpment. The seeps are located at the abrupt boundary between the abrupt abyssal plain and the steep-sloped escarpment (arrow). Vertical exaggeration is ×8. (Both figures were provided through the courtesy of C.K. Paull, Scripps Institution of Oceanography, UCSD.)

nity of animals. The community includes holothurians, sea stars, shrimp, snails, limpets, brittle stars, anemones, vestimentiferan worms, galatheid crabs, clams, mussels, and zoarcid fish.

Drilling into the limestone rocks of the platform east of the escarpment revealed fluids with temperatures of up to 115°C (239°F) and salinities of 250‰. This dense water apparently seeps deep into the jointed rocks of the platform and is trapped by the impermeable clay deposited on the ocean floor at the base of the escarpment. It then flows out on to the surface clays, mixes with sea water, and produces the spring water with salinity intermediate between that of the platform fluids and ocean water.

Also observed in 1984 were dense biological communities associated with oil and gas seeps on the Gulf of Mexico continental slope (Fig. 11.40). Trawls at depths of between 600–700 m (1,968–2,296 ft) recovered epifauna and infauna similar to those observed at the hydrothermal vents and the hypersaline seep at the base of the Florida Escarpment. Carbon isotope analysis indicates the hydrocarbon-seep fauna is based on a chemosynthetic productivity that derives its energy from hydrogen sulfide and methane. It is certainly possible that such assemblages will be found throughout the oceans in areas where hydrocarbons are being generated in the underlying sediment.

Finally, a third environment of seep communities was observed from *Alvin* during the summer of 1984. It is located in the subduction zone of the Juan de Fuca Plate at the base of the continental slope off the coast of Oregon (Fig. 11.41). Here, the trench is filled with sediments. At the seaward edge of the slope, clastic sediments are folded into a ridge. At the crest of the ridge, pore water escapes from the 2-million-year-old folded sedimentary rocks into a thin overlying layer of soft sediment. Occurring at a depth of 2,036 m (6,678 ft), the seeps produce water that is only slightly warmer (about 0.3°C, or 0.5°F) than ambient conditions. The seep water contains CH_4 (methane) that is probably produced by decomposition of organic material in the sedimentary rocks. The methane serves as the source of energy for bacteria that oxidize it and chemosynthetically produce food for themselves and the rest of the

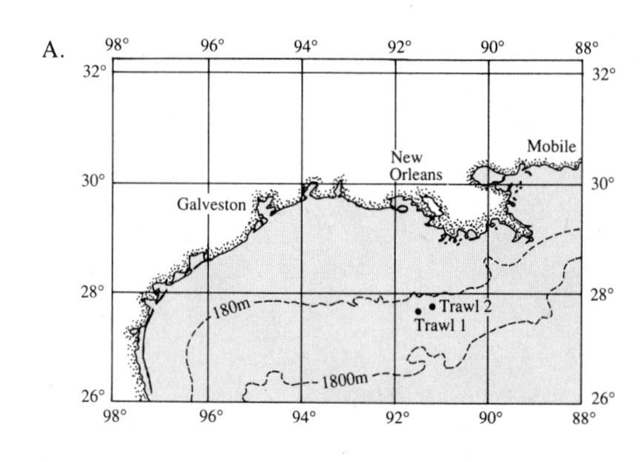

FIGURE 11.40
Hydrocarbon Seep Biocommunities–Gulf of Mexico

(A) Location of trawls that recovered hydrocarbon seep fauna. **(B)** North-south acoustic profile (25 kHz) across a brine pool located 285 km southwest of the Mississippi delta. The pool is 22 m long and 11 m wide, and ringed by a large bed of mussels, *Bathymodiolus* n. sp. Part A of this figure shows the corrected topography for the acoustical survey shown in Part B. The topography of the Part B profile had to be corrected because the depth of submarine *NR-1*, which conducted the survey, changed along the transect. (From I. R. MacDonald et al., 1990. Chemosynthetic Mussels at a Brine-Filled Pockmark in the Northern Gulf of Mexico. *Science,* 248(4959):1096–1099. Copyright 1990 by the AAAS.)

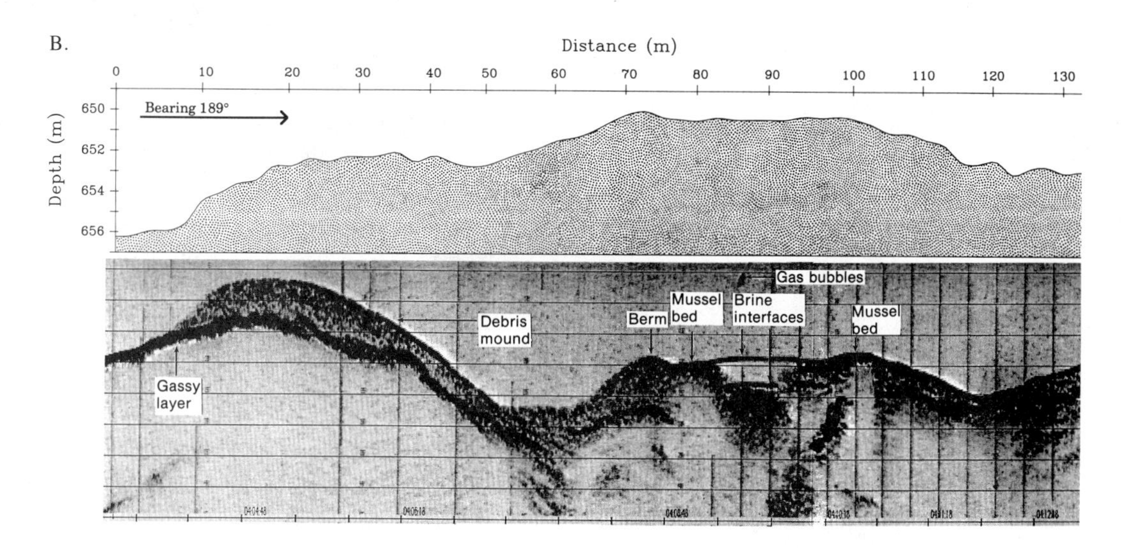

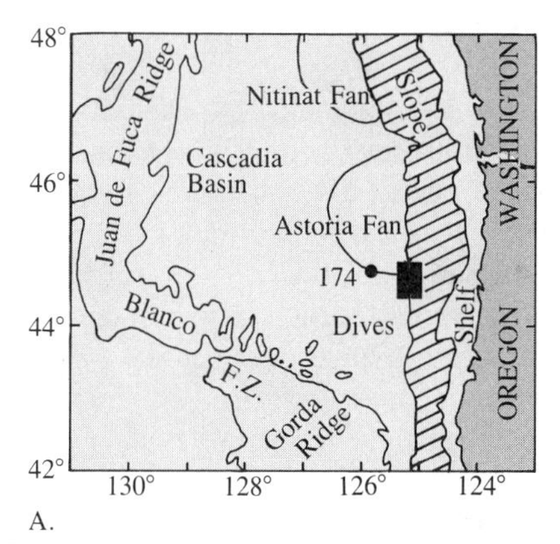

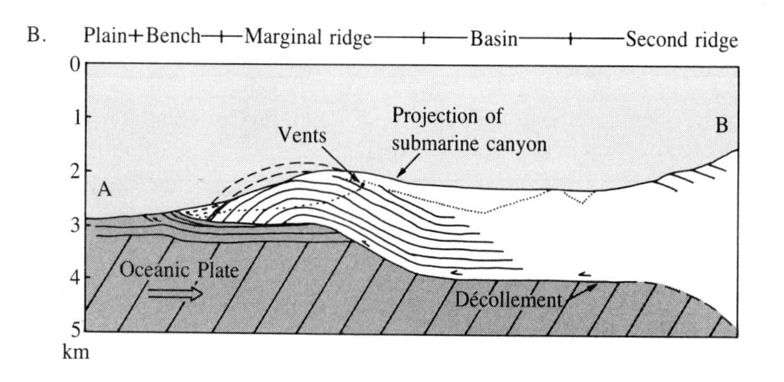

FIGURE 11.41
Subduction Vents off the Oregon Coast

(A) Structural cross section running east-west across the center of the vent area indicated by the black rectangle in A. Compression of the folded sedimentary rock units by the subduction process squeezes pore water out of the rocks near the crest of the fold-producing the vents. **(B)** Location of subduction zone methane-enriched vents at base of Oregon continental slope. The number, 174, is a Deep Sea Drilling Project location. (Courtesy of L.D. Kulm, Oregon State University.)

community, which contains many of the same genera found at other vent sites. During 1985, similar communities were located in subduction zones of the Japan Trench and the Peru-Chile Trench. All the seeps are located on the landward side of the trenches at depths from 1,300–5,640 m (4,264–18,499 ft).

SUMMARY

The part of the ocean more than 1,000 m (3,280 ft) below the surface (both pelagic and benthic components) is defined as the deep sea. It is the largest marine habitat, encompassing 80 percent of the ocean bottom and over 75 percent of the water. It is impoverished in animal life, however, accounting for only 10 percent of the ocean's total biomass of organisms. Its dimensions are immense and environmental characteristics strikingly uniform—no sunlight, low uniform temperature, moderate to high levels of dissolved oxygen, and almost constant salinity. Sediment composition of the abyss varies, as does the amount of food reaching the bottom from the surface. Food for deep-sea animals consists primarily of copepod fecal pellets as well as detritus brought by turbidity slumps and the occasional large carcass.

To fully appreciate the biology of the deep ocean, one must understand the fundamental nature of the physical marine environment. We discussed the bathymetric features of the deep-ocean basin in Chapter 2. The sediments that accumulate in those basins and the global plate tectonics process that created them are discussed here. The particles that make up this sediment are of four types: lithogenous (of rock origin), biogenous (of biological origin), hydrogenous (of water origin), and cosmogenous (of outer-space origin). Sediments accumulating on the continental margin are composed primarily of lithogenous particles and are called neritic. The sediments deposited slowly in the deep-ocean basins are called oceanic and contain particles that are primarily lithogenous or biogenous. If the biogenous component is less than 30 percent, the sediment is called an abyssal clay. If it exceeds 30 percent, the sediment is an ooze. Biogenous particles derived from diatoms and radiolarians are composed of silica (SiO_2), while those derived from Coccolithophoridae and Foraminifera are composed of calcium carbonate ($CaCO_3$). Whether or not enough biogenous particles accumulate on the ocean floor to produce an ooze depends on three processes: production, solution, and dilution of biogenous particles.

Although silica is dissolved at the same rate at any ocean depth, the rate of dissolving of $CaCO_3$ increases with depth. This is because of the increase in the amount of carbon dioxide (CO_2) with increased depth. The depth at which $CaCO_3$ is dissolved into the water at a rate equal to that which it arrives from sinking is the calcium carbonate compensation depth. Although $CaCO_3$ ooze covers more of the deep-ocean floor than abyssal clay or siliceous ooze, it is not found where the ocean floor lies below an average depth of 4,000 m (13,120 ft). Abyssal clay is typically found on the floor of the deepest ocean basins.

Global plate tectonics is the process by which new ocean floor is created and continents move across the earth's surface. New ocean floor is created at the axes of mid-ocean ridges and rises as molten material rises and hardens into rigid lithosphere. Beneath the lithosphere is the plastic asthenosphere, which can deform slowly and allow the lithosphere to move off down the slopes of the ridges and rises. As it moves away from the high heat-flow region of the spreading centers, the lithosphere cools and thickens. Since there are spreading centers in all the oceans, the lithosphere is broken into at least thirteen lithospheric plates that collide at convergent boundaries to produce ocean trenches and folded mountain ranges. They also slide past one another along transform faults such as the San Andreas Fault.

Beneath ocean trenches where one plate subducts beneath another, earthquakes may be generated at depths of up to 650 km (403 mi), indicating that the lithosphere does not remelt until it reaches that depth. At spreading centers where the lithosphere is thin, earthquakes rarely occur at depths greater than 10 km (6.2 mi).

There are oases of high biological productivity in the deep sea—the hydrothermal vents. Where magma from the earth's mantle pushes to the ocean floor, high concentrations of hydrogen- and sulphur-rich gases dissolve in the much warmer water. Bacteria use the gases to produce organic molecules through chemosynthesis. A fascinating group of invertebrates and fish are found surrounding these deep-ocean vents.

The rest of the ocean floor is like a desert in the amount of life present. The benthic invertebrates for the most part look similar to taxonomically related shallow-water invertebrates. The same is generally true for other zooplankton. Many fish, however, have highly modified morphology. The deep-sea pelagic fish have huge mouths and generally reduced body musculature.

The deepest parts of the abyss are the deep-ocean trenches. These trenches range from 8,000 to over 10,000 m beneath the surface and are physically separated from each other. Each has a distinct group of zooplankton, fish, and benthic invertebrates. The differences in fauna can be explained in part by varying concentrations of food-reaching trenches. Those close to shore receive a relatively large amount of food. Those in oceanic areas are impoverished. In part, however, the differences in species found is due to evolution of endemic forms. Since there is little contact of populations between trenches, species specific to each trench have the opportunity to evolve.

Three vent-type communities first observed in 1984 are the hypersaline seep community at the base of the Florida Escarpment, the hydrocarbon seep communities observed on the continental slope of the Gulf of Mexico, and subduction zone methane vent communities of the coasts of Oregon, Japan and, Peru. Their discovery opens the possibility of a much wider distribution of chemosynthetically supported communities of the deep ocean floor.

QUESTIONS AND EXERCISES

1. Discuss the roles of productivity, solution, and dilution in determining whether or not an ooze deposit will form on the floor of the deep ocean. Include a discussion of the calcium carbonate compensation depth.
2. Describe the relative motion of lithospheric plates that meet at (1) spreading centers, (2) ocean trenches, and (3) transform faults.
3. Discuss the distribution of species and biomass from the top of the abyssal zone to the lowest depths. Include both the pelagic and benthic communities.
4. Discuss the food sources important to the deep-sea benthos.
5. Discuss the food source of the hydrothermal vent communities. Compare this with the food chains of the rest of the deep-sea benthos.
6. Given the absence of sunlight in the abyss, discuss how pelagic animals perform the functions of feeding and reproduction.
7. Discuss the adaptations of deep-sea pelagic animals to the problem of maintaining neutral bouyancy.

REFERENCES

Cushing, D.H., and **Walsh, J.J.** 1976. *The ecology of the seas.* London:Blackwell Scientific.

George, R.V. 1981. Functional adaptation of deep-sea organisms. In Vernburg, F.J., and Vernburg, W.B., eds. *Functional adaptation of marine organisms.* New York:Academic Press.

Heezen, B.C., and **Hollister, C.D.** 1971. *The face of the deep.* New York:Oxford University Press.

Herring, P.J., and **Clarke, M.R.** 1971. *Deep oceans.* New York:Praeger.

Kennicutt, M.C. II, Brooks, J.M., Bidigare, R.R., Fay, R.R., Wade, T.L., and **McDonald, T.J.** 1985. Vent-type taxa in a hydrocarbon seep region on the Louisiana slope. *Nature* 317(6035):351–352.

Kulm, L.D., et al. 1986. Oregon subduction zone: Venting, fauna and carbonates. *Science* 231(4738):561–566.

Longhurst, A.R. 1981. *Analysis of marine ecosystems.* New York:Academic Press.

Marshall, N.B. 1979. *Deep sea biology.* London:Blanford University Press.

Menzies, R.J., George, R.Y., and **Rowe, G.T.** 1973. *Abyssal environment and ecology of the world's oceans.* New York:Wiley.

Paull, C.K., Hecker, B., Commeau, R., Freeman-Lynde, R.P., Neumann, C., Corso, W.P., Golubic, S., Hook, J.E., Sides, E., and **Curray J.** 1984. Biological communities at the Florida Escarpment resemble hydrothermal vent taxa. *Science* 223(4677):965–967.

Shepard, F.P. 1977. *Geological oceanography.* New York:Crane, Russak.

Smith, R.V., and **Kinsey, D.W.** 1976. Calcium carbonate production, coral reef growth, and sea level change. *Science* 194:937–938.

Spencer, D.W., Honjo, S., and **Brewer, P.G.** 1978. Particles and particle fluxes in the ocean. *Oceanus* 21(1):20–26.

Sverdrup, H., Johnson, M., and **Fleming, R.** 1942. *The oceans: Their physics, chemistry and general biology.* Englewood Cliffs, NJ:Prentice-Hall.

Tarling, D., and **Tarling, M.** 1971. *Continental drift: A study of the earth's moving surface.* Garden City, NY:Doubleday.

Thurman, H.V. 1988. *Introductory oceanography.* 5th ed. Columbus, OH:Merrill.

SUGGESTED READING

OCEANS

The abyss, the pelagos, and benthos of the deep ocean. 1979. 12(6):20–45.

Cone, J. 1988. Prospecting—two miles down. 21(1):16–22.

OCEANUS

Ballard, R.D. 1977. Notes on a major oceanographic find. 20(3):35–44.

Deep-sea hot springs and cold seeps. 1984. 27(3):2–78.

Edmond, J.M. 1982. Ocean hot springs: A status report. 25(2):22–27.

Grassle, J.F. 1978. Diversity and population dynamics of benthic organisms. 21(1):42–49.

Jannasch, H.W. 1978. Experiments in deep-sea microbiology. 21(1):50–59.

SCIENCE 80, 81, 82

West, S. 1982. A patchwork earth. (Some problems related to plate tectonics) 3(5):46–53.

SCIENTIFIC AMERICAN

Bonatti, E. 1987. The rifting of continents. 256(3):96–103.

Childress, J., Felback, H., and **Somero, G.** 1987. Symbiosis in the deep sea. 256(5):114–121.

Edmond, J., and **Von Damm, K.** 1983. Hot springs on the ocean floor. 248(4):30–41.

Hallam, A. 1975. Alfred Wegener and continental drift. 232(2):88–97.

Heezen, B.C. 1956. The origin of submarine canyons. 195(2):36–41.

Isaacs, J.D., and **Schwartzlose, R.A.** 1975. Active animals of the deep-sea floor. 233(4):84–91.

Macdonald, K.C., and **Luyendyk, B.P.** 1981. The crest of the East Pacific Rise. 244(5):100–117.

Tokosoz, M.N. 1975. The subduction of the lithosphere. 233(5):88–101.

SEA FRONTIERS

Palmer, R. 1986. Life in the sunless sea. 32(4):269–277.

DEEP-OCEAN BENTHOS

The following is an interview with Dr. Robert R. Hessler, Professor of Oceanography, Scripps Institute of Oceanography, University of California, San Diego.

Q: Dr. Hessler, when did life first appear on the deep-ocean floor?

A: There is no absolute evidence for an answer to that question. The conditions for life on the deep-sea floor have probably existed for as long as the deep sea has existed. More than likely, there always has been oxygen and food. The present ocean sediments only give us evidence back to the Cretaceous (65–136 million years ago).

Q: How about the fossil record? Are there any fossils in rocks that appear to have been deposited in an ancient deep-ocean environment?

A: I have never pursued this much myself, I think there are some. Some of the black shales are thought to have accumulated in deep-basin conditions.

Q: Do you have any ideas about the evolution of life in the deep ocean?

A: My guess would be that ultimately life must have begun in shallow water, and penetrated the deep sea from there. One would expect animal life to evolve closest to the source of primary production, and that can only be in shallow water. Once you get deeper than a few hundred meters there isn't enough light to support primary productivity. Furthermore, considering the variety of physical circumstances in shallow water, the multitude of biogeographic barriers and so on, I would say that shallow water would be the area where evolution was the most active. Influx into the deep sea probably has been continuous.

Q: What groups seem to be most effective in moving from the shallow water into the deep benthic environment?

A: Some groups have done particularly well. All the phyla make it into the deep sea. As you go to lower taxonomic levels, the deep-sea fauna is more and more unique, so that at the level of species, there are hardly any species at all that are common to both the deep-sea and shallow water. The percarid crustaceans, and macrofauna that brood their young—such as amphipods, isopods, tanaids and cumaceans—have been extraordinarily successful in the deep sea. You can get as many as one hundred species of isopod in a single sample. Bivalve molluscs also are very abundant and diverse. In certain environments, other groups tend to be important. Sometimes you find large quantities of sipunculids, sometimes large quantities of pogonophorans.

Among the very small animals, another group that is very important is the nematodes. There are incredible quantities of nematodes in the deep sea. The tiny harpacticoid copepods (deposit-feeding copepods less than 2 mm in length) also are very abundant. Foraminifera is an interesting group. One doesn't think of the Foraminifera as being an important group of benthic organisms, but in the deep sea they are dominant, certainly in terms of biomass and probably also in terms of species diversity.

If you look at the megafauna (the animals that can be photographed), holothurians are fairly important. Glass sponges, tunicates, gorgonians, and certain kinds of decapods also are common. There are a variety of deep-sea families of fishes that are fairly characteristic. The grenadiers are almost always seen.

Q: Which of the physical factors such as pressure, temperature, and the absence of light do you think is most important in limiting the vertical migration of life in the deep ocean?

A: Temperature seems to be the most important factor that prevents deep-sea animals from migrating upward. The deep sea is isothermal; that is, at any one spot the temperature hardly varies by a measurable amount, and deep-sea animals have become accustomed to this. This means that even a slight change in temperature is likely to have an enormous effect. Organisms can't break through this temperature barrier. At north and south high latitudes, however, you find portions of the deep-sea fauna in shallow water. I think the reason for this is that the temperature barrier isn't there; it's equally cold in the surface waters.

With animals penetrating the deep sea from shallow water, the issue is more complex. My guess is that the most important factor preventing animals from penetrating the deep sea is not physical, but food. In order to survive in the deep sea, animals have to be adapted to conditions in which both the rate of supply and the concentration of food is very low. An organism that does not have a life style that can cope with this is not going to make it in the deep sea. Temperature adaptation for a shallow-water animal isn't that difficult. Of the shallow water species that live through the summer and winter, some are capable of adapting to 20–25° C temperature changes on an annual basis. Shallow-water animals could become, or may already be, used to low temperature, but getting used to low food supply requires major changes in behavior, morphology, and physiology. That's much harder to achieve.

Q: The overall consensus is that food is in low supply in general on the deep-ocean floor?

A: That's correct. The reason is that we find so little life there. Any single sample has less life in it by several orders of magnitude than does a sample of shallow water. There is only one sensible explanation for that: There's not much food.

Q Are we talking here in terms of biomass?

A: Total biomass, but the comparison holds in terms of numerical abundance, too. The two factors, of course, tend to be correlated.

Q: How about the occasional availability of large supplies of food? Does that have any significant bearing on the distribution of life in the deep ocean?

A: This is called the dead whale hypothesis. The first thing I have to mention is that I don't believe that anyone has written a completely convincing budget for the rate of food supply to the deep sea. So I don't think anyone knows, with any degree of confidence, the rate at which small particles of food are descending to the deep sea, and nobody knows anything at all about the big ones. What we do know about large organic falls is that there seems to be a group of animals that are very well adapted for taking advantage of them—fish and amphipods. When a large organic fall hits the bottom, it isn't long before these animals arrive in large quantities, consuming it at a fairly rapid rate and secondarily dispersing it to the rest of the fauna. This dispersal is achieved through their feces, through breaking the food parcel into small particles which are scattered about, and by decomposition of the organisms themselves when they die elsewhere.

Q: What evidence is there that deep-ocean life is moving out of the deep sea and back to the shallow shelf environment from which you think it came?

A: Not much. You have to ask yourself, What kind of evidence can I rely on to detect such a phenomenon? It's difficult to come up with really strong criteria, but there is one that isn't bad—the absence of eyes. In a group I work on, the isopods, there's a large variety of families that live in the deep sea but are also found in shallow water at high latitudes. What's interesting about them is that they lack eyes. I won't go into all the details of the argument, but it looks as though the reason they lack eyes in shallow water is because they originally lost them in the deep sea. When they reinvaded shallow water, they couldn't evolve them back. So, some groups apparently have come back into shallow water in high latitudes.

Q: What are some other important general issues in your field of research?

A: Well, there are several very important issues. First of all, everything that has been done in the deep sea has been done on soft, mud bottoms. The reason for that is simply that oceanographers have gotten tired of losing their gear while sampling rocky bottoms. But now, with the discovery of the deep-sea hydrothermal vents and their wonderful fauna, more and more people are getting interested in studying the hard-bottom environment. The fauna of the hydrothermal vents is fascinating, and really very different from the normal deep-sea fauna. For one thing, it's so abundant. This is a nice thing to have found because it seems to be good documentation of the fact that food really is the limiting factor. Here is a source of large quantities of food in the deep sea and, lo and behold, there is a large amount of biomass taking advantage of it. And these individuals are relatively large, as opposed to most of the animals of the deep sea.

Another point about the deep sea that deserves mention is that almost all deep-sea animals are deposit feeders. Here, I am talking about mud bottom, where suspension feeders and carnivores are rare. For years now, there has been a debate about how there can be so many deposit feeders in the same space at the same time. It is a very difficult issue, and it probably never will be solved in the deep sea.

Q: With the discovery of large biomass concentrations at the vents, is there any hope that further study of the deep-ocean biology will lead to an increase in the food that will be available to us from the ocean?

A: Only to the most limited extent. Except for the vent community, which is inedible because of ambient water chemistry, life is too sparse on the ocean floor to support a fishery. There is a limited fishery for deep, bottom-associated fish, but the turnover rate is so slow in the deep sea that such areas would be easily fished out.

RUTH D. TURNER—INVESTIGATOR OF DEEP-SEA BENTHOS

Although her initial interest was in birds, Ruth Dixon Turner has gained international eminence as a benthic marine biologist specializing in molluscan anatomy and systematics. Dr. Turner authored definitive monographs on the clam families Teredinidae (shipworms) and Pholadidae (includes deep-sea wood borers), and she is the first woman to dive in the DSRV Alvin (see Fig. 1).

At an age of 70 plus, Ruth Turner still talks excitedly as she details anecdotes from her experiences—ones usually including the contributions of her colleagues who were involved directly or indirectly in her investigations. One of her major interests is the comparison of the ecosystems that develop around wood islands (artificially deposited masses of wood on the deep-ocean floor) with those of

FIGURE 1

Dr. Ruth Turner is shown preparing to enter the deep-sea research vessel, *Alvin*. Dr. Turner was the first woman to dive in this marvelous tool, which has allowed so many deep-sea researchers to see firsthand the realm they wish to learn more about. Many experiments that helped Dr. Turner research her deep-sea wood borers were made possible only because of the availability of *Alvin,* in which she has made approximately 50 dives. (Photo courtesy of the Museum of Comparative Zoology, Harvard University.)

hydrothermal vents. Dr. Turner is particularly interested in securing data related to feeding types, life histories, and growth rates. Both ecosystems are located in the deep ocean, where the physical environment has long-term stability, and each is supported by a stable high level of primary productivity (wood and chemosynthetic bacteria). Long-term observations of the wood islands and hydrothermal vents should help test the hypothesis that stable environments with adequate primary productivity will produce complex, diverse communities, with the greatest diversity occurring at the predator level.

Ruth Turner's view of science is that it should be fun, it can't be fully scheduled, and it is wonderful. It requires the cooperative effort of many individuals and, above all, "Nothing replaces knowing your animals."

12

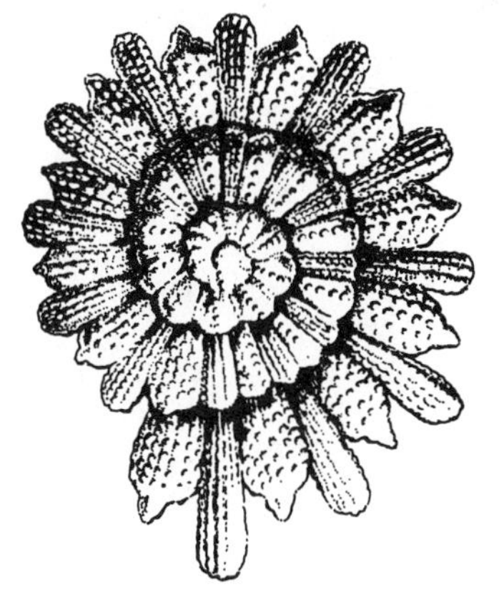

THE INTERTIDAL ZONE

Nature is kind to those of us who are curious about life in the oceans. At most shorelines of the world, the water level drops and rises due to gravitationally driven tides. For those few hours that the tide recedes, we can examine the edge of the marine world. This is the **intertidal zone.**

The intertidal zone is a benthic habitat. The benthic organisms that make their home in the intertidal can be categorized according to the substrates they occupy. Algae and benthic diatoms are referred to as **epiflora,** and attached animals as **epifauna.** Animals that live burrowed in the sediments are **infauna.** All epiflora are fixed to the substrate and cannot move (i.e., they are sessile). Epifauna such as barnacles, sponges, mussels, bryozoans, and tunicates are sessile. Other benthos such as sea stars, snails, chitons, limpets, and flatworms have limited movement. A few animals—crabs, lobsters, isopods, and other crustaceans—can move more quickly. The infauna move very little. Many clams, polychaetes, and burrowing crustaceans can tunnel slowly in the sediments, but their movement is limited to inside the burrows. The intertidal benthos, then, are marine organisms that, because of their limited ability to move, are exposed when the tide recedes.

The intertidal zone is small in area when compared with the pelagic zone, or the bottom of the sea. There are 450,000 km (270,000 miles) of coastline on earth. If we assume that the average width of the intertidal zone is 80 m (262 ft), then the intertidal habitat of the world includes 36,000 km^2 (13,900 mi^2), or less than 0.002 percent of the ocean surface. Intertidal communities are most extensive in temperate latitudes. Because of its permanent ice cap, the Antarctic has no intertidal zone. In the Arctic, tidal amplitude is less than a meter, and because winter ice freezes to a depth of 2 m (6.6 ft), the intertidal zone is frozen most of the year, greatly restricting the intertidal flora and fauna. In the tropics, tidal amplitude is generally less than a meter, and so intertidal communities are not extensive. In temperate latitudes, well-developed tides combined with moderate environmental conditions give rise to rich and varied biological communities.

The flora and fauna of the intertidal zone are exposed to a much wider range of environmental conditions than is generally found in the oceans (Fig. 12.1). For example, temperature rarely changes more than 1°C (1.8°F) in any 24-hour period in the open ocean. In the intertidal zone, however, temperature may vary 10–20°C over a single tidal period of exposure and immersion. Heat from the sun and drying from wind also contribute to rapid changes in the intertidal zone. The extent to which a particular organism is exposed to the varied environmental conditions depends on its location in the intertidal zone. At the lower limit of the intertidal zone, the period of exposure may be only minutes, while at the upper limit, exposure is virtually continuous.

The most important physical factors affecting the intertidal zone are tides and waves. Tides establish the extent of the intertidal zone. Waves affect the spacing and types of organisms found. On rocky shores, the extent of wave action determines how securely organisms

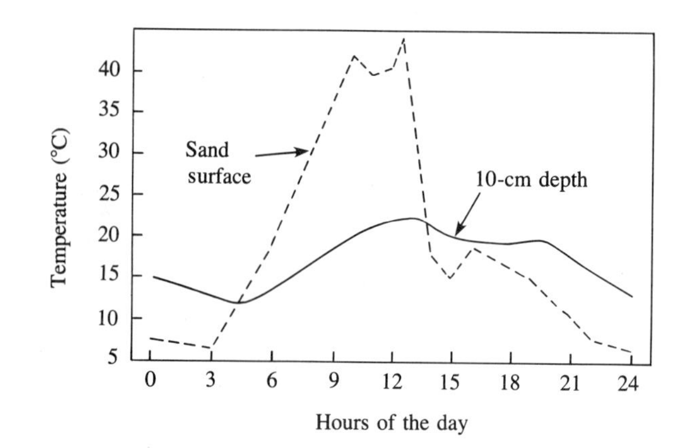

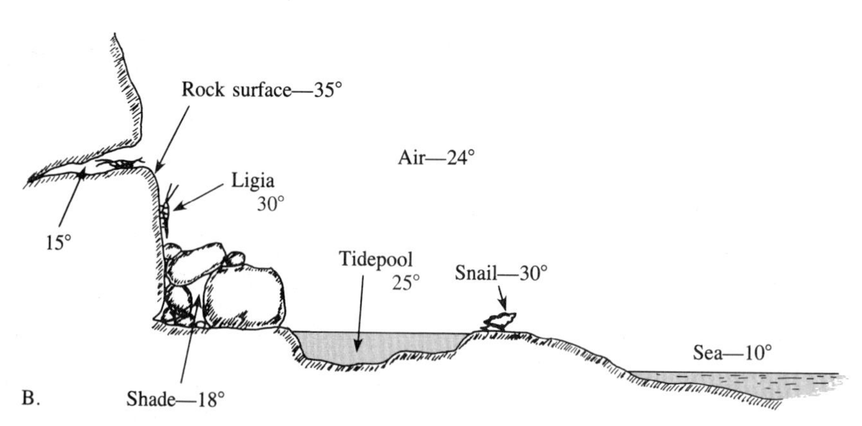

FIGURE 12.1 Environmental Conditions in the Intertidal Zone

Temperatures (in degrees Celsius) at the intertidal surface change rapidly and widely in a 24-hour period. **(A)** Organisms that burrow in sand or mud find more uniform conditions than those at the surface or in the atmosphere. **(B)** The rocky intertidal may have microhabitats with conditions considerably different than sun-exposed surfaces. (After Carefoot, 1977.)

must attach to the rocks. On shores made of particulate material, the decreasing degree of wave action determines if the shore will be gravel, sand, or mud.

In order to understand the ecology of the intertidal zone it is important to understand the physical basis of waves and tides. In this chapter we will consider waves and tides first, and then will examine the types of organisms found in various intertidal communities and the major ecological factors involved with the adaptation of life to the stresses of the intertidal zone.

WAVES

Both ocean waves and tides are basically wave phenomena, the former driven by atmospheric winds and the latter by the gravitational attractions of the moon and sun. As we are more familiar with the wave forms of wind-generated waves, we will begin our discussion with them.

Winds produce the most readily observed form of ocean water motion: waves. These wind-generated waves advance across the ocean's surface, so they are called **progressive waves.** The wave form is composed of a crest, the highest part of the wave, and a trough, the lowest part of the wave, a series of which move across the ocean surface as a wave train (Fig. 12.2). The horizontal distance between corresponding points on successive wave forms in a wave train is the wavelength *(L)*, and the vertical distance from trough to crest is the wave height *(H)*. The time required for one wavelength to pass a fixed point is the period *(T)*. The speed of waves is determined by use of the formula

$$S = L/T$$

(See Fig. 12.3 for a graphic depiction of this relationship.)

As the wave form moves across the ocean surface, the water particles do not travel with the wave but in-

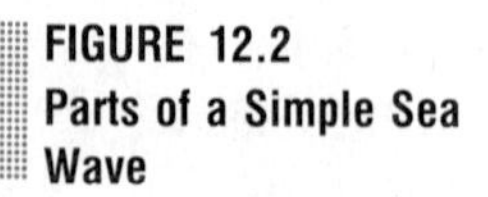

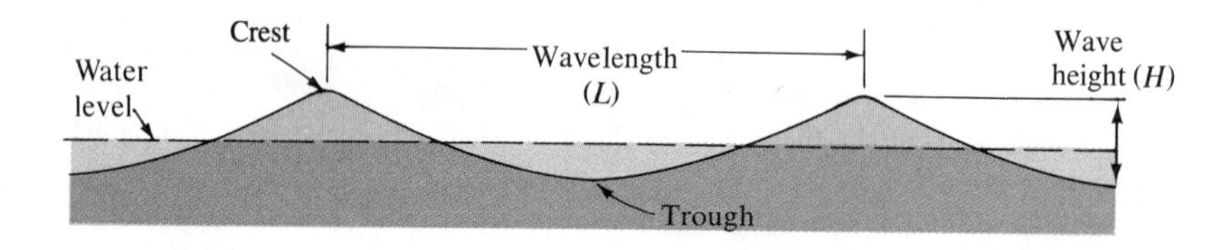

FIGURE 12.2 Parts of a Simple Sea Wave

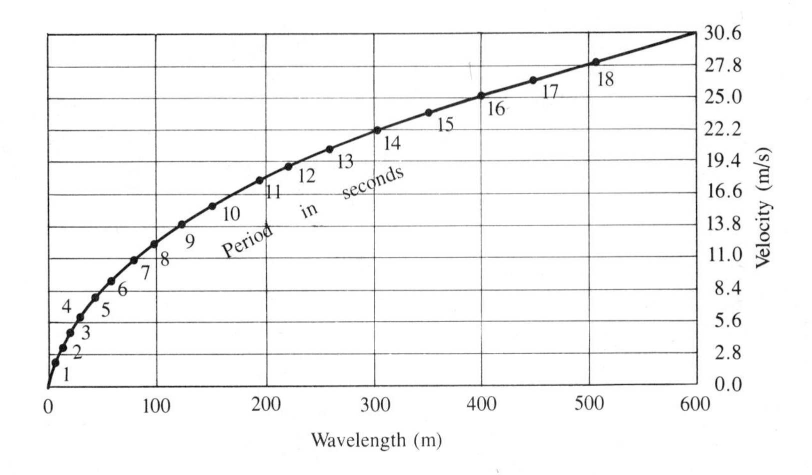

FIGURE 12.3
Speed of Deep-Water Waves

The theoretical relationships between wave speed, wavelength, and period for deep-water waves. Wave speed is equal to wavelength divided by period.

stead move in an orbital path (Fig. 12.4). This orbital motion extends to a depth of one-half wavelength below the surface, and particles move through the top half of their orbit in the direction the wave is traveling.

As light winds blow over the water, tiny ripples (less than 1.74 cm in length) are formed. These are called **capillary waves,** and disappear as soon as the wind stops. As the wind increases above 6 km/hr (4 mph), larger waves, called **gravity waves,** are formed. These gravity waves, unlike capillary waves, do not stop when the wind stops. Gravity waves continue until the water becomes too shallow, and they die as surf. Gravity waves formed by storms in the South Pacific off Antarctica have been known to travel the whole length of the Pacific and crash ashore on the coast of Alaska. Gravity waves more than 30 m (98 ft) high and 300 m long have been measured.

With increased wind energy, waves become longer, higher, and move at greater speeds. Since wave height increases faster than the length, the steepness *(H/L)* also increases. If the steepness reaches a ratio of the wavelength divided by seven (*L*/7), then the wave becomes unstable and breaks as a whitecap (Fig. 12.4).

Most wind-generated waves form in regions of the ocean where storm winds—called **seas**—blow. What factors affect the size of waves? There are three: wind speed, the time the wind blows in the same direction, and the **fetch,** that is, the distance over the water that the wind can blow uninterrupted. Once waves leave the area of the sea, they take the form of long, rounded waves called **swells.** Swells lose very little energy as they move over the ocean.

The energy of swell and other stable waves is released in the surf zone of the shallow water at the shoreline. When the water depth becomes less than one-half the wavelength (*L*/2), the shallowness interferes with the circular particle motion associated with the wave. Wave energy is lost due to frictional resistance,

FIGURE 12.4
Movement and Forms of Wind-Generated Waves

(A) Water particles within a deep-water wave move in orbital paths. The size of the orbit diminishes with increasing depth. **(B)** As energy is put into the ocean surface by wind, small rounded waves with V-shaped troughs develop (capillary waves). As the water gains energy, the waves increase in height and length. When they exceed 1.74 cm in length, they take on the shape of the sine curve and become gravity waves. Increased energy increases the steepness of the waves. The crests become pointed and the troughs rounded. As the steepness reaches 1/7, the waves become unstable, and whitecaps form as they 'break'.

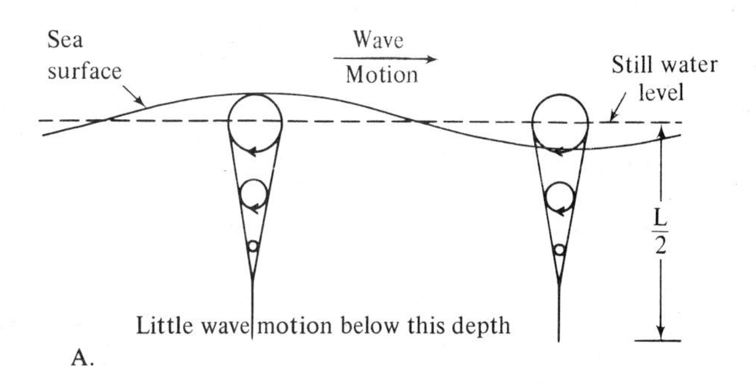

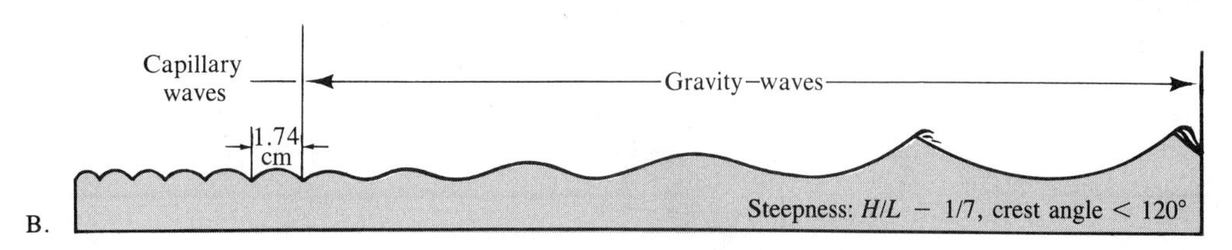

FIGURE 12.5
Release of Wave Energy in the Surf Zone

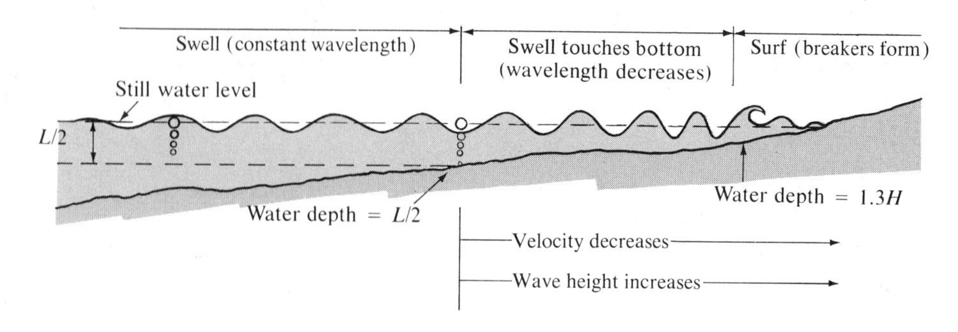

and the base of the wave slows down, becoming shorter, higher, and steeper. The wave breaks when the steepness reaches $L/7$. This usually occurs when the water depth has decreased to $1.3H$ (Fig. 12.5).

Since different segments of a wave are in different depths as it approaches shore, the wave does not slow down at the same rate. As a result, waves bend or refract, and the concentration of energy released along the shore will vary. Orthogonal lines drawn perpendicular to a wave identify segments with the same energy. Observing how these orthogonal lines bend as a wave approaches shore over uneven depth helps to understand wave refraction (Fig. 12.6). Due to refraction, the release of energy along the shore is higher on headlands jutting out into the ocean, where orthogonal lines converge, and lower in bays, where the orthogonal lines diverge. Erosion is greatest along high-energy segments of the shoreline, while deposition occurs along low-energy segments and results in the formation of beaches. Very different biological communities are found on high- and low-energy shorelines.

TIDES

The second wave phenomenon in the ocean that profoundly affects the intertidal habitat is the **tide.** Tides are generated by the gravitational attractions between the earth and moon and the earth and sun. Examined as a wave, tides have a wavelength that is half the circumference of the earth and a height that averages 1 m (3.3 ft), although they may reach heights of greater than 15 m as in the Bay of Fundy in Nova Scotia. High tide marks the passing of the tide wave crest, and low tide the passing of the tide wave trough. The tidal range is the vertical difference between high tide and low tide. Obviously, plants and animals that live in the intertidal zone—where they are alternately covered and uncovered—have adapted to an environment that is much more variable than is typical throughout most of the ocean. The adaptation of the reproductive behavior of a fish to tides is discussed in the special feature, "Grunions and the Tides", at the end of this chapter.

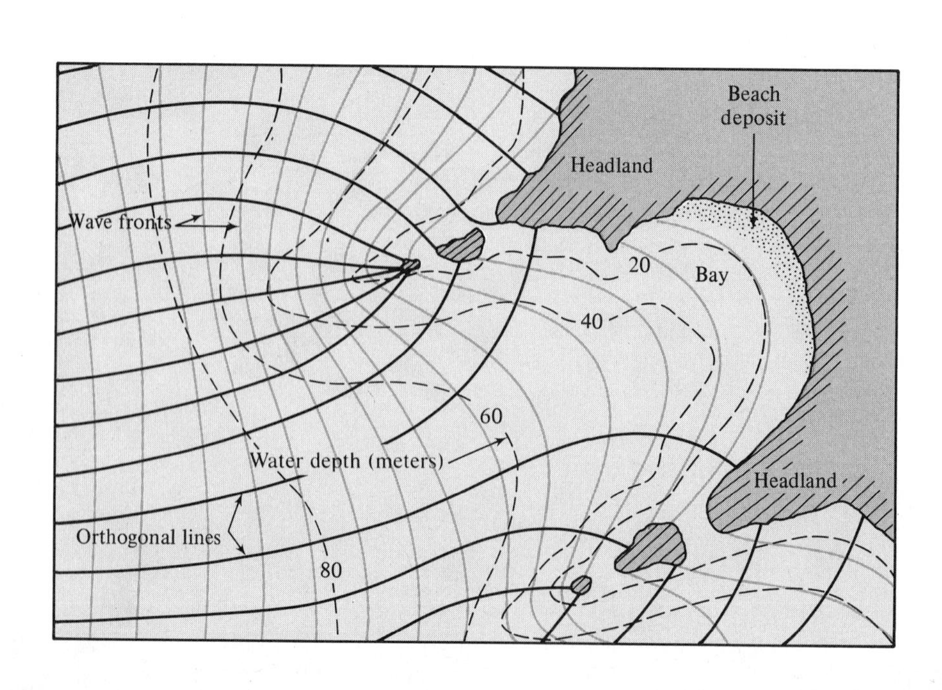

FIGURE 12.6
Refraction of Waves Along Shoreline

As the waves 'feel bottom' first in the shallow areas off the headlands, they are slowed. The segments of the waves that move through the deeper water leading into the bay are not slowed until they are well into the bay. As a result, the waves are *refracted* (bent), and energy is concentrated on the stacks and headlands. Spaces between the orthogonal lines represent equal amounts of energy.

The tide-generating force of the sun is only 46 percent that of the moon because of its greater distance from the earth. Thus, the moon dominates the earth's tides. Every 29.5 days the moon completes a synodic month, or cycle of phase changes—new moon, first quarter, full moon, third quarter, and new moon again (Fig. 12.7). The tides are greatly affected by these phase changes because the tides we experience on the earth are the combined effect of the relative positions of the solar tide (with crests pointed toward and away from the sun), and the larger lunar tide (with crests pointing toward and away from the moon).

During the new moon phase—when the moon is on the same side of the earth as the sun (i.e., is in conjunction), and is nearly aligned with a line running through the center of the earth and sun—and about two weeks later, during full moon, when the moon is in a similar alignment but on the opposite side of the earth from the sun (in opposition), the crests and troughs of the solar tide and the crests and troughs of the lunar tide are at about the same position on the earth's surface. Adding together the displacements of the ocean above and below the mean tide level caused by each of the tides produces higher high tides and lower low tides, and a maximum tidal range called the **spring tide.** Occurring halfway between the new and full moon phases are the **quadratures**—first-quarter and third-quarter phases—when the lunar tide crests align with the solar tide troughs, and vice versa. This phenomenon occurs because the moon and sun are at right angles to one another, relative to the earth. The lunar crest is reduced by an amount equal to the solar tide trough, and the lunar trough is reduced by the amount of the solar crest. This produces a high tide that is not very high, and a low tide that is not very low. This condition of minimum tidal range associated with the first and third quarters is called the **neap tide.** Every month we experience two spring tides, separated by neap tides. Figure 12.8 shows how the effects of the lunar and solar tides combine during the period of one day to produce significantly different tidal ranges during a new-moon spring tide and a third-quarter neap tide.

Because the moon may move from as much as 28.5° north of the equator to 28.5° south of the equator, and back again, in one month, the two high tides or low tides that occur during a day may not be of the same height. The difference in these heights is called **diurnal inequality.** This diurnal equality of the moon is complicated over a year's time by the declination of the earth relative to the sun, which results in the sun traveling a course from 23.5° north to 23.5° south of the earth's equator and back again. Along with these changes in declination, the distance between the earth and the moon changes over a period of one year. The combined effects of all these variables, plus varying ocean size and depth and friction, produce three fundamental types of tides throughout the world. Each type requires some discussion.

The classification of tides observed around the world is based on tidal period. The **diurnal tide** has

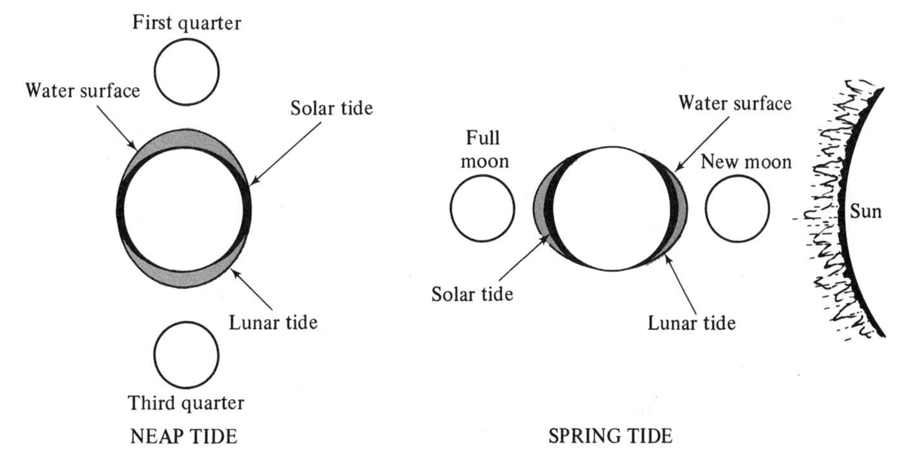

FIGURE 12.7
The Effect of Lunar Phase Changes on Earth's Tides

When the moon is in the new or full position, the tidal bulges created by the sun and moon are aligned, producing a larger bulge. When the moon is in the positions halfway between the new and full phases—the first and third quarters—the tidal bulge produced by the moon is at right angles to the bulge created by the sun. The bulges tend to 'cancel' each other, and the resulting bulge is smaller. New moon and full moon phases produce spring tides with maximum tidal ranges, while the first and third quarter phases of the moon produce neap tides with minimal tidal ranges.

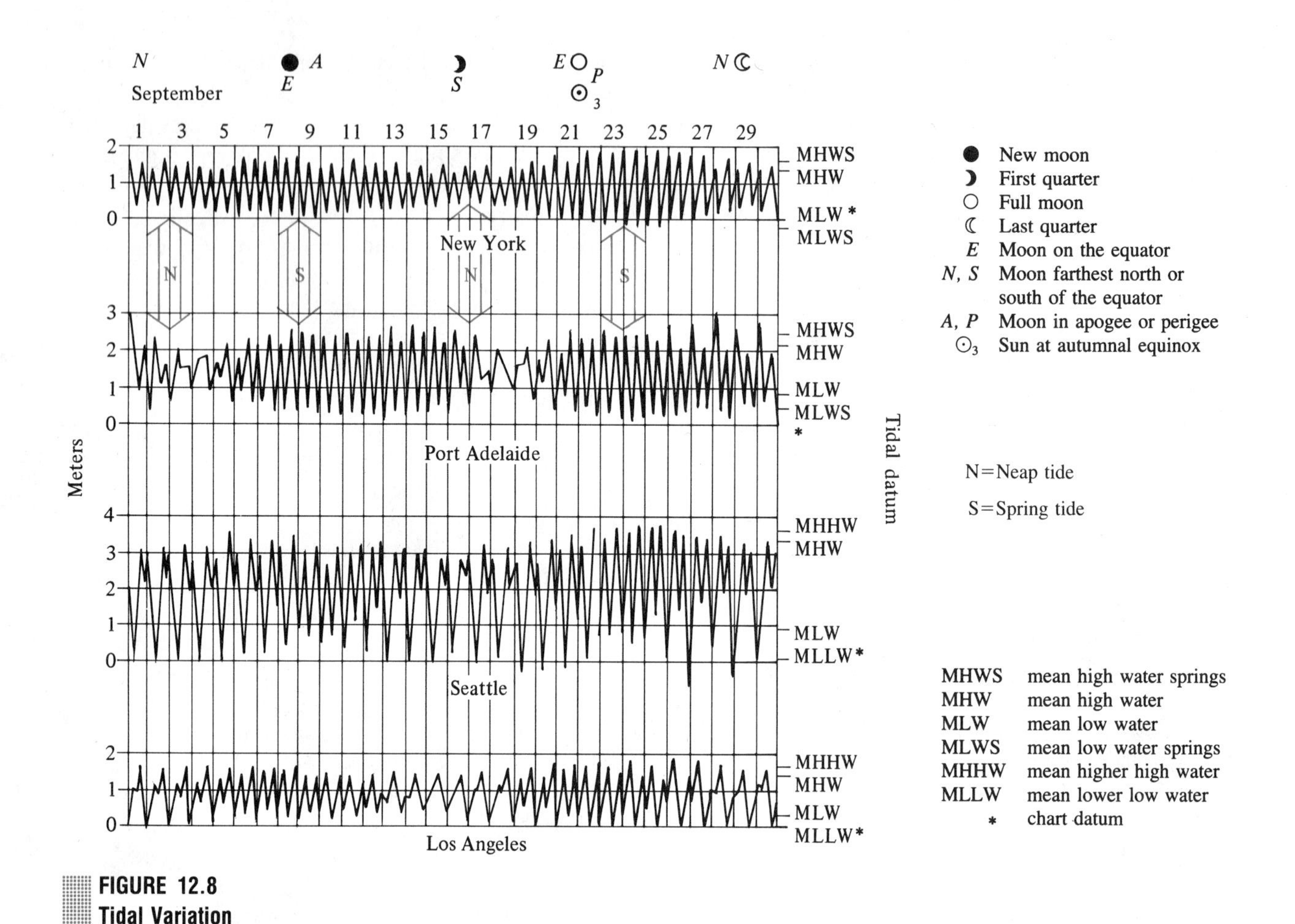

FIGURE 12.8
Tidal Variation

The declinations of the moon and the sun, the phase of the moon, and the position of the moon in its orbit all contribute to the tidal variations. The position of the earth in its orbit around the sun and the configuration of the sea bottom and basin boundaries are also factors. (From C. Hauge. 1972. Tides, currents, and waves. *California Geology,* July. Reprinted by permission of the author.)

one high and one low per tidal day. Since the tides are dominated by the moon rather than the sun, the tidal day is a lunar day of 24 hours and 50 minutes, rather than the 24-hour solar day. Thus, the period of the diurnal (daily) tide is 24 h 50 min. Such tides are the dominant tide found in the Gulf of Mexico, Aleutian Islands, and along the Pacific coast of southeast Asia. **Semidiurnal** (semidaily) **tides** have a period of 12 h 25 min, and are characterized by two high and low waters per day, with little or no diurnal inequality. This is the most common type of tide occurring throughout the world, and is found at most of the open coastal regions of the Atlantic Ocean. **Mixed tides** display both diurnal and semidiurnal periods (Fig. 12.9), and are common along the west coast of North America. When semidiurnal conditions prevail, there is a significant diurnal inequality. Even though a tide at a place can be identified as one of these types, it may pass through stages of one or both of the other types. The tides at four different locations during the month of September are compared in Figure 12.9.

A significant feature of the tides is the tidal current that develops as water moves in and out between shores, or restrictions that intensify the flow of water during flooding and ebbing of the tides. The speed of tidal currents varies from zero at tidal extremes—when water is about to change its direction of flow—to maximum values halfway between tidal extremes. Although tidal currents are of great concern to navigators, they are much less important biologically than the breaking of ocean waves and the physical changes of sea level that the tides bring.

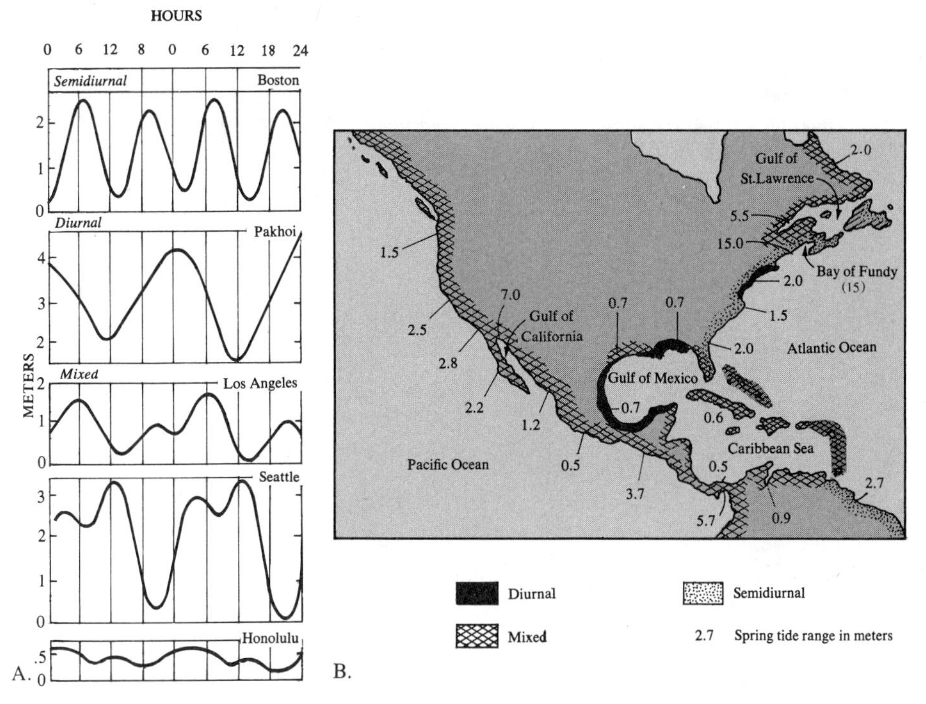

FIGURE 12.9
Types of Tides and Their Locations in North America

(A) In a semidiurnal (twice daily) type of tide, there are two highs and lows during each tidal day, and the heights of each successive high and low are about the same. In the diurnal (daily) type of tide, there is only one high and one low each tidal day. In the mixed type of tide, both diurnal and semidiurnal effects are detectable, and the tide is characterized by a large difference in the high water heights, the low water heights, or both, during one tidal day. **(B)** Maxim tidal range for each type of tide varies at different locations. The numbers give the spring tide range in meters (the maximum tidal range that can be expected). Storm waves, lower barometric pressure, ocean currents, and the alignment of additional celestial bodies could increase the range. (After C. Hauge. 1972. Tides, currents, and waves. *California Geology,* July.)

THE INTERTIDAL COMMUNITY

Since the intertidal zone is a transitional zone between the terrestrial and the marine environments, so is sometimes exposed to the atmosphere and sometimes covered by seawater, we might expect to find in it a mixture of marine and terrestrial organisms. To some extent, we do. Salt marshes and mangroves, both found in the mid- to upper part of the intertidal zone, are dominated by terrestrial plants that can tolerate periods of immersion in salt water. Salt marshes and mangroves, however, are an exception to the general pattern. The plants and insects that dominate terrestrial environments are generally absent in the marine environment, and the intertidal flora and fauna belong to taxonomic groups that are distinctly marine. Intertidal organisms often show striking adaptations to atmospheric exposure caused by the tides, but their evolutionary affinities are with the organisms of the sea.

Although restricted in distribution to the edge of the sea, the biological communities of the intertidal zone are very diverse and rich. Most phyla have intertidal benthic representatives, and the diversity of marine life in the intertidal zone reflects a wide range of morphological and functional adaptations based on the particular characteristics of the various phyla. Low tide, because it offers a brief opportunity to examine marine organisms under near-natural conditions, has been the most important laboratory of marine biologists. Much of what we know about the physiology and ecology of marine organisms has come from studies in the intertidal zone. This zone includes some of the most biologically productive communities known on earth. Many salt marshes, for example, produce more biomass per year than any crop from energy-subsidized human farming.

Several patterns of distribution of organisms are apparent in the intertidal zone. The most obvious one is

FIGURE 12.10
Species Richness in the Intertidal Zone

In general the number of species increases toward the low tide level. This graph is for the littoral zone of a mixed rocky and soft substrate shoreline at Partridge Point in Puget Sound.

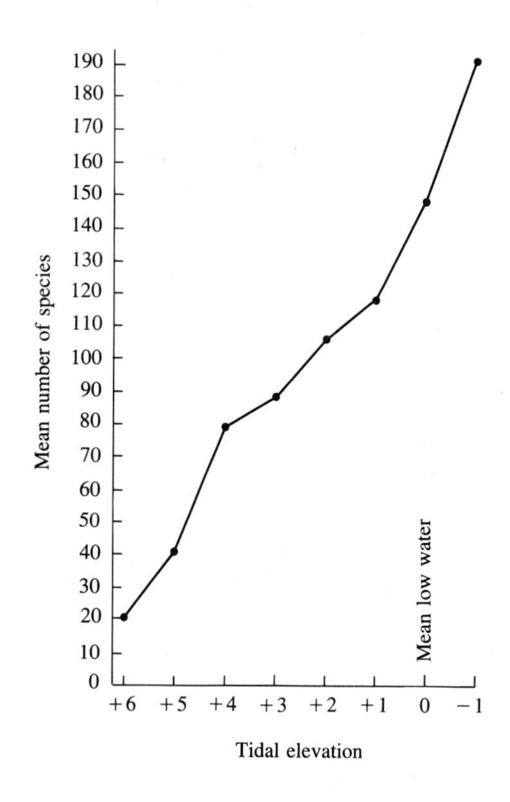

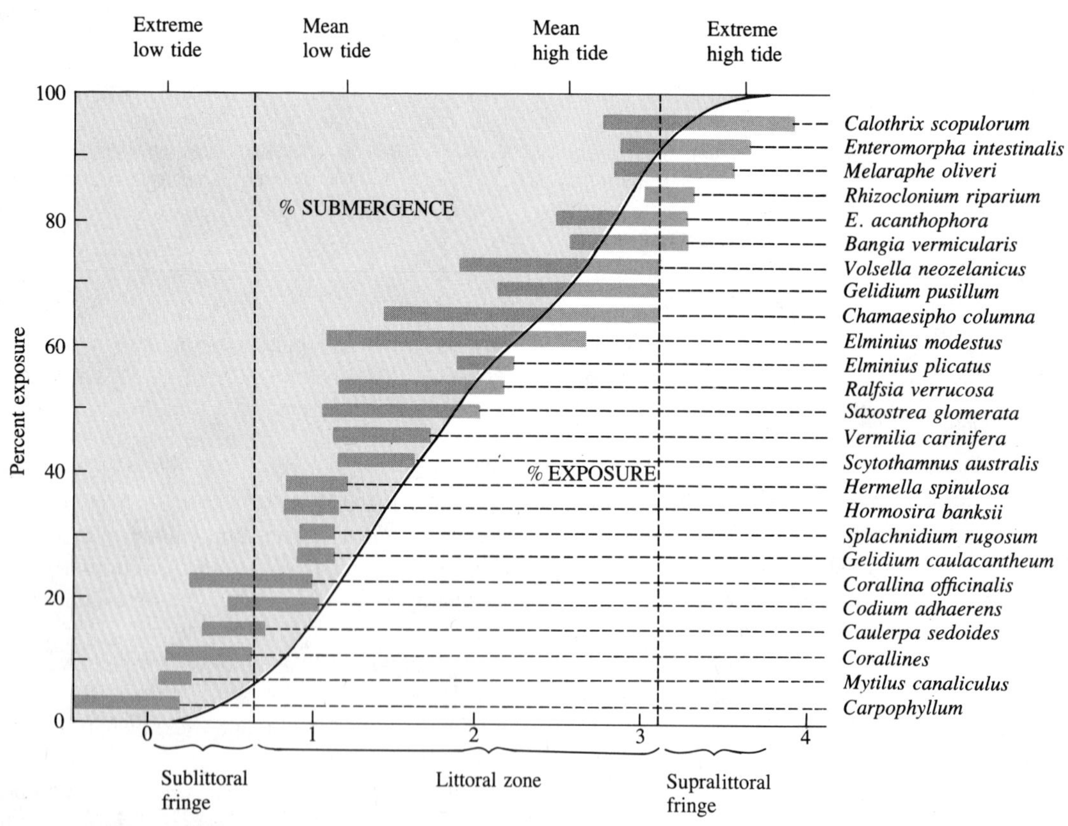

FIGURE 12.11A
Zonation of Intertidal Organisms

Organisms in the intertidal zone are generally found in only a narrow range of tidal elevation. This graph shows the zonation of intertidal algae and animals from a rocky shore in New Zealand.

FIGURE 12.11B
Zonation of Major Species on Rocky Shores

This figure is a general scheme of common animals and algae found in eastern North America. Details will differ for specific locations. (After Amos, W.H. 1966.)

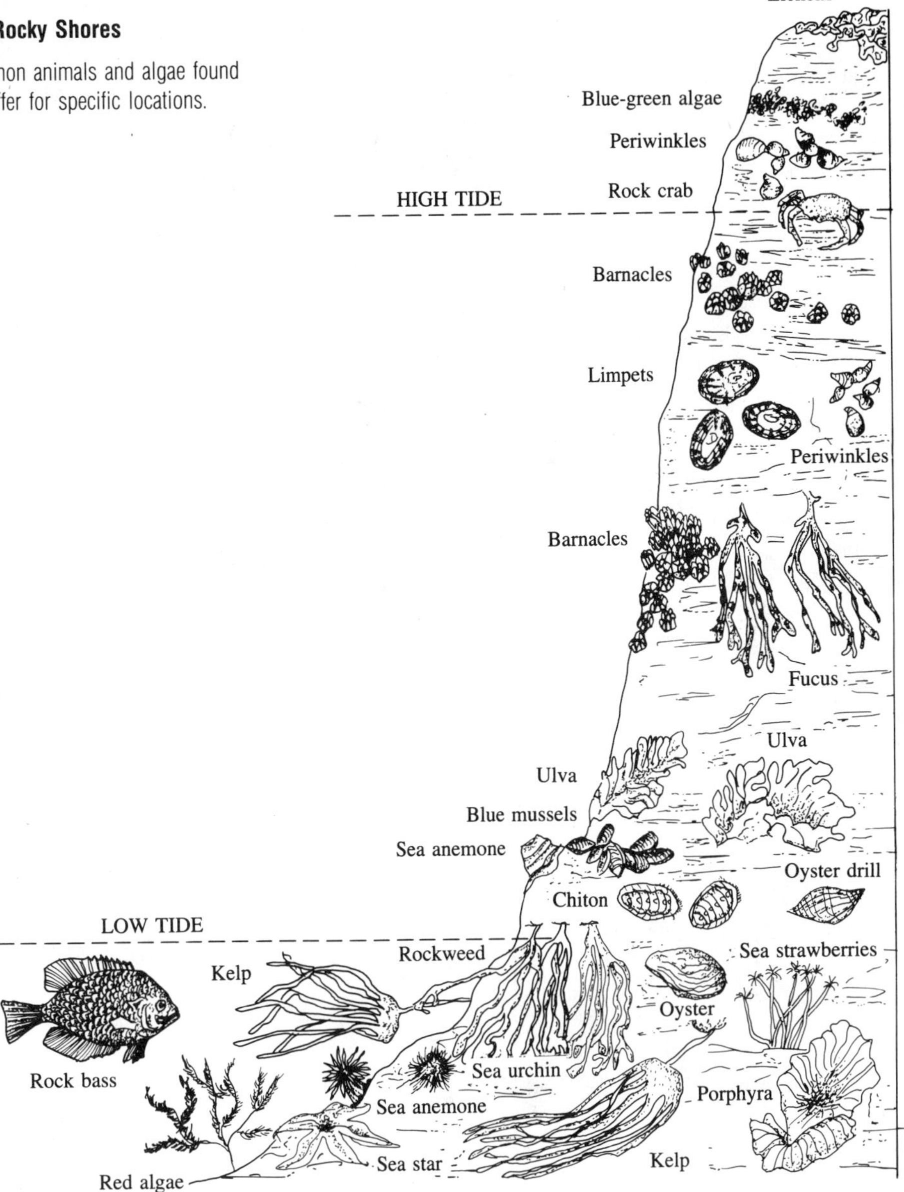

the steady increase in species richness (the number of different flora and fauna) from the high tide level to the low tide level. The higher the level in the intertidal zone, the greater the exposure time to the atmosphere and the greater the impact of the harsh environmental conditions associated with this exposure. Only a few hardy organisms can tolerate the exposure of the high tide level. The variation in the number of species from a rocky shore in Puget Sound is shown in Figure 12.10. Because of the complexity of environmental factors found in the intertidal zone, few species are adapted to the full range of physical and biological conditions encountered throughout the tidal cycle. In fact, most species are restricted to only some part of the intertidal zone.

Zonation is one of the most interesting biological features of the intertidal region, and has been studied most extensively on rocky shores, where the distribution of large organisms can be conveniently observed. Zonation, however, is characteristic of all intertidal habitats. The zonation of intertidal algae and animals of a rocky shore is shown in Figure 12.11A; that of common animals and algae of a rocky shore is shown in Figure 12.11B.

The intertidal zone can be divided into two major habitats on the basis of substrate type. Rocky shores are characterized by a solid substrate. Particulate shores consist of sediment particles ranging in size from clay through sands and gravels to cobbles. Particulate shores include cobble, gravel, and mud beaches.

Rocky Shores

The rocky intertidal zone is one of the most luxuriant marine environments. The combination of solid substrate for attachment and the frequent wave action and clean water create a very favorable habitat for marine organisms. Hundreds of species representing most phyla are found crowded on the rocky surfaces. Because of the abundance of easy-to-observe epibenthos and the relative ease of access, rocky shores are the favorite intertidal habitat for study by marine biologists.

The patterns of distribution of organisms on rocky shores are readily seen. Typically, a band of barnacles or sometimes mussels is located in the midtide zone. Snails of the genus *Littorina* are found near the high-tide line. Large concentrations of macroalgae are typically found only at or near the low-tide line.

One of the important questions of marine biology in the late nineteenth and early part of the twentieth century was relating the zonation of flora and fauna of rocky shores to specific tide heights. The most extensive work was done by a British husband-and-wife team of marine biologists, T.A. and Anne Stephenson. Examining rocky shores world wide, they found and described a common pattern of three zones: the supralittoral fringe, the eulittoral zone, and the sublittoral fringe (see Fig. 12.12). Each of the three zones corresponds to a general tidal elevation. (See a discussion of spring and neap tides, above). Note that the supralittoral fringe generally extends somewhat landward of the height of the highest spring tides, and that there are some days (during neap tides) when this zone is not wetted at all by the tides. The eulittoral zone is the part of the intertidal that is covered and exposed in each 24-hour period, although the relative times of exposure and immersion differ from spring to neap tide. The sublittoral fringe has as its lower boundary the extreme low water of spring tides. Because the lower level of the spring tide varies, most of this area is exposed only briefly every two weeks. In fact, the lower limit of the sublittoral fringe may be exposed for only a few hours once or twice a year, at the time of the most extreme spring tides.

The Stephensons' extensive examination of rocky shores led to two important conclusions. First, the three intertidal zones are found world wide, and the same groups of organisms can be expected in each of the zones. Second, the upper and lower boundaries of these three zones cannot be set at precise tidal elevations because of the action of waves.

Surf action, as well as tides, affects the distribution of marine life in the rocky intertidal. At exposed locations where waves are large, the surf may regularly wet the intertidal zone to a level well above the high-tide line. In such locations, the distribution of organisms is shifted upward relative to tide heights (Fig. 12.13). In very exposed locations, the supralittoral fringe and much of the eulittoral zone may be elevated entirely above the level of the extreme spring tides.

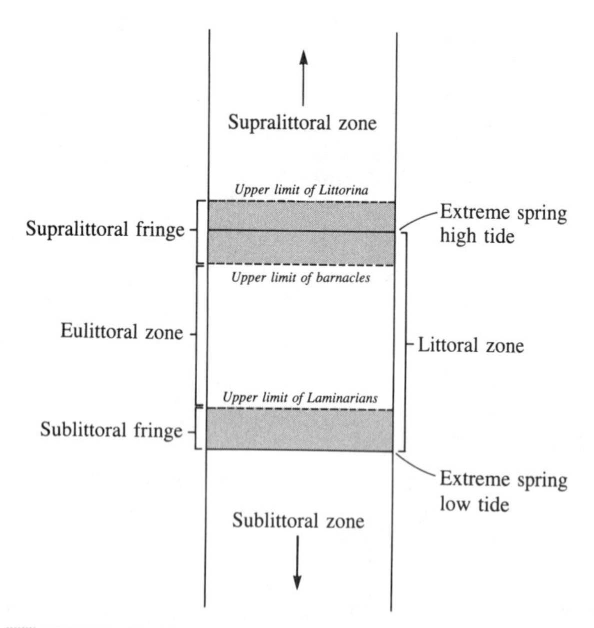

FIGURE 12.12
General Scheme of Intertidal Zonation

The intertidal zone is divided into three segments. The supralittoral fringe is wetted by the ocean but is for the most part above the level of high tides. The eulittoral zone is generally covered and uncovered by the tides each day. The sublittoral fringe is only infrequently exposed.

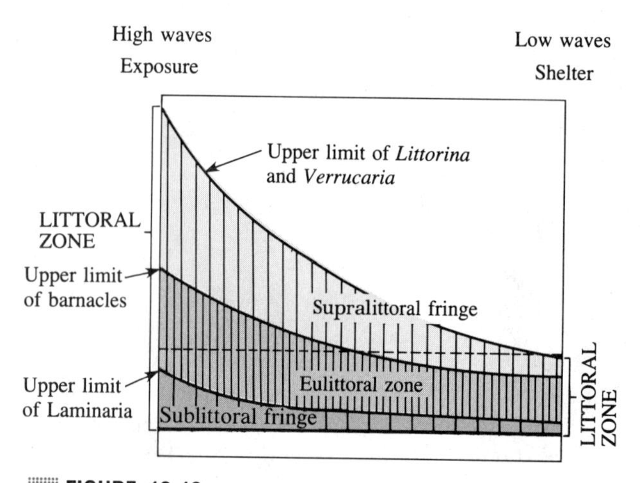

FIGURE 12.13
Surf Action in the Intertidal Zone

Wave action, in addition to the tides, affects the distribution of benthic organisms in the intertidal zone. In general, the boundaries of the zones shift upward with greater surf action.

Other environmental factors affect the position of the three zones relative to the tide height. On southern-facing shores (in the Northern Hemisphere), the three zones occur at lower tidal elevations than they do on shores with a northern or shadier exposure. The kind of rock also can affect the height of the zones. Black basalt rock absorbs more heat and becomes hotter than lighter-colored limestone or sandstone. The three zones are found at lower tidal elevations on basalt shores.

Clearly a number of interrelated environmental factors influence the distribution of marine life on rocky shores. Despite this complexity, however, three general zones can be described for most rocky shores, and a group of organisms is characteristic of each zone.

The Supralittoral Fringe. The supralittoral fringe is the part of the intertidal zone that is generally above the height covered by the tide each day. This highest of the zones on rocky shores has a number of indicator species. Periwinkle snails (*Littorina;* see Fig. 12.14) and isopods *(Ligia)* are the most common animals. Vegetation is dominated by the black band. This black band, which appears as a series of blotches, consists of lichen and/or cyanobacteria. Lichens are mostly the genus *Verrucaria,* and the cyanobacteria is commonly of the genus *Calothrix*. Occasionally, red algae *(Porphyra, Hildenbrandia)* or brown algae *(Fucus)* are found in the lower part of this zone, particularly during seasons when exposure is minimal.

At rocky shores with high wave action, the supralittoral fringe may be well above the high-tide level and is dependent on wave spray for periodic wetting. Marine life here is particularly well adapted to the long exposure of the tidal cycle. The cyanobacteria *Calothrix* has adapted to the high-tide zone by secreting a mucilaginous sheath around each filament. This sheath decreases the loss of cell moisture during drying conditions, and protects filaments against fresh water inundation during periods of rain. *Calothrix* is one of the groups of cyanobacteria that can convert atmospheric nitrogen into nitrate molecules, which can be used as a nutrient for photosynthesis.

Verrucaria, which looks like black rubber tar patches, is a lichen common to both the east and west coasts of North America and Europe. The fungus component has thick resistant cell walls that can absorb up to 35 times their weight in water, and act like a sponge holding moisture during drying exposures. The other component of the lichen, a green alga, is particularly interesting because it is also found as a free-living filamentous form. When the reproductive spores of the alga land on the fungus, they are absorbed into it, but if the spores land on bare rock they grow into free-living filaments.

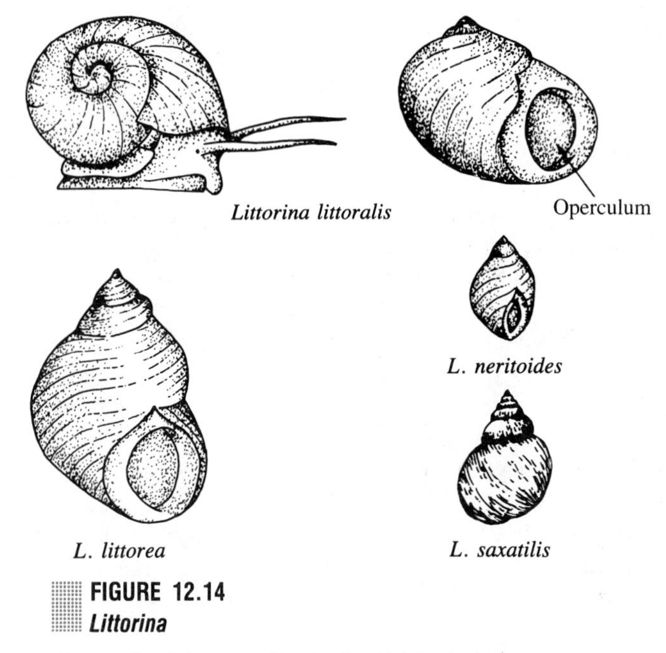

FIGURE 12.14
Littorina

Many snails of the genus *Littorina* live high in the intertidal zone. When exposed, the snail protects itself from desiccation by pulling back into the shell and covering the opening with the operculum. First it secretes a mucous thread that attaches the shell to the rock.

The littorinid snails found in the supralittoral fringe are also adapted to long periods of tidal exposure. They can survive out of seawater easily for over a week. These snails can retract completely into the shell and yet remain firmly attached to the substrate by a mucous thread. A mucous film also seals the operculum against the shell, and the mantle cavity holds considerable seawater, which provides moisture during exposure. Littorinid snails also avoid hot sun by seeking crevices or the shady side of rocks. While exposed to the atmosphere, littorinids are able to obtain oxygen directly across the surface of the mantle cavity. In this way, the mantle cavity acts as a lung, similar to that of land snails.

Another adaptation of littorinid snails to tidal exposure is the production of insoluble uric acid, rather than the more soluble urea or ammonia, as a waste product of metabolism. The water needed for excretion is thereby reduced, further helping the snails to withstand drying. Littorinid snails can survive heating to 47°C, a much hotter temperature than other marine snails can tolerate. Littorinids are also tolerant of low temperature, although some high-tide littorinids crawl to lower parts of the intertidal zone when the temperature is lower than 2°C.

The species of littorinid snails found in the upper tide zone apparently actively choose this highly exposed environment. When dislodged to lower tidal elevations by waves (or by experimenters), the snails move back to the high tide zone. Littorinids are herbivores and feed on microalgae. Although high-tide littorinids are tolerant of long periods of exposure, they are still dependent on water for feeding and reproduction.

The second animal characteristic of the supralittoral fringe is the isopod *Ligia,* also called the sea slater. Sea slaters are relatively large (up to 5 cm long) and are fast, active runners. During the day they generally seek out caves and crevices for protection from both predators and exposure, and they are only occasionally seen on exposed rocky surfaces. In periods of hot weather, however, European sea slaters purposely come into the sun to evaporate body water which, in turn, cools the isopod. They are not dependent on being covered by the tide to feed, and during the night they scavenge the drift of the high-tide line. They feed on drift algae or on the attached seaweeds found in the supralittoral fringe. *Ligia* is almost totally free of the ocean, even for reproduction. Following the general pattern of isopods, the female broods her eggs in a pouch on the body and only occasionally dampens the eggs by dipping her abdomen in the water. The eggs hatch into young, fully formed isopods that stay in the supralittoral fringe. In most respects, sea slaters have adapted to life in the atmosphere so successfully that they can almost be considered terrestrial animals.

The Eulittoral Zone. The eulittoral zone roughly corresponds to the part of the intertidal zone that is covered and exposed by the tides each day. Rocky shores exposed to high wave action have an elevated eulittoral zone compared to more protected shores. Marine organisms in the eulittoral zone frequently show adaptations to conditions of tidal exposure, but since these conditions are not as severe as at higher tidal levels, species diversity in the eulittoral zone is much greater than in the supralittoral fringe.

Acorn barnacles of the genera *Balanus* and *Chthamalus* are usually the most obvious species found in the midtide zone and are typically found as a distinct band starting at the top of the zone. Two other species are key indicators of the eulittoral zone: the blue mussel, *Mytilus,* and the alga *Fucus* or *Ascophyllum.* Often the mussel and alga are found together. Other organisms characteristic of the eulittoral zone are gastropod snails (particularly limpets), shore crabs *(Hemigrapsus, Carcinus),* and sea urchins. Toward the lower part of the eulittoral zone, many different organisms are found, including polychaete worms, algae, coelenterates, sponges, and other sessile animals. In the following paragraphs we will look more closely at the indicator species and other common organisms of the eulittoral zone.

Acorn barnacles produce a strong calcium carbonate box that protects them against mechanical abrasion, desiccation, and other environmental stresses of atmospheric exposure. They firmly attach to rocks by means of a tightly bonding glue. The glue is a mix of calcium carbonate and protein that hardens underwater. Barnacles are tolerant of a wide range of temperatures (from below freezing to around 44°C) and can tightly close their shell plates for periods of two to three days. During periods of exposure, barnacles can directly absorb adequate atmospheric oxygen through the soft tissues lining the shell. Even when the shell plates are tightly closed, a small opening, the micropyle, allows fresh air to enter the shell cavity. If exposure is prolonged, even the micropyle closes and the barnacles switch to anaerobic respiration.

One of the most interesting aspects of the biology of acorn barnacles is their selection of a place to live by pelagic larvae. The decision to settle is irreversible; once cemented in place, the barnacle remains in that spot for the rest of its life. Acorn barnacles have a specialized larvae—the cypris—that chooses a site for attachment. Cypris larvae (Fig. 12.15) have no digestive system and cannot feed. They rely on food reserves of the egg to provide the energy required for the larval period. The best-developed appendages of the larvae are a pair of antennules which they use to walk across the substrate, testing the surface for a suitable site. Once the site is selected, the cypris larva attaches by secreting cement from the antennules and then metamorphoses into a juvenile barnacle.

The second indicator species of the eulittoral zone is the blue mussel, *Mytilus edulis* (Fig. 12.16). At some rocky shores the mussel may replace acorn barnacles as the dominant species. At others there may be a codominance of mussels and barnacles. As adults, mussels are sessile and rely on their tight-fitting shell for protection from exposure. Mussels attach to the substrate by byssal threads that originate in a gland in the foot. This byssal gland consists of three parts: one secretes a protein, the second an organic molecule (a kind of phenolic substance), and the third an enzyme. The thread is a sort of epoxy resin. In the presence of the enzyme, the phenolic substance links with the protein to form a strong water-resistant strand. The foot is used to attach the strand to the substrate, and forms a thin, strong thread. A number of strands securely anchors the mussel to the substrate. Young mussels, freshly settled from the plankton, can use the foot to move over the rock surface, but once the byssal threads are attached, the mussel becomes sessile.

The third important indicator of the eulittoral zone of rocky shores is the brown alga rockweed *(Fucus* or *Ascophyllum).* Like mussels, rockweed may dominate

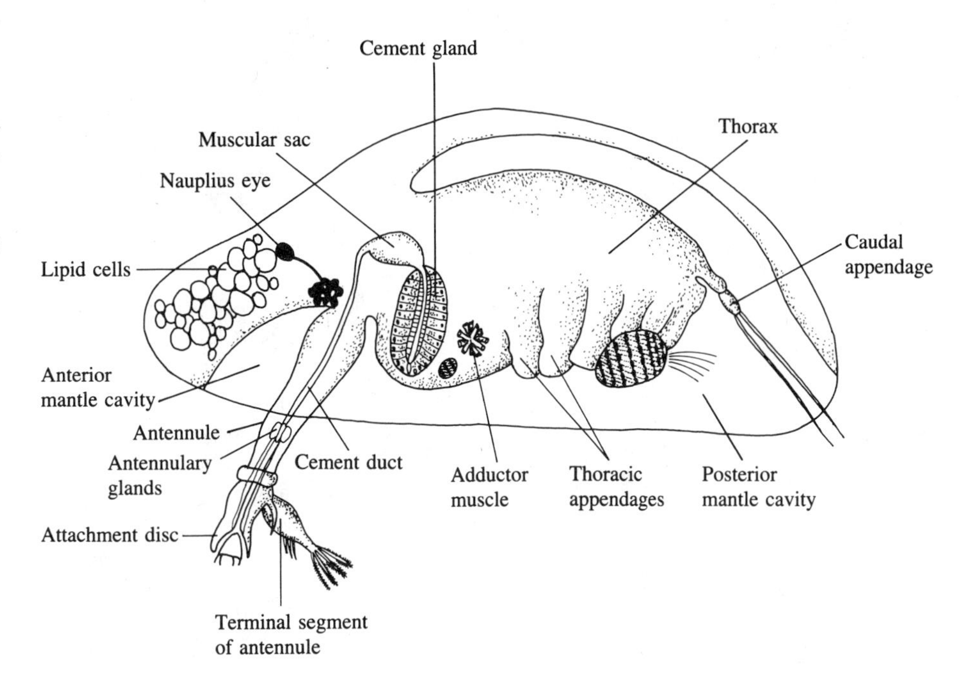

FIGURE 12.15
Cypris Larva

The cypris larva of the acorn barnacle is adapted to the task of choosing a suitable substrate. Note that there is no digestive system—metabolic energy is stored as lipids in specialized cells. Cypris larvae are around 0.5 mm in length.

the shore or may be mixed with barnacles. Rockweed is uncommon at locations with high surf action. *Fucus* or *Ascophyllum* can tolerate extreme drying, and experiments have shown it can recover even after losing 91 percent of its water.

On the rocky shores of the Pacific Ocean that are exposed to the full force of the waves, two additional animals are indicator species of the eulittoral zone: stalked (or goose-necked) barnacles and sea urchins. Stalked barnacles are larger than acorn barnacles, and have a fleshy stalk. Unlike acorn barnacles, they are not as resistant to desiccation and rely on spray from the nearly continuous surf for adequate wetting. Sea urchins—*Paracentrotus* in the European North Atlantic, and *Strongylocentrotus* in the northeastern Pacific—prefer rocky depressions. Where the rock is soft, they are typically found in burrows. The opening of these burrows is often smaller than the urchin, making these individuals prisoners of their own excavations. The urchins carve the burrows with the Aristotle's lantern (see Ch. 5) by continuously scraping at the surface of the rock. In optimal habitats, these urchins riddle the surface of the rock with burrows almost touching each other.

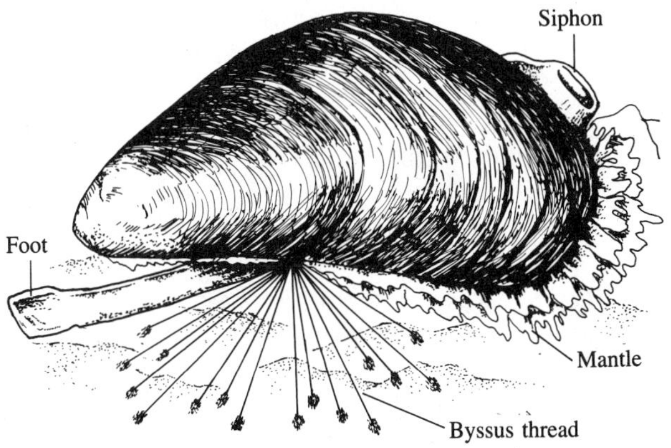

On the wave-exposed rocky shores of the Pacific Ocean, the small blue mussel, *Mytilus edulis*, is replaced by the much larger *Mytilus californianus*. This mussel reaches a length of 10 cm (4 in) and may be up to 10 times the size of *M. edulis*. *M. californianus* occurs in dense clumps that may be three or four animals deep. The densely packed mussels break the force of the waves, and the byssal threads trap sediment and create a moist microhabitat in which burrowing animals such as bivalve clams and polychaete worms, as well as epifauna such as sea anemones, limpets, and snails are common.

FIGURE 12.16
Blue Mussel

The blue mussel *Mytilus edulis* is commonly found in the eulittoral zone of rocky shores (or on pilings). A series of byssal threads from the base of the foot are used for attachment.

Other indicator animals of the eulittoral zone are not as visible. Among them are the gastropod snails (particularly limpets) and rock crabs. Limpets (Fig. 12.17) have a cone-shaped shell and cling to the substrate by means of a large ventral foot. The foot is mucous covered and attaches like a suction disc. The conical shell fits closely to the rocky surface, and the water held in the mantle cavity helps to prevent desiccation during exposure. Some species secrete a mucous

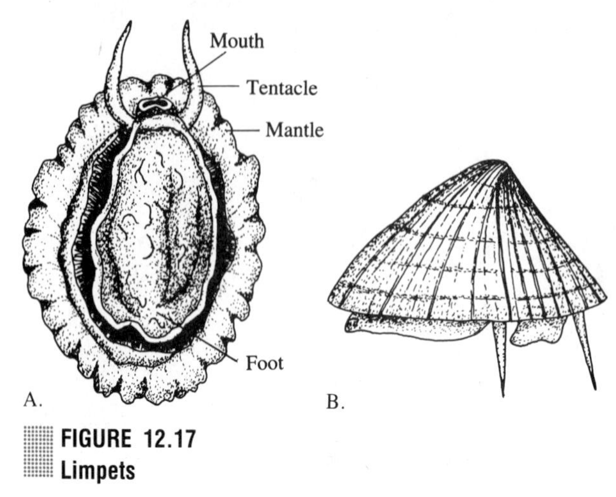

FIGURE 12.17
Limpets

Limpets are common animals throughout the full range of the littoral zone. Protection from exposure come from the conical shell and from water that the limpets hold in the space between the foot and the shell during periods of exposure. **(A)** Undersurface. **(B)** Side view.

band around the edge of the shell, which further retards evaporation. (A California species—*Acmaea limatula*—rocks its shell when stressed by high temperature, to encourage evaporation and promote cooling.) In general, limpets living at higher levels in the eulittoral zone have relatively taller shells. This shell shape increases the relative size of the mantle cavity for storing water and decreases the area from which evaporation can take place.

Perhaps the most active animal of the eulittoral zone are the shore crabs *(Hemigrapsus, Pachygrapsus, Carcinus)*. During tidal exposure, this crab seeks out crevices and the undersides of rocks, where it congregates in large numbers. During tidal immersion, shore crabs range widely over the intertidal in search of food. Shore crabs lack specific morphological adaptations to tidal exposure. They rely, instead, on their agility to avoid extreme conditions. They do share with other intertidal life, however, an increased tolerance to high temperatures and desiccation. During periods of tidal exposure, shore crabs retain seawater in their gill cavities and rely on the diffusion of atmospheric oxygen into the water of the gill cavities and then to the gills.

The Sublittoral Fringe. In general, the sublittoral fringe is exposed to the atmosphere for only a few hours during spring tides. The flora and fauna of this zone seldom have any special morphological adaptations to atmospheric exposure and show only slight physiological adaptations. The diversity of species and abundance of individuals is greater in the sublittoral fringe than at higher levels. In fact, a visit to the sublittoral fringe of a rocky shore will show you more kinds of algae and animals than any other intertidal area.

Probably the most consistent visible feature of this low zone is encrusting red algae. These pink, red, and purple-colored coralline algae attach directly to the rocky substrate, mostly in sheets or a short turf. In warm regions, the low-tide zone also may have dense growths of tunicates. In colder latitudes, extensive growths of large brown-algae kelps (particularly *Laminaria* and *Alaria*) may be prominent. These stands of brown algae mark the upward limit of the more extensive algae beds found in the shallow subtidal zone of temperate rocky shores (see Ch. 13). The algae bed, normally a three-dimensional community, collapses at low tide into two dimensions. When the tide is in, most of the brown algae float near the surface, forming a canopy that intercepts most of the light. The encrusting red algae seen at low tide are shade-tolerant species that receive only low light intensities through the algae canopy. The presence of the three-dimensional algae contributes greatly to the species diversity of animals. A wide range of animals are found as epifauna on the algae fronds, and many more are found in the holdfasts of the algae.

Intertidal Rock Pools. Rock pools occur in all three subzones of the intertidal zone. As the rocky shore is eroded, depressions form that collect water, and depending on where they are located relative to the tide, they may have a diverse assemblage of marine flora and fauna. Rock pools are fascinating laboratories for marine biologists. The life forms can be closely observed under the most natural conditions possible since the organisms are both supported and bathed by seawater. Rock pools, however, do have their own stresses. Pools located in the supralittoral fringe may be diluted by rainwater or concentrated by evaporation. Often, drift algae accumulate in these high pools and the decomposition products are an additional stress. Few marine organisms inhabit these high rock pools. The lichens, periwinkles, and isopods characteristic of the supralittoral fringe prefer the exposed surfaces to the pools.

Rock pools of the eulittoral zone often have a diverse and abundant assemblage of marine life. Environmental conditions in these rock pools are not as severe as those of higher pools, although they are more extreme than in marine waters. Temperatures of rock pools tend to follow those of the atmosphere and may exceed air temperatures during the summer. The oxygen content and pH of the water of eulittoral rock pools is likely to be more variable. In rock pools with algae, oxygen released from photosynthesis may, during the day, add to that already in the water. In rock pools dominated by animals, the combination of elevated temperatures and metabolic oxygen demand can reduce the dissolved oxygen to stressful levels. Still, rock pools have much less extreme environmental conditions than adjacent exposed surfaces.

Individual rock pools are highly variable in the organisms they contain. Two generalizations can be made. First, the most prominent indicator species of the eulittoral part of the intertidal zone—the acorn barnacles—tend to avoid rock pools. Second, many species more characteristic of the sublittoral fringe subzone—anemones, sponges, hydroids, and tunicates—are likely to be found in eulittoral pools. Two of the most important common animals in the rock pools are the hermit crab and the rock pool fish.

If you examine a rock pool for a few minutes, one of the first signs of movement you are likely to see is a snail shell moving rapidly in a most uncharacteristic manner. The shell contains a hermit crab, a decapod crab of the family Paguridae that has a large, fleshy abdomen housed in the snail shell. If threatened, the crab can withdraw totally into the shell and cover the opening with its claws. Hermit crabs are the clowns of rock pools. They are continually examining empty shells, for as they outgrow their shell they must locate a larger one. If no suitable shell is available, the hermit crab will aggressively attack another crab with a shell the correct size. Hermit crabs only take over empty gastropod shells and do not attack live snails.

Rock pool fish are of two main types—blenny (eel-shaped) and flattened. These sculpin-shaped fish often have a ventral sucker for attaching to rocks. Most rock pool fish stay in a particular pool during periods of tidal exposure. When the tide is in, they may range widely for feeding, but they know when and where to return to their pool for the next period of low tide.

Because of the relatively short period of tidal exposure, the rock pools of the sublittoral fringe subzone of the intertidal zone have virtually the same organisms as adjacent rocky surfaces.

Food Habits in the Rocky Intertidal Zone

The sources of food and feeding relationships in the intertidal zone are complex. Some of the food is provided by intertidal flora and some is imported from the pelagic environment during periods of tidal immersion.

The dominant visible flora of the intertidal zone are the macroalgae, particularly the midtide-level *Fucus* or *Ascophyllum*. Because these large macroalgae are not eaten directly by herbivores to any great extent, they eventually fragment and the pieces of drift algae are carried first to the high-tide drift line, and then to the bottom of either fine sand and mud flats or the subtidal zone. Here, where currents are slow, the food enters the detritus food chain and the energy becomes available to other organisms, primarily as products of bacterial decomposition.

The intertidal zone does have a number of important herbivores, including the littornid snails and limpets. If these animals do not eat macroalgae, what is their food? The apparently bare rocky surfaces of the high- and midtide zones are covered with an **algal film.** This film consists mostly of benthic diatoms but also includes bacteria, small algal species, and juvenile algal forms. Algal films are tolerant of drying during tidal exposure, and have their greatest productivity where tidal exposure occurs in cool, moist atmospheric conditions.

The two dominant indicator animals of the eulittoral zone (barnacles and mussels) are filter feeders that take plankton during periods of tidal immersion. Barnacles feed mostly on zooplankton, while mussels filter the smaller phytoplankton.

Another important method of feeding in the intertidal zone is scavenging. Scavengers include isopods, amphipods, and shore crabs. The shore crabs are of particular interest. As adults, they feed mostly on animal remains but may also chip mussels, barnacles, and limpets off rocks with their claws. As juveniles, shore crabs may feed on algae, particularly leafy green algae such as *Ulva* and *Enteromorpha*.

Two major predators of the intertidal zone are the whelks (a type of snail) and sea stars. Whelks *(Nucella)* feed mostly on acorn barnacles of the eulittoral zone. In fact, experiments on whelk predation of acorn barnacles have demonstrated that the lower limit of distribution of barnacles is set by predation. Whelks are found only on the lower part of the eulittoral zone, probably because of a lack of tolerance to extended tidal exposure. Since barnacles are more tolerant of tidal exposure, they flourish only above the levels that whelks can tolerate.

Unlike whelks, sea stars prey on a wide range of intertidal animals. Large stars favor bivalves and large barnacles. Sea stars can evert their stomachs through the mouth and either surround their prey whole, as with limpets, or, as with bivalves, insert the stomach into the shell and digest the prey in place. Experiments have shown that sea stars can insert the stomach through an opening only 0.1 mm in width, so they do not have to pull the bivalve shells apart to eat them. Studies on the predation of one intertidal sea star found on the west coast of North America—*Pisaster ochraceus*—have shown that its predatory activities have an important impact on the presence or absence of a number of species of algae and animals. On wave-exposed shores, *Pisaster* prefers mussels, and its numbers tend to set the lower limits of the mussels. When *Pisaster* is experimentally removed, the mussels move lower in the tide zone and force out a wide diversity of flora and fauna normally found in association with the sea stars. In this part of the intertidal zone, the community structure is greatly influenced by the feeding activities of this one predator. These experiments have led to the concept of a keystone species, that is, a species whose presence greatly influences the species composition of the community.

Predation is an important factor that can affect the kinds of organisms found in the lower part of the intertidal zone. Competition and larval recruitment, however, also can have an affect. Most animals in the intertidal zone have planktonic larvae, which maintain a pelagic life for usually a few weeks. There are two major advantages to a pelagic larvae. The first is dispersion. Since benthic organisms have a limited ability to move, the larval phase is a distinct advantage for expanding a species range or recolonizing areas. The second advantage of a planktonic phase is the ability to exploit rich food sources. Most larvae feed on phytoplankton, and larvae generally occur at times of high phytoplankton concentrations. When larvae first enter the plankton, they are attracted to the light (i.e., are positively phototrophic) and use their rudimentary swimming ability to stay near the surface. This keeps the larvae in areas of high food concentration and strong surface currents.

One disadvantage of a planktonic larval phase is having to find a suitable space at the correct level of the intertidal zone when it is time to settle. Most larvae have an active site selection phase before metamorphosis. A larva can postpone settlement for a limited time until a suitable substrate is located. The choice to settle is influenced by many factors, the most important being presence of adults of the same species. The texture, contour, angle, color, and light-reflecting properties of the substrate also influence settlement. Strength of the current, size of sediment particles, and presence of bacteria also may be important.

For some rocky intertidal species (limpets, mussels), the metamorphosed juveniles can move a little to choose their final location. Sessile species such as barnacles, however, must make an irreversible decision at metamorphosis. Another sessile group are the polychaetes with calcareous tubes (serpulid polychaetes). Figure 12.18 shows the major factors in site selection by the North Atlantic serpulid *Spirorbis borealis,* and sequence of events leading to settlement. A number of factors—a photonegative phase when settling, the microtopography of the surface, biotic features of the substrate, and the presence of adults of the species—are important in the decision to settle. Clearly, the settlement of benthic larvae, an important factor in determining the distribution of species on rocky shores, is not a random process.

Patterns of Distribution in the Rocky Intertidal Zone. Competition, seasonal distribution, and migrations, in addition to exposure to the atmosphere, affect the distribution of marine life in the intertidal zone. **Competition** refers to the attempt of two or more species to utilize a resource when there is not enough for all users. The main resources competed for are space, food, and, in the case of algae, light. In the mid- to low levels of the intertidal zone, space is the most limiting resource, and the effects of space competition on species distribution is clearly understood. One of the best-studied examples of competition for space is that found in two species of intertidal acorn barnacles: *Chthamalus* and *Balanus*. *Chthamalus* is smaller and, because it is more tolerant of tidal exposure, lives higher in the tide zone than *Balanus*. In some cases, *Balanus* simply overgrows and smothers *Chthamalus*. In others, a *Balanus* individual forces its shell underneath a *Chthamalus* shell and lifts it off the rock. Finally, two *Balanus* may grow together and slowly crush the *Chthamalus*. The end result is that *Balanus* flourishes at lower levels of the intertidal and *Chthamalus* is found only at higher, more exposed levels.

The timing of low tides can produce a seasonal distribution of some species. At rocky shores on the northwest coast of North America, the lower low tides generally occur in the night during the winter and in the day during the summer. During the early spring (February and March), conditions of temperature and moisture in the midtide zone are suitable to promote rapid growth, and a rich stand of brown algae characteristic of lower tidal levels becomes established. In late spring, the low tides start to occur during the midday. By the end of spring, the combination of higher temperatures and desiccation from direct sun cause a die-off of the macroalgae in the midtide zone, and they are found only in their regular low zone. A similar distribution is found in a species of red algae *(Porphyra)* occurring on exposed shores. During the winter, with generally high surf conditions, the *Porphyra* becomes well established in the supralittoral fringe. With the onset of summer and lower surf conditions, desiccation causes this high-level *Porphyra* to die back.

Particulate Shores

Although rocky shores have been studied in more detail by marine biologists, particulate shores, which include gravel, sand, and muddy beaches, are more common. All these particulate shores are composed of sediments

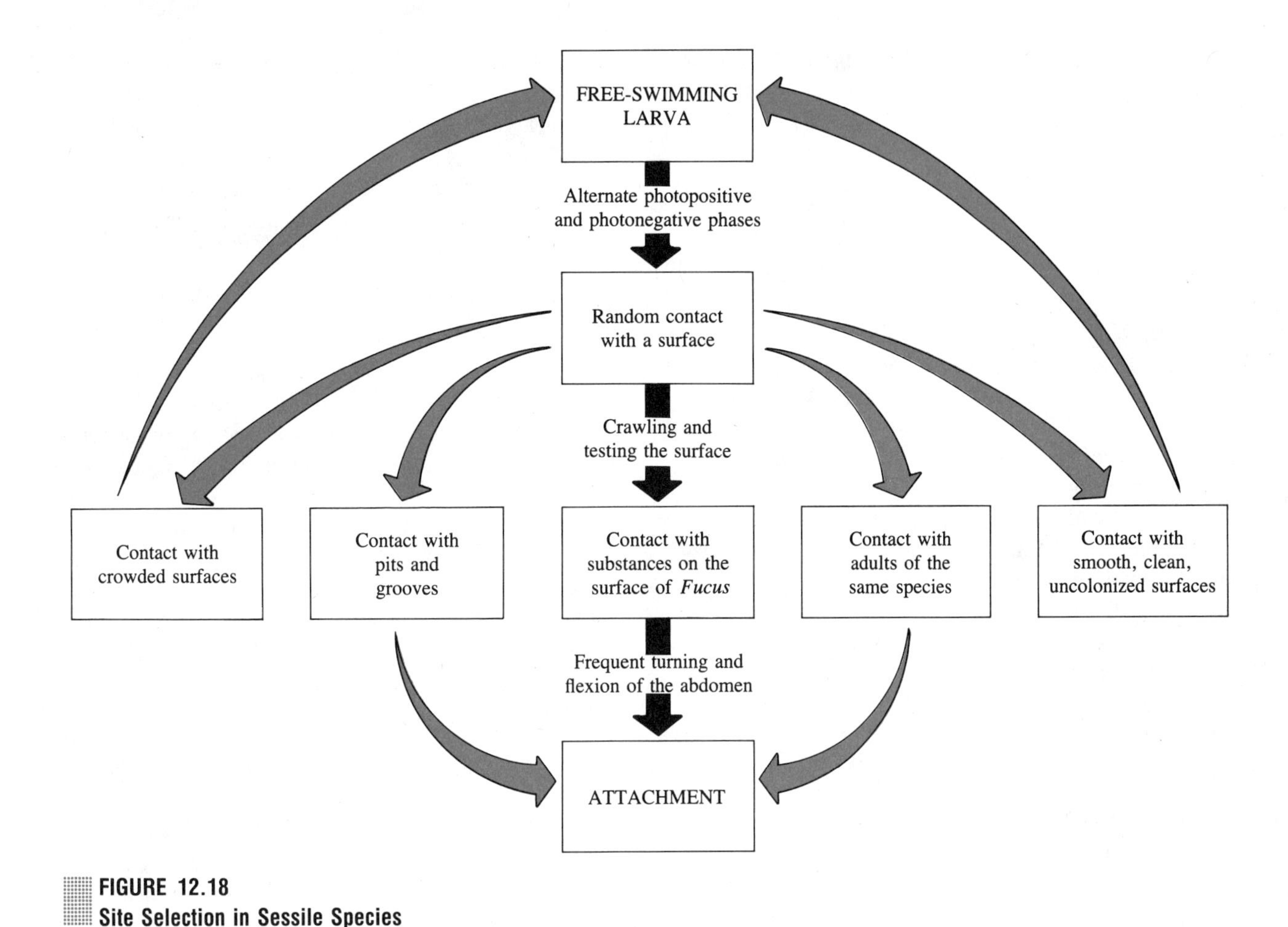

FIGURE 12.18
Site Selection in Sessile Species

The process of selection of a site for attachment by larvae is far from random in sessile species. Shown is the process of choice and the range of possible behavior for the polychaete worm *Spirobis*.

that are sorted by wave action. The most important influence on marine life of these shores is the instability of the sediments. A lack of solid substrate virtually eliminates colonizing of the sediment by macroalgae and epifauna. The organisms of particulate shores are very different from those of rocky shores. The flora is restricted to benthic diatoms which can attach to the surface of sediment particles and, on mud flats, to a few species of green algae or cyanobacteria that can anchor in the mud. The animals either live in the interstitial spaces between the particles (meiofauna) or burrow into the sediment (infauna).

It is difficult to make any further generalizations about life on particulate shores because of the wide range of environmental conditions associated with different sediment types. The fauna of sandy beaches, for example, differs greatly from that of mud flats. Beaches are categorized on the basis of particle size. Gravel beaches have the coarsest particles and mud flats the finest. Generally, mudflats occur in areas where wave action is very low, and gravel and sand beaches where wave action is high; but sediment size cannot be predicted from wave energy alone. There is, however, a close relationship between sediment size and beach slope. Beaches with coarse deposits are steeper, and those with fine deposits are flatter (Tab. 12.1)—hence the term *mud flat*. Beaches of fine deposits are well

TABLE 12.1
Relationship of Particle Size to Beach Slope

Particle Size	Mean Slope of Beach
Cobble	24°
Pebble	17°
Granule	11°
Very coarse sand	9°
Coarse sand	7°
Medium sand	5°
Fine sand	3°
Very fine sand	1°

sorted; that is, all the particles tend to be the same size. Coarse deposits are poorly sorted; that is, particle sizes are variable. Wave action affects not only sorting but the stability of the surface sediments (Fig. 12.19). In conditions of high surf, the upper sediments are suspended by wave action. In surf 1 m (3.3 ft) high, sediments are suspended to a depth of 8 cm (3 in).

The water content of the sediment varies with particle size, beach slope, and degree of sorting. Coarse sediments with steep slopes drain quickly and the surface may become quite dry. As particle size decreases, the capillary attraction to water increases. This, coupled with the gentle slope of fine-particle beaches, promotes water retention. Mud flats, for example, lose little interstitial water during tidal exposure. Depending on water content, agitation of sediments may cause them to become more firm. Perhaps you have noticed when walking on a sandy beach that some areas are more firm and that the sand looks lighter around your footprint. This firmness under pressure, or **dilatancy,** is dependent on the water content of the sand. The opposite condition is when with pressure the sediments become more fluid. This response, called **thixotropy,** is more characteristic of muddy sediments. Thixotropic sediments are much easier for burrowing animals to penetrate.

In addition to greater water retention, fine sediments have higher organic content. Organic particles are about the same size as fine sand and mud particles, and the same factors that cause accumulation of fine sediments (see wave refraction, above) also cause organic material to collect. Because the high organic content promotes rapid bacterial growth, poorly drained

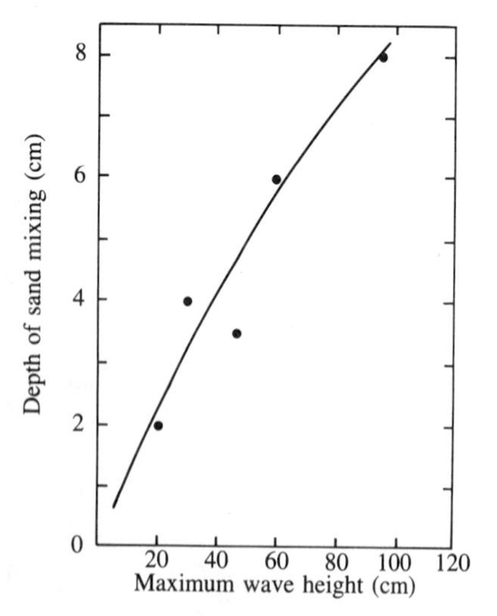

FIGURE 12.19
Surface Stability of Particulate Shores

Surf causes a suspension of the particles. Waves 1 m in height disturb the sediments to a depth of 8 cm. Burrowing in this shifting substrate is difficult.

sediments are oxygen deficient. The oxygen-deficient sediments become anaerobic, attracting anaerobic bacteria that, because of their metabolism, turn the sediments black just below the surface. These black sediments are characteristic of mud flats.

The interstitial water of sediments resists changes in salinity and temperature. Dry conditions and rainfall affect salinity in only the top few centimeters of a mud flat. Temperatures of surface sediments are moderate compared to the atmosphere. Below 10 cm (4 in), sediment temperature does not respond at all to changes of air temperature. In the top 10 cm sediment, temperatures vary only 2–3°C (3.6–5.4°F), while air temperature may have a diurnal variation of 10–15°C.

The major environmental stresses of living in sediments, then, are different from those of a rocky shore, where large temperature ranges and desiccation limit tolerance of atmospheric exposure. Marine life in the sediments is affected instead by shifting sediments and oxygen availability. The combination of different particle sizes and varied environmental factors in particulate shores creates a range of habitats (Fig. 12.20) that are colonized by quite different organisms. At one end of the spectrum are gravel beaches, or shingle beaches as they are known in Europe. Because of the great wave action that turns the sediments, and high porosity that promotes water drainage, this habitat is almost devoid of marine life. On the other end of the spectrum are fine mud sediments, where anaerobic conditions are found just a few centimeters below the surface because of poor drainage and high organic content. Burrowers in the sediments must maintain contact with the surface to guarantee a supply of oxygen-rich water, and organisms living in interstitial spaces are restricted to a thin surface layer that has high-temperature stresses. To see how marine life has adapted to conditions of life in the sediments, we will look at exposed sandy beaches and mud flats.

Sandy Beaches. Sandy beaches offer no solid substrate for the attachment of organisms because wave action continuously turns the substrate particles. Two basic sizes of marine organisms are adapted to life in this constantly shifting substrate: large infaunal organisms (1–10 cm) capable of burrowing in the sand, and small meiofauna (0.1–1 mm) that live between the sand grains in interstitial spaces.

Zonation, so conspicuous on rocky shores, does occur on sandy beaches, but it is much less obvious since it is related to the moisture retention capabilities of the sand (Fig. 12.21). The area above mean high tide is generally dry, wetted only occasionally by very high tides or large waves. In the portion of the beach that is covered and exposed by the tide each day there are two

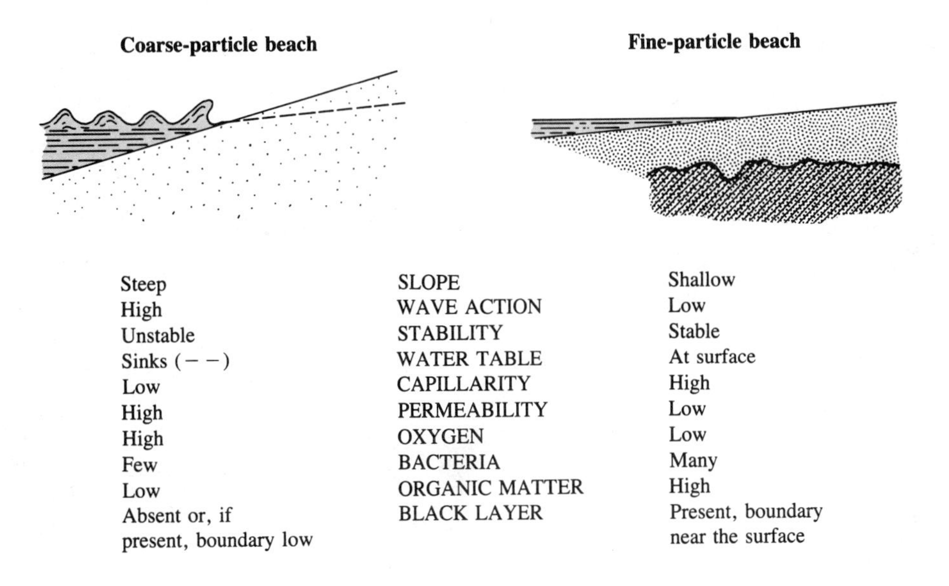

Coarse-particle beach		Fine-particle beach
Steep	SLOPE	Shallow
High	WAVE ACTION	Low
Unstable	STABILITY	Stable
Sinks (− −)	WATER TABLE	At surface
Low	CAPILLARITY	High
High	PERMEABILITY	Low
High	OXYGEN	Low
Few	BACTERIA	Many
Low	ORGANIC MATTER	High
Absent or, if present, boundary low	BLACK LAYER	Present, boundary near the surface

FIGURE 12.20
Environmental Characteristics of Coarse- and Fine-Particle Beaches

zones: the zone of retention and the zone of resurgence. The **zone of retention** is the part of the beach where water drains seaward. The finer the particles, the greater the retention through capillary action. The lowest intertidal zone on sandy beaches is the **zone of resurgence.** This zone, even during periods of tidal exposure, is always saturated by water draining from higher elevation.

Waves often bring up masses of algae and attached animals, called **drift,** that pile up at the high-tide line. This drift is a source of food to a number of scavengers, the most important of which are amphipods in temperate latitudes, and ghost crabs (Fig. 12.22) in the tropics. Both are almost terrestrial; they do not swim or require water for oxygen exchange, only needing the ocean for reproduction. They do, however, require the dampness of night to be fully active. They use the algal drift for food or explore the intertidal zone during ebb tide for both algal and animal remains. During daylight hours, they burrow into the sand above the high-tide mark.

The dominant macroorganisms of the intertidal zone of sandy beaches are mostly bivalve molluscs and crustaceans. Some bivalve species, such as the razor clams (*Ensis* on the Atlantic coast and *Siliqua* on the Pacific), burrow deeply into the sand to avoid the unstable sediments, while others live closer to the surface. The pismo clam, found on California beaches, lies just below the surface, often with part of the shell exposed. The massive valves of this clam help to maintain its position in the shifting sand. Another clam that lives close to the surface is the bean clam *Donax*. *Donax* occurs by the thousands on beaches of the Atlantic and Pacific Oceans. Waves wash them out of the sand and they burrow back to a depth of 1–2 cm (0.4–0.8 in). *Donax*

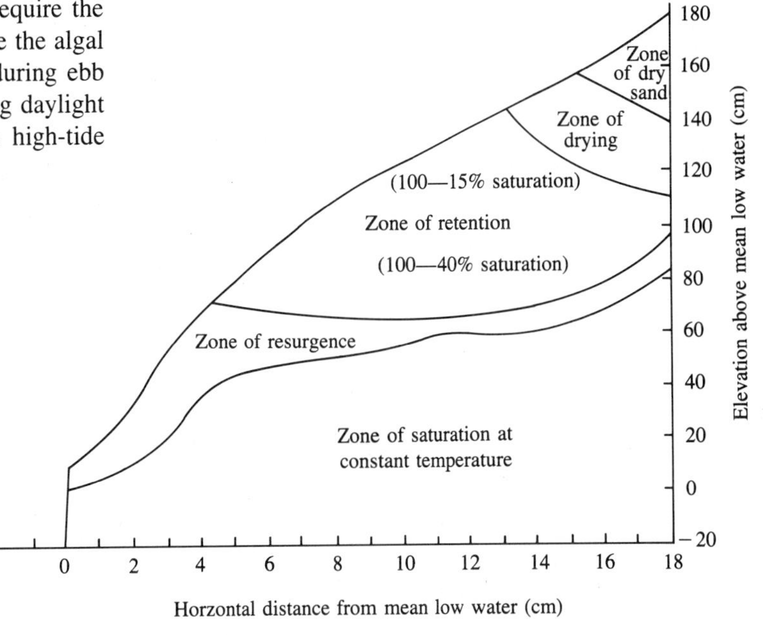

FIGURE 12.21
Zones of Water Content on Sandy Beaches

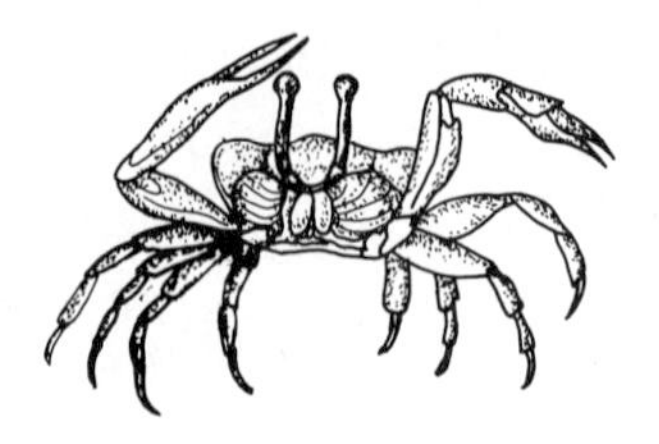

FIGURE 12.22
Ghost Crabs

Ghost crabs are found in burrows above the high tide of tropical sandy beaches.

moves up and down the intertidal zone with the tides (Fig. 12.23). They use incoming waves on flooding tides to carry them up the beach, and outgoing waves to carry them down the beach. In this manner they remain in the wash zone of the waves, which is their preferred habitat.

Sandy beaches are an important habitat for meiofauna, which live in the interstitial spaces between sand grains (Fig. 12.24). They include harpactacoid copepods, nematodes, flatworms, small molluscs and polychaetes, rotifers, gastrotrichs, tardigrades, and some foraminiferans. The food available for meiofauna includes bacteria, detritus, and diatoms. Some meiofauna have high metabolic activity and short reproductive times, and may be much more important in the overall productivity of sandy beaches than their small size and biomass indicate. Marine biologists still are not sure how important meiofauna are as a source of food to larger animals. Their important ecological role is be-

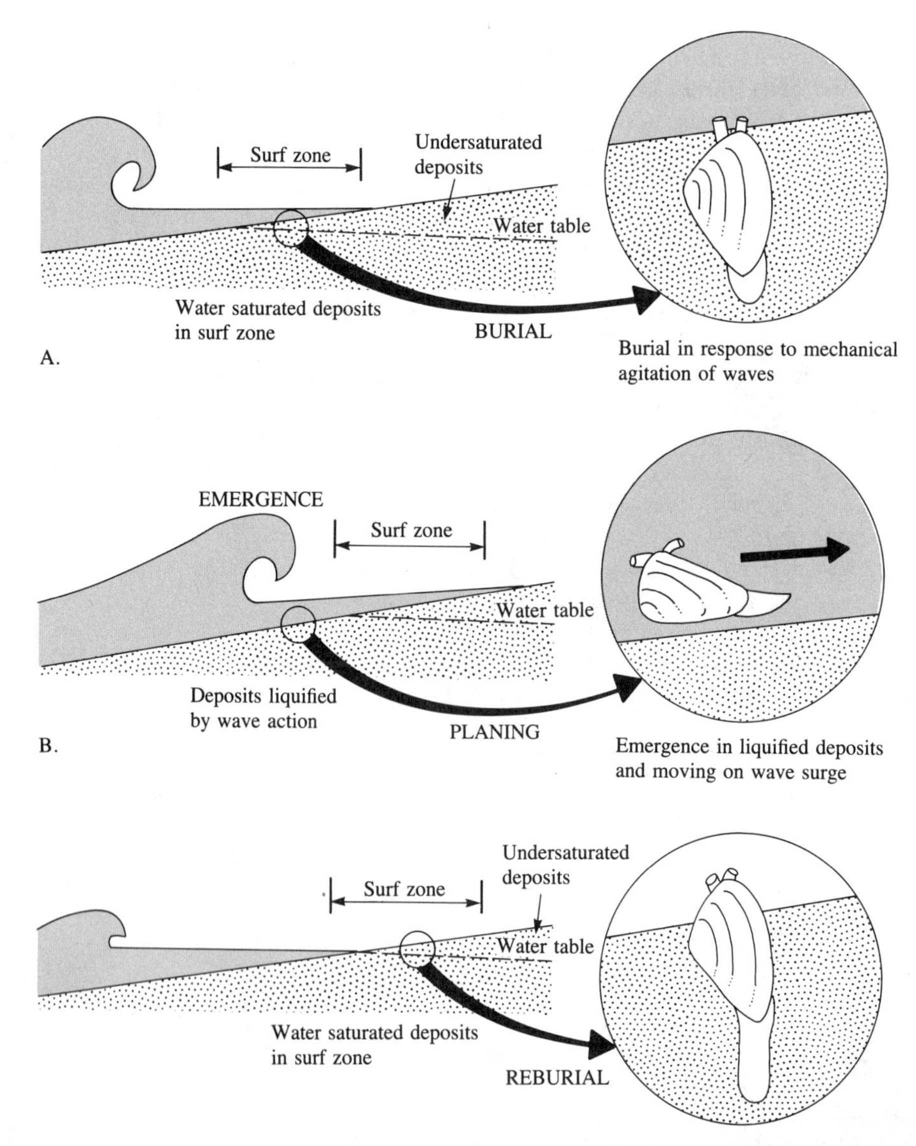

FIGURE 12.23

The clam *Donax* moves to stay in the surf zone. **(A)** Burial. **(B)** Moving. **(C)** Reburial.

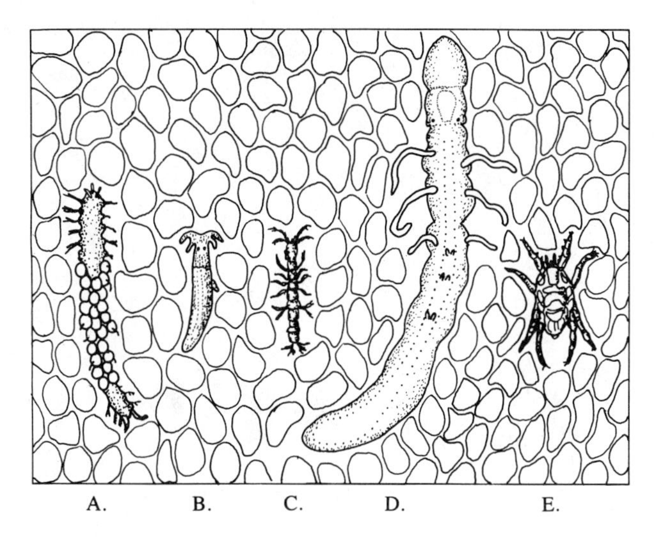

FIGURE 12.24
Meiofauna of Sandy Beaches

Meiofauna are small, generally elongated and thin, or flattened vertically or laterally. **(A)** Polychaete worm. **(B)** Mollusc. **(C)** Arthropod. **(D)** Polychaete. **(E)** Arthropod mite.

coming clear, however. Even though they have much greater numbers and lower biomass than the macrofauna, they may be equally important in the cycling of energy and nutrients through the community of sandy beaches.

The sandy beach habitat is important for the reproduction of a number of species that otherwise spend their lives at sea or in the subtidal benthos. Three such species are the horseshoe crab, the sea turtle, and the grunion. (See Ch. 7 for the reproduction of sea turtles, and the special feature at the end of this chapter for the grunion). Horseshoe crabs are considered to be living fossils (see Ch. 5) because their body form has not changed in over 360 million years. Horseshoe crabs normally live in the deeper water benthos of the Atlantic and Asian Pacific shores. In early summer, however, they come into shallow water to breed. On a spring tide of a full or new moon, the crabs come ashore and climb to the high-tide mark. The female excavates a nest and deposits eggs, which are fertilized by the male. The nest is then left for waves to cover. Two weeks later, with the return of the high spring tide, the eggs hatch and the young immediately return through the beach to the subtidal mud.

The exposed sand habitat is one of the harshest of intertidal environments. The effects of the shifting substrate, low moisture due to the porous nature of the sediments, and the force of the waves combine to exclude all but a few species. For those that can successfully adapt to this habitat, however, there is little competition for the resources of food and space.

Mud Flats. Intertidal areas of fine sand and mud are found where soft sediments are protected from wave action. Mud flats are areas of deposition, both of fine sediments and organic detritus. In addition to having finer particles, mud flats are flatter and therefore have a greater area than rocky or sandy beaches. The fine sediment and flatness create a high degree of water retention. The fine particles hold the water tightly, allowing little movement. This, coupled with a high concentration of organic matter, creates an optimal habitat for bacteria. The bacteria quickly deplete the oxygen, because of the slow movement of water through the sediments. Anaerobic conditions follow, and the sediments of most mud flats lack oxygen a few millimeters below the surface.

The transition from anaerobic to aerobic conditions is marked by a color change of the sediments. Surface-oxygenated sediments are gray, yellow, or brown, while the anaerobic sediments are uniformly black. The black color is caused by anaerobic bacteria that break down organic molecules in a process similar to the anaerobic fermentation of sugars by yeasts. These anaerobic bacteria are relatively inefficient and produce, as the end product of metabolism, molecules such as alcohols and fatty acids. The anaerobic bacteria use inorganic molecules such as sulphate and nitrate ions as part of their metabolism. One common anaerobic bacterium—*Desulfovibrio*—converts sulphate (SO_4^{2-}) to hydrogen sulphide gas. The hydrogen sulphide gas then combines with iron molecules in the sediments to produce the characteristic black iron sulphides. Often, the hydrogen sulphide gas is produced in excess and gives a distinct odor to the sediments when they are disturbed. More important is the toxic effects of the gas. Hydrogen sulphide is very toxic to all aerobic organisms. Hence, virtually no metazoans are able to enter the black zone of the sediments unless in burrows that maintain contact with water from the surface.

Specific adaptations to the mud environment are found in those species that burrow. Since there is no disturbing wave action, burrows in mud can be quite elaborate and relatively permanent. The important burrowing fauna in mud flats are bivalve molluscs, polychaetes, and crustaceans (Fig. 12.25). One of the most common clams in mud is the soft shell *Mya*. This clam predominates on mud flats in temperate latitudes and may be found to a depth of 30 cm (11.8 in), where the sediments are anaerobic. The required supply of fresh water for oxygen must be brought from the surface. A pair of siphons extend from the clam to the water surface: One to bring fresh water to the mantle cavity where oxygen is removed by the gills, and the second to return water to the surface. The siphons can withdraw into the mud and close quite tightly, allowing the clam to tolerate periods of tidal exposure.

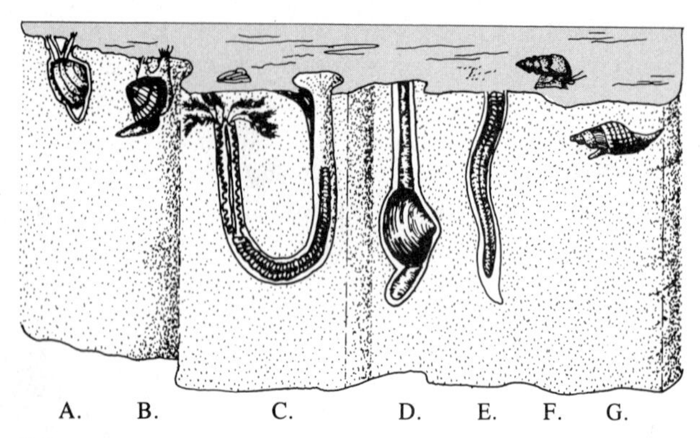

FIGURE 12.25
Infauna of Mud Flats

(A) *Macoma.* **(B)** Cockle *Clinocardium.* **(C)** Polychaete worm. **(D)** The soft shell clam *Mya.* **(E)** Polychaete worm. **(F)** The gastropod snail *Hydrobia.* **(G)** Gastropod. Most benthic infauna (A–E) are dependent on the surface for food. A few species (G) are predators that can move through the mud beneath the surface. Fewer animals (F) live on the surface.

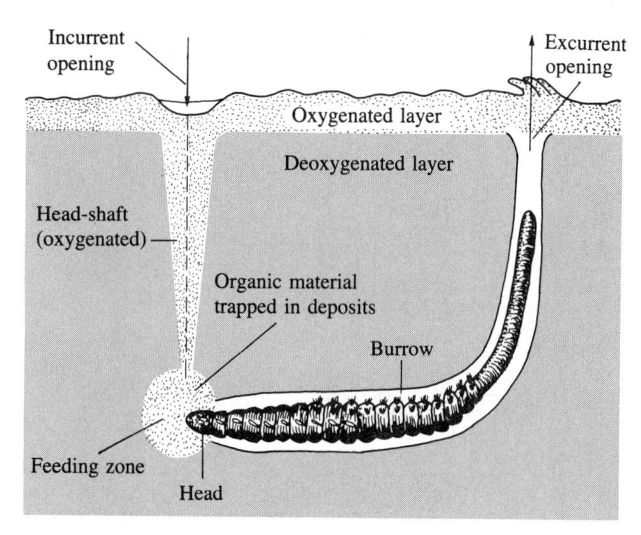

FIGURE 12.26
Arenicola

The polychaete *Arenicola* can live in anaerobic sediments by continually pumping oxygenated water through its burrow. Food is mostly the organic material in the sediments.

Most clams on mud flats are not active diggers. When threatened, they withdraw their siphons, often sending a jet of water into the air. The depth of a clam is limited by the length of its siphons. On the northwest coast of the Pacific, two genera of large clams are found at depths of a meter or more. The horse clam *(Tresus)* and the goeduck *(Panope)* have siphons that far exceed the size of the shell. The goeduck weighs up to 5 kg (11 lbs) and may be buried more than 50 cm (20 in) below the surface.

Polychaetes that build burrows on mud flats include the lugworms (genera *Arenicola* and *Abarenicola*). Lugworms build U-shaped burrows open to the surface at the exhalant end, while a depression in the surface of the mud marks the inhalant area. Detritus with some sand and mud particles are drawn into the incurrent opening (Fig. 12.26). Lugworms swallow all the particles and expel the inorganic portions as a coiled fecal casting. Oxygen in the incurrent flow keeps the water in the tube and the sediments in the feeding zone from becoming anaerobic. Other polychaetes are more selective in their food. The tube-dwelling *Chaetopterus* (Fig. 12.27) secretes a mucous bag at the anterior end that catches food. The posterior segments of the body are modified into fans that pump water through the burrow. Small particles of detritus and plankton are trapped in the mucus, which is periodically eaten. Yet another tube-dwelling polychaete feeding pattern is to use the tentacles to gather detritus particles from the surface. Figure 12.28 shows how the tentacles of the polychaete *Amphitrite* are extended across the surface of the sediments. Small detrital particles adhere to the tentacles and are carried by cilia back to the mouth.

Some crustaceans also may be common on mud flats. In many subtropical and temperate mud flats, areas are riddled with burrows of the ghost shrimp *Callianassa*. Ghost shrimp build U-shaped burrows with two or more openings that may be up to 2 m in total length. Pumping with their abdominal appendages, ghost shrimp circulate water through the burrow. In addition to providing food and oxygen to the ghost shrimp, the burrow is a home for other animals including goby fish, pea crabs, and copepods.

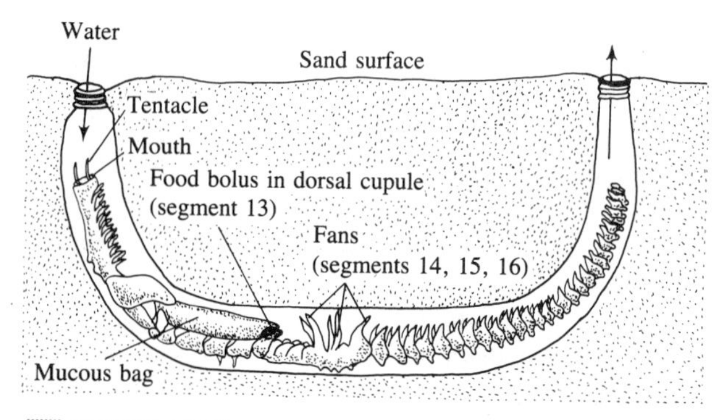

FIGURE 12.27
Chaetopterus

The polychaete *Chaetopterus* also burrows in anaerobic sediments. Its burrow opens on both ends to oxygen-rich surface water. Suspended particles in the water are used as food.

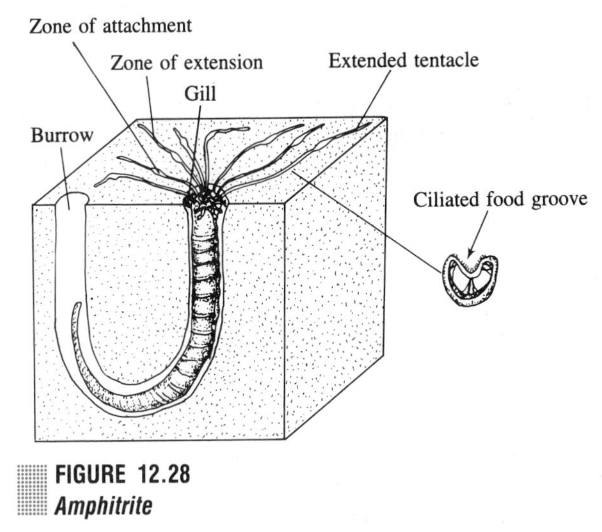

FIGURE 12.28
Amphitrite

The polychaete *Amphitrite* uses its tentacles to draw food from the surrounding surface deposits to the mouth.

The pattern of zonation obvious on rocky shores and evident on sandy beaches is not apparent on mud flats. At the mid- to high-tide zone there often is a salt marsh (temperate and polar latitudes) or mangrove swamp (tropical latitudes). These systems are unique in their own right and are discussed in the following sections. Near the low-tide mark there often are patches of sea grasses. These patches are generally the upward most extension of a more widespread subtidal meadow.

Living in the mud flats of the intertidal zone is difficult because of the soft sediments and lack of oxygen below the surface. As with the sandy beach, those organisms that solve these problems find a rich habitat in which to live.

Salt Marshes. A salt marsh is one of the few examples of a community of true plants that can tolerate salt water and thrive in the marine environment. Many species are perennial grasses that can withstand periods of immersion in seawater. Salt marshes are found on the mud flats of protected bays, but only in the mid- to upper-tide zone.

Salt marsh communities are found only in temperate and polar latitudes (Fig. 12.29). In tropical areas, they are replaced by mangrove swamps. Many marine biologists consider the two communities to be ecological equivalents because both are found in muddy sediments at higher tidal elevations of protected water. Three environmental factors are required for extensive salt marsh formation: protection from wave action, shallow intertidal slopes, and some freezing temperatures during the year. Salt marshes are widely distributed on the east coast of North America (Fig. 12.30). In total there are some 500 species of plants belonging to

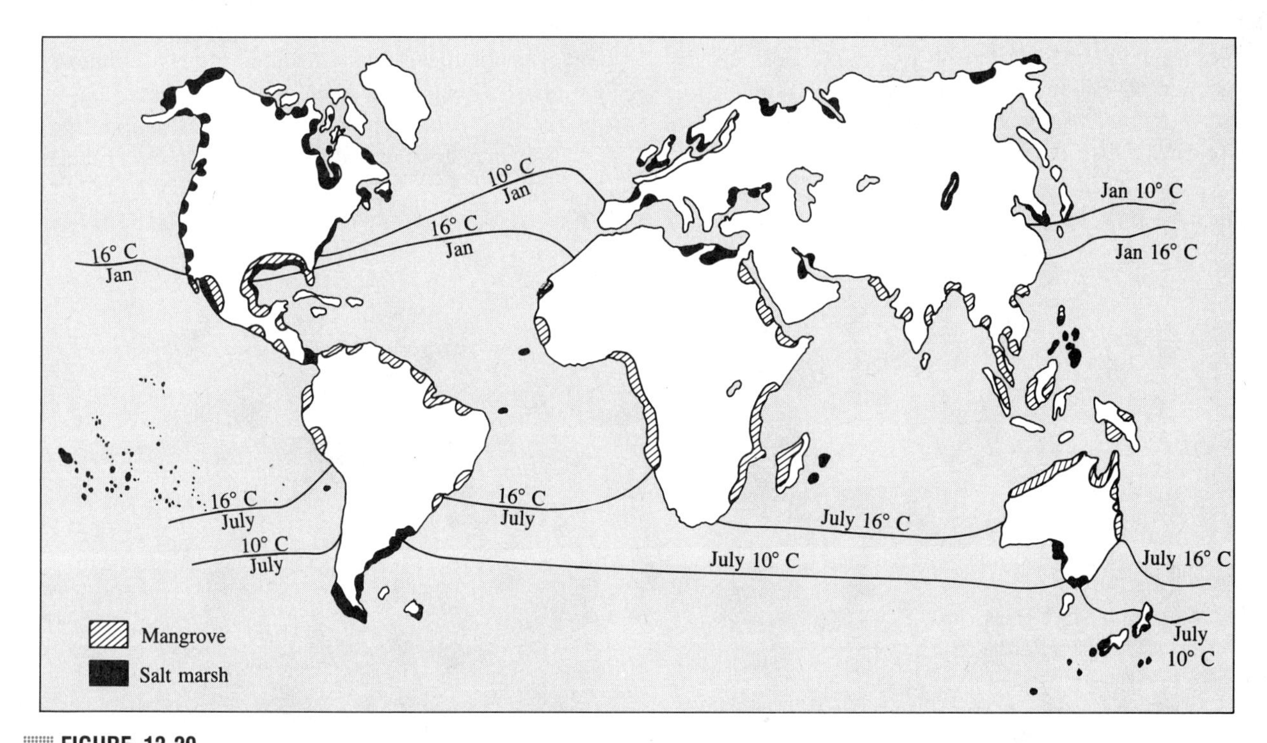

FIGURE 12.29
Worldwide Distribution of Salt Marshes and Mangrove Swamps

(From Chapman, Y.J. 1977. *Ecosystems of the world,* Vol. 1, *Wet coastal ecosystems.* New York:Elsevier.)

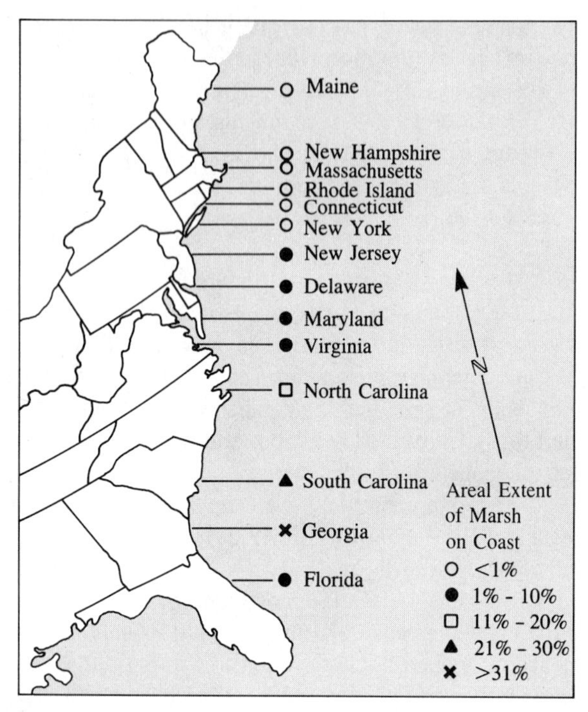

FIGURE 12.30
Distribution of Salt Marshes on the Atlantic Coast of North America

(From Chapman, 1977.)

18 families of angiosperms that are found in salt marshes world wide. Although species composition varies from marsh to marsh, a number are widespread. *Spartina, Salicornia, Juncus,* and *Puccinellia* have a world-wide distribution.

As in other intertidal habitats, plant species in salt marshes show a zonation that is related to the degree of exposure and inundation during the tidal cycle (Fig. 12.31). Salt marshes generally start at the level of the average neap tide and extend upward to and beyond the height of the highest tides. These marshes often are complex, having drainage creeks, mud flat areas (called pans), and tidal flats (Fig. 12.32). Salt marsh plants have special mechanisms to tolerate the salts in seawater. Special cells in the roots pump salts out so that the plants have an adequate supply of fresh water to serve the needs of its metabolism, particularly transpiration.

Although salt marsh plants can tolerate full-strength seawater, they grow better where salinity is reduced. The salts of seawater are a stress, and under experimentation salt marsh vegetation grows better in fresh water, and yet these plants are found only in locations periodically inundated by salt water. In conditions of optimal growth, other plants out-compete the salt marsh plants. Since other vegetation cannot tolerate periodic immersion in salt water, biological competition is not a problem in the salt marsh habitat. The stress of salt water imposes a metabolic cost on salt marsh vegetation. Almost 50 percent of the energy derived from photosynthesis is required to pump the salts out of the plant tissues.

Salt marshes stabilize the sediments, thereby promoting their own growth. The roots and stems tend to capture the fine sediments that are deposited by currents and waves (Fig. 12.33). Although large waves have enough energy to wash out the marsh plants, small waves continuously bring a source of new sediments to the marsh. Rates of accumulation of up to 15 cm (5.9 in) per year have been observed in some salt marshes.

Salt marshes have very high biological productivity. Some of the highest rates of net primary productivity of any photosynthetic organism in the world have

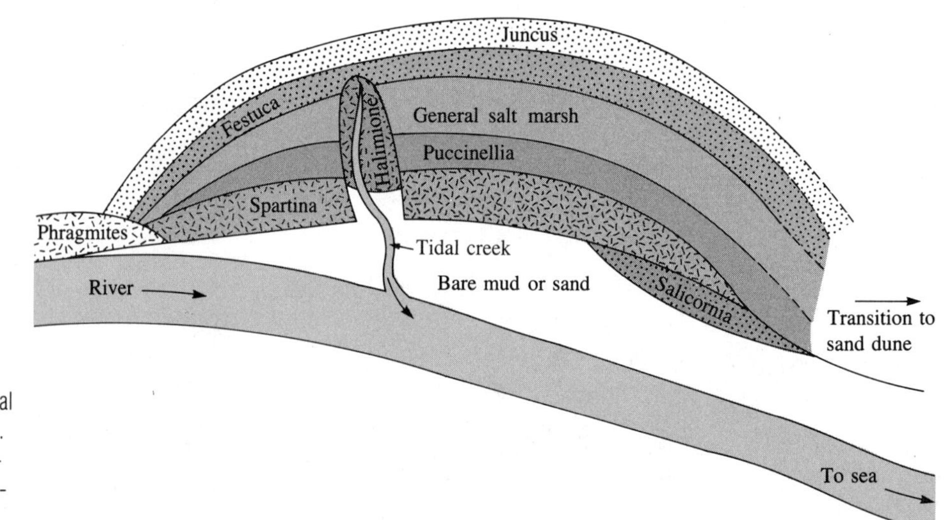

FIGURE 12.31
Salt Marsh Zonation

Salt marsh vegetation occurs in zones related to tidal elevation. Shown is a tidal transect of a typical English salt marsh. (From Green, J. 1968. *The biology of estuarine animals.* Seattle:University of Washington Press.)

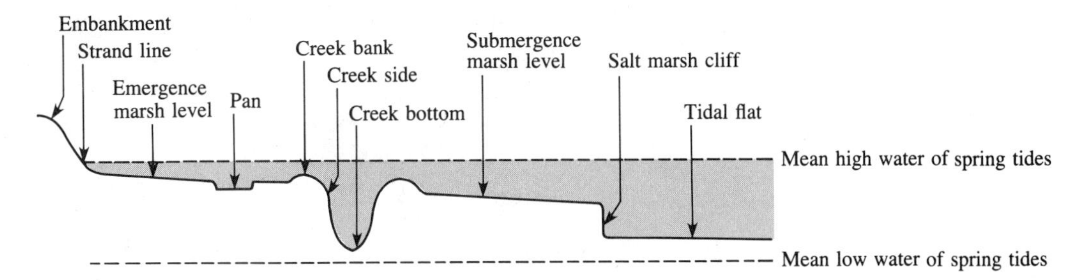

FIGURE 12.32
Subhabitats of Salt Marsh Communities

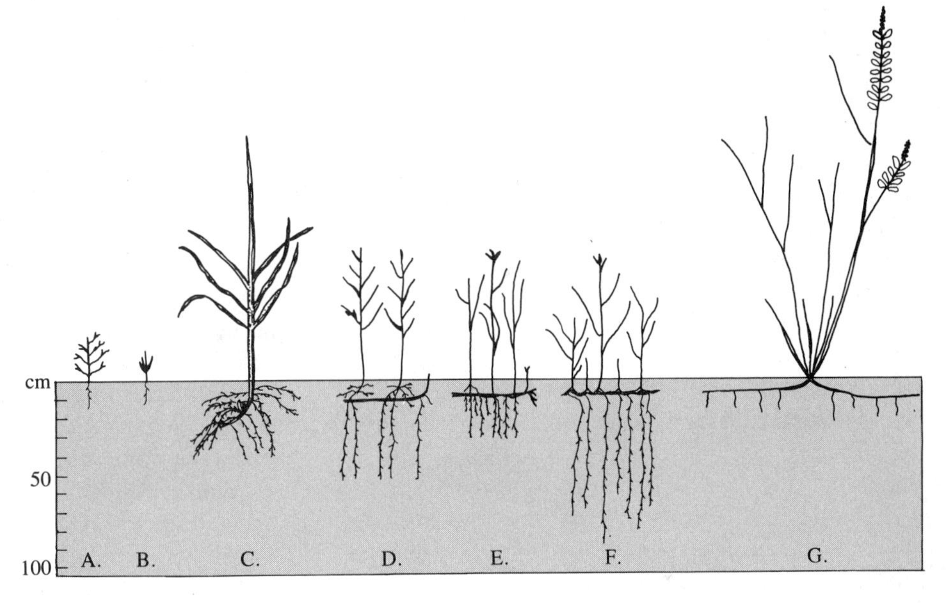

FIGURE 12.33
Root Systems of Salt Marsh Plants

The roots of salt marsh plants penetrate deeply into the sediments, binding them together so they are less susceptible to the erosion action of waves. **(A** and **B)** *Salicornia spp.* **(C** and **E)** *Spartina spp.* **(D)** *Distichlis.* **(F)** *Juncus.* **(G)** *Iva.*

been observed in salt marshes. Few animals, however, eat the salt marsh vegetation directly. Most of the energy captured by the salt marsh in photosynthesis is slowly released to the adjacent water and sediments as the vegetation decays. This energy source is very important to many marine organisms, a number of which are commercially significant. One study showed that 95 percent of all sport and commercial fish caught off the coast of Georgia were nurtured in the early part of their life cycle by energy from the salt marsh.

The animals found in the salt marsh are a mix of terrestrial and marine. About 50 percent of the animal species are terrestrial. They include insects, birds, and mammals. These species are found in the marsh primarily when the tide is out, as they do not have any special adaptations for marsh life. The marsh may be very important to the biology of these tidal visitors, however. Some insects feed on the sap of marsh vegetation, and mosquitoes use marsh pools for development of their larvae. Ducks and geese use the marsh for roosting and feeding, and many other birds use various areas of the marsh as nesting sites. Mammals, including mice, rabbits, and deer, forage in the marsh. Often, marshes are managed for agriculture. For example, they typically are used for grazing cattle and as a source of hay.

Of the marine invertebrates found in salt marshes, bivalves, gastropod snails, and crabs are the most common. The ribbed mussel *Geukensia* is the most common of the bivalves. This mussel attaches to the rhizome and stalks of salt marsh plants. It is particularly tolerant of the wide ranges of temperature and salinity found in the marshes. When exposed to the atmosphere, it can obtain oxygen by air gaping, or periodically sucking air into the mantle cavity. A number of gastropod snails including periwinkles *(Littorina)* and the coffee bean snail *(Melampus)* are also common in salt marshes. These snails are able to move and respire while exposed, and only retreat into their shells under extreme desiccation. In fact, the coffee bean snail has been observed climbing up the marsh grass stems to avoid tidal submergence.

FIGURE 12.34
Fiddler Crabs

The fiddler crab *Uca* is common in salt marshes. They build extensive burrows, and the males have a large claw.

The most obvious crabs in salt marshes are the pugnacious fiddler crabs, *Uca* (Fig. 12.34). The males have an oversized right pincer that is almost as large as the body itself. *Uca* lives in burrows in the mud. The species perhaps most characteristic of the marsh proper is *Uca pugilator*. These fiddler crabs are nocturnal and feed only when the tide is ebbing. They retreat into their burrows on incoming tides, and plug the openings of their burrow with mud.

The ribbed mussel, the snail, and fiddler crab illustrate the various feeding habits of marine animals in the salt marshes. The ribbed mussel is a filter feeder and strains its food (mostly phytoplankton and microzooplankton) from the water at high tide. The snails are grazers. They scrape epiphytic algae (mostly diatoms) from the surface of the salt marsh stems and leaves. Finally, the fiddler crabs are detritus eaters, feeding on the rich organic surface deposits of the marsh. Their food is mostly bacteria that in turn get energy by decomposing salt marsh vegetation.

Mangrove Swamps. The salt marshes found in the mid- to upper-intertidal area of temperate and polar latitudes give way to mangrove swamps in tropical latitudes. Mangroves are trees and shrubs that may rise to 30 m or more. These trees are found in muddy bays with protected waters (Fig. 12.35). Mangrove trees can tolerate full-strength seawater, but also are found in estuarine conditions. As with salt marsh plants, mangroves are **facultative halophytes;** that is, they tolerate salt water but grow better in fresh water. Other species, however, are more competitive, and it is only in salt conditions that mangroves are not out-competed. Immersion of roots in seawater up to 1 m (3.3 ft) in depth is common.

There are some 80 species of mangrove trees found world wide. By far the greatest diversity is in the Indo-Pacific region (65 species). Mangroves are a heterogeneous group of plants and come from a number of taxonomic groupings. One of the most widely distributed genera of mangrove is the red mangrove, *Rhizophora*. In many ways, this genus epitomizes mangrove swamps. It generally is found on the seaward edge of the swamp and is characterized by its stiltlike prop roots, which give extra support to the tree. *Rhizophora* also may send out 'drop roots' from branches for further support. The roots of mangroves are morphologically specialized for anchoring and nutrient transport. Roots that penetrate the anaerobic zone of the mud are heavily protected, and do not exchange nutrients or gases. Feeding root hairs are found only in the aerobic top centimeter of the mud. A further adaptation is viviparous reproduction: The embryo, or propagule, starts to develop while still on the tree, and has an anchorlike attachment. When the propagle drops from the tree, it may fall directly into the mud and commence growth as a new tree. If the propagule falls into the water, it will float until it touches and embeds in mud.

Mangrove swamps, like salt marshes, tend to accumulate sediment. Fine particles trapped in the root systems become permanently deposited. Mangroves also are very productive communities, ranking as high as estuaries, evergreen forests, or farmland. As in salt marshes, most of the plant material is not eaten directly, but decays and enriches the adjacent waters through detritus food chains.

The fauna of mangroves is a mixture of terrestrial and marine organisms. Numerous birds, such as kingfishers, cormorants, and herons, use the upper parts of the trees as habitats. Monkeys may frequent the tree

FIGURE 12.35
Mangrove Zonation

The distribution of plants in a mangrove community is affected by the extent of tidal immersion. Shown is mangrove zonation in southern Florida.

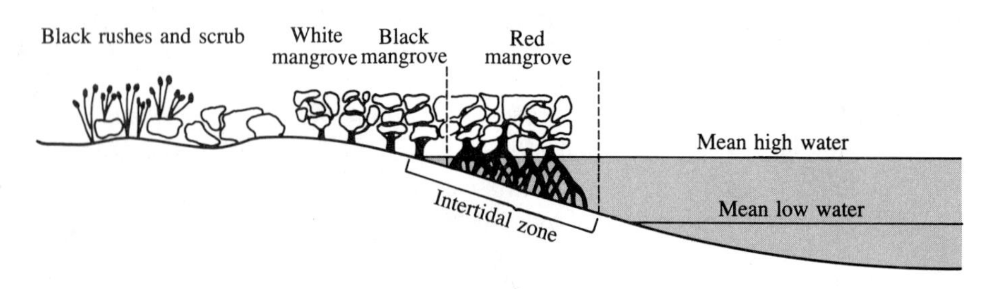

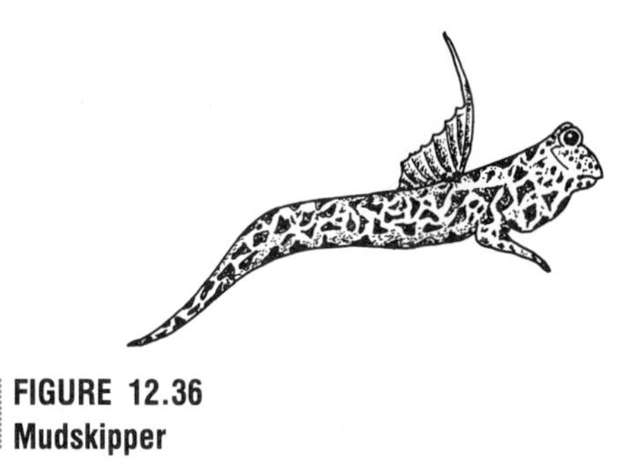

FIGURE 12.36
Mudskipper

One of the important fish of mangroves is the mudskipper. The male raises its dorsal fin in courtship display.

canopy, and snakes, frogs, and insects are common inhabitants.

Marine animals of the mangrove swamp are for the most part similar to those of the adjacent mud flats. Some species, however, are associated primarily with the mangrove. Perhaps the most interesting is the mudskipper (Fig. 12.36). This fish is partly amphibious and uses its pectoral fins to climb on the mangrove roots or hop across the mud flats. They periodically moisten their gills to prevent drying.

Oysters are commonly found in mangrove swamps with the shell firmly attached to roots. A few species of the mangrove community have become nearly totally terrestrial. A littoral snail, which on rocky shores tend toward a terrestrial existence, can live in the roots and branches above the highest water in mangroves. A number of crabs also have adapted to life in the branches above the high tide. These crabs return to the sea only to breed.

SUMMARY

The intertidal zone is the transitional region between land and sea. In general, it is covered and uncovered by the tides each day. Compared with the total area of the earth's surface covered by the ocean, the intertidal zone is minuscule, less than 0.002 percent of the ocean's surface. Intertidal zones are greatest in extent in temperate latitudes, less extensive in the tropics, and restricted in polar regions.

The two most important physical factors that shape the intertidal zone are waves and tides. Wind-generated waves have their origin in storm areas called seas. The energy put into waves by wind increases with increases in wind velocity, duration of wind blowing in the same direction, and distance over which the wind blows freely in one direction (fetch). When waves leave the sea area, they transfer the energy put into them as swell, which can travel over great stretches of ocean with little energy loss. The energy is released primarily in the surf zone, where the shallow bottom near the continents interferes with wave motion. Refraction causes this energy release to be concentrated on headlands, and diffused in bays.

The tide grows out of the gravitational attraction between the earth, sun, and moon. The sun, which has about half the tidal effect of the moon, modifies the lunar tide. During the synodic month, the moon goes through a complete phase cycle producing spring tides during new moon and full moon, and neap tides during the first and third quarters. Tides also are modified by the constantly changing distance between the earth, sun, and moon, and the position of the moon and sun relative to the earth's equatorial plane (declination). The changing depth and size of ocean basins produces three identifiable types of tidal patterns: diurnal, semidiurnal, and mixed. The changing level of the ocean, because of tidal motion, produces the most varied of marine environments, the intertidal zone.

The intertidal zone is divided into rocky shores and particulate shores. Rocky shores are solid substrates, and particulate shores consist of sediment particles ranging in size from clay through cobbles. Six distinct biological communities are found in the intertidal zone. Four are named for the dominant substrate in which the flora and fauna live: rock, sand, mud, and gravel. The remaining two are salt marshes and mangroves. The rocky intertidal consists mostly of epibenthic organisms attached to the rock surface. The communities of mud flats, sand beaches, and gravel beaches consist primarily of infauna, organisms that burrow into the sediments. Salt marshes and mangroves are found in the upper levels of the intertidal zone, and are characterized by grasses and shrubs (salt marsh) or trees (mangrove).

Although the intertidal zone is the transition from the terrestrial to the marine environment, the majority of organisms are marine. Only in salt marshes and mangroves are terrestrial plants and animals found. These communities are truly transitional.

The characteristic flora and fauna of rocky shores, sand beaches, gravel shores, and mud flats all belong to taxonomic groups that are widely distributed in the marine environment and have little or no distribution on land. In tidal marshes and mangroves, the dominant organisms are land plants, and the flora and fauna consist of both marine and terrestrial taxonomic groups.

The flora and fauna of the intertidal zone are not uniformly distributed throughout the zone. The complex and rapidly changing environmental conditions found from high tide to low tide create a series of subzones. There are three generally recognized: the supralittoral fringe (that part covered by only higher than normal tides), the eulittoral zone (that part exposed and submerged every day), and the sublittoral fringe (that part exposed only on lower-than-normal tides).

Organisms of the intertidal zone show a number of adaptations to the harsh environmental conditions. Many organisms have heavy shells or exoskeletons for protection from atmospheric drying or drenching from rain. More common, however, are physiological adaptations. Many flora and fauna are more tolerant of extremes of environmental conditions, such as drying and high and low temperatures, than organisms found below the level of the low tide.

QUESTIONS AND EXERCISES

1. How are the sea, swell, and surf zone related to energy flow and transfer?
2. Based on the comparative energy levels present on the shores of headlands and bays, on which would you more likely find a rocky intertidal and sandy beach intertidal biocommunity?
3. Why is a spring tide condition associated with new and full moon phases, and a neap tide condition associated with the first and third quarters of the moon?
4. How do diurnal, semidiurnal, and mixed tides differ?
5. Discuss the relationship between velocity of tidal currents and tidal extremes.
6. Discuss the adaptations of intertidal organisms to the stress of desiccation.
7. Discuss the role and importance of bacteria in the muddy shore environment.
8. The upper limits of distribution of organisms in the intertidal are set by physical factors, while the lower limits are set by biological factors. For what marine intertidal communities is this statement valid?
9. Discuss the environmental factors that are associated with salt marsh and mangrove communities.
10. The tidal cycle affects physical factors which, in turn, influence the distribution of organisms in the intertidal zone. What are the major environmental factors that are affected by the tides?
11. Discuss the kinds and number of species of marine organisms of the intertidal zone, compared to the following other habitats: pelagic zone, shallow-water benthos, deep-ocean benthos, and deep-sea pelagic zone.
12. Define benthic, epibenthic, epifauna, meiofauna, epiflora, and infauna. Give a range of examples for each.

REFERENCES

Amos, W.H. 1966. *The life of the sea shore.* New York:McGraw-Hill.

Carefoot, T. 1977. *Pacific seashores.* Seattle:University of Washington Press.

Chapman, J., ed. 1977. *Ecosystems of the world: Volume 1, Wet coastal ecosystems.* New York: Elsevier.

Clancy, E.P. 1969. *The tides: Pulse of the earth.* Garden City, NY:Doubleday.

Denny, M.W. 1988. *Biology and the mechanics of the wave-swept environment.* Princeton, NJ: Princeton University Press.

Daiber, F.C. 1982. *Animals of the tidal marsh.* New York:Van Nostrand, Reinhold.

Green, J. 1968. *The biology of estuarine animals.* Seattle:University of Washington Press.

Higgens, R.P. and **Thiel, H.,** eds. 1988. *Introduction to the study of meiofauna.* Washington, DC: Smithsonian Institution Press.

Lear, R.R., and **Turner, T.** 1977. *Mangroves of Australia.* Brisbane, Australia:University of Queensland Press.

Lewis, J.R. 1964. *The ecology of rocky shores.* Agincourt, Ontario:Hodder and Stoughton.

Moore, P.G. and **Seeds, R.** 1986. *Ecology of rocky coasts.* New York:Columbia University Press.

Pickard, G.L. 1975. *Descriptive physical oceanography: An introduction,* 2nd ed. New York:Pergamon.

Ricketts, E.G., and **Calvin, J.** 1962. *Between Pacific tides.* Stanford, CA:Stanford University Press.

Thurman, H.V. 1985. *Introductory oceanography.* Columbus, OH:Merrill.

Tomlinson, P.B. 1986. *The botany of mangroves.* New York:Cambridge University Press.

SUGGESTED READING

OCEANS

Jerome, L.E. 1977. Mangroves. 10(5):38–47.

Jerome, L.E. 1979. Marsh restoration. 12(1):57–61.

Schafer, K. 1988. Mangroves. 21(6):44–49.

Wertheim, A. 1976. Tidepools. 9(5):16–23.

Wertheim, A. 1984. Tidepool gardens. (Tidal rockpools) 17(6):14–16.

OCEANUS

Phillips, R.C. 1978. Seagrasses and the coastal marine environment. 21(3):30–40.

Rutzler, K., and **Feller, C.** 1987. Mangrove swamp communities. 30(4):16–22.

Van Raalte, C.D. 1977. Nitrogen fixation in salt marshes: A built-in plant fertilizer. 20(3):58–63.

SCIENTIFIC AMERICAN

Bascom, W. 1959. Ocean waves. 201(1):89–97.

Cohen, I. 1981. Newton's discovery of gravity. 244(3):166–179.

Goldreich, P. 1972. Tides and the earth-moon system. 226(3):42–57.

Lynch, D.K. 1982. Tidal bores. 247(4):146–157.

SEA FRONTIERS

Moring, J. 1985. New trouble in the tide pools. 31(4):196–204.

GRUNION AND THE TIDES

Along the beaches of southern California and Baja California from March through September, a very unusual fish-spawning behavior can be observed. Shortly after the maximum spring tidal range has occurred, small silvery fish come ashore to leave their fertilized eggs buried in the sand. They are the grunion *(Leuresthes tenuis)*, slender little fish 12–15 cm in length. The name grunion comes from the Spanish *grunon*, which means grunter. The early Spanish settlers gave the fish this name because of the faint noise they make during spawning.

The type of tide that occurs along southern California and Baja California beaches is a mixed tide. On most tidal days (24 h, 50 min), there are two high and two low tides. There usually is a significant difference in the heights of the two high tides that occur each day. During the summer months, this higher high tide occurs at night. As the higher high tides become higher each night—as the maximum spring tide range is approached—sand is eroded from the beach (Fig. 1). After the

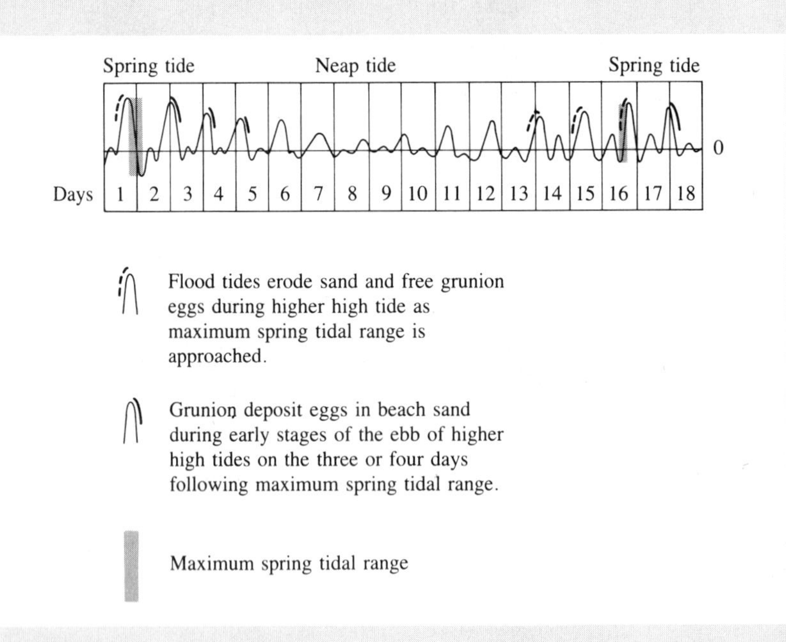

FIGURE 1
Tidal Conditions for Grunion Spawning

Eggs deposited in beach during ebb tides on days 3, 4, and 5 will be ready to hatch when eroded by the flood tides of days 13, 14, 15, and 16. The lines separating days on the tide curve chart represent midnight.

maximum height of the spring tide has occurred, the higher high tide that occurs each night will be a little lower than the one of the previous night. During this sequence of decreasing heights of the tides, as neap tide conditions approach, sand is deposited on the beach. The grunion depend greatly on this pattern of beach sand deposition and erosion in their spawning process.

Grunion spawn only after each night's higher high tide has peaked on the three or four nights following the night of the occurrence of the highest spring high tide. This behavior assures that the eggs will be covered deeper by sand deposited by the receding high tides. The fertilized eggs buried in the sand are ready to hatch nine days after spawning. By this time, another spring tide is approaching, and the higher high tide that occurs each night will be higher than that of the previous night. This condition causes the beach sand to erode, exposing the eggs to agitation by waves that break ever higher on the beach. The eggs hatch about three minutes after being freed into the water. Tests done in laboratories have shown that the eggs will not hatch until agitated in a manner that simulates the agitation of the eroding waves in nature.

The spawning begins as the grunion come ashore immediately following an appropriate high tide, and it may last from one to three hours. The peak of the spawning activity usually occurs about an hour after the start, and may last from 30 minutes to an hour. Thousands of fish may be on the beach at this time. During a run, females move high on the beach. If no males are near, a female may return to the water without depositing her eggs. In the presence of males, she will drill her tail into the semifluid sand until only her head is visible. The female continues to twist, depositing her eggs 5–7 cm below the surface. The male curls around the female's body, and deposits his milt against it (Fig. 2). The milt run down the body of the female to fertilize the eggs. The spawning completed, both fish return to the water with the next wave.

As soon as the eggs are deposited, another group of eggs begins to form within the female. They will be deposited during the next spring tide run. Larger females are capable of producing up to 3,000 eggs for each series of spawning runs that are separated by the two-week period between spring tides. Early in the spawning season, only older fish spawn, but by May even the one-year-old females are in spawning condition.

Young grunion grow rapidly, and are about 12 cm long when they are a year old and ready for their first spawning. They usually live to the age of two or three years, but four-year-olds have been recovered. The age of grunion can be determined by the scales. After growing rapidly during the first year, they grow very slowly. There is no growth at all during the six-month spawning season, which causes growth marks to form on each scale.

How the grunion are able to time their spawning behavior so precisely is not known. Some investigators believe that grunion are able to sense very small changes in the hydrostatic pressure caused by the changing level of the water associated with the tidal ebb and flow. Certainly some very dependable detection mechanism keeps the grunion accurately informed concerning the tidal conditions, because their survival depends on a spawning behavior precisely tuned to tidal motions.

FIGURE 2
Grunion Spawning

A male grunion wraps around the female that has burrowed into the sand to deposit her eggs beneath the surface. The male releases sperm-laden milt, which runs down the female's body to fertilize the eggs. (Photo by Bill Beebe.)

13

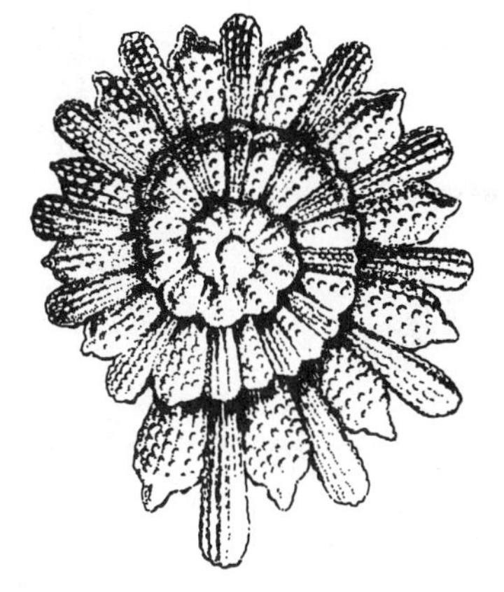

THE COASTAL ZONE

The level of the lowest tide to the deep edge of the continental shelf is the coastal zone (Fig. 13.1). Although the coastal zone is a relatively small area compared with the rest of the ocean (less than 10% of the total ocean area), it is biologically very diverse and rich. The benthos of the coastal zone is quite different from that of the deeper ocean. The physical features of the coastal zone differ from those of the offshore ocean. The edge of the continental shelf is, for the most part, a dividing line between the currents of the great oceanic gyres and coastal currents. These coastal currents are driven by local winds, tides, and fresh water runoff. The result is variations of temperature and salinity in the coastal zone that are more extreme than those of the open ocean.

The biology of the coastal zone is influenced most by high concentrations of nutrients (nitrates and phosphates) in surface waters. Coastal winds, estuarine circulation resulting from fresh water runoff, and tidal action all contribute to a continued replenishing of the nutrients to the surface water from deeper areas (Fig. 13.2). This means that the shallow plants, algae, and phytoplankton are supplied with a rich nutrient source that supports high levels of photosynthesis. These high levels of photosynthesis in turn support rich and diverse biological communities.

To a large extent, the coastal zone benthic communities rely on a food web that is based on macroalgae and plants rooted in the sediments of shallow water. In deeper waters of the open ocean, the relatively impoverished benthos relies, for the most part, on organic material raining down from the plankton-based community of surface waters.

There are a number of distinct biological communities found in the coastal zone. In shallow water close to shore, sea-grass meadows, kelp forests, and seaweed beds can be found. In warmer waters, coral reefs are found adjacent to land masses. Although these biological communities are highly visible, they occupy only a small fraction of the coastal zone. The bulk of the continental shelf consists of sand and mud bottoms, and the benthic organisms of these areas are the most common biological communities of the coastal zone. In this chapter we examine the physical features that are important in supporting the rich biological communities of the coastal zone and, in turn, consider the structure and function of these communities.

PHYSICAL FEATURES

Temperature

In coastal regions of the ocean where the water is relatively shallow, very great ranges in temperature may occur on a yearly basis. Sea ice forms in many of the high-latitude coastal areas, where temperatures are determined by the freezing point of the water, which generally will be above −2°C (28.4°F). Maximum surface temperature in low-latitude coastal water may approach 45°C (113°F) in areas where the coastal water is restricted in its circulation with the open ocean, and pro-

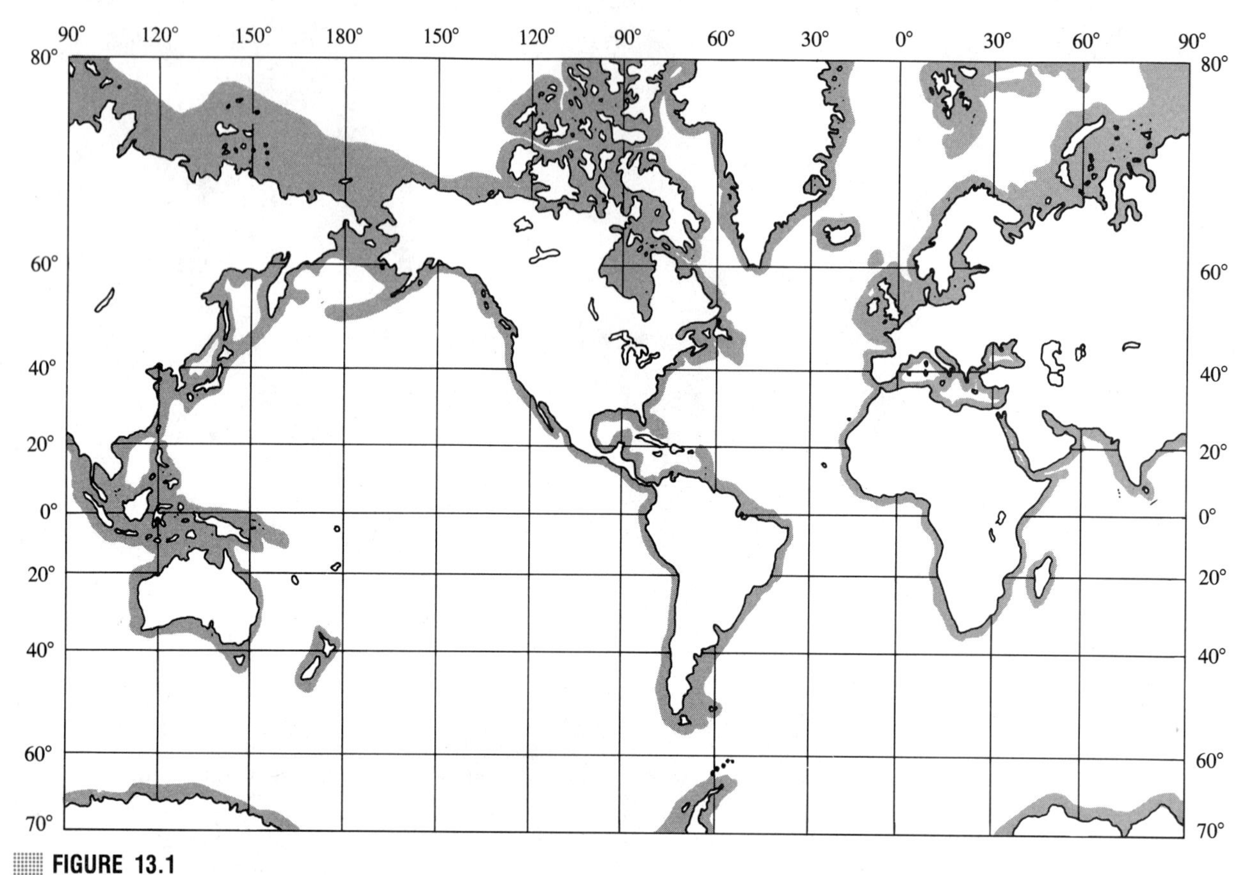

FIGURE 13.1
The Coastal Zone Benthic Habitat

The coastal zone benthic habitat is defined as ocean floor shallower than a depth of 200 m (656 ft), the approximate seaward boundary of the continental shelf.

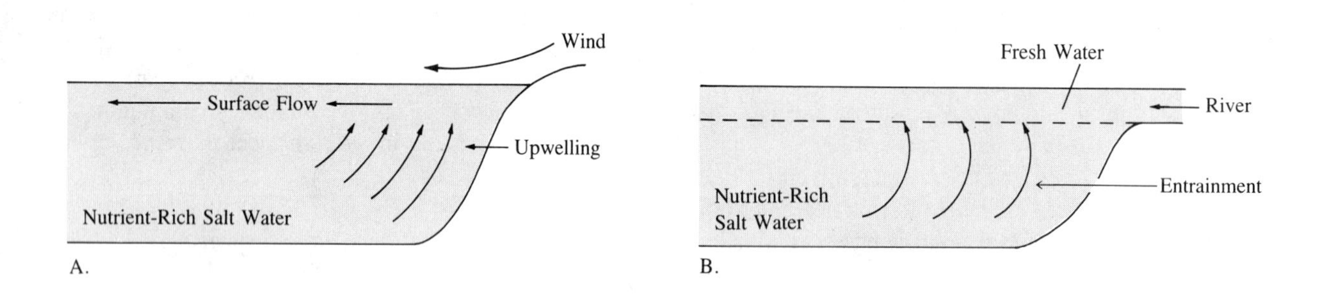

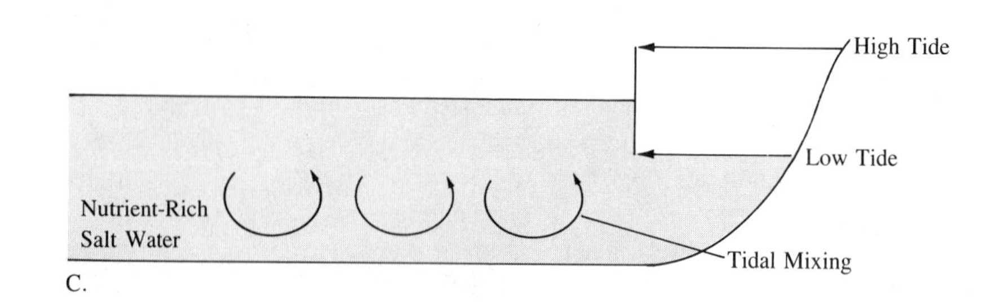

FIGURE 13.2

The three major mechanisms responsible for bringing deeper nutrient-rich water to surface waters of the coastal zone. **(A)** Offshore wind. **(B)** River runoff. **(C)** Tidal mixing.

tected from strong mixing. The seasonal change in temperature can be most easily detected in the coastal regions of the midlatitudes, where surface temperatures are at a minimum in the winter and reach maximum values in the late summer.

Figure 13.3 shows how strong thermoclines may develop in areas where mixing does not occur. High-temperature surface water may form a relatively thin layer. Mixing reduces the surface temperature by distributing the heat through a greater vertical column of water, thus pushing the thermocline deeper and making it less pronounced. One major factor in mixing coastal water is the effect of tidal currents, which can have a considerable influence on the vertical mixing of shallow water near the coast (Fig. 13.2C). Prevailing winds also can have a significant effect on surface mixing.

Coastal Currents

The currents of the coastal ocean are, like those of the open ocean, caused by wind. Winds blowing parallel to the coast in one direction cause water to pile up along the shore, under the influence of the Coriolis effect. The water, under the influence of gravity, runs back down the slope toward the open ocean (geostrophic currents). As the water runs down the slope away from the shore, the Coriolis effect causes it to veer to the north on the western coast, and to the south on the eastern coast of continents in the Northern Hemisphere. When winds blow along the coast in such a direction that Coriolis effect causes water to move offshore, the result is coastal upwelling. Coastal upwelling brings deep, cold, nutrient-rich water to the surface.

Another condition that produces geostrophic flow along the margins of continents is the runoff of large quantities of fresh water that gradually mix with the oceanic water. This produces a surface slope of water away from the shore. This seaward slope is associated with salinity and density gradients, as both increase seaward (Fig. 13.4). These variable currents, which depend upon the wind and the amount of runoff for their strength, are bounded on the ocean side by the more steady boundary currents of the open-ocean gyres.

These local geostrophic currents frequently flow in the opposite direction of the boundary current, as is the case with the Davidson Current that develops along the coast of Washington and Oregon during the winter. A very great amount of precipitation occurs in the Pacific

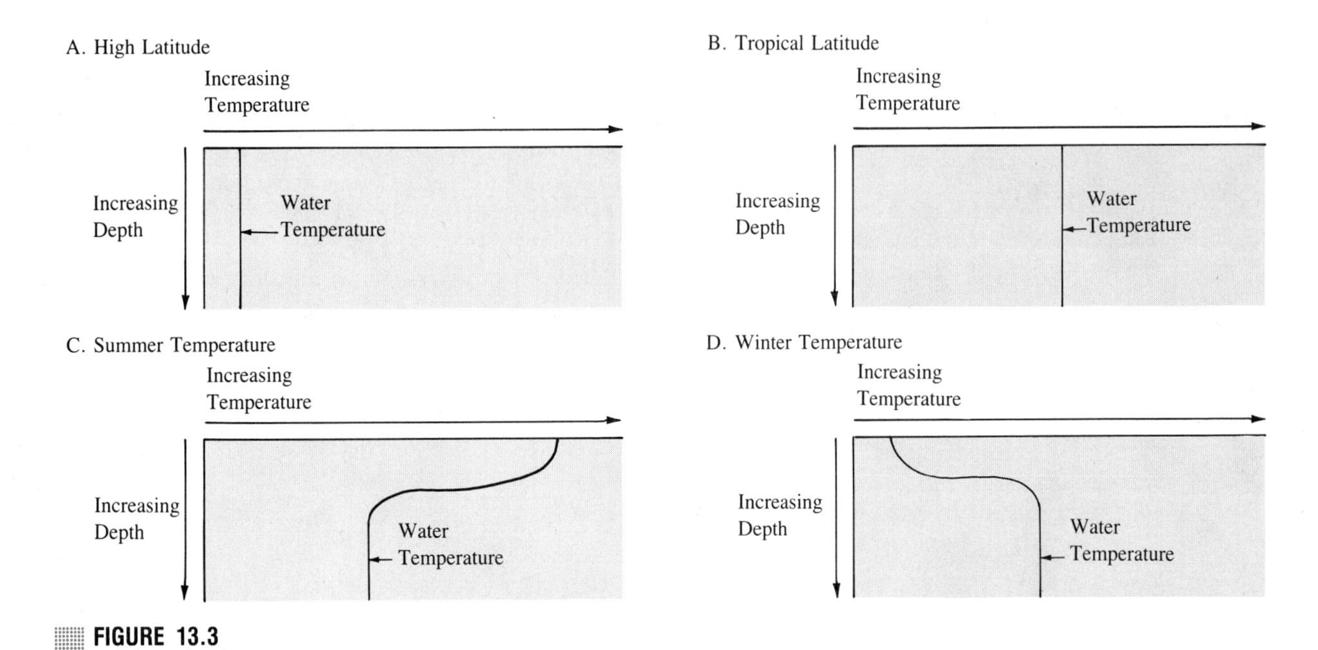

FIGURE 13.3
Temperature Relations in the Coastal Ocean

In arctic and antarctic areas **(A)**, sea ice keeps coastal waters uniformly cold with depth. Uniform temperatures with depth also are found in tropical areas that are protected from free circulation with the open ocean **(B)**. Warm water heated by the atmosphere extends from surface to bottom. In temperate latitudes, there generally is a clear stratification of temperature with depth. In summer, warmer surface water tends to float on the surface **(C)**. In winter, a layer of low-temperature water may develop **(D)**. When cold water persists, the higher density causes it to sink and the water becomes uniformly cold as in **(A)**.

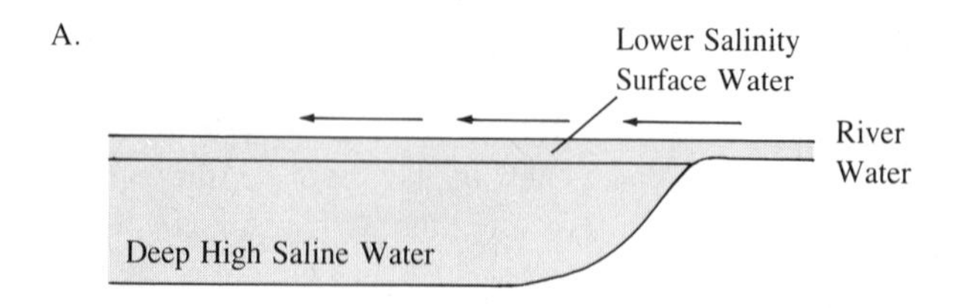

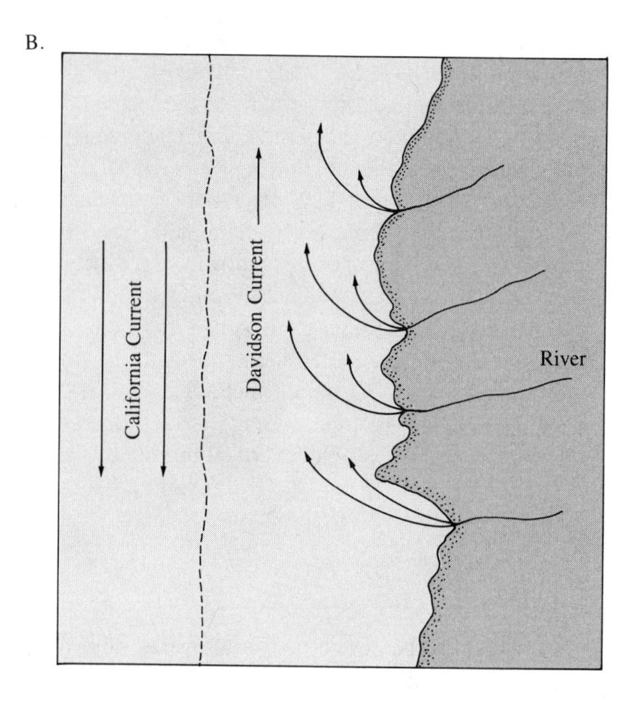

FIGURE 13.4

Along the coast of Oregon and Washington, the coastal current flows north in the winter. This northerly flowing Davidson Current runs in an opposite direction to the southerly flowing offshore California Current. The Davidson Current, for the most part, is driven by runoff from coastal rivers. The fresh-water runoff produces a seaward slope away from the shore **(A)**. As the flow runs seaward, the Coriolis effect turns the water northward **(B)**, creating the Davidson Current.

Northwest during the winter months, and the winds that generally blow out of the southwest are the strongest during these same months. Their combined effect produces a relatively strong northward-flowing geostrophic current between the southward-flowing California Current, which is part of the open-ocean circulation, and the continent of North America.

Salinity and Estuaries

In general, salinity is lower in the coastal ocean because of fresh water runoff from the continents (Fig. 13.5A). In some coastal regions such as the Mediterranean Sea, however, prevailing winds blowing from the land have lost their moisture over the continent. As these winds blow across the coastal ocean, they evaporate considerable quantities of water, resulting in increased surface salinity (Fig. 13.5C).

In the vicinity of rivers where fresh water runoff mixes with the sea, **estuaries** are formed. Technically, estuaries are defined as semienclosed bodies of water in which ocean water is significantly diluted by fresh water from land runoff. In some cases, estuaries are formed where rivers empty directly into the coastal

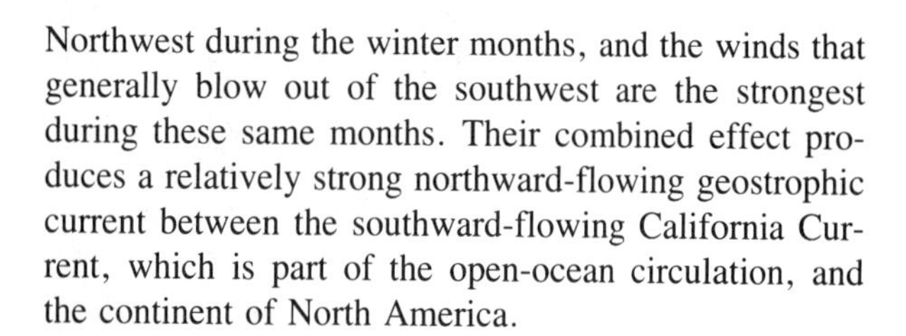

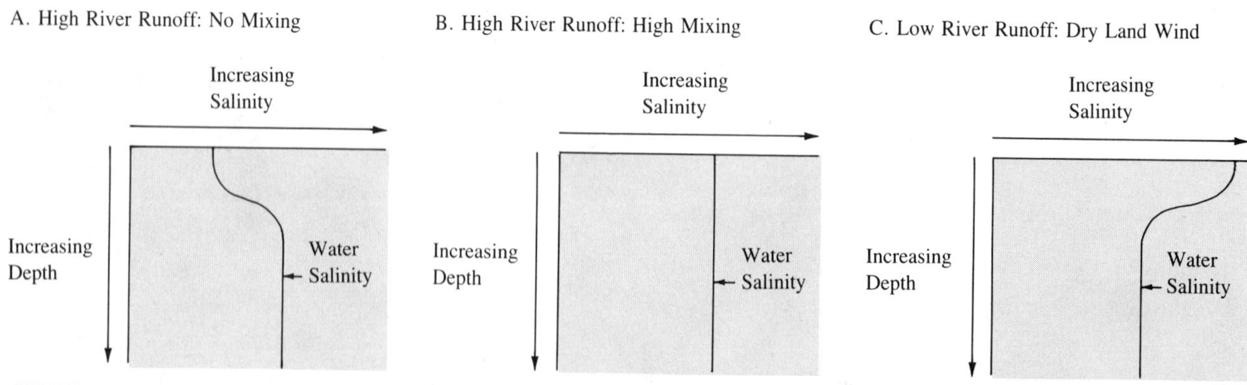

FIGURE 13.5

The salinity of the coastal ocean is affected by processes on land. When river runoff is not completely mixed with the sea water, there is a layer of fresher water on the surface and a halocline is formed **(A)**. When lower-salinity surface water is mixed well **(B)**, the salinity is a little lower than the open ocean but is uniform from surface to bottom. In areas where a dry land wind blows over the coastal water **(C)**, a layer of higher-salinity water forms on the surface and, when there is little mixing, a halocline forms.

ocean. For example, the Amazon River pushes a layer of brackish water almost 330 km (200 mi) offshore. In most cases, however, estuaries are associated with physical features of the continental margin. For example, bays, fjords, and barrier islands all form areas of protected water in the region of river discharges.

Origin of Estuaries

Essentially, all estuaries in existence today owe their origin to the fact that in the last 18,000 years, sea level has been raised approximately 120 m (394 ft), owing to the melting of much of the major continental glaciers that covered portions of North America, Europe, and Asia during the last Ice Age. Four major classes of estuaries can be identified on the basis of their geological origin (Fig. 13.6):

Coastal plain estuaries were formed as the rising sea level caused the oceans to invade the existing river valleys. These estuaries are sometimes referred to as drowned river valleys. Chesapeake Bay is an example of this estuary type.

Fjords are glaciated valleys that are U-shaped, with steep walls. They usually have a glacial deposit forming a sill near the ocean entrance. Fjords are common along Canadian coasts.

Bar-built estuaries are shallow estuaries separated from the open ocean by bars composed of sand deposited parallel to the coast by wave action. Lagoons separating barrier islands from the mainland are bar-built estuaries.

Tectonic estuaries are produced by geologic faulting or folding, which causes a restricted down-dropped area into which rivers flow. San Francisco Bay is in part a tectonic estuary.

Water Mixing in Estuaries

Generally, the fresh-water runoff that flows into an estuary moves as an upper layer of low-density water across the estuary toward the open ocean. An inflow from the ocean takes place below the upper layer, and mixing takes place at the contact between these water masses. Estuaries have been classified into the following categories, on the basis of the distribution of water properties (Fig. 13.7).

Vertically mixed These are shallow, low-volume estuaries where the net flow always proceeds from

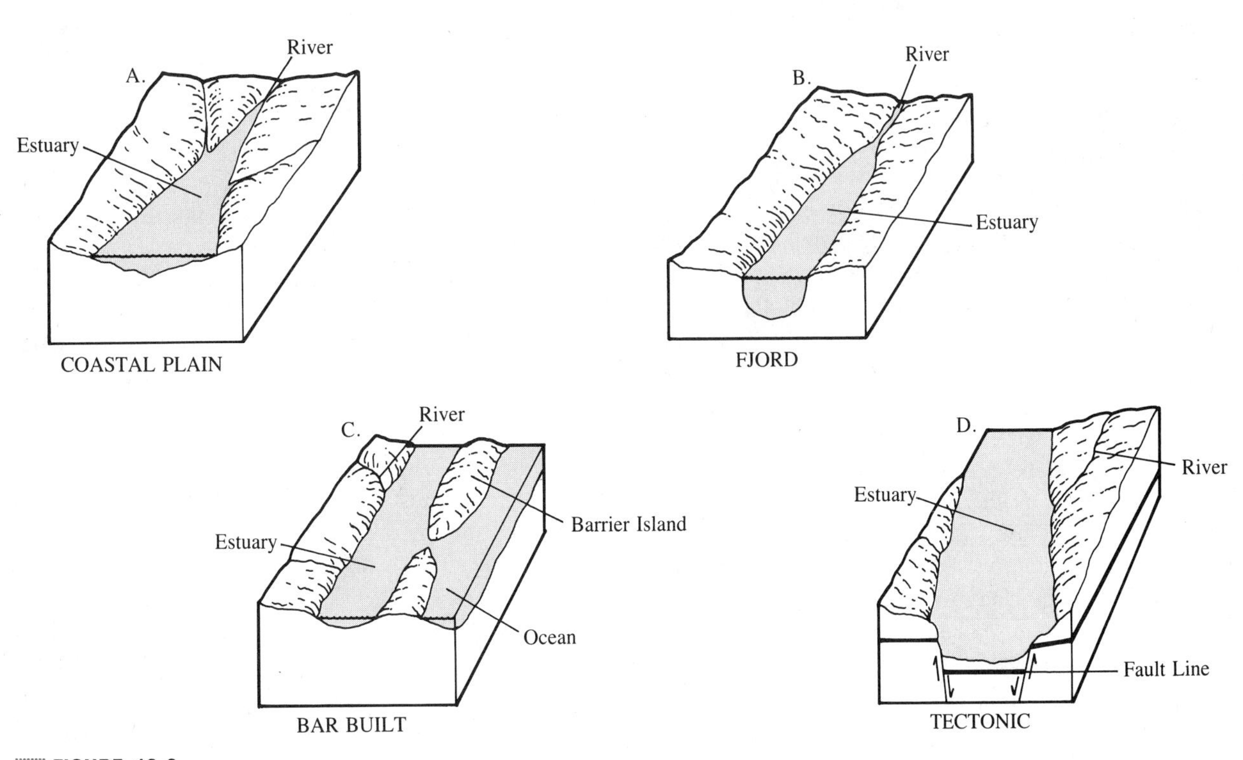

FIGURE 13.6
Classification of Estuaries on the Basis of Their Origin

Coastal plain estuaries **(A)**, such as Chesapeake Bay, are formed from 'drowned' river valleys. Fjords **(B)**, such as Puget Sound, are formed by glaciation. Bar-built estuaries **(C)**, such as the Outer Banks of North Carolina, are formed by longshore transport of sediment that forms barrier beaches. Tectonic estuaries **(D)**, such as San Francisco Bay, are formed by earthquakes or continental drift.

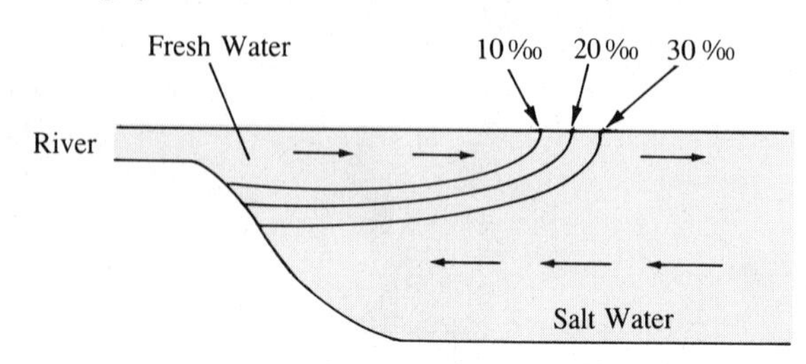

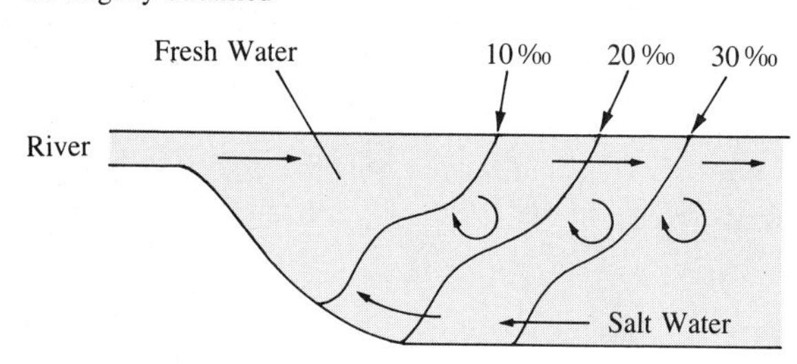

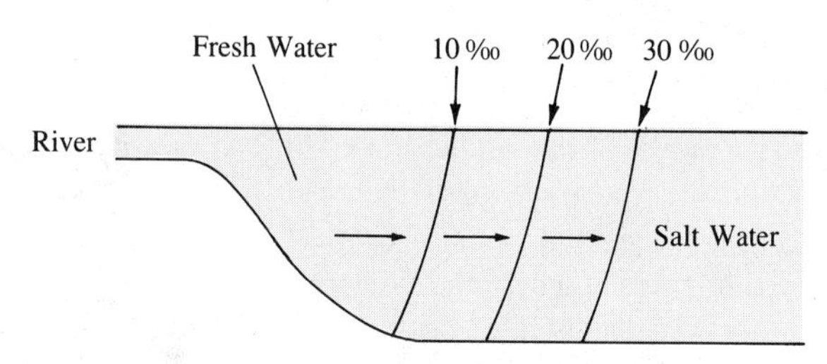

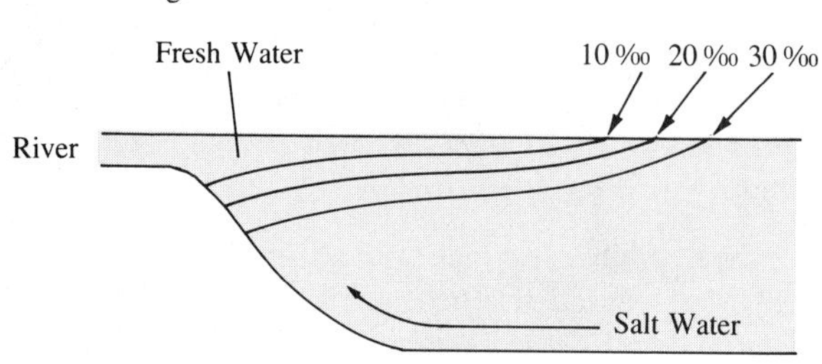

FIGURE 13.7
Classification of Estuaries on the Basis of Mixing of Salt and Fresh Water

When mixing is low, the fresh river water flows on top of the saltier water below **(A)**, forming a well-stratified estuary. When moderate mixing energy is present **(B)**, the estuary is slightly stratified. When mixing energy is high **(C)**, the estuary is vertically mixed. When river water flow is low to moderate, the salt water can intrude as a salt wedge **(D)**.

the river at the head of the estuary toward the mouth. Salinity at any point in the estuary will be uniform from the surface to the bottom, owing to the even mixing of the river water with ocean water by diffusion at all depths. Salinity increases from the head to the mouth of the estuary.

Slightly stratified Relatively shallow estuaries in which the salinity increases from the head to the mouth at any depth, they contain two basic water layers that can be identified: the less saline upper water provided by the river, and the deeper marine water separated by a zone of mixing. It is in this type of estuary that we begin to see the typical estuarine circulation pattern develop, as there is a net surface flow of low-salinity water into the ocean, and a net subsurface flow of marine water toward the head of the estuary.

Highly stratified This kind is typical of deep estuaries, where the salinity in the upper layer increases from the head of the estuary to the mouth, at which point it reaches a value close to that of open-ocean water. The deep-water layer, however, has a rather uniform marine salinity at any depth throughout the length of the estuary. The net flow of the two layers is similar to that described for the slightly stratified estuary, with one exception: The mixing that occurs at the interface of the upper water and the lower water is such that the net movement is from the deep-water mass into the upper water. The less saline surface water does not seem to dilute the deep-water mass, and simply moves from the head toward the mouth of the estuary, having its salinity increased as water from the deep mass joins it. Relatively strong **haloclines** (the water depth at which there is a sharp increase in salinity) develop in such estuaries at the contact between the upper- and lower-water masses.

Salt wedge Salt wedges are a wedge of salt water intruding from the ocean below that of the river water. Salt wedges are typical of the mouths of deep, large-volume rivers. There is no horizontal salinity gradient at the surface in these deep estuaries; the surface water is essentially fresh throughout the length of, and even beyond, the estuary. There is, however, a horizontal salinity gradient at depth, and a very pronounced vertical salinity gradient manifested as a strong halocline at any location throughout the length of the estuary. This halocline will be shallower and more highly developed near the mouth of the estuary.

The mixing patterns described above often cannot be applied to an estuary as a whole. Mixing within an estuary may change with location, season, or tidal conditions.

That much is yet to be learned about estuarine circulation is illustrated by the following observations made over an extended period in Chesapeake Bay. Just the opposite of the expected pattern—intervals of upstream surface flow accompanied by downstream deep flow—have been recorded. Also, periods when all flow was upstream, as well as periods of total downstream flow, have been observed. Even more complex patterns known to develop involve a surface and bottom flow in one direction separated by a middepth flow in the opposite direction, or landward flows along the shores and seaward flows in the central portions of the estuary.

Understanding the dynamics of estuary circulation is essential if desired advances are to be made in our ability to meaningfully describe: (a) parameters of water quality such as dissolved oxygen and coliform bacteria; (b) suspended particulate matter and bottom sediment transport; and (c) biological activity, particularly in relation to microscopic plants and animals, fish eggs, and larvae.

Estuarine Flow and Nutrient Enrichment. One of the reasons that estuaries support productive biological communities is that estuarine surface waters contain high levels of nutrients (nitrates and phosphates). In part, these nutrients come from the fresh water. However, a more important source is the nutrient enrichment resulting from estuarine flow. As the river water flows over the heavier salt water below, friction (or entrainment) pulls the salt water along with it, and gradually the two mix. Because the fresh water pulls the underlying salt water seaward, it is replaced by deeper nutrient-rich salt water that is pulled in from offshore. This circulation is diagrammed in Figure 13.8. The amounts of water involved are remarkable. In general, the total volume of water moving seaward from the estuary is ten times the volume of the river. In some cases, it is higher. The total seaward estuarine flow of the Saint Lawrence River is around 25 times the volume of the river itself. The nutrients in this entrained sea water are very important in promoting the high levels of photosynthesis found in estuaries.

BIOLOGICAL COMMUNITIES OF THE COASTAL ZONE

The biological communities of the coastal zone include coral reefs, sea-grass meadows, and kelp and seaweed beds, as well as the benthos of the wide expanse of sand and mud bottoms that dominate the continental shelves.

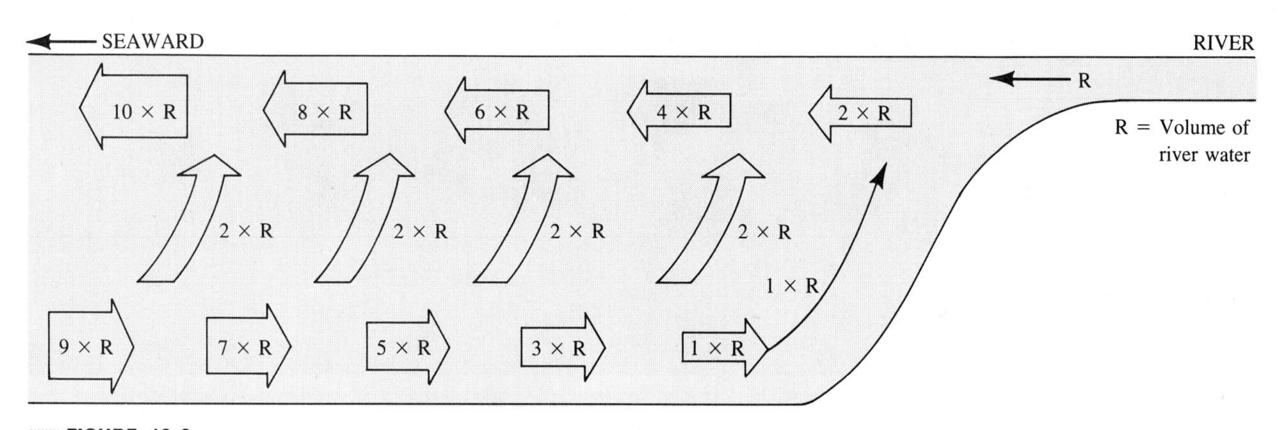

FIGURE 13.8
Schematic Diagram of Estuarine Circulation

The total volume of water that flows seaward is ten times the volume of river water entering the estuary. As the fresh water spreads across salt water below, friction pulls the salt water seaward, where it mixes with the fresh water. Deep nutrient-rich seawater is pulled to the surface.

Coral Reefs

It is difficult to describe with words the beauty, complexity, and majesty of the coral reef, one of the most fascinating marine communities. The physical structure of the coral reef is produced by the growth of both reef corals and algae. Continual growth creates a three-dimensional environment that is home to hundreds of species of marine organisms, many of them brilliantly colored. (The factors that influence the distribution of coral reef sponges in the Caribbean and the Pacific Ocean are discussed in a special feature at the end of this chapter.) Much of the biology of coral reefs remains mysterious, but the major mechanisms of coral reef structure and function are now well known.

Coral reefs are found in shallow waters surrounding tropical land masses. They are restricted to waters warmer than 18°C (64°F), generally in a band that lies between the Tropic of Cancer and the Tropic of Capricorn (Fig. 13.9). The exact boundaries of the coral reef band depend on ocean currents. Oceanic surface currents may carry warm water to higher latitudes, as in the western Atlantic, and coral reefs may follow. In other areas, such as the west coast of South America, colder water may be carried into the tropical zone, and so coral distribution is restricted. In addition to warm water, corals require high-salinity and silt-free water. In the area of the Amazon River, water temperature is suitable for coral growth, but silt and low-salinity water from the river prevent corals from thriving.

Coral reefs are of three main types: atoll reefs, barrier reefs, and fringing reefs (Fig. 13.10). Atoll and barrier reefs are fully developed and considered to be mature. Both are characterized by a lagoon that is protected on the seaward side by the reef. Barrier reefs occur in association with continental land masses. The largest single coral reef in the world is the Great Barrier Reef off eastern Australia, which measures up to 100 km (160 mi) across and nearly 2,000 km in length. Atolls, conversely, are associated with islands, particularly oceanic volcanic islands (Fig. 13.11). On atolls, the volcanic mountain has subsided beneath the ocean surface, leaving only the coral reef. Fringing reefs are considered immature because they have not yet produced a lagoon between the reef and its associated land mass. It is important to note that all types of coral reefs are constantly subjected to turnover by death and new growth.

Coral reef morphology is complex, and those studying particular reefs recognize many specialized aspects. The degree of exposure and the amount of wave

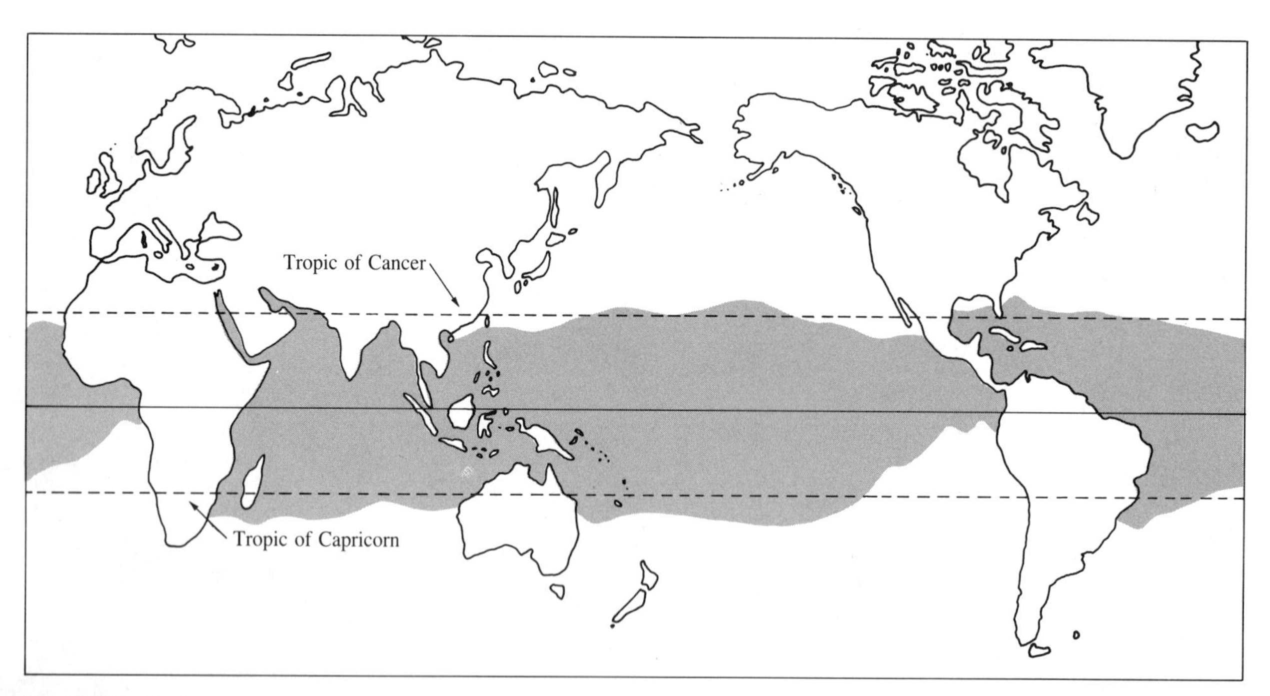

FIGURE 13.9
The Coral Reef Band

Coral reefs are generally restricted to a band between the Tropic of Cancer and the Tropic of Capricorn. Water temperature limits coral distribution. Where waters of 18°C (64°F) or cooler invade the tropics, coral reefs retreat toward the equator; where water of 18°C or warmer is found in higher latitudes, coral reefs follow.

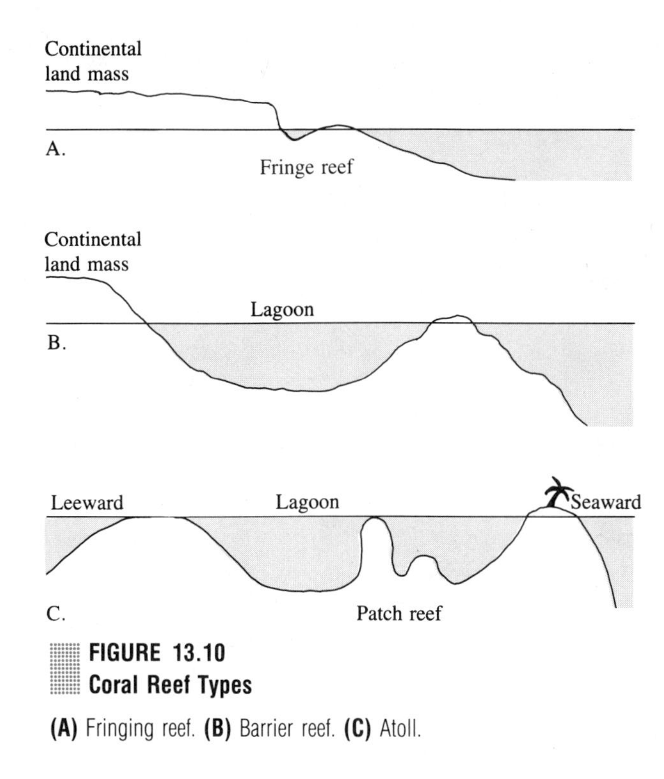

FIGURE 13.10
Coral Reef Types

(A) Fringing reef. **(B)** Barrier reef. **(C)** Atoll.

action is of particular importance in determining reef structure. Many reefs have a number of general features including the reef terrace, or flat, the algal ridge, the buttress zone, and reef face, or fore reef (Fig. 13.12). The surface of the reef is the **terrace,** or reef flat. Its elevation is at approximately average low water so the coral reef community is periodically exposed during periods of lower tides. The seaward edge of the coral reef generally has an **algal ridge** that is at a slightly higher elevation than the reef flat. The algal ridge takes the force of the breaking waves and is constantly awash from the movements of surf and surge.

Just seaward and below the algal ridge is the **buttress zone,** which reduces the force of the breaking waves on the algal ridge. The buttress zone is a series of massive fingers of coral alternating with surge channels. The buttress zone deflects and channels waves so that they tend to be directed upward over the algal ridge. In this way, the destructive energy of waves is dissipated over the seaward edge of the reef.

Below the buttress zone is the fore reef, or **reef face,** which extends to deep water. The maximum depth of actively growing corals and algae is approximately 70 m (230 ft). On atolls, the vertical face may continue deeper, but the reef face at depths greater than

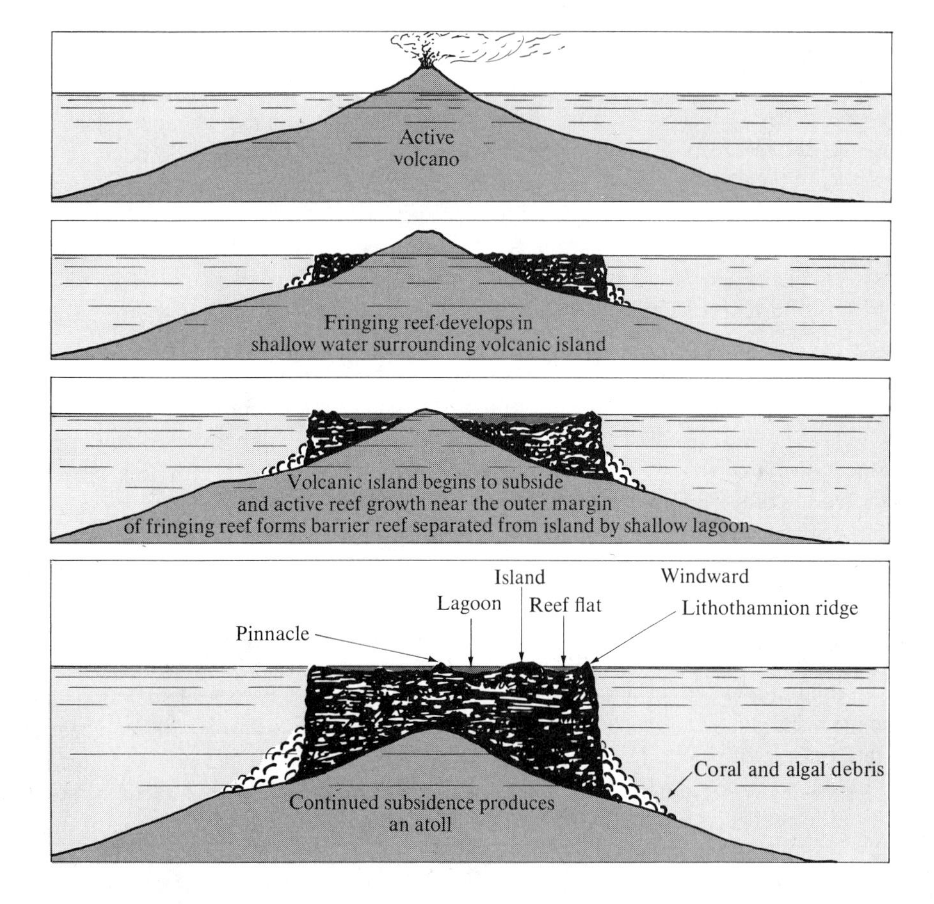

FIGURE 13.11
Darwin's Theory of Reef Development

Charles Darwin first postulated that coral atolls were the tops of volcanoes that had subsided beneath the water's surface.

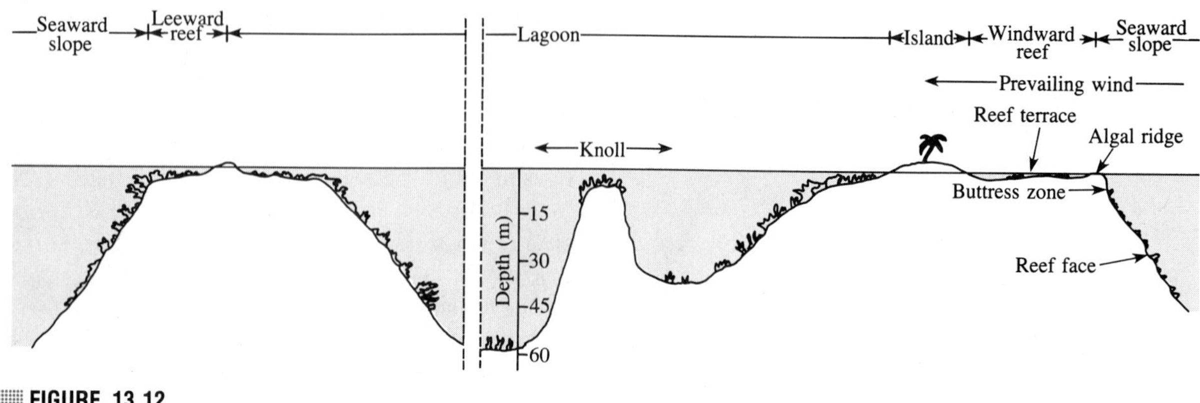

FIGURE 13.12
Influence of Surf on Features of a Coral Reef

High surf on the windward side of an atoll creates a well-developed algal ridge and reef flat. In these areas, coral development is minimal. On the more protected leeward side, the algal ridge is reduced and larger corals predominate.

70 m was deposited in past geological time and is no longer growing.

Growth of the coral reef is greatest on the algal ridge, the buttress zone, and the seaward portion of the reef flat. The farther toward land from the outer reef edge, the less favorable the conditions for reef growth. Why reef formation decreases with distance from the reef edge is not clear. Some argue that the amount of oxygen decreases and the amount of waste product metabolites increases. There also is a decrease in the amount of zooplankton inward from the reef edge. Perhaps the quieter water away from the reef edge favors those organisms responsible for reef erosion and destruction. Whatever the reason, the coral reef is a product of the actions of those organisms that produce and cement into place the calcium carbonate of the reef, and those organisms and physical factors that erode and reduce the solid reef material to sand.

Reef Formation. Coral reefs are named for those species of cnidarians that secrete an external skeleton of calcium carbonate that becomes part of the reef. The production of a calcium carbonate reef, however, is not restricted to corals. Some species of algae (both red and green) also secrete calcium carbonate skeletons. In fact, the majority of the limestone of many coral reefs is produced by the algae. Algae not only secrete limestone, but also cement loose sand into the reef. Pieces of coral, and other shells and sea urchins, bivalves, snails, and foraminiferans, contribute to the sand found in and near the reef. Sand that accumulates among the corals on the reef top or on the reef face is cemented into the reef by the action of the algae. Both algae and coral are essential to reef formation. Any threat to either component is a threat to the total reef.

Corals are cnidarians of the class Anthozoa (see Ch. 5). Like anemones, corals have no pelagic medusae stage in the life cycle. Not all corals produce calcium carbonate skeletons. Two types of corals are found on a reef: stony corals, which have calcium carbonate skeletons (Fig. 13.13), and soft corals, which have a protein-like skeleton. Only the stony corals contribute to the production of reef limestone. Typically, stony corals are colonial. Each polyp produces a calcium carbonate cup in which the polyp tissue sits. Polyps can retract into the cup and generally extend only during the darkness of night. Each polyp is attached by tissue to the other. Reproduction is primarily by asexual budding, and colonies once established can quickly grow (approximately 1 cm/yr). Corals also reproduce sexually and produce a planktonic larval form, which is the means by which corals disperse to new areas.

Corals are carnivorous and feed on zooplankton. This carnivorous feeding, however, does not account for the massive growth of corals on a reef. Although corals feed only at night, experiments have shown that they grow more during the day than at night. The explanation for this growth pattern lies in the presence of zooxanthellae, small rounded dinoflagellates that live inside the cells of corals. These dinoflagellates are primarily of two genera *(Gymnodinium* and *Amphidinium),* and both can be found as **endosymbionts,** or as free-living forms.

The zooxanthellae live in the cells of the inner gut tissues of the coral, and are found in concentrations of up to 30,000 cells per cubic cm of coral tissue. The corals provide a protected place for the zooxanthellae to live, and adequate sun for photosynthesis penetrates the mostly transparent middle and outer layers of the coral tissue. The metabolic wastes of the coral (CO_2 and ni-

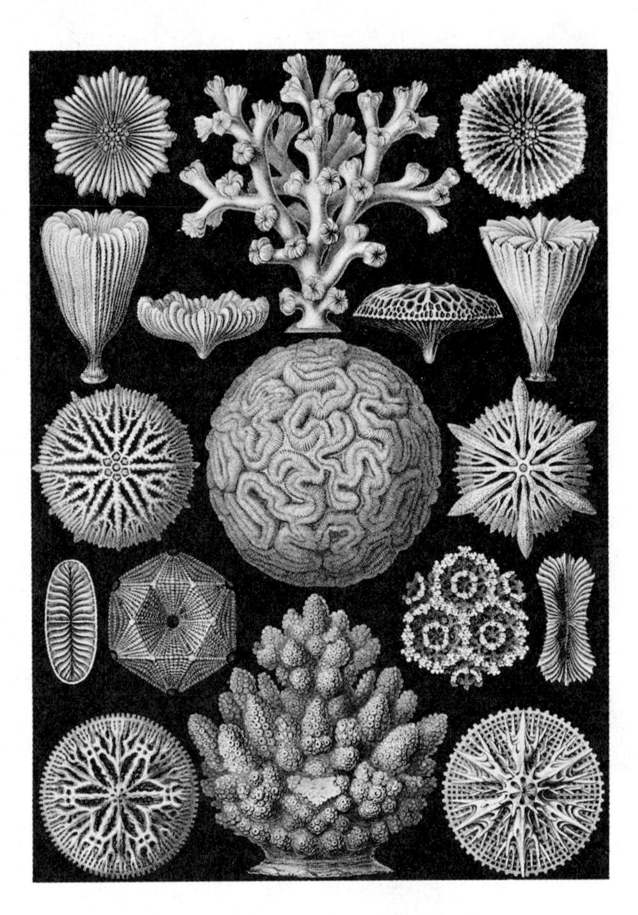

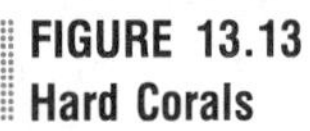

FIGURE 13.13
Hard Corals

(From Haeckel, E. 1974. *Art forms in nature.* New York:Dover.)

trogen molecules) are used by the zooxanthellae for photosynthesis. The benefits to the coral are equally important. Oxygen (as a product of photosynthesis) is made available to the coral, and up to 60 percent of the organic molecules produced by the dinoflagellates during photosynthesis escape the membranes of the cell, and are used by the coral as food. Up to 75 percent of the coral tissue may be zooxanthellae.

Soft corals differ from stony corals in the detail of their anatomy (the polyps have eight tentacles, instead of the six of stony corals), in their skeletons (soft corals have a central, flexible, proteinlike core), and in their distribution and abundance. Soft corals are found primarily on the reef face in deeper waters, but are similar to stony corals in having zooxanthellae in the tissues.

Although coral reefs are named for the characteristic stony corals, there are certain species of green and red algae that are particularly important to reef formation. Two groups of algae are important: the red coralline algae, and some green algae species. Red algae that contribute to reef formation are encrusting forms found primarily on the reef top and the wavebreaking ridge. These red algae help the reef grow rapidly by trapping any loose sand and cementing it to the solid reef. The most common green algae found on coral reefs are members of the genus *Halimeda*. This algae looks somewhat like a small underwater cactus; its photosynthetic filament consists of calcified segments arranged in a branching pattern.

Halimeda is often one of the most conspicuous organisms of the coral reef. It is found in the deep water of the outer reef slopes, in reef passages, on coral, and on the adjacent floor of reef lagoons. This green algae is important for contributing both biomass and calcium carbonate to the coral reef. Productivity rates of *Halimeda* may be as great as the red algae or the zooxanthellae of the corals. *Halimeda* is eaten directly, or may enter into the detritus food chain. The outer segments of the *Halimeda* filament easily break off, and the small discs of calcium carbonate are moved by waves and currents. Much of this calcium carbonate is washed to the lagoon floor, where it adds to the sand. Much of it, however, is carried to the fore reef slope or is trapped in crevices on the reef by the action of other organisms.

Up to 25 percent of the limestone of the coral reef may be produced by *Halimeda*.

A coral reef is the result of the dynamic interplay of the production of calcium carbonate by algae and corals and the erosion of the reef by physical and biological factors. Although coral reefs have structural characteristics that reduce the shock of wave surf (buttresses and the algal ridge), there still is a loss of coral to the waves, particularly during storms. Hurricanes in the Caribbean and cyclones on the Great Barrier Reef have caused severe localized damage, breaking off large coral colonies from reef ridges and flats. Some of the coral may tumble down the reef face to be recemented at lower depths. Most, however, is broken into finer and finer pieces until it finally becomes sand. As dramatic as storm damage is, however, the erosion by waves is minor compared to the erosion of coral reefs by biological factors.

Many varied organisms contribute to reef erosion. Fungi, sponges, bivalves, snails, polychaetes, sipunculids, and barnacles are capable of boring into the limestone of coral reefs. Some borers are even able to penetrate the outer living tissue layer of stony corals and invade the skeleton. One small clam found on many coral reefs burrows through the coral tissue as a larvae, and excavates a burrow into the skeleton. Boring sponges can penetrate older nonliving parts, and spread throughout the coral skeleton. These boring sponges can weaken the skeleton of the coral head to the point that the coral is unable to support itself and falls over. Sea urchins are particularly active burrowers in the coral, and eat material as they excavate.

Coral is also eroded by attacks on the surface. Parrot fish, whose strong jaws are capable of breaking off pieces of coral, feed directly on the polyps. Both the polyps and limestone are chewed into a fine sand that passes through the digestive tract. By eating coral (and coralline red algae), a single parrot fish may produce up to 90 kg (198 lbs) of sand per year.

The forces of erosion and growth on a coral reef are in dynamic balance. The optimal environmental conditions for reef formation are found at the reef edge. Inward from the reef edge, physical factors and organisms favor erosion. Erosion forms the lagoon of coral atolls, the open water between barrier reefs and land.

Corals as Food. Numerous predators, besides parrot fish, feed on the polyps of stony corals, but few cause any damage. Among these more or less harmless feeders are many fish, crabs, polychaete worms, snails, and sea stars. Nudibranch snails cruise over the coral surface, biting off individual polyps. Some butterfly fish 'farm' an area of coral, defending a territory of coral, and eating polyps, which are quickly regenerated.

One sea star that feeds on corals can cause damage. Since 1957, the crown-of-thorns sea star *(Acanthaster planci)* has become a menace to coral reefs. This species, previously rare, has undergone a population explosion on many coral reefs throughout the west Pacific Ocean. The crown-of-thorns uses spines on its legs to climb over the coral, protecting its tube feet, which appear to be sensitive to the nematocysts of the coral polyp. The stomach of the sea star is not, however, sensitive to nematocysts and readily digests the coral polyps. The damage to the coral is extensive and regeneration is slow. The crown-of-thorns has killed large patches of coral from many reefs through the western Indo-Pacific area. On the Great Barrier Reef, no corals reappeared for four years after attack, and it has been estimated that 20–40 years will be required for full recovery. Crown-of-thorn attacks affect more than just coral species. The whole reef community breaks down as corals are replaced by species of algae, causing many of the invertebrate and fish species associated with corals to be displaced. Other factors may cause widespread reef damage. Coral reef bleaching, thought to be related to increased water temperatures, is discussed in a special feature at the end of this chapter.

The most visible group of animals associated with coral reefs are the reef fish. To the visitor of a coral reef, the clouds of brightly colored fish are probably the most exciting feature. No marine habitat is richer in fish species, and thousands of individuals belonging to hundreds of species may be found on even a relatively small coral reef.

The species of fish observed at a particular coral reef change from day to night (Fig. 13.14). Daytime fish are characteristically brightly colored and patterned, and are found in schools. Most of the schooling fish feed on zooplankton that is constantly swept over the coral reef by waves and currents. By congregating, schooling fish are able to detect an approaching predator more effectively. Movement of the school tends to confuse the predator and allows time for the fish to flash back to the protection of the reef.

Besides schooling, many fish that are prey species use camouflage or protective coloration to confuse predators. Sharp, bold patterns on a fish break up the outline of the body shape, making it less obvious. Brightly colored spots on prey (often on the tail) look like large eyes that likely confuse predators. (Often, predators aim for the eye of the prey.) If a predator attacks the tail spot, the prey has a better chance of escape.

Other fish that are poisonous advertise their presence with bold coloration. Lion fish have bright stripes across the body. However, not all toxic fish advertise their presence. The deadly stone fish is the color of the

FIGURE 13.14

The fish population of a coral reef changes from day to night. Plankton-eating fish that form clouds around the reef by day take refuge in the reef crevices at night. Night-feeding fish for the most part feed on the benthos.

surrounding coral, and is very difficult to see when perched on a piece of coral.

Some fish, especially algae feeders or grazers on sessile animals, establish territories on coral reefs. The occupant of a territory vigorously defends its space against others of the same species. Some territorial fish even 'farm' their territory. For example, some damsel fish, which feed on filamentous algae, weed out sponges, hydroids, and polychaete worms that try to settle in their territory, creating a greater space for the food algae to grow on.

At nighttime, the fish of the day period disappear into the protection of the crevices and shelter of the reef. The nighttime fish are different from the day fish: Night fish are both fewer in number of species and individuals. The nocturnal appearance of these fish is probably linked to the nocturnal nature of the polychaetes, molluscs, and crustaceans on which they feed.

An important feature of the biological community of coral reefs is the range of symbiotic relationships, such as that between zooxanthellae and corals. Many other species live together in obligatory or facultative relationships. Crabs and shrimp are typical commensal species on coral reefs. Although not prominent and visible, shrimp are particularly abundant on coral reefs. Some 98 genera have been collected from all parts of the reef. These shrimp are found as commensals with sponges, soft and hard corals, sea anemones, annelids, bivalve and gastropod molluscs, sea urchins, brittle stars, tunicates, and fish.

The best-known commensal relationship in coral reefs is between a large predatory fish and a smaller cleaning fish or shrimp. The large predator allows the small cleaning shrimp or fish to nip ectoparasites off its gills, or even the insides of its jaws. The cleaning fish and shrimp also eat away dead or diseased tissue. They have a specific station where they receive the large predators, and often advertise their readiness to clean with a dance. The predator generally assumes a nonthreatening posture (lying down, floating head down near the bottom, or displaying a color change) while the cleaners do their job. Cleaners are effective in maintaining the good health of the predators on a coral reef. Experiments in which cleaners were removed from a section of a reef showed an increase in the incidence of parasitism, and a decline in the health and vigor of the predators.

As for biological productivity, there are many more species on a coral reef than the surrounding water or sediments. Productivity, as measured by photosynthesis, is some ten times greater on the reef than in the surrounding ocean waters. This high primary productivity has been a puzzle because most tropical waters are nutrient poor and have relatively low rates of photosynthesis. The higher productivity of coral reefs is related to rapid recycling of nutrients and nutrient enrichment from benthic cyanobacteria that can convert nitrogen gas to nitrate ions. The presence of the underwater mountain on which many coral reefs are located interferes with oceanic circulation, and causes a form of island upwelling (Fig. 13.15). Coral reefs truly are islands of intense biological productivity in the relative desert of the surrounding water.

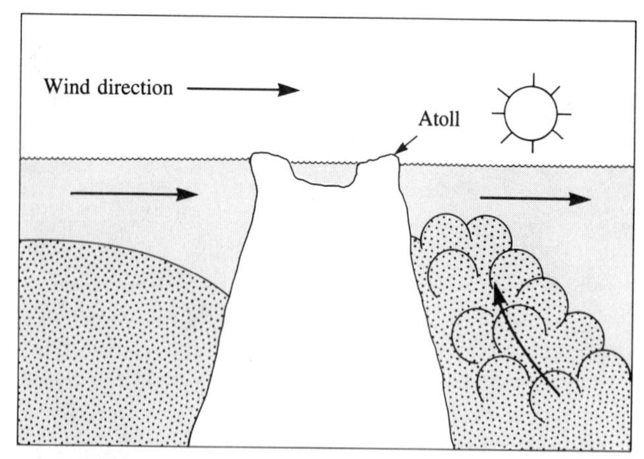

FIGURE 13.15
Upwelling in Coral Reefs

Where persistent winds blow across a coral atoll, a localized upwelling may occur on the leeward shore, bringing nutrient-rich water to the surface. This is only one of a number of explanations for the higher biological productivity of coral reefs compared to the surrounding oceanic water.

Sea-Grass Meadows

The lagoons associated with many coral reefs are often carpeted with a meadow of sea grass. The sea grasses are among the very few true plants that are totally adapted to the marine environment. There are about 50 species of these flowering plants that can live totally submerged in seawater. Sea grasses are only distantly related to the familiar grasses of land. Although a few species are found in the full surf of exposed rocky shores in temperate latitudes *(Phyllospadix),* sea grasses are most abundant in shallow, quiet waters of temperate and tropical latitudes. Sea grasses are not found in polar zones. In temperate latitudes, the most common genus is *Zostera* (eelgrass; Fig. 13.16A), while in tropical waters the most common genus is *Thalassia* (turtle grass).

Although sea grasses are true vascular plants that flower and set seeds, they have made a number of specific adaptations to their life in seawater. The leaves have no **stomata.** Instead, the carbon dioxide required for photosynthesis is absorbed across the leaf epidermis. In the flowering process, pollen is dispersed by water and fertilization is by a chance encounter of the pollen grain with a flower. To protect the pollen grain from the seawater, they are encapsulated in a mucous sheath. The vascular system of sea grasses is poorly developed. Nutrients required for plant growth (NO_4, PO_4) can be absorbed from the water directly across the leaf surface. Nutrients also may be obtained from the roots. In fact, in some cases nitrates and phosphates may travel from sediments through the roots and vascular system to the leaves and then escape to the surrounding seawater. In this way, seagrass meadows can increase the dissolved nutrient level of the water.

Sea-grass meadows are rich biological communities with rates of primary production comparable to high-quality, intensively farmed agricultural crops. Despite this rich production of plant material, few animals eat sea grasses directly. In the tropics, manatee, green turtles, parrot fish, and surgeon fish are the principal vertebrate herbivores. Among the invertebrates, only sea urchins eat sea grasses directly. In coral reef lagoons there often is a well-defined strip of bare sand between coral heads and the sea-grass meadow. The extent of the bare sand indicates the distance that sea urchins range away from the safety of the coral head as they graze.

In temperate sea-grass meadows, the major herbivores are ducks and geese that feed on the fringes of sea-grass meadows exposed at low tide. Less than 5 percent of the plant material of sea-grass meadows is

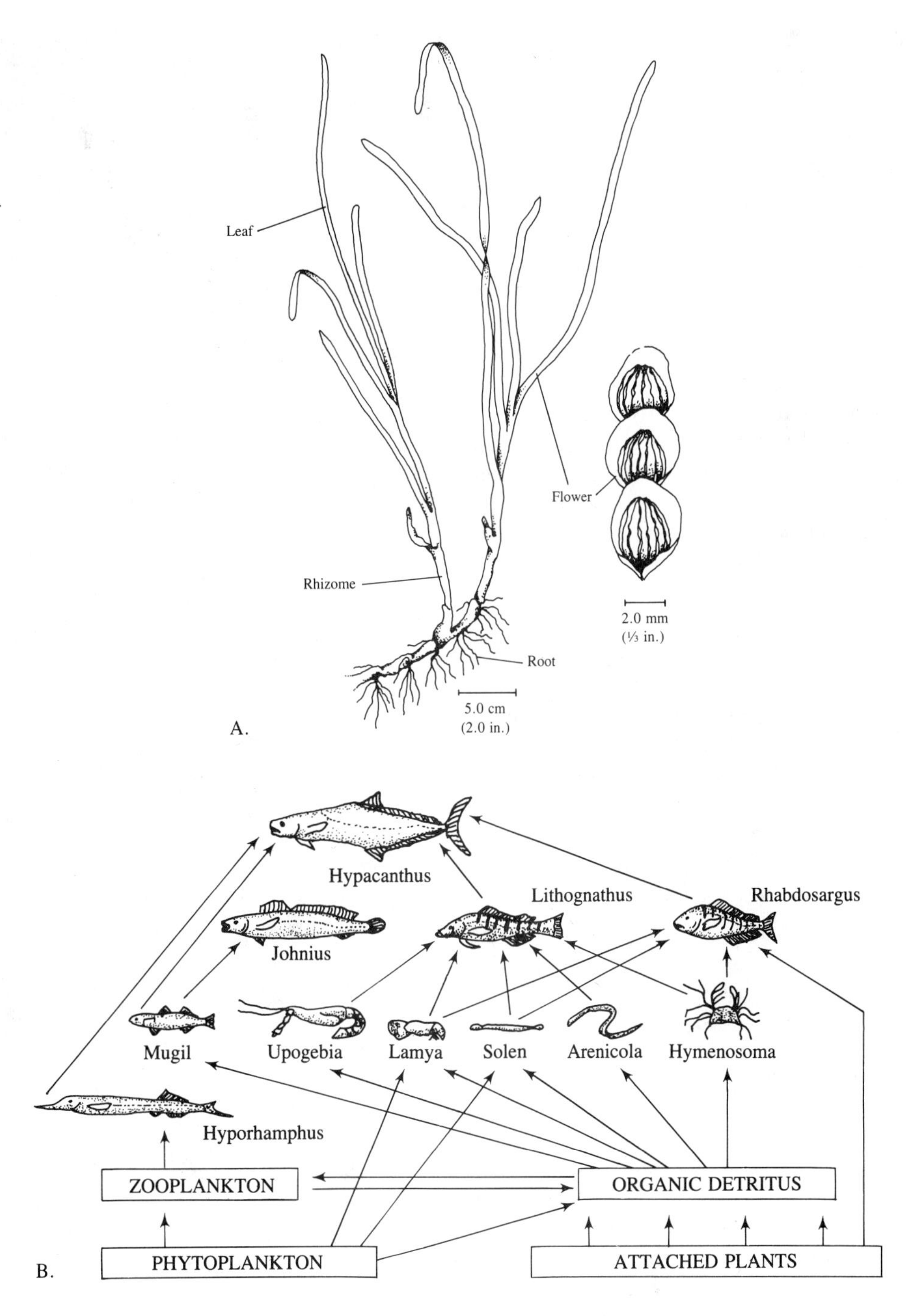

FIGURE 13.16

(A) The eelgrass *Zostera.* **(B)** Food chain in a seagrass meadow. Although phytoplankton is a source of food for the community of eelgrass organisms, it is not nearly as important as the sea grass. Little sea grass is eaten directly; most of it enters into a detritus food chain. Shown are the food pathways in a South African sea-grass bed.

eaten by herbivores. The remainder eventually enters the detritus food chain. Individual leaves of sea-grass plants have natural cycles of growth and senescence. As the leaves age, organic molecules escape into the surrounding water. These organic molecules are probably taken up by animals that live as epiphytes on the leaves. Ultimately, sea-grass leaves break free from the plant and may float quite a distance away from the point of attachment. The leaves gradually break down into smaller particles that sink and become part of the sediments of the bottom. During this process of breakdown, the leaf serves as a physical substrate and provides energy for bacterial growth. The bacteria in turn are used as a source of food by a wide range of filter and deposit feeders in the benthos (Fig. 13.16).

The leaves of sea-grass meadows may be exported great distances from the point of growth. Sea-grass blades have been photographed on the sediments of the Puerto Rico Trench at depths of 7,860 m (25,780 ft). Off the coast of North Carolina, sea-grass leaves are quite common. Of some 5,300 photographs taken at an average depth of 3,900 m in the vicinity of the Virgin Islands, sea-grass leaves were observed in virtually every picture.

Many more species of algae and animals are found in a sea-grass community than in adjacent areas of sandy or muddy bottoms. This is because the sea-grass meadow provides both a solid substrate on which organisms can settle, and a refuge where animals can hide and avoid predation.

Sea-grass meadows serve as a host to a wide variety of epiphytes. Microalgae (particularly diatoms), macroalgae (particularly filamentous red algae), and many invertebrates use the leaves as a substrate for attachment. The epiphytic algae may make a significant contribution to the total photosynthesis of sea-grass meadows. Between 25–30 percent of total photosynthesis may be due to epiphytic algae, which serve as a source of food for many invertebrate species found in the sea-grass community. Gastropod snails, nudibranchs, polychaete worms, isopods, and amphipods are commonly found crawling on sea-grass leaves and among the roots. Other animals swim in and out of sea-grass meadows, feeding selectively. Mysids, chaetognaths, shrimp, squid, and a number of fish use the sea-grass meadows in this way.

Since the early 1900s it has been argued that sea grasses are important to the biology of the organisms of all areas of the continental shelf. The most dramatic evidence for this view came from the devastation of eelgrass meadows on the east coast of North America between 1931 and 1933. Between 90–95 percent of all eelgrass meadows suddenly died off. This was caused by a fungus attack called **wasting disease.** The reason for the rapid spread of the disease is still not well understood, although it may have been related to changes in water temperature. The disease affected a wide range of species. Numbers of the black brant, a goose that feeds on eelgrass during annual migrations, declined sharply. The Canada goose population decreased by 80 percent in affected areas. Many invertebrate species not obviously connected with eelgrass beds disappeared or were greatly reduced in numbers. Scallops, clams, crabs, lobsters, cod, and flounder all showed severe decreases when the eelgrass disappeared.

Sea-grass meadows stabilize the sediments in which they grow. Sea-grass leaves act as baffles that slow water movements from waves and currents. Suspended material tends to accumulate in the quiet waters of the meadow and is trapped by the network of rhizomes and roots. As the roots and rhizomes bind the sediment together, surface erosion is decreased.

Kelp Forests and Seaweed Beds

Shallow waters of rocky shores in temperate latitudes (where water temperatures seldom exceed 20°C) of both the Northern and Southern Hemispheres are characterized by rich growths of macroalgae (Fig. 13.17). The prominent algae that give the seaweed beds and kelp forests their shape are brown algae, although numerous red and green algae also are found.

The largest of the brown algae belong to the Laminariales and are commonly called kelp. They have a holdfast for attachment to the bottom, and a stemlike stipe that permits the algae to be anchored in deeper water, but allow the fronds, which provide the major photosynthetic surface, to lie near the water surface. In the North Atlantic, the most common kelps are of the genus *Laminaria*. The stipes of *Laminaria* are relatively short and the depth to which they can grow successfully is limited. In temperate regions of both the North and

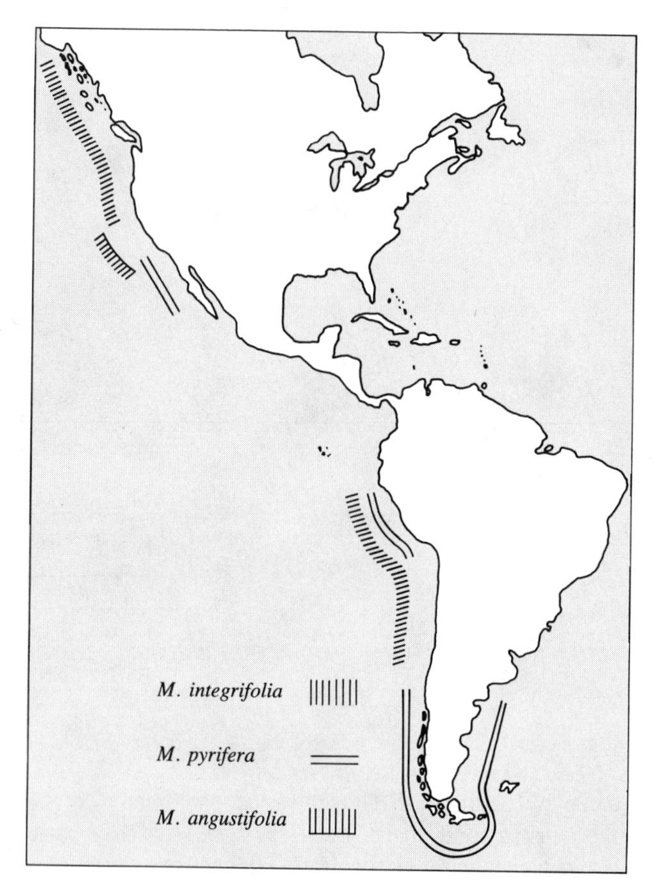

FIGURE 13.17
Distribution of Macroalgae

Large macroalgae are for the most part restricted to temperate latitudes. Shown is the distribution of the giant kelp *Macrocystis* in North and South America.

South Pacific Oceans and the South Atlantic, the most conspicuous kelps are those of the genera *Macrocystis, Nereocystis,* and *Eklonia*. These kelps are characterized by long stipes that permit them to grow to depths of around 25 m (82 ft). The maximum depth recorded for a kelp is 65 m, off the Pacific coast of South America. Kelps with long stipes often are found in dense masses referred to as **kelp forests** (Fig. 13.18).

Kelp forests and seaweed beds are rich in both algae and animal species. Thick mats of algae, exposed during the lowest tides, offer protection to organisms from the harsh conditions of the atmosphere. When the tide is in, the now upright algae create a complex habitat for other species. The spaces between and below the algae fronds provide protected areas for fish and swimming crustaceans. This refuge is of particular importance for juvenile fish. The surface of the larger algal blades offers a substrate for many epiphytic animals and other algae. Of the epiphytic algae, reds are most common, and of the sessile epiphytic fauna, hydroids, cnidarians, and bryozoans are common. The algae blades also provide a habitat for many motile crawling animals such as isopods, copepods, and snails. The surface of the bottom is another optimal habitat for shade-tolerant algae species and many sessile and motile invertebrates, and the holdfasts of the algae provide yet another habitat for a range of animals (Fig. 13.19).

Growth of most species of kelp and seaweed is greatest during the winter months and early spring. *Laminaria* on the Atlantic coast of Canada has its maximum growth in February and March, when water temperature hovers around 0°C (32°F). In the Arctic, *Laminaria* has been found to grow in complete darkness under the ice pack. This early growth is apparently due to the mobilization of food reserves rather than photosynthesis. Not all kelps and seaweeds show this seasonal growth. The giant kelp of California (*Macrocystis*) grows more or less evenly all year round.

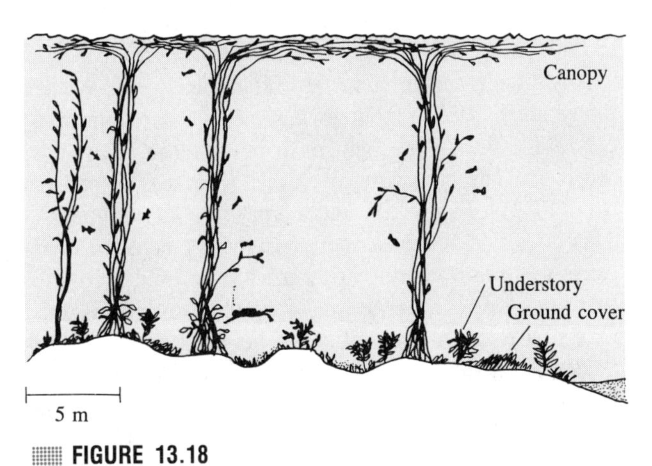

FIGURE 13.18
Kelp Forest

The giant kelp *Macrocystis* forms forestlike communities.

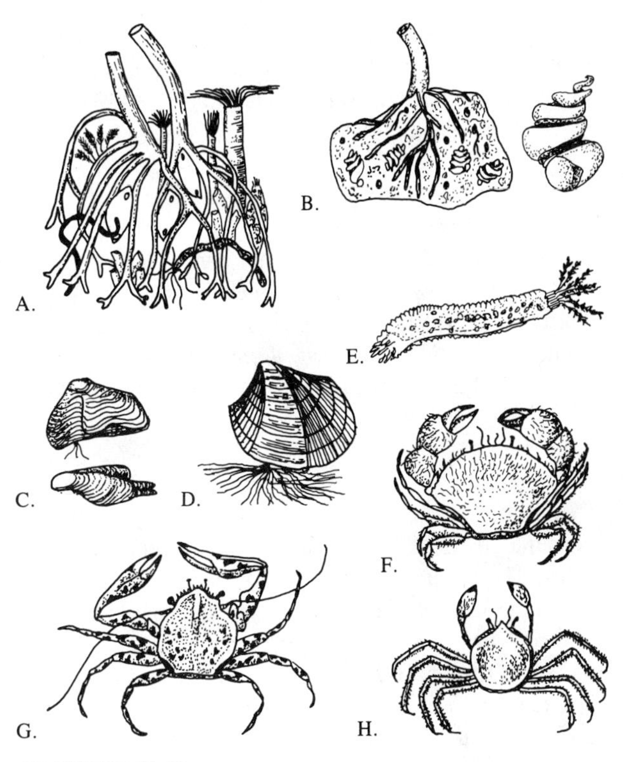

FIGURE 13.19

Kelp holdfasts often serve as a place to live for a variety of animals. Shown are characteristic fauna of the giant kelp *Eklonia* of southern latitudes. **(A)** *Eklonia* holdfast with tube worms and sponges. **(B)** *Eklonia* holdfast surrounded by a sponge. Note the coiled gastropod shells. **(C)** The clam *Hiatella.* **(D)** The clam *Ryenella.* **(E)** The sea cucumber *Ocnus.* **(F)** The crab *Pilumnus.* **(G)** The crab *Petrocheles.* **(H)** The crab *Hymenicus.*

Rates of growth of seaweeds and kelps are impressive. Fronds of *Macrocystis* may grow up to 30 cm (12 in) per day, and in Puget Sound the stipe of the bull kelp *Nereocystis* may grow up to 25 cm per day. Annual rates of production of kelps and seaweeds are equally impressive. Production values of 2–3 $kg/m^2/yr$ (dry weight) have been measured for *Macrocystis,* and *Laminaria* on the east coast of Canada has an annual productivity of between 0.5–4 $kg/m^2/yr$ (1–7 $lb/yd^2/yr$). These values equal or exceed the production values of tropical rain forests, long considered the most productive biological communities on earth.

The color of seaweeds and kelp is caused primarily by accessory photosynthetic pigments. These pigments are known to increase the efficiency of photosynthesis by capturing wavelengths of light not absorbed by chlorophyll. Red algae have the accessory photosynthetic pigment phycocyanin. Phycocyanin absorbs the shorter (blue) wavelengths of visible light. Most red algae are found in shady conditions and tend to be deeper in their distribution than brown and green algae. At these depths, the longer wavelengths of light (reds) have been mostly absorbed by the water and there is a higher proportion of blue wavelengths. Other factors also may influence the distribution of red algae. In temperate climates, some red algae are found in the supralittoral zone in full exposure to light, and in the tropics red algae often predominate in the well-lighted algal ridge of coral reefs.

There are only a few grazing animals that eat seaweeds and kelp of temperate latitudes. These herbivores consume a small portion of the algae produced each year. The large accumulation of seaweeds and kelp found on beaches at certain times of the year is a reflection of their limited use as a direct food source. Most of the algae from seaweed beds and kelp forests enters the detritus food chain. Some kelp *(Nereocystis)* are annuals and break free during fall storms. Others such as *Laminaria* and *Macrocystis* are constantly eroding material from the ends of their blades. This material may sink to the bottom in the vicinity of the algae. The larger pieces of broken algae are called **drift.** Drift is an important food source for a number of animals including spider crabs, snails, sea urchins, abalone, and some sea stars. The drift is broken into smaller pieces (by physical abrasion or by being passed through the gut of drift-eaters), and these pieces are broken down further primarily through the action of bacteria. These bacteria are then a source of food for detritus suspension feeders including sponges, tunicates, polychaetes, bivalves, and some snails.

Benthos of the Continental Shelves

As we have seen, the ocean bottom immediately adjacent to land but below the influence of tides may host rich and intensely productive biological communities. Coral reefs, kelp forests, and sea-grass meadows all rely on high concentrations of light for their rich diversity and productivity. As the ocean bottom recedes from the edge of the land, the conditions of increasing depth and decreasing light limit photosynthesis. No large primary producers are found. Bottom conditions become more uniform on the great plains of the continental shelves. Generally, the bottom slopes gradually to the 200 m (650 ft) depth (the average depth of the outer edge of the shelf). The substrate on continental shelves ranges from muddy gravel through fine sand to large areas of mud. In general, the sediments are much coarser than those of the deeper parts of the ocean.

The continental shelf at all depths generally has some measurable light during the day. At deeper depths there may not be enough light to support photosynthesis or color vision, but even at 200 m there usually is enough light for those animals with eyes to identify shapes and forms.

Compared to the extent of the coastal fringe of the littoral zone or the nearshore benthic communities of coral reefs, kelp forests, and sea-grass meadows, the extent of the continental shelves is immense. Approximately 8 percent of the total area of the oceans is continental shelf.

Although the animal life on the continental shelf is not as diverse or abundant as that of the fringe of the land and the sea, it still is relatively rich compared with the rest of the ocean. Many more species are found in and on the sediments of the continental shelf than in the pelagic neritic zone above the sediments. In the deeper parts of the ocean beyond the shelf, the diversity of both planktonic and benthic species decreases further.

Infauna and Epifauna. The animals of the continental shelf live in the sediment (infauna) or on its surface (epifauna). The infauna are burrowers. For the most part, they are sessile and stay in one location. Animals from many phyla have adapted to an infaunal way of life. The most common infauna are the polychaete worms (phylum Annelida), the bivalve clams (phylum Mollusca), and isopod crustaceans. Anemones (phylum Cnidaria), gastropod molluscs, some crustacea (amphipods and crabs), sea cucumbers, and brittle stars (Echinodermata) also are common (Fig. 13.20). A number of the smaller phyla, including the phoronids, brachiopods, priapulids, and echiurids also are found in the infauna.

Although the infauna generally stay in one place, many are capable of burrowing. Most burrowing species use a similar mechanism to move through the soft sediment. The burrowing cycle has a number of phases (Fig. 13.21). First is the initial entry into the sediments. This does not utilize the active digging mechanism because there is no **penetration anchor** to hold the body in place. Once the body has started to enter the sediments, it forms a penetration anchor by localized muscular contraction. When the penetration anchor is set, the organism can actively push the leading edge of the foot or head into the sediment. Often water is pumped to this area to liquify the sediments. Once the foot or head has been extended into the sediments, it forms a terminal anchor. By setting the terminal anchor and pulling against it, the rest of the body is pulled into the

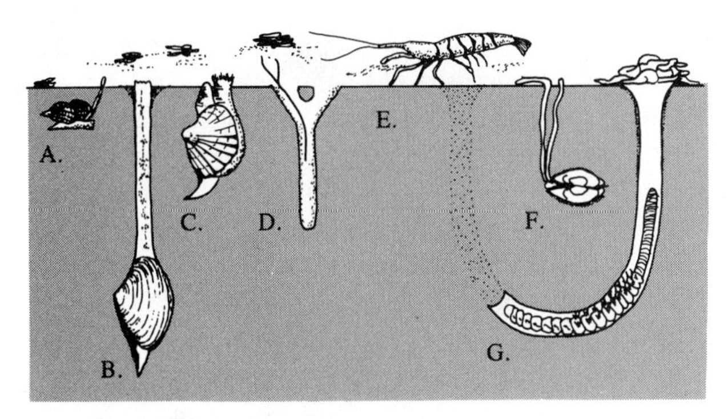

FIGURE 13.20
Fauna of the Soft Sediments

Polychaetes and bivalves are the most common organisms living in the sediments. Some gastropod snails are able to burrow and move beneath the surface. **(A)** The mudsnail *Nassarius.* **(B)** The clam *Mya.* **(C)** The cockle *Cardium.* **(D)** The bristle worm *Pygospio.* **(E)** The shrimp *Crangon.* **(F)** The clam *Macoma.* **(G)** The lug worm *Arenicola.*

sediments. The effectiveness of this basic burrowing mechanism differs among species. One of the most adept burrowers, the razor clam, can burrow through sand faster than a person can dig. Razor clams can develop fluid pressures in the anchors that are ten times that of the normal body fluid pressure.

Epifauna, for the most part, move slowly over the surface of the soft sediments or are firmly attached to the occasionally solid substrate found on the bottom. Epifaunal organisms are more diverse than infauna. Of the some 125,000 species found in the subtidal benthos of the ocean, about 75 percent are epifauna, and only 25 percent are infauna. Animals living on the substrate surface are generally capable of slow movement. Sea urchins, sea stars, brittle stars, isopods, and amphipods all have limited movement.

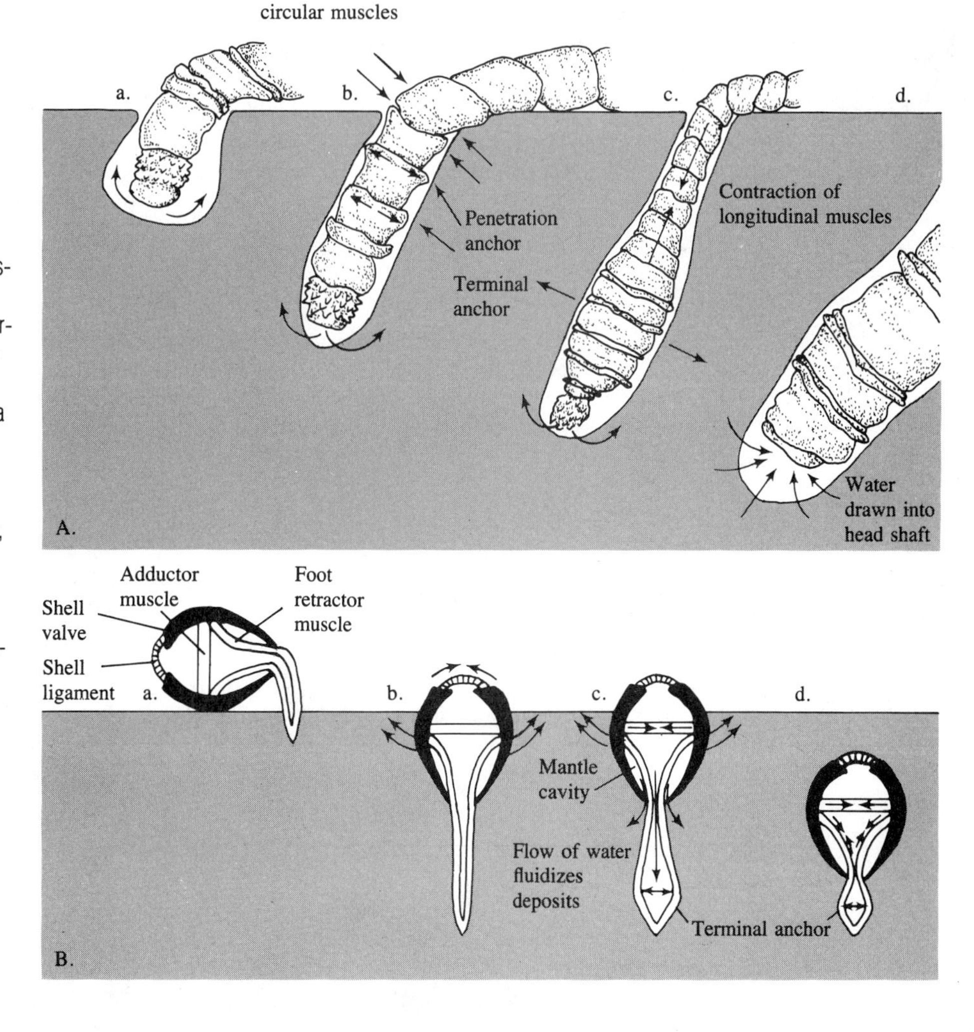

FIGURE 13.21
Burrowing Mechanisms

(A) Burrowing in the polychaete *Arenicola.* As soon as the head reaches an adequate depth (*a*), muscle contraction forms a penetration anchor (*b*). This permits more vigorous formation of the burrow by the head. When the burrow is deep enough, muscle contraction forms a terminal anchor (*c*) and the body can be pulled into the hole. Rapid movement of the head end (*d*) causes liquefaction of the sediment, which makes burrowing easier. **(B)** Burrowing in bivalves. When the foot has extended far enough into the sediment (*a*), the shell and ligament form the penetration anchor (*b*). The foot liquifies the sediment as it probes (*c*), allowing easier digging. The foot can expand and form a terminal anchor (*d*), which permits pulling the shell and the rest of the body deeper into the sediment. (After Newell, R.C. 1979. *Biology of intertidal animals.* Kent, United Kingdom:Marine Ecological Surveys.)

A number of common animals of the benthos are not easily characterized as either epifauna or infauna. Examples of these include American lobsters, which can walk quickly over the surface; scallops and cockles, which use jet propulsion from snapping their shells shut to swim awkwardly for short distances; cancer crabs, which burrow into the mud during the day and actively range during the night; and mysids, which are planktonic during the night but hover just above the bottom during daylight. These and many other animals demonstrate the difficulty of classifying the wide range of life styles into simple categories.

Although the movement of most benthic infauna and epifauna is relatively limited, the geographical range of a species over the wide plains of the continental shelf is relatively great. This raises the question of how species disperse and expand their range. This problem is similar to that of the benthos of the intertidal zone. Most benthic species of the continental shelf have a planktonic larval form as part of the life cycle. Characteristic of this method of reproduction is the production of many small eggs, with the expectation of high larval mortality. Because each egg has a small amount of food reserves (yolk), larval forms must actively feed on the plankton. Reproduction of the benthic forms is synchronized to times of adequate food supply for the larvae in the plankton. Some 80 percent of continental shelf benthic animals show this type of reproductive pattern.

The meiofauna found in the sediment of the continental shelf consist of organisms similar to those of the littoral sand habitat. Most continental shelf sediments have adequate water exchange to insure aerobic conditions in at least the top few centimeters of sediment. Meiofauna are well represented in these surface sediments, and they feed on the bacteria that use detritus as a source of energy.

The richness of the biological community of the continental shelf is reflected in the numerous commercially important species of fish and shellfish harvested there. Crustaceans (crabs, shrimps, lobsters), scallops, clams, and a number of bottom-dwelling fish all are important participants in the continental shelf ecosystem. This continental shelf ecosystem is not uniformly distributed at all depths. Numbers of organisms and biomass decrease with increasing depth. In the shallower waters of the continental shelf, up to 1,000 g (2.2 lb) of animal weight can be expected for each square meter, but at the outer edge of the shelf at depths of around 200 m (650 ft), biomass values fall to around 50 g/m^2 (0.1 lb/yd^2). (Values for the deep-sea benthos are even less, around 1 g/m^2.) This range of biomass is similar for both tropical and temperate latitudes.

Food Sources. Three sources of food are available for benthic organisms of the continental shelf: first, the limited supply of photosynthetic diatoms at the sediment surface; second, the zooplankton of the neritic water column above the continental shelf; and third, detritus brought to the continental shelf from primary producers located in shallower waters (e.g., kelps, sea grasses, salt marshes, mangrove plants, and phytoplankton).

Although benthic diatoms are not numerous enough to contribute a large share of the food eaten by continental shelf animals, their biology is interesting (Fig. 13.22). They have been found actively photosynthesizing at remarkable depths (depths that receive only 1% of maximum surface daylight). On the continental shelf on the North Atlantic, diatoms have been taken at a depth of 110 m (360 ft); off the California and Florida coasts, to a depth of 40 m; and in the Bay of Bengal, to 200 m. Diatoms at these great depths have much higher concentrations of chlorophyll than those in areas of higher light intensity.

The second source of food for the benthos is the phytoplankton and zooplankton of the neritic water column above the shelf. In the shallower areas of the continental shelf, light is adequate to promote phytoplankton growth just above the sediments. Benthic filter feeders can feed directly on both phytoplankton and zooplankton (Fig. 13.23). They too, however, are a minor source of food for the continental shelf benthos. Generally, less than 10 percent of the food requirement of the benthos comes from living plankton.

More important as a food source is the detritus derived from the plankton above. The planktonic detritus is a constant rain of dead or fragmented zooplankton and phytoplankton, as well as the sometimes abundant zooplankton feces. Over the continental shelf, this rain of food is greater than in the deeper oceanic areas. There are two reasons for this. First, the detrital plank-

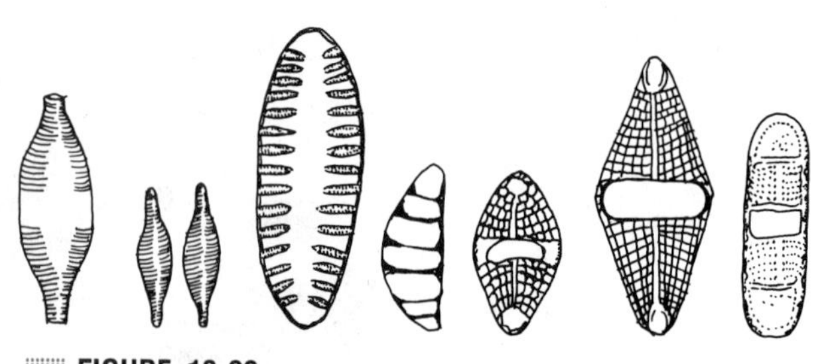

FIGURE 13.22
Benthic Diatoms

A variety of benthic diatoms live in the few millimeters of sediment at the surface where light is sufficient to allow photosynthesis.

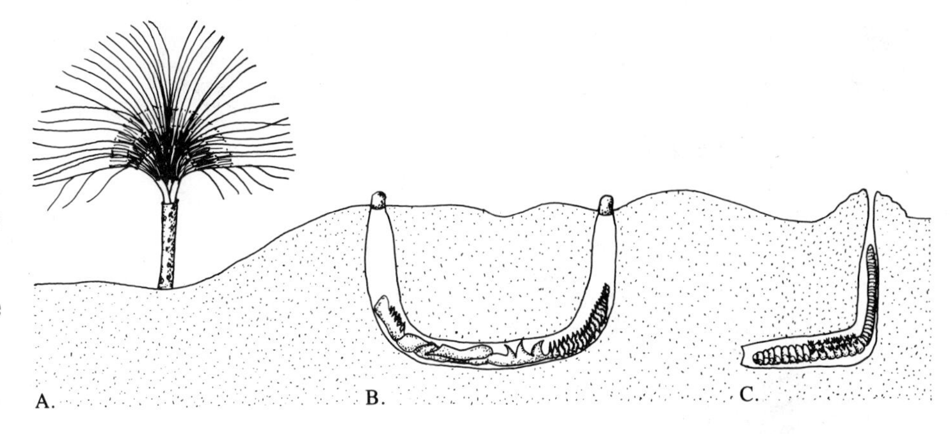

FIGURE 13.23
Filter Feeders of the Continental Shelf

A number of polychaete worms filter the water for food. **(A)** The sabellid polychaete extends a plume of tentacles into the water. **(B)** *Chaetopterus* pumps water through a burrow with two openings. **(C)** *Arenicola* pumps water through a burrow with a single opening.

tonic particles don't have as far to sink to the bottom, and there is not as much time for bacterial breakdown along the way. Second, the planktonic productivity (of both zooplankton and phytoplankton) is greater in coastal areas than in comparable oceanic areas. Consequently, there is more planktonic detritus formed over the shelf. Up to approximately 30–40 percent of the productivity of the water column finds its way to the bottom as detritus.

Another important source of detritus to the continental shelf is sea-grass meadows, salt marshes, mangrove swamps, and the algae of the littoral zone and kelp forests. As we noted earlier, only a small part of the plants or algae in these communities is eaten directly by herbivores. The majority of these plants and algae die or are broken off and wash away. Some of this material may reach deep water, but most of it settles on the continental shelf, where it is continuously broken down by physical and biological activity into finer and finer particles. Even as these particles become smaller and smaller, it is unlikely that they are used directly as a source of food. Instead, bacteria are the energy link between the detritus and the benthos of the continental shelf.

The amount of detritus available to the benthos is large. Some 200 g/m^2 (0.37 lb/yd^2) of carbon per year finds its way to the shallow continental region. This value decreases with depth, dropping to around 30 g/m^2 of carbon per year near the outer edge of the continental shelf. In the deep ocean basins, detrital input is only 1–2 g/m^2 of carbon per year.

Detritus and associated bacteria are used as food by the benthos in a number of ways. Some animals are filter feeders at the surface, taking in detrital particles that become suspended by current or wave action. Others scrape the organic-rich layer off the surface, ingesting detritus and some surface sediments. Yet others, in the manner of the terrestrial earthworm, eat the subsurface deposits directly. Filter feeders include mussels, scallops, and clams, and subsurface deposit feeders are mostly polychaete worms. Most benthic animals are selective deposit feeders, taking just the surface layer. These include most epifauna (mysids, shrimp, amphipods, isopods, gastropods, and brittle stars) and many infauna (polychaete worms and bivalves).

Meiofauna and microfauna also use the organic detritus of the sediments as a source of food. Their exact role is uncertain, however. Because of their small size, they produce a relatively small amount of biomass compared with the larger benthos, but they do have rapid turnover times with high growth rates and short life spans. It is not clear whether the meiofauna themselves are an important source of food to the benthos. Some studies conclude that meiofauna are not eaten by larger organisms but are important in the process of detritus breakdown, while others show that some meiofauna (particularly harpactacoid copepods) are important food for nekton organisms.

The detritus-eating benthos are an important source of food to other animals. These predators include fish, carnivorous snails, and crustaceans such as crabs and lobsters. Often the benthos is an important source of food for only part of the life cycle of an animal. One interesting example is the plaice, a commercially important flatfish found in the North Atlantic. Juvenile plaice feed exclusively on the siphons of tellinid clams, nipping off just the ends of the siphon, which then regenerate. Adult plaice are more general feeders of the benthos, taking a range of animals.

The pattern of distribution of the organisms found on the continental shelf is not uniform. Marine biologists working in the North Atlantic during the early part of the twentieth century noticed a similarity in the groupings of benthic organisms in various regions. In the North Atlantic and arctic areas, and in other cold temperate and some warm temperate areas, the biologi-

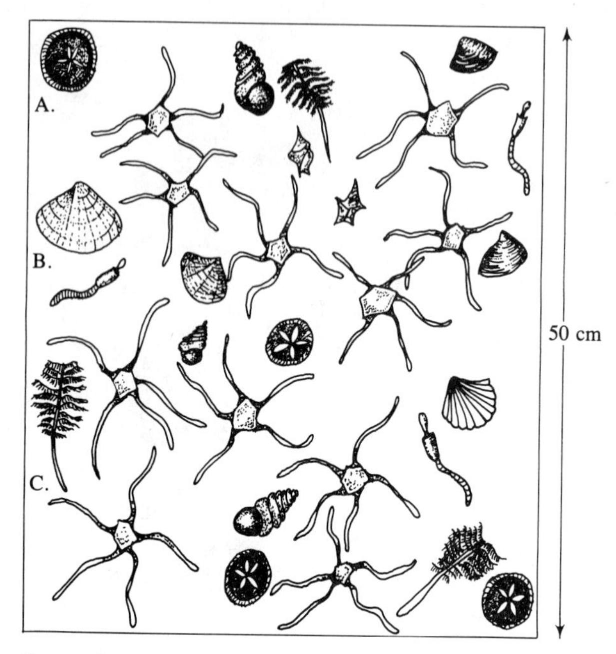

Community 1

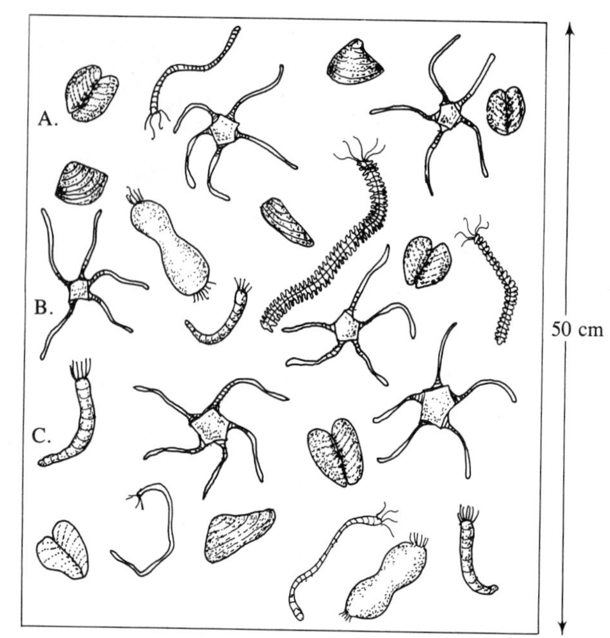

Community 2

FIGURE 13.24
Soft Sediment Communities

Animals of soft sediments can be classified into distinct communities. Shown here are two mud sediments of Norwegian waters. *Community 1*: **A**. The brittle star *(Amphiura)*. **B**. The bivalve *Arctica*. **C**. The sea pen *(Virgularia)*. [Also various snails and worms.] *Community 2*: **A**. The heart urchin *(Brissopsis)*. **B**. The brittle star *(Ophiura)*. **C**. The bristle worm *(Maldane)*. [Also other bivalves and polychaetes.]

cal communities on bottoms of similar substrate and at similar depths are characterized by similar species (Fig. 13.24). Where the bottom is more sand than mud, and at a depth of between 10–20 m (33–66 ft), bivalves of the genus *Venus* are dominant. In shallow sandy mud bottoms in the North Sea, the Greenland Ocean, and the north Pacific, the clam *Macoma* is the dominant bivalve. On muddy bottoms between 15–20 m depth, the brittle star *Amphiora* is dominant. These various biological communities, referred to as **parallel benthic communities,** have not only a particular dominant form but a predictable assemblage of other benthic species. For example, the *Macoma* community commonly also has cockles *(Cardium)*, the polychaete *Arenicola*, shrimp *(Crangon)*, and amphipods *(Corophium)*.

In tropical latitudes there are no similar parallel benthic communities. At lower latitudes, the diversity of benthic organisms is greater and the patterns of distribution are not as clear-cut.

SUMMARY

The benthic habitat that stretches from the limit of the low tide to the edge of the continental shelf (approximate depth of 200 m) is less than 10 percent of the total area of the ocean floor. The continental shelf sediments range from muddy gravel through fine sand, and generally are more coarse than deep-sea sediments.

The benthic community of the continental shelf is characterized by relatively high species diversity. There are nearly ten times more species on the shelf than in the sediments of adjacent deep-sea areas. Many are infauna. Others are invertebrates and fish that either walk over the substrate or swim close to it.

For most of the continental shelf, food is imported. Except for the very shallow waters close to shore, there is no significant primary productivity on the surface of continental shelf sediments. Food in the form of detritus comes from two sources: phytoplankton and zooplankton in the water column, and algae and plants that grow in shallow water close to shore. This detritus is most common in shallow waters (200 g/m^2 of carbon per year), and diminishes with depth (30 g/m^2 of carbon per year at 200 m depth). Food supply to the continental shelf is much greater than that to the deep-sea sediments (which is around 1.2 g/m^2 of carbon per year).

Three biological communities found in shallow water close to shore are of particular importance: coral reefs, sea-grass meadows, and kelp forests and seaweed beds.

Coral reefs are cemented limestone structures made of the skeletons of stony corals (cnidarians), calcareous algae, and other organisms that produce calcium carbonate. The hard surface is covered with a great diversity of epifauna and epiflora as well as many mobile invertebrates and fish. Coral reefs are found in shallow water from the level of mean low tide to depths of 70 m (230 ft). Their distribution is restricted to water with a mean annual temperature of 20°C (68°F) or warmer, and generally are found between the Tropics of Cancer and Capricorn.

There are three classifications of coral reefs: fringing reefs, which are attached to the land mass; barrier reefs, in which there is a water channel between the land and the coral reef; and atolls, which are reefs surrounding a lagoon that are not associated with a land mass.

Coral reefs have an especially high productivity and species diversity compared to surrounding areas. They are some ten times more productive than adjacent benthic or pelagic communities. The high productivity is due primarily to an enrichment of nutrients on the reef. Cyanobacteria, which have the ability to fix atmospheric nitrogen into nitrates, and localized upwelling caused by the reefs, contribute to higher levels of nutrients. Free-living algae and symbiotic protists found in the tissue of hard corals, sponges, and some clams are the main sources of the high rate of primary productivity.

The species diversity of coral reefs also is greater than adjacent benthic or pelagic communities. The high productivity and habitat complexity of coral reefs support a rich variety of benthic algae, invertebrates, burrowing forms, mobile crustacea, and many species of fish.

Sea grasses are true plants that complete their total life cycle (including flowering and setting seeds) under water. The 50 or so species are only distantly related to the terrestrial grasses. Sea grasses are found in dense growths, called meadows, in shallow waters of temperate and tropical latitudes. Their distribution generally is from the level of low tide to no more than a depth of 10 m (33 ft). Sea grasses require relatively high light intensities to support photosynthesis.

The sea-grass community has a much greater species diversity than adjacent areas with no grasses. The leaves serve as a substrate for many epiflora and epifauna species. They also create a three-dimensional habitat that attracts many mobile invertebrates and fish. The primary productivity of sea-grass meadows is high, although few animals eat the leaves directly. Most vegetation breaks off, decays, and enters detritus-based food webs.

Kelp forests and seaweed beds are extensive growths of macroalgae found primarily in the shallow waters of temperate latitudes. Kelps and sea weeds require rocky bottoms. They require high light intensities, and maximum growth is found only in the top 1–2 m of water. Kelps are able to grow in deeper water by using a long, trunklike stipe to connect the surface blades to the bottom. Stipe lengths of 10–20 m are common. Kelp forests and seaweed beds have a higher species diversity than adjacent areas. The three-dimensional habitat offers protection from predation to a wide range of mobile invertebrates and fish. Juvenile fish, in particular, use kelp forests and seaweeds to hide from predators.

Few animals feed directly on kelps and seaweeds, although free-floating pieces of kelp and algae, called drift, are eaten by a number of invertebrates. Most of the food energy of kelps and seaweeds decays and enters the detritus-based food web.

QUESTIONS AND EXERCISES

1. Discuss the importance of the sea-grass community to other parts of the marine environment, particularly the coastal zone and the deep sea.
2. Why do coral reefs have such a high biological productivity?
3. Discuss the organisms that are responsible for the secretion of the limestone in coral reefs.
4. What are the major types of coral reefs? Under what conditions does each occur?
5. Discuss the relative importance of the primary productivity of the shallow-water benthos to total ocean productivity.

6. The closest thing to trees in the oceans is kelp. How do trees and kelp differ? How are they similar?

7. Compare the species diversity and biomass of the various communities of the shallow benthos on the continental shelf.

REFERENCES

Emery, A. 1981. *The coral reef.* Ottawa, Canada:Canadian Broadcasting Company.

Haeckel, E. 1973. *Art forms in nature.* New York:Dover.

Newell, R.C. 1979. *Biology of intertidal animals.* Kent, United Kingdom:Marine Ecological Surveys.

Odum, H.T., and **Copeland, B.J.,** eds. 1974. *Coastal ecosystems of the United States, Vol. 1.* Washington, DC:Conservation Foundation.

Phillips, R.C., and **McRoy, C.P.** 1980. *Handbook of seagrass biology: An ecosystem perspective.* New York:Garland.

Roessler, C. 1986. *Coral kingdoms.* New York:Abrams.

Wilkinson, C.R. 1987. Interocean differences in size and nutrition of coral reef sponge population. *Science*, 236 (4809):1654–1656.

SUGGESTED READING

OCEANS

Alevizion, W.S. 1976. Fisher of the Santa Cruz Island kelp forests. 9(5):44–49.
1979. Australian coral reefs. 12(3):18–31.

Burrell, J. 1976. The reef. 9(3):48–55.

Carlson, B.A. 1979. Cups, plates, boulders, and branches. (Form and function of a coral reef) 12(1):46–53.

Faulkner, D. 1981. Palau coral reefs. 14(4):36–43.

Genthe, H.C. 1975. The sea forest. 8(5):50.

Kerstitch, A. 1982. Cleaning shrimp. 15(5):41.

Mandojana, R. 1983. Cleaning station. (Caribbean coral reef) 16(1):38–40.

Martin, R.A. 1975. Reef life in the Bahamas. 8(6):44.

Morrissey, S. 1988. Estuaries: Concern over troubled waters. 21(3):22–27.

Phillips, R.C., and **Seaborn, C.** 1981. Eelgrass. 14(4):18–25.

Reed, J.K., and **Jones, R.B.** 1982. Deep water coral. 15(1):38–41.

OCEANUS

Borowitzka, M.A., and **Larkum, A.W.** 1986. Reef algae. 29(2):49–54.

Drew, E.A. 1986. *Halimeda*—the sand-producing alga. 29(2):45–54.

Hopely, D., and **Davis, P.J.** 1986. Evolution of the Great Barrier Reef. 29(2):37–44.

Leschine, T.M. 1981. The Panamanian sea-level canal. 24(2):20–30.

Lucas, J. 1986. Crown-of-thorns starfish. 29(2):55–69.

SEA FRONTIERS

Logan, A. 1986. Aggressive behavior in reef corals: A strategy for survival. 32(5):347–350.

Thomas, L.P. 1984. Coral landscapes. 30(2):76–81.

BLEACHING OF CORAL REEF COMMUNITIES

The bleaching or loss of color by coral reef organisms has occurred on a local basis numerous times in the past. However, a mass mortality of at least 70 percent of the corals along the Pacific Central American coast occurred as a result of a bleaching episode associated with the severe El Niño of 1982–1983. Although the cause is not known for certain, this eastern Pacific bleaching may have been caused by the increased water temperatures associated with the El Niño event. Whatever the cause, it is clear that it will take many years for these reefs to recover.

One thing is clear: The direct cause of the bleaching is the expulsion of symbiont algae called zooxanthellae. Essentially, all reef-building corals and some other reef animals are nourished by these algae that live within their tissues. The loss of this source of nourishment can be life threatening to the reef dwellers.

Figure 1 shows the bleaching of a round starlet coral, *Siderastrea siderea,* on Enrique Reef, Puerto Rico. The photograph was taken in November 1987. Normally a rusty brown color, this coral has been bleached on its lower left half.

A husband-and-wife team of coral reef biologists—Drs. Lucy Bunkley Williams and Ernest H. Williams, Jr. of the University of Puerto Rico—have provided the

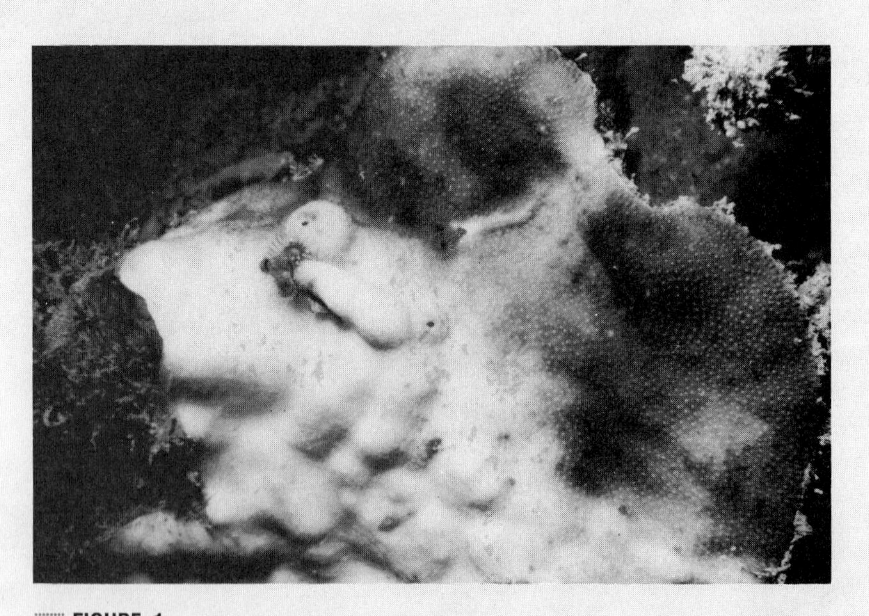

FIGURE 1
Coral Bleaching on Enrique Reef, Puerto Rico, November 1987

(Photo by Lucy Bunkley Williams.)

following statement to give us an expert view concerning the possible causes of coral reef bleaching and its implications for the future (Fig. 2):

> Extensive bleaching (loss of zooxanthellae) of 60 species of stoney corals, fire corals, stylaster corals, sea anemones, coral-like anemones, zoanthids, gorgonians, and at least three orders of sponges, began in the northern Caribbean, western Colombia, Bahamas, Florida, and Texas from summer to early fall of 1987. Recovery of many animals began in late fall to early winter, while some bleaching continued and new bleaching began in other parts of the Caribbean at that time. The disturbance occurred from near the surface down to the approximate limit where zooxanthellae occur. Similar events occurred in the Pacific in Australia (January–June), the Galapagos Islands (February–April), and possibly Hawaii, Thailand, and Indonesia. Some animals have died, but no massive die-offs [like those of] the eastern Pacific during the 1982–1983 bleaching event have occurred. Higher-than-normal water temperatures, light-related effects, or a combination of both have been suggested as the most likely causes of the event. Bleaching of coral reef animals, while very important, may have a larger meaning as an indicator of a more widespread and enduring problem.

FIGURE 2
Drs. Lucy Bunkley Williams and Ernest Williams, Jr.

The Williams are involved in coral reef research at the University of Puerto Rico at Mayaguez. Lucy Bunkley Williams also is a prize-winning underwater photographer. Ernest Williams, Jr. is Professor of Marine Parasitology, and is director of the Caribbean Aquatic Animal Health Laboratory. Both have conducted five week-long saturation dives in the undersea habitat, *Hydrolab*. Each has logged over 1,500 hours of research dives, and they have coauthored over 80 scientific publications.

WHY ARE THE CORAL REEF SPONGE POPULATIONS OF THE GREAT BARRIER REEF SO DIFFERENT FROM THOSE OF THE CARIBBEAN SEA?

A comparison of coral reef communities in the western Pacific Ocean and the Caribbean Sea shows that great differences exist. Very few species are common to both regions. Too, there are 85 percent more coral species in the western Pacific. A recent study (Wilkinson, 1987) of the sponge populations on coral reefs produced some interesting findings regarding the distribution patterns and sources of nutrition for sponges in the two areas.

Along a 200-mile transect from the Australian shore to the Coral Sea across the Great Barrier Reef (GBR), the biomass of sponges decreased seaward. Although we normally think of sponges as heterotrophs feeding on particulate organic matter, the sponges on the transect became increasingly phototrophic (Fig. 1) with increasing distance from shore. The phototrophic sponges are flattened, and receive at least 50 percent of their energy requirements from photosynthetic symbionts, usually cyanobacteria.

A similar study of reef sponges in the Caribbean Sea indicated that no sponge species were common to both areas. Caribbean sponge populations at depths of

FIGURE 1

This new species of the genus *Carteriospongia* is a flat-spreading phototroph found in 2 m-deep water in the Whitsunday region of the Great Barrier Reef. (Photo courtesy of C.R. Wilkinson, Australian Institute of Marine Science.)

15–20 m possess up to six times more biomass than those of the GBR, and individual sponges are larger in the Caribbean. While the mean sponge weight of individuals decreased seaward on the GBR, it increased in the Caribbean Sea. The greatest difference was the near absence of phototrophic sponges on Caribbean reefs.

Although few Caribbean sponges were phototrophic, 40 percent of the species (Fig. 2) did have some cyanobacteria in their outer layer of cells. Caribbean sponges depend overwhelmingly on heterotrophism, and consume by an order of magnitude more particulate and dissolved organic matter than do GBR sponges found on the middle and outer reefs.

FIGURE 2

The sponge on the left is the species *Niphates digitalis,* which is a true heterotroph. The sponge on the right, *Agelas sp.,* contains very few cyanobacterial symbionts. Both are found in Carrie Bow Cay, Belize at 20-m depth in the Caribbean Sea. (Photo courtesy of C.R. Wilkinson, Australian Institute of Marine Science.)

The greater biomass and lower phototrophic feeding rate of Caribbean sponges may result from the availability of greater food supply. Phytoplankton productivity is reportedly higher in the western Atlantic than in the western Pacific. Possibly due to the low productivity of the Coral Sea waters, sponges that evolved as flattened phototrophs survived on the outer GBR more readily than did the heterotrophic species. It also may be possible that the distribution pattern of phototrophic sponges mirrors that of the hermatypic corals that build reefs. Reef-building corals appear to have spread throughout the oceans from their evolutionary focus in the western Pacific, where their generic diversity is greatest.

PART FOUR

THE OCEAN UNDER STRESS

Having described the marine organisms and their ecology, in this final unit we will concentrate our attention on the ways in which humans have inserted themselves into the ecology of the oceans. Since these considerations often are not part of a course in marine biology, and are certainly not the focus of any such course (usually due to time constraints), they often must be addressed informally. For those who wish to have some formal coverage of these important problems, we include the following chapters.

Historically, fishing has been, and still is, the area in which humans have most strongly interfered with the ecology of the oceans. From the very beginning, those who fished knew that the objects of their attention were living in concert with the physical resources and the other populations of organisms within the sea. However, due to primary interest in fish populations, fisheries biologists began to focus more on fishing statistics than on the study of ecological relationships in a fishery. A fishing limit called the maximum sustainable yield was developed, in large part on the basis of these statistics. Although it did provide a basis for regulating fisheries, it did not reflect an understanding of the ecological system of which the fishery was a part. In Chapter 14, Food from the Sea, we discuss the development of a new term—total allowable catch—which serves the same purpose as the maximum sustainable yield, but is determined primarily from ecological data. We present the underlying principles important to the study of fishery ecology, and discuss fishery management and the present state of mariculture.

Chapter 15, Marine Pollution, begins with considerations relative to the capacity of the oceans to accept the wastes of human civilization, and the levels at which the effects of pollution may be detected in marine organisms. These levels range from those that are detectable within minutes of exposure to those that may not show up for decades. After a brief look at the national legislation directed at prevention of marine pollution, we examine some of the existing problems relative to plastics, petroleum, sewage, persistent compounds, and mercury. The chapter concludes on the positive note of discussing the international efforts that are being developed in the interest of improving the health of the world's coastal waters, where the major problems presently exist.

14

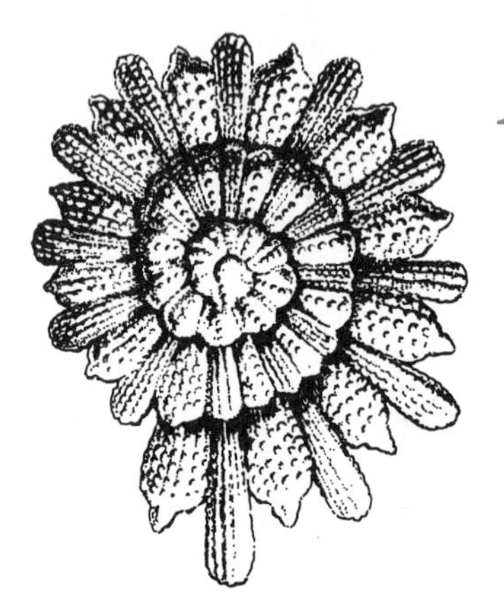

Food from the Sea

Since well before the beginning of recorded history, humans have used the sea as a source of food. Although food from the sea has in recent years provided only about 5 percent of the protein consumed by the world's population, this seemingly small percentage has meant the difference between starvation and an adequate diet for millions of the world's inhabitants.

In this chapter, the problems and prospects for the future of the world's natural marine fisheries and **mariculture** (marine aquaculture) industries will be discussed.

FISHERIES

Not only is it important that we understand the feeding and other ecological relationships that exist among the ocean's organisms, but also that we understand the effects of our attempts to exploit the biological resources of the oceans as a resource of human food. While we seek to increase the food resources we remove from the oceans, it is clear that the greatest negative impact humans have had on the ecology of the oceans is through overfishing.

It is very disturbing that the fisheries of fish, shellfish, and seaweeds from the ocean has only increased from 59.7 metric tonnes in 1970 to 76.9 metric tonnes in 1985 (the latest year for which complete information is available), despite a significantly increased fishery effort. It is also certain that the cost per unit of production is rapidly increasing (Tab. 14.1).

Fishing peoples have always understood that fish must be permitted to reproduce if a fishery is to be maintained. However, it has become increasingly difficult to regulate fisheries in a manner that would assure sufficient reproduction to maintain the fishery. Primarily, this is the result of the great demand that accompanies an increasing human population. Fisheries such as the anchovy, cod, flounder, haddock, herring, and sardine apparently are suffering from overfishing; that is, they are being harvested at a rate above the natural rate of production, in such numbers that their population is driven below the size necessary to produce maximum yield. International efforts are now being made to provide worldwide ocean fisheries management.

Early History of Fisheries Assessment and Management in the United States

Federal research into the nature of fisheries has been significant since the founding of the U.S. Fish Commission in 1871. However, regulation of North American fisheries dates from the prohibition of seining for mackerel by the Plymouth Colony in 1684.

The first major U.S. fishing grounds was Georges Bank, where a cod fishery was conducted by 1721 (Fig. 14.1). Mackerel and halibut fisheries were added by 1831. Although cod held up under heavy fishing, halibut began to decline in 1850. The **purse seine** was introduced in Gloucester, Massachusetts in 1825, and by 1850 the mackerel fishery there began to decline. Mackerel, which salted well and were fished only dur-

TABLE 14.1
World Commercial Fish Catch, 1951–1985 (million metric tonnes)

Year	Fresh Water	Peruvian Anchovy	Other Marine	Total Marine	Total
1951	2.6	NA	20.9	20.9	23.5
1952	2.8	NA	22.3	22.3	25.1
1953	3.0	NA	22.9	22.9	25.9
1954	3.2	NA	24.4	24.4	27.6
1955	3.4	NA	25.5	25.5	28.9
1956	3.5	.1	27.2	27.3	30.8
1957	3.9	.3	27.5	27.8	31.7
1958	4.5	.8	28.0	28.8	33.3
1959	5.1	2.0	29.8	31.8	36.9
1960	5.6	3.5	31.1	34.6	40.2
1961	5.7	5.3	32.6	37.9	43.6
1962	5.8	7.1	31.9	39.0	44.8
1963	5.9	7.2	33.5	40.7	46.6
1964	6.2	9.8	35.9	45.7	51.9
1965	7.0	7.7	38.5	46.2	53.2
1966	7.3	9.6	40.4	50.0	57.3
1967	7.2	10.5	42.7	53.2	60.4
1968	7.4	11.3	45.2	56.5	63.9
1969	7.6	9.7	45.4	55.1	62.7
1970	8.4	13.1	46.6	59.7	68.1
1971	9.0	11.2	48.3	59.5	68.5
1972	5.7	4.8	53.7	58.5	64.2
1973	5.8	1.7	55.3	57.0	62.8
1974	5.8	4.0	56.8	60.7	66.5
1975	6.2	3.3	57.0	60.2	66.4
1976	5.9	4.3	59.7	63.9	69.8
1977	6.1	.8	62.3	62.8	68.9
1978	5.8	1.4	63.3	64.6	70.4
1979	5.9	1.4	63.8	65.2	71.1
1980	6.2	.8	65.4	65.8	72.0
1981	6.6	1.5	66.7	68.2	74.8
1982	6.8	1.8	67.9	69.7	76.5
1983	7.2	.1	69.2	69.3	76.5
1984	9.7	.1	73.0	73.1	82.8
1985	8.0	.1	76.8	76.9	84.9

NA—Not available

Source: U.S. Department of Commerce, N.O.A.A. Publication: Fisheries of the United States, 1985.

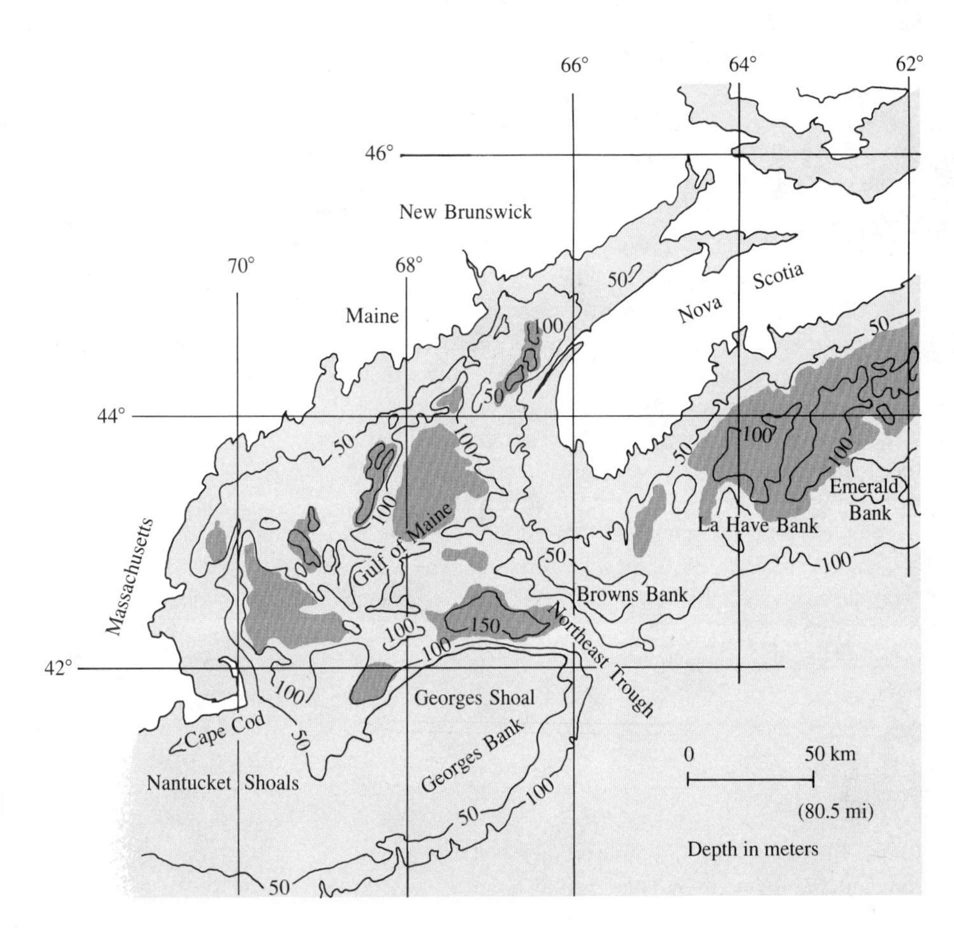

FIGURE 14.1
Georges Bank

Georges Bank, located off Cape Cod in Massachusetts, was the first major U.S. offshore fishing grounds. It is still a major fishing area.

ing spring and summer, were protected in 1887 by the U.S. Congress. They prohibited the landing of purse-seined mackerel from March to May, but this action did not improve the state of the mackerel population. The salt mackerel fishery continued to decline, and within a few years it collapsed.

This early failure in fishery management is but the first in a series of failures. The details may vary, but the results are always similar. Of all the human endeavors, few activities have been so consistently fruitless as fisheries management.

Fisheries and Marine Biology

Over the years, huge fluctuations in fish stocks that were not clearly related to fishing pressures convinced many fisheries biologists that a detailed study of fisheries ecology was needed.

Recruitment, the addition of young adult fish to the fishery, was found to depend on survival at critical stages in the life of a fish. The mortality rate is very high for eggs and larvae. Mortality decreases in juvenile fish, and reaches its lowest level in adults. The following discusses in detail the mortality issues of each stage.

Larval survival—After the larva uses up its yolk reserves, it must find appropriate food or die. For instance, anchovy larvae require at least 20 phytoplankton cells/ml of water within 2-1/2 days, and the minimum cell size must be at least 40 μm in diameter. Once a fish survives the larval stage, it still must survive a juvenile existence before it can be recruited into the fishery.

Juvenile survival—Juvenile fish may die from predation, disease, parasitism, genetic defects, and the effects of pollution. Availability of adequate food is still important because rapid growth rate is the best insurance against predators. If the juvenile fish grows faster than its predator, it quickly will become too large to be suitable prey. Assuming the availability of suitable food, the juvenile fish will grow more rapidly in warmer water. The feeding history of fish at this stage is complex, as they may feed on a variety of plankton of various size ranges as they grow larger. Even at a small size, juvenile fish may feed on a large size range of plankton. However, there usually is a narrow size range of plankton on which they can feed with high efficiency.

The Ecosystem Approach to Fishery Assessment

In an effort to understand the role of the physical environment and the many populations of organisms that affect fishery recruitment, fishery scientists have adopted an ecosystem approach to fishery assessment.

In most ecosystems, nitrogen is the nutrient that becomes so limited that it curtails primary productivity. Assessing the (a) cycling of nitrogen within an ecosystem, and (b) the influx of new nitrogen into the system can lead to an estimate of the maximum amount of protein that can be removed from the ecosystem by fishing, yet maintain a healthy fishery.

Less than 10 percent of the primary productivity in oligotrophic waters results from an influx of new nitrogen, while about 50 percent of the primary production results from the influx of new nitrogen in areas of upwelling such as those that occur in coastal Peruvian waters (Fig. 14.2). In both the examples cited, the balance of the production is supported by nitrogen recycled within the ecosystem.

If, on a worldwide basis, slightly less than 20 percent of primary production results from the flux of new nitrogen into the ecosystem, we can estimate the **potential world fishery** on this basis. Such estimates must be considered tentative, but if 0.1 percent of this primary productivity sustained by new nitrogen found its way into the world fishery, the potential world fishery would be about 100 million metric tonnes. The present catch is just over 70 million metric tonnes. The key point here is that the fishery must not remove more fish than are supported by the influx of new nitrogen, or the ecosystem supporting the fishery will be damaged.

What fishery managers want to know is the **standing stock,** the total mass present at a given time in the fishery. Previous methods of fishery assessment were directed at making this determination based on statistics of landings and scientific sampling of fish populations. It was discovered that these methods were inaccurate. It is much easier to sample the phytoplankton and zooplankton populations, and the productivity of these populations can be determined with a greater degree of accuracy. Despite this improved accuracy, there still are many problems in moving from phytoplankton standing stock to fishery standing stock.

The major areas of concern in estimating the potential fishery are knowledge of size grades, standing stocks, and size doubling times of each component of the food chain, from the phytoplankton to the fishery species. Where ecological methods have been applied to a number of fisheries, the conclusion was that the fisheries were already taking an amount equal to the potential fishery. It has become clear that the physical resources of the marine environment determine the level of biological productivity, and therefore the fishery standing stock. Any significant change in the physical resources will have a corresponding effect on the total fishery.

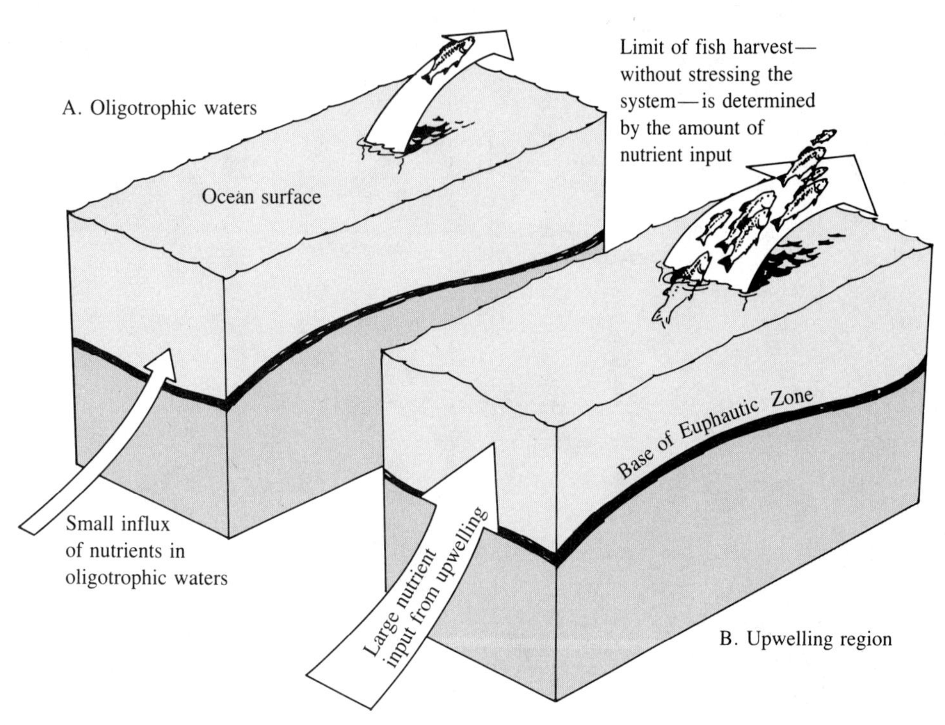

FIGURE 14.2
Potential Fishery and Influx of Nitrogen

All marine ecosystems are supported by nutrients that are recycled within the euphotic zone that contains their plant population and nutrients added by influx from surrounding and deeper waters. The biomass of fish taken from the ecosystem cannot exceed that which is supported by the influx of nutrients, or the ecosystem will become stressed. The small arrows in **(A)** represent a small amount of nutrient flux into an oligotrophic open-ocean ecosystem. The much larger arrows in **(B)** represent a large influx of nutrients into the euphotic zone of an upwelling region. This greater influx of nutrients makes possible a much greater fishery that can be extracted from the upwelling region than from the oligotrophic ecosystem.

Upwelling and Fisheries

Because so much of the productivity of upwelling areas is supported by the influx of nutrients, it is not surprising to find that these areas, which represent about 0.1 percent of the ocean surface area, account for around 50 percent of the world fishery. Within upwelling areas, research has identified two variables that significantly affect the fisheries associated with them. These variables are the duration and rate of upwelling (Fig. 14.3). Generally, upwelling of long duration (more than 250 days) produces ideal conditions for fishery development. However, even in areas of shorter duration, high levels of productivity still may result if the rate of upwelling is moderate. Regardless of the duration, extremely low or high rates of upwelling produce a decreased fishery, compared to moderate rates. Waters that are too stratified or too well mixed, where upwelling rates are too low or high, respectively, tend to develop smaller phytoplankton species that cannot be grazed efficiently by fishes such as sardine and anchovy. This decrease in phytoplankton size is thought to result from the decreased concentration of nutrients that results from either condition. Under conditions of small phytoplankton, herbivorous zooplankton must first feed on the tiny plant cells and subsequently be grazed by the fish.

Where upwelling rates are moderate, and the water column less turbulent, higher concentrations of nutrients support large species of diatoms that can be grazed directly by the fish. This elimination of the herbivorous zooplankton greatly increases the energy available to the fish, and thus increases the potential fishery. Therefore, it appears that areas of upwelling of long duration and moderate rates produce the optimum environment to produce a large fishery.

A Global Approach to the Study of Upwelling Ecology

Because of the great potential of these areas, the International Oceanographic Congress (IOC) and the United Nations Food and Agriculture Organization (FAO) are at present conducting a joint research project. Specifically, they are testing Rubin Lasker's stable ocean hypothesis in the four major eastern boundary current upwellings of the California, Peru, Canary, and Benguela current systems (Fig. 14.4). Lasker's hypothesis states that the upper mixed layer of the ocean must be in a stable (nonturbulent) state to provide a sufficient concentration of food to ensure the survival of enough first-feeding fish larvae to produce a good **year class** of juvenile fish. This phenomenon is closely related to the same physical processes that account for the varied nutrient concentrations discussed above—duration and rate of upwelling.

The physical relationships behind this hypothesis are tied to the role of wind in producing upwelling. In

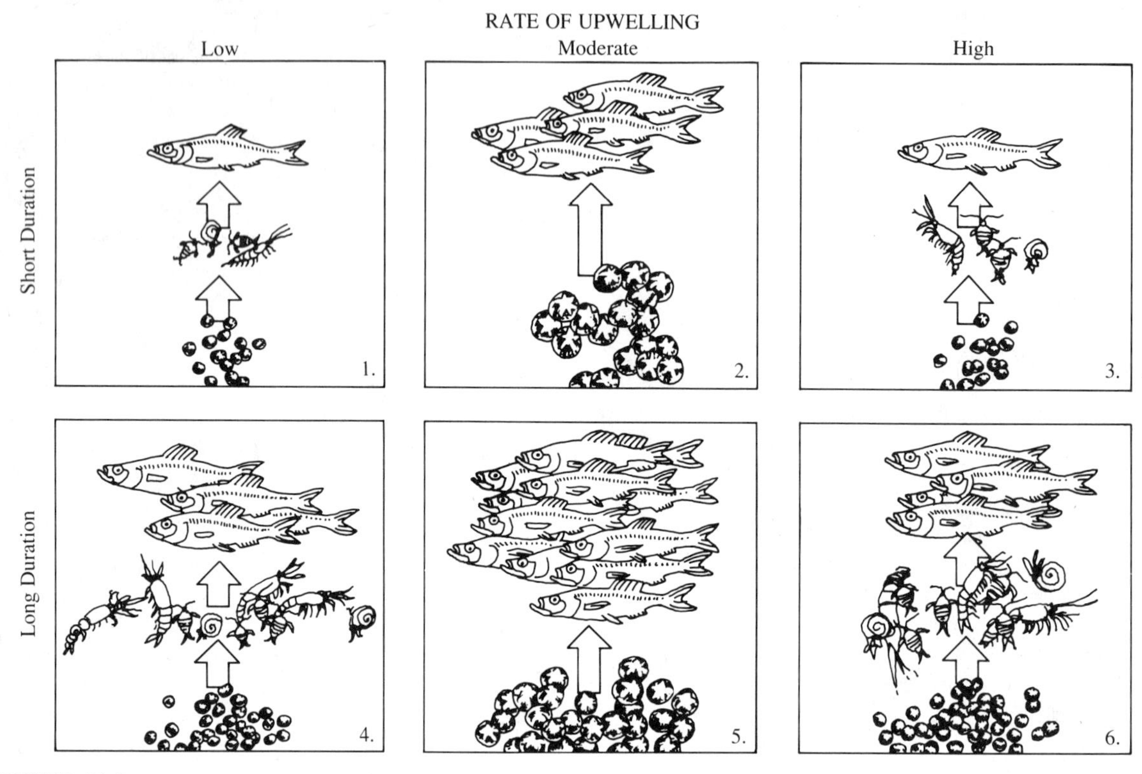

FIGURE 14.3
Classification of Upwelling Systems in Terms of Duration and Rate of Upwelling

Duration determines the absolute amount of production, and rate of upwelling determines the algal cell size. Moderate rates of upwelling for short or long duration (blocks 2 and 5) produce large enough concentrations of nutrients in nonturbulent water to support large algal cells that can be grazed directly by sardines and anchovies. These conditions will support relatively large fisheries. Whether the duration is short or long, low or high rates of upwelling will not allow high enough concentrations of nutrients to produce large algal cells for the fish to graze directly. Therefore, zooplankton must graze the tiny algal cells and be grazed in turn by the fish. This increases the length of the food chain and greatly reduces the amount of energy available to the fish. The result is a much smaller fishery biomass. (After Wyatt, 1980.)

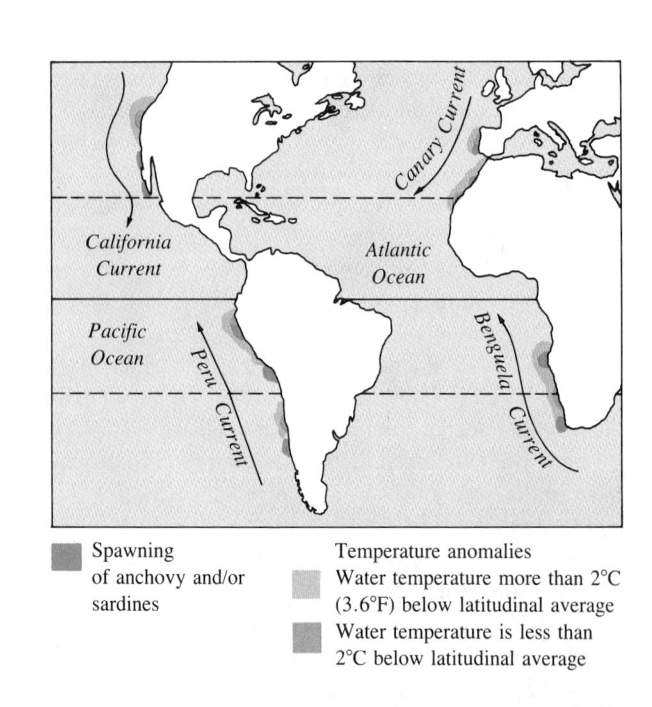

FIGURE 14.4
Eastern Boundary Current Upwelling Systems

The eastern boundary currents of the subtropical gyres in the Pacific and Atlantic oceans are associated with wind systems that produce coastal upwelling. The map labels the boundary current systems and shows the degree of temperature anomalies associated with the upwellings. Note that spawning of sardines and anchovies usually occurs in areas of moderate cooling, where the upwelling rates are moderate. The only exception is the Peruvian anchoveta, which spawn in very cold water, where the upwelling rate is high and the water turbulent. The reason for this Peruvian exception to the norm has not been determined. (After Parrish et al., 1983.)

each of these areas, winds usually blow in a direction that moves surface water offshore to enhance upwelling. The stronger the winds blow, the higher the rate of upwelling and the greater the turbulence. If the winds are too strong, the following conditions will result: (a) Turbulent mixing will reduce the concentration of food particles required by fish larvae, and (b) strong offshore water transport may carry eggs and larvae away from the favorable coastal region.

Thus, it would be expected that spawning of anchovies and sardines in these systems would occur in areas where winds are weaker and upwelling is nonturbulent. Since upwelling centers are characteristically colder, the spawning would be found to occur in warmer, less-turbulent waters, away from upwelling centers.

The study reveals that all four coastal upwelling areas have a number of features in common. All regions have equatorward current flow and narrow continental shelves. They are dominated by a small number of pelagic species of fishes that can achieve very large numbers of individuals. In addition to sardines *(Sardina* and *Sardinops)* and anchovy *(Engraulis),* there are jack mackerel *(Trachurus)* and hake *(Merluccius),* as well as smaller populations of bonito *(Sarda)* and mackerel *(Scomber)* (Fig. 14.5).

The California and Benguela systems each have one upwelling center (see Fig. 14.4), and the Canary and Peru systems each have two. Only in the California Current is there a pronounced weakening of upwelling during winter. Preferred spawning areas are along coastal indentations, where waters are protected from wind-driven turbulence and offshore transport. Also, the continental shelves are wider in the spawning regions. The greater spawning populations are found equatorward of the upwelling centers, with smaller spawnings occurring poleward.

As long as other physical conditions are favorable, temperature does not seem to influence spawning habits. Evidence shows that sardines spawn in waters with temperatures ranging from 13–21°C (55–75°F).

The Peruvian Anomaly

The Peruvian anchoveta stock has been the largest in the world ocean. Spawning of this population of anchoveta occurs with a spring peak in the Chimbote region of maximum upwelling. This is not the expected behavior, and its cause is not understood. There is a broader continental shelf in this region compared to other upwelling centers, and this unusual spawning behavior may be effective in helping the anchoveta cope with the periodic negative effects of the El Niño events. Similar phenomena affect the other eastern boundary upwelling regions, but they do not reach the magnitude of the El Niño along the Peruvian coast. (See the special feature at the end of Chapter 2 for further discussion of this phenomenon.)

Much remains to be learned about the ecology of the four eastern boundary current systems. An ongoing study of the four areas should produce a rapid increase in knowledge. Dr. Paul Smith of the National Marine Fisheries Service is participating in such a study, and believes that observing the variables that occur in the four areas over a five-year span will allow as much new knowledge to be gained as would be obtained from a 20-year study of only one upwelling area.

New Technologies Useful in Studying Fisheries Ecology

The use of **side-directed sonar** has become increasingly effective in assessing the masses of populations within the ecosystem associated with fisheries. When these signals are reflected off schools of fish, squid, or

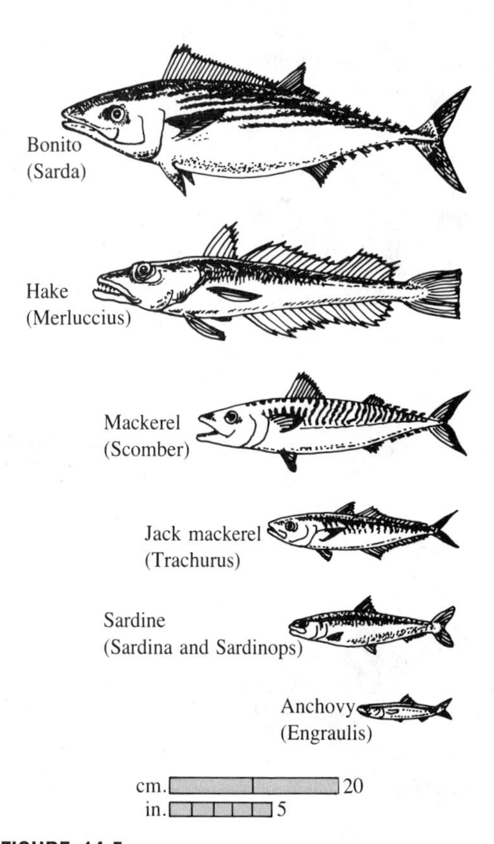

FIGURE 14.5
Fishes Common to All Four Subtropical Eastern Boundary Current Upwelling Ecosystems

(Courtesy of National Marine Fishery Service.)

other animals, the return signals can be analyzed to determine the volume and mass of the aggregations. Careful analysis of the frequencies that return with the greater resonance can tell investigators the species and even the ages of organisms in the school. Using a version of this sonar equipment attached to the manned submersible the *Johnson Sea Link*, investigators added significantly to the knowledge of the ecosystem that supports the important fisheries of Georges Bank.

It long had been a problem to explain how the benthic community on Georges Bank could support the large demersal fish (fish that live on or near the bottom) and squid fisheries there. The benthic and demersal zooplankton communities did not appear to be large enough to support these fisheries.

Sonar investigations in the submarine canyons along the southeast flank of Georges Bank (see Fig. 14.1) revealed that large masses of the krill *Meganyctiphanes norvegica* (Fig. 14.6) are present on or near the bottom day and night. In the 50 m adjacent to the bottom, densities of krill averaged 813 animals/m^3. It is known that these krill are an important component of the diet of squid, cod, flounder, haddock, hake, pollock, and redfish. It is clear that if they descend into the canyons and deep waters surrounding Georges Bank to feed on this previously unknown food supply, the krill easily could provide the needed support for the existence of the large fisheries represented by these animals.

Fishermen and Marine Science

Such activities as we have discussed above fall into the realm of the fisheries scientist. The fisherman, however, is much less concerned with these findings. For instance, fishermen always have fished where the greatest concentrations of saleable fish could be found. They began by finding fish in the nearshore waters. Improved technology allowed them to travel greater distances to catch fish and still get them back to market in good-enough shape to sell. The distant-water fleets of the 1960s and 1970s steadily increased the world catch. Now that the world fishery has leveled off at more than 70 million metric tonnes, the economy of the distant-water fleets is failing. They are becoming the victims of their own success—a common occurrence. As fishery scientists learn more about fishery ecology, this information will be of great interest to those assigned the task of managing the fisheries. This increased knowledge probably will continue to be of little interest to fishermen, who are primarily concerned with finding and catching the fish. Although many of those involved in fisheries management are pessimistic about the chances of fishermen developing an interest in fishery ecology, let us hope they are wrong.

The Future of Fisheries Management

A disturbing trend in fisheries management developed during the recent past: The reduced priority of preserving the ecology of the fishery in making management decisions. Increased emphasis was given to social benefits, such as occupational opportunity. Since it is the ecosystem that makes the social benefits possible, it surely was shortsighted to relegate the welfare of the ecosystem to any position less than the number-one priority in fisheries management. The fact that this philosophy is being abandoned surely will increase the likelihood of successful fisheries management in the future.

A.

B.

FIGURE 14.6
High-Density Near-Bottom Krill Aggregations in Submarine Canyons off Georges Bank

(A) Close-up view of the krill *Meganyctiphanes norvegica.* This species is about 3.8 cm (1.5 in) long. (Photo by Pam Blades-Eckelbarger, Harbor Branch Oceanographic Institution, Inc.) **(B)** A large aggregation of the krill observed in the submarine canyons off the southeast flank of Georges Bank (see Fig. 14.1). (Photo by Marsh Youngbluth, Harbor Branch Oceanographic Institution, Inc.)

With all the effort being directed at gathering data upon which to base the management of fisheries, it still is often said that the fisheries of the world are unmanaged. One reason for this assessment of fisheries management is the method commonly used to control catches. Most regulating bodies now set a **total allowable catch (TAC)** for each species. However, with the fishing methods in use, many species are recovered as a 'by catch' along with the fish at which the fishery is primarily directed. To cope with this inevitable occurrence, a total allowable catch is established to cover all species affected by the fishery. This value is always less than the total of catch allowances for all the species that were set on an individual species basis. The setting of these total allowable catch values is in itself difficult, but not nearly so difficult as assuring accurate reporting of catches.

Fisheries scientists are an optimistic lot, and in spite of the dismal record of management in the past, most believe fishery managers have at last begun to appreciate the nature of the problem. What scientists need now is experimental evidence that they are correct in their understanding. However, this evidence may be many years in coming. Future success will require large-scale studies involving both physical oceanographers and marine biologists engaged in an ecosystems approach to understanding fisheries. Studies like that of the eastern boundary current system upwellings discussed above will be the types of studies fishery scientists will be conducting in the future. Although the world catch may not be much larger than it is today, the technology and knowledge that will be available to fisheries managers of the future will make it possible to set and enforce catch limits that will assure the sustained yields of most fisheries.

Regulation

The effects of inadequate management can be seen in the history of the fisheries in the northwest Atlantic. In this area, regulated by the International Commission for the Northwest Atlantic Fisheries, the fishing capacity of the international fleet increased 500 percent from 1966 to 1976. The total catch, however, rose by only 15 percent. This indicated a significant decrease in the catch per unit of effort, a good indication that the fishing stocks of the Newfoundland–Grand Banks areas are being overexploited. Biologists who had been setting quotas for the major species within this region complained that enforcement was extremely difficult, and that the quotas had become international currency used in political bargaining. Politicians were accused of allocating quotas for fish that exceeded the total allowable catch set by the commission, thus depleting the reserve and preventing stock recovery.

Occurrences like these make it seem that the only answer for regulation with adequate enforcement may be to have fisheries controlled by the nation in whose coastal waters they exist. The difficulty of enforcing regulations by the international commission was largely responsible for Canada's unilateral decision to extend its right to control fish stocks for a distance of 200 mi (322 km) from its shores beginning January 1, 1977. The United States followed with a similar action on March 1, 1977. These 200-mile strips of coastal water over which coastal nations assume authority for controlling resources and pollution are called **Exclusive Economic Zones (EEZ),** and were included in the United Nations Law of the Sea Convention, completed in 1982. Figure 14.7 shows the area of the ocean covered by the U.S. EEZ.

FIGURE 14.7
Exclusive Economic Zone (EEZ)

Area included in the exclusive economic zone established by the United States. With the establishment of such zones by all nations, 35 percent of the oceans will be included. (Courtesy of Southwest Fisheries Center, National Marine Fisheries Service, La Jolla, CA.)

In April 1990, the tuna canning industry responded to a boycott of tuna products initiated by the environmental group Earth Island Institute by declaring it would not buy or sell tuna caught by methods that kill or injure dolphins. The nature of the problem of dolphin deaths was graphically presented to the world in video footage taken by biologist Samuel F. La Budde. Undoubtedly, it was the impact of this footage that turned the tide. It was aired first in March, 1988, and in just over two years accomplished what the U.S. government was unwilling to do since the Marine Mammal Protection Act was passed in 1972: It will reduce greatly the practice of using dolphins to spot yellowtail tuna in the eastern Pacific Ocean. Spotted (*Stenella attenuata*; see Fig. 14.8) and spinner *(S. longirostris)* dolphins have been used by the industry to spot yellowfin tuna schools that are for unknown reasons commonly found beneath the dolphins.

A measure designed in the 1970s to reduce the number of deaths was the development of purse seines of the type shown in Figure 14.9. Along with the net change, a technique for its use—**backdown**—helped dolphins escape over a fine-meshed apron as the net is pulled in. It was this process that was the focal point of La Budde's video, which showed dolphins being caught in the net and pulled over and crushed by the block pulley used to land the net.

Another means of netting tuna that will be affected by this decision is the use of **drift nets** or **gill nets,** which are made of monofilament line that is virtually

FIGURE 14.8

Spotted dolphin, photographed in the Pacific Ocean by W. High. (Courtesy of the Southwest Fisheries Center, National Marine Fisheries Service.)

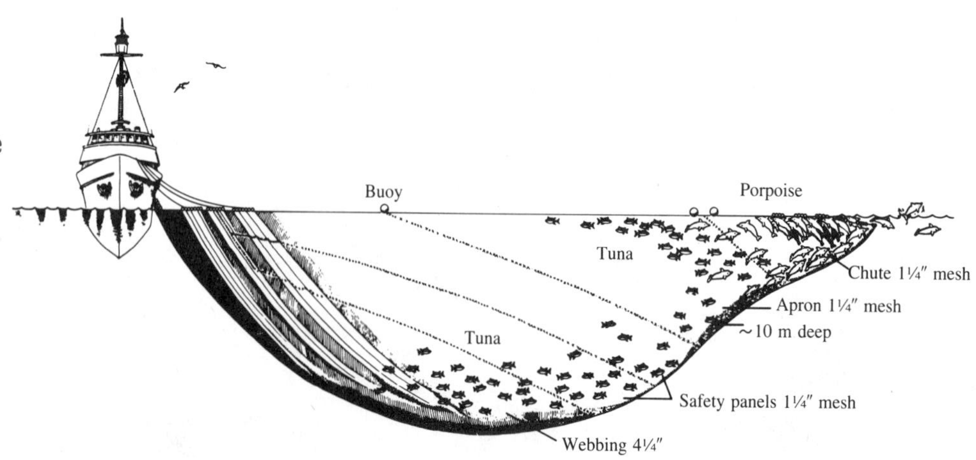

FIGURE 14.9 Backdown Using Apron-Chute Rescue System

A technique used to reduce the number of porpoises killed by U.S. tuna purse seiners of the East Pacific tuna fleet is referred to as *backdown.* Nets with small mesh in an apron-chute area prevent dolphins from being entangled as the net is pulled under them. (Courtesy of National Marine Fisheries Service.)

FIGURE 14.10
Dolphin Caught in Drift Net

Although not the target kill, this dolphin became ensnared in a net positioned off the northeast coast of the United States. (Photo: Cannon/ ©Greenpeace.)

invisible and cannot be detected by dolphin echolocation (Fig. 14.10). Japan, Korea, and Taiwan are the nations with the largest drift net fleets. At present, they are sending 1,500 fishing vessels into the North Pacific, and setting over 37,000 km (23,000 mi) of nets every day. They are said to be fishing for squid, but are illegally taking large quantities of salmon and steelhead trout. Drift netters are moving into the South Pacific and targeting immature 2-year-old tuna. There is considerable concern that this practice will destroy the South Pacific tuna fishery in a very short time. In addition to dolphins, birds and turtles are killed in these nets. At least 20 percent of drift net catches is thrown away because it contains unsaleable or badly damaged species. This most efficient means of stripping animals from the sea is an extremely inefficient method of maintaining the health of the world's fisheries for future years, and it has become a focus of international negotiations. The United States and Japan already have agreed to put a program of observers on Japanese vessels to monitor their location and type of catch.

Fishing fleets from Russia, Poland, Japan, and Italy fish off the U.S. coasts, interfering at times with the U.S. lobster fishery. Occasionally, fishing vessels are boarded, seized, and escorted to port for violation of the fishing regulations established by the United States within their 322 km (200-mi) limit. Such violations can result in severe fines, imprisonment of the captain, and the confiscation of the ship. The U.S. fishing vessels, particularly from the tuna fleet, occasionally are seized for fishing within the limit of Latin American nations, where they also have been subjected to extremely large fines. In 1988, Japanese fishing fleets were accused of illegally taking at least 300,000 metric tonnes of pollock and cod from U.S. EEZ waters in the Bering Sea. It is a great irritation to many that the U.S. imports two-thirds of its fish products each year, and yet most of these imports are caught in U.S. waters. Although establishment of the EEZs throughout the world may aid in protecting fisheries, it certainly has created new political and economic problems. The U.S. fishing industry, happy with the establishment of an EEZ by our government, has found it, too, is limited by the accompanying regulations.

THE PERUVIAN ANCHOVETA FISHERY

The magnificent anchoveta fishery that existed at the north end of the Peru Current can be used as a case study to show the importance of placing the understanding of the ecology of a fishery above all other considerations in fisheries management. The unfortunate expansion of this fishery was based in large part on the consideration of the jobs that were provided. Many of those jobs, today, are gone forever.

Physical Conditions

Southeasterly winds blow along the coast of Peru and drive the surface waters away from the coast. The surface water is replaced by an upwelling of cold water from depths of 200–300 m (660–990 ft).

The Peru Current (Fig. 14.11A) is the northward-flowing eastern boundary of the South Pacific current gyre. Speed seldom exceeds 0.5 km/h (0.3 mi/h), and the current extends to a depth of 700 m (2,300 ft) near its seaward edge, some 200 km (124 mi) from the coast. The Peru Countercurrent knifes its way between the coastal and oceanic regions of the Peru Current near the equator. This southbound surface flow of warm equatorial water seldom reaches more than 2–3° south of the equator, except during the summer when it reaches its maximum strength. Beneath the surface current is the southward-flowing Peru Undercurrent.

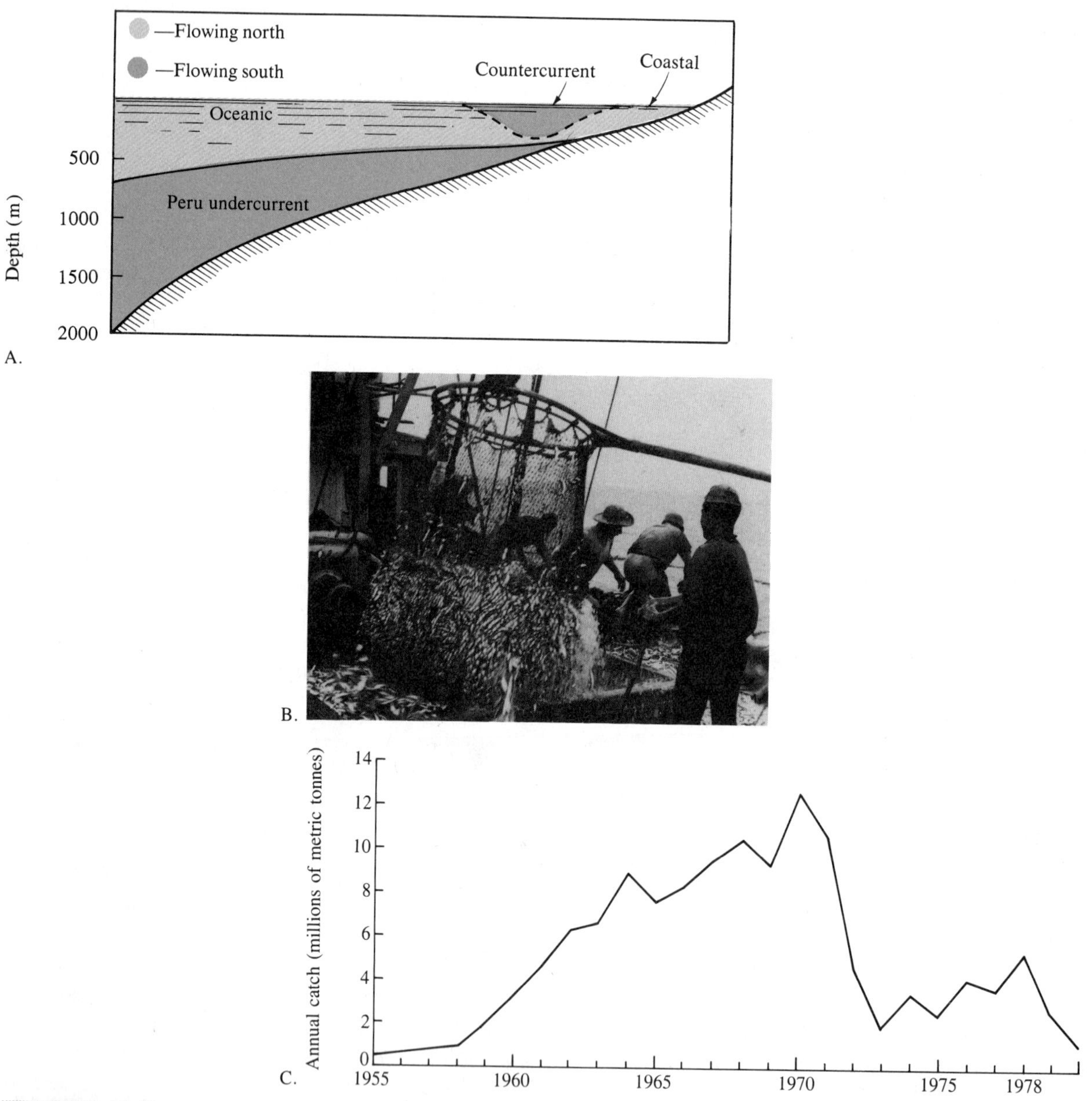

FIGURE 14.11
The Peruvian Anchoveta Fishery—The Loss of a Fishery

(A) As the Peru Current flows north along the Peruvian coast, it is periodically split into oceanic and coastal branches by the Peru Countercurrent. The Peru Countercurrent is a warm water mass that moves south from the equator during the Southern Hemisphere summer season. Although it usually moves south no more than 2° or 3° below the equator, it periodically moves well down the Peruvian coast in an event called the El Niño. **(B)** Peruvian Anchoveta. A catch of Peruvian anchovy during the early 1960s, when the fishery was undergoing rapid development. These small fish represented one of the world's major fisheries until its collapse in 1972. An active fish meal industry based on anchoveta made Peru, for a while, the world's leading fishing nation. (Photo courtesy of FAO.) **(C)** Annual Anchoveta Catch, 1955–1980. After a peak catch of 12.3 million metric tonnes in 1970, the fishery rapidly declined, and collapsed in 1972.

Biological Conditions

Biologically, the significance of the cold water and nutrients provided by upwelling is great. During the years 1964–1970, fully 20 percent of the world fishery was represented by anchoveta taken off the coast of Peru.

The anchoveta food web starts with diatoms and other phytoplankton that support a vast zooplankton population of copepods, arrowworms, fish larvae, and other small animals. Energy is passed on through a series of predators, but most stops with the anchoveta, which once attained a biomass of 20 million metric tonnes (Fig. 14.11B). Natural predators that feed on the anchoveta are marine birds—boobies, cormorants, and pelicans—and large fishes and squid.

A Natural Hazard

Periodically during the summer, the southeasterly trade winds weaken. The northward-flowing current and upwelling slows down, and the Peru Countercurrent may move as far as 1,000 km (621 mi) south of the equator, laying down a warm, low-salinity layer over the coastal water. This event is associated with the development of the El Niño–Southern Oscillation (discussed in the special feature at the end of Chapter 2).

The increase in water temperature and decrease in nutrients brought about by the slowdown or complete halt of upwelling cause a severe decrease in the anchoveta population, the remnants of which seek deeper, cooler water. Marine birds that feed primarily on the anchoveta and produce the enormous **guano** (excrement rich in phosphorus and nitrogen compounds) deposits on the islands off the Peruvian coast have had their populations cut severely by the decrease in the anchoveta.

In 1972, a severe El Niño occurred. The total biomass of anchoveta decreased from an estimated 20 million metric tonnes in 1971 to an estimated 2 million metric tonnes in 1973. But before blaming this decrease solely on the natural event of El Niño—with which the anchoveta have contended for thousands of years—another factor must be evaluated.

An Artificial Hazard

The Peruvian anchoveta fishery began to develop into a major industry in 1957. In 1960, the Instituto del Mar del Peru was established with the aid of the United Nations. Its purpose was to study the anchoveta fishery, and recommend a management program to the Peruvian government. It was hoped that such an early study would prevent the fishery from going the way of the Japan herring and California sardine fisheries, which had collapsed as a result of overexploitation.

Biologists estimated that 10 million metric tonnes was the maximum annual take anchoveta could sustain. Figure 14.11C shows that this limit was exceeded in 1968, 1970, and 1971. The 1970 catch of 12.3 million metric tonnes probably would reach 14 million tonnes if processing losses and spoilage were included.

Dr. Paul Smith, who is conducting a study of the anchoveta fishery, as mentioned above, believes that the yearly anchoveta catches may have been much larger than the official figures indicate. Other biologists familiar with the problem agree. Their belief grows out of the fact that many catches contained large numbers of **pelladilla,** young anchoveta that are low in protein. This underreporting would result from buyers reducing the total tonnage of a catch they are purchasing by a percentage determined by the proportion of the catch that was accounted for by high water-content and low oil-content pelladilla. For example, a 100 metric tonne catch may be recorded as a 70 tonne catch because it contained a large number of pelladilla.

Dr. Smith thinks the crisis may have resulted from a combination of the effects of El Niño and the industry's lack of understanding of the anchoveta fishery. This is reflected in the fact that the January 1972 allotment was set at 1.2 million metric tonnes. This figure was reached easily by the industry with about half the expected effort. Because the fishing vessels were able to meet their allotment so easily without going far from shore, the fishery was thought to have developed handsomely, and the February allotment was raised to 1.8 million metric tonnes. Much more effort was required to meet this allotment, and in March the fishery collapsed. What had happened?

Apparently, El Niño, or the Peru Countercurrent, had come south in January, but it remained offshore. The anchovy were concentrated in the nearshore portion of the Peru Current that had not been affected (Fig. 14.11A). In January, the nets were filled easily near shore, and it was natural to think the high concentration of anchovy extended over the normal fishery area as well. This one event cannot explain the collapse of the fishery, but it is a symptom common to fisheries that are poorly understood.

It may be that even with upwelling, ammonia (NH_3) from anchoveta excrement is required to support enough phytoplankton to allow the anchoveta population to expand. Since anchoveta populations are now small and scattered, they may swim through great expanses of water without adding much nutrient enrichment. When the population was larger, the whole region received it. It may require a long time, with a gradual increase in the population, until this nutrient supplement is made available to the phytoplankton throughout the entire region.

The anchoveta are gradually being replaced by sardines, a more valuable variety of fish. The entire anchoveta catch was converted to fish meal and exported as poultry and hog feed. The sardine fishery, instead, will be directed at producing canned and frozen products for direct human consumption.

With anchoveta fishing banned since 1980, and an emphasis on species conservation in Peru's future fishing activity, the anchoveta fish meal fishery may be viable in the future. However, it never will reach its previous levels. Although it is much smaller in biomass, the sardine fishery eventually may develop into a major source of revenue, due to the higher value of the products that can be produced from it. To improve the present state of affairs, the Peruvian Ministry of Fisheries will have to conduct a prudent program of fishery assessment and management.

MARICULTURE

Marine aquaculture, or mariculture, has been conducted for years throughout the Far East, making major contributions to the available food supply in that part of the world. Most attempts at commercial mariculture in the United States have not been successful because most of the projects were based on intensive production, tainted with overambition and impatience.

Organisms chosen for mariculture should be popular marine products that command a high price, are easy and inexpensive to grow, and will reach marketable size within a year or less. Candidates should be hardy and resistant to disease and parasites. They also should be capable of feeding with a high degree of growth per unit of intake. Severe economic problems can result if the chosen organism is not able to reproduce or cannot be brought to sexual maturity in captivity.

Algae

Algae, in the form of certain seaweeds, are grown and marketed as luxury foods. The life cycle of most of these algae is very complicated, and most of the successful operations are conducted in Japan, where spore-producing plants are cultivated in laboratories. When the spore-producing plants release their spores into the water, local growers come to the laboratories to dip their nets, ropes, or whatever devices they use for attachment, into the water and allow the spores to attach themselves to these devices. They next are submerged into estuarine waters, where the spores grow into the desired seaweed form using nutrients that are naturally available.

Bivalves

Cultivation of oysters and mussels is probably the most successful form of mariculture. Commercially, no bivalves have been reared from the larval stage to a marketing size in a controlled environment because of the problems related to producing sufficient phytoplankton to feed them. Usually, hatchery-produced juveniles are reared to maturity in natural environments such as bays, estuaries, and other protected coastal waters. They may be set on the bottom or in trays or other suspension devices above the bottom. This simple procedure results in yields ranging from 10 to over 1,000 metric tonnes per acre of edible meat per year. There is no attempt at artificial feeding, and the molluscs are fed by the phytoplankton carried to them by natural water movements.

An interesting outcome of placing oil platforms in the coastal ocean has developed off the Santa Barbara, California coast. Mussels grow so rapidly on the substructure of the platforms that it had cost thousands of dollars per year to remove them from each platform. In a deal that has worked out well for Bob Meek and his Ecomar Marine Consulting firm, the oil companies now get their platforms cleaned for free, and Ecomar harvests 3,500 pounds of mussels per day (Fig. 14.12). The mussels are sold along the west coast and as far inland as Chicago. Their main market is the restaurant industry, which claims these mussels are as fine as any

FIGURE 14.12
Harvesting an Oil Platform

Using a pneumatically powered scraper/hoe, these divers loosen the mussels, scallops, and other sea creatures that have adhered themselves to this oil rig leg. All organisms are sucked up a tube to the surface for further processing. (Photo courtesy of Ecomar Marine Consulting.)

that can be found. Ecomar also cultures up to 50,000 scallops and oysters annually on trays that are suspended from the platforms.

Crustaceans

Crustaceans such as shrimp and lobster have not yet been raised in the United States with any great success, even though the price for such organisms is relatively high (Fig. 14.13). The problem with most shrimp operations is that they must be fed extraneously, and this adds greatly to the cost of production. World-wide, the production of shrimp exceeds 450,000 metric tonnes per year, which is about 22 percent of the world supply of this delicacy. Despite the modest state of development in the United States, shrimp are, on a world-wide basis, the most valuable commodity produced by mariculture. Asian nations lead the way, with China being responsible for 22 percent of world production. Other major producing nations are Ecuador, Taiwan, Indonesia, Thailand, and the Philippines. Mexico is beginning to develop projects as well. Shrimp-rearing projects in the United States are found in South Carolina, Texas, and Hawaii.

The brightest spot in shrimp mariculture in the western hemisphere is Ecuador, where a very profitable operation is producing 45,000 metric tonnes per year. This success in Ecuador, where the cost of labor and dealing with government regulation is low, contrasts with high costs in the United States, where a similar operation may have to cope with over 35 regulating agencies, along with high labor costs. However, it is clear that shrimp can be produced in mariculture operations at much lower cost than they can be provided by fishing. This fact is beginning to be felt by the world shrimp fishing fleet. The largest problem the industry faces is the production of larvae that are susceptible to bacterial and viral infections.

American lobsters *(Homarus americanus)* can easily be carried through the complete life cycle in captivity, but they have a cannibalistic nature. To be successfully grown, these lobsters would have to be compartmentalized and maintained at an optimum temperature of about 20°C (68°F) for two years. Artificial feeding would also be necessary. Encouraging results have been achieved in rearing American lobster in the warm effluent from an electrical generating plant in Bodega Bay, California. Similar projects will be tried in the coastal waters of Maine. The spiny lobster is not so easily reared through its long complicated larva development, and less effort has been exerted toward developing commercial mariculture with these animals.

Fish

Mariculture with fish is also difficult, and only in Japan has a significant marine program been developed. Salmon, yellowtail tuna, puffers, and a few other fish

A.

B.

FIGURE 14.13
Crustacean Mariculture

(A) The University of Texas uses a hot water well to raise shrimp in its marine laboratories. (Photo by Lowell Georgia/Science Source.) **(B)** These European lobsters *(Homarus gammarus)* are being laboratory raised by the University of California to determine their commercial feasibility. (Photo by William E. Townsend, Jr./Photo Researchers.)

are being raised there on an experimental basis. Some estuary fish farming involving mullet and milkfish has been practiced successfully in the Far East for centuries. These fishes are hardy and tolerate salinity ranging from that of fresh water to full seawater. They do not have to be artificially fed. However, since they cannot be spawned and raised to sexual maturity in an artificial environment, the fry must be collected from a natural nursery environment. The costs of operating the farms are relatively low, and yields of about 1 metric tonne per acre are common (Fig. 14.14).

Salmon ranching operations in the Pacific Northwest have cost tens of millions of dollars, and it is still questionable as to when the operations may begin to make a profit. However, good returns of salmon released by the Alaska Fish and Game hatcheries are being reported. In Canada, the addition of nitrates and phosphates to lakes provided a larger biomass of plankton for young salmon to feed on before going to sea. This rather simple modification, which had no ill effects on the ecology of the lake, increased salmon production in the lake and salmon returns from the open ocean. Hopefully, when the management of commercial operations is refined as knowledge is gained, we will be hearing success stories from the Pacific Northwest.

It is projected that Norway will produce 80,000 metric tonnes of salmon from Atlantic salmon *(Salmo salar)* mariculture operations in which the fish are maintained in pens and fed automatically by computerized processes. The success of Norway in this type of mariculture greatly exceeds that of any other nation. Because the national government limits the production in Norway, Norwegians have transported their techniques to Scotland, Canada, and the United States. If these projects are only partially as successful as the Norwegian operations, they dramatically will change the industry of salmon production.

WHAT LIES AHEAD?

Mariculture requiring intensive artificial feeding and high labor costs cannot be expected to help alleviate the world shortage of animal protein. Actually, more food is consumed than is produced by such systems, which are typical of the enterprises that have been attempted in the United States. However, those farming enterprises that use natural feeding in estuarine environments have good potential for helping increase the food supply. Most of these efforts require juvenile animals that are obtained from natural nursery areas, which represent the greatest costs to the marine farmer. Potential for increased productivity using the 1 billion acres or so of coastal wetlands in the world would be quite significant. Assuming only 10 percent of the wetlands would be put into production, it could result in 100 million metric tonnes of fish based on the productivity that has been established for mullet and milkfish.

Many supporters of mariculture believe that this industry, which now accounts for less than 10 percent of the present food taken from the sea, can grow to a 60 million metric tonnes per year business. Many scientists who have studied the field of mariculture are not as optimistic about the ability of the industry to significantly increase the world food supply. The new projects being developed in the United States are directed at luxury foods that will certainly be too expensive for the world's hungry poor. They think these projects may generate a profit, but not much food.

The Food and Agriculture Organization (FAO) of the United Nations has estimated that, based on tradi-

FIGURE 14.14
Tilapia Fish Farm

The vats at this Caldwell, ID farm are used to raise species of the popular food fish *Tilapia.* (Photo by David R. Frazier.)

tional fisheries, the maximum catch that might be expected from the oceans is about 120 million metric tonnes. This estimate may be considered conservative, and could well be exceeded, particularly with the addition of new species to the significant fisheries. A candidate for addition to the list of major fisheries might be the Antarctic krill. Grenadiers and lantern fishes found in deeper waters and the pelagic red crab of the Eastern Pacific and subantarctic regions also may be added in the near future. Another significant addition would be cephalopods such as squid, cuttlefish, and octopus, which yield about 80 percent edible flesh—compared to 20–50 percent for the most common representatives of our present fishery.

Surely, improved management of existing fisheries, progress in mariculture, and the addition of new species to the world fishery will lead to a much-needed increase in the supply of marine food.

SUMMARY

Increased effort in some of the world's fisheries has not significantly increased catches. In some cases, the result has been a dramatic decrease in takes, indicating overexploitation of these fisheries has resulted in decreased catch per unit of effort.

Fisheries management in coastal U.S. waters began with the prohibition against seining for mackerel imposed by the Plymouth Colony in 1684. Attempts at regulation of fisheries have not been successful right up to the present. This failure may be in large part due to the fact that the ecology of most fisheries is not well understood. Present studies of the ecology of marine fisheries attempts to assess the factors that affect survival of fish eggs, larvae, and juveniles. Understanding these important factors helps fishery scientists make better predictions of recruitment of young adults to the fishery. A promising method of fishery assessment involves an ecosystem approach. This method requires knowledge of the annual nutrient flux into the ecosystem, size grades of organisms, standing stocks, and size doubling times of each component of the food chain.

About 50 percent of the world fishery comes from about 0.1 percent of the ocean surface area where upwelling occurs. Therefore, studies of duration and rate of upwelling in these regions have been undertaken. They reveal that moderate rates of upwelling over long periods of time—or even short durations—produce larger fisheries than high or low rates of upwelling.

Studies of the four major subtropical eastern boundary current fisheries supported by upwelling reveal that they have much in common. They are dominated by four species of pelagic fishes: sardine, anchovy, jack mackerel, and hake. Spawning occurs mostly in areas of moderate upwelling, where food conditions are optimum for fish larvae. The Peruvian anchoveta are an exception; they spawn where upwelling rates are high. There is as yet no explanation for this behavior.

Side-directed sonar used in conjunction with a manned submersible has proven effective in identifying a previously unknown krill population that inhabits the submarine canyons off the slopes of Georges Bank. These krill may provide the answer to the previously unexplained high levels of fish production on Georges Bank.

The incidental killing of dolphins while conducting the tuna fishery in the eastern Pacific has produced serious problems for the U.S. tuna fleet, and in the field of international relations. The decision in 1990 by U.S. tuna canners to not purchase tuna caught beneath dolphin schools should go a long way toward bringing this problem to a manageable solution.

An example of the combined effects of natural phenomena and human activities on a marine population is the sharp decline in the population of Peruvian anchoveta that occurred in 1972. This reduction seems to have been the result of a severe El Niño condition, and a poor understanding of the fishery in the coastal waters of Peru. A less euphemistic statement of the cause would be that the fishery was destroyed by the failure of regulators to enforce recommended procedures.

Mariculture projects around the world have, in some cases, indicated potential for increasing the food taken from the ocean. The most successful mariculture to date is directed at algae, oysters, mussels, mullet, milkfish, and salmon. These activities, combined with adding new species such as krill, grenadiers, lantern fish, red crabs, and cephalopods to our open-ocean fisheries, may help increase the amount of food we take from the ocean.

QUESTIONS AND EXERCISES

1. Describe the nature of and the results of the first attempt at fisheries management in U.S. coastal waters.
2. What are the two critical survival stages that must occur before young adults can be recruited to a fishery? What factor is most important in increasing the chance of survival at each stage?
3. Explain the relationship that exists between the influx of nutrients (nitrogen) into an ecosystem, and the amount of the fishery that can be removed safely from it each year.
4. Which two characteristics of an upwelling system are related to the magnitude of the fishery it can produce? What combination of these factors appears to produce the maximum environment for the development of a large fishery?
5. What are the characteristics that the fisheries of the four major eastern boundary current upwellings of the subtropical gyres have in common?
6. Discuss how the discovery of large demersal aggregations of krill found in submarine canyons off the flank of Georges Bank relate to the ecology of the fisheries on the bank.
7. List and discuss some positive and negative effects that you think may result from the establishment of the 322-km (200-m) Exclusive Economic Zones (EEZ).
8. Discuss the natural and artificial conditions that may have combined to cause the catastrophic decline of the Peruvian anchoveta fishery in 1972.
9. What characteristics are desirable in an organism chosen for a mariculture project?
10. What do the cultivation of algae and bivalves have in common that helps make them good choices for mariculture operations?
11. Discuss the positive and troublesome aspects of the mariculture of crustaceans.
12. Discuss how the mariculture of mullet and milkfish is similar to that of bivalves.
13. From which mariculture and natural fisheries do we have the greatest hope of obtaining an increased quantity of food from the sea?

REFERENCES

Backus, R.H., and **Bourne, D.W.**, Eds. 1987. *Georges Bank.* Cambridge, MA:MIT Press.

Borgese, E.M., and **Ginsburg, N.** 1986. *Ocean yearbook 6.* Chicago:University of Chicago Press.

Greene, C.H., **Wiebe, P.H.**, **Burczynxki, J.**, and **Youngbluth, M.J.** 1988. Acoustical detection of high-density krill demersal layers in the submarine canyons off Georges Bank. *Science* 241(4863):359–361.

Knauss, J.A. 1974. Marine science and the 1974 Law of the Sea Conference—science faces a difficult future in changing Law of the Sea. *Science* 184:1335–1341.

Longhurst, A.R., and **Pauly, D.** 1987. *Ecology of tropical oceans.* San Diego:Academic Press.

Parrish, R., **Bakun, A.**, **Husby, D.M.**, and **Nelson, C.S.** 1983. Comparative climatology of selected environmental processes in relation to eastern boundary current pelagic fish reproduction. Unpublished FAO study.

Parsons, T.R., **Takahashi, M.**, and **Hargrave, B.** 1984. *Biological oceanographic processes, 3rd ed.* New York:Pergamon Press.

Royce, W.F. 1987. *Fishery development.* San Diego:Academic Press.

Smith, P.E. 1969. The horizontal dimensions and abundance of fish schools in the upper mixed layer as measured by sonar. La Jolla, CA:Bureau of Commercial Fisheries.

Wyatt, T. 1980. The growth season in the sea. *J. Plankton Research* (2):81–97.

SUGGESTED READING

SCIENTIFIC AMERICAN

Holt, S.J. 1969. The food resources of the ocean. 221(3):178–197.

Pinchot, G.B. 1970. Marine farming. 223(6):14–21.

SEA FRONTIERS

Brown, M. 1984. Counting dolphins: New answers to an important question. 30(2):68–75.

Cooper, L. 1981. Seaweed with potential. 27(1):24–27.

Cuyvers, L. 1984. Milkfish: Southeast Asia's protein machine. 30(3):173–179.

Iverson, E.S. 1984. Mariculture: Promise made, debt unpaid. 30(1):20–29.

Iverson, E.S., and **Iverson, J.Z.** 1987. Salmon-farming success in Norway. 33(5):354–361.

Iverson, E.S., and **Jory, D.E.** 1986. Shrimp culture in Ecuador: Farmers without seed. 32(6):442–453.

Katz, A.S. 1984. Herring fisheries: Comparing the two stories. 30:335–341.

Maranto, G. 1988. Caught in conflict: Managers trapped in fisheries dilemma. 34(3):144–151.

Nicol, S. 1987. Krill: Food of the future? 33(1):12–17.

15

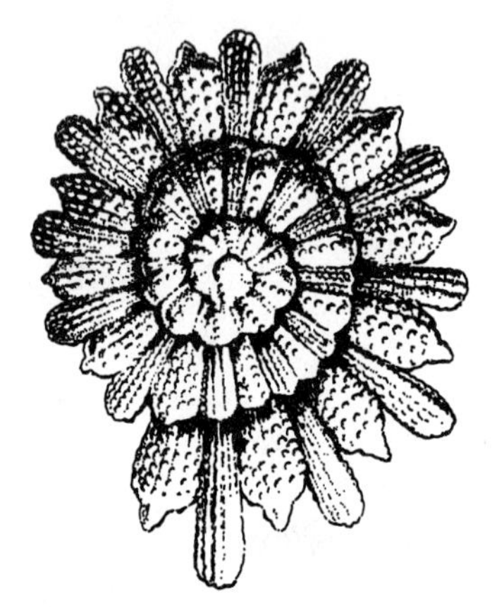

MARINE POLLUTION

As society provides for an ever-increasing population, the role played by the oceans in meeting these needs also is increased. One of these roles has become a threat: The site for disposal of a growing amount of waste materials, from sewage sludge to radioactive waste. An additional threat faced by the marine environment is its contamination from transportation and production accidents associated with the efforts to meet present petroleum energy needs.

Since 80 percent of its population lives within easy access of its oceanic coasts and the Great Lakes, the ocean is an important recreational resource in the United States. Well over 100 million Americans participate in marine recreational activities, spending almost $150 billion on them annually. In the past, it has not been unusual to see local areas of the shore closed to recreational activities such as fishing, boating, and swimming because of pollution.

We have little difficulty understanding the meaning of pollution. The World Health Organization (WHO) defines **pollution** of the marine environment as

> The introduction by man, directly or indirectly, of substances or energy into the marine environment, including estuaries, which results or is likely to result in such deleterious effects as harm to living resources and marine life, hazards to human health, hindrance to marine activities, including fishing and other legitimate uses of the sea, impairment of quality for use of sea water and reduction of amenities.

But as far as the marine environment is concerned, there is great difficulty in establishing the degree to which pollution is occurring. In most cases, the ocean has not been studied sufficiently prior to society's introduction of pollutants. Therefore, we cannot tell how the marine environment has been altered by these activities.

CAPACITY OF THE OCEANS FOR ACCEPTING SOCIETY'S WASTES

At present, there is a strong consensus developing in the United States (and in many other nations) that we should not dispose of any waste materials or energy in the ocean. A more realistic view of the problem, however, states that we will have to consider the oceans as an option in disposing of some waste materials. An even more realistic view of this problem is that regardless of what we say we are going to do, polluting materials will continue to enter the oceans in the foreseeable future. In light of this probability, research into the capacity of the oceans to absorb various potentially polluting substances continues.

One of the present goals of marine pollution research is to determine an elusive entity—the **assimilative capacity** of the oceans. This is the amount of a given material that can be contained within the ocean without causing an unacceptable degree of harm to its living or nonliving resources. Determining at what concentrations an unacceptable degree of harm is done to the ocean's resources requires long-term monitoring of disposal activities. Although some researchers believe we have enough data to establish values for the assimilative capacity of the coastal ocean for certain substances, all agree that what we do not know greatly exceeds what we do know.

PREDICTING THE EFFECTS OF POLLUTION ON MARINE ORGANISMS

To date, the most widely used technique for determining the level of pollutant concentration that negatively affects the living resources of the ocean is the **standard laboratory bioassay.** Regulatory agencies such as the Environmental Protection Agency (EPA) use a bioassay that determines the concentration of a pollutant that causes 50 percent mortality among the test organisms in order to set allowable concentration limits of pollutants in coastal water effluents. The shortcoming of the bioassay is that it does not predict the long-term effects of sublethal doses on the organisms.

It is desirable to be able to predict the effects of a given concentration of pollutants at four different levels of biological organization. Some of these effects can be detected in a few minutes, whereas others may require decades of observation. These levels of biological response, the time spans over which they occur, and examples of each follow (see also Tab. 15.1).

Biochemical-cellular: *Minutes to hours* Exposure to organic pollutants such as hydrocarbons and PCBs produce dysfunction in metabolic processes, causing death.

Organismal: *Hours to months* Pollution stress may create changes in physiological processes, such as metabolism or digestive efficiency, that will severely reduce the energy available for growth and reproduction.

TABLE 15.1
Biological Responses to Pollutant Stress

Level	Adaptive Response	Destructive Response	Result at Next Level
Biochemical-cellular	Detoxification		Adaptation of organism
		Membrane disruption Energy imbalance	Reduction in condition of organism
Organismal	Disease defense Adjustment in rate functions Avoidance		Regulation and adaptation of populations
		Metabolic changes	Reduction in performance of populations
		Behavior aberrations Increased incidence of disease Reduction in growth and reproduction rates	
Population	Adaptation of organism to stress No change in population dynamics		No change at community level
		Changes in population dynamics	Effects on coexisting organisms and communities
Community	Adaptation of populations to stress		No change in community diversity or stability
			Ecosystem adaptation
		Changes in species composition and diversity	Deterioration of community
		Reduction in energy flow	Change in ecosystem structure and function

Adapted with permission of Judith M. Capuzzo, Woods Hole Oceanographic Institution.

Population dynamics: *Months to decades* Although it is difficult to tie long-term effects to a certain pollutant, short-term or chronic effects may be tied to events such as oil spills or malfunctions at sewage plants. These responses occur as changes in the abundance and distribution of a species, population structure on an age-class basis, growth rates within age classes, fecundity, and incidence of disease.

Community dynamics and structure: *Years to decades* Communities, as assemblages of populations, become less diverse, and more resistant populations increase in numbers under the stress of pollutants.

Much progress is needed before the identification of early warning signs of stress at each level of biological response is possible. Since the degree of system complexity and the time needed to measure a response increase exponentially from the biochemical-cellular level to the community level, the predictive difficulties also increase at each level.

CEPEX

The Controlled Ecosystem Pollution Experiment (CEPEX) used plastic experimental bags (Fig. 15.1) to simulate communities of plankton ranging in size from 68–1,700 m^3 (2,400–60,000 ft^3) to investigate the effects of pollutants. Observations of the recovery rates of bacteria, phytoplankton, and zooplankton to pollution by heavy metals and hydrocarbons showed that the bacteria and phytoplankton, with their short generation times, recovered much more rapidly than did the larger and more complex zooplankton.

MARINE POLLUTION CONTROL IN THE UNITED STATES

The United States has long been concerned with the legal protection of the marine environment. The first Rivers and Harbors Act of 1899 provided a straightforward

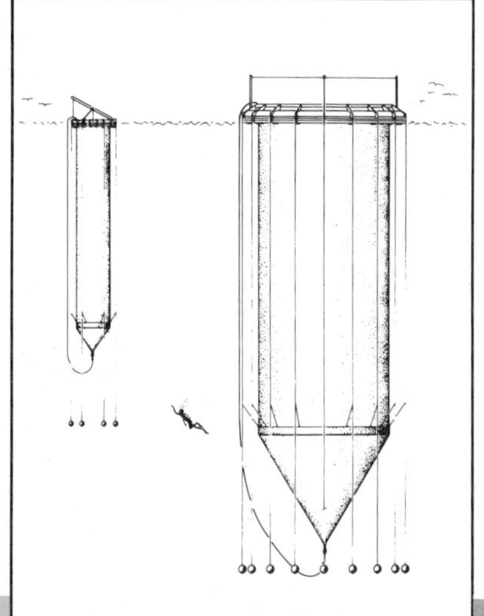

FIGURE 15.1 Controlled Ecosystem Pollution Experiment (CEPEX)

Views of CEPEX bags used in experiments to investigate the effects of pollutants on plankton communities. (Photo courtesy of the National Science Foundation.)

prohibition against dumping any kind of refuse into the navigable waters of the United States. Most current regulations on waste disposal in the coastal ocean, the 332-km(200-mi) Exclusive Economic Zone (EEZ), are contained in two major statutes. The Federal Water Pollution Control Act (Clean Water Act) of 1948 and its amendments deal primarily with problems of point sources of municipal and industrial waste, plus spills of oil and hazardous materials. The Marine Protection, Research, and Sanctuaries Act (Ocean Dumping Act) of 1972 controls dumping of wastes at sea, research, and the establishment of marine sanctuaries.

Although these regulations are quite complex, these laws basically (a) prohibit the dumping of materials known to be harmful, and (b) specify the criteria under which other materials may be dumped. The basic premise of each seems to be that land disposal is preferable to marine disposal. If it cannot be proved to the satisfaction of the regulating agencies that dumping of a material will not adversely affect human health, it will not be allowed in the ocean.

Despite the strong language of the laws, the coastal ocean has been adversely affected by the dumping of waste materials. Conversely, some waste materials that could possibly have been safely disposed of in the coastal ocean were not because of the difficulty in proving it would not have been harmful to dump them. It is very easy to prove the harmful effects of dangerous materials, but it is almost impossible to prove that other seemingly benign substances would not create a long-term harmful condition.

Because some scientists determined that we probably can better predict the paths of materials dumped into the ocean than those dumped on land, the option of ocean disposal received renewed attention. In the past, uncertainty led to a decision to prohibit dumping. Soon, there was growing support for placing the obligation of proving harmful effect on the regulating agency, instead of requiring the dumping applicant to prove that no harmful effect will occur. With the public outcry over the littering of northeastern U.S. beaches with medical waste during Summer 1988, legislation to prohibit marine dumping was quickly passed.

In light of recent advances in scientific knowledge, it is frustrating to have to admit the broad scope of our ignorance of the marine environment. Considering the great cost and amount of time required to remove uncertainty from the field of marine waste disposal in the near future, it would seem prudent to continue exercising a great deal of caution in pursuit of this option. Considering the rapid escalation of the cost of land-based disposal, short-term political priorities point to stopping ocean dumping. Economic reality, however, tends to make ocean dumping more attractive. Clearly, politicians and their constituents need to have the best possible data available to proceed rationally. Some very difficult decisions will have to be made in the near future.

AREAS OF CONCERN

Plastics

During the last 30 years, the use of plastics has increased at a tremendous rate. The problem of disposing of this throw-away component of western culture has already strained the capacity of our land-based disposal systems. It is also an increasingly abundant component of oceanic flotsum.

Small pellets that are used in the production of essentially all plastic products are transported in bulk aboard commercial vessels. Found throughout the oceans, the pellets probably find their way into the oceans as a result of spillage at loading terminals. In coastal waters, plastic products used in fishing and thrown overboard by recreational and commercial vessels are common. They also find their way into the open-ocean waters via careless dumping by commercial vessels.

The best documentation of negative effects of plastics in the ocean on marine organisms is found in the strangulation of seals and birds caught in plastic netting and packing straps (Fig. 15.2). Marine turtles are known to mistake plastic bags for jellyfish or other transparent plankton on which they typically feed.

It is believed that all plastics that enter the ocean may eventually be removed by the shorelines of islands and continents that filter out the particles as ocean currents wash against their shores. Beaches throughout the world are probably increasing their plastic pellet content as a result of the filtering process. Some Bermuda beaches have up to 10,000 pellets/m^2. The beaches of Menemsha Harbor on Martha's Vineyard, Massachusetts yielded 16,000 plastic spherules/m^2. Thin plastic film products eventually will be broken into smaller particles by photochemical degradation, and others may sink as they become denser because of photochemical degradation and encrustation by epifauna such as bryozoans and hydroids. These data are from a survey conducted between 1984 and 1987, which also found more than 10,000 plastic pieces and 1,500 pellets/km^2 in the northern Sargasso Sea, between 28–40° N latitude. The 1987 survey found that the concentration of plastic pellets had doubled since 1972.

The U.S. Congress is considering a bill to ban disposal of plastics in the 200-mile-wide EEZ. Internationally, the Convention for the Prevention of Pollution

FIGURE 15.2
Seal Trapped by Plastic Packing Strap

This animal was found on a beach of San Clemente Island off the southern California coast. It was saved from strangulation by the removal of the strap by scientists from the National Marine Fisheries Service. (Photo by Wayne Perryman, NMFS.)

from Ships prevents disposal of all plastics into the oceans. The acceptance of both these regulations would aid greatly in turning around the increasing concentration of plastics in the world's oceans.

Petroleum

There have been a number of major oil spills since the *Torrey Canyon* ran aground at Seven Stones Rocks off the Cornwall, England coast on March 18, 1967. They have resulted from additional tanker accidents and the blowout of oil wells being drilled or produced in the coastal ocean. Although it is important to know how much oil enters the oceans and from what sources, our primary concern is the effect of hydrocarbon pollution on marine organisms and the marine environment in general.

Since hydrocarbons are organic substances that are biodegradable by microorganisms, they are considered by some to be among the least damaging pollutants of the ocean. Yet they are complex, containing molecules ranging in molecular mass from 16–20,000. Most crude oils also will contain, in addition to their hydrocarbon components, chemicals containing oxygen, nitrogen, sulfur, and trace metals. When this complex chemical mixture is combined with the ocean—another complex chemical mixture containing organisms—the result can indeed be negative to the organisms.

West Falmouth Harbor. Because of the complexity of petroleum, it is difficult to answer the question, *How long does it take for a shore to recover from an oil spill?* The oldest well-studied oil spill in the United States occurred on September 16, 1969, near West Falmouth Harbor in Buzzards Bay, Massachusetts. When the barge *Florida* came ashore and ruptured, currents carried the Number 2 fuel oil north into Wild Harbor, where the most severe damage occurred (Fig. 15.3). The initial kill was almost total for intertidal and subtidal animals in the most severely oiled area. A severe reduction in species diversity was accompanied by rapid increases in the population of polychaete worms that were resistant to the oil. The species, *Capitella capitata,* accounted for up to 99.9 percent of the individuals taken in samples of the most severely oiled locations during the first year. Species diversity did not appreciably increase until well into the third year after the spill. Although conditions have returned to near normal, oil can still be found in the fine sediment of areas that underwent the most severe contamination.

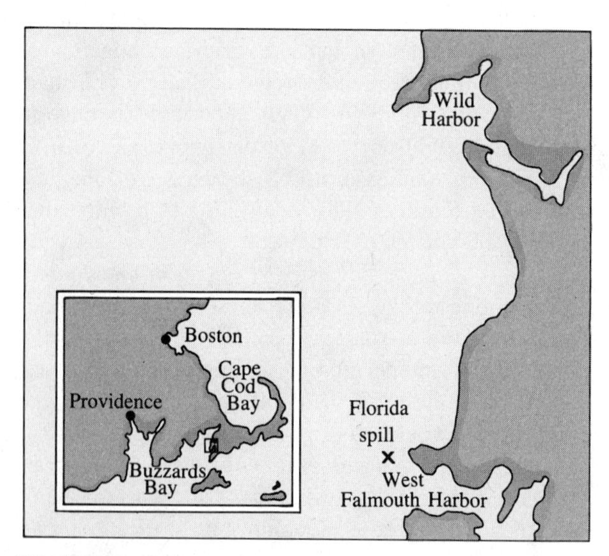

FIGURE 15.3
***Florida* Oil Spill at West Falmouth Harbor**

Chedabucto Bay, Canada. A spill of heavy Bunker C fuel oil in Chedabucto Bay, Nova Scotia on February 4, 1970, has provided some indication of the recovery time. Based on changes in visible contamination, about one-third of the initially oiled 200 km (124 mi) of shore was clear after three months. This cleaning was primarily the result of wave action on the rocky shore. By 1973, mechanical cleaning by waves, evaporation, photochemical decomposition, and bacterial decomposition had cleaned all but the restricted low-energy estaurine marshes. Fully 75 percent of the shore was clean. Visible contamination had been reduced to 15 percent by 1976 and is rarely observable beneath some of the finest sediments (Fig. 15.4).

Although the oil disappeared rapidly from the rocky shore, the low-energy sediments of the marshes will not be cleaned so quickly. Although the half-life (the time required for half the hydrocarbon concentra-

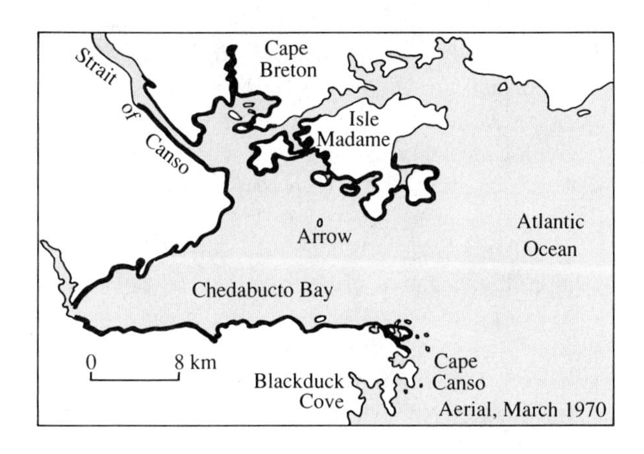

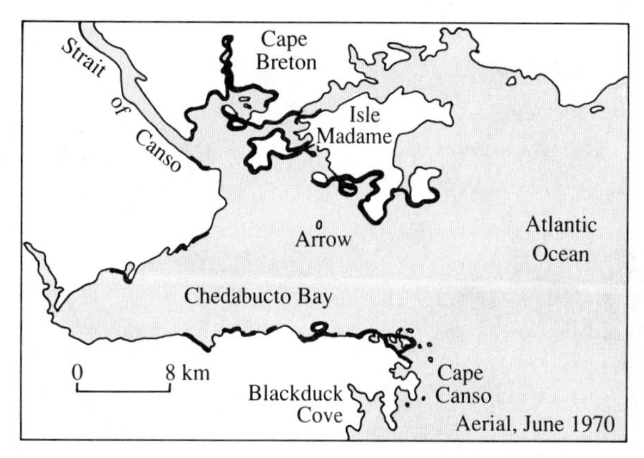

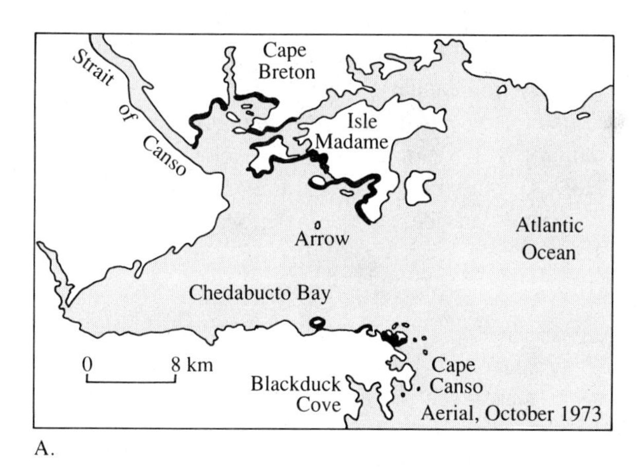

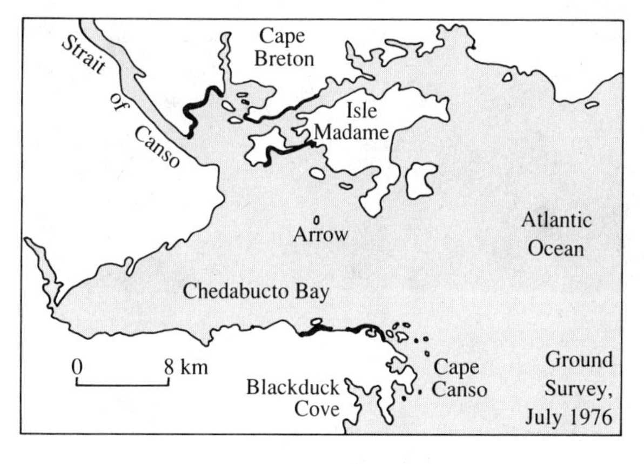

A.

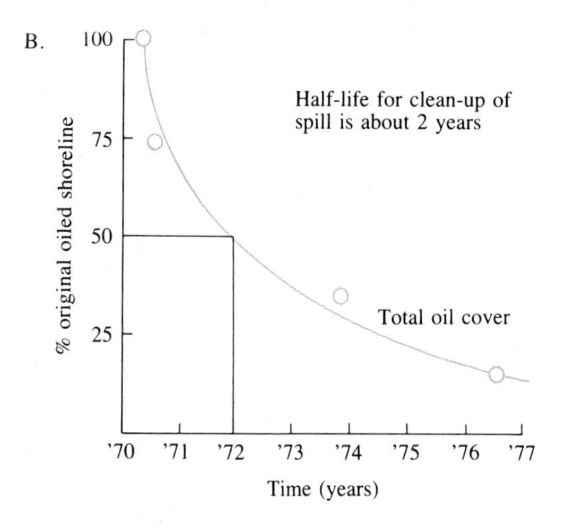

FIGURE 15.4
Oil Spill in Chedabucto Bay

(A) Distribution of oil residues in the shore zone of Chedabucto Bay, Nova Scotia, Canada, determined by aerial inspection in 1970 and 1973, and by ground survey in 1976. **(B)** Oil removal pattern of stranded Bunker C oil from shorelines, 1970–1976. No oil is visible at the surface now, but can be found by removing the surface sediments in some marshes.

tion to be degraded) for removal of the visible oiling in the bay as a whole is about two years, it appears that the half-life for removal from the fine sediment will be up to 25 years.

Dr. John H. Vandermeulen of the Bedford Institute of Oceanography in Nova Scotia, Canada returned to the bay in 1989 to examine the long-term effects of the spill. He says that sandy and muddy sediments still have detectable hydrocarbons, and boulder beach deposits still are cemented with heavy tar deposits. Oil released from local commercial and recreational fishing activities over the past 18 years has made it impossible to tie the present contamination to the *Arrow* spill in 1970. In terms of total biomass, the area seems to be fully recovered.

The biological consequences of the Chedabucto Bay spill have been indicated by the recovery of kelp, *Fucus vesiculosus,* the salt marsh grass, *Spartina alterniflora,* and the soft-shell clam, *Mya arenaria*. Over half the *Fucus* was destroyed, and a recovery half-life of four years is indicated. *Spartina* suffered a similar reduction, but was much slower to start recovery. Whereas *Fucus* made a gradual recovery that began immediately, *Spartina* showed no signs of recovery until two years after the spill. Once recovery started, however, it was rapid, and an estimate of the recovery half-life is about three years. The clam, *Mya,* does not show such an encouraging pattern of recovery. Severely stressed by the contamination, *Mya* has experienced a serious reduction in efficiency of food utilization. Tissue and shell growth were greatly reduced, and the age-class structure has been altered because of the high degree of mortality that occurred with the spill. The recovery half-life of the *Mya* was in excess of ten years.

Argo Merchant. When oil spills do not come ashore, the negative effects are not so obvious. The sinking of the *Argo Merchant* after it ran aground on Fishing Rip Shoals off Massachusetts on December 15, 1976, provides the best-studied effects of such a spill. A full load of Number 6 fuel oil was dumped into the ocean 40 km (25 mi) southeast of Nantucket Island. Fortunately, the winds were such that no oil came ashore. The surface slick moved east out to sea and was gone by mid-January. Even though the oil was not visible for long, it did a significant amount of biological damage that was investigated by scientists from Woods Hole Oceanographic Institution, the National Marine Fisheries Service, and other local institutions (Fig. 15.5).

A.

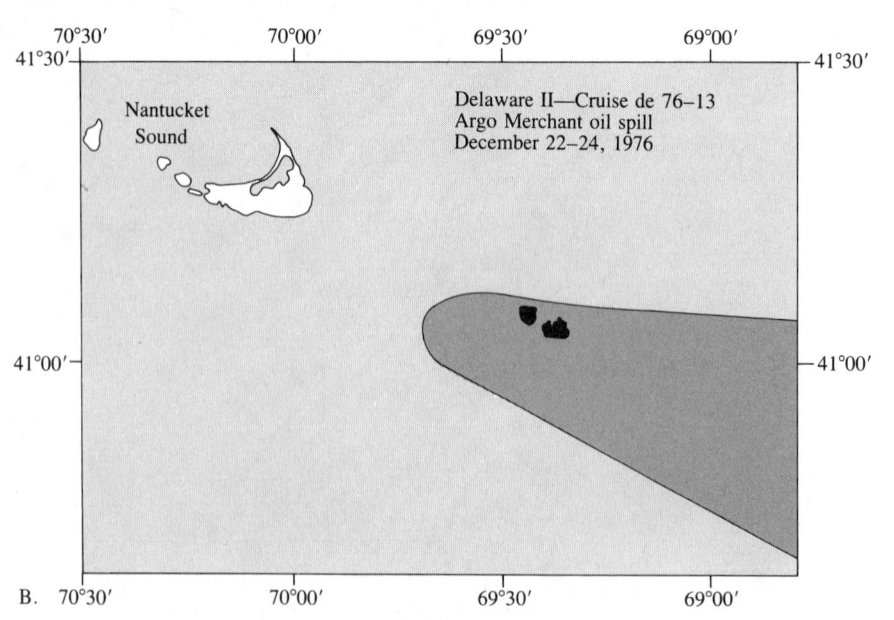

B.

FIGURE 15.5
The *Argo Merchant* Oil Spill

(A) The *Argo Merchant* after breaking up and spilling much of its cargo of oil into the Atlantic Ocean southeast of Nantucket. **(B)** Location of grounding site and oiled area of the ocean. (A: Photo courtesy of U.S. Environmental Protection Agency. B: From *The Argo Merchant oil spill: A preliminary scientific report.* 1977. Washington, DC:Dept. of Commerce.)

The primary observable effect of the *Argo Merchant* spill on marine organisms resulted from the recovery of pelagic fish eggs, pollock, and cod, in plankton samples taken shortly after the spill. Of the 49 pollock eggs recovered, 94 percent were oil-fouled, while 60 percent of the 60 cod eggs were in that state. A total of 20 percent of the cod eggs and 46 percent of the pollock eggs were dead or dying (Fig. 15.6). Although little is known about the natural mortality rate of pelagic fish eggs, only 4 percent of a sample of cod eggs spawned in a laboratory were dead or dying at the same developmental stage. Based on the available knowledge, hydrocarbon contamination appears capable of causing death to fish eggs.

Because most of the oil floated in a surface slick, contamination in subsurface water samples did not exceed more than 250 parts per billion. Other than the described damage to fish eggs and a significant amount of damage to other members of the plankton, little direct evidence of major biological damage was recovered. There were, however, numerous oiled birds that washed ashore at Nantucket and Martha's Vineyard.

Because pollock females spawn about 225,000, and cod about 1 million eggs each season over a range extending from New Jersey to Greenland, it is unlikely that this single occurrence will have a major effect on those fisheries. However, it is easy to see why the fishing communities of the Northwest Atlantic are so happy about the marginal results of oil exploration on Georges Bank. It may be that a fishery already severely stressed by overexploitation could not survive even the lowest level of pollution brought about by the establishment of oil production on the Bank.

Exxon Valdez. The first major oil spill resulting from the development of the North Slope of Alaska petroleum reserves occurred on March 23, 1989 when the California-bound *Exxon Valdez* went aground on rocks 40 km (25 mi) out of Valdez, Alaska (Fig. 15.7). The 32,000 metric tonnes of oil spilled makes this the largest spill to occur in U.S. waters. This spill in the pristine waters of Prince William Sound spread to the Gulf of Alaska, and 1,775 km (1,100 mi) of shoreline were damaged. The U.S. Fish and Wildlife Service reported that at least 994 sea otters and 34,434 birds were killed by the spill. The actual kill could be ten times that amount, according to some official estimates. The total death toll will never be known, but it is predicted that

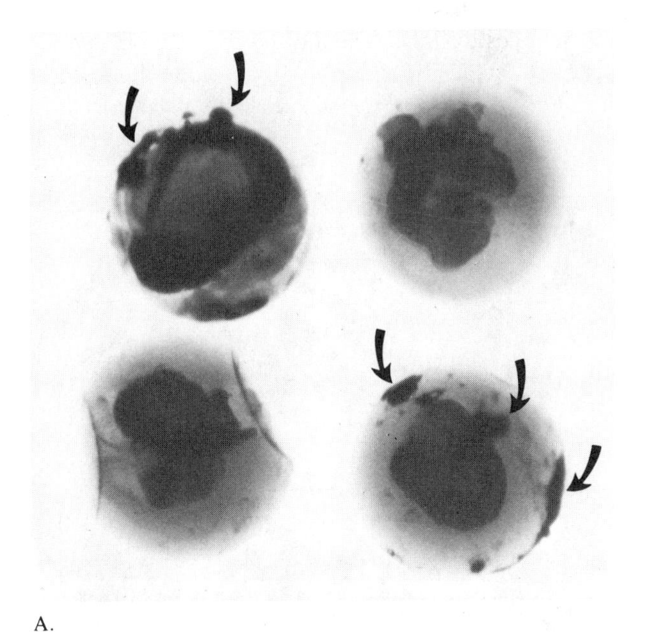

A.

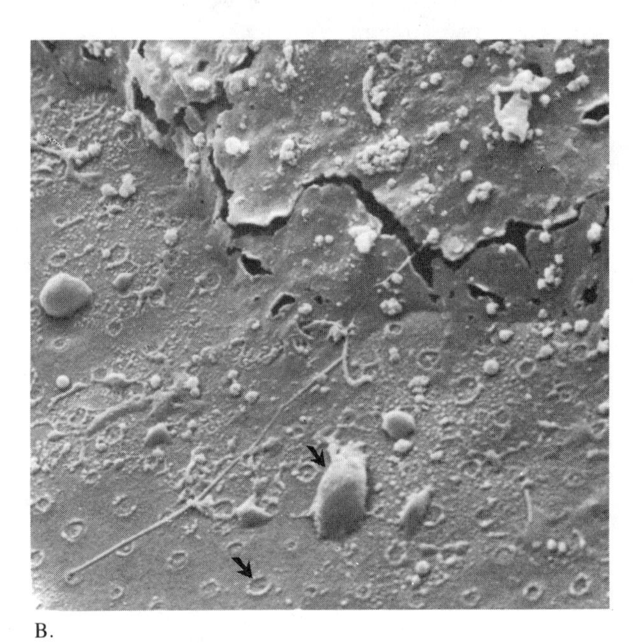

B.

FIGURE 15.6
Pollock Eggs Damaged by the *Argo Merchant* Spill

(A) These eggs, in the tail-bud and tail-free embryo stages, were taken from the edge of the oil slick. The outer membranes of the eggs at the upper left and lower right (see arrows) are contaminated with a tarlike oil. The uncontaminated egg at the upper right has a malformed embryo; the lower left egg is collapsed and also has an abnormal embryo. The actual size of pollock eggs is about 1 mm. **(B)** Surface of an oil-contaminated egg. The upper arrow points to one of many oil droplets, the lower arrow to one of the membrane pores. (×5000) (Photo courtesy of A. Crosby Longwell.)

FIGURE 15.7
Oil-Covered Bird, Prince William Sound, Alaska

This victim of the *Exxon Valdez* oil spill was found during the cleanup. (Photo by Ron Levy/Gamma-Liaison.)

these Alaskan waters may have a long, slow recovery as a result of the low water temperatures that will impede natural cleanup processes.

Encouraged by good results on small test plots, Exxon is spending $10 million to spread phosphorus- and nitrogen-rich fertilizers on Alaskan shorelines to boost the development of indigenous oil-eating bacteria. It is hoped that this will result in a cleanup rate that will produce in two to three years what would require up to seven years under natural conditions.

Oil spills are a problem that will be with us for many years, and they may become more common as the petroleum reserves underlying the continental shelves of the world are increasingly exploited. The largest oil spill on record occurred June 3, 1979, as a result of this exploitation. The *Petroleus Mexicanos Ixtoc 1* blew out and caught fire. Before it was capped on March 24, 1980, it spewed 468,000 tons of oil into the Gulf of Mexico (Fig. 15.8).

Oil can be broken down by microorganisms such as bacteria, fungi, and yeast. Only certain of these organisms are effective in breaking down a particular variety of hydrocarbon, and none are effective against all forms. Dr. A.M. Chakrabarty, a microbiologist at the University of Illinois, produced a microorganism capable of breaking down nearly two-thirds of the hydrocarbons in most crude oil spills. By manipulating special rings of DNA called plasmids, he was able to combine the consumptive characteristics of numerous natural strains of bacteria into what might be considered a 'superbug' that could be very effective in cleaning oil spills. However, before this superstrain can be used against actual oil spills, much testing is needed to ensure that it does not have adverse effects on the environment.

Sewage

Over 500,000 metric tonnes per year of sewage sludge containing a mixture of human waste, oil, zinc, copper, lead, silver, mercury, PCBs, and pesticides have been dumped through sewage outfalls into the coastal waters of southern California. A more massive 8 million metric tonnes have been dumped into the New York Bight each year. Although the Clean Water Act of 1972 prohibited dumping of sewage into the ocean after 1981, the high cost of treating and disposing of it on land resulted in extended waivers being granted to these areas. Ironically, the nonbiodegradable debris that washed up

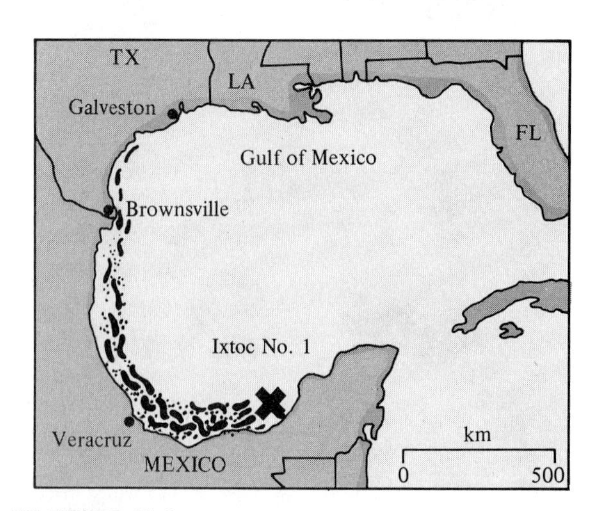

FIGURE 15.8
Oil Spill from the *Ixtoc 1* Blowout

Location of the Gulf of Campeche blowout and oil slick that threatened the Texas coast during the summer of 1979.

on Atlantic coast beaches during the summer of 1988 and hurt the tourist business probably was responsible for legislation again being passed to terminate disposing of sewage into the ocean. (The cutoff date is now January, 1992.) There is little likelihood that this debris is associated with sewage. It is more likely the result of plastic debris being carried to the ocean by storm drain systems after heavy rains.

It now appears that the political pressure to clean up the coastal water is sufficient to assure that the more expensive land treatment and disposal will be the primary disposal procedure of the future. For the present, Los Angeles has stopped pumping sewage into the sea. If it is allowed to resume this practice, the sewage sludge pumped through the lines will have had its toxic chemicals and pathogens removed.

Off the east coast of the United States, over 8 million metric tonnes of sewage sludge was dumped by barges over dump sites totalling 150 km^2 (58 mi^2) each year during the past few years. These dumps occurred at the New York Bight Sludge Site and the Philadelphia Sludge Site shown on Figure 15.9. The New York Bight water depth is about 29 m (95 ft), and Philadelphia's site is 40 m (130 ft) deep. In such shallow water, the water column displays relatively uniform characteristics from top to bottom; even the smallest sludge particles reach the bottom without having undergone much horizontal transport, and the ecology of the dump site can be totally destroyed. At the very least, the concentration of organic and inorganic nutrients seriously disrupt the biogeochemical cycling. Greatly reduced species diversity results, and in some locations the environment becomes anoxic. The use of such shallow water sites has been abandoned, however, and sewage is now being transported to a deep water site 106 miles out (Fig. 15.9). At the deep-water site beyond the shelf break, there is usually a well-developed density gradient separating low-density, warmer surface water from high-density, colder deep water. Internal waves moving along this density gradient can retard the sinking rate of particles, and may allow a horizontal transport rate 100 times greater than the sinking rate.

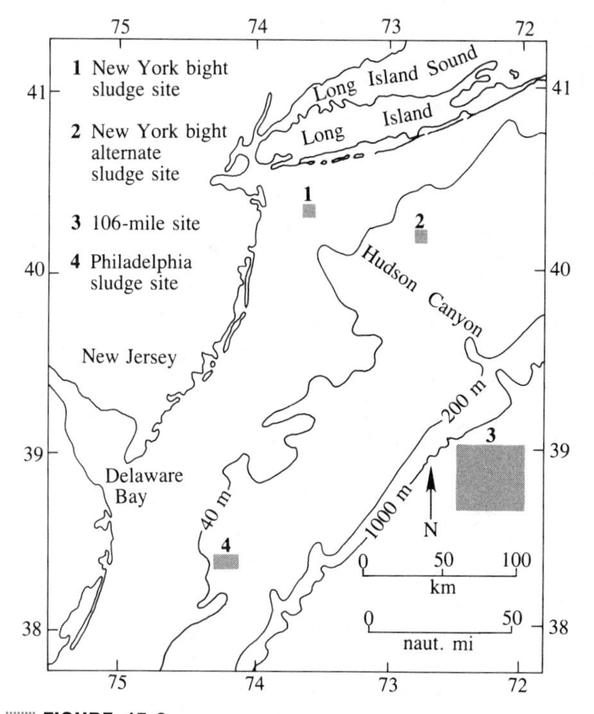

FIGURE 15.9
Operative and Alternative Sewage Sludge Disposal Sites in the Northeastern Atlantic

Local fishermen are enjoying none of the rosy scenario described above. Rather, they have been complaining about the adverse effects of the deep water dumping on their fisheries, soon after it began. As it stands now, this east coast sewage dumping program is doomed, and municipalities will have to find the money for land-based processing.

Radioactive Waste

As with other waste materials, a case can be made that oceans can accept a given amount of radioactive waste without doing significant harm to the marine environment. Since 1944, artificial **radionuclides** (radioactive atoms) have been reaching the oceans as **fallout** from the production and testing of nuclear weapons. This input has been about four times greater in the Northern Hemisphere than in the Southern Hemisphere. Because of the nature of the fallout process, no high concentrations of any radionuclides has resulted. Table 15.2 shows the amount of radiation in thousands of curies that has reached the oceans via fallout. (A **curie** is a quantity of any radioactive nuclide in which exactly 3.7 $\times 10^{10}$ disintegrations occur per second.)

The second largest source of radionuclides is the nuclear fuel cycle. The release from this source occurs in local concentrations. Nuclear power-generating plants release little radiation to the oceans, but nuclear fuel reprocessing plants contribute a significant quantity. Such plants are found at Sellafield and Dounreay in Great Britain, and at Cape de la Hague, France.

Another significant source of radioactive waste is a dumping operation being carried out over the past three decades under international agreement at various worldwide locations. Since 1967, the only significant dumping operation has been conducted in the Northeast

TABLE 15.2
Amounts of Radionuclides in the Ocean

	Plutonium-239, -240 (kCi)	Cesium-137 (kCi)	Strontium-90 (kCi)	Carbon-14 (kCi)	Tritium (kCi)
Total worldwide fallout by early 1970s	320	16,700	11,500	6,000	3,000,000
Sellafield plant discharge (1957–1978)	14	830	130		370
	Total α-emitters		Total β/γ-emitters (other than tritium)		
NEA dumpsite (1967–1979)	8.3		258		262

NOTE: kCi = 1,000 curies
SOURCE: Nuclear Energy Agency

Atlantic under the coordination of the Nuclear Energy Agency of the Organization for Economic Cooperation and Development. The input of this operation is shown in Table 15.2.

In order to determine safe input rates for various radionuclides, oceanographers must determine the average concentration of a radionuclide if mixed totally in the ocean. This is of primary importance for radionuclides with half-lives much greater than the mixing time of the ocean. It is a major consideration for materials like plutonium, which has a half-life of 24,400 years. **Mixing time** is the time required to evenly distribute a substance added to the ocean. The scientific community's best guess is that the mixing time of the world ocean is several thousand years; that of the Atlantic Ocean may be less than 1,000 years.

A second consideration that must be made is to determine the transfer mechanisms to critical populations. This is important for all radionuclides in areas near a dumping locality, regardless of the length of their half-lives.

The International Atomic Energy Agency has determined maximum dumping rates on the basis of these two considerations. As an example, the dumping rate for one of the most dangerous radionuclides—plutonium^{239}—is placed at 10,000 curies/yr. This limit would result in the plutonium concentration reaching the safe dose limit in 40,000 years.

Although research has been done on deep sea disposal of high-level nuclear waste, the research has been terminated. The U.S. Nuclear Energy Agency has concluded that if this waste is to be disposed of in the thick sediments of abyssal plains, it will not occur for at least 50 years. Land-based disposal sites are now the top priority of research toward the solution of this problem. (See the special feature, "Subsea Disposal of High-Level Nuclear Waste"—at the end of this chapter—for a discussion of the Organization for Economic Cooperation and Development's findings.)

Halogenated Hydrocarbons (DDT and PCBs)

DDT (dichlorodiphenyltrichlorethane) and **PCBs** (polychlorinated biphenols) are found throughout the marine environment. They are persistent, biologically active chemicals that have been put into the oceans entirely as a result of human activities.

At the time of a near-total ban on U.S. production of the pesticide DDT in 1971, 2.0×10^{12} g had been manufactured, most of it by the United States. Its use in the Northern Hemisphere since 1972 has virtually ceased. The danger of excessive use of DDT and other similar pesticides was first manifested in the marine environment in bird populations. During the 1960s, there was a serious decline in the brown pelican population of Anacapa Island off the coast of California. High concentrations of DDT in the fish eaten by the birds had caused the production of excessively thin egg shells. A decline in the osprey population of Long Island Sound that began in the late 1950s and continued throughout the 1960s also was caused by thinning egg shells brought on by DDT contamination. Studies showed a 1 percent increase in brown pelican and osprey egg shell thickness from 1970 to 1976, while the concentration of pesticide residue decreased. Since there was an increased egg hatching rate associated with these changes, it was concluded that DDT was the cause of the decline in the bird populations during the 1960s (Fig. 15.10).

PCBs are industrial chemicals found in a variety of products from paints to plastics. They have been indicated as causes of spontaneous abortions in sea lions, and the death of shrimp in Escambia Bay, Florida.

FIGURE 15.10
Brown Pelican

Pelican populations in the Channel Islands off southern California are increasing after a serious decline in the 1960s caused by DDT toxicity.

The main route by which DDT and PCBs enter the ocean is the atmosphere. They are concentrated initially in the thin surface slick of organic chemicals at the ocean surface, until they gradually sink to the bottom, attached to sinking particles. A study off the coast of Scotland indicated that open-ocean concentrations of DDT and PCBs are 10 and 12 times less, respectively, than in coastal waters. Long-term studies have shown that DDT residue content in molluscs along the U.S. coasts reached a peak in 1968. The pervasiveness of DDT and PCBs in the marine environment can best be demonstrated by the fact that antarctic marine organisms contain measurable quantities of halogenated hydrocarbons. Obviously there has been no agriculture or industry on that continent to explain this presence. These substances have been transported from far, distant sources by winds and ocean currents.

Mercury

The stage was set for the first tragic occurrence of mercury poisoning with the establishment of a chemical factory on Minamata Bay, Japan, in 1938. One of the products of this plant was acetaldehyde, the production of which requires mercury as a catalyst. The first ecological changes in Minamata Bay were reported in 1950; human effects were noted in 1953; and the disease became epidemic in 1956. It was not until 1968 that the Japanese government declared mercury the cause of the disease, which involves a breakdown of the nervous system. The plant was immediately shut down, but by 1969 over 100 people were known to suffer from mercury poisoning. Almost half of these victims died. A second occurrence of mercury poisoning resulted from pollution by an acetaldehyde factory, shut down in 1965, that was located in Niigata, Japan. Between 1965 and 1970, 47 fishing families contracted the disease.

During the 1960s and 1970s, much attention was given to the problem of mercury contamination in seafood. Studies done on the amount of seafood consumed by various populations of humans have led to the establishment of safe levels of mercury content in fish to be marketed. To establish these levels for a given population, three variables must be considered:

Fish consumption rate of the human population under consideration;

Mercury concentration of the fish being consumed by that population; and

Minimum ingestion rate of mercury that induces symptoms of disease.

If dependable data on these variables are available, and a safety factor of 10 is applied to the determination, a maximum concentration level allowable for the contaminant (in this case, mercury) can be established with a high degree of confidence that it will protect the health of the human population.

Figure 15.11 shows how such data can be used to arrive at a safe level of mercury concentration in fish to be consumed by a population. For the general populations of Japan, Sweden, and the United States, the average individual fish consumption rates are 84, 56, and 17 grams per day, respectively. By comparison, the members of the Minamata fishing community averaged 286 and 410 g/day during winter and summer seasons, respectively. Based on the minimum level of mercury consumption that is known to bring on symptoms of poisoning as determined by Swedish scientists using Japanese data (0.3 mg/day over a 200-day period), the three populations shown in Figure 15.11 would begin to show symptoms if they ate fish with the following mercury concentrations:

Japan	4 ppm
Sweden	6 ppm
United States	20 ppm

If a safety factor of 10 is applied, the maximum concentration of mercury concentration in fish that could be safely consumed by these populations would be:

Japan	0.4 ppm
Sweden	0.6 ppm
United States	2.0 ppm

Although the U.S. Food and Drug Administration (FDA) initially established an extremely cautious limit of 0.5 mg/kg (ppm), the present limit of 1 mg/kg ade-

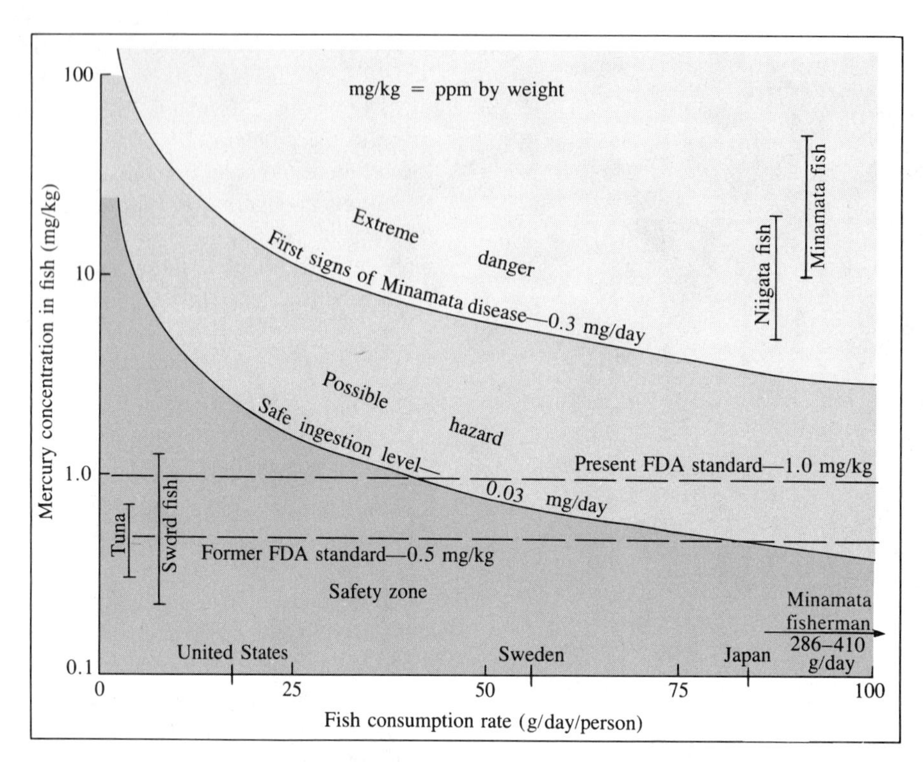

FIGURE 15.11
Fish Consumption Rates and Mercury Safety Levels

Plotted curves show the safe ingestion level of mercury until the first signs of Minamata disease can be expected (0.3 mg/d), and the safe ingestion level of mercury after the safety factor of 10 is included (0.03 mg/d). For the United States, it appears that tuna and swordfish are safe to consume, although some swordfish will be banned because their mercury concentrations are above the present 1.0 mg/kg (ppm) FDA limit.

quately protects the health of U.S. citizens. Essentially all tuna falls below this concentration, and most swordfish will be acceptable. The present limit amounts to the use of a safety factor of 20 instead of 10, so unless one eats an unusually large amount of tuna or swordfish, we should have little concern about consuming these fishes.

INTERNATIONAL EFFORTS TO PROTECT THE MARINE ENVIRONMENT

Beginning with the Action Plan for the Human Environment established by the United Nations in 1972, a comprehensive plan for international environmental protection has evolved. It involves procedures for assessment of problem areas and monitoring activities that are potential sources of pollution.

One of the most critically polluted areas identified was the Mediterranean Sea, polluted in many coastal regions from essentially every conceivable source—domestic sewage, industrial discharge, pesticides, and petroleum. To aid its recovery, an action plan was initiated in 1975, completed in 1976, and has been ratified by 13 of the 18 nations with Mediterranean coasts.

Other regions identified as environmental units with action plans in effect are the Red Sea, Gulf of Aden, East Asian Seas, East Africa, and Kuwait (Fig. 15.12). Areas with action plans in preparation are the Caribbean, South Asian Seas, West and Central Africa, Southeast Pacific, and Southwest Pacific.

If such efforts proceed effectively, they can reverse the trend toward progressive degradation of the environment that accelerated throughout the 1960s. Recent efforts appear to have slowed this process and may well have reversed it in local coastal areas. The prognosis for the future health of the marine environment is much improved from a decade ago. This change serves as testimony to the human will to preserve the quality of the earth's environment. There is still much to do, and a continuing effort shall be required if we are to succeed.

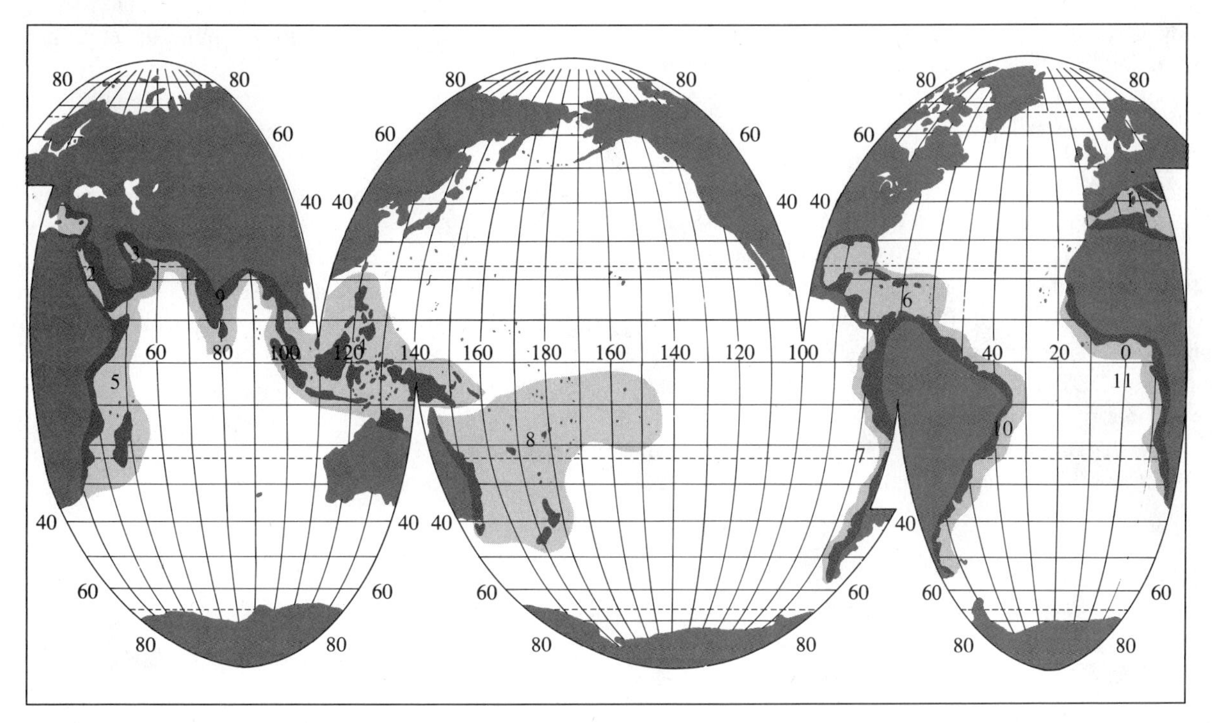

FIGURE 15.12
Areas Covered by UNEP's Regional Seas Program

According to the United Nations Environment Programme, the definition of boundaries is the responsibility of the governments concerned; those shown here are only illustrative. (1) Mediterranean, (2) Red Sea and Gulf of Aden, (3) Kuwait region, (4) East Asian Seas, (5) East Africa, (6) Caribbean Sea, (7) Southeast Pacific, (8) Southwest Pacific, (9) South Asian Seas, (10) Southwest Atlantic, and (11) West and Central Africa.

SUMMARY

Pollution is the introduction of substances or energy into the marine environment that results in harm to the living resources of the oceans, or to humans that use these resources. Scientists are trying to determine the ocean's assimilative capacity, the amount of a substance it can absorb without becoming polluted.

The U.S. government has been concerned about coastal pollution since it passed the Rivers and Harbors Act in 1899. Later laws have been passed to prohibit the dumping of materials known to be harmful, and specify criteria under which other materials may be dumped.

The amount of plastic accumulating in the oceans has increased dramatically. Certain forms of plastic are known to be lethal to marine mammals and turtles, and national and international legislation is being considered to ban the disposal of plastic in the oceans.

Study of areas such as Wild Harbor, Massachusetts and Chedabucto Bay, Nova Scotia have shown that oil pollution reduces species diversity and persists longer on muddy bottoms than on sandy or rocky bottoms. Oil spills that do not come ashore, like that of the *Argo Merchant,* do considerably less environmental damage than those that do. The greatest damage resulting from the *Argo Merchant* spill was probably to plankton, especially fish eggs.

Although 1972 legislation required an end to dumping of sewage in the coastal ocean by 1981, exceptions continued to be made. Especially off the southern California coast and in New York Bight, large amounts of sewage and sewage sludge continued to be dumped. Increased public concern resulted in new legislation that sets January 1992 as the deadline for stopping sewage dumping in the ocean. Some scientists expected that the transfer of dump sites along the northeast Atlantic coast to deeper water farther offshore would make it safe to continue to dump sewage from Philadelphia and the New York area into the coastal ocean. However, fishers have complained that the deep water dumping has damaged fisheries less than one year after it began.

Most of the radioactive waste in the ocean comes from fallout resulting from the testing of nuclear explosives since 1944. However, the areas where the greatest concentration of radionuclides is entering the ocean are near the fuel reprocessing plants in Great Britain and France.

DDT pollution produced a decline in the Long Island osprey population in the 1950s, and the brown pelican population of the California coast in the 1960s. Virtual cessation of DDT use in the Northern Hemisphere in 1972 allowed the recovery of both populations. The DDT had thinned the egg shells and reduced the number of successful hatchings.

The first major human disaster resulting from ocean pollution reached epidemic proportions in 1956 within Minamata Bay, Japan. More than 50 people died, and another 100 were affected by mercury poisoning. A similar episode occurred during the late 1960s at Niigata, Japan. Mercury contamination levels of mercury have been set based on the fish consumption of populations throughout the world, and they appear to have been effective in preventing further poisonings.

With its Action Plan for the Human Environment initiated in 1972, the United Nations Environment Program has established eleven environmental units within which assessment and monitoring have begun or are planned. They are the Mediterranean Sea, Red Sea, Kuwait, Caribbean, West and Central Africa, East Asian Seas, Southeast Pacific, Southwest Pacific, East Africa, South Asian Seas, and Southwest Atlantic.

QUESTIONS AND EXERCISES

1. Discuss why it is difficult to determine the assimilative capacity of the oceans for accepting our waste materials.
2. List the acts of the U.S. Congress that have protected water quality, and describe restrictions contained in each.
3. Do you think it is a better policy (a) not to allow dumping into the ocean any substance that can't be proven to be harmless to the ocean, or (b) to consider the effects of disposing of the substance on land and in the ocean before deciding where to dispose of it? Discuss the reasons for your choice.
4. Discuss the means by which plastics are thought to enter the ocean, their known negative effects on organisms, and legislation being considered to outlaw dumping of plastics in the ocean.
5. Describe the effect of oil spills on species diversity and recovery in the benthos of Wild Harbor and Chedabucto Bay.
6. Discuss whether oil spills that wash ashore or those that do not—like that of the *Argo Merchant*—are more destructive to marine life.
7. How might dumping sewage in deeper water off the eastern U.S. coast help reduce the degradation of the ocean bottom?
8. What components of the nuclear energy generation industry introduce the most radioactive wastes into the oceans?
9. Which group of animals, from what we have been able to determine, suffered the greatest threat from DDT?
10. Look again at Figure 15.11. Discuss the desirability of setting a mercury contamination level for fish at the 1.0 mg/kg level for the citizens of the United States, Sweden, and Japan.

REFERENCES

Bascom, W. 1978. Quantifying man's influence on coastal waters. *Journal of Ocean Engineering* 3:4.

Borgese, E.M., and **Ginsburg, N.**, eds. 1988 *Ocean yearbook 7.* Chicago:University of Chicago Press.

Clark, W.C. 1989. Managing planet earth. *Scientific American* 261:3,47–54.

Comptroller General of the United States. 1977. Problems and progress in regulating ocean dumping of sewage sludge and industrial wastes. Washington, DC:Government Printing Office.

International Atomic Energy Agency. 1980. International symposium on the impacts of radionuclide releases into the marine environment. Paris:International Atomic Energy Agency Press.

Krendle, P.A., ed. 1975. Heavy metals in the aquatic environment. New York:Pergamon Press.

Lawrence, W.W. 1979. Of acceptable risk: Science and the determination of safety. Los Altos, CA: William Kaufman, Inc.

Mearns, A.J., and **Word, J.Q.** 1981. Forecasting effects of sewage solids on marine benthic communities. In: *Ecological effects of environmental stress,* J. O'Connor and G. Mayer, eds. Estuarine Research Foundation, special publication. Lexington, MA:Estuarine Research Foundation.

National Academy of Sciences, National Research Council. 1979. Report on a multimedium management of municipal sludge, vol. 9. Washington, DC: National Academy of Sciences.

National Advisory Committee on Oceans and Atmosphere. 1980. The role of the ocean in a waste management strategy: A special report to the President and the Congress. Washington, DC: National Academy of Sciences.

National Oceanic and Atmospheric Administration. 1979. Assimilative capacity of U.S. coastal waters for pollutants. Proceedings of Workshop at Crystal Mountain, Washington, July 29–August 4. Oil spills and spills of hazardous substances. 1975. Washington, DC:U.S. Environmental Protection Agency.

Orlando, S.P., Shirzad, F., Schuerholz, J.M., Mathieux, D.P., and **Strassner, S.S.** 1988. Shoreline Modification, Dredged Channels, and Dredged Material Disposal Areas in the Nation's Estuaries. Rockville, MD: National Oceanic and Atmospheric Administration.

Wilbur, R.J. 1987. Plastic in the North Atlantic. *Oceanus* 30:3, 61–68.

SUGGESTED READING

SCIENTIFIC AMERICAN

Bascom, W. 1974. Disposal of waste in the ocean. 231:2,16–25.

Butler, J.N. 1975. Pelagic tar. 232:6,90–97.

Houghton, R.A., and **Woodwell, G.M.** 1989. Global climatic change. 260:4,36–44.

SEA FRONTIERS

Driessen, P.K. 1987. Oil rigs and sea life: A shotgun marriage that works. 33:5,362–372.

Gruber, M. 1971. The great ocean sweepstakes. 17:3,146–159.

Hersch, S.L. 1988. Death of the dolphins: Investigating the east coast die-off. 34:4,200–207.

Idyll, C.P. 1971. Mercury and fish. 17:4,230–240.

Nowadnick, J. 1977. There is but one ocean. 23:3,130–140.

Parry, J. 1983. Nations unite to fight pollution. 29:3,143–150.

Reynolds, J.E., and **Wilcox, J.R.** 1987. People, power plants and manatees. 33(4):263–269.

White, I.C. 1985. An organization to help combat oil spills. 31:1,15–21.

MODIFICATION OF UNITED STATES ESTUARIES

Over 92 distinct estuaries have been described by the National Oceanic and Atmospheric Administration (NOAA) for the shorelines of the 48 contiguous United States. The value of these estuaries to fish and wildlife is threatened by modification of both shoreline and bottom features. Shoreline modification often affects intertidal areas of the estuary. These intertidal areas include wetlands and tidal flats and are integral in providing protection from storms, trapping pollutants, and acting as nursery grounds for commercially important fisheries. Despite the ecological importance of estuarine shorelines, they frequently are modified for industrial and/or residential development. Modification includes dredging and filling, or implanting bulkheads, groins, and jetties. Generally, these modifications permanently destroy the ecological importance of the shorelines.

The Strategic Assessment Branch of NOAA recently has determined the extent of the modification of estuarine shorelines of the continental United States. The results of this study (Orlando et al., 1988) are summarized in Figure 1. Of the 68,150 km of estuarine shoreline, 4,876 km, or more than 7 percent, have been modified. Modification is greatest in those estuaries associated with major urban areas. Table 1 lists the estuaries with the greatest degree of shoreline modification. Along the Gulf Coast, however, shoreline modification is not from urban development, but instead results from extensive residential development of the Florida coast.

Damaging modification of estuaries is not restricted to shoreline change. The dredging of channels and the disposal of dredge material also degrades estuaries by disturbing submerged aquatic vegetation and benthic communities; increasing turbidity and decreasing light penetration; and resuspending sediment-bound toxins.

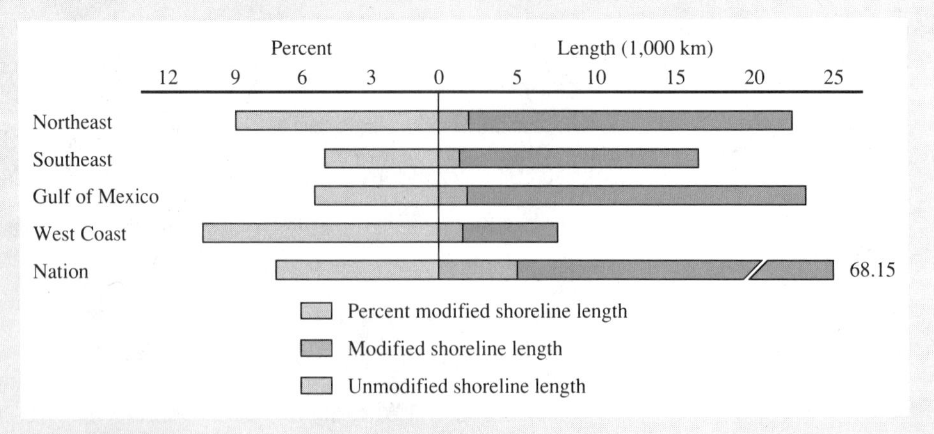

FIGURE 1
Modification of Estuarine Shoreline by Region

TABLE 1
Top-Ten Rankings for Selected Shoreline and Modification Features

(From Orlando et al., 1988.)

	Modified Shoreline Length			Dredged Material Disposal Area	
	Estuary	km		Estuary	km
1	Chesapeake Bay	480	1	Mississippi Delta	190
2	Hudson/Raritan Bay	340	2	Mobile Bay	120
3	Tampa Bay	300	3	Long Island Sound	80
4	Indian River	230	4	Mississippi Sound	70
5	Galveston Bay	190	5	Laguna Madre	70
6	San Francisco Bay	180	6	Chesapeake Bay	70
7	Long Island Sound	180	7	Galveston Bay	50
8	St. Johns River	160	8	Tampa Bay	40
9	San Pedro Bay	160	9	Matagorda Bay	40
10	Great South Bay	150	10	Sabine Lake	20

The modification of estuarine bottom surface area by dredging is shown in Figure 2. Of the 66,720 sq km of estuarine bottom, 1,387 sq km (2.1%) have been modified by dreding or disposal. Modification of the bottom by dredging is most pronounced on the West Coast of the United States, whereas the impact of dredge disposal is greatest in the Gulf of Mexico.

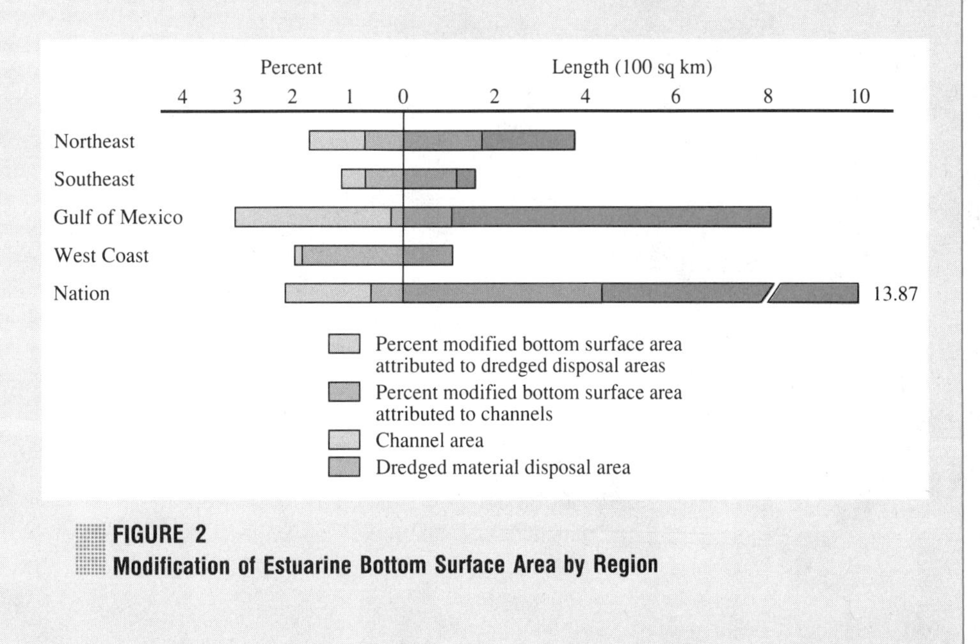

FIGURE 2
Modification of Estuarine Bottom Surface Area by Region

SUBSEA DISPOSAL OF HIGH-LEVEL NUCLEAR WASTE

For four decades, high-level nuclear waste has been accumulating from the production of nuclear weapons and commercial power generation. The United States alone has more than 284 million liters of waste from weapons production, and 12,000 metric tonnes of spent reactor fuel that must be safely disposed of until it is no longer dangerous.

This material will be radioactive for more than 1 million years, but a safe disposal system probably would require a shorter time of high secure confinement. The wastes are composed of more than 50 isotopes, each of which has different chemical, half-life, abundance, and radioactive emission characteristics. Considering these factors, investigators believe confinement that allows release of radiation into the atmosphere no greater than that of the natural uranium ore from which it was generated may be considered safe, and is possible with today's technology (Fig. 1).

The major emphasis at present is on disposal at land sites, because the materials remain accessible. However, since 1974, research has been conducted into the possibility of disposal of high-level nuclear waste in the deep sea. This research was terminated by the United States in 1986 due to budget constraints, and on an international basis in 1987 when international agreement was reached to pursue land-based disposal. However, the sea-bed disposal research did produce encouraging results, and given the political realities that may be faced as work

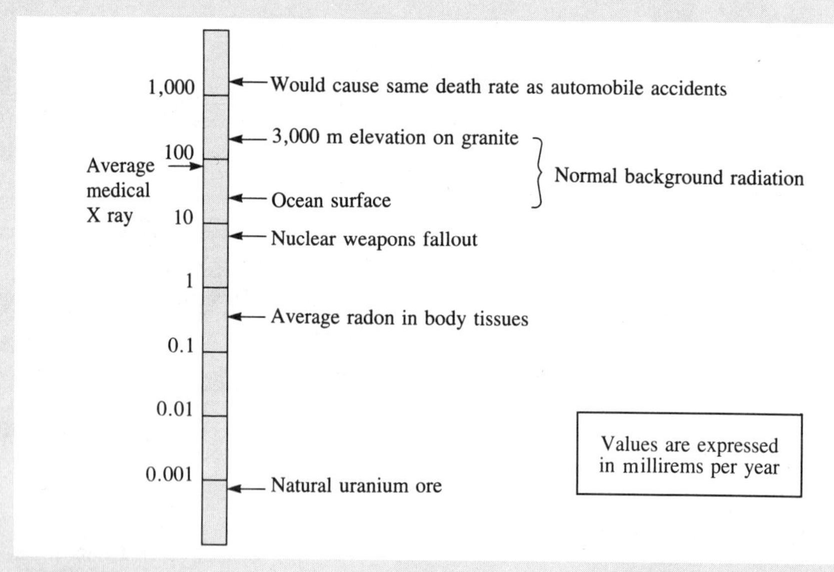

FIGURE 1
Radiation Doses from Various Sources

proceeds toward land-based disposal, it may be considered again in the future. The major features of the program are described below.

Given the requirements that the waste must be kept out of the way of human activities, safe from exposure by natural erosion, and away from seismically active regions, the centers of oceanic lithospheric plates might be ideal disposal sites. Such regions are beneath at least 5,000 m of water—far from the marine activities of humans. Fishing, petroleum production, and mining activities now are conducted primarily on the continental shelves. Only a potential area of manganese nodule mining between Mexico and the Hawaiian Islands would need to be avoided.

The lithospheric plates are covered with up to 1,000 m of fine sediment that has accumulated uninterrupted for over 100 million years in some regions. This pattern of accumulation indicates long periods of stability as the plates move across the earth's surface, and the pattern can be expected to continue for millions of years. The sediment could serve as an ideal medium in which to place the high-level nuclear waste. Many of the radionuclides, after being released from the canisters (they may fail within 1,000 years), would naturally adhere to the clay particles in the sediment. This would slow diffusion of radioactivity away from the burial site.

A number of midplate, midgyre sites (MPGs) have been identified in the North Atlantic and North Pacific oceans (Fig. 2). The initial plan is to place the canisters

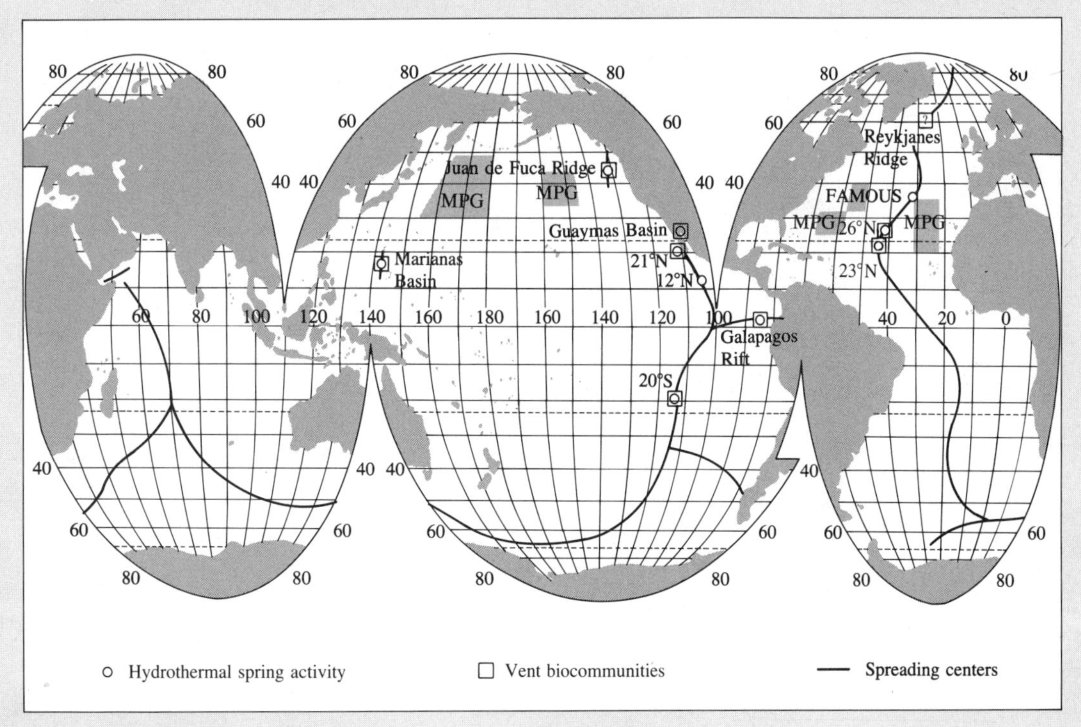

FIGURE 2

Midplate, midgyre regions (MPGs) in northern oceans can be expected to be stable for millions of years in the future.

under 30–100 m of sediment at intervals of at least 100 m. How to emplace the canisters has not been decided yet, but it may be possible to simply drop them from an appropriate distance above the ocean floor (Fig. 3).

Models indicate that heat release from the waste will be transferred almost entirely by conduction, so upward convection of radiation carried by sediment pore water will be negligible. This is a crucial consideration. Although these results are promising, yet to be completed are models of how well the burial holes will reseal themselves, and the means by which radiation will be transported in the water column once it reaches the sediment surface.

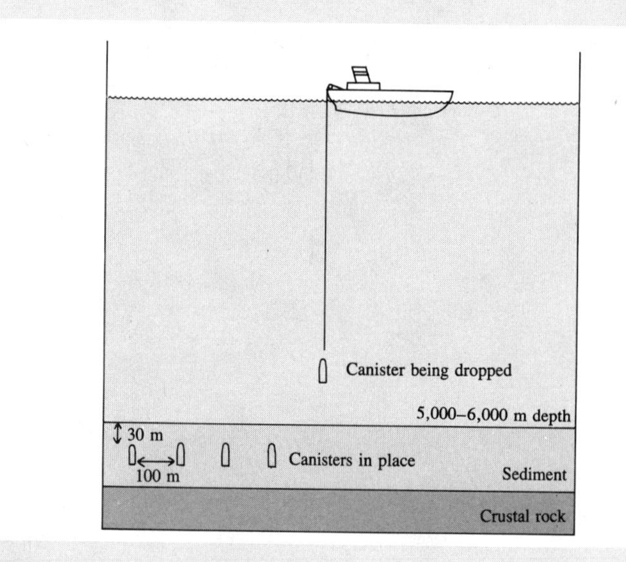

FIGURE 3
Hypothetical Emplacement of Waste Canisters

A
THE METRIC SYSTEM AND CONVERSION FACTORS

Units of Length	1 micrometer (μm) = 10^{-6} m = 0.000394 in 1 millimeter (mm) = 10^{-3} m = 0.0394 in = 10^3 μm 1 centimeter (cm) = 10^{-2} m = 0.394 in = 10^4 μm 1 meter (m) = 10^2 cm = 39.4 in = 3.28 ft = 1.09 yd = 0.547 fath 1 kilometer (km) = 10^3 m = 0.621 statute mi = 0.540 nautical mi
Units of Volume	1 liter = 10^3 cm^3 = 1.0567 liquid qt = 0.264 U.S. gal 1 cubic meter (m^3) = 10^6 cm^3 = 10^3 l = 35.3 ft^3 = 264 U.S. gal 1 cubic kilometer (km^3) = 10^9 m^3 = 10^{15} cm^3 = 0.24 statute mi^3
Units of Area	1 square centimeter (cm^2) = 0.155 in^2 1 square meter (m^2) = 10.7 ft^2 1 square kilometer (km^2) = 0.292 nautical mi^2 = 0.386 statute mi^2 1 hectare (ha) = 10^3 m^2 = 2.47 acres
Units of Time	1 day = 8.64×10^4 s (mean solar day) 1 year = 8,765.8 h = 3.156×10^7 s
Units of Mass	1 gram (g) = 0.035 oz 1 kilogram (kg) = 10^3 g = 2.205 lb 1 metric tonne = 10^6 g = 1,000 kg = 2,205 lb = 1.1 ton
Units of Speed	1 centimeter per second (cm/s) = 0.0328 ft/s 1 meter per second (m/s) = 2.24 statute mi/h = 1.94 kt 1 kilometer per hour (km/h) = 27.8 cm/s = 0.55 kt 1 knot (1 nautical mile per hour, kt) = 1.15 statute mi/h = 0.51 m/s

Units of Temperature

Celsius (°C)	Fahrenheit (°F)	Kelvin (°K)	
−273.2	−459.7	0	Absolute zero (lowest possible temperature)
0	32	273.2	Freezing point of water
100	212	373.2	Boiling point of water

Conversions: °F = (1.8 × °C) + 32 °C = (°F−32)/1.8

B
PREFIXES AND SUFFIXES

A

a- not, without
ab-, abs- off, away, from
abysso- deep
acanth- spine
acro- the top
aigial- beach, shore
aero- air, atmosphere
albi- white
alga- seaweed
alti- high, tall
alve- cavity, pit
amoebe- change
annel- ring
annu- year
anomal- irregular, uneven
antho- flower
apex- tip
aplysio- sponge, filthiness
aqua- water
arachno- spider
arena- sand
arthro- joint
astheno- weak, feeble
astro- star
auto- self
avi- bird

B

bacterio- bacteria
balaeno- whale
balano- acorn
barnaco- goose
batho- deep
bentho- the deep sea
bio- living
botryo- bunch of grapes
brachio- arm
branchio- gill
broncho- windpipe
bryo- moss
bysso- a fine thread

C

calci- limestone ($CaCO_3$)
calori- heat
capill- hair
cara- head
carno- flesh
cartilagi- gristle
caryo- nucleus
cen-, ceno- recent
cephalo- head
chaeto- bristle
chiton- tunic
chlor- green
choano- funnel, collar
chondri- cartilage
cilio- small hair
cirri- hair
clino- slope
cnido- nettle
cocco- berry
coelo- hollow
cope- oar
crusta- rind
cteno- comb
cyano- dark blue
cypri- Venus, lovely
-cyte cell
-cyst bladder, bag

D

deca- ten
delphi- dolphin
di- two, double
dino- whirling
diplo- double, two
dolio- barrel
dors- back

E

echino- spiny
eco- house, abode
ecto- outside, outer
edrio- a seat
en- in, into
endo- inner, within
entero- gut
epi- upon, above
estuar- the sea
eu- good, well
exo- out, without

F

fec- dregs
fecund- fruitful
flagelli- whip
flacci- flake
fluvi- river

fossili- dug up
fuco- red

G

geno- birth, race
geo- the earth
giga- very large
globo- ball, globe
gnatho- jaw
guano- dung
gymno- naked, bare

H

halo- salt
haplo- single
helio- the sun
helminth- worm
hemi- half
herbi- plant
herpeto- creeping
hetero- different
hexa- six
holo- whole
homo- alike
hydro- water
hygro- wet
hyper- over, above, excess
hypo- under, beneath

I

ichthyo- fish
-idae- members of the animal family of
infra- below, beneath
insecti- cut into
insula- an island
inter- between
involute- intricate
iso- equal
-ite rock

J

juven- young
juxta- near to

K

kera- horn
kilo- one thousand

L

lacto- milk
lamina- layer
larvi- ghost
lati- broad, wide
latent hidden
lateral at a side
lemni- water plant
limno- marshy lake
lipo- fat
litho- stone
-littorial near the seashore
lopho- tuft
lorica- armor
luci- light
luna moon
-lux light

M

macro- large
mala- jaw, cheek
mamilla- teat
mandibulo- jaw
mantle cloak
mari- the sea
masti- chewing
masto- breast, nipple
mastigo- whip
maxillo- jaw
medi- middle
mega- great, large
meio- less
meridio- noon
meso- middle
meta- after
meteor- in the sky
-meter measure
-metry science of measuring
mid- middle
milli- thousandth
mio- less
moll- soft
mono- one, single
-morph form

N

nano- dwarf
necto- swimming
nemato- thread
neo- new
nerito- sea nymph
noct- night
-nomy the science of
nucleo- nucleus
nutri- nourishing

O

o-, oo- egg
ob- reversed
octa- eight
oculo- eye
odonto- teeth
oiko- house, dwelling
oligo- few, scant
-ology science of
omni- all
ophi- a serpent
opto- the eye, vision
orni- a bird
-osis condition
ovo- egg

P

pan- all
para- beside, near
pari- equal
pecti- comb
pedi- foot
penta- five
peri- all around
phaeo- dusky
pholado- lurking in a hole
-phore carrier
photo- light
phyto- plant
pinnati- feather
pisci- fish
plani- flat, level
plankto- wandering
pleisto- most
pleuro- side
plio- more
pluri- several
pneuma- air, breath
pod- foot
poikilo- varigated
poly- many
poro- channel

post- behind, after
-pous foot
pre- before
pro- before, forward
procto- anus
proto- first
pseudo- false
ptero- wing
pulmo- lung
pycno- dense

Q

quadra- four
quasi- almost

R

radi- radial
rhizo- root
rhodo- rose-colored

S

sali- salt
schizo- split, division
scyphi- cup
semi- half
septi- partition
siphono- tube
-sis process
spiro- spiral, coil
stoma- mouth
strati- layer
sub- below
supra- above
symbio- living together

T

taxo- arrangement
tecto- covering
terra- earth
terti- third
thalasso- the sea
trocho- wheel
tropho- nourishment
turbi- disturbed

U

un- not

V

vel- veil
ventro- underside

X

xantho- yellow
xipho- sword

Z

zoo- animal

C

TAXONOMIC CLASSIFICATION OF COMMON MARINE ORGANISMS

Kingdom Monera

Organisms without nuclear membranes; nuclear material is spread throughout cell; predominantly unicellular

Saprobic bacteria Break down organic material, both aerobic and anaerobic forms

Autotrophic bacteria Both photosynthetic (cyanobacteria, formerly called blue-green algae) and chemosynthetic species

Parasitic/Pathogenic bacteria

Kingdom Protista

Organisms with nuclear material confined to nucleus by a membrane

Phylum Chrysophyta Golden-brown algae; includes diatoms, coccolithophores, and silicoflagellates; chlorophyll *a* and *c,* xanthophyll, and carotene pigments ($6{,}000^+$ species)

Phylum Pyrrophyta Dinoflagellate algae; chlorophyll *a* and *c,* xanthophyll, and carotene pigments (1,100 species)

Phylum Chlorophyta Green algae, chlorophyll *a* and *b,* and carotene pigments (7,000 species)

Phylum Phaeophyta Brown algae; chlorophyll *a* and *c,* xanthophyll, and carotene pigments (1,500 species)

Phylum Rhodophyta Red algae; chlorophyll *a,* carotene, and phycobilin pigments (4,000 species)

Phylum Protozoa Nonphotosynthetic, heterotrophic protists (27,400 species)

Phylum Sarcodina Ameboid; radiolarians; and foraminiferans (11,500 species)

Phylum Ciliophora Ciliated protozoa (6,000 species)

Kingdom Fungi

Phylum Ascomycota Yeasts, molds

Phylum Basidiomycota Rusts, mushrooms

Kingdom Metaphyta

Multicellular, complex plants

Phylum Tracheophyta Vascular plants with roots, stems, and leaves that are serviced by special cells that carry food and fluids (287,200 species)

Class Angiospermae Flowering plants with seeds contained in a closed vessel (275,000 species)

Kingdom Metazoa

Multicellular animals

Phylum Porifera Sponges; spicules are the only hard parts in these sessile animals that do not possess tissue (10,000 species)

Class Calcarea Calcium carbonate spicules (50 species)

Class Desmospongiae Skeleton may be composed of siliceous spicules or spongin fibers, or be nonexistent (9,500 species)

Class Sclerospongiae Coralline sponges; massive skeleton composed of calcium carbonate, siliceous spicules, and organic fibers (7 species)

Class Hexactinellida Glass sponges; six-rayed with siliceous spicules (450 species)

Phylum Cnidaria Radially symmetrical, two-cell, layered body wall with one opening to gut cavity; polyp (asexual, sexual, benthic) and medusa (sexual, pelagic) body forms (9,000 species)

Class Hydrozoa Polypoid colonies such as pelagic Portuguese man-of-war and benthic Obelia are common; medusa present in reproductive cycle but reduced in size (3,000 species)

Class Scyphozoa Jellyfish; medusa up to 1 m in diameter is dominant form; polyp is small if present (250 species)

Class Anthozoa Corals and anemones possessing only polypoid body form and reproducing asexually and sexually (6,000 species)

Phylum Ctenophora Predominantly planktonic comb jellies; basic eight-sided radial symmetry modified by secondary bilateral symmetry (50 species)

Phylum Platyhelminthes Flatworms; bilateral symmetry; hermaphroditic (13,000 species)

Phylum Nemertea Ribbon worms; as long as 30 m; benthic and pelagic (650 species)

Phylum Nematoda Roundworms; marine forms are primarily free-living and benthic; most 1–3 mm in length (10,000 marine species)

Phylum Rotifera Ciliated, unsegmented forms less than 2 mm in length (400 species, only a few marine)

Phylum Bryozoa Moss animals; benthic, branching, or encrusting colonies; lophophore feeding structure (4,000 species)

Phylum Branchiopoda Lamp shells; lophophorate benthic bivalves (280 species)

Phylum Phoronida Horseshoe worms; 24 cm-long lophorate tubeworms that live in sediment of shallow and temperate shallow waters (10 species)

Phylum Sipuncula Peanut worms, benthic (300 species)

Phylum Echiura Spoon worms; sausage shaped with spoon-shaped proboscis; burrow in sediment or live under rocks (100 species)

Phylum Pogonophora Tube-dwelling, gutless worms 5–80 cm long; absorb organic matter through body wall (100 species)

Phylum Tardigrada Marine meiofauna that have the ability to survive long periods in a cryptobiotic state (400 species)

Phylum Mollusca Soft bodies possessing a muscular foot and mantle that usually secretes calcium carbonate shell (110,000 species)

Class Monoplacophora Rare trench-dwelling forms with segmented bodies and limpetlike shells (10 species)

Class Polyplacophora Chintons; oval, flattened body covered by eight overlapping plates (600 species)

Class Gastropoda Large, diverse group of snails and their relatives; shell spiral if present (64,500 species)

Class Bivalvia Bivalves; includes mostly filter-feeding clams, mussels, oysters, and scallops (7,500 species)

Class Aplacophora Tusk shells; sand-burrowing organisms that feed on small animals living in sand deposits (350 species)

Class Cephalopoda Octopus, squid, and cuttlefish that possess no external shell except in the genus *Nautilus* (600 species)

Phylum Annelida Segmented worms in which musculature, circulatory, nervous, excretory, and reproductive systems may be repeated in many segments; mostly benthic (8,700 marine species)

Phylum Arthropoda Joint-legged animals with segmented body covered by an exoskeleton (925,000 species)

Subphylum Crustacea Calcareous exoskeleton, two pairs of antennae; cephlon, thorax, and abdomen body parts; includes copepods, ostracods, barnacles, shrimp, lobsters, and crabs (31,300 species)

Subphylum Chelicerata Horseshoe crabs (4 species)

Class Merostomata

Class Pycnogonida Sea spiders

Subphylum Uniramia Insects; genus *Halobites* is the only truly marine insect

Phylum Chaetognatha Arrow worms; mostly planktonic, transparent, and slender; up to 10 cm long (50 species)

Phylum Hemichordata Acorn worms and pterobranchs; primitive nerve chord; gill slits; benthic (90 species)

Phylum Echinodermata Spiny-skinned animals; benthic animals with secondary radial symmetry and water vascular system (6,000 species)

Class Asteroidea Starfishes; free-living, flattened body with five or more rays with tube feet used for locomotion; mouth down (1,600 species)

Class Ophiuroidea Brittle stars and basket stars; prominent central disc with slender rays; tube feet used for feeding; mouth down (200 species)

Class Echinoidea Sea urchins, sand dollars, heart urchins; free-living forms without rays; calcium carbonate test; mouth down or forward (860 species)

Class Holothuroidea Sea cucumbers; soft bodies with radial symmetry obscured; mouth forward (900 species)

Class Crinoidea Sea lilies, feather stars; cup-shaped body attached to bottom by a jointed stalk or appendages; mouth up (630 species)

Phylum Chordata Notochord; dorsal nerve chord and gills or gill slits (39,000 species)

Subphylum Urochordata Tunicates; chordate characteristics in larval stage only; benthic sea squirts and planktonic thaliaceans and larvaceans (1,375 species)

Subphylum Cephalochordata Amphioxus or lancelets; live in coarse temperate and tropical sediment (25 species)

Subphylum Vertebrata Internal skeleton; spinal column of vertebrae; brain (37,700 species)

Class Agnatha Lampreys and hagfishes; most primitive vertebrates with cartilaginous skeleton, no jaws and no scales (50 species)

Class Chondrichthyes Sharks, skates, and rays; cartilaginous skeleton; 5–7 gill openings; placoid scales (625 species)

Class Osteichthyes Bony fishes; cycloid scales; covered gill opening; swim bladder common (20,000 species)

Class Amphibia Frogs, toads, and salamanders; Asian mud flat frogs are the only amphibians that tolerate marine water (2,600 species)

Class Reptilia Snakes, turtles, lizards, and alligators; orders Squamata (snakes) and Chelonia (turtles) are major marine groups (6,500 species)

Class Aves Birds; many live on and in the ocean but all must return to land to breed (8,600 species)

Class Mammalia Warm-blooded; hair; mammary glands; bear live young; marine representatives found in the orders Sirenia (sea cows, dugong), Cetacea (whales), and Carnivora (sea otter, pinnipeds) (4,100 species)

GLOSSARY

Abiotic environment The nonliving components of an ecosystem: soil, air, water.

Abyssal *See* Abyssal zone.

Abyssal clay Deep-ocean (oceanic) deposits containing less than 30 percent biogenous sediment.

Abyssal plain A flat depositional surface extending seaward from the continental rise or oceanic trenches.

Abyssal zone The benthic environment between 4,000–6,000 m (13,120–19,680 ft).

Abyssopelagic zone Open-ocean (oceanic) environment below 4,000 m depth.

Acrorhagus The swollen tissue of aggressive sea anemones that contains large concentrations of stinging nematocysts.

Adaptation A genetically based characteristic that allows an organism to survive and flourish in a particular environment.

Aerobic Metabolic respiration that requires oxygen (O_2); most organisms carry on respiration this way.

Algal film A growth of microscopic algae and diatoms on bare substrates (mostly on rocks in the intertidal zone).

Algal ridge The seaward edge of a coral reef, it consists mostly of algal species, and absorbs most of the wave energy that hits a reef.

Ambulacral The water vascular system of the phylum Echinodermata; a fluid system with tube feet that uses hydraulic energy to create movement.

Ampullae of Lorenzini A system of pores on the snout of sharks that can detect small electrical charges, used for locating prey at short distances.

Anadromous The migration pattern of fish where reproduction is in fresh water, yet the majority of life is spent in salt water (e.g., salmon live this way).

Anaerobic Respiration carried on in the absence of free oxygen (O_2); some bacteria and protozoans carry on respiration this way.

Antarctic Bottom Water (AABW) Antarctic surface water that cools and sinks near the Weddell Sea; it spreads across the deep ocean floor because it is the densest water in the ocean.

Aphotic zone Basically, 'without light'; the ocean is generally in this state below 1,000 m (3,280 ft).

Assimilative capacity The level of concentration to which the ocean can accumulate foreign substances without harming marine life, or humans who use the ocean or its resources.

Asthenosphere A plastic layer in the upper mantle between the depths 80–200 km that may allow lateral movement of lithospheric plates and isostatic adjustments.

Atrich nematocyst Type of stinging nematocyst used by aggressive sea anemones to attack other sea anemones during territorial conflicts.

Autolytic decomposition Decomposition of organic matter achieved by enzymes present in the tissue; the enzymes are triggered to begin their work upon tissue death.

Autotroph Plants and bacteria that can synthesize organic compounds from inorganic nutrients.

Autotrophic chemosynthetic bacteria Bacteria that synthesize organic molecules from inorganic nutrients using chemical energy released from the bonds of some chemical compounds by oxidation.

Autumn bloom An increase in phytoplankton growth that occurs in temperate regions due to decreased thermal stratification and increased mixing of nutrients in deep water into the sunlit surface waters.

Backdown A system of using purse seines for tuna capture where the net is 'backed down' (allowed to sink) so that dolphins can escape.

Bacteriovore Any organism that consumes bacteria.

Bathyal *See* Bathyal zone.

Bathyal zone The benthic environment between the depths of 200–4,000 m (656–13,120 ft). It includes mainly the continental slope and the oceanic ridges and rises.

Bathypelagic zone The pelagic environment between the depths of 1,000–4,000 m (3,280–13,120 ft).

Benthic environment The ocean floor.

Benthos The forms of marine life that live on the ocean bottom.

Binomial nomenclature System of naming used by biologists to identify organisms. Each name consists of two parts—a genus and a species designation. For example, the scientific name of the bottle-nosed dolphin is *Tursiops truncatus*.

Biogenous Ocean floor sediment derived from biological organisms.

Biogeochemical cycles The natural cycling of compounds among the living and nonliving components of an ecosystem.

Biomass The total mass of a defined organism or group of organisms in a particular community or the ocean as a whole.

Biomass pyramid The trophic structure of a biological community, where each level is represented by the total standing stock biomass (usually represented as a chart).

Biotic community The living organisms that inhabit an ecosystem.

Boundary current The northward- or southward-flowing currents that form the western and eastern boundaries, respectively, of the subtropical circulation gyres.

Buttress zone The part of a coral reef that first meets the breaking waves of the surf.

Calcium carbonate $CaCO_3$; a chalklike substance secreted by many organisms in the form of shell or skeletal structures.

Calcium carbonate compensation depth The depth of the ocean where the mineral calcium carbonate spontaneously dissolves because of high concentrations of carbon dioxide in the water. This depth ranges from 3,500 to over 6,000 m.

Capillary wave Ocean wave whose wavelength is less than 1.74 cm; the dominant restoring force for such waves is surface tension.

Carapace The dorsal (back) shell of a turtle.

Carnivore An animal with a diet consisting primarily of other animals.

Catadromous The migration pattern of fish where reproduction is in the ocean, yet the majority of adult life is spent in fresh water (e.g., the Atlantic eel lives this way).

Caudal fin Fin located at the tail end.

Caudal peduncle The part of the fish tail just in front of the caudal fin.

Centric diatom Member of the class Bacillariophyceae of algae that possesses a wall of silica valves.

Cephalization The tendency in animal evolution to form a head end.

Chemosynthesis Metabolism that synthesizes organic molecules from inorganic nutrients using chemical energy released from the bonds of some chemical compound by oxidation.

Chlorinity The amount of chloride ion and ions of other halogens in ocean water, expressed in parts per thousand by weight (‰).

Choanocyte The cell type of sponges that has a collar and a flagellum. It usually is used to create a current.

Circumpolar current Eastward-flowing current that extends from the surface to the ocean floor and encircles Antarctica.

Cnidocytes Stinging cell of phylum Cnidaria that contains the stinging mechanism (i.e., the nematocyst) used in defense and capturing prey.

Coastal upwelling The movement of deeper nutrient-rich water into the surface water mass as a result of wind-blown surface water moving offshore.

Commensalism A symbiotic relationship between organisms in which both benefit.

Competition The demand of two different species for the same resource.

Consumers Those organisms that are heterotrophs (i.e., they cannot produce their own food through photosynthesis).

Continental rise A gently sloping depositional surface at the base of the continental slope.

Continental shelf A gently sloping depositional surface extending from the low-water line to the depth of a marked increase in slope around the margin of a continent or island.

Continental slope A relatively steeply sloping surface lying seaward of the continental shelf.

Coriolis effect A force resulting from the earth's rotation which causes particles in motion to be deflected to the right in the Northern Hemisphere, and to the left in the Southern Hemisphere.

Cosmogenous Term used to describe ocean floor sediments that have their origin outside the earth.

Counterlighting The color pattern of pelagic animals that is light on the bottom and dark on the top, it helps reduce the contrast of the organism with the light in the water.

Curie The unit of measure of radioactivity.

Cyanobacteria The major group of bacteria that is capable of photosynthesis (formerly called blue-green algae).

Decomposers Bacteria and fungi that decompose dead organic material and convert it to inorganic nutrients that can be used by the producers of an ecosystem.

Deep scattering layer (DSL) A layer of marine organisms in the open ocean that scatters signals from an echo sounder. The layer migrates daily from depths of slightly over 100 m at night to more than 800 m during the day.

Deep water The water beneath the permanent thermocline (or pycnocline) that has a uniformly low temperature.

Demersal fish Fish that live close to the bottom (e.g., halibut).

Denitrifying bacteria Bacteria that reduce oxides of nitrogen to produce free nitrogen (N_2).

Density-driven circulation Ocean currents that are caused by differences in the density of two water masses.

Detritus Dead and decaying organic material.

Deuterostome A group of invertebrate phyla that have certain similar features of embryonic development; the blastophore of deuterostomes does not contribute to the formation of the mouth.

Dichlorodiphenyltrichloroethane The pesticide DDT.

Diffusion A process by which fluids move through other fluids from areas of high concentration to areas in which they are in lower concentrations by random molecular movement.

Dilatancy The measure of firmness of sand and mud shores.

Diploid Having a double set of chromosomes; 2n. *See also* Haploid.

Diurnal inequality The difference in height between the two high tides or the two low tides in any given 24-hour and 50-min daily cycle.

Diurnal tide A tide with one high water and one low water during a tidal day.

Drift Algae (or pieces of algae) that have been dislodged and are free-floating or washed up on the high-tide line of the shore.

Drift nets Long monofilament gill nets set at sea to catch fish or squid; these nets have a high incidental catch of birds and mammals (hence their nickname—walls of death).

Drowned river valley *See* Coastal plain estuary.

Dysphotic zone The depth of the water column that has low light intensities permitting some vision but not photosynthesis; generally from 100–1,000 m.

Ecdysis The skeletal molting, or shedding, process of arthropods.

Ecological efficiency Ratio of the amount of energy transferred between trophic levels to the amount of food required to produce the transfer.

Ecology The science pertaining to the study of organisms in relation to one another and the environment.

Ecosystem All the organisms in a biotic community, and the abiotic environment factors with which they interact.

Ekman spiral A theoretical explanation of the effect of a steady wind blowing over an ocean of unlimited depth and breadth and of uniform viscosity. The result is a surface flow at 45° to the right of the wind in the Northern Hemisphere. Water at increasing depth drifts in directions increasingly to the right until, at about 100 m depth, it is moving in a direction opposite that of the wind. The net water transport is 90° to the wind, and velocity decreases with depth.

Ekman transport The net transport of surface water set in motion by wind; due to the Ekman spiral phenomenon, transport is theoretically in a direction 90° to the right and 90° to the left of the wind direction in the Northern Hemisphere and Southern Hemisphere, respectively.

Electromagnetic spectrum The entire range of electric and magnetic radiation. (*See* Infrared radiation, Ultraviolet radiation, and Visible spectrum for further discussion.)

Endemic Refers to an area where a particular species underwent its evolution.

Endogenous community A unique biological community found only in certain places and consisting of species not found in other communities (e.g., hydrothermal vent communities).

Endogenous rhythm A behavior that seems to be initiated in an organism by an internal biological clock.

Endosymbionts A symbiotic relationship where one organism lives inside the cell of the other (e.g., coral cells).

Entropy A thermodynamic quantity that reflects the degree of randomness in a system; it increases in all natural systems.

Epifauna Animals that live on the ocean bottom, either attached or moving freely over it.

Epiflora Photosynthetic organisms attached to a solid substrate (e.g., kelp).

Epipelagic zone A subdivision of the oceanic province that extends from the surface to a depth of 200 m (656 ft).

Equatorial submergence The change in distribution of species with depth over a latitude gradient. In temperate and polar latitudes, a particular species will be found close to the surface. In equatorial latitudes, the same species will be found at much deeper depths where temperatures are similar to those where the species is found at temperate and polar latitudes.

Equatorial upwelling The surfacing of deeper, nutrient-rich water into the equatorial surface mass that results from wind-driven surface water moving to higher latitudes.

Estuary The mouth of a river valley where marine influence is manifested as tidal effects and increased salinity of the river water.

Eulittoral zone The benthic zone between the highest and lowest spring tide on shorelines; the intertidal zone. *See also* Sublittoral fringe, Supralittoral fringe.

Euphotic zone The surface layer of the ocean that receives enough light to support photosynthesis; the bottom of this zone, which marks the compensation depth, varies and reaches a maximum value of around 150 m in the very clearest open ocean water.

Euryhaline Able to tolerate a wide range of salinity.

Eurythermal Able to tolerate a wide range of temperature.

Eutrophic A condition in which waters are nutrient-rich and support a high level of biological productivity.

Evolution The permanent change in the genetic material of a species in response to changes in the environment.

Exclusive economic zone (EEZ) The area adjacent to the shoreline (200 mi in the United States) that is not claimed as sovereign land; all resources in the water and on the bottom are claimed for the exclusive use of the country.

Fallout The deposition on land or water of radioactive particles that are in the atmosphere.

Facultative halophyte A plant that, although not requiring immersion in salt water for growth, can tolerate at least short periods of immersion in salt water.

Fetch (a) Pertaining to the area of the open ocean over which the wind blows with constant speed and direction, thereby creating a wave system. (b) The distance across the fetch (wave-generating area), measured in a direction parallel to the direction of the wind.

Fin rays The skeletal supports of fins in bony fish.

Fjord A long, narrow, deep, U-shaped inlet that usually represents the seaward end of a continental glacial valley that has become partially submerged after the melting of the glacier.

Food chains The passage of energy materials from producers through a sequence of herbivores and a number of carnivores.

Food web A group of interrelated food chains.

Frond The leaflike blade of algae.

Frustule The siliceous covering of a diatom consisting of two halves (epitheca and hypotheca).

Geostrophic current A current that grows out of the earth's rotation and represents a near balance of gravitational force and Coriolis effect.

Gill net A net type used in commercial fishing where the fish's gills are trapped by webbing when the fish tries to swim through.

Gill rakers Skeletal supports for the gills of fish that support and protect the gills, used by some fish to filter zooplankton from the water.

Global plate tectonics The process by which lithospheric plates are moved across the earth's surface to collide, slide by one another, or diverge to produce the topographic configuration of the earth.

Globigerinous ooze A deep ocean sediment that consists mostly of foraminifera shells.

Gran Method A system involving the observation of the change in dissolved oxygen within paired transparent and opaque bottles containing phytoplankton, suspended in the ocean surface water to determine the base of the euphotic zone.

Gravity waves A wave for which the dominant restoring force is gravity; such waves have a wavelength of more than 1.74 cm, and their velocity of propagation is controlled mainly by gravity.

Guano The fecal material of sea birds.

Gular pouch A pouch formed by tissues between the edges of the lower beak of certain sea birds, such as the pelican.

Gulf Stream current The high intensity western boundary current of the North Atlantic Ocean subtropical gyre that flows north off the east coast of the United States.

Gyre A circular spiral form, used mainly in reference to the circular motion of water in each of the major ocean basins centered in subtropical high-pressure regions.

Hadal *See* Hadal zone.

Hadal zone Pertaining to the deepest ocean benthic environment, specifically that of ocean trenches deeper than 6 km (3.7 mi).

Haloclines A layer of water exhibiting a high rate of change in salinity in the vertical dimension.

Haploid Having a single set of chromosomes. 1n. *See also* Diploid.

Herbivore An animal that relies chiefly or solely on plants for its food.

Heterocercal caudal fin The tail fin of sharks and rays; the lower lobe of these tail fins is larger than the upper.

Heterotrophic *See* Heterotrophs.

Heterotrophs Animals and bacteria that depend on the organic compounds produced by other animals and plants as food; organisms not capable of producing their own food by photosynthesis.

Higher phyla Those phyla that have evolved more recently than the older, lower phyla.

Holdfast The rootlike base of marine algae that anchors the organism to the substrate.

Holoplankton Organisms that spend their entire life as members of the plankton.

Homocercal The tail fin of fish where the lower lobe of the tail is the same size as the upper lobe.

Host An organism that has a symbiotic relationship with a symbiont or parasite.

Hydrogenous sediment Sediment that forms from ocean water precipitation or ion exchange between existing sediment and ocean water, such as manganese nodules, phosphorite, glauconite, phillipsite, and montmorillonite.

Hyperosmotic Pertaining to an aqueous solution having a lower osmotic pressure (salinity) than another aqueous solution from which it is separated by a semipermeable membrane that allows osmosis to occur; the hypotonic fluid loses water molecules through the membrane to the other solution.

Hypo-osmotic Pertaining to aqueous solution having a lower osmotic pressure (salinity) than another aqueous solution from which it is separated by a semipermeable membrane that allows osmosis to occur; the hypotonic fluid loses water molecules through the membrane to the hypertonic solution.

Iceberg A massive piece of glacier that has broken from the front of the glacier (calved) into a body of water; it floats with its tip at least 5 m above the water's surface and at least four-fifths of its mass submerged.

Infauna Animals that live buried in the soft substrate (sand or mud).

Infrared radiation Electromagnetic radiation lying between the wavelengths of 0.8–1,000 μm; this wave band is bounded on the shorter wavelength side by the visible spectrum and on the long side by microwave radiation.

Insolation The rate at which solar radiation is received per unit of surface area at any point at or above the earth's surface.

Intertidal zone The ocean floor covered by the highest normal tides and exposed by the lowest normal tides and the water environment of the tide pools within this region; the littoral zone, or foreshore.

Iso-osmotic Having equal osmotic pressure; if two iso-osmotic fluids are separated by a semipermeable membrane that allows osmosis to occur, there is no net transfer of water molecules across the membrane.

Kelp forest Large varieties of Phaeophyta (brown algae).

Kinetic energy Energy of motion; it increases as the mass or velocity of the object in motion increases.

Krill A common name frequently applied to members of the crustacean order Euphausiacea (euphausiids).

Langmuir circulation A cellular circulation set up by winds that blow consistently in one direction with velocities in excess of 12 km/h (7 mph); helical spirals running parallel to the wind direction are alternately clockwise and counterclockwise.

Lateral line The sensory organ of fish, used for detecting changes in pressure, that consists of a series of pores in the lateral line running down the side of the body.

Lithogenous sediment Sediment composed of mineral grains derived from the rocks of continents and islands and transported to the ocean by wind or running water.

Lithosphere The outer layer of the earth's structure including the crust and the upper mantle, to a depth of about 200 km; this layer breaks into the plates that move according to the theory of plate tectonics.

Lithospheric plates *See* Lithosphere.

Lower phyla Those phyla that evolved early in the evolution of life.

Macroplankton Plankton with maximum dimensions falling between 2–20 cm (0.8–8 in); composed primarily of metazoans.

Mariculture The application of the principles of agriculture to the production of marine organisms.

Megaplankton Plankton with maximum dimensions that exceed 20 cm (8 in); composed entirely of metazoans.

Meiosis Cell division in eukaryotic cells in which chromosomal numbers are reduced to one half. *See* Mitosis.

Meroplankton Planktonic larval forms of organisms that are members of the benthos or nekton as adults.

Mesoglea Jellylike substance found between endoderm and ectoderm in cnidarians and ctenophores; ranges from noncellular and nonliving to highly cellular nature, which leads some to believe it represents a stage in the development of mesoderm.

Mesopelagic zone That portion of the oceanic province from about 200–1,000 m depth; corresponds approximately with the dysphotic (twilight) zone.

Mesoplankton Zooplankton species that are between 0.2–20 mm in size.

Metamere The individual segment of an annelid worm.

Microplankton Plankton not easily seen with the unaided eye but easily recovered from the ocean with the aid of a fine mesh plankton net.

Mid-ocean ridge A linear, seismic mountain range that extends through all the major oceans rising from 1–3 km above the deep-ocean basins; averaging 1,500 km in width, rift valleys are common along the central axis and are a source of new oceanic crustal material.

Migration Predictable movements of animals for reproduction, feeding, or protection.

Mitosis Cell division in eukaryotic cells in which chromosomal numbers remain constant. *See* Meiosis.

Mixed layer The surface layer of the ocean water mixed by wave and tide motions to produce relatively isothermal and isohaline conditions.

Mixed tide A tide having two high waters and two low waters per tidal day with a marked diurnal inequality. This tide type also may show alternating periods of diurnal and semidiurnal components.

Mixing time The time it takes for the fresh water from a river to mix with the salt water of an estuary.

Mutualism A symbiotic relationship in which both participants benefit.

Nacreous The innermost layer of the mollusc shell; the shell layer that makes pearls.

Nanoplankton Plankton with maximum dimensions falling between 2–20 μm (0.008–0.08 in); consists of fungi, phytoplankton, and protozoans.

Natural selection The increased reproduction of those individuals in a species that are best adapted to environmental conditions.

Neap tide Tides of minimal range occurring when the moon is in quadrature (first and third quarters).

Nektobenthos Those members of the benthos that can actively swim and spend much time off the bottom; can determine their position in the ocean by swimming.

Nekton Large animals in the water column of the ocean that are free swimming (e.g., tuna).

Neritic province That portion of the pelagic environment from the shoreline to the point where the depth reaches 200 m.

Neritic sediments Sediments composed primarily of lithogenous particles, deposited relatively rapidly on the continental shelf, continental slope, and continental rise.

Neuromast cell The individual sensory cell of the lateral line organ of fish. Neuromast cells are deformed by changes in pressure. The deformation causes an electrical discharge in the nervous system.

Neuston Planktonic organisms of the pelagic zone that live at or very near the surface.

Nitrogen-fixing bacteria Those bacteria capable of converting nitrogen gas into nitrate salts.

North Atlantic Deep Water (NADW) A deep water mass that forms primarily at the surface of the Norwegian Sea and moves south along the floor of the North Atlantic Ocean.

Oceanic province The portion of the pelagic environment where the water depth is greater than 200 m.

Oceanic sediments The inorganic abyssal clays and the organic oozes that accumulate on the deep ocean floor slowly, particle by particle.

Oceanic spreading center The axes of oceanic ridges and rises that are the locations at which new lithosphere is added to lithospheric plates. The plates move away from these axes in the process of sea-floor spreading.

Oligotrophic Referring to areas of low biological productivity.

Ommatidia The individual light collection and image creation units of the arthropod compound eye.

Omnivores Animals that use a wide range of organisms including algae, plants, phytoplankton, and animals as food.

Oozes A pelagic sediment containing at least 30 percent skeletal remains of pelagic organisms, the balance being clay minerals. Oozes are further defined by the chemical composition of the organic remains (siliceous or calcareous) and by their characteristic organisms (diatoms, foraminiferans, radiolarians, pteropods).

Operculum A flap of proteinaceous or calcium carbonate attached to the foot of snails that covers the shell opening when the foot is withdrawn into the shell.

Orthophosphates Phosphoric oxide (P_2O^5) can combine with water to produce orthophosphates ($3H_2O \cdot P_2O_5$ or H_3PO_4) that may be used by plants as nutrients.

Oscula The large openings of the sponge body that allow water to exit.

Osmosis Passage of water molecules through a semipermeable membrane separating two aqueous solutions of different solute concentrations; the water molecules pass from the solutions of lower solute concentration into the other.

Osmotic pressure The pressure that must be applied to the more concentrated solution to prevent the passage of water molecules into it (through a semipermeable membrane) from the less concentrated solution; a measure of the tendency for osmosis to occur.

Ossicles The individual unit of the echinoderm skeleton.

Ostia The microscopic openings of the sponge body that allow water to enter.

Oxygen compensation depth The depth in the ocean at which marine plants receive just enough solar radiation to meet their basic metabolic needs; it marks the maximum depth of the euphotic zone.

Oxygen utilization rates (OUR) The rate at which oxygen is used in the decomposition of organic material that originates in the surface euphotic zone.

Parallel benthic communities The similarity of species composition in temperate and arctic bottom communities of widely separated locations.

Parasite *See* Parasitism.

Parasitism A symbiotic relationship between two organisms in which one benefits at the expense of the other.

Parenchyma A soft cellular tissue that fills the space between the outer layer and the inner organs of flat worms.

Patchiness Term used in marine biology to describe the nonrandom or clumpy distribution of organisms.

Pathogenic/parasitic bacteria Those bacteria that either cause disease or parasitize other organisms.

Pelagic environment The open-ocean environment which is divided into the neritic province (water depth 0–200 m) and the oceanic province (water depth greater than 200 m).

Pelladilla The juvenile stage of the Peruvian anchovy.

Penetration anchor The part of the body of burrowing organisms that swells and locks into place, allowing the animal to pull down.

Pennate diatoms The major group of diatoms that have elongated shells and that are generally found on the bottom.

Percoform Referring to the fish family Perciformes.

Photic zone The upper ocean in which the presence of solar radiation is detectable; it includes the euphotic and dysphotic zones.

Photophores One of several types of light-producing organs found primarily on fishes and squids inhabiting the mesopelagic and upper-bathypelagic zone.

Photosynthesis The process by which plants produce carbohydrate food from carbon dioxide and water in the presence of chlorophyll, using light energy and releasing oxygen.

Phototrophs Those organisms that are capable of photosynthesis.

Phycocyanin An accessory photosynthetic pigment found in red algae.

Phycoerythrin A red pigment characteristic of the Rhodophyta (red algae).

Phylogeny The evolutionary relationships of higher taxonomic groups such as kingdoms and phyla.

Phytoplankton The most important community of primary producers in the ocean; photosynthetic plankton.

Planktivores A heterotroph that consumes plankton.

Plankton Passively drifting or weakly swimming organisms that do not move independently of currents; includes mostly microscopic algae, protozoans, and larval forms of higher animals.

Pneumatocysts The float structures of algae, particularly the Sargasso seaweed.

Polar easterlies The polar winds caused by the high atmospheric pressures of high latitudes.

Polar emergence The emergence of low- and midlatitude temperature-sensitive deep ocean benthos onto the shallow shelves of the polar regions, where temperatures similar to that of their deep ocean habitat exist.

Pollution (marine) The introduction of substances or energy into the marine environment that results in harm to the living resources of the ocean or humans who use these resources.

Polychlorinated biphenol The toxic chemical PCB.

Polymorphic Colonial organisms that have individually distinct shapes (e.g., the Portuguese man-of-war).

Population A group of individuals of one species living in an area.

Potential world fishery The mass of fish that are supported annually by nutrients that originate outside an ecosystem (world fishery ecosystems) and flow into it through upwelling or other processes. This amount of fish can be removed on an annual basis without degrading the ecosystem.

Primary production The amount of organic matter synthesized by organisms from inorganic substances within a given volume of water or habitat in a unit of time.

Primitive species Those species of a phylum that have a number of features that characterize the early evolution of the group.

Producers Autotrophic organisms that synthesize organic molecules to support an ecosystem.

Progressive wave A wave in which the wave form progressively moves.

Protoplasm The complicated self-perpetuating living material that makes up all organisms. The elements carbon, hydrogen, and oxygen constitute more than 95 percent of protoplasm; water and dissolved salts make up from 50–97 percent of most plants and animals; carbohydrates, lipids (fats), and proteins constitute the remainder.

Protostome An animal in which the blastopore of the embryo persists and forms the mouth.

Protozoa Phylum of one-celled animals in which nuclear material is confined within a nuclear sheath.

Pseudopodia An extension of protoplasm in a broad, flat, or long needlelike projection used for locomotion or feeding; typical of amoeboid forms such as Foraminifera and Radiolaria.

Purse seine A curtainlike net that can be used to encircle a school of fish. The bottom is pulled tight much the way a string is used to close a baglike purse.

Pycnocline A layer of water exhibiting a high rate of change in density in its vertical dimension.

Quadrature The position of the moon where it is at right angles to the sun, relative to the earth.

Radionuclides Those chemical substances that have radioactivity.

Radula The mouth structure of gastropod snails that is used for scraping food off a surface.

Raptorial feeders Animals that selectively remove appropriate food particles from ocean water.

Recruitment (fishery) The year class (number of fish or mass of fish) of young adults added to a fishery following each spawning season.

Red muscle mass The muscle tissue of fish that is used for short spurts of rapid swimming.

Red tide A reddish-brown discoloration of surface water, usually in coastal areas, caused by high concentrations of microscopic organisms, usually dinoflagellates. It probably results from increased availability of certain nutrients for various reasons. Toxins produced by the dinoflagellates may kill fish directly, or large populations of animal remains may use up the oxygen in the surface water to asphyxiate many animals.

Reef face The upper seaward face of a reef from the reef edge (seaward margin of reef flat) to the depth at which living coral and coralline algae become rare (16–30 m).

Rete mirabile A complex of blood vessels that function as a countercurrent heat exchange system.

Salt wedge The intrusion into a river of a lens of salt water on the bottom.

Salinity A measure of the quantity of dissolved solids in ocean water. Formally, it is the total amount of dissolved solids in ocean water in parts per thousand by weight after all carbonate has been converted to oxide, the bromide and iodide to chloride, and all the organic matter oxidized. Normally, it is computed from conductivity, refractive index, or chlorinity.

Saltationist An evolutionary scientist who argues that there are time periods of very rapid evolutionary change.

Saprobic bacteria Bacteria that decompose dead organisms.

Sargasso Sea water A region of convergence in the North Atlantic lying south and east of Bermuda, where the water is very clear deep blue in color and contains large quantities of floating Sargassum.

Sea (a) The subdivision of an ocean. Two types of seas are identifiable and defined: the Mediterranean seas, where a number of seas are grouped together collectively as one sea, and adjacent seas, which are connected individually to the ocean. (b) A portion of the ocean where waves are being generated by wind.

Sea-floor spreading A process producing the lithosphere when convective upwelling of magma along the oceanic ridges moves the ocean floor away from the ridge axes at rates of from 1–10 cm (0.4–4 in) per year.

Semidiurnal tide Semidaily tide having two high and two low waters per tidal day, with small inequalities between successive highs and successive lows; tidal period is about 12 h and 25 min solar time.

Semipermeable A material that has the property of being permeable to some substances but not to others.

Sessile Permanently attached to the substrate and not free to move about.

Shelf break The depth on the outer edge of the continental shelf where depth starts to rapidly increase, usually at 200 m.

Side-directed sonar A type of underwater radar that can be directed to the side of a ship instead of straight down.

Silica Silicon dioxide (SiO_2).

Slumping The sliding of underwater sediments on an unstable slope.

Sound Fixing and Ranging (SOFAR) Channel A low-velocity sound channel near the base of the thermocline. Sound can be transmitted for great distances through the ocean within this channel.

Specialized species Those species of a phylum that have lost the characteristics of the early evolution of the group.

Species Those individuals that can potentially reproduce with each other but not with other individuals; the basic unit of taxonomy and evolution.

Sporangia Cells on the blades of algae that produce the reproductive spores.

Spring bloom A rapid increase in phytoplankton growth that occurs in temperate regions of the ocean because of increased availability of solar radiation.

Spring tide Tide of maximum range occurring every fortnight when the moon is new and full.

Standard laboratory bioassay Tests conducted in a laboratory setting to determine the percentage of dying caused within a popualtion by a given concentration of foreign material.

Standing crop (stock) The biomass of a population present at any given time.

Stenohaline Able to withstand only a small range of salinity change.

Stenothermal Those species that can tolerate only a narrow range of environmental temperatures.

Stipe Part of an algae plant that is similar to the stem of true plants.

Stokes Law For particles less than 60 μm (0.024 in), the smaller they are the slower they fall in a fluid medium. For plankton, this upper size limit can be extended to 500 μm (0.2 in).

Stratified estuaries Those estuaries that are characterized by a distinct layer of low-salinity water on top of a layer of high-salinity water.

Subducts The diving melting of lithospheric plates under other plates.

Sublittoral fringe That portion of the benthic environment extending from low tide to a depth of 200 m; considered by some to be the surface of the continental shelf. *See also* Eulittoral zone, Supralittoral fringe.

Submarine canyon A steep, V-shaped canyon cut into the continental shelf or slope.

Subneritic province Benthic environments from the high-tide shoreline to the seaward edge of the continental slope.

Suboceanic province Benthic environments seaward of the continental shelf.

Substrate The base on which an organism lives and grows.

Subsurface chlorophyll maximum (SCM) The concentration of chlorophyll in many places in the ocean, usually highest immediately below the oxygen compensation depth.

Subsurface oxygen maximum (SOM) Generally observed within the lower euphotic zone, the SOM may represent supersaturation of oxygen of up to 120 percent as a result of photosynthetically produced oxygen.

Supralittoral fringe The splash or spray zone above the high tide mark. *See also* Eulittoral zone, Sublittoral fringe.

Swell One of a series of regular, long-period waves that travels out of a wave-generating area.

Swim bladder An elongated, gas-filled sack, dorsal to the digestive tract of most bony fishes, which serves chiefly as a hydrostatic organ to help achieve neutral buoyancy.

Symbiont In commensalism and mutualism, the organism that benefits from its relationship with its host.

Symbiosis A mutual relationship between two species in which one or both benefit.

Systematist A scientist who is concerned with the naming of species and the evolutionary relationships of one species with others.

Taxonomy A system of identification, naming, and classification of organisms.

Terrace A flattened part of the nearshore land that has been formed by higher sea levels of the past.

Test The shell of an organism.

Thermocline A layer of water in which a rapid change in temperature can be measured in the vertical dimension.

Thixotropy The tendency of sands and muds to become fluid when pressure is applied.

Tide Periodic rise and fall of the surface of the ocean and connected bodies of water resulting from the gravitational attraction of the moon and sun acting unequally on different parts of the earth.

Total allowable catch The permissible annual catch of a given species of fish that will not degrade the fishery given the knowledge of the ecosystem from which the fishery is being removed.

Trade winds The air masses moving from subtropical high-pressure belts toward the equator, northeasterly in the Northern Hemisphere, and southeasterly in the Southern Hemisphere.

Transform fault A fault characteristic of mid-ocean ridges along which they are offset.

Trophic level A nourishment level in a food chain. Plant producers constitute the lowest level, followed by herbivores and a series of carnivores at the higher levels.

Turbidity current Underwater avalanche where slumping of sediments on an unstable slope creates a movement of both sediment and water that forms turbidity currents.

Ultraviolet radiation Electromagnetic radiation of wavelengths shorter than visible radiation and longer than X-rays. The approximate frequency range is from 1–20,000 Hz.

Upwelling The process by which deep, cold, nutrient-laden water is brought to the surface, usually by diverging equatorial currents or coastal currents that pull water away from the coast.

Viscosity The property of offering resistance to flow; internal friction.

Visible spectrum The segment of the electromagnetic spectrum that includes visible light, it falls between the wavelengths of 0.38 μm and 0.76 μm.

Viviparous Bearing live young—rather than eggs—at sea.

Wasting disease A disease caused by a fungus that destroyed the sea-grass meadows of the North Atlantic in the 1930s.

Westerlies Winds moving away from the subtropical high pressure belts toward higher latitudes; they are southwesterly in the Northern Hemisphere, and northwesterly in the Southern Hemisphere.

Year class (cohorts) Those fish recruited into a fishery during a given year.

Zone of resurgence The part of the intertidal zone of sandy beaches that is saturated with water during tidal exposure.

Zone of retention That part of the intertidal zone of sandy beaches in which water drains during tidal exposure and sediments are not saturated.

Zooplankton Animal plankton.

Zooxanthellae A modified form of dinoflagellate that lives inside other organisms, as found in some clams and corals.

Acknowledgments

Insert: Color in the Marine Environment
Page 2, Both: Marty Snyderman
Pages 3–4, All: H. H. Webber
Page 5, Both: Norbert Wu
Page 6, Top: Charles Seaborn/Odyssey Productions, Chicago; Bottom: H. H. Webber
Page 7, Top: Marty Snyderman; Center: H. H. Webber; Bottom: Norbert Wu
Page 8, Top: R. J. Boldstein/Visuals Unlimited; Center: Robert Frerck/Odyssey Productions, Chicago; Bottom: Norbert Wu
Page 9, All: Norbert Wu
Page 10, Top: Norbert Wu; Bottom: Wm. G. Jorgensen/Visuals Unlimited
Page 11, Top & Center: Norbert Wu; Bottom: H. V. Thurman
Page 12, Top: Norbert Wu; Bottom: Alex Kerstitch/Sea of Cortez Enterprises
Page 13, Both: Marty Snyderman
Pages 14–15: Otis Brown and Robert Evans, University of Miami, REMAS
Page 16: Jane A. Elrod and Gene Feldman, NASA, Goddard Space Flight Center
Page 85, Figure 5.4: Alex Kerstitch/Sea of Cortez Enterprises
Page 144, Figure 1: Charles Seaborn/Odyssey Productions, Chicago
Page 145, Figure 2: Marty Snyderman
Page 162, Figure 1: Robert L. Pitman
Page 191, Figure 9.2: Alex Kerstitch/Sea of Cortez Enterprises

Insert: Habitats in Color
Page 1: Marty Snyderman
Page 2, Top: Marty Snyderman; Bottom: H. V. Thurman
Page 3: H. V. Thurman
Pages 4–5: Laura Ovresat, Proof Positive/Farrowlyne Associates, Inc.
Page 6, Both: Dr. Fred N. Spiess, Scripps Oceanographic Institute, University of California, San Diego
Page 7, Top: Scripps Oceanographic Institute, University of California, San Diego; Bottom: Dr. Robert N. Hessler, Stereo Camera System, Scripps Oceanographic Institute, University of California, San Diego
Pages 8–11, All: Marty Snyderman
Page 12, Top: Norbert Wu; Bottom: Marty Snyderman
Page 13, Top: Marty Snyderman; Bottom: Alex Kerstitch/Sea of Cortez Enterprises
Page 14, Top: H. V. Thurman; Bottom: H. H. Webber
Page 15, Top: Sea Studios, Inc.; Bottom: Marty Snyderman

Page 16, Top: H. H. Webber; Bottom: Sea Studios, Inc.
Page 359, Figure 14.10: Cannon/Greenpeace
Page 363, Figure 14.13: Lowell Georgia/Science Source/Photo Researchers
Page 363, Figure 14.13: William E. Townsend, Jr./Photo Researchers
Page 364, Figure 14.14: David R. Frazier Photolibrary
Page 376, Figure 15.7: Ron Levy/Gamma-Liaison

INDEX

A

D

G

H

I

T

U

V

W

X

Y

Z